Tools for Learning

Tools for Learning is an on-line review area that provides a variety of activities designed to help students study for their class. Students will find chapter outlines, review questions (written by the author), flash cards, figure labeling exercises, and links to a variety of tutorials and other useful learning tools to help students master the basic science material.

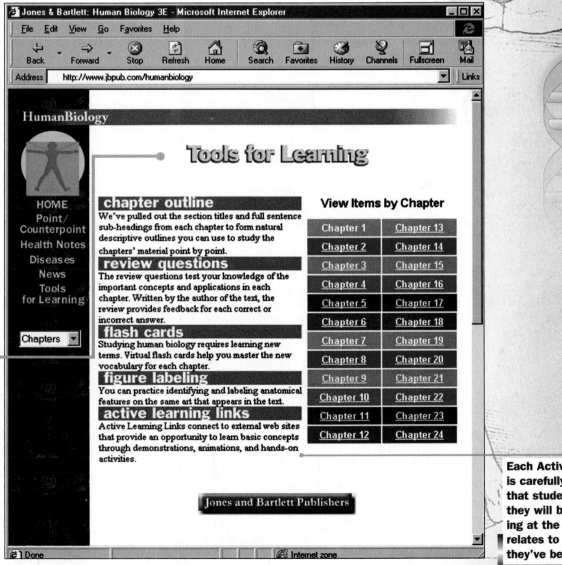

Each Active Learning Link is carefully introduced so that students know what they will be doing and seeing at the site and how it relates to the chapter they've been reading.

To find out more about HumanBiology, please e-mail info@jbpub.com, or call your Jones and Bartlett sales representative at 800-832-0034.

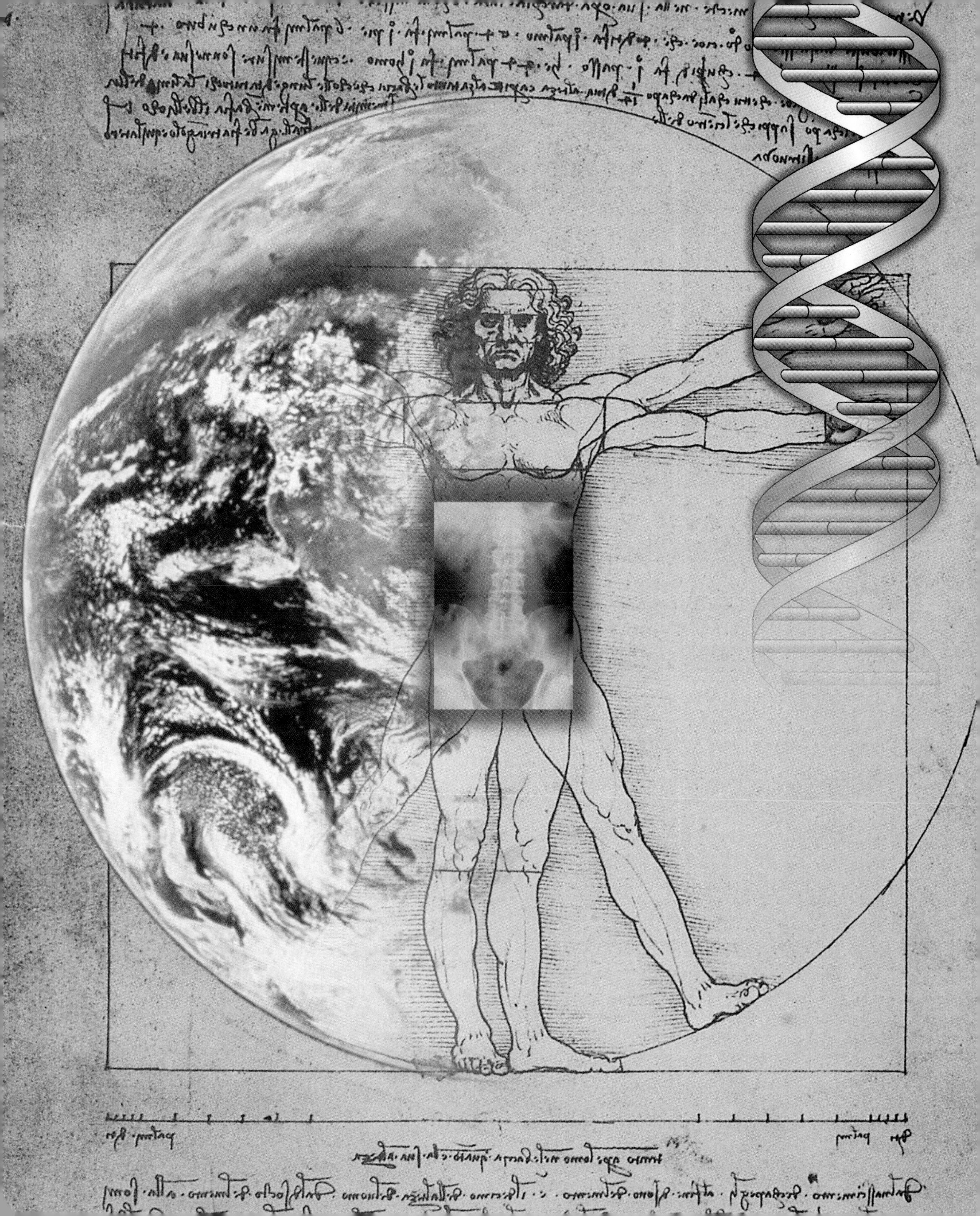

Human Biology

HEALTH, HOMEOSTASIS, AND THE ENVIRONMENT

THIRD EDITION

Daniel D. Chiras
University of Denver

JONES AND BARTLETT PUBLISHERS
Sudbury, Massachusetts
BOSTON   TORONTO   LONDON   SINGAPORE

World Headquarters
Jones and Bartlett Publishers
40 Tall Pine Drive
Sudbury, MA 01776
978-443-5000
info@jbpub.com
www.jbpub.com

Jones and Bartlett Canada
2100 Bloor St. West, Suite 6-272
Toronto, ON M6S 5A5
CANADA

Jones and Bartlett Publishers International
Barb House, Barb Mews
London W6 7PA
UK

■ JONES AND BARTLETT'S COMMITMENT TO THE ENVIRONMENT
As a book publisher, Jones and Bartlett is committed to reducing its impact on the environment. This and many other Jones and Bartlett titles are printed using recycled post-consumer paper. We purchase our paper from manufacturers committed to sustainable, environmentally sensitive processes. We employ the Internet and office computer network technology in our effort toward sustainable solutions. New communication technology provides opportunities to reduce our use of paper and other resources through on-line delivery of instructors' materials and educational information—and of course, we recycle in the office.

■ COVER CREDITS: Vitruvian Man by Leonardo daVinci, courtesy of Corbis; X-ray courtesy of Custom Medical Stock; Earth image courtesy of Photodisc, Inc. The Virtuvian Man by Leonardo daVinci is used throughout this book courtesy of Corbis.

PRODUCTION CREDITS
CHIEF EXECUTIVE OFFICER: Clayton Jones
CHIEF OPERATING OFFICER: Don Jones, Jr.
PUBLISHER: Tom Walker
V.P., SALES AND MARKETING: Tom Manning
V.P., MANAGING DIRECTOR: Judith H. Hauck
V.P., COLLEGE EDITORIAL DIRECTOR: Brian L. McKean
V.P., DIRECTOR OF INTERACTIVE TECHNOLOGY: Mike Campbell
MARKETING DIRECTOR: Rich Pirozzi
DIRECTOR OF DESIGN AND PRODUCTION: Anne Spencer
DIRECTOR OF MANUFACTURING AND INVENTORY: Therese Bräuer
SENIOR DEVELOPMENTAL EDITOR: Dean W. DeChambeau
ASSISTANT EDITOR: Karen McClure
PRODUCTION EDITOR: Rebecca S. Marks

DESIGN: Anne Spencer and Graphic World Publishing Services
MEDICO-ANATOMICAL ART: Imagineering
TYPESETTING AND COMPOSITION: Graphic World, Inc.
COVER DESIGN: Anne Spencer
WEB DESIGNER: Mike DeFronzo
INTERACTIVE TECNOLOGY PROJECT EDITOR: Scott Smith
PRINTING AND BINDING: Courier Company
COVER PRINTING: John Pow Company
EDITORIAL PRODUCTION SERVICE AND PACKAGING: Kathy Smith

Library of Congress Cataloging-in-Publication Data

Chiras, Daniel D.
    Human biology : health, homeostasis, and the environment / Daniel Chiras.—3rd ed.
      p.   cm.
    Includes index.
    ISBN 0-7637-0808-9
    1. Human biology.   I. Title.
    QP36.C46   1999
    612—dc21
                                98-49008
                                CIP

Printed in the United States of America
03 02 01 00 99   10 9 8 7 6 5 4 3 2 1

Printed on Recycled Paper

This book is dedicated to my family:
my mother and father, for their continual love and support;
my sons, Skyler and Forrest, for their joyous laughter and bright smiles;
and Linda, for her patience, kindness, and unwavering love.

———————

ABOUT THE AUTHOR

DR. CHIRAS received his Ph.D. in reproductive physiology from the University of Kansas Medical School in 1976 where his research on ovarian physiology earned him the Latimer Award. In September 1976, Dr. Chiras joined the Biology Department at the University of Colorado in Denver in a teaching and research position. Since then, he has taught numerous undergraduate and graduate courses, including general biology, cell biology, histology, endocrinology, and reproductive biology. Dr. Chiras also has a strong interest in environmental issues and has taught a variety of courses on the subject.

Currently an adjunct professor at the University of Colorado in Denver and at the University of Denver, Dr. Chiras has also been a visiting professor at the University of Washington, where he taught environmental science. Most of his time is spent writing books and articles and lecturing on a variety of topics, including ways to build a sustainable society.

Dr. Chiras is the author of numerous technical publications on ovarian physiology, critical thinking, sustainability, and environmental education, which have appeared in the *American Biology Teacher* and other journals. He has also written numerous articles for newspapers and magazines on environmental issues. He is the author of the environment section for World Book Encyclopedia's annual publication, *Science Year*, and wrote the environmental issues and air pollution articles in *Encyclopedia Americana*, as well as dozens of articles on mammals.

Dr. Chiras has published five college and high school textbooks, including *Environmental Science: Action for a Sustainable Future* and *Natural Resource Conservation: An Ecological Approach* (with John P. Reganold and the late Oliver S. Owen). Dr. Chiras's high school textbook, *Environmental Science: A Framework for Decision Making,* was selected as the official book of the U.S. Academic Decathlon, a nationwide competition involving thousands of American high school students in over 3000 schools.

Dr. Chiras's books for general audiences include *Beyond the Fray: Reshaping America's Environmental Response* and *Lessons from Nature: Learning to Live Sustainably on the Earth,* which outlines ways to apply ecological principles to create a sustainable society. He has also published *Study Skills for Science Students.*

In addition to writing, teaching, and lecturing, Dr. Chiras plays an active role in the environmental movement. He is founder and president of the Sustainable Futures Society in Evergreen, Colorado. Dr. Chiras currently serves as an editor of *Environmental Carcinogenesis and Ecotoxicology Reviews.* Besides his active scientific and environmental pursuits, Dr. Chiras is an avid bicyclist, organic gardener, and musician. He plays guitar, saxophone, and flute, and has written numerous songs in the past 20 years. He and his two sons live in Evergreen, Colorado, in a passive solar home supplied by solar and wind power and constructed from recycled materials, including 800 used automobile tires.

Finished product.

Work in progess.

BRIEF CONTENTS

CONTENTS

ix

SPECIAL FEATURES

END OF CHAPTER MATERIAL

CHAPTER 3

THE LIFE OF THE CELL 39

SPECIAL FEATURES

END OF CHAPTER MATERIAL

II. The Human Organism: Structure and Function of the Human Body

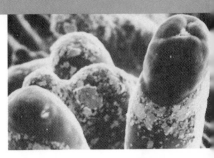

CHAPTER 6

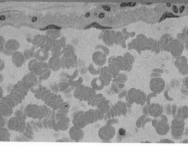

CHAPTER 7

THE BLOOD 155

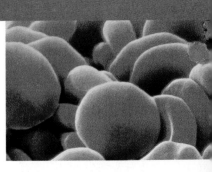

CHAPTER 8

THE IMMUNE SYSTEM 168

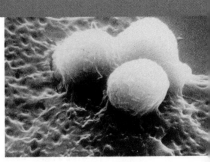

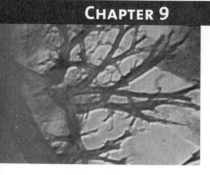
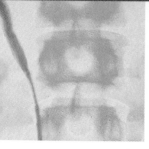

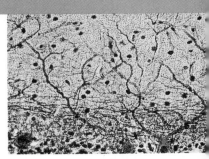

CHAPTER 13

CHAPTER 14

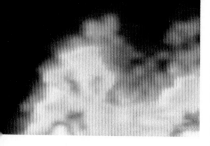

III. Heredity and Reproduction

CHAPTER 15

CHROMOSOMES, CELL DIVISION, AND CANCER 337

CHAPTER 16

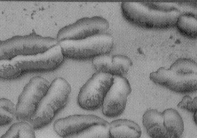

PRINCIPLES OF HUMAN HEREDITY 359

CHAPTER 17

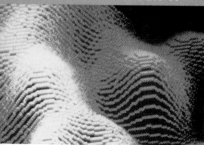

MOLECULAR GENETICS: HOW GENES WORK AND HOW GENES ARE CONTROLLED 387

CHAPTER 18

GENETIC ENGINEERING AND BIOTECHNOLOGY: SCIENCE, ETHICS, AND SOCIETY 408

CHAPTER 19

HUMAN REPRODUCTION 422

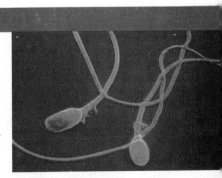

CHAPTER 20

HUMAN DEVELOPMENT AND AGING 453

Student Detail Schedule

Days = (**M**)onday, (**T**)uesday, (**W**)ednesday, Thu(**R**)sday, (**F**)riday, (**S**)aturday, S(**U**)nday
Note: **TR** means class meets on **Tuesday and Thursday**

For information regarding Yavapai College policies, click the HELP link in the upper right corner of the screen.

My Account Balance and Pay

View/Buy Textbooks

Total Credit Hours: 3.000

Drawing I - ART 110 - 103

Associated Term:	Spring 2012
CRN:	13891
Status:	**Web Registered** on Nov 21, 2011
Assigned Instructor:	Harvel C. Wardwell
Grade Mode:	Standard (A, B, C, D, F, S, U)
Credits:	3.000
Level:	Credit
Campus:	Prescott

Scheduled Meeting Times

Type	Time	Days	Where	Date Range	Schedule Type	Instructors
Class	5:30 pm - 8:15 pm	MW	Center for the Arts (Bldg. 15) 211	Jan 18, 2012 - May 08, 2012	Lecture/Lab	Harvel C. Wardwell (P)

Return to Previous

release: 8.3.0.3

IV. Evolution and Ecology

CHAPTER 21

EVOLUTION: FIVE BILLION YEARS OF CHANGE 477

CHAPTER 22

TRACING OUR ROOTS: THE STORY OF HUMAN EVOLUTION 502

CHAPTER 23

CHAPTER 24

ENVIRONMENTAL ISSUES: POPULATION, POLLUTION, AND RESOURCES 538

PREFACE

Human Biology: Health, Homeostasis, and the Environment is written for the introductory human biology course. This book explores the biology of human beings. It examines the structure and function of cells, tissues, and organs that make up our bodies. The central theme of this book is homeostasis—a kind of internal balancing act that is vital for our survival. Through numerous examples, I show that human health is dependent on properly functioning homeostatic mechanisms. These, in turn, are dependent on living in a "healthy" environment.

In writing *Human Biology*, I had several goals in mind. First, this book was written to teach students about the human body. This knowledge, in turn, helps students understand and take care of themselves and make informed health decisions both now and in the future. Second, this book promotes scientific literacy, that is, an understanding of how scientific information is gathered and the way it influences our lives. Furthermore, it provides background material so that students can understand new scientific discoveries and the way they will influence our lives. This will aid students in understanding some of the thorniest political issues of our times. Third, this book promotes critical thinking skills, which will be valuable to students throughout their lives. Fourth, the discussions of the environment assist students in understanding our relationship to the larger whole, the natural environment, and ways we affect it, sometimes to our own detriment. Fifth, this book helps students begin to learn systems thinking—that is, thinking in terms of whole systems and how the parts interact.

Organization

Human Biology is divided into four parts. Part I outlines basic biological and chemical principles vital to your understanding of the human organism. Part I contains a discussion of science and the scientific method and introduces critical thinking skills that are applied throughout the book. It closes with an overview of the cell, the basic building block of all organisms.

Part II outlines the structure and function of human beings. The chapters in this section describe how the major organ systems operate. Homeostasis is emphasized in these chapters as a unifying principle of biology.

Part III discusses cell division, heredity, and genetic engineering. It also includes the discussion of reproduction and development. These chapters outline the evolution, structure, and function of reproductive systems in animals and portray the dramatic events that lead to the formation of new individuals.

Finally, Part IV focuses on the big picture. It looks at evolution—how we got here—and basic principles of ecology, the study of ecosystems. The final chapter surveys the problems modern society has created in the natural world and offers solutions for redirecting human society onto a sustainable course—one that ensures well-functioning homeostatic mechanisms.

New to the Third Edition

This text has been thoroughly revised and trimmed of extraneous material that reviewers have pointed out to me. Previous users of the text will notice the larger page size, which allows for a more dynamic presentation of art and text. Major pieces of art, notably the human systems illustrations, have been redrawn to improve their pedagogical value. Major sections in each chapter are now numbered to help instructors assign readings.

Perhaps the most significant change to this new edition is its organization. As outlined above, I have reorganized the chapters so that the more demanding genetics material comes later in the text. Furthermore, the publisher and I have taken a bold new approach to teaching chemistry to human biology students. Instead of discussing all the chemistry students need to know in one chapter early on in the book, I have introduced just the basics of chemistry in Chapter 2, with a brief overview of the major biological molecules. Specific information on the important biological molecules—carbohydrates, lipids, proteins, amino acids, and nucleic acids—is provided as students need the information. For example, students learn about proteins and lipids in the section on plasma membranes in Chapter 3, "The Life of the Cell," just before they learn about the structure of the plasma membrane. They then learn a little about carbohydrates in the energetics section of that chapter. More information on carbohydrates, lipids, and proteins is presented in the nutrition chapter. DNA and RNA are presented in detail in the discussion of molecular genetics. We have placed an icon ◯ in the table of contents so instructors can see where the basic chemistry has been moved.

This approach has at least two benefits. First, it helps prevent chemistry overload—the deluge of seemingly isolated facts about chemistry that so many students find difficult to grasp and impossible to remember. Second, it offers information on chemistry in context, so that it is easier to learn and retain. It shows the relevancy of chemistry to students' understanding of important topics.

Web Enhancement. This third edition is now linked to a extensive web site, *HumanBiology*, developed exclusively for this edition by the author and Jones and Bartlett Publishers. Icons in the book identify material that is matched to relevant Internet sites both students and instructors can visit through *HumanBiology* (www.jbpub.com/humanbiology). These independent sites provide supplementary information on the diseases introduced in the book, topics debated in the Point/Counterpoints, the subjects discussed in the book's health notes, and the latest biology news. We've provided a brief description to place each link in context *before* the student connects to the site. Jones and Bartlett and the author continually monitor the links to ensure there will always be a working and appropriate site on line.

HumanBiology also contains Tools for Learning, an online student review area that provides a variety of activities designed to help students study for their class. Students will find chapter outlines, review questions (written by the author), flash cards, figure labeling exercises, and links to a variety of tutorials and other useful learning tools to help them master the basic science material. An illustrated description of *HumanBiology* is provided just inside the front cover of this book.

A Note to Instructors about Using *HumanBiology* in Their Course. Instructors often find the Internet useful for managing their courses. A tour of the Internet shows that instructors are using web pages to post their course syllabi and assignments, add their favorite hot links, test on-line, receive and respond to e-mail from their students, and provide a bulletin board for their students. In many cases, however, the technology or resources to do this within an instructor's school network aren't available. Jones and Bartlett Publishers offers this service on-line at the *HumanBiology* site in a format that allows you to easily and quickly add material specific to your course that's available to your students only.

You can have your students use the web in a variety of ways. For example, it serves as a source of background material and resources for the term papers you assign. The students can use the links we have created to other web sites to further their understanding of human biology—to increase their breadth and depth of knowledge. This information helps them become more knowledgeable citizens, more health conscious, and better informed when it comes to making health-care decisions.

Students can also use the web to find information to help them in class discussions and class debates. Many instructors use the Point/Counterpoints in the book to stimulate lively debate. Because of space limitations, however, most Point/Counterpoint authors cannot fully develop their arguments. A trip to the web can help students uncover information that will expand their knowledge and assist them in class debates.

Special Features

Human Biology is a user-friendly book. The material is presented in a friendly style. Complex subjects are simplified somewhat, and numerous analogies are used to make material more meaningful. For the most part, this book concentrates on basic information—key facts and concepts essential to students of human biology. You will find numerous examples that are not only relevant, but also fascinating. I've attempted to hold terminology down wherever possible and provided pronunciations for virtually all terms. This book is enhanced by numerous features listed below.

Scientific Discoveries That Changed the World

Science is as much a body of facts as it is a process of discovery. Many great discoveries have been made over the years. To highlight some of these discoveries, I've included numerous essays covering such topics as the discovery of the structure of DNA. These features highlight the work of some of the world's most important scientists and illustrate how scientific discoveries have changed our view of the world. They also further students' understanding of the scientific method and illustrate the fact that scientific advances usually require the work of many scientists, sometimes working in seemingly unrelated areas.

Health Notes

I have updated the Health Notes and added a few new ones. All Health Notes emphasize information on ways of maintaining or improving your health. They offer practical advice on proper diet, exercise, stress management, and cancer prevention.

Point/Counterpoints

Many discoveries in biology have had profound impacts on our lives. Today, however, new discoveries often result in controversial applications such as genetic engineering or fetal cell transplantation. This book presents a number of modern-day controversies in Point/Counterpoints. Some debate social and political issues that require a good biological background and others focus on scientific debates.

Each Point/Counterpoint consists of two brief essays written by distinguished writers and thinkers. These essays present opposing views on important issues of our times such as genetic engineering, fetal cell transplantation, cancer, and global warming. Point/Counterpoints also offer students a chance to practice critical thinking skills. Additional discussions, questions, and links to web pages supporting differing views on the issues are available at the Jones and Bartlett web site *HumanBiology.*

Controversary over the Use of Animals in Laboratory Research (Chapter 1)
Controversy over Food Irradiation (Chapter 2)
Fetal Cell Transplantation (Chapter 3)
Tracking People with AIDS (Chapter 8)
Prioritizing Medical Expenditures (Chapter 10)
Are We in a Cancer Epidemic? (Chapter 15)
Are Current Procedures for Determining Carcinogens Valid? (Chapter 17)
Controversy over Herbicide Resistance Through Genetic Engineering (Chapter 18)
The Medical Debate over Circumcision (Chapter 19)
Physician-Administered Euthanasia (Chapter 20)
Why Worry About Extinction? (Chapter 23)

Critical Thinking Skills

As noted earlier, Chapter 1 presents a number of "rules" for improving critical thinking skills. These guidelines will help students become more discerning thinkers, a skill that could prove useful in this and many other college courses—not to mention the benefit it will have in later life.

With the new book design, additional emphasis is placed on critical thinking throughout the text. Each chapter, for example, contains a Thinking Critically exercise. This exercise at the beginning of the chapter outlines a problem or presents the results of a study and asks students to apply their critical thinking skills. A brief analysis is offered at the end of the chapter.

As in the last edition, I've also included an Exercising Your Critical Thinking Skills at the end of each chapter. These exercises also call on students to use their critical thinking skills and include case studies, hypothetical scenarios, or summaries of news or scientific reports. Each exercise emphasizes one or two of the critical thinking rules presented in the first chapter. Critical thinking questions are also included after each Point/Counterpoint and in the Test of Concepts questions.

Health, Homeostasis, and the Environment Sections

The health of the Earth's organisms and the environment in which they live are closely connected. To illustrate these connections, each chapter concludes with a Health, Homeostasis, and the Environment section. These describe the importance of homeostasis and demonstrate some of the ways in which the physical, chemical, social, and psychological environments affect our health by upsetting homeostasis.

In-Text Summaries

To help students learn key concepts, virtually all chapter section heads are written as summary statements. These statements capture key concepts presented in the material that follows. These in-text summaries provide students with a way to review major concepts as they prepare for exams.

End-of-Chapter Summaries

At the end of each chapter is a summary of the material covered in the chapter. Students can use the chapter headings to glean key concepts and the End-of-Chapter summaries to review the most important factual information presented in the chapter.

Test of Concepts

Each chapter contains a number of brief essay questions that enable students to assess their understanding of the material. These questions go beyond the regurgitation of facts in the test of terms. The in-text summaries, detailed end-of-chapter summaries, and test of concepts should provide an excellent study guide.

Art Program

This book contains a remarkable collection of drawings and photographs, many of which are new in this edition. These colorful illustrations supplement the text and make the more complex concepts and processes understandable. In this edition, more of the art contains explanatory boxes describing the key processes illustrated.

Ancillary Materials

Jones and Bartlett Publishers offers an impressive variety of traditional print and interactive multimedia supplements to assist instructors and aid students in mastering human biology. Additional information and review copies of any of the following items are available through your Jones and Bartlett Sales Representative.

Instructor's ToolKit CD-ROM

This CD-ROM contains full-color art files of the text's illustrations for computer projection, labeled and unlabeled transparency masters, and Microsoft PowerPoint™ Lecture Outline Slides. Also included are the complete text files for the instructor's manual and test bank, an electronic test bank and test-generating software, and concept maps of key systems.

Instructor's Manual and Test Bank

Written by Jay Templin of Montgomery Community College, Pottstown, PA, the instructor's manual contains lecture outlines, learning objectives, key terms, concept questions, teaching tips—including ideas for using the Point/Counterpoints in class—and concept maps. The test bank contains over two thousand questions in a variety of formats.

Study Guide

Written by Jim Blahnik of Lorain County Community College, the study guide contains a section on developing good study skills, chapter overviews, chapter outlines, learning objectives, study tips pertinent to each chapter, art labeling exercises, concept map exercises, and practice tests.

HumanBiology Web Site (www.jbpub.com/humanbiology)

The *HumanBiology* web site offers students an unprecedented degree of integration between their text and the on-line world. The site contains Tools for Learning, an on-line student review area to help students study for their class. Students will find complete chapter outlines, review questions with feedback (written by the author), hundreds of flash cards, figure labeling exercises, and links to a variety of interactive tutorials. The site also introduces students to independent sites that directly relate to diseases mentioned in the chapters and the text's Point/Counterpoint and Health Notes features. This close linkage means the text always remains current. An illustrated description of *HumanBiology* is provided just inside the front cover of this book.

CyberClass

CyberClass is a customizable, web-based teaching and learning environment. It allows instructors to easily and quickly post material specific to their course, such as a syllabus, assignments, and favorite hot links. Instructors can also administer tests on-line, receive and respond to e-mail from students in the course, and maintain a secure on-line grade book. Students can use CyberClass to study flash cards, use on-line practice exams, and post messages to a class bulletin board. Please contact your Jones and Bartlett representative for more information and a demonstration of this exciting technology.

Acknowledgments

A project of this magnitude is the fruit of a great many people. I wish to thank the thousands of scientists and teachers who have contributed to our understanding of human biology. A special thanks to the extraordinary teachers who have made tremendous contributions to my education, especially the late Weldon Spross, Edward Evans, Dr. H. T. Gier, Dr. Gilbert Greenwald, Dr. Floyd Foltz, Dr. Howard Matzke, and Dr. Douglas Poorman.

I am also deeply indebted to many people for their assistance during the writing of this book. A special thanks to my editor, Brian McKean, for his patience, guidance, and inspiration. Brian's a true delight to work with. A great debt of gratitude goes to Dean DeChambeau, my extrordinary developmental editor. His efforts in all areas of the book are greatly appreciated. Many thanks to my production editors, Anne Spencer and Kathy Smith. Much appreciation for their calmness amidst the turmoil of a rapid production schedule, cordiality, attention to detail, and diligence. Additional thanks must be extended to Karen McClure who coordinated the supplements and provided invaluable assistance to the production editors. A word of thanks also to my colleague Dr. John Cunningham of Visuals Unlimited, who supplied the excellent photographs. Thanks also go to my team of artists: Wayne Clark, Darwen and Vally Hennings, Imagineering, Carlyn Iverson, Georg Klatt, Elizabeth Morales, Precision Graphics, Publication Services, Pat Rossi, John and Judy Waller, Cyndie C.H.-Wooley, and J.B. Woolsey and Associates. I greatly appreciate the efforts of the many sales representatives who have helped make this book a success. It has been a pleasure and an honor to have worked with such a fine and talented group of people.

Thanks also to the many authors who contributed the Point/Counterpoints in this book. Your work will make this a more exciting journey for students as they begin to appreciate different perspectives of crucial issues.

Throughout this time, my two delightful sons, Skyler and Forrest, have offered considerable support and a counterbalance to the stresses and strains of a project of this magnitude. You're the light of my life. Thanks, too, to my partner and part-time research assistant Linda Stuart for her friendship and inspiration. Many thanks to Scott Reuman, my other research assistant, for giving 150 percent and cheerfully.

Finally, a special thanks to all the reviewers who offered many useful comments throughout this project. Their insights and attention to detail have been greatly appreciated. Below is a list of those who have reviewed the manuscript.

First and Second Edition Reviewers

D. Darryl Adams
Mankato State University

Donald K. Alford
Metropolitan State College

David R. Anderson
Pennsylvania State University—Fayette Campus

Jack Bennett
Northern Illinois University

Charles E. Booth
Eastern Connecticut State University

J. D. Brammer
North Dakota State University

Vic Chow
City College of San Francisco

Ann Christensen
Pima Community College

Francoise Cossette
University of Ottowa

Peter Colverson
Mohawk Valley Community College

John D. Cowlishaw
Oakland University

Richard Crosby
Treasure Valley Community College

Stephen Freedman
Loyola University of Chicago

Martin Hahn
William Paterson College

John P. Harley
Eastern Kentucky University

Robert R. Hollenbeck
Metropolitan State College

Carl Johnson
Vanderbilt University

Florence Juillerat
Indiana University—Purdue University

Ruth Logan
Santa Monica College

Charles Mays
DePauw University

David Mork
St. Cloud State University

Donald J. Nash
Colorado State University

Emily C. Oaks
State University of New York—Oswego

Lewis Peters
Northern Michigan University

Richard E. Richards
University of Colorado at Colorado Springs

Miriam Schocken
Empire State College

Richard Shippee
Vincennes University

Beverly Silver
James Madison University

David Weisbrot
William Paterson College

Terrance O. Weitzel, Jr.
Jacksonville University

Roberta Williams
University of Nevada—Las Vegas

Tommy Wynn
North Carolina State University

Third Edition Reviewers

Felix Baerlocher
Mount Allison University

Judith Byrnes-Enoch
Empire State College

John Cummings
Waynesburg College

Jeffrey Dean
Vanderbilt University

Debby Dempsey
Northern Kentucky University

Penni Croot
SUNY Potsdam

Melanie DeVores
Sam Houston State University

Sheldon R. Gordon
Oakland University

Wendel J. Johnson
University of Wisconsin, Marinette

David Mork
St. Cloud State University

Donald J. Nash
Colorado State University

Lynette Rushton
South Puget Sound Community College

Richard Weisenberg
Temple University

Roberta Williams
University of Nevada—Las Vegas

Mary Vetter
University of Regina

LIFE IN THE BALANCE

AN INTRODUCTION TO HUMAN BIOLOGY

Human health is dependent on a healthy lifestyle and a healthy environment.

Thinking Critically

Your local newspaper reports the results of an experiment a student in one of the local high schools performed to test the effects of a special diet on cholesterol content in chicken eggs. He obtained 20 chickens from two different breeders. Half of the chickens were fed his special diet; the other half were fed a diet of standard chicken feed, which he purchased at the local feed store. The boy found that cholesterol levels in the eggs from the group fed his special diet were lower than the other group. The story created quite a stir in the local media—so much so that the boy's father is trying to acquire funding to market his son's new feed. What problems do you see with this study? ■

TO BEGIN OUR EXPLORATION OF HUMAN BIOL-
ogy and gain some perspective on our species,
we will travel back in time about 3.5 million
years to the grasslands of Africa where the first recognizable humanlike organism roamed (**FIGURE 1-1**). Scientists dubbed this creature *Australopithecus afarensis*
(aus-TRAL-owe-PITH-a-CUSS A-far-EN-suss).

Standing only 3 feet tall and walking upright, our
earliest ancestors subsisted in large part on a diet of roots,
seeds, nuts, and fruits. They supplemented their primarily vegetarian diet with **carrion** (CARE-ee-on), animals
that had been killed by predators or that had died from
other causes. They may also have captured
and killed other animals for meat.

Weak and slow compared to other large animals, our
earliest ancestors could have easily ended up as an evolutionary dead-end. Fortunately, though, they possessed
several anatomical features that tipped the scales heavily
in their favor. Undoubtedly, one of the most important
characteristics of all was their brain, a feature that would
thrust humans to a position of evolutionary dominance.

Although, as one humorist quipped, to an insect a
human is just something good to eat, our preeminence in
the biological world is indisputable. Today, as testimony to
the success of our journey, human beings inhabit a world
that is markedly different from that of our ancestors.
Rather than collecting nuts
and berries from the

FIGURE 1-1 Australopithecus afarensis Current scientific
evidence suggests that *Australopithecus afarensis* was the
first humanlike ape. Its skeletal remains indicate that it
walked upright.

(a)

(b)

FIGURE 1-2 What Price Progress? Technological development and economic progress are a double-edged sword, providing many benefits, but not without a huge price as witnessed in *(a)* polluted skies and *(b)* denuded hillsides cleared to provide wood for human use.

plants around us, today many people purchase their food from grocery stores supplied by farms throughout the world. Rather than roaming in small bands, the vast majority of the world's people live in cities and towns that offer amenities our ancestors would have never dreamed possible. Instead of being restricted to whimsical gazing into space, we travel there ourselves. And, in recent years, no longer content with what nature provides, our scientists have begun to tinker with the genetic material of the cell to alter plants and animals to increase food production. Some scientists have even begun to tinker with our hereditary material.

For better or for worse, humans have become a major player in **evolution** itself, a process of biological change that occurs in distinct groups of organisms, known as **populations.** Evolution results in structural, functional, and behavioral changes in populations. These changes, in turn, result in organisms better equipped to cope with their environment—that is, better able to survive to reproduce.

From most perspectives, the human experiment has been an overwhelming success. However, growing evidence shows that our victory has not occurred without a huge price tag. Examples of the cost of our success are many: overpopulation, polluted skies, vanishing forests, and lost species (**FIGURE 1-2**). Growing evidence, in fact, suggests that human civilization could destroy millions of years of evolutionary history in a relatively brief time span if we persist along our current path. It is

at this point in our evolutionary history that we take up our study of human biology.

Like other texts, the bulk of this book will take you on a journey through the human body. On this journey, you will learn a great deal about yourself—how you got here, how you inherited certain characteristics from your parents, and how your body functions. You will learn the basics of nutrition and how broken bones mend. You will discover how your immune and nervous systems operate. You will explore many common diseases and—perhaps more important—how to prevent them. You will find that many of the diseases afflicting modern humans result from the environment we live in—the stressful, polluted, human-engineered environs that reflect a lifestyle that often puts convenience and efficiency ahead of healthy patterns of existence. In essence, then, you will be looking at human beings in the modern context.

The information you learn from this book will prove useful to you in many ways. It will also help you understand important political debates over issues such as genetic engineering, pollution, and vaccination. As you proceed through this book, though, be sure to take some time to marvel at the wonders of the human body—the intricate details of the cell, the fascinating structure and function of the body's organs, and the intriguing manner in which the various parts work together.

Health and Homeostasis: Life's Essential Balancing Act

In this book, you will see how human health depends on numerous internal checks and balances that have evolved over many millions of years. These internal processes help to maintain a fairly constant internal condition, a state often referred to as *homeostasis* (home-e-oh-STAY-siss).

Homeostasis Is a State of Relative Constancy

The term **homeostasis** comes from two Greek words, *homeo,* which means "the same," and *stasis,* which means "standing." Literally translated, homeostasis means "staying the same." Thus, many people refer to homeostasis as a state of internal constancy. Unfortunately, this view is not entirely accurate. In reality, homeostasis is not a static state; rather, it is a dynamic (ever-changing) state. Let me explain what this means by using an example: body temperature.

Humans are warm-blooded creatures. We generate body heat internally and maintain body temperature at a fairly constant level—about 98.2°F according to recent studies with modern temperature-measurement devices, not the 98.6°F determined long ago (in 1868) with less precise instruments (thermometers).

In reality, though, body temperature varies during the day, falling slightly at night when you sleep and rising during the daylight hours. It increases even more when you participate in strenuous physical activity.

Like many other internal conditions, then, body temperature fluctuates within a range. This is what is meant by a *dynamic state.* Homeostasis is therefore defined as a condition of relative constancy, which is achieved through a variety of automatic mechanisms that compensate for internal and external changes. As Chapter 4 illustrates, homeostatic mechanisms require **sensors,** structures that detect internal and external change—for example, changes in external temperature. Sensors then elicit a response that offsets the change, helping to maintain a fairly constant state. Shivering, for example, is a rhythmic contraction of muscles that generates body heat when the outside temperature drops. It is one of several homeostatic mechanisms in warm-blooded animals triggered by cold.

Homeostatic mechanisms also maintain fairly constant levels of nutrients in the blood, which is essential for normal body function. If nutrient levels fall too low or climb too high, serious problems can arise. For ex-ample, sugar (glucose) is essential to the function of body cells, especially brain cells. If blood glucose falls too low, brain cells are deprived of energy, and a person may go into shock. If levels go too high, other problems arise. Consequently, the blood glucose concentration must be maintained at a fairly constant level. The concentration of blood glucose is regulated by two hormones you will study in Chapter 3.

Homeostatic mechanisms also exist in larger biological systems or ecosystems. An **ecosystem** is a biological system consisting of organisms and their environment. Homeostatic mechanisms help achieve balance in ecosystems, a phenomenon referred to in this book as **environmental homeostasis.**

A highly simplified example illustrates the point. In the grasslands of Kansas, rodent populations generally remain fairly constant from one year to the next. This phenomenon results, in part, from **predators** (animals that hunt and kill other organisms). Coyotes, hawks, and other predators feed on rodents, helping to hold rodent populations in check.

Although predators are a crucial element in maintaining environmental homeostasis in the grasslands of Kansas and virtually all other natural systems, weather, food supplies, and a host of other factors also contribute. In short, it is the net effect of these factors that determines population sizes. The balance can be quite complex. Excess rain may result in a bumper crop of seeds, upon which rodents feed. This, in turn, may lead to a rise in the rodent population. Predator populations may increase in response to the expanding rodent population. Drought in succeeding years, however, may trim rodent populations, with their decline resulting in a fall in predator populations. Homeostatic mechanisms in ecosystems such as these maintain a balance between food supply and populations of organisms.

Some homeostatic mechanisms can also be thought of as ways in which ecosystems recover from changes. Landslides in the Pacific Northwest, for instance, sometimes occur after heavy rains and can deposit huge quantities of sediment in salmon streams. This, in turn, buries spawning grounds and kills fish. Over time, however, sediment is usually purged by flowing water in streams, which restores gravel beds where fish spawn and allows populations to recover. On the hillside, grasses, shrubs, and trees may take root, stabilizing the slope and preventing future landslides.

Because of these and other similar mechanisms, ecosystems undisturbed by human activity tend to persist for thousands of years—barring natural geological and climactic changes. However, when human disturbances occur frequently or are severe, the changes may overwhelm the recuperative capacity of the environment. In such instances, disruption may lead to destruction. Today, pollution, mining, overgrazing, deforestation, and many other human activities threaten the environment in this manner.

In this book, the term *homeostasis* is used to refer to the balance that occurs at all levels of biological organization—from cells to organisms to entire ecosystems. Although many of the details still need to be worked out, the abundance of homeostatic mechanisms in nature suggests their evolutionary significance. Maintaining "balance" is essential to the continuation of life. Without it, life would be a chancy proposition. Cells would fall into disarray. Organisms would perish and ecosystems would be wiped out.

Human Health and the Health of Ecosystems Are Closely Tied

Scientists have found that the health of the environment and the health of organisms, including human beings, are interdependent. Alterations in one system—for example, adverse changes in the chemical composition of the air—can have impacts on the other. Thus, polluted air, water, and soils take a toll on humans and other species. According to a recent study by the Environmental Protection Agency, over 76 million Americans (nearly one of every three people) live in cities where the ground-level pollutant ozone poses a threat to health. Ozone is a chemical pollutant indirectly derived from automobiles that irritates the eyes and the lungs. Ozone also destroys the tiny air sacs in the lung, resulting in **emphysema,** a disease that kills its victims slowly and painfully (**FIGURE 1-3**).

Ozone is just one of many pollutants in our environment. Dozens of cancer-causing chemicals are also found in our drinking water, food, and air; these chemicals can alter cells in the body, causing some to proliferate uncontrollably. Unthwarted growth forms cancerous tumors that are often fatal. From a strictly human perspective, it should become clear that planet care is the ultimate form of self-care.

Emphysema
www.jbpub.com/humanbiology

Human Health also Depends on a Healthy Psychological and Social Environment.

The health of organisms requires more than a clean environment. It requires social and psychological conditions conducive to mental health. Stressful environments can lead to serious ailments in those individuals unfortunate enough to be stuck in them. Health, Homeostasis, and Environment sections at the end of each chapter illustrate the connection between our social/psychological environment and our health—a connection, until recently, rarely mentioned in discussions about human health.

Although humans are the central focus of this book, it is important to pause for a moment to note that many other species share this world with us. They, too, are profoundly affected by the condition of the environment. In Colorado, for example, construction activity and road traffic generated in the building of a dam southwest of Denver caused stress in a bighorn sheep population nearby that wiped out half of the herd. Countless other examples show the connection between the condition of the environment and the health of species that suffer our environmental insults.

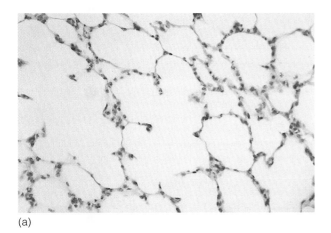

(a)

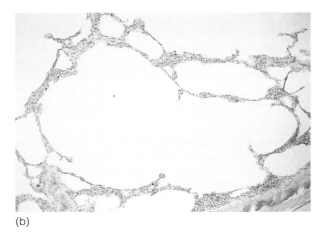

(b)

FIGURE 1-3 Emphysema Pollutants in cigarette smoke and air pollution destroy the tiny air sacs in the lung, which are necessary for the absorption of oxygen into the blood and the release of carbon dioxide. As the air sacs break down, the surface area of the lung decreases, making it harder for victims of this progressive disease to breathe. *(a)* Section through a normal lung; *(b)* section through an emphysemic lung.

Health Is a State of Physical and Mental Well-Being

Before we move on, it is important to understand what health is. For many years, human health was defined as the absence of disease (**FIGURE 1-4A**). As long as a person had no obvious symptoms of a disease, that person was considered healthy. Although such a person may have had clogged arteries from a lifetime of eggs-and-bacon breakfasts, it wasn't until symptoms of heart disease—for example, chest pain—became apparent that the patient was considered unhealthy.

In recent years, many health experts have sought to create a much broader and more useful definition of the term *health*. Today, human health considerations take into account two broader categories: physical *and* emotional well-being.

Physical health refers to the state of the body—how well it is working. Physical health can be measured by checking temperature, blood pressure, blood sugar levels, and a host of other variables. Medical scientists use the term **risk factor** to indicate abnormal conditions such as high blood pressure that put a person at risk for disease. The presence of one or more risk factors is a sign of less-than-perfect health. Obviously, the more risk factors there are, the worse one's physical health (**FIGURE 1-4B**).

The new concept of health is very different from the old one because it says that even though a person feels healthy and doesn't exhibit signs of disease, such as a failing heart, the presence of risk factors must be taken into account when categorizing the person's health. As shown in **FIGURE 1-4B**, the absence of risk factors results in the best health. A few risk factors mean health is only good.

More risk factors mean health must be considered poor, even though the individual may not have had a heart attack or some other problem—yet.

Consider the example of high blood pressure. A person with high blood pressure often feels fine, especially early on. In the long run, though, a person with high blood pressure is much more likely to suffer a heart attack than a person with normal blood pressure. Therefore, all other things being equal, a person with high blood pressure is less healthy than one with normal blood pressure.

Some additional risk factors include: high cholesterol levels, abnormal hormonal levels, and high or low blood sugar. Scientists also use the term *risk factor* to refer to activities that predispose someone to disease. Smoking and a fatty diet, for example, are risk factors for strokes and heart attacks.

Physical health is also measured by level of physical fitness. If you can't walk up a set of stairs without gasping for air, you're not considered very physically fit. You're more likely to have other problems later in life—for example, heart disease.

Along with physical health goes emotional well-being. Especially relevant today is the ability to cope with the stresses and strains of modern life. Inability to cope may lead to physical ill health, such as high blood pressure, heart disease, and stroke.

In essence, then, mental and physical fitness are measures of our psychological and physical abilities to meet the demands of everyday life. Fit people are able to cope with day-to-day psychological stresses and are able to walk up a flight of stairs without becoming

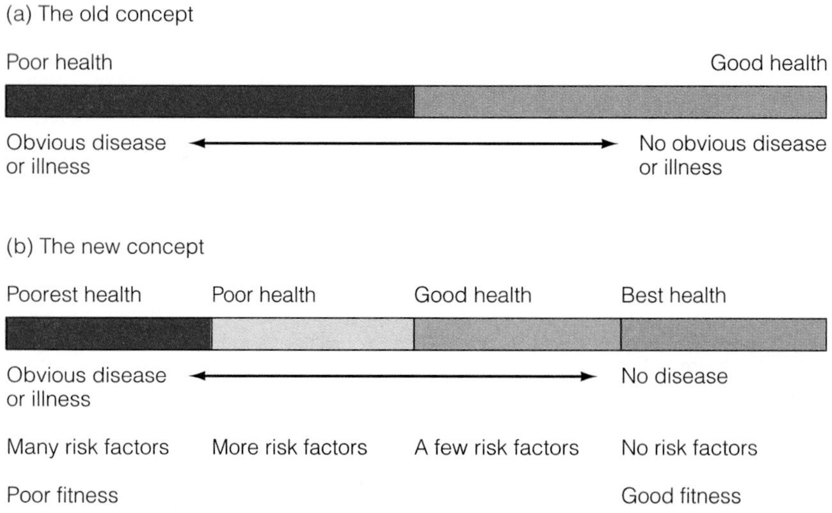

(a) The old concept

Poor health   Good health

Obvious disease ⟵————————————⟶ No obvious disease
or illness   or illness

(b) The new concept

Poorest health   Poor health   Good health   Best health

Obvious disease ⟵————————————⟶ No disease
or illness

Many risk factors   More risk factors   A few risk factors   No risk factors

Poor fitness   Good fitness

FIGURE 1-4 Old and New Concepts of Health

TABLE 1-1

Healthy Habits

Sleep seven to eight hours per day

Eat breakfast regularly

Eat a balanced diet

Avoid snacking on junk food (sweets or fatty foods) between meals

Maintain ideal weight

Do not smoke

Avoid alcohol or use it moderately

Exercise regularly

Manage stress in your life

short of breath. (If you can't, it may be time to seek counseling and start an exercise program.)

Maintaining good health is a lifelong job. TABLE 1-1 lists numerous healthy habits. By incorporating these habits into your lifestyle, you can increase your chances of living a long, healthy life.

Health Is Dependent on Properly Functioning Homeostatic Mechanisms

As pointed out earlier, physical health depends on properly functioning homeostatic mechanisms—that is, regulatory controls in the body that help maintain homeostasis. When these controls function improperly or break down completely, illness results. Persistent stress, for example, can disrupt several of the body's homeostatic mechanisms, leading to disease.

Stress Results in Disease by Disrupting Homeostatic Mechanisms

Stress is prevalent in many of our lives, so let's examine it more carefully. Most people cope with infrequent or short-term stress fairly well. Stress causes little harm when it occurs infrequently. If it is prolonged, however, stress can increase the risk of cardiovascular disease (diseases of the heart and arteries). Persistent stress may also increase the risk of ulcers and weaken the immune system. In addition, it may increase the likelihood of developing mental disorders. Fortunately for us, stress can be alleviated by exercise, relaxation training, massage, acupuncture, and other measures discussed in Health Note 1-1.

This discussion is not meant to imply that all diseases result from homeostatic imbalance. Some are produced by genetic defects; others are caused by bacteria or viruses. But even in these instances, diseases often result from an upset in regulatory mechanisms of the body that disrupt homeostasis. An excellent example is acquired immune deficiency syndrome, or AIDS. AIDS is caused by a virus that attacks certain cells of the immune system. This, in turn, results in a reduction in a key protective mechanism of the body, which is vital to homeostasis. In other diseases, temporary upsets in homeostasis may make us more susceptible to infectious agents. According to a recent study, people under stress are twice as likely to suffer from colds and the flu as those who are not.

Section 1-2

Evolution: The Unity and Diversity of Life

Homeostasis is a central theme of this book because it is so essential to life and is now threatened by modern culture. Another key concept of biology and a subtheme of this book is **evolution.** A few words on the subject are essential to your understanding of human biology and the unique evolutionary predicament we're in.

All life-forms alive today exist because of evolution. In fact, every cell and every organ in the human body is a product of millions of years of evolutionary refinement. Even the intricate homeostatic mechanisms described in the previous section evolved over long periods.

FIGURE 1-5 shows the five major groups or **kingdoms** of organisms that exist today. This simple diagram also provides an overview of evolution. The simplest, bacterialike organisms, belonging to a group called the **monerans,** were the first to evolve. They gave rise to a more complex set of organisms, known as the **protistans.** The protistans consist of single-celled organisms such as amoebas and paramecia. During the course of evolution, the protistans gave rise to three additional groups: plants, fungi, and animals, the kingdom to which we humans belong.

This common lineage is responsible for the striking similarities among the Earth's organisms, even in remarkably different organisms. For example, all organisms rely on the same type of genetic material. Similarities also exist on other levels besides the biochemical one. A comparison of certain anatomical features, such as the bones in a person's arm and the wings of birds, reveals an eerie resemblance.

Humans Are Similar to Other Organisms in Many Ways

An analysis of living things turns up seven common features, typically referred to as the characteristics of life.

Health Note 1-1

Maintaining Balance: Reducing Stress in Your Life

Heart disease

www.jbpub.com/humanbiology

Hard as you try, you can't avoid stress; it's a normal occurrence in everyday life.

Most writings on the subject define **stress** as a mental or physical tension. It's what we feel when we're exposed to stressful situations. Physical danger, for instance, may result in a feeling of stress. Humans have evolved mechanisms that deal very successfully with danger. But non-lifethreatening situations, like a blind date or a final exam, can also evoke feelings of stress.

Stress can be real or imagined. Either way, it elicits the same response in the body: an increase in heart rate, an increase in blood flow to the muscles and a decrease in blood flow to the digestive system, a rise in blood glucose levels, and a dilation of the pupils. All of these physical changes in the body help prepare us to respond to the stress. Once the stimulus is gone, though, the body returns to normal. That is, it is said to recover.

So the first thing we must recognize about stress is that the body doesn't discriminate between stressors—the sources of the stress. The response is always the same.

In pondering the effects of stress on the human body, two considerations are paramount: the duration of the **stressor** (the source of the stress response) and our ability to cope with stress.

In general, the source, be it physical or mental, may be short-lived (acute) or long-term (chronic)—or somewhere in between. Long-term exposure is of greatest concern, for it prevents recovery.

A prolonged period of stress may lead to disease, for during the state of stress, the body's immune system—the system that protects us from bacteria and other microorganisms that lead to disease—is depressed. Thus, prolonged immune-system suppression may make a person more susceptible to infectious disease—bacteria and viruses that cause colds and flu and other diseases. In addition, during a prolonged period of stress, changes occur in the blood vessels that accelerate the accumulation of cholesterol, which clogs the arteries and may eventually result in strokes and heart attacks.

Studies show that severe stress, if prolonged, can cause great physical harm. It may interfere with performance at home, at work, and at school.

Some argue that a little stress may actually help improve a person's performance. Healthy people recognize the stress they're feeling and channel its energies into productive work. This leads us to the second consideration: how people cope with stress.

A study of upper-echelon executives in American corporations shows that, despite the daily stresses and strains of the corporate pressure cooker, these people live longer, on average, than people of the same sex in the general population. Why? The answer is that they handle stress well. Some psychologists believe that these people have a sense of being in control, despite the stress of their work. They have clear objectives and meet them with a strong sense of purpose. Ultimately, such people view their jobs as a challenge, not a threat.

Unfortunately, not all people are so lucky. Studies suggest that many people are not in a position of control; they feel expendable and often view themselves as victims. For them, stress is a bad thing and may be associated with many severe and chronic diseases.

To a large degree, then, the effect stress has on us depends on the duration of exposure and our attitudes: what we think about ourselves and our environment. Researchers are finding that stress turns against us when we feel insignificant, powerless, and overwhelmed.

What can be done to deal with stress? Two basic strategies exist. First, it's important to select an environment and create a lifestyle that minimize stress. For some people, that means living in the country, rather than a crowded city. For others, it may mean choosing a major in college that suits their personality. The possibilities are limitless.

But this strategy may not be possible. You may have to be in a stressful environment. In this case, it's up to you to take direct action to learn to cope with stress or lessen its impact. Coping with stress requires mental strategies. Lessening its impact requires physical strategies. Let's consider the physical strategies first.

One of the easiest routes is to lessen the impact of stress through exercise. Studies show that a single workout at the gym, a ride on your bicycle, a vigorous hike, or a day of cross-country skiing reduces tension for two to five hours. Even better is a regular exercise program, which reduces the overall stress in your life. An individual who is easily stressed may find that stress levels decline after several weeks of exercise.

Exercise can be supplemented by relaxation training, another physical strategy. As you prepare for a difficult test or get ready for a date that you

are nervous about, tension often builds in your muscles. Periodically stopping to release that tension helps you reduce physical stress. For some people, getting up and stretching or taking a walk can reduce the tension. Others find it useful to tighten their muscles forcefully, then let them relax. Massage therapy and acupuncture have also been used to successfully reduce stress. Cassette tapes can teach you relaxation methods. If that's not for you, consider signing up for stress-reduction classes from a trained therapist. The more you practice relaxation, the better you will become at relieving tension.

Stress from within is often blown out of proportion. By dealing with the thoughts that exaggerate your stress, you can help eliminate stress before it begins. This is a mental strategy. Start paying attention to the thoughts that provoke anxiety in your life. Are they exaggerated? If so, why? For example, are you nervous before exams? Why? Do you always study adequately? Could you prepare better? Would better preparation reduce your anxiety? Finding the source of your anxiety and taking positive action to alleviate it are helpful ways of reducing stress.

Stress reduction is not always so easy. Test anxiety, for example, may be deeply rooted in feelings of insecurity and inadequacy. Many people struggle with low self-esteem their entire life. A trained psychologist can help you find the roots of your problem and assist you in learning to feel better about yourself. Psychological help is as important as medical help these days. Given the complexity and pace of our society, there is no shame in seeking counseling.

Biofeedback is another form of stress relief. A trained health care worker hooks you up to a machine that monitors heart rate, breathing, muscle tension, or some other physiological indicator of stress (**FIGURE 1**). During a biofeedback session, your trainer will help you relax, and then perhaps discuss a stressful situation. When one of the indicators shows that you are suffering from stress, a signal is given off. Your goal is to consciously reduce the frequency of the signal. For example, if your heart started beating faster when you thought about taking an exam, a clicking sound might be heard. By deep breathing and relaxing, you consciously slow down your heart rate, and the clicking sound slows down, and then disappears. Learning to recognize the symptoms of stress and to counter them is the goal of biofeedback. Eventually, you should be able to do it without the aid of a machine.

You can also reduce tension by managing your time and your workload efficiently. Numerous books on this subject can help you learn to budget your time more effectively. Efficient time management alone can often keep you from feeling stressed.

It is important to challenge yourself in college, but be realistic in what you expect of yourself. If you must work, for example, sign up for a class load that you can handle. Even if you are not working, take a reasonable class load, and be sure to exercise regularly and get plenty of sleep.

Relieving stress in our lives helps us reduce the risk of cardiovascular disease. It helps us relax and enjoy life. Lest we forget, it also makes us more pleasant to be around. All in all, it is best to start learning early in life how to reduce or cope with stress. Lessons learned now will be useful for years to come.

Visit *Human Biology's* Internet site, www.jbpub.com/humanbiology, for links to web sites offering more information on this topic.

www.jbpub.com/humanbiology

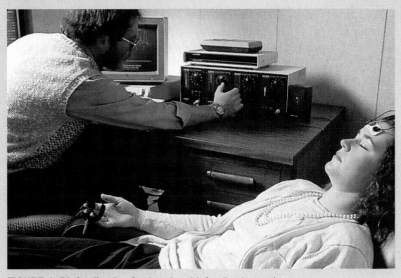

FIGURE 1 Biofeedback Student in a biofeedback session.

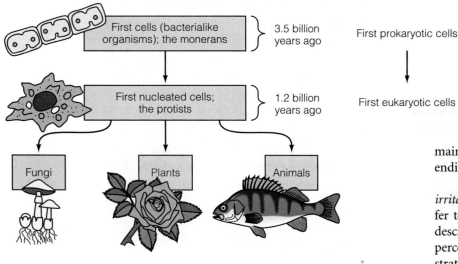

First cells (bacterialike organisms); the monerans — 3.5 billion years ago — First prokaryotic cells

First nucleated cells; the protists — 1.2 billion years ago — First eukaryotic cells

Fungi   Plants   Animals

FIGURE 1-5 Evolution of the Prokaryotic and Eukaryotic Cells This diagram illustrates the evolutionary history of life. Organisms fall into one of five major groupings, or kingdoms. The first life-forms were the monerans, which are single-celled prokaryotic organisms. They gave rise to the protistans, single-celled eukaryotic organisms. Eukaryotic protistans, in turn, gave rise to plants, fungi, and animals.

A brief look at these not only shows our evolutionary connection with other organisms, but helps us understand people better.

The first characteristic of life is that all organisms, including humans, are made of cells, tiny structures that are the fundamental building block of living things. Cells, in turn, consist of molecules, nonliving particles composed of smaller units called **atoms.** Glucose molecules, for instance, contain carbon, hydrogen, and oxygen atoms. Molecules, in turn, combine to form the parts of the cell.

The second characteristic of life is that all organisms grow and maintain their complex organization by taking in molecules and energy from their surroundings. As you will see in subsequent chapters, humans must expend considerable amounts of energy to maintain their complex internal organization. In fact, 70%–80% of the energy that adult humans need is used just to maintain our bodies—to transport molecules in and out of cells, to synthesize proteins, and to perform other basic body functions. The rest is used for activity—walking, running, talking, and so on.

The third characteristic of life is that all living things exhibit a feature called metabolism. **Metabolism** refers to the chemical reactions in organisms. These reactions consist of two types—those in which food substances are built up into living tissue (anabolic reactions) and those in which food is broken down into simpler substances, often releasing energy (catabolic reactions). In human body cells, thousands of reactions occur each second to maintain life.

The fourth characteristic of life is homeostasis. All organisms rely on numerous homeostatic mechanisms. Vital in the constantly shifting world of many organisms,

maintaining a constant internal environment is a never-ending task.

The fifth characteristic common to all forms of life is irritability. In biological parlance, irritability does not refer to grouchiness, although some of us clearly fit this description. Rather, **irritability** refers to the ability to perceive and respond to stimuli. A young girl demonstrates this vital property when she withdraws her hand from a hot stove or closes her eyes in response to bright light.

The ability to perceive and respond to stimuli allows all organisms to respond to their environment and is, therefore, an important survival tool. Not surprisingly, evolution has resulted in some rather remarkable ways by which organisms both detect and respond to stimuli. Some of the most impressive ones occur in human beings.

The sixth characteristic of life is reproduction and growth. All organisms are capable of reproduction and growth. Two types of reproduction are known to exist: sexual and asexual. **Sexual reproduction** occurs in organisms that produce offspring by combining sex cells, sperm and eggs. In vertebrates such as birds and mammals, these cells come from male and female individuals.

Asexual reproduction is common in single-celled organisms such as the amoeba (**FIGURE 1-6A**). In these and other single-celled organisms, reproduction generally occurs by simple cell division of a parent cell. When it divides, it forms two identical offspring.

Some multicellular organisms reproduce asexually by budding. **Budding** is a process in which new individuals are produced from small outgrowths called **buds.** Tiny buds enlarge and soon develop into a fully formed organism. **FIGURE 1-6B** shows how an aquatic organism known as a hydra buds. Like cell division in unicellular organisms, budding produces generation after generation of identical offspring.

The seventh characteristic of life is that humans and all other organisms are the products of evolution. Evolution is a process that leads to structural, functional, and behavioral changes in species, known as **adaptations.** Favorable adaptations increase an organism's chances of survival and reproduction. Evolutionary change occurs

(a)

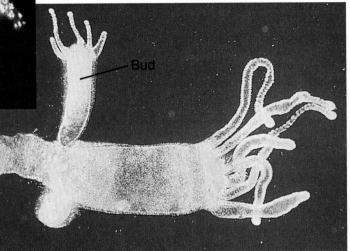

Bud

(b)

FIGURE 1-6 Asexual Reproduction *(a)* Asexual reproduction takes place when an amoeba divides, producing identical offspring. *(b)* Small cellular buds in some organisms like this hydra enlarge to produce genetically identical offspring, which eventually break free from the parent.

through the interplay of genetic variation and environment. Permit me to explain what this means.

Genetic variation refers to naturally occurring genetic differences in various members of a population, groups of genetically similar organisms. These differences result in variations in structure, function, and behavior of organisms—that is, adaptations. Those that give members of a population an advantage over others in survival and reproduction tend to persist.

Charles Darwin, the nineteenth century British naturalist who spent much of his life studying evolution, proposed a **theory of natural selection,** which explains the role of the environment in evolution. In essence, he found that the environment acts as a selective agent, weeding out organisms less able to survive.

Darwin described natural selection as a process in which slight variations, if useful, are preserved. As a result, natural selection is a process by which organisms become better adapted to their environment. Because organisms endowed with beneficial variations are more likely to survive and reproduce, they pass on the favorable genetic material. Over time, the genetic composition of the species may change. Individuals of a species may become better camouflaged or faster and thus better able to escape being eaten or to capture prey.

The details of evolution are discussed in Chapters 21 and 22. It is important to point out here, however, that individuals do not evolve, only populations do. Put another way, although evolution results from the emergence of adaptations in individuals, it occurs only when there is a shift in the genetic composition of a population. Profound shifts can result in the formation of entirely new species.

Humans Are One Form of Life on Earth and Have Many Unique Characteristics

Sobering as it may seem, humans are just one of evolution's many delightfully interesting products. Nonetheless, humans are also a unique form of life. Several features distinguish us from other species.

One of the key differences between humans and other animals is our ability to acquire and use complex languages. Another distinguishing feature is our culture. Culture has been defined in many ways. Humorist Will Cuppy remarked that culture is anything we do that monkeys don't. **Culture** can be defined as the ideas, values, customs, skills, and arts of a given people in a given time. While other species may have some rudiments of culture, for instance, some communication skills, humans are unique in the biological world because of the complexity and splendor of our cultural achievements.

Humans differ in other ways as well. One noteworthy difference is our ability to plan for the future. Although a few other animals seem to share this ability, most of what appears to be planning is probably the result of instinct. A bird's nest-building activities, for example, are programmed by its genetic material and are probably not the result of conscious planning. In contrast, building skyscrapers and launching rockets require an extraordinary amount of forethought.

Another unique characteristic of humans is our unrivaled ability to reshape the environment. The ability of our minds to shape images and the ability of our hands to translate those images into reality have given us a power almost beyond our control. Despite the benefits of our remarkable technologies, attempts to control nature sometimes backfire, creating larger problems. Efforts to control upstream flooding on the Mississippi River, for instance, have led to more frequent floods and more damage downstream, as witnessed by the floods of

1993 (**FIGURE 1-7**). By building levees along the river, upstream communities prevent water from spilling over their banks, but this protection delivers a larger slug of water to downstream sections.

The marvelous human achievements over the past 200 years have led many to think of our species as the crowning accomplishment of nature. This view, however, can be dangerously misleading. It makes us think of ourselves as separate from nature and immune to its laws, which we are not. It also leads many to believe that nature (the environment) is at our disposal and that we can do with it as we like.

To live on the planet without destroying the natural world around us, disrupting evolution, and ruining the prospects of the human race requires a new way of thinking and acting, one that fosters a cooperative relationship between humans and nature. Barry Commoner, a biologist, put it best when he said that "nothing can survive on the planet unless it is a cooperative part of a larger, global life." It is our challenge to find a way to build such a way of life. Although not the main focus of this book, a few seminal ideas are presented here, especially in the Health, Homeostasis, and the Environment sections and in Chapters 23 and 24.

FIGURE 1-7 Human Control of Nature   The use of earthen walls, or levees, along the Mississippi River has increased considerably in the past hundred years in an effort to stop flooding. Levees and other measures to control the river have led to more frequent flooding, as explained in the text.

Science, Critical Thinking, and Social Responsibility

Over the millenia, human societies have accumulated enormous amounts of information about the world around us. The systematic study of the universe and its many parts today falls into the realm of science.

The term *science* comes from the Latin word *scientia*, which means "to know" or "to discern." Today, **science** is defined as knowledge derived from observation, study, and experimentation. It is also a method of accumulating knowledge. In other words, it involves ways of learning facts as well as an immense body of knowledge.

Many people I talk to seem to view science as a dull and uninteresting endeavor best left in the hands of a select few. In reality, science is an exciting endeavor that often involves enormous creative energy. Because it teaches us about the workings of the world around us, it can be a source of great fascination. As the sometimes zany paleontologist Robert Bakker, a consultant to the company that made the dinosaurs for the movie *Jurassic Park,* once noted, science is "fun for the mind."

Science also has a practical side, as pointed out earlier in this chapter. It provides information that improves our lives in many ways. It helps us understand fascinating and important phenomena such as the weather and the spread of disease. A knowledge of science makes us better voters, better able to discern fact from fiction in a political debate. An understanding of science can help us decide on the safest forms of energy to meet future demands. And an understanding of human biology can help us make informed health care decisions.

The Scientific Method Generally Involves Observations, Hypotheses, and Experiments

The methods of science are not as foreign as you might think. In fact, most of us use the scientific method nearly every day of our lives. When your car fails to start, for example, or when your computer acts up, you very likely engage in the type of thinking that scientists use to gather information and test ideas. This process, called the **scientific method,** is summarized in **FIGURE 1-8**. As shown, the scientific method generally begins with observations and measurements. In some cases, these may be part of carefully conducted experiments. Others may occur more haphazardly. A scientist on vacation in the tropics, for instance, may notice a phenomenon that sparks her curiosity and leads to in-depth studies—and an excuse to return to the tropics for a research expedition!

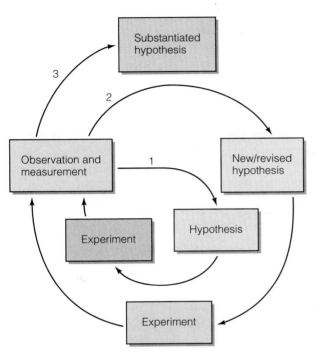

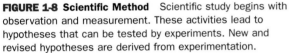

FIGURE 1-8 Scientific Method Scientific study begins with observation and measurement. These activities lead to hypotheses that can be tested by experiments. New and revised hypotheses are derived from experimentation.

To see how the scientific method works, let us look at a familiar example. Suppose you sat down at a computer, turned on the switch, but nothing happened. You might also have noticed that the lights in the room hadn't come on. These two observations would lead to a **hypothesis** (pronounced high-POTH-eh-siss), a tentative explanation of the phenomenon. From your observations, you might hypothesize that the electricity in your house was cut off.

You would then test your hypothesis by performing an **experiment.** In this case, you wouldn't need a $500,000 research grant; all you would need to do would be to traipse into the kitchen and try the light switch. If the lights in the kitchen worked, you would reject your original hypothesis and form a new one. Perhaps, you hypothesize, the circuit breaker to your study had been tripped. To test this idea, you would "perform" another experiment, locating the circuit breaker to see if it was turned off. If the circuit breaker was off, you would conclude that your second hypothesis was valid. To substantiate your conclusion, you would throw the switch, and then test your computer and the room light.

This simple process involving observations and measurements, hypothesis, and experimentation forms the foundation of the scientific method. Although scientific experimentation may be much more complicated

than discovering the roots of a computer failure, the process itself is the same.

■ Theories Are Broad Generalizations Based on Many Experimental Observations

Scientific method leads to the accumulation of scientific facts (really, tested hypotheses). Over time, as our knowledge accumulates, scientists are able to assemble knowledge and gain broader understanding of the way the world works. These broad generalizations are known as **theories.**

Theories are supported by numerous facts that have been established by careful observation, measurement, and experimentation. Unlike hypotheses, theories cannot be tested by single experiments because they encompass many bits of information. Atomic theory, for instance, explains the structure of the atom and fits numerous observations made in different ways over many decades.

A theory commands respect in science because it has stood the test of time and survived. This does not mean, however, that theories are always correct. As history bears out, numerous theories have been modified or discarded as new scientific evidence was gained. Even widely held theories that have persisted for hundreds of years have been overturned. In 140 A.D., for example, the Greek astronomer Ptolemy (TAL-eh-me) proposed a theory that placed Earth at the center of our solar system. This was called the **geocentric theory.** For nearly 1500 years, the geocentric view held sway. Many astronomers vigorously defended this position while ignoring observations that did not fit the theory. In 1580, the astronomer Nicolaus Copernicus dared to expound a new theory—the **heliocentric view,** which places the sun at the center of the solar system. His work stimulated considerable controversy, but it eventually prevailed because it better fit the observations.

Because theories may require modifications or rejection, scientists must be open-minded and willing to analyze new evidence that throws into question their most cherished beliefs. For the most part, though, theories are talked about as if they were fact. Some people even object to calling a theory a "theory" for fear that it sounds tentative.

Another word about theories before we move on. The word *theory* is commonly misused in everyday conversation. A friend, for instance, might say, "My theory about why Jane missed the party is that she didn't want to see her ex-boyfriend." Jane's feelings aside, this is hardly worthy of the status of a theory. What your friend really meant was his "hypothesis," for his explanation was truly a tentative explanation that could be tested by experimentation.

Science Helps Shape Our Lives and Our Values

Science and the scientific process are essential to modern existence. We wouldn't have the microwave oven or compact disc if it weren't for science. Science can also influence political decisions regarding health care, environmental protection, and a host of other issues.

Many decisions in the public-policy arena, however, are not made solely on the basis of scientific facts. Rather, they are heavily influenced by values—what we view as right or wrong—and economic needs. When human values are framed in the absence of scientific knowledge, however, they can lead to a lopsided way of viewing the world. The political and economic decisions that emerge may be fundamentally flawed.

Although many people do not realize it, science can even influence human values. Environmental values, for instance, are influenced by information gained from the study of ecology. Ecology helps us understand the interconnectedness of living things. It helps uncover relationships that are not obvious to most people, such as the role of bacteria in recycling nutrients. In widening our understanding of the relationships among living organisms and the environment, science helps us understand our dependence on other species and thus helps us act more thoughtfully and compassionately. It widens our ethical and cognitive boundaries, assisting us in living sustainably on the Earth.

Critical Thinking Helps Us Analyze Problems, Issues, and Information More Clearly

Another benefit of your study of science is that it can help you learn to think more critically. **Critical thinking** is not being "critical" or judgmental. Rather, it's a process that allows one to objectively analyze problems, facts, issues, and information. Ultimately, critical thinking permits people to distinguish between beliefs (what we believe to be true) and knowledge (facts that are well supported by research). In other words, critical thinking is a process by which we separate judgment from facts. It is our most ordered kind of thinking. It is not just thinking deeply about a subject, although that is necessary. Critical thinking subjects facts and conclusions to careful analysis, looking for weaknesses in logic and other errors of reasoning. Critical thinking skills, therefore, are essential to analyzing a wide range of problems, issues, and information.

TABLE 1-2 summarizes 11 critical thinking rules that will come in handy as you read the newspaper, watch the news, listen to speeches, and study new subjects in school. Here is a brief description of each rule.

TABLE 1-2
Critical Thinking Rules

When analyzing an issue or fact, you may find it useful to employ these rules:

1. Gather complete information.

2. Understand and define all terms.

3. Question the methods by which data and information were derived:
   Were the facts derived from experiments?
   Were the experiments well executed?
   Did the experiment include a control group and an experimental group?
   Did the experiment include a sufficient number of subjects?
   Has the experiment been repeated?

4. Question the conclusions:
   Are the conclusions appropriate?
   Was there enough information on which to base the conclusions?

5. Uncover assumptions and biases:
   Was the experimental design biased?
   Are there underlying assumptions that affect the conclusions?

6. Question the source of the information:
   Is the source reliable?
   Is the source an expert or supposed expert?

7. Don't expect all of the answers or complete information.

8. Examine the big picture.

9. Look for multiple causes and effects.

10. Watch for thought stoppers.

11. Understand your own biases and values.

The first rule of critical thinking is: gather complete information. Critical thinking requires facts. Don't sit back and accept everything you read and hear. If you do, you may at times run the risk of learning only what the media or biased special-interest groups want you to think. Question what you learn, and seek other sources of information. Gathering lots of information keeps you from falling into the trap of mistaking ignorance for perspective. By keeping an open mind to alternative views and continually being on the lookout for new facts, you can develop an enlightened viewpoint.

The second rule of critical thinking is: understand and define all terms. Critical thinking requires a clear understanding of all terms. As you study biology, or most any subject, you will encounter hordes of new terms. Although sometimes tedious, learning new terms is essential. Without a clear understanding of the terminology, you will not be able to master this or any other subject.

Understanding terms and making sure that others define them in discussions also bring clarity to issues and debates. The Greek philosopher Socrates, in fact, destroyed many an argument in his time by insisting on clear, concise definitions of terms. As you analyze any information or issue, always be certain that you understand the terms, and make sure that others define their terms as well.

The third rule of critical thinking is: question the methods by which the facts are derived. In science, many debates over controversial topics hinge on the methods used to discover new information. The first question you should ask is: was the information gained from careful experimentation, or is it an offshoot of casual, unscientific observations?

Proper experimentation in biology usually requires two groups: experimental and control. The **experimental group** is the one that is tested or manipulated in some way. The **control group** is not. Valid conclusions come from such comparisons because, in a properly run experiment, both groups are treated identically except in one way. The difference in treatment is technically known as the *experimental variable.* Consider an example to illustrate this point. In order to test the effect of excess vitamin ingestion on laboratory mice, a good scientist would start with a group of mice of the same age, sex, weight, genetic composition, and so on. These animals would be divided into two groups, the experimental and control groups. Both groups would be treated the same throughout the experiment, receiving the same diet and being housed in the same type of cage at the same temperature. The only difference between the two should be the vitamin supplements given to the experimental group. Consequently, any observed differences between the groups could be attributed to the treatment (the experimental variable).

Besides having an experimental group and a control group, good experimentation requires an adequate number of subjects to ensure that any observed differences are real. Individual variation is natural. As a rule, the smaller the number of animals in each group, the less reliable the data because of variation. In laboratory experiments, at least 10 test animals are required for reliable statistical analysis. Groups larger than 10 are even better. For human health studies, much larger groups are generally used.

As you analyze new facts, first check to see how they were derived. If the results were obtained from experimentation, were the experiments well planned and executed? Did the experiment have a control group? Were the control and experimental groups treated identically except for the experimental variable? Did the experi-

menters use an adequate number of subjects? Even if all of these conditions are met, beware. In science, one experiment is rarely adequate to permit one to draw firm conclusions. Careful scrutiny, for example, may show small but significant design flaws: perhaps the mice being tested were resistant to the drug under study or were hypersensitive to it. As a rule of thumb, then, wait for scientific verification of the results. A second researcher may repeat the experiment with similar results. In some cases, a new researcher may find different results.

Nowhere is caution more necessary than when one encounters announcements of scientific breakthroughs on the television news and in magazine and newspaper articles. Ever eager to showcase new scientific studies *before* they have been verified by others, the media sometimes does a grave disservice to the advancement of scientific knowledge because further study shows earlier results to be invalid. The cold fusion debacle in the 1980s is a classic example. Unfortunately, the media often fail to publish the results from follow-up studies that contradict early results. Ultimately, the public is left with a false impression of reality.

The fourth rule of critical thinking is: question the conclusions derived from facts. Surprisingly, even if an experiment is run correctly, there's no guarantee that the conclusions drawn from the results will be correct. How can that be? The answer may lie in bias, ignorance, and error. Bias refers to beliefs that taint the interpretation of results. Ignorance is a lack of full knowledge. This, in turn, may lead a scientist to misinterpret his or her results. Finally, error does occur, in spite of our best efforts.

Two questions should be asked when one analyzes the conclusions of an experiment: (1) Do the facts support the conclusions, and (2) are there alternative conclusions? An example will illustrate the importance of these questions.

One of the earliest studies on lung cancer showed that people who consumed large quantities of table sugar (sucrose) had a higher incidence of lung cancer than those who ate moderate amounts. The researchers concluded that lung cancer was caused by sugar. Did the facts support this conclusion? A reexamination of the patients showed that the group with lung cancer had a noticeably higher percentage of smokers. It turns out that smokers consume more sugar. Thus, the association between sugar and lung cancer is probably false.

This example illustrates an important principle of biological and medical research: correlation does not necessarily mean causation. In other words, two factors that appear to be related may, in fact, not be linked at all.

The fifth rule of critical thinking is: look for assumptions and biases. This rule is related to the previous rule,

Scientific Discoveries that Changed the World 1-1

Debunking the Theory of Spontaneous Generation

FEATURING THE WORK OF ARISTOTLE, REDI, AND PASTEUR

The Greek philosopher and scientist Aristotle, who lived from 384 to 322 B.C., proposed a theory to explain the origin of living things. It was called the **theory of spontaneous generation.** It asserted that living things arose spontaneously from innate matter. Mice, he believed, arose from a pile of hay and rags placed in the corner of a barn. Flies could be produced by first killing a bull, then burying it with its horns protruding from the ground. After several days, one of the horns would be sawed off and flies would emerge. People, it was said, emerged from a worm that developed from the slime in the bottom of a mud puddle.

As absurd as the idea of spontaneous generation sounds to us today, the view remained compelling to many scientists well into the nineteenth century, despite observations that contradicted the theory, such as the phenomenon of childbirth itself.

Debunking the theory of spontaneous generation engaged some of the best scientific minds of the day. Although many scientists were involved in gunning down the theory of spontaneous generation, two promi-

nent scientists, Francesco Redi and Louis Pasteur, played pivotal roles.

Redi, an Italian naturalist and physician, was one of the first scientists to refute spontaneous generation through experimentation. Around 1665, Redi performed a simple but compelling experiment to determine whether ordinary houseflies were spontaneously generated in rotting meat. He began by placing three small pieces of meat in three separate glass containers. The first one was covered with paper. The second was left open, and the third was covered with gauze. Left at room temperature, the meat quickly began to rot and attract flies. Soon, the meat in the open container began to seethe with maggots, larvae hatched from fly eggs. The paper-covered container showed no evidence of maggots, nor did the meat in the gauze-covered container, although maggots did appear in the gauze itself. Redi's conclusion from this experiment was that maggots (which give rise to flies) do not come from the meat itself but from the eggs deposited by flies.

Redi's experiments convinced many people that flies and other organisms did not arise by spontaneous generation, but this did not put the debate to rest. Soon after Redi's

now-famous experiment, in fact, a Dutch linen merchant by the name of Anton Leeuwenhoek discovered bacteria using simple microscopes he had built. This discovery revived arguments for spontaneous generation on the microscopic level. Many scientists asserted that although flies and other organisms did not arise spontaneously, microorganisms probably did—testifying to the allure of ingrained ideas. As evidence, they cited studies showing that microorganisms could arise from boiled extracts of hay or meat. According to die-hard proponents of spontaneous generation, nonliving plant and animal matter was thought to possess a vital, or life-generating, force that could give rise to microorganisms.

The debate over spontaneous generation persisted for the next 200 years even though a number of studies showed that a sterilized medium sealed from the outside air remained free of bacteria, disproving spontaneous generation. But supporters of spontaneous generation claimed that these experiments were flawed. They argued that experimenters had eliminated some vital force needed to give rise to new life.

In 1861, Louis Pasteur published the results of an experiment that

questioning conclusions, but is so important that it warrants closer examination. Biases and hidden assumptions are to thinking what cyanide is to food, a poison. Unfortunately, biases and hidden assumptions run rampant in the media.

In many contemporary debates over a wide range of issues, proponents often present information that supports their point of view. This selective inclusion of supportive data and exclusion of contradictory information is often an expression of a hidden agenda. What

often happens is that people make up their mind about an issue, then seek out information that supports their point of view. Several of the Scientific Discoveries boxes presented in this book illustrate the pervasive nature of biases and hidden assumptions.

The sixth rule of critical thinking is: question the source of the facts—that is, who is telling them. Closely related to the rule on assumptions and biases, this rule calls on us to question the people behind various research studies or various positions, in part, because they

helped put this debate to rest. He placed sterilized broth in a sterilized swan-necked flask, a flask with a long curved neck (**FIGURE 1**). The design of the flask permitted air to enter, eliminating criticism that he had destroyed any vital forces necessary for spontaneous generation, but it blocked airborne bacteria from entering. (Airborne bacteria probably deposited on the tube leading to the flask and were prevented from entering the broth.) In his experiments, Pasteur clearly showed that bacteria could not arise spontaneously. Only when the broth was open to the air did bacteria emerge.

Besides helping to put an end to the debate about the origin of living organisms, Pasteur's work helped lay the foundation for the modern understanding of infectious disease. Before his simple but convincing experiments, some scientists argued that the diseases we now call infectious were caused by a malfunction in one or more body parts. This malfunction, they claimed, resulted in the production of poisons that caused the illness. The microbes present in the diseased individuals, they went on to say, arose spontaneously as a result of the malfunction and were not the cause of the disease per se.

This brief history points out three important lessons. First, scientific discovery is usually the result of many scientists, each working on different parts of the puzzle. Often, though, one scientist is credited for an important discovery that is, in reality, built on a foundation of scientific research by many others. Second, this brief discussion also shows how discoveries open up new ways of thinking. Third, it illustrates the persistence of ideas that shape the way we think and the resistance people often exhibit to new ideas even in the face of contradictory evidence.

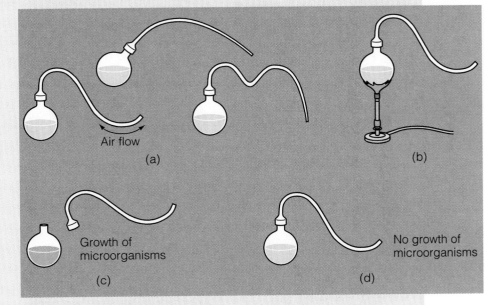

FIGURE 1 Pasteur's Experiment Pasteur's simple but elegant experiment helped debunk the theory of spontaneous generation. *(a)* These specially designed swan flasks allowed air to enter, but prevented bacteria from entering the broth. *(b)* The broth was boiled to kill microorganisms. *(c)* Microorganisms appeared only if the flasks' necks were removed. *(d)* No microbes grew if the neck remained intact.

may have performed their experiments incorrectly or they may have drawn erroneous conclusions because they were influenced by bias.

Sometimes a study of the biographies of the people delivering the information is as instructive as an examination of their conclusions. An association with a partisan group may be a red flag, warning that bias may have influenced their conclusions. So, beware of "experts" who have a hidden agenda or such a narrow focus that they miss the big picture. So-called experts from industry who swear under oath to the safety of their product are likely to be biased or even deceitful.

Also beware of people who may not know as much as you think they should. Although we think of physicians as experts on human health, most of them received little or no training in nutrition in medical school. Many medical students still graduate without a full understanding of the role of nutrition in preventing disease and promoting good health.

Point/Counterpoint

Controversy over the Use of Animals in Laboratory Research

ANIMAL RESEARCH IS ESSENTIAL TO HUMAN HEALTH
Frankie L. Trull

Frankie L. Trull *is president of the Foundation for Biomedical Research, a nonprofit educational organization dedicated to informing the public about the importance of humane animal research.*

Virtually every major medical advance of the last 100 years—from chemotherapy to by-pass surgery, from organ transplantation to joint replacement—has depended on research with animals. Animal studies have provided the scientific knowledge that allows health care providers to improve the quality of life for humans and animals by preventing and treating diseases and disorders and by easing pain and suffering.

Some people question animal research on the ground that data from animals cannot be extrapolated to humans. But physicians and scientists agree that the many similarities that exist provide the best insights into the complex systems of both humans and animals. Knowledge gained from animal research has contributed to a dramatically increased human life span, which has increased from 47 years in 1900 to more than 76.3 years in 1995. Much of this increase can be attributed to improved sanitation and better hygiene; the rest of this increased longevity is a result of health and medical advances made possible in part through animal research.

Research on animals has also led to countless treatments, techniques, and medical technologies. Animal research was indispensable in the development of immunization against many diseases, including polio, mumps, measles, diphtheria, rubella, and hepatitis. One million insulin-dependent diabetics survive today because of the discovery of insulin

and the study of diabetes using dogs, rabbits, rats, and mice. Organ transplantation, considered a dubious proposition just a few decades ago, has become commonplace because of research on mice, rats, rabbits, and dogs.

Animal research has contributed immeasurably to our understanding of tumors and has led to the discoveries of most cancer treatments and therapies. Virtually all cardiovascular advances, including the heart-lung machine, the cardiac pacemaker, and the coronary bypass, could not have been possible without the use of animals. Other discoveries made possible through animal research include an understanding of DNA; X-rays; radiation therapy; hypertension; artificial hips, joints, and limbs; monoclonal antibodies; surgical dressings; ultrasound; the artificial heart; and the CAT scan.

Animal research will be essential to medical progress in the future as well. With the use of animals, researchers are gaining understanding into the cause of—and treatments for—AIDS, Alzheimer's disease, cystic fibrosis, sudden infant death syndrome, and cancer in the hopes that these problems can be eliminated. Although many nonanimal research models have been developed, no responsible scientist believes that the technology exists today or in the foreseeable future to conduct biological research without using animals.

Despite distortions and exaggerations put forth by those opposed to animal research, occurrences of poor animal care are extremely rare. Researchers care about the welfare of laboratory animals. Like everybody else, scientists don't want to see animals suffer or die. In fact, treating animals humanely is good science. Animals that are in poor health or under stress will provide inaccurate data.

Many people are under the false impression that laboratory animals are not protected by laws and regulations. In fact, many safeguards are in place to guarantee the welfare of animals used in research. A federal law, entitled the Animal Welfare Act, stipulates standards for care and treatment of laboratory animals, and the U.S. Public Health Service (PHS), the country's major source of funding for biomedical research, sets forth requirements with which research institutions must comply in order to qualify for grants for any biomedical research involving *any* kind of animal.

Both the Animal Welfare Act and the PHS animal welfare policy mandate review of all research by an animal-care committee set up in each institution to ensure that laboratory animals are being used responsibly and cared for humanely. The committee, which must include one veterinarian and one person unaffiliated with the institution, has the power to reject any research proposal and stop projects if it believes proper standards are not being met.

Although animal research opponents portray the medical community as deeply divided over the merits of animal experimentation, the percentage of physicians opposed to animal research remains very small. A 1989 survey by the American Medical Association of a representative sample of all active physicians found that 99% believed animal research had contributed to medical progress, and 97% supported the continued use of animals for basic and clinical research.

The general public, when presented with the facts, has also been supportive of animal research. This support must not be allowed to erode through apathy or misconceptions. Should animal research be lost to the scientific community, the victims would be all people: our families, our neighbors, our fellow humans.

VIVISECTION: A MEDIEVAL LEGACY
Elliot M. Katz

Vivisection, an outdated and extremely cruel form of biomedical research, is the purposeful burning, drugging, blinding, infecting, irradiating, poisoning, shocking, addicting, shooting, freezing, and traumatizing of healthy animals. In psychological studies, baby monkeys are separated from their mothers and driven insane; in smoking research, dogs have tobacco smoke forced into their lungs; in addiction studies, chimpanzees, monkeys, and dogs are addicted to cocaine, heroin, and amphetamines, then forced into convulsions and painful withdrawal symptoms; in vision research, kittens and monkeys are blinded; in spinal cord studies, kittens and cats have their spinal cords severed; in military research, cats, dogs, monkeys, goats, pigs, mice, and rats die slow, agonizing deaths after being exposed to deadly radiation, chemical, and biological agents.

Started at a time when the scientific community did not believe animals felt pain, vivisection has left a legacy of animal suffering of unimaginable proportions. Descartes, the father of vivisection, asserted that the cries of a laboratory animal had no more meaning than the metallic squeak of an overwound clock spring. Though the research community considers vivisection a "necessary evil," a growing number of scientists and health professionals see vivisection as simply *evil*.

As a veterinarian, I was taught that vivisection was essential to human health. My eyes were finally opened to the full horrors and futility of vivisection when years later, faculty members and campus veterinarians at the University of California informed me that animals were dying by the thousands from severe neglect and abuse; that vivisectors and campus officials were denying and concealing the abuses; and that experiments were conducted whose only benefit was to the school's finances and researchers' careers. I discovered that animal "care" committees, typically cited as assurance that animals are used responsibly, are in fact "rubber stamp" committees composed mostly of vivisectors who routinely approve each other's projects. Over the years, I have witnessed an ongoing pattern of university officials denying documented charges of misconduct, attempting instead to discredit critics of vivisection, ultimately defending even the most ludicrous and cruel experiments as necessary and humane.

I discovered that assertions by the biomedical community that vivisection is an essential and indispensable part of protecting the public's health are simply untrue. Vivisection can and should be ended. It is scientifically outdated and morally wrong. There is a plethora of modern biomedical technology that can be used to improve society's health without harming animals. The advent of sophisticated scanning technologies, including computerized tomography (CT), positron emission tomography (PET), and magnetic resonance imaging (MRI), has given scientists the ability to examine people and animals noninvasively. This technology has isolated abnormalities in the brains of patients with Alzheimer's disease, epilepsy, and autism, revolutionizing diagnosis and treatment of these diseases. Tissue and cell cultures are being increasingly used to screen cancer and AIDS drugs. Progress with AIDS has come from areas entirely unrelated to animal experimentation. Human skin cell cultures are used to test new products and drugs for toxicity and irritancy.

Why, then, is vivisection so entrenched and defended with an almost religious fervor? Dr. Murry Cohen summed it up when he stated, "Change is difficult for most people, but it is particularly painful for scientific and medical bureaucracies, which fight to maintain the status quo, especially if required change might imply admission of previously held incorrect ideas or flawed axioms." Vivisection continues today because of vested interests, habit, economics, and legal considerations, not for the real advancement of science and public health.

When presented with the facts, members of the public almost unanimously express their desire to see an end to the horrors of vivisection. Thousands of professionals like myself have reevaluated the sense, efficacy, and worth of vivisection and have formed or joined organizations working to end this outdated and cruel form of research. The ending of vivisection will lead to improved public health and restore to medicine and science much needed excellence and compassion for all beings, human and nonhuman alike.

Elliot M. Katz, *DVM (here with companion, Manco) is a graduate of the Cornell University School of Veterinary Medicine. He is president and founder of In Defense of Animals, a national animal rights organization.*

SHARPENING YOUR CRITICAL THINKING SKILLS

1. Summarize the main points of each author.
2. Do these authors use data or ethical, anecdote (stories or experiences) arguments to make their cases?
3. Do you have a view on this issue? What factors weigh most heavily in making up your mind?

Visit *Human Biology's* Internet site, www.jbpub.com/humanbiology, to research opposing web sites and respond to questions that will help you clarify your own opinion. (See Point/Counterpoint: Furthering the Debate.)

The seventh rule of critical thinking is: do not expect all of the answers. Although this rule may seem contradictory at first, and maybe a little absurd in a section that seeks to sharpen your thinking, remember that hard and fast answers are not always available in science. As scientists work out the details, we must be patient and therefore comfortable with incomplete knowledge that produces uncertainty. Consider an example.

Many atmospheric scientists believe that the temperature of the Earth is warming as a result of carbon dioxide and other gaseous pollutants produced by modern society. These pollutants trap heat in the atmosphere, much like a blanket or the glass in a greenhouse. Many of these scientists have staked their reputation on global warming and cite an impressive body of information in support of their view. A few scientists disagree, however. They think that too many uncertainties exist to allow a firm conclusion. Thus, global warming may be an issue on which critical thinkers might reserve opinion, although evidence continues to mount in favor of those who claim that global warming truly is occurring.

The eighth rule of critical thinking requires us to examine the big picture. Many examples of this critical thinking rule exist. Health is one of those areas where a "big-picture" view is essential. Many illnesses, for instance, result from lifestyle. Stress, diet, exercise, smoking, and drinking, for instance, all impact our health. An individual suffering from constant colds may need to look at the big picture—his or her lifestyle—to find the cause. While it may be expedient to go to a doctor to get an antibiotic to combat a bacterial infection or a flu shot to prevent getting the flu, if one doesn't take steps to reduce stress and other unhealthy lifestyle factors, persistent illness could plague that person for years. The rule, then, is to be aware of events in their context.

The ninth rule of critical thinking is: look for multiple causes and effects. Be wary of dualistic thinking. The self-proclaimed "thinking man," *Homo sapiens,* exhibits remarkable intelligence. But as intelligent as our species is, we frequently succumb to simplistic thinking when analyzing problems and devising solutions.

In the 1970s, for example, one noted ecologist argued vigorously that the world's environmental problems stemmed from overpopulation—too many people

for the available resources. Another equally notable scientist argued that the problems were due to technology and its by-product, pollution.

A more critical analysis of the environmental crisis shows that our problems are the result of many factors. Overpopulation and technology are but two of many. Inadequate laws and education must be factored into the equation. So must various psychological and cultural elements—for instance, our view of nature as something to overcome. Many more could be added to the list.

Critical thinking demands a broader view of cause and effect. Look at a variety of parameters when you assess problems. Avoid simplistic thinking by considering as many contributing factors as you can.

The tenth rule of critical thinking reminds us to watch for thought stoppers. Professor Ann Causey, a colleague of mine, reminds us of the ability of thought stoppers to inhibit critical thinking. Thought stoppers are words or phrases that cause us to temporarily suspend our critical faculties. They elicit an overwhelming emotional response or call upon myths and other half-truths that we've heard so often that we accept them blindly.

If someone derails your thinking, stop for a moment or two to assess the situation. Have you been derailed by a half-truth, a myth, or an emotional roadblock that will fall apart upon closer scrutiny?

Finally, the eleventh rule of critical thinking is: understand your own biases, hidden assumptions, and areas of ignorance. So far, this discussion on critical thinking has concentrated on ways you can uncover mistakes in reasoning that other people make. But what about your own mistakes? What about your biases, hidden assumptions, and areas of ignorance that affect your ability to think critically? Are you blissfully going through life thinking everyone else has it wrong? Uncomfortable as it may be, it's essential to grapple with your own weaknesses. Only then can you become a truly critical thinker.

As you read this text, you will be presented with examples to help you sharpen your critical thinking skills. The Point/Counterpoints will help you practice these rules.

SUMMARY

HEALTH AND HOMEOSTASIS: LIFE'S ESSENTIAL BALANCING ACT

1. Humans, like all other organisms, have evolved mechanisms that ensure relative internal constancy *(homeostasis)*. These homeostatic mechanisms are vital to survival and reproduction.

2. Homeostatic mechanisms exist at all levels of biological organization, from cells to organisms to ecosystems.

3. The health of all species and ecosystems is dependent on the functioning of homeostatic mechanisms. When these mechanisms break down, illnesses often result.

4. Human *health* has traditionally been defined as the absence of disease, but a broader definition of health is now emerging. Under this definition, good health implies a state of physical and mental well-being.

5. Physical well-being is characterized by an absence of disease or symptoms of disease, a lack of risk factors that lead to disease, and good physical fitness.

6. Mental health is also characterized by a lack of mental illness and a capacity to deal effectively with the normal stresses and strains of life.

7. Human health and the health of the many species that share this planet with us depend on a properly functioning, healthy ecosystem. Thus, alterations of the environment can have severe repercussions for all species, including humans.

EVOLUTION: THE UNITY AND DIVERSITY OF LIFE

8. The *theory of evolution* says that all life evolved from earlier forms. The process of evolution results in improvements in existing species—that is, modifications that make a species better suited to its environment. It may also lead to the formation of new species.

9. Evolution is responsible for the great diversity of life-forms. However, because the Earth's organisms evolved from early cells that arose over 3.5 billion years ago, all organisms, including humans, share many common characteristics. Thus, evolution is responsible for the unity of life as well. Items 10–16 list the common characteristics of all life-forms.

10. All organisms, including humans, are made up of cells.

11. All other organisms grow and maintain their complex organization by taking in chemicals and energy from the surroundings.

12. All living things house many chemical reactions. These reactions are collectively referred to as *metabolism.*

13. All organisms possess homeostatic mechanisms.

14. All organisms exhibit *irritability*—the capacity to perceive and respond to stimuli.

15. All organisms are capable of reproduction and growth.

16. All organisms are the product of evolutionary development and are subject to evolutionary change.

17. Although humans are similar to many other organisms, we also possess many unique abilities. One of the key differences is culture. In addition, not only can we humans plan for the future, but we also possess enormous abilities to reshape the Earth through ingenuity and technology. This, however, does not imply that humans are in any way superior to other forms, nor that we are separate from nature—just that we are unique.

SCIENCE, CRITICAL THINKING, AND SOCIAL RESPONSIBILITY

18. *Science* is both a systematic method of discovery and a body of information about the world around us.

19. Scientists gather information and test ideas through the *scientific method*. The scientific method begins with observations and measurements, often made during experiments. Observations and measurements often lead to *hypotheses*, explanations of natural phenomena that can be tested in *experiments*. The results of experiments help scientists support or refute their hypotheses.

20. The body of scientific knowledge also contains *theories* or broad generalizations about the way the world works. Theories can change over time as new information becomes available.

21. Scientific discovery can influence ethics. New knowledge about our place in the biosphere, for example, may temper our current notions of human dominance and help people the world over to build a more sustainable relationship with nature.

22. *Critical thinking* is a useful tool in science and is best defined as careful analysis that helps us distinguish knowledge from beliefs or judgments.

23. Critical thinking provides a way to analyze issues and information. It requires that you first define an issue, then study the evidence.

24. Table 1-2 summarizes the critical thinking rules.

Critical Thinking

THINKING CRITICALLY— ANALYSIS

This analysis corresponds to the Thinking Critically scenario that was presented at the beginning of this chapter.

This experiment has several major flaws. First, the boy used chickens from two different farms, so the chickens could have been genetically dissimilar. Differences in genetics could have been responsible for the differences in cholesterol content in the eggs. Second, although differences were found in the cholesterol content of the eggs of the two groups, we don't know if they were statistically significant. Good statistical analysis is necessary to determine whether measured differences are substantial enough to be attributable to the treatment. Another problem is the small sample size. Before I donated any money to this new venture, I'd want to see it performed on a larger group of genetically similar chickens. The fourth and final problem is that no mention was made of the differences between the two feeds. A careful analysis is essential to solidify one's confidence.

EXERCISING YOUR CRITICAL THINKING SKILLS

Read an article on a current medical or environmental problem in a magazine or newspaper. Then, study the facts presented by the author. Does the report offer enough information to allow you to evaluate the assertions made? Do you see places where conclusions are not supported by the facts? Do you see places where bias has entered into the author's writing? Do you see citations to support issues from authorities who may not be as knowledgeable as you would like or who may be biased? Do you see faulty experiments?

TEST OF CONCEPTS

1. How would you define life?
2. In what ways are a rock and a living organism (for example, a bird) similar, and in what ways are they different?
3. Describe the concept of homeostasis. How does it apply to humans? How does it apply to ecosystems? Give examples from your experiences.
4. Using your critical thinking skills debate the statement "planet care is a form of self-care."
5. Using the definition of health and the list of healthy habits in Table 1-1, assess your own health. What areas need improvement?
6. In what ways are humans different from other animals? In what ways are they similar?
7. Describe the scientific method, and give some examples of how you have used it recently in your own life.
8. How do a hypothesis and a theory differ?
9. Can you think of an example in which a single group of animals (including humans) serves as an experimental group and a control group in the same experiment? If you can't, ask one of your campus biologists for some ideas.
10. List and discuss the critical thinking skills presented in this chapter. Which skills seem to be most important for the kind of thinking you normally do?
11. A graduate student injects 10 mice with a chemical commonly found in the environment and finds that all of his animals die within a few days. Eager to publish his results, the student comes to you, his adviser. What would you suggest the student do before publishing his results?
12. Given your knowledge of scientific method and critical thinking, make a list of reasons why scientists might disagree on a particular issue or research finding.

TOOLS FOR LEARNING

www.jbpub.com/humanbiology

Tools for Learning is an on-line student review area located at this book's web site HumanBiology (www.jbpub.com/humanbiology). The review area provides a variety of activities designed to help you study for your class:

Chapter Outlines. We've pulled out the section titles and full sentence sub-headings from each chapter to form natural descriptive outlines you can use to study the chapters' material point by point.

Review Questions. The review questions test your knowledge of the important concepts and applications in each chapter. Written by the author of the text, the review provides feedback for each correct or incorrect answer. This is an excellent test preparation tool.

Flash Cards. Studying human biology requires learning new terms. Virtual flash cards help you master the new vocabulary for each chapter.

Figure Labeling. You can practice identifying and labeling anatomical features on the same art content that appears in the text.

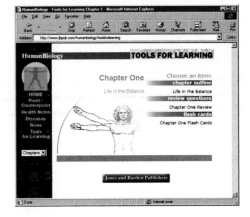

Active Learning Links. Active Learning Links connect to external web sites that provide an opportunity to learn basic concepts through demonstrations, animations, and hands-on activities.

Light micrograph of crystals of ATP molecules, which serve as energy carriers in all organisms.

Thinking Critically

In 1971, Joseph Chatt and his colleagues at the University of Sussex in England synthesized two compounds with the same chemical formula. However, he found that the crystals formed by one compound were sapphire-green; the crystals formed by the other were bright green.

Chatt hypothesized that he had created two compounds that have the same molecular weight and atomic composition but differ in chemical or physical properties because the atoms are linked in different ways. But, when Chatt subjected his compounds to X-ray analysis, which helps determine the structure of molecules, he found that the atoms of his compounds were in the same relative positions. The only difference appeared to be in the chemical bond between two atoms (oxygen and molybdenum), which was longer in the bright green compound than in the sapphire-green one.

Assuming that impurities are rare in single, well-formed crystals, Chatt's group concluded that the color difference in the two compounds resulted from differences in the length of a single bond, a phenomenon not previously reported.

At first, Chatt's findings were discounted, because chemists believe that molecular bonds have characteristic lengths. Most scientists concluded that his experiments were flawed. Can you find any reason(s) in the above information that might have led Chatt to an incorrect conclusion? ■

SURVEYS OF PUBLIC KNOWLEDGE OFTEN PROVE to be embarrassing. In a poll of world geography a few years ago, for example, Americans from 18 to 24 years of age only scored tenth among industrial nations in their knowledge of simple geographic facts. Many Americans couldn't locate Egypt or Germany on a world map, and 14% of those polled couldn't locate the United States!

Nowhere is public lack of knowledge more blatant than in the field of chemistry. Considering the importance of chemistry to our lives, it is surprising how little most people know about it. Why is chemistry so important to us?

Chemistry is important, in part, because it has so many useful applications. In recent decades, thousands of new drugs have been developed by scientists. These chemicals have helped humankind combat a wide range of often deadly or debilitating diseases. One of the most notable examples is antibiotics. Not only have these chemicals reduced suffering, but they have also greatly reduced mortality, especially among infants.

Advances in chemistry have also led to countless improvements in household products that make our lives safer. House paints are a case in point. At one time, most paints contained lead-based pigments. Because lead is toxic to the nervous system, chips of paint flaking off walls and ceilings pose a health hazard to young children, who routinely put foreign objects in their mouths as they explore their new world. Modern paints lacking lead-based pigments are a definite improvement.

The truth of the matter is that almost everything we touch or look at in our daily lives has been brought to us or improved by chemists. But chemistry is also important in other ways. A little knowledge of chemistry, for example, can help us understand environmental problems such as ozone depletion, urban air pollution, and acid rain. Closer to home, it can help us determine which solvent to use to remove dirt or stains from carpets. On a more personal level, a knowledge of chemistry helps people who want to live long, healthy lives make informed decisions on nutrition.

But why study chemistry at the beginning of a biology course? The answer is simple. First, cells and organisms are made up of chemicals. These chemicals undergo many reactions that are the basis of life. In order to understand the structure and function of cells and organisms, you must understand some basic principles of chemistry.

This chapter introduces you to several key principles of chemistry, most of which form the foundation for the study of modern biology. We begin with a discussion of atoms and molecules, then look at water,

acids, bases, and buffers. This information could prove useful to you throughout your life. The major biological molecules encountered in the study of biology are discussed in subsequent chapters.

Atoms and Subatomic Particles

The world in which we live is composed of a wide variety of physical matter such as wood, water, plastic, metal, and air. **Matter** is defined as anything that has mass and occupies space. Technically, **mass** is a measure of the amount of matter in an object. Thus, a water balloon has a greater mass than an air-filled balloon, as anyone who has ever been hit by the former will tell you.

Atoms Are the Fundamental Unit of All Matter

Matter is composed of tiny particles called **atoms.** The word *atom* comes from the Greek *atomos,* which means "incapable of division." The ancient Greek philosophers who proposed the existence of atoms believed that all matter consisted of small particles that lacked any internal structure. Over time, though, evidence has shown that atoms consist of smaller particles, called **subatomic particles.** In this chapter, we will examine three of the most important subatomic particles: electrons, protons, and neutrons (**FIGURE 2-1**). **Protons** (PRO-tauns) have the greatest mass of all and are located in the center of the atom, the **nucleus** (**FIGURE 2-1**). Each proton has a positive charge. **Neutrons** (NEW-trauns) are uncharged particles with slightly less mass than protons. Like protons they are also found in the nucleus. Because protons are positively charged and neutrons have no charge, the nucleus of an atom is positively charged.

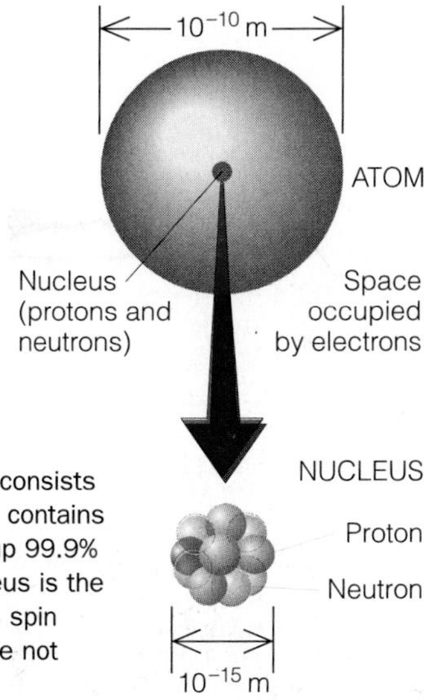

FIGURE 2-1 The Atom The atom consists of two regions. The central nucleus contains protons and neutrons and makes up 99.9% of the mass. Surrounding the nucleus is the electron cloud, where the electrons spin furiously around the nucleus. (Figure not drawn to scale.)

Surrounding the positively charged nucleus is a region called the **electron cloud.** It contains tiny, negatively charged particles, **electrons** (ee-LECK-trauns), that move rapidly through a large volume of space.

Most of the mass of an atom can be attributed to the protons and neutrons, which are packed into a tiny space. These particles are so heavy that a single cubic centimeter (about $\frac{1}{3}$ teaspoon) would weigh 100 million tons. However, most of the volume of an atom is composed of the electron cloud. Consequently, the late Carl Sagan, one of America's leading astronomers, remarked that atoms are mostly space, or "electron fluff."

Atoms are electrically neutral because they contain the same number of protons and electrons. Thus, the positive charges cancel out the negative ones.

The Elements Are the Purest Form of Matter

Early scientists accounted for the diversity of matter by assuming that there must exist a number of pure substances and that these substances could be combined into an infinite variety of different forms. They called these pure substances *elements.* Today, modern chemists define an **element** as a pure substance (for example, gold or lead) that contains only one type of atom. Put another way, elements are substances that cannot be separated into different substances by ordinary chemical means.

At this writing, 92 naturally occurring elements are known. Another dozen or so can be made in the laboratory. Of the 92 naturally occurring elements, only about 20 are found in organisms. Four elements in this group: carbon, oxygen, hydrogen, and nitrogen (remember: COHN) comprise 98% of the atoms of all living things.

Elements are listed on a chart called the **periodic table of elements** (**TABLE 2-1**; Appendix A). The periodic table is a handy summary of vital statistics on each element. As illustrated in **TABLE 2-1**, most elements are represented on the table by a one- or two-letter symbol. The element hydrogen, for example, is indicated by H. Oxygen is designated by O, and carbon is represented by C.

Above each symbol on the periodic table is its atomic number. The **atomic number** of an element is the number of protons found in the nuclei of the element's atoms. Because all atoms are electrically neutral, the atomic number also equals the number of electrons in the atoms.

As you can see, below the chemical symbol for each element is another number, called the mass number. The **mass number** is the average mass of the atoms of

---

TABLE 2-1

A Simplified Periodic Table

I	II	III	IV	V	VI	VII	VIII
1 ← Atomic number: PROTONS H ← Atomic symbol Hydrogen 1.0 ← Mass number (amu) = PROTONS + NEUTRONS							2 He Helium 4.0
3 Li Lithium 7.0	4 Be Beryllium 9.0	5 B Boron 11.0	6 C Carbon 12.0	7 N Nitrogen 14.0	8 O Oxygen 16.0	9 F Fluorine 19.0	10 Ne Neon 20.2
11 Na Sodium 23.0	12 Mg Magnesium 24.3	13 Al Aluminum 27.0	14 Si Silicon 28.1	15 P Phosphorus 31.0	16 S Sulfur 32.1	17 Cl Chlorine 35.5	18 Ar Argon 40.0
19 K Potassium 39.1	20 Ca Calcium 40.1						

Note that each element is represented by a one- or two-letter symbol. Elements are listed according to their atomic number (the number of protons). Their atomic mass is also shown.

each element. The mass number is approximately equal to the number of protons and neutrons in a given atom. Because electrons have virtually no mass, they contribute very little to the mass of an atom. As shown in **TABLE 2-1**, carbon has an atomic mass of 12.

Isotopes Are Alternative Forms of Atoms, Differing in the Number of Neutrons They Contain

Although I just noted that elements are made of identical atoms, most elements are actually made up of mixtures of two or more forms of the same atom, called **isotopes** (EYE-so-topes). These forms differ in the number of neutrons they contain. Hydrogen, for example, has three slightly different atomic forms. The most common hydrogen atom contains one proton, one electron, and no neutrons. The next most common form has one proton, one electron, and one neutron. The least common hydrogen atom has one proton, one electron, but two neutrons. The three forms, or isotopes, of hydrogen are chemically identical—that is, they react the same.

Additional neutrons not only make some isotopes heavier than others, they often make isotopes unstable. In order to achieve a more stable state, many isotopes release **radiation**—small bursts of energy, or tiny energetic particles—from their nuclei. Radiation carries away energy and mass from an atom's nucleus and helps atoms achieve more stable states. Radioactive isotopes are called **radionuclides** (ray-dee-oh-NEW-klides).

Radiation has provided numerous benefits to humankind. For example, scientists have discovered a wide variety of uses for radiation in medicine and research, including X-rays and radioactive chemicals used in medical tests. One of the most important advancements is the use of radioactive isotopes as markers in experiments (Scientific Discoveries 2-1). Another extremely important application is the use of naturally occurring radioactive isotopes to date rocks and fossils. Information from this technique has helped scientists pinpoint important evolutionary events and has given us important insights into the history of life on Earth.

Despite the many benefits of radiation, its use has not occurred without some risks. For example, radiation exposure damages molecules in body cells and causes changes (mutations) in the genetic material that can result in birth defects and cancer. Despite these and other problems, radiation is widely used. It is now even being used to sterilize foods, thus eliminating the need to refrigerate foods. The Point/Counterpoint in this chapter examines the pros and cons of this procedure.

Section 2-2

The Making of a Molecule

The diversity of matter in the world around us is, as noted above, a result of the presence of a wide range of elements, each with its own unique properties. It also arises from the fact that atoms of various elements can combine to form compounds and molecules. The term **compound** refers to a substance made up of two or more atoms. Each compound has a unique chemical formula that indicates the ratio of atoms found in it. For instance, HCl is the chemical formula of hydrochloric acid. It contains one hydrogen atom and one chlorine atom. The chemical formula of the flammable gas methane is CH_4; it contains one carbon atom for every four hydrogen atoms.

The term **molecule** refers to the smallest particle of a compound that still retains the properties of that compound. Methane gas given off by rotting vegetation, for instance, is a chemical compound consisting of methane molecules. Hydrogen gas is a chemical compound consisting of hydrogen molecules, each one of which contains two hydrogen atoms.

Atoms Bond to Form More Stable Configurations

Atoms in molecules are joined together by chemical bonds. There are two types of bonds that form *between* atoms: ionic and covalent. In both instances, the electrons are responsible for creating the bonds that hold atoms together.

Ionic Bonds Are Electrostatic Attractions Between Two Oppositely Charged Particles

Ionic bonds are formed when two atoms react in such a way that electrons are transferred from one atom to another. This creates two charged particles, or **ions** (EYE-ons). The atom that loses its electron becomes positively charged, and the one that gains an electron is negatively charged. This exchange results in the formation of a weak electrostatic force between the two charged particles. This force holds them together and is referred to as an *ionic bond*. **FIGURE 2-2** shows how an ionic bond forms between a chlorine atom and a sodium atom.

Covalent Bonds Are Formed by the Sharing of Electrons Between Atoms

In contrast to ionic bonds, covalent bonds occur when two atoms "share" electrons. This arrangement usually produces a much stronger bond. When atoms are bound by covalent bonds, the electrons they share

Scientific Discoveries that Changed the World 2-1

The Discovery of Radioactive Chemical Markers

Featuring the Work of Schoenheimer

Biochemist Rudolf Schoenheimer (1898–1941) and his colleagues offered science a chemical tool that would yield important new information and result in significant scientific and medical advances: radioactively labeled molecules—that is, common biological molecules containing one or more radioactive atoms (isotopes).

A radioactive atom incorporated in a biological molecule is akin to a radio-transmitter collar on a grizzly bear. The atom emits tiny bursts of radioactivity, which permit scientists to track molecules on their course through the

body, in much the same way that the collar emits radio signals that allow biologists to electronically track the bear through its forbidding terrain.

Marking common molecules like amino acids with radioactive atoms permits scientists not just to track molecules but to follow what happens to them along the way. Before the emergence of this technique, substances such as fatty acids or amino acids administered to an animal were lost almost immediately, because they were indistinguishable from molecules in the body.

Radioactive markers permit scientists to follow marked molecules as they progress through the body's maze of biochemical reactions. Radioactive markers also permit scientists to determine where chemicals are stored and how they are removed from the body. In fact, many of the physiological processes you will read about in this book were investigated by using radioactively labeled molecules.

Before Schoenheimer's discovery, the inability to track molecules resulted in a huge gap in our knowl-

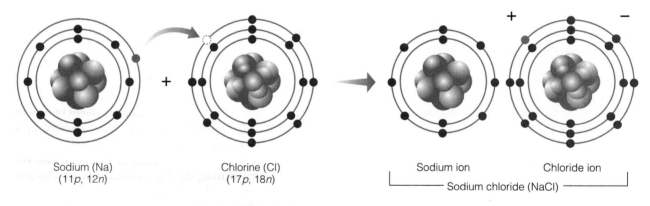

Sodium (Na)
(11*p*, 12*n*)

Chlorine (Cl)
(17*p*, 18*n*)

Sodium ion

Chloride ion

Sodium chloride (NaCl)

FIGURE 2-2 Ions and Ionic Bonds   Sodium (Na) and chlorine (Cl) atoms differ in their electron affinity. Sodium has a weaker affinity and tends to give up an electron to atoms like chlorine. As a result, sodium becomes a positively charged ion and chlorine a negatively charged ion. The oppositely charged ions attract each other, forming an ionic bond.

actually orbit around both atoms. This holds the atoms together. **FIGURE 2-3** illustrates the formation of hydrogen gas, H_2, from two hydrogen atoms.

Consider methane (CH_4), a gas that consists of four hydrogen atoms covalently bonded to one carbon atom. To understand how this molecule forms, we turn first to the central atom, carbon.

Carbon atoms have six electrons, two in a region or *shell* close to the nucleus and four in an outer shell. Carbon reacts with four hydrogen atoms to form methane. This arrangement results in a full outer shell containing eight electrons.

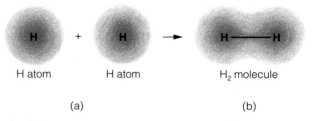

H atom   H atom H_2 molecule

(a)   (b)

FIGURE 2-3 Covalent Bond   Nonpolar covalent bonds are characterized by a sharing of electrons. *(a)* The shaded regions represent the hydrogen atoms' electron clouds. *(b)* Two atoms bound together by a covalent bond. The electrons orbit around both nuclei, holding them together to create a molecule.

edge about metabolism and other body functions. Futile attempts to "tag" molecules with nonradioactive molecular markers generally failed, because the altered molecules differed so much from their natural substances that they were treated differently by the body.

Through experimentation, Schoenheimer showed that radioactively labeled molecules were so similar to the unlabeled molecule that they reacted identically. This made them ideal markers and opened the door to many additional discoveries.

From Schoenheimer's brilliant experiments, a new picture of life has emerged. He and his colleagues showed that molecules of the body such as DNA, which had previously been considered stable, are actually in a continuous state of flux. Even apparently dormant fat deposits in the body were ever-changing. Additional studies eventually showed that amino acids were rapidly broken down and rebuilt into new ones.

Since their introduction in the 1930s, radioactive markers have also enjoyed great popularity in medicine where they're often used to determine the functional status of various organs. Radioactive iodide ions, for instance, can be administered to a patient to assess the function of the thyroid gland, a hormone-producing gland in the neck.

Radioactively labeled molecules are helping scientists unravel the mysteries of body functions. Labeled glucose molecules, for instance, can be used to determine the relative activity of different parts of the brain, permitting scientists to map brain functions accurately.

These are but a few examples of the tremendous contribution of radioactively labeled molecules to medicine and science.

Many other atoms react similarly, filling the outer shell during covalent bonding. Thus, chemists have coined the term *octet rule* to explain this behavior.

Covalent bonding can be represented by using dots and Xs for the outer-shell electrons of an atom. **FIGURE 2-4** uses this technique to show the electron sharing in a molecule of methane. In this example, carbon is represented by the letter C surrounded by four outer-shell electrons, indicated by dots. Each hydrogen atom is represented by H, and the outer-shell electron of each hydrogen atom is indicated by an X. As illustrated, each hydrogen atom shares its electron with the carbon atom, giving the carbon atom four additional electrons and thus creating a full outer shell for carbon.

Another important molecule is water, H_2O. As the formula indicates, a water molecule consists of a single atom of oxygen bonded to two atoms of hydrogen. Oxygen atoms contain eight electrons, two in the inner shell and six in the outer shell. By combining with two hydrogen atoms, oxygen fills its outer shell.

The bonds between carbon and hydrogen in methane and oxygen and hydrogen in water involve the sharing of one pair of electrons—one from each atom. These bonds are referred to as **single covalent bonds.** In chemical nomenclature, they are often indicated by a single line joining two atoms, as in H–H.

Some atoms share two pairs of electrons with another atom, forming **double covalent bonds.** Oxygen, for instance, can share two of its outer-shell electrons with a carbon atom, forming a double covalent bond (C=O), as shown in **FIGURE 2-5**. Other atoms can share three pairs of electrons, forming triple covalent bonds (**FIGURE 2-5**).

Polar Covalent Bonds Occur Any Time There Is an Unequal Sharing of Electrons by Two Atoms

The electrons shared between two such atoms are not always shared equally. Unequal sharing occurs when one of the two atoms has a slightly higher affinity (attraction) for electrons than the other. For example, oxygen and hydrogen join to form a single covalent bond, but oxygen has a slightly higher affinity for electrons than hydrogen. Consequently, the electron of the hydrogen atom tends to spend more time around the oxygen atom than around the hydrogen atom. Because of this, the

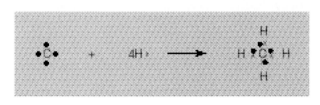

FIGURE 2-4 Covalent Bonding Simplified You can keep track of the electrons being shared in covalent bonds by representing the outer-shell electrons as dots or Xs. See the text for explanation.

Point/Counterpoint

Controversy over Food Irradiation

FOOD IRRADIATION: TOO MANY QUESTIONS

Donald B. Louria

Donald B. Louria *teaches at the New Jersey Medical School, where he is chairman of the Department of Preventive Medicine and Community Health.*

The debate about the safety of irradiating foods raises six issues:

1. The safety issue. Irradiated food does not become radioactive, but the radiation does cause chemical changes in food molecules. The major concerns are possible induction of cancer and genetic damage. A highly controversial but flawed study on small numbers of malnourished children in India suggested that consumption of freshly irradiated wheat could produce chromosomal abnormalities. However, a study of a much larger group of healthy adults showed no such abnormalities. The proponents of food irradiation have criticized the first Indian study and pointed to the larger study that they feel showed no chromosome damage. But that does not answer the concerns. If food irradiation becomes an accepted technique, millions of malnourished people will eat irradiated foods. What is needed is a careful study with adequate numbers of malnourished children and adults. I am prepared to believe that consumption of irradiated foods is unlikely to produce significant harm, but such a belief must be supported by proper data.

2. The nutrition issue. Irradiated food loses some of its nutritional value; the extent of loss of vitamin content depends on the type of food and the dose of radiation—the higher the radiation dose, the greater the loss. Furthermore, some evidence suggests that when irradiated foods are processed (frozen, thawed, heated), there is an accelerated loss of vitamins. Irradiation proponents suggest that the effects of irradiation on vitamin content are not different from that of conventional food processing (heating, freezing, and the like). They also maintain that the diet in the United States contains redundant vitamins, so some loss through irradiation would not be of concern. Of course, this would not be true for millions of people in other countries, people over 60 in the United States, and people with various diseases.

   Arguing that our diet contains adequate vitamins and that destroying the vitamin content of foods by irradiation is of no importance is not likely to be viewed well by many Americans.

3. The necessity of this technology. Proponents say that food irradiation will reduce diarrheal illness from infected poultry. However, proper cooking procedures are just as effective. Proponents say irradiating meats will reduce the dangers of trichinosis, a parasitic disease, but trichinosis occurs infrequently.

4. Helping to solve world hunger. Will irradiating foods prolong their shelf lives and thus feed the world? Shelf lives will definitely be increased; that is an important benefit for the less-developed world, but the proponents have provided no adequate data on the extent of the benefit. It is likely that the foods will be sold primarily in affluent countries where the shelf life issue is of less concern.

5. The issue of safer competing technologies. Food irradiation could reduce the use of toxic chemicals applied to foods in storage. During the next decades, scientists will develop food crops that resist pests, grains that do not require chemicals to protect them in storage, and foods that have longer shelf lives. Advances in biotechnology are likely to give us much safer alternatives to food irradiation, if we will only have a little patience.

6. The environmental pollution issue. Food irradiation is a technology that will result in numerous food irradiation plants in the United States. The few irradiation plants operating in the United States have contaminated their workers and the environment. Imagine the potential for contamination if there were hundreds of such plants using radioactive materials.

Food irradiation does have potential benefits, but it also raises substantial concerns. Whether we should adopt the technology is obviously a matter for continuing debate. Certainly the technology should not be adopted until the issues have been resolved.

FOOD IRRADIATION: SAFE AND SOUND
George G. Giddings

Irradiated foods are safe and wholesome. The ionizing radiation process applied to foods offers certain proven public health and economic benefits without significant public health risks when carried out according to well-established principles and procedures. Decades of worldwide research and testing by competent, knowledgeable, objective, and responsible scientists have led to this conclusion. This conclusion is also supported by some 30 years of experience in the radiation sterilization of medical devices and other health care products to prevent infections, plus a growing list of industrial and consumer products, including foods and their raw materials, ingredients, and packaging materials. The technology is so pollution-free that the EPA exempts it from environmental impact statements.

There has been organized political opposition to food irradiation by a network of special-interest activists serving various political agendas, notably the antinuclear/anti-irradiation one. This network includes a handful of scientists and medical professionals from other fields who act as "expert witnesses" against food irradiation to serve their hidden agendas. For example, they still point to an old Indian study on irradiated wheat and a few children as suggestive of risk even though it was dismissed in the mid-1970s. Opponents point to minor losses of certain vitamins where in fact irradiation is gentler towards vitamins and other nutrients than

comparable food processes and even cooking. Their campaign is doomed to failure in the face of the unshakable facts, including a growing appreciation for public health and other proven benefits of food irradiation, and its growing worldwide regulatory approval, industrial usage, and public acceptance. The American Medical Association and World Health Organization are among the many professional bodies that have endorsed food irradiation.

Food irradiation is undoubtedly the most versatile physical process yet applied to food substances in terms of the range and variety of objectives it can accomplish. Radiation can

- inhibit the sprouting of foods, such as potatoes and onions, and delay spoilage
- rid fruits, vegetables, and grains of insect pests
- prevent parasites from infecting consumers of fish and meats
- rid foods of microbial pathogens such as the salmonellae
- delay microbial spoilage of a wide variety of animal and plant products by reducing microbe levels
- sterilize packaging materials, eliminating microorganisms that would otherwise contaminate products
- sterilize a wide variety of prepared or cooked foods such as meat, poultry, and fishery products, which have already been used to feed astronauts in space and immune-compromised patients

All of these beneficial effects and more can, and are, being readily accomplished by the application of

ionizing (gamma, electron, and X-ray) radiation according to well-established principles and procedures. Nevertheless, irradiation must compete with a number of other new technologies. As a result, it is not likely to be applied to all, or even a high percentage, of the national and world food supply. It will therefore be used in cases in which it is clearly the best all-around choice.

George G. Giddings, Ph. D., *has been involved in food irradiation research and development since 1963 and has written and spoken extensively on the subject.*

SHARPENING YOUR CRITICAL THINKING SKILLS

1. What is food irradiation? Why is it used?
2. List and summarize the key points of each author.
3. Using your critical thinking skills, analyze each author's position. Are you inclined to agree with either one on all issues? Why or why not?
4. In your opinion, what is the best course for the development and implementation of this technology?
5. What facts would you need to know to form a personal conclusion about this issue?

Visit *Human Biology's* Internet site, www.jbpub.com/humanbiology, to research opposing web sites and respond to questions that will help you clarify your own opinion. (See Point/Counterpoint: Furthering the Debate.)

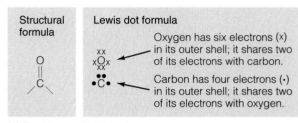

Structural formula	Lewis dot formula
	Oxygen has six electrons (x) in its outer shell; it shares two of its electrons with carbon.
	Carbon has four electrons (·) in its outer shell; it shares two of its electrons with oxygen.

(a) Double covalent bond

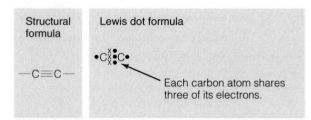

Structural formula	Lewis dot formula
—C≡C—	Each carbon atom shares three of its electrons.

(b) Triple covalent bond

FIGURE 2-5 Double and Triple Covalent Bonds   *(a)* When atoms share two pairs of electrons, a double covalent bond is formed. Chemists have devised several ways to draw these bonds. In the formula on the left, each line represents a pair of shared electrons. On the right, electrons are indicated by Xs and dots. *(b)* On rare occasions, atoms share three pairs of electrons, creating triple covalent bonds.

oxygen atom has a slightly negative charge. The hydrogen atom is visited less frequently by the electron and has a slightly positive charge (**FIGURE 2-6A**). The result is a **polar covalent bond**. A polar covalent bond is simply a covalent bond whose atoms bear a slight charge—either positive or negative.

The presence of polar covalent bonds in molecules makes entire molecules polar. This, in turn, has profound implications to life on Earth, as you will soon see. Consider one of the most important of all polar molecules, water.

As **FIGURE 2-6A** illustrates, water contains two polar covalent bonds and, therefore, has three slightly charged atoms. The slightly charged atoms are often attracted to oppositely charged atoms on *neighboring* molecules (**FIGURE 2-6B**).

FIGURE 2-6 Hydrogen Bonding   *(a)* Slightly unequal sharing of electrons in the water molecule creates a polar molecule. Electrons tend to spend more time around the oxygen nucleus, making it slightly negative. The hydrogen atoms are, therefore, slightly positive, as indicated by the symbols δ+. *(b)* Because of the polarity, hydrogen atoms of one molecule are attracted to oxygen atoms of another. The attraction is called a hydrogen bond.

Polar molecules are also formed when hydrogen atoms covalently bond to atoms such as nitrogen and fluorine, which, like oxygen, have a slightly higher affinity for electrons than hydrogen.

Hydrogen Bonds Form Between Slightly Charged Atoms Usually on Different Molecules

Covalent bonds occur between atoms of a molecule. They are the bonds that hold the atoms together. But neighboring molecules can also be attracted to one another, thanks to the presence of polar covalent bonds in them. Take water for an example. In a glass of water, positively charged hydrogen atoms of water molecules attract negatively charged oxygen atoms of other water molecules. The electrostatic attractions *between* the positively charged hydrogen atoms of *one* water molecule and the negatively charged oxygen atoms of *another* are called **hydrogen bonds**. This is an electrostatic attraction responsible for many of the unique properties of water and absolutely essential to life on Earth (discussed later). Hydrogen bonds are also found in other biologically important molecules that will be discussed in subsequent chapters.

Chemical Compounds Fall into Two Broad Groups: Organic and Inorganic

Chemists classify compounds as either organic or inorganic. **Organic compounds** contain molecules that are made primarily of carbon atoms. Many organic mole-

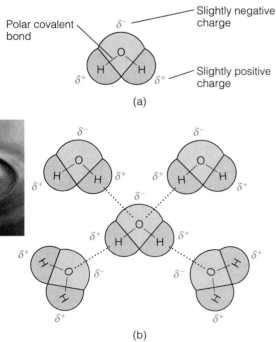

TABLE 2-2

Comparison of Organic and Inorganic Compounds

Organic	Inorganic
Consists primarily of molecules containing carbon and hydrogen	Usually consists of positively and negatively charged ions
Atoms linked by covalent bonds	Usually consists of atoms joined by ionic bonds
Often consists of large molecules that contain many atoms	Always contains small numbers of atoms

cules are quite large. That's because carbon atoms can be joined to one another like the beads of a necklace. These long chains of carbon atoms form the "backbone" of many organic compounds. Attached to the backbone are a number of other atoms, most often hydrogen, oxygen, and nitrogen. In organic molecules, all atoms are covalently linked.

Inorganic compounds are defined primarily by exclusion—that is, they are compounds that are *not* organic. Inorganic molecules are generally small molecules. They usually consist of atoms joined by ionic bonds, as opposed to organic compounds, which are always covalently bonded. Sodium chloride and magnesium chloride are examples of inorganic compounds.

As you will see in the study of biology, not all rules apply 100% of the time. In this instance, not all inorganic compounds contain ionic bonds. Water, for example, is an inorganic compound whose atoms are joined by covalent bonds. **TABLE 2-2** summarizes the main features of each type of compound.

FIGURE 2-7 Perspiration Cools the Body Perspiration is an automatic homeostatic response that helps rid the body of heat.

Water, Acids, Bases, and Buffers

Now that you've learned a few basics of chemistry, let's look at water, a biologically important molecule.

Water Is Vital to Life

Scientific historian and naturalist Loren Eiseley once wrote that "if there is magic on the planet, it has to be water." This remarkable substance produces billowy clouds, icicles that hang from the limbs of trees, and elegant waterfalls. Aesthetically pleasing as water is, though, it is also of great practical importance to humans and all other living organisms.

Water is a major component of all cells and organisms. In fact, nearly two-thirds of the human body is water. Thus, if you weigh 100 pounds, nearly 70 pounds of your body weight is water. Water is an important biological **solvent,** a fluid that dissolves other chemical substances. Human blood, for example, contains large amounts of water that dissolves and transports nutrients, hormones, and wastes throughout our bodies.

Water also participates in many chemical reactions in the body—for example, in the breakdown of protein (described later). In addition, water serves as a lubricant. Saliva, which is largely water, lubricates food in the mouth and esophagus, the tube that transports food from our mouths to our stomachs. A watery fluid in the joints called *synovial fluid* enables bones to slide over one another, thus facilitating body motion.

Finally, water helps us regulate body temperature. Perspiration, for example, rids our bodies of heat, for when it evaporates, water draws off heat (**FIGURE 2-7**). People sometimes suffer heatstroke in hot humid climates because the excess water in the atmosphere reduces evaporation. This causes heat to build up, which can cause a person to collapse.

Water Molecules Dissociate into Hydrogen and Hydroxide Ions

Water molecules are fairly stable; nonetheless, some molecules dissociate, or break apart, forming two charged units known as ions as shown in the following reaction:

$$H_2O \rightleftharpoons H^+ + OH^-$$

water   hydrogen ion   hydroxide ion

Because the hydrogen and hydroxide ions formed in this reaction can react with one another to re-form water molecules, this reaction is said to be *reversible.*

The double arrow in the reaction formula indicates that it is reversible.

Although water can dissociate, the ratio of water molecules to the ions, H^+ and OH^-, in the human body is about 500 million:1. Nevertheless, even the slightest change in the hydrogen ion concentration can alter cells and organisms, shutting down biochemical pathways and sometimes killing organisms. It is not surprising, then, that the human body contains a number of homeostatic mechanisms to ensure a constant level of these ions.

Acids Are Substances That Add Hydrogen Ions to Solution; Bases Remove Hydrogen Ions

In pure water, the concentrations of H^+ and OH^- ions are equal.[1] A solution containing an equal number of these ions is said to be *chemically neutral.*

Chemical neutrality can be upset by adding or removing H^+ or OH^-. For example, hydrochloric acid (HCl) dissociates into hydrogen ions (H^+) and chloride ions (Cl^-) when added to water. This substance, therefore, increases the hydrogen ion concentration. Scientists call a substance that adds hydrogen ions to a solution an **acid.** Solutions with proportionately more H^+ than OH^- are said to be acidic. **Bases** are substances that remove H^+ from solution. Sodium hydroxide, for example, dissociates when added to aqueous solutions, forming Na^+ and OH^-. The hydroxide ions react with hydrogen ions in the solution, forming water molecules. As sodium hydroxide is added to a solution of pure water, the hydrogen ion concentration declines, and the hydroxide ion concentration increases. A solution with a greater concentration of OH^- than H^+ ions is said to be *basic.*

Acidity is measured on the **pH scale** (**FIGURE 2-8**). As illustrated, the pH scale ranges from 0 to 14. Neutral substances are assigned a pH of 7. Basic substances have a pH greater than 7, and acidic substances have a pH less than 7.

The pH scale is logarithmic (law-gah-RITH-mick). This means that the pH scale mathematically condenses a wide range of numbers onto a fairly small scale. The lesson in all of this is: Don't be fooled by a small change in pH. One pH unit represents a tenfold change in acidity. Accordingly, a solution with a pH of 3 is 10 times more acidic than one with a pH of 4 and 100 times more acidic than one with a pH of 5.

Most biochemical reactions occur at pH values between 6 and 8. Human blood, for example, has a pH of 7.4. So important is this to normal body function that a slight shift in the pH of the blood for even a short period can be fatal. Fortunately, evolution has "provided" us with numerous homeostatic systems that maintain required pHs in the body.

Homeostasis Is Ensured in Part by Buffers, Molecules That Help Maintain pH within a Narrow Range

Biological systems operate within a narrow pH range maintained by a relatively simple homeostatic mechanism created by buffers. **Buffers** (BUFF-firs) are chemicals found in organisms and ecosystems, particularly water, that protect against drastic shifts in pH. How do they work?

In many ways, buffers are "hydrogen-ion sponges." They help maintain a constant pH by removing hydrogen ions from solution when levels increase. Buffers give back the hydrogen ions when levels fall.

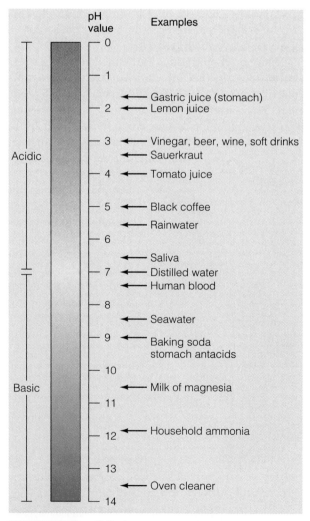

FIGURE 2-8 The pH Scale

---

[1]Concentration is the amount of solute (dissolved substance) in a given amount of solvent (usually water).

One of the most prevalent of all buffers is carbonic acid. Found in the blood of animals, carbonic acid is formed from water and carbon dioxide, a waste product of cellular energy production. In blood, carbonic acid dissociates into bicarbonate and hydrogen ions:

$$H_2CO_3 \;\rightleftharpoons\; H^+ + HCO_3^-$$

carbonic acid   hydrogen ion   bicarbonate
(weak acid)   (weak base)

In our blood and in lakes and rivers where carbonic acid is also present, this chemical reaction shifts back and forth in response to changing levels of hydrogen ions. Thus, when hydrogen ions are added to water, they combine with bicarbonate ions, driving the reaction to the left. But when the hydrogen-ion concentration falls, the reaction is driven to the right. In either case, the pH of the blood remains constant.

Overview of Other Biologically Important Molecules

This introduction to chemistry is intended to give you a foundation upon which to add to your chemical literacy. In your study of biology, you will encounter chemistry terms and many chemical compounds. Your studies will reveal four major groups of biological molecules: (1) carbohydrates; (2) lipids; (3) amino acids, peptides, and proteins; and (4) nucleic acids. Rather than describe each one in detail here, I will present a few important facts. These molecules are discussed in more detail in subsequent chapters.

Carbohydrates are organic molecules that range in size from very small such as the blood sugar glucose, which cells use to generate energy, to very large ones such as starch. The large molecules are actually composed of many small molecules. Carbohydrates supply energy to cells. Pastas and cereals are a great source of carbohydrates.

Lipids (LIP-ids) are a rather diverse group of molecules that includes fats and steroids. Some lipids are a source of energy and others are important structural components of cells. Lipids form a layer of insulation beneath the skin of many animals, including whales and humans. Lipids are found in many foods, especially meat, and fried foods such as french fries.

Most readers have heard the terms *amino acids* and *proteins*. As you probably already know, **proteins** are long molecules (polymers) consisting of many smaller molecules, the **amino acids.** Found in many foods, especially meats and milk products, proteins are important

structural elements of cells, as you will see in the next chapter. Many proteins are **enzymes,** special molecules that speed up chemical reactions in the body. For now, suffice it to say that **peptides** are very small proteins—chains of amino acids.

The fourth group of biologically important molecules is the **nucleic** (new-CLAY-ick) **acids.** This structurally complex group includes DNA, the genetic material found in cells. These molecules are long chains of smaller molecules called *nucleotides* (NEW-klee-oh-tides). Another nucleic acid is RNA. Its structure and function are discussed in Chapter 17.

Health, Homeostasis, and the Environment
Tracking a Chemical Killer

Cancer is the ultimate symbol of homeostatic imbalance. That is, it results from a failure of the body's homeostatic system. Cancer occurs when body cells lose control and begin to divide unremittingly, producing tumors. These tumors eventually kill the person unless they're destroyed. Cancer sometimes results from environmental imbalances, excesses of chemical toxins, or some other cancer-causing agents in our environment as the following story reveals.

For 2000 years, the people in Lin Xian, China, 250 miles south of Beijing, have been dying in record numbers from cancer of the esophagus, the muscular tube that transports food to the stomach. So prevalent is the disease that one of every four persons once succumbed to this ruthless killer, whose incidence is higher there than anywhere else in the world.

In 1959, scientists began a systematic study of 70,000 people in the valley around Lin Xian in an attempt to discover origins of the disease and put an end to the scourge. This disease would lead scientists on a lengthy journey of investigation that would show how minute imbalances in the environment can produce internal chemical imbalances that result in cancer.

Scientists first found that esophageal cancer in Lin Xian was the result of a group of chemicals called *nitrosamines* (NYE-trose-ah-MEANS). Nitrosamines, they discovered, were being produced in the stomachs of the residents from two other chemical compounds: nitrites and amines. But where did they come from?

Research showed that residents had abnormally high levels of nitrites, which came from nitrates in the vegetables that residents ate (**FIGURE 2-9**). The amines were present in moldy bread, a delicacy in the region. But why the excess of nitrates in vegetables? And why the higher-than-normal levels of nitrites in people?

FIGURE 2-9 Tracking a Deadly Killer   In Lin Xian, China, the link between esophageal cancer and the soil was uncovered by the diligent work of medical researchers.

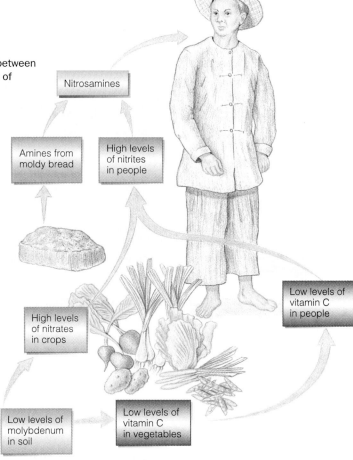

Chemical analyses of the soil in which the region's crops were grown showed that the topsoil around Lin Xian was deficient in molybdenum (meh-LIB-deh-num), an element required by plants in minute quantities. This deficiency caused the plants to concentrate nitrates. It also reduced vitamin C production by plants. Nitrates from plants were apparently being converted to nitrites in the stomachs of the residents. Low levels of vitamin C promoted this conversion. When the residents were given vitamin C tablets, their nitrite levels dropped.

Ultimately, then, a soil deficiency, caused either by intensive agriculture, which depletes the soil, or from a natural deficiency, put the local residents who ate moldy bread rich in amines at high risk for esophageal cancer.

To reduce the risk of esophageal cancer, the villagers now coat wheat and corn seeds with molybdenum. As a result, nitrite levels in vegetables have dropped 40%, and vitamin C levels have increased 25%. Although it is still too early to tell whether restoring the soil's nutrient balance will put a halt to this deadly killer, scientists are optimistic.

SUMMARY

1. An understanding of chemistry is essential to understanding biology because all cells and organisms are composed of molecules and many life processes are nothing more than chemical reactions.

ATOMS AND SUBATOMIC PARTICLES

2. The physical world we live in is composed of matter. Matter is anything that has mass and occupies space.
3. All matter is composed of tiny particles called *atoms*.
4. Atoms contain many *subatomic particles*, three of the most important being electrons, protons, and neutrons.
5. Negatively charged particles, *electrons*, are the smallest of these three subatomic particles; they are found in the electron cloud around the dense, central *nucleus*.
6. *Protons* are positively charged particles that are located in the nucleus with *neutrons*, which are noncharged. Neutrons and protons are large and massive compared to electrons.

7. Atoms are electrically neutral because the number of electrons always equals the number of protons.
8. A pure substance such as gold or lead containing only one type of atom is known as an *element*. The elements are listed on the *periodic table* by *atomic number*, which is the number of protons in the nucleus.
9. Most elements are made up of mixtures of two or more *isotopes*, atoms with slightly different atomic masses resulting from the presence of additional neutrons.
10. Additional neutrons often make isotopes unstable. To achieve a more stable state, many isotopes release *radiation*—small bursts of energy, or tiny energetic particles—from their nuclei.
11. Radiation has proved to be a useful tool in medicine and science, but it can also damage cells of organisms, leading to birth defects and cancer.

THE MAKING OF A MOLECULE

12. Atoms react with one another to form organic and inorganic compounds.

Their reactivity is a result of the interaction of electrons.

13. Two types of bonds are found between atoms, ionic and covalent.
14. In general, atoms tend to react with others to form complete outer shells, creating more stable atomic configurations.
15. During some atomic reactions, electrons are transferred from one atom to another. Transferring an outer-shell electron from one atom to another creates two oppositely charged ions. An electrostatic attraction forms between them and is referred to as an *ionic bond*.
16. *Covalent bonds* form between atoms that share one or more pairs of electrons. This sharing of electrons holds the atoms together.
17. Unequal sharing of electrons results in the formation of a *polar covalent bond*, which is often sufficient to result in the formation of *polar molecules*. Polar molecules are frequently attracted to one another by *hydrogen bonds*, weak electrostatic attractions between the oppositely charged atoms of neighboring molecules.

WATER, ACIDS, BASES, AND BUFFERS

18. *Organic molecules* are compounds made up primarily of carbon and hydrogen. The atoms of these generally large molecules are held together by covalent bonds.

19. *Inorganic molecules* are much smaller molecules whose atoms are often linked by ionic bonds.

20. Water is an important inorganic molecule and a major component of all body cells. It serves as a solvent, which transports many substances in the blood, body tissues, and cells. It also participates in many chemical reactions, serves as a lubricant, and helps regulate body temperature.

21. Although fairly stable, water molecules can dissociate, forming hydrogen and hydroxide ions.

22. In pure water, the concentration of these ions is equal and water has a pH of 7. Adding *acidic substances* increases the concentration of hydrogen ions, causing the pH to fall.

23. Substances that remove hydrogen ions from solution cause the pH to climb and are called *bases*.

24. A solution with a pH less than 7 is acidic; a solution with a pH greater than 7 is basic.

25. Biological systems contain *buffers*, chemical compounds that offset changes in the concentration of hydrogen and hydroxide ions. Buffers are therefore an important component of chemical homeostasis in organisms and the environment.

OVERVIEW OF OTHER BIOLOGICALLY IMPORTANT MOLECULES

26. Carbohydrates, lipids, amino acids, proteins, and nucleic acids are biologically important molecules. They vary considerably in structure and function. Details of their structure and function are covered in later chapters.

Critical Thinking

THINKING CRITICALLY— ANALYSIS

This analysis corresponds to the Thinking Critically scenario that was presented at the beginning of this chapter.

The phrase that may have raised a red flag in your mind while you were reading this exercise is "Assuming that impurities are rare in single, well-formed crystals." Beware of conclusions based on assumptions. They may prove to be wrong.

But this may not be readily apparent at first. In 1985, in fact, chemists at the Ruhr University in Germany reported the synthesis of another pair of isomers like Chatt's, but from tungsten-containing compounds. These isomers were, according to the scientists, also chemically and structurally identical except for the length of one metal-oxygen bond. Several other laboratories also reported similar findings, creating a great deal of excitement among scientists, for it appeared that a long-held belief of the constancy of bond length was about to topple.

Theoretical chemists began proposing explanations for this exciting phenomenon. In 1988, Nobel Prize-winning chemist Roald Hoffman and his colleagues published one explanation. However, as chemists began to repeat the experiments (a process essential to good science) to learn more about the isomers, enthusiasm began to fade. In one replication of Chatt's experiment, scientists found that impurities in samples resulted in different colored crystals. After the German experiment was repeated, it was shown that those results had also resulted from contamination.

"Bond-stretch isomerism," as it was called, nearly became accepted theory based on the widely accepted, but faulty, assumption that impurities are rare in single, well-formed crystals. Obviously, this was not the case.

EXERCISING YOUR CRITICAL THINKING SKILLS

Smart pharmaceuticals—chemicals that supposedly boost intelligence—have grown in popularity in certain subcultures in the United States. According to one estimate, 100,000 people in the United States now use them. Several major pharmaceutical firms are also developing brain-enhancing drugs, primarily to treat stroke victims and people with Alzheimer's disease. Despite all the hoopla and the billions being spent on these chemicals, Gary Wenk, a professor of neurology and psychology at the University of Arizona, who's been testing various brain-enhancing chemicals for nearly 20 years, says he's never tested one that proved to be more than minimally effective. Yet users of the drugs make bold claims about their effectiveness. They say they increase brain cell metabolism to improve memory, concentration, alertness, and problem-solving ability. Using your knowledge of critical thinking and experimental design, why do you think this discrepancy exists? Which side of the debate would you side with? Or do you need more information? If you were thinking about using some of these drugs, what questions would you want answered about them?

TEST OF CONCEPTS

1. Why is the matter on Earth so varied in its appearance?
2. Describe the structure of an atom, using a diagram to illustrate your answer.
3. Define the following terms: atomic number, mass number, and isotope.
4. How are ionic and covalent bonds different? Describe what holds the atoms together in both cases. In what ways are the bonds similar?
5. Describe how polar covalent bonds form and why they result in the formation of hydrogen bonds.
6. Temperature is a measure of the speed of molecules: the higher the temperature, the higher the speed. With this information and your knowledge of water and hydrogen bonds, why does water have a higher boiling point than a nonpolar liquid like alcohol?
7. Describe the biological significance of water.
8. How does the body regulate H^+ levels in the blood? Describe the process.

www.jbpub.com/humanbiology

TOOLS FOR LEARNING

Tools for Learning is an on-line student review area located at this book's web site HumanBiology (www.jbpub.com/humanbiology). The review area provides a variety of activities designed to help you study for your class:

Chapter Outlines. We've pulled out the section titles and full sentence sub-headings from each chapter to form natural descriptive outlines you can use to study the chapters' material point by point.

Review Questions. The review questions test your knowledge of the important concepts and applications in each chapter. Written by the author of the text, the review provides feedback for each correct or incorrect answer. This is an excellent test preparation tool.

Flash Cards. Studying human biology requires learning new terms. Virtual flash cards help you master the new vocabulary for each chapter.

Figure Labeling. You can practice identifying and labeling anatomical features on the same art content that appears in the text.

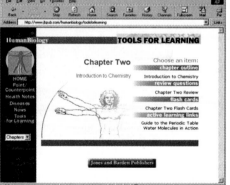

Active Learning Links. Active Learning Links connect to external web sites that provide an opportunity to learn basic concepts through demonstrations, animations, and hands-on activities.

THE LIFE OF THE CELL

Thinking Critically

Researchers at a major medical college in your area have discovered that cells live longer when cultured (grown) in petri dishes containing a certain vitamin. They suggest that this vitamin might help people live longer. A local newspaper runs with the story, and your friends are thinking of taking the vitamin supplement in hopes of living longer lives. What advice would you give your friends before they embark on this course of action? ■

Mitochondria, energy factories of human cells.

IN SWEDEN AND MEXICO, MEDICAL RESEARCHERS are experimenting with a controversial procedure called fetal cell transplantation. By transplanting normal, healthy cells from aborted human fetuses into adults, medical scientists hope to be able to cure a number of chronic, debilitating diseases. Scientists believe that fetal cells can replace defective cells in the organs and tissues of adults. This procedure may someday make a significant contribution to medical science, reducing pain and suffering in tens of thousands of people worldwide—people who have little hope of a cure.

Although promising, fetal cell transplantation poses a number of technical challenges. It also poses several significant moral and political dilemmas, which are discussed in the Point/Counterpoint in this chapter.

This chapter focuses on the cell and some key biological molecules. The information presented here forms a foundation on which a good understanding of human biology can be built. This material will also help you understand the details of fetal cell transplantation. Before we examine the structure and function of cells, though, we take a journey back in time to trace the evolution of the cell.

Cellular Evolution and Homeostasis

Two Types of Cells Exist: Prokaryotes and Eukaryotes

All organisms are composed of one or more cells. Scientists believe that the very first cells appeared on Earth approximately 3.5 billion years ago. Resembling modern-day bacteria, these single-celled organisms inhabited the oceans and were the dominant form of life for over 2 billion years.

These first cells and their modern relatives (the bacteria) are known as prokaryotes (*pro* = before; *karyon* = nucleus). **Prokaryotes** (pro-CARE-ee-oats) are single-celled organisms that contain a single circular strand of DNA (**FIGURE 3-1A**). Prokaryotic organisms formed the Earth's first major group (kingdom) of organisms, the **monerans** (moe-NARE-ens), briefly introduced in Chapter 1. Unlike the more complex cells that make up humans and other multicellular organisms, prokaryotic cells possess few of the internal structures found in the cells of more recent organisms.

Over time, the prokaryotes gave rise to more complex cells that we know as **eukaryotes** (*eu* = true; *karyon* = nucleus; pronounced you-CARE-ee-oats). These cells first appeared on Earth about 1.2 billion years ago. The eukaryotes contain DNA within a membrane-bound structure, the **nucleus** (**FIGURE 3-1B**)—a structure not found in monerans. The earliest eukaryotic cells were also single-celled organisms that lived in the world's oceans. They were members of the Earth's second kingdom, the protistans (pro-TEES-tans).

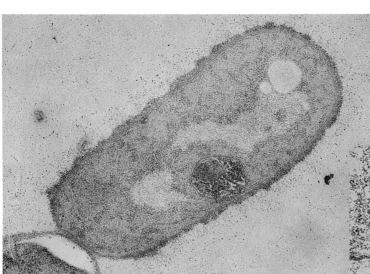

(a)

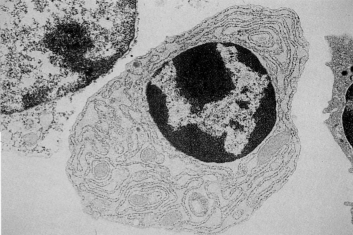

(b)

FIGURE 3-1 Prokaryotes and Eukaryotes *(a)* Magnified many thousands of times, this bacterium is a typical prokaryote. It contains DNA, but the DNA is not located within a nucleus. *(b)* Also greatly magnified, this cell is a typical eukaryote. Its DNA is membrane-bound, and it contains numerous cellular organelles.

The Formation of Complex, Multicellular Organisms Resulted from Evolution

During evolution, eukaryotic organisms gave rise to three additional groupings, or kingdoms: plants, fungi, and animals. The evolution of the three new kingdoms was characterized by two major developments: (1) multicellularity, the development of organisms composed of many cells; and (2) cellular specialization, the emergence of cells that perform specific tasks.

Multicellularity and cellular specialization increased the complexity of life on Earth because they opened up new avenues for existence and new evolutionary challenges. For example, the emergence of cells that perform specialized functions resulted in the evolution of complex motor and sensory systems of numerous animal species, enabling these organisms to move about in their environment and to respond to a wide range of external stimuli. But growing complexity also required control mechanisms to ensure that the parts worked in unison. Over time, distinct body systems of multicellular animals, such as the nervous and endocrine systems, evolved and took control over a wide variety of body functions, most of which contribute to the maintenance of homeostasis. As pointed out in Chapter 1, homeostasis is necessary for the proper functioning of the cells and, thus, of the body as a whole.

Despite dramatic evolutionary changes that have led to a diverse array of life-forms, the cell's basic features have endured, reinforcing the principle of evolution that successful patterns tend to persist. This chapter focuses on general cell characteristics. Although this knowledge is applicable to a wide range of organisms, our focus will remain on humans. Before you peer into the internal architecture of the human cell, however, you might want to take a look at the microscope, an important tool that makes our journey possible.

Microscopes: Illuminating the Structure of Cells

Cells are exceedingly small and, with few exceptions, cannot be seen with the naked eye. Our knowledge of them has depended on the **microscope,** a device consisting of a lens or combination of lenses that enlarges tiny objects. (For a discussion of the discovery of cells, see Scientific Discoveries 3-1.)

Microscopes fit into two broad categories: (1) **light microscopes,** which use ordinary visible light to illuminate the specimen under study; and (2) **electron microscopes,** which use a beam of electrons to create a visual image of the specimen.

Both types of microscopes enlarge minute images and allow us to see details of the cell. Light microscopes provide less magnification and allow us to see less detail than the electron microscope. Light microscopes magnify objects from 100 to 400 times their original size. Electron microscopes enlarge objects 100,000 times their original size. For a comparison, see **FIGURES 3-2** and **3-3.** One type of electron microscope, known as a *scanning electron microscope,* allows scientists to see a three-dimensional view of cells. Photographs from both light and electron microscopes are used in this book to elucidate the structure of important cells, tissues, and organs.

(a)

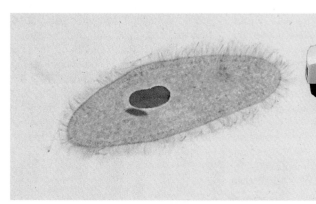

(b)

FIGURE 3-2 The Light Microscope *(a)* In the ordinary light microscope, the image is illuminated by light that is cast from a mirror or (in this case) a light bulb located below the object. Two lenses (one in the eyepiece and one just above the object) magnify the object. *(b)* Light micrograph of a section through a paramecium.

Point/Counterpoint

Fetal Cell Transplantation

FETAL TISSUE TRANSPLANTS: AUSCHWITZ REVISITED

Thomas J. Longua

Thomas J. Longua *has been an educator for 30 years and is currently instructor of anatomy and physiology at the Denver Academy of Court Reporting. He has been active in the pro-life movement for 17 years and is vice-president of the Colorado Right to Life Committee.*

Opposition to the practice of using tissue from aborted human babies for medical purposes starts with opposition to abortion itself.

In the mid-1960s, Western society decided that it needed an "acceptable" way to rid itself of the many unwanted babies it was creating in its newfound sexual permissiveness. Therefore, abortion was legalized, even though it was recognized—as it had been for over 2000 years—as an act of killing an innocent human life.

Pro-life citizens predicted that legalization would lead to other evils, including infanticide and euthanasia. What most of us did not foresee was the use of aborted human fetuses as medical "spare parts." Even when this practice began to be discussed, we could not believe that a "civilized" society could take it seriously. After all, Auschwitz was still a recent memory.

After World War II, the Nuremberg Tribunal was convened for the Nazi war criminals, including the physicians who had experimented on concentration camp victims. In his opening statement, the prosecutor said, "The defendants in this case are charged with . . . atrocities committed in the name of medical science. . . . The wrongs which we seek to condemn and punish have been so calculated, so malignant, and so devastating that civilization cannot tolerate their being ignored, because it cannot survive their being repeated."

Yet with fetal tissue transplantation, we are again committing the same "malignant wrongs" as the Nazi doctors committed. Even the arguments are the same, including the pretense that the victims are "subhuman." (It is ironic that abortion proponents who once denied that the fetus was human now endorse fetal transplants precisely because the tissue *is* human.)

Compare a Nazi doctor to a modern-day advocate:

"Wouldn't it be ridiculous to send the bodies to the crematory oven without giving them an opportunity to contribute to the progress of society?" (Nazi Doctor August Hirt, 1942)

"We are simply using something which is destined for the incinerator to benefit mankind." (Transplant Doctor Lawrence Lawn, 1970)

Advocates of such transplants argue that the procedure is acceptable if it is "separated" from the abortion itself. But even this argument echoes the Nazi doctors' defense: "We caused no deaths. They were all consigned to death by legal authorities; with . . . professional correctness we tried to salvage some good from their plight." Furthermore, new abortion techniques are being developed solely for the purpose of ensuring that the tissue is more usable. This is hardly "separation."

The fact is, the entire practice of fetal tissue transplants rests on the "acceptability" of killing some humans in the first place. (Certainly there is nothing wrong with using tissue from ectopic pregnancies or miscarriages, but the debate revolves around babies that are purposely killed.)

Besides the Nazi-like perverseness of the practice itself, it will certainly have other detrimental effects: It will further legitimize—and even

sanctify—the original killing. What woman contemplating abortion will not be swayed by the argument that the decision to kill her baby will help others? Dr. James J. Parks, a Denver abortionist, bragged, "[My patients] say, 'Thank God, some good is going to come out of this.'" (*New York Times,* November 19, 1989.)

It will certainly increase the number of abortions. Some medical journals claim that fetal tissue could be used to treat a vast array of diseases. In a guest editorial in the November 7, 1988, *Wall Street Journal,* Dr. Emanuel D. Thorne estimated that the current 1.6 million abortions in America every year will not be enough to keep up with the demand for such tissue!

It will increase the pressure to legalize euthanasia. If harvesting unwanted humans becomes standard practice, who can doubt that there will be further demand for organs from older humans whose lives, like those of aborted babies, are deemed "meaningless"?

It will lead to trafficking in human "spare parts"—even to pregnancy planned specifically for that purpose. Indeed, this has already happened in both America and Europe.

It will decrease research that would otherwise be pursued to find alternative methods of dealing with disease, such as the use of autografts, transgene cell lines, animal donors, and synthetic and neurotropic drugs. These methods hold great promise, yet acceptance of fetal tissue as the "easy way out" is already curtailing such research.

Advocates of fetal tissue transplants depict themselves as "humanitarians." But then, so did the doctors at Nuremberg.

HUMAN FETAL TISSUE SHOULD BE USED TO TREAT HUMAN DISEASE

Curt R. Freed

Despite its legalization, abortion remains a controversial issue and will continue to stir debate in the future. In the United States, the debate centers on whether a woman has the right to control her reproduction. The future developments of this political debate are uncertain, but recent elections suggest that pro-choice candidates have been victorious when elections are based on the abortion issue.

Currently, over 1 million legal abortions are performed in the United States each year; most abortions are performed in the first trimester. For nearly all women, having an abortion is an anguishing choice filled with regret and ambivalence. Nonetheless, the difficult personal decision to terminate a pregnancy is made. After the abortion, fetal tissue is usually discarded.

As an alternative to throwing this tissue away, research has shown that fetal tissue may be useful for treating patients with disabling diseases. For over 50 years, research in animals has demonstrated that fetal tissue has a unique capacity to replace certain cellular deficiencies and so may be useful for treating some chronic diseases of humans. These diseases include Parkinson's disease, diabetes, and some immune system disorders.

Cadaver fetal tissue offers the promise of helping large numbers of Americans with crippling diseases. Parkinson's disease, for example, affects hundreds of thousands of Americans. By reducing the ability to move, the disease can end careers and turn people into invalids. The disease is caused by the death of a small number of critically important

nerve cells that produce a chemical called dopamine. Experiments in animals and early experiments in humans indicate that fetal dopamine cells transplanted into the brains of these patients may restore a patient's capacity to move and may even eliminate the disease. Patients whose minds work perfectly well and whose bodies are otherwise normal may become healthy and productive citizens once again.

Using cadaver tissue to treat humans has been debated for nearly 40 years. As kidney, cornea, and other organ transplants were developed in the 1950s, many objected to recovering organs from cadavers. In the intervening decades, opinion has changed so that the practice of recovering these organs from cadavers has gone from a provocative and controversial practice to an accepted policy endorsed by most states. In fact, in most states, a check-off box on the back of driver's licenses is used to give permission for organ donation in the event of the death of the driver. Because abortions are induced, some argue that fetal tissue should be regarded differently from other cadaver tissue. Given the facts that abortion is legal and that fetal tissue is ordinarily discarded, there should be no moral dilemma in using fetal tissue for therapeutic purposes. As with the use of all human tissue for transplant, specific informed consent by the woman donating the tissue must be obtained.

Some have proposed that using fetal tissue for therapeutic purposes will increase the number of abortions. This is preposterous. It strains the imagination to think that a woman would get pregnant and have an abortion simply on the chance that the aborted fetal tissue might be used to treat a patient unknown to her. An un-

wanted pregnancy is an intimate and deeply personal crisis; it is inconceivable that the pregnancy would be seen primarily as a philanthropic opportunity.

Politics and medicine have frequently mixed in the past and will continue to do so in the future. As a physician, I think it is important to try to improve the health of patients with serious diseases. Legally acquired fetal cadaver tissue that would otherwise be discarded should be used to treat humans with disabling diseases.

Curt R. Freed, M.D., *is a professor of medicine and pharmacology at the University of Colorado School of Medicine in Denver. He has written some 70 articles on medical topics as well as numerous abstracts, chapters, and reviews.*

▌SHARPENING YOUR CRITICAL THINKING SKILLS

1. Summarize the positions of each author.
2. Using your critical thinking skills, analyze the view of each author. Is each stand well substantiated? Do the author's biases play a role in each argument?
3. Do you see this debate as scientific or ethical? Explain your answer.
4. Each position is based on at least one key argument. Can you pinpoint them?
5. Which viewpoint do you agree with? Why? What factors (biases) affect your decision?

Visit *Human Biology's* Internet site, www.jbpub.com/humanbiology, to research opposing web sites and respond to questions that will help you clarify your own opinion. (See Point/Counterpoint: Furthering the Debate.)

www.jbpub.com/humanbiology

Scientific Discoveries that Changed the World 3-1

The Discovery of Cells
Featuring the Work of Hooke, Leeuwenhoek, Brown, Schleiden, Schwann, and Virchow

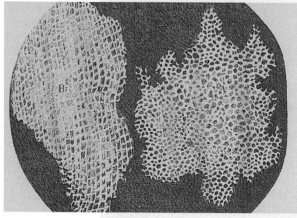

FIGURE 1 Thin Slice of Cork Hooke discovered tiny, boxlike compartments that reminded him of a honeycomb. He called them cellulae, meaning "little rooms."

One of the fundamental principles of biology is known as the **cell theory.** The cell theory consists of three parts: (1) all organisms consist of one or more cells, (2) the cell is the basic unit of structure of all organisms, and (3) all cells arise from pre-existing cells. Although this may seem rather elementary, it was not so obvious to early scientists, who labored with relatively crude instruments and without the benefit of many facts we now take for granted.

One of those scientific pioneers who opened our eyes to the world of cells was Robert Hooke, a seventeenth-century British mathematician, inventor, and scientist. Equipped with a relatively crude microscope, Hooke observed just about everything he

could lay his hands on—which he described in his book *Micrographia,* published in 1665.

One especially useful description was that made on a thin slice of cork (**FIGURE 1**). Peering through his microscope, Hooke beheld a network of tiny, boxlike compartments that reminded him of a honeycomb. He called these compartments cellulae, meaning "little rooms." Today, we know them as cells.

Hooke did not really see cells but, rather, cell walls, the structures that surround the plasma membranes of plant cells. The cytoplasm and cellular organelles had disappeared.

Hooke's work was complemented a few years later by Antony van Leeuwenhoek, a Dutch shopkeeper

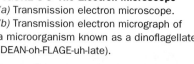

FIGURE 3-3 The Electron Microscope
(a) Transmission electron microscope.
(b) Transmission electron micrograph of a microorganism known as a dinoflagellate (DEAN-oh-FLAGE-uh-late).

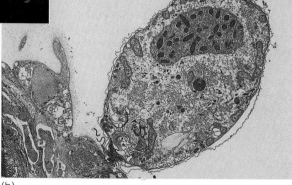

(a)

(b)

Section 3-3

An Overview of Cell Structure

In humans, most cells exhibit a high degree of structural and functional specialization. Very few look alike. But we won't worry about the differences. Instead, we'll focus on generalities as we examine a fictional entity, the typical animal cell. This structure, shown in **FIGURE 3-4**, is a potpourri of structures found in the various body cells. Don't look for it under the microscope, because it only exists on the pages of biology textbooks. You may want to refer to **TABLE 3-1** as you study the following material and as you review your knowledge later.

The Cell Consists of Two Main Compartments

Human cells are the product of millions of years of evolution. As illustrated in **FIGURE 3-4**, human cells—like those of other eukaryotes—consist of two basic compartments, the nuclear and the cytoplasmic. The nuclear compartment, or **nucleus,** is the control center of the cell because it contains the genetic information that regulates the structure and function of the cell. As shown

who spent much of his free time designing simple microscopes. Like Hooke, Leeuwenhoek examined just about everything he could find and wrote extensively on his observations. Wayne Becker, a cell biologist at the University of Wisconsin writes, "His detailed reports attest to both the high quality of his lenses and his keen powers of observation." Becker continues: "They also reveal an active imagination, since at one point he reported seeing a 'homunculus' (little man) in the nucleus of a human sperm cell."

Leeuwenhoek discovered bacteria and protozoans (single-celled eukaryotic organisms). Although his microscopes were superior to any others around at that time, they were still crude by modern standards. This fact and the tendency of the scientific community back then to focus only on the observation and description of life-forms stalled progress in our understanding of cells for some time. In fact, more than a century passed before cell biology moved significantly forward.

Aided by improved lenses, the eighteenth-century English botanist Robert Brown noted that every plant cell he studied contained a centrally placed structure, now called a nucleus. A German colleague, Matthias Schleiden, concluded that all plant tissues consisted of cells. One year later, Theodor Schwann, a German scientist, arrived at a similar conclusion regarding animal cells. This discovery laid to rest an earlier hypothesis that plants and animals were structurally different.

Based on earlier research and his own work, Schwann proposed the first two parts of the cell theory—that all organisms consist of one or more cells and that the cell is the basic unit of structure for all organisms. Less than 20 years after the discovery of cell division, the German physiologist Rudolf Virchow added the third tenet of the cell theory, that all cells arise from preexisting cells.

Like other discoveries, the cell theory is the work of many people over many years. Although only a few people are named as having made possible this important discovery, the credit really belongs to an entire line of scientists who, through careful observation and experimentation, have changed our view of the world.

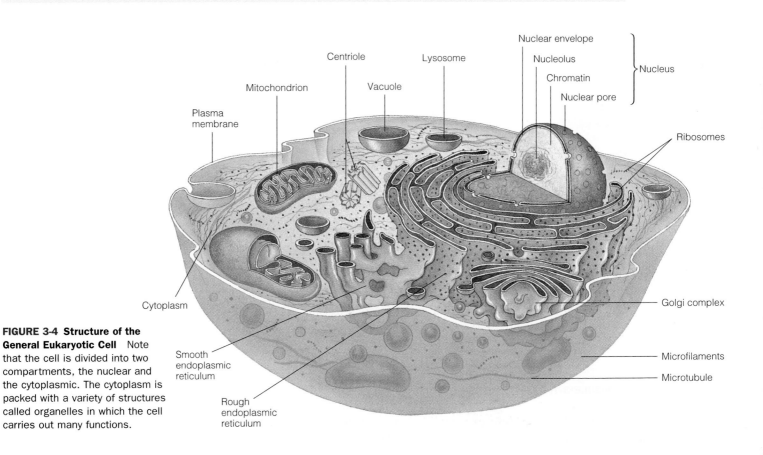

FIGURE 3-4 Structure of the General Eukaryotic Cell Note that the cell is divided into two compartments, the nuclear and the cytoplasmic. The cytoplasm is packed with a variety of structures called organelles in which the cell carries out many functions.

TABLE 3-1

Overview of Cell Organelles

Organelle	Structure	Function
Nucleus	Round or oval body; surrounded by nuclear envelope	Contains the genetic information necessary for control of cell structure and function; DNA contains hereditary information.
Nucleolus	Round or oval body in the nucleus consisting of DNA and RNA	Produces ribosomal RNA
Endoplasmic reticulum	Network of membranous tubules in the cytoplasm of the cell. Smooth endoplasmic reticulum contains no ribosomes. Rough endoplasmic reticulum is studded with ribosomes.	Smooth endoplasmic reticulum (SER) is involved in the production of phospholipids and has many different functions in different cells; round endoplasmic reticulum (RER) is the site of the synthesis of lysosomal enzymes and proteins for extracellular use.
Ribosomes	Small particles found in the cytoplasm; made of RNA and protein	Aid in the production of proteins on the RER and polysomes
Polysome	Molecule of mRNA bound to ribosomes	Site of protein synthesis
Golgi complex	Series of flattened sacs usually located near the nucleus	Sorts, chemically modifies, and packages proteins produced on the RER
Secretory vesicles	Membrane-bound vesicles containing proteins produced by the RER and repackaged by the Golgi complex; contain protein hormones or enzymes	Store protein hormones or enzymes in the cytoplasm awaiting a signal for release
Food vacuole	Membrane-bound vesicle containing material engulfed by the cell	Stores ingested material and combines with lysosome
Lysosome	Round, membrane-bound structure containing digestive enzymes	Combines with food vacuoles and digests materials engulfed by cells
Mitochondria	Round, oval, or elongated structures with a double membrane. The inner membrane is thrown into folds.	Complete the breakdown of glucose, producing NADH and ATP
Cytoskeleton	Network of microtubules and microfilaments in the cell	Gives the cell internal support, helps transport molecules and some organelles inside the cell, and binds to enzymes of metabolic pathways
Cilia	Small projections of the cell membrane containing microtubules; found on a limited number of cells	Propel materials along the surface of certain cells
Flagella	Large projections of the cell membrane containing microtubules; found in humans only on sperm cells	Provide motive force for sperm cells
Centrioles	Small cylindrical bodies composed of microtubules arranged in nine sets of triplets; found in animal cells, not plants.	Help organize spindle apparatus necessary for cell division

in **FIGURE 3-4**, the nucleus is surrounded by a double membrane.

The cytoplasmic compartment lies between the nucleus and the **plasma membrane,** the outermost structure of the cell. It contains a substance known as cytoplasm. **Cytoplasm** is a semifluid material that consists of numerous molecules, including water, protein, ions, nutrients, vitamins, dissolved gases, and waste products.

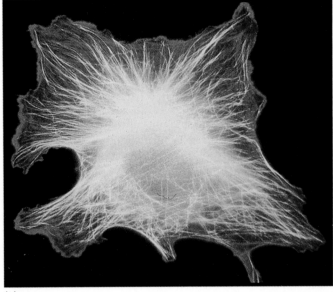

(a)

FIGURE 3-5 The Cytoskeleton *(a)* Photomicrograph of the cytoskeleton of a human fibroblast (connective tissue cell). The microtubules are yellow, and the microfilaments are red. *(b)* Artist's rendition of the cytoskeleton, showing its two major components: microtubules (tubes made of the protein tubulin) and smaller microfilaments (solid fibers made of actin and myosin proteins).

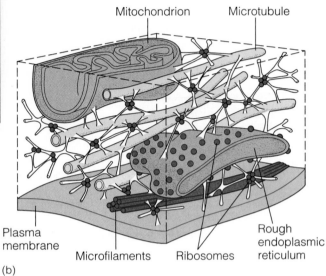

(b)

It forms a nutrient pool from which the cell draws chemicals it needs for metabolism. It is also a dumping ground for wastes.

As shown in **FIGURE 3-4**, the cell contains numerous structures called organelles. **Organelles** ("little organs") carry out specific functions (**TABLE 3-1**). Many organelles are bounded by membranes and form subcompartments within the cytoplasm.

Compartmentalization is essential to the smooth functioning of complex cells because it permits cells to segregate many of their functions. Thus, it increases the efficiency of cells and helps complex organisms perform many functions simultaneously. In a way, a cell is much like a modern factory with different departments in different parts of the building, each performing specific functions, but all working together toward one end.

The cytoplasm of the cell also contains organelles that are not membrane-bound. These organelles perform many important functions (outlined shortly).

Giving shape to the cell is a network of protein tubules and filaments found in the cytoplasm and known as the **cytoskeleton** (**FIGURE 3-5**). Besides giving support, the cytoskeleton helps organize the cell's activities, greatly increasing cellular efficiency.

The cytoskeleton facilitates this vital function by binding with enzymes. **Enzymes** are proteins that increase the rate of chemical reactions in cells. As you will see later in the chapter, many chemical reactions in cells of the body occur in series, with one reaction giving rise to a substance that's used in the next. A series of linked reactions is called a **metabolic pathway** and is the cellular equivalent of an assembly line in a factory. The enzymes of the metabolic pathway are the cell's equivalent to factory workers that work on an assembly line.

Molecules enter a metabolic pathway at one end and are modified along their course by enzymes (**FIGURE 3-6**). Eventually, a finished product comes off the line. Because enzymes are arranged along the cytoskeleton in proper order, the product of one reaction is conveniently situated for the next reaction. Without the cytoskeleton, chaos might erupt as in a factory that haphazardly assembles its products.

FIGURE 3-6 Metabolic Pathway Reactions in the cell occur as parts of larger pathways where the product of one reaction becomes the reactant of another, as in this biochemical pathway. This illustrates the early steps in glycolysis, the breakdown of glucose in the cytoplasm of eukaryotic cells. Note that each reaction has its own enzyme (labeled in red). Also note that the names of these, and all other enzymes, end in *-ase*.

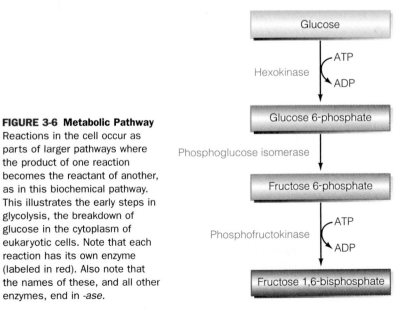

Section 3-4

The Structure and Function of the Plasma Membrane

The **plasma membrane** is a thin layer of lipids and proteins. The plasma membrane controls what goes in and out of the cell thereby controlling the cell's internal chemical environment—a function essential to cellular homeostasis and survival (**TABLE 3-2**).

TABLE 3-2

Overview of Plasma Membrane Functions

Ensures the cell's structural integrity

Regulates the flow of molecules and ions into and out of the cell

Maintains the chemical composition of cytoplasm and extracellular fluid

Participates in cellular communication

Forms a cellular identification system

The Lipids of the Plasma Membrane Form a Double Layer in Which Many of the Proteins Float Freely

The plasma membrane and all of the other internal membranes of the cell consist of lipids, proteins, and a small amount of carbohydrate (**FIGURE 3-7**). **Lipids** or fats are a chemically diverse group of biochemicals that form waxes, grease, and oils. All are all insoluble in water. Lipids serve a variety of functions. Some provide energy. Others serve as insulation and energy storage. Still others play a structural role. Three different lipids are discussed in this book. The first, which is found in the membranes of cells, is the **phospholipid** (FOSS-foh-lipid). Shown in **FIGURE 3-8**, phospholipids form lollypop-shaped molecules with a polar (charged) head and a nonpolar (uncharged) tail. Most of the lipid molecules in the plasma membrane are phospholipids.

Experimental studies strongly suggest that the phospholipid molecules of the plasma membrane are arranged in a double layer. **FIGURE 3-7** shows the most widely accepted model (theory) of the plasma membrane's structure, the **fluid mosaic model.** As shown, the polar heads of the outer layer of phospholipid mol-

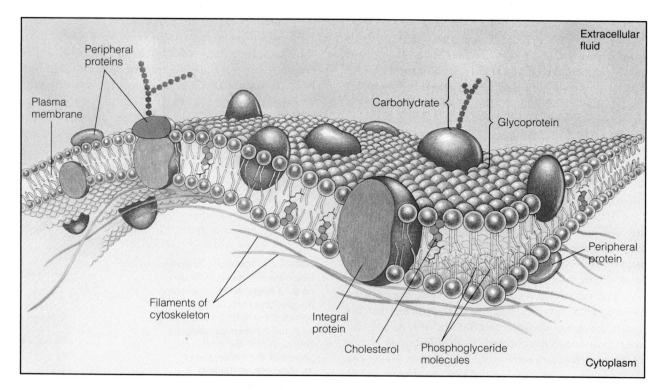

FIGURE 3-7 Fluid Mosaic Model of the Plasma Membrane of Animal Cells   The fluid mosaic model is the most widely accepted theory of the plasma membrane. Phospholipids are the chief lipid component. They are arranged in a bilayer. Integral proteins float like icebergs in a sea of lipids.

FIGURE 3-8 Phospholipids (a) Each phospholipid consists of a glycerol backbone, two fatty acids, a phosphate, and a variable group generally designated by the letter R. In this case, the R group is choline, which is polar. (b) Because of the R group, the molecule has a polar head. The nonpolar tail region is formed by the two fatty acid chains.

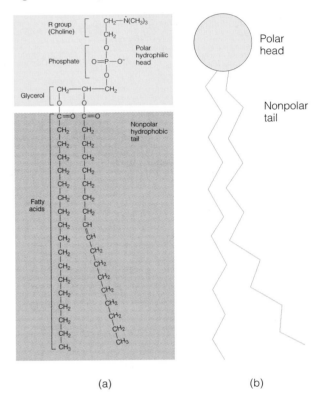

(a)

(b)

FIGURE 3-9 Structure of Amino Acids (a) The amino acid is a small organic molecule with four groups, one of which is variable and is designated by the letter R. (b) Some representative R groups.

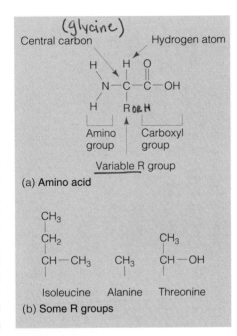

(a) Amino acid

(b) Some R groups

ecules face the watery extracellular fluid, and the polar heads of the inner layer of molecules are directed inward toward the watery cytoplasm.

Interspersed in the lipid bilayer of human cells are large protein molecules known as integral proteins. **Proteins** are large molecular weight compounds made of many smaller molecules, known as amino acids. **Amino acids** contain a central carbon atom, attached to which are an amino group (NH_2) and a carboxylic acid group (COOH)—from which they derive their name (**FIGURE 3-9**). As shown in the figure, the central carbon atom also bonds to a hydrogen and a variable group (denoted by the letter R). To date, approximately 250 amino acids have been discovered in plants and animals or synthesized in the laboratory. However, only 20 of them are found in natural proteins. From this seemingly limited pool of amino acids, cells build thousands of different proteins.

Proteins serve many different functions. One important class of proteins, the **enzymes**, accelerates chemical reactions in the body. Without enzymes, few reactions would occur. Other proteins play a structural role.

Keratin (CARE-ah-tin), for example, is a protein in human hair and nails. Collagen (coll-AH-gin), the most abundant protein in the human body, forms fibers that are found in ligaments, tendons, and bones.

Proteins are synthesized in the body one amino acid at a time, but a large protein containing hundreds of amino acids is assembled in under a minute. **FIGURE 3-10** shows how amino acids bind. The bond between two amino acids is called a **peptide bond.** As a protein is synthesized, the chain of amino acids begins to twist and bend (**FIGURE 3-11**). Further along in the process, the amino acid structure forms a three-dimensional structure. It is often globular in nature. The integral proteins of the plasma membrane are globular structures that float freely like giant icebergs in their sea of lipid. As illustrated, some of them completely penetrate the plasma membrane; others penetrate only partway. Several integral proteins may join to form pores that

FIGURE 3-10 Formation of the Peptide Bond The carboxyl and amino groups react in such a way that a covalent bond is formed between the carbon of the carboxyl group of one amino acid and the nitrogen of the amino group of another. This bond is called a peptide bond.

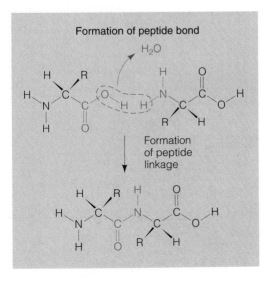

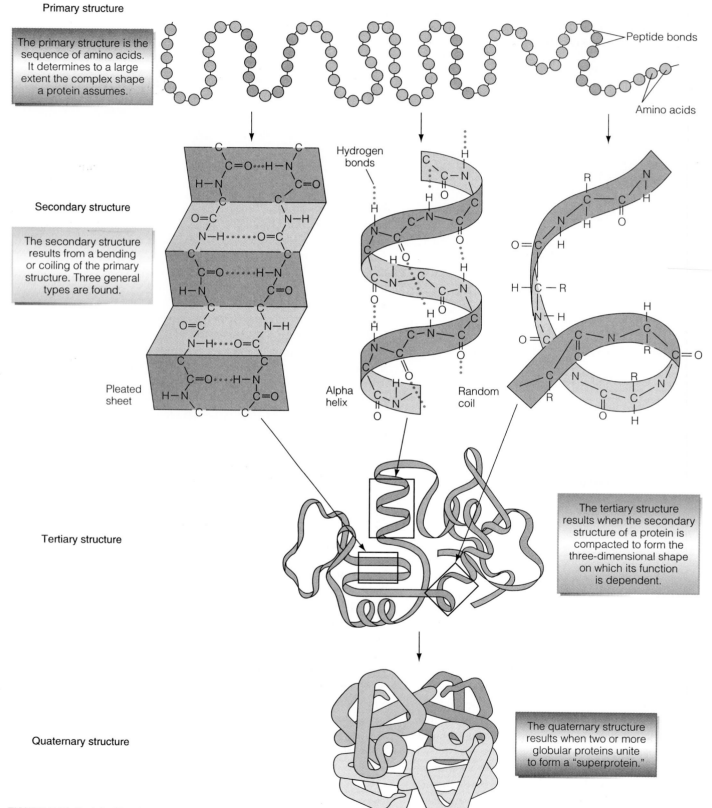

Primary structure

The primary structure is the sequence of amino acids. It determines to a large extent the complex shape a protein assumes.

Peptide bonds

Amino acids

Secondary structure

The secondary structure results from a bending or coiling of the primary structure. Three general types are found.

Pleated sheet

Hydrogen bonds

Alpha helix

Random coil

Tertiary structure

The tertiary structure results when the secondary structure of a protein is compacted to form the three-dimensional shape on which its function is dependent.

Quaternary structure

The quaternary structure results when two or more globular proteins unite to form a "superprotein."

FIGURE 3-11 Protein Structure

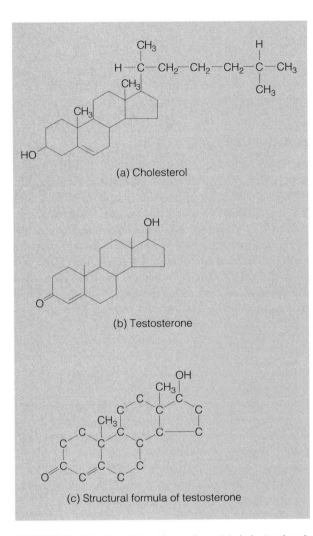

(a) Cholesterol

(b) Testosterone

(c) Structural formula of testosterone

FIGURE 3-12 Steroids Steroids, such as *(a)* cholesterol and *(b)* testosterone, consist of four rings joined together. Parts *(a)* and *(b)* are a shorthand way of drawing *(c)* the structural formula.

permit the movement of molecules into and out of the cell. Others may attach to the underlying cytoskeleton, anchoring the plasma membrane in place.

Another group of proteins, the **peripheral proteins,** is found in the plasma membrane. On the cytoplasmic side of the membrane, peripheral proteins often attach to the exposed surfaces of integral proteins. They may also bind to the cytoskeleton.

Plasma membranes contain small but significant amounts of carbohydrate. Most carbohydrates are oligosaccharides (short chains of monosaccharides). Shown in **FIGURE 3-7**, these segments attach to integral proteins that protrude into the extracellular fluid of cells. A protein combined with a carbohydrate is a **glycoprotein.** The glycoproteins of the plasma membrane are vital to cell function.

Another type of lipid is also found in the plasma membrane. This lipid is cholesterol. Cholesterol mole-cules are found wedged between the phospholipids. They appear to make the membrane more elastic. Cholesterol is a **steroid,** another type of lipid. Its chemical formula is shown in **FIGURE 3-12**. Cholesterol and other steroids consist of four ring structures that are joined together. Cholesterol molecules are found in the plasma membranes of cells, wedged in between the phospholipid molecules (**FIGURE 3-8**). Their importance is not well understood.

The Plasma Membrane Serves Many Functions and Is Essential to Cellular Homeostasis

The plasma membrane is a vital component of homeostasis. The plasma membrane regulates the flow of molecules and ions into and out of the cell. This helps maintain the precise chemical concentrations inside and outside the cell—both essential for optimal cellular function. (For a review of plasma membrane functions, see **TABLE 3-2.**)

Because the plasma membrane regulates the molecular traffic, it is said to be **selectively permeable.** Put another way, the plasma membrane selects or controls what enters and leaves the cell. As you will learn in later chapters, the plasma membrane also plays a vital role in cellular communication. Communication is defined very broadly here to include the transmission or receipt of any kind of message. Some hormones, for example, attach to specific integral proteins (membrane receptors) in the plasma membranes of cells. Like the ringing of a doorbell, this sends a message to the cell's interior that triggers important changes in the cell's structure and function. Consequently, hormones are said to "communicate" with the cell through its membrane.

The plasma membrane is also part of an elaborate cellular identification system. The cells in each of us have a unique protein "cellular fingerprint."[1] Because the protein composition of the plasma membrane of the cells of each individual is unique, a person's immune system can recognize its own cells and determine when foreign cells—such as bacteria—have invaded. The immune system also identifies cancer cells, whose cellular fingerprints have become altered. Once it has identified microbial invaders or tumor cells, the body mounts an attack, often successfully eliminating them. Without this marvelous ability to recognize foreign invaders or altered cells, we would very likely perish in rapid order from infections or tumors.

---

[1]These proteins are called glycoproteins because they contain smaller carbohydrate molecules.

As noted in Chapter 1, human actions often disrupt homeostasis. Perhaps one of the most blatant examples is an organ transplant. During this procedure, surgeons remove damaged organs and replace them with healthy organs from donors. Kidneys, hearts, and skin are the most commonly transplanted organs and tissues. However, unless the tissue or organ comes from an identical twin, who's genetically identical to the recipient, the immune system of the recipient recognizes the cells of a transplanted organ as foreign and attempts to destroy them. To prevent rejection, physicians prescribe drugs that suppress the immune system. Unfortunately, this treatment makes a patient vulnerable to bacterial and viral infections.

In a sense, then, a change in the body's delicate balance (caused by a transplant) results in a grave imbalance that must be counterbalanced by drugs. As evidence of the seriousness of this tinkering, many early heart transplants failed as patients died from massive infections. Not to be deterred, scientists have developed new drugs that suppress the immune system and prevent the rejection of transplanted tissues and organs without completely eliminating the body's protection from bacteria and viruses. In such instances, the body maintains some ability to control its internal environment and to keep itself free of bacteria and other infectious agents.

Fetal cell transplantation, discussed at the opening of this chapter, represents another means of trying to dupe the immune system. Fetal cells have no protein fingerprint and can therefore be transplanted without triggering an immune reaction.[2]

Molecules Move Through the Plasma Membrane in Five Ways

Cells are bathed in a liquid that fills the tiny spaces between them. This is called **interstitial fluid** (in-ter-STICH-al). This liquid and the cytoplasm of cells contain a great deal of water and a variety of chemical substances that are dissolved or suspended in them. Some of these substances pass through the membrane with ease; others must be transported across the membrane with the aid of special molecules. All told, five basic mechanisms exist for moving molecules and ions across plasma membranes (**TABLE 3-3**).

The Movement of Molecules from High to Low Concentrations Either Directly Through the Membrane Lipid or Through Pores Is Called Simple Diffusion.

Lipid-soluble substances pass directly through the membrane via diffusion. Why? Because the plasma membrane is principally made of lipid, and chemical compounds readily dissolve in chemically similar solvents. Steroid hormones, oxygen, and carbon dioxide are all lipid-soluble and therefore pass directly through the lipid bilayer of the plasma membrane with ease (**FIGURE 3-13A**).

---

[2]The immunologic immaturity of fetal cells is probably an evolutionary adaptation that protects fetuses from attack by the mother's immune system.

TABLE 3-3	
Overview of Plasma Membrane Transport	
Process	**Description**
Simple diffusion	Flow of ions and molecules from high concentrations to low. Water-soluble ions and molecules probably pass through pores; water-insoluble molecules pass directly through the lipid layer.
Facilitated diffusion	Flow of ions and molecules from high concentrations to low concentrations with the aid of protein carrier molecules in the membrane.
Active transport	Transport of molecules from regions of low concentration to regions of high concentration with the aid of transport proteins in the cell membrane and ATP
Endocytosis	Active incorporation of liquid and solid materials outside the cell by the plasma membrane. Materials are engulfed by the cell and become surrounded in a membrane.
Exocytosis	Release of materials packaged in secretory vesicles
Osmosis	Diffusion of water molecules from regions of high water (low solute) concentration to regions of low water (high solute) concentrations

Diffusion refers to any movement of molecules or ions down a concentration gradient—that is, from high to low concentration. As you will soon see, two types of diffusion exist, simple and facilitated. Movement of lipid-soluble chemicals through the membrane without assistance, as described above, is **simple diffusion.**

Water-soluble materials cannot pass through the lipid bilayer of the plasma membrane and must travel via other routes. Scientists believe that water and many other small water-soluble molecules and ions pass through pores in the plasma membrane and that these pores are formed by integral proteins (**FIGURE 3-13B**). Movement through membrane pores is another form of simple diffusion. Interestingly, no one has ever even seen a pore in electron micrographs of cell membranes, and their existence is based entirely on physiological research. Nonetheless, researchers believe that some protein pores remain open all the time, thus permitting small molecules and ions to cross the plasma membrane.

Water-Soluble Molecules Can Also Diffuse Through Membranes with the Assistance of Protein Carrier Molecules.

Carrier proteins are molecules that transport small water-soluble molecules and ions across the membrane in ways not yet completely understood. Carrier proteins "shuttle" molecules across membranes from regions of high concentration to regions of low concentration. This process is called **facilitated diffusion** to distinguish it from simple diffusion through pores or through the lipid bilayer (**FIGURE 3-13C**).

Molecules Are Also Actively Transported Across the Membrane.

Another transport mechanism found in plasma membranes is known as active transport. **Active transport** is the movement of molecules across membranes with the aid of protein carrier molecules in the plasma membrane and with energy supplied by a special molecule called ATP (**FIGURE 3-13D**). ATP or **adenosine triphosphate** (ah-DEN-oh-seen try-FOSS-fate) is a molecule with an important job. ATP consists of three smaller molecules, shown in **FIGURE 3-14**: a nitrogen-containing organic base, adenine; a five-carbon sugar, ribose; and three phosphate groups.

ATP is often likened to a form of energy currency. When the cell needs energy, say to move molecules across its membrane, ATP splits off a phosphate, forming ADP (adenosine diphosphate). The breaking of the phosphate bond yields energy that the cell can use directly. The reaction is:

$$ATP \rightarrow ADP + P_i + Energy$$

Reactions in the body that give off energy often turn the energy over to ADP. That way, new ATP can be formed.

$$ADP + P_i + Energy \rightarrow ATP$$

ATP therefore also serves as a kind of an energy shuttle, picking up energy from reactions that release it and shuttling it to reactions that require it. The energy is stored in the covalent bond between ADP and the last phosphate. During active transport, molecules and ions are transported from regions of low concentration to regions of high concentration—movement "up," or against, the concentration gradient.

Active transport occurs in cells that must concentrate chemical substances to function properly and, like the various forms of diffusion, it is essential for maintaining homeostasis. For example, cells in the thyroid glands in the neck require large quantities of the iodide ion (I^-) to manufacture the hormone thyroxine (thigh-ROX-in). But since levels of I^- in the bloodstream are fairly low, thyroid cells must actively transport them into the cytoplasm, where the iodide concentration is many times higher. Ultimately, if thyroid cells had to rely on diffusion, they would probably not be able to produce enough hormone to meet the body's needs. Cells also use active transport to move materials out of their cytoplasm against concentration gradients.

Large Molecules Such as Proteins and Cells Are Ingested by Endocytosis.

Endocytosis (en-doe-sigh-TOE-siss; literally, "into the cell") is illustrated in **FIGURE 3-13E**. This process requires ATP and consists of two related activities: phagocytosis and pinocytosis. **Phagocytosis** ("cell eating") occurs when cells engulf larger particles, such as bacteria and viruses. In humans, phagocytosis is limited to relatively few types of cells—those involved in protecting the body against foreign invaders. **Pinocytosis** (cell drinking) occurs when cells engulf extracellular fluids and dissolved materials. Most, if not all, cells are capable of pinocytosis.

Cells Also Regurgitate Materials, Releasing Large Molecules, Such as Hormones.

This process is called **exocytosis** ("out of the cell") and is essentially the reverse of endocytosis (**FIGURE 3-13F**). For example, in the cells of the pituitary gland, which is located under the brain, protein hormones are packaged internally into tiny membrane-bound vesicles. These vesicles migrate to the plasma membranes and fuse with them. At the point of fusion, the membranes break down, and the protein hormone is released into the extracellular fluid. This process is exocytosis.

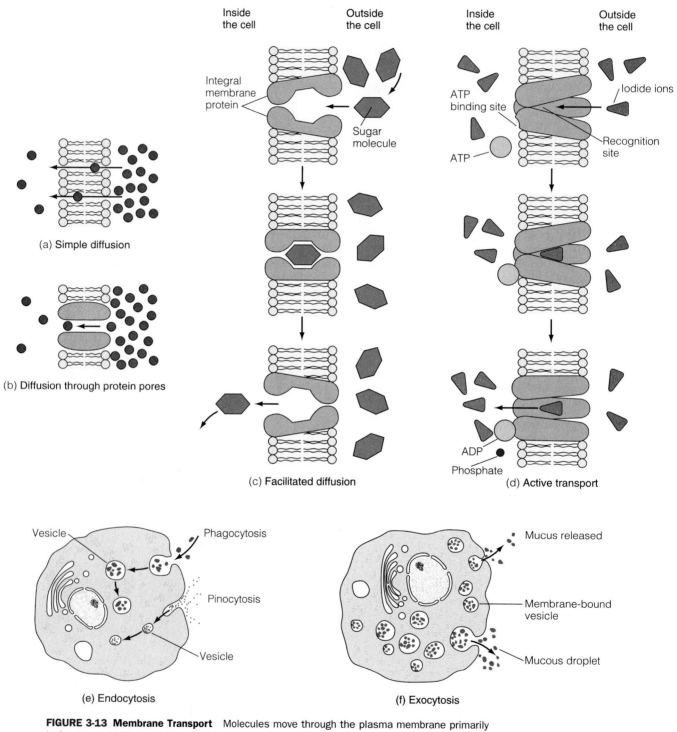

Inside the cell **Outside the cell** **Inside the cell** **Outside the cell**

Integral membrane protein

Sugar molecule

(a) Simple diffusion

(b) Diffusion through protein pores

(c) Facilitated diffusion

ATP binding site

Iodide ions

ATP

Recognition site

ADP

Phosphate

(d) Active transport

Vesicle

Phagocytosis

Pinocytosis

Vesicle

(e) Endocytosis

Mucus released

Membrane-bound vesicle

Mucous droplet

(f) Exocytosis

FIGURE 3-13 Membrane Transport Molecules move through the plasma membrane primarily in five ways. *(a)* Lipid-soluble substances pass through the membrane directly via simple diffusion. *(b)* Water-soluble molecules can diffuse passively through pores formed by protein molecules. *(c)* Water-soluble molecules can also diffuse through membranes with the assistance of proteins in facilitated diffusion. *(d)* Other proteins use energy from ATP to move against concentration gradients in a process called active transport. *(e)* Finally, cells can engulf large particles, cell fragments, and even entire cells via endocytosis. *(f)* Exocytosis, the reverse process, rids the cell of large particles.

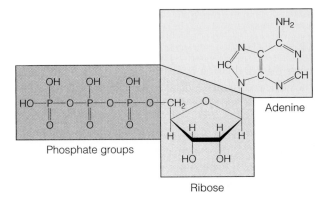

FIGURE 3-14 Molecular Structure of the Nucleotide ATP
ATP is often called the energy currency of the cell.

■ The Diffusion of Water Across the Plasma Membrane Is Known as Osmosis

Like any other small molecule, water moves from one side of a plasma membrane to the other by diffusion. The diffusion of water across a selectively permeable membrane, however, is given a special name, **osmosis** (oss-MOE-siss; this word comes from a Greek word meaning "to push"). To understand osmosis, consider a simplified example in **FIGURE 3-15**.

In this example, we begin with a large bag made of a special plastic that's selectively permeable. We fill the bag with water and table sugar (sucrose) and then submerge it in a large beaker of distilled water (**FIGURE 3-15**). Because the bag permits the water (but not the sucrose) to move in and out, the water molecules diffuse down the concentration gradient, and the bag begins to swell. This swelling results from the inward movement of water molecules shown by the arrows in **FIGURE 3-15B**. In this system, water moves from regions of high water concentration (distilled water) to regions of low water concentration (inside the bag).

The concentration difference across the membrane "drives" the water across a selectively permeable membrane. Therefore, whenever two fluids with different concentrations of solute (dissolved substance) are separated by a selectively permeable membrane, water will flow from one to the other, moving down the concentration gradient; the driving force is called **osmotic pressure**. The greater the difference in water concentration, the greater the osmotic pressure, and the more quickly the water moves.[3]

In humans and other animals, osmotic pressure is responsible for the movement of water across many membranes. Combined with blood pressure and other forces, osmotic pressure plays an important role in the filtering of blood in the kidney—a function essential to homeostasis (Chapter 10).

Because osmosis helps equalize water concentrations on opposite sides of membranes, it is an important homeostatic mechanism in all body cells. In other words, osmosis helps regulate the concentration of the fluid surrounding the cells of multicellular organisms (also called **extracellular fluid**), keeping it at the same concentration as that of the cells' cytoplasm. In such cases, the extracellular fluid is said to be **isotonic** (having the same strength).

If the fluid surrounding a cell is not isotonic, serious problems can be expected. For example, if blood cells are immersed in a solution that is more concentrated than their cytoplasm, water will move out of the cells, and the cells will shrink. A solution with a solute concentration higher than the cell's cytoplasm is said to be **hypertonic** (having a greater strength). A solution with a solute concentration lower than the cell's cytoplasm is said to be **hypotonic** (having a lesser strength). If red blood cells are placed in such a solution, water will rush in, causing the cells to swell and burst.

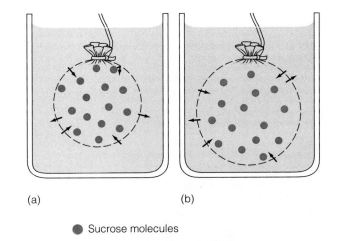

(a)   (b)

● Sucrose molecules

FIGURE 3-15 Osmosis Osmosis is the diffusion of water molecules from a region of higher water concentration (or low solute concentration) to one of lower water concentration (or high solute concentration) across a selectively permeable membrane. *(a)* To demonstrate the process, immerse a bag of sugar water in a solution of pure water. *(b)* Water diffuses into the bag (toward the lower concentration), causing it to swell.

---

[3]Water concentration is measured by the number of water molecules per milliliter. Pure water has a higher water concentration than salt water; that is, there are more water molecules per unit volume in pure water than in salt water.

Cellular Compartmentalization: Organelles

We next turn our attention to the **organelles**—structures that carry out many of the key functions of the cell. We begin with the nucleus.

The Nucleus Houses the DNA and Is the Cell's Command Center

As noted earlier, the nucleus is one of the cell's two major compartments. It is also one of the cell's most conspicuous and most important organelles (**FIGURE 3-16**). It is often likened to a cellular command center. The nucleus of the cell consists of (1) the nuclear envelope, (2) the chromatin, (3) the nucleoplasm, and (4) the nucleolus.

The **nuclear envelope** is a double membrane that isolates the nuclear material from the cytoplasm. Minute channels in the envelope, the **nuclear pores,** allow materials to pass into and out of the nucleus (**FIGURE 3-16B**).

The bulk of the nucleus contains long, threadlike fibers of DNA and protein, called **chromosomes** (CHROME-ah-zomes). The DNA-protein material, known as **chromatin** (CHROME-eh-tin), appears as fine granules in transmission electron micrographs of nuclei (**FIGURE 3-16A**). The DNA in the nucleus is the genetic material. It houses all of the information a cell needs for proper development and functioning—hence the description of the nucleus as a central command center. Just before cell division, the chromosomes coil to form short, compact bodies, highly visible in the light microscopic preparations (**FIGURE 3-17**). This compaction greatly facil-

itates the cell division by making it much easier to divide the nuclear contents. A detailed discussion of DNA can be found in Chapter 17. Proteins, water, and other small molecules and ions are also found in the nucleus, forming a semifluid material known as the **nucleoplasm** (new-KLEE-oh-plazem)—the nuclear equivalent of cytoplasm.

Nucleoli ("little nuclei"; singular, nucleolus) are temporary structures found in the nuclei of cells between cell divisions (**FIGURE 3-4**). They appear as small, clear, oval structures in light micrographs and as dense bodies in transmission electron micrographs. Nucleoli are regions of the DNA actively engaged in the production of **ribosomal RNA** (**rRNA;** pronounced RYE-bow-zomal), so named because it combines with certain proteins to form ribosomes. **Ribosomes** are nonmembranous organelles that appear as small, dark granules in electron micrographs and consist of two subunits. Produced in the nucleus, the subunits enter the cytoplasm through the nuclear pores. In the cytoplasm, they combine to play an important part in protein synthesis, described in Chapter 17. A detailed discussion of RNA is also presented in that chapter.

The Mitochondrion Is the Site of Cellular Energy Production in Animal Cells

No factory would operate without a source of energy and a place where that energy is generated. In most industrial nations, a factory's energy comes from power plants fueled by coal or oil or nuclear fuels. Like the factory, the cell also requires energy and a special structure to generate it. This structure is the mitochondrion (MY-toe-CON-dree-on).

The **mitochondrion,** shown in **FIGURE 3-18**, is an organelle that liberates energy from organic molecules,

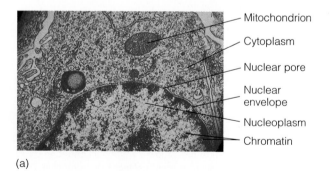

(a)

(b)

FIGURE 3-16 The Nucleus *(a)* The nucleus houses the genetic information that controls the structure and function of the cell. The nuclear envelope, made of lipids and proteins like the plasma membrane, actually consists of two membranes, separated by a space. Pores in the membrane allow the movement of molecules into the nucleus, providing raw

materials for the synthesis of DNA and RNA. They also allow RNA molecules to travel into the cytoplasm, where they participate in the production of protein. Chromatin may be densely packed or loosely arranged in the nucleus. *(b)* Colorized scanning electron micrograph of the nuclear membrane, showing numerous pores.

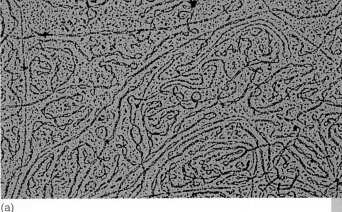

(a)

FIGURE 3-17 Chromosomes (a) The threadlike chromosomes, made of chromatin fibers consisting of protein and DNA, must condense before the nucleus can divide. (b) Condensed chromosomes. Can you think of any advantages to this strategy?

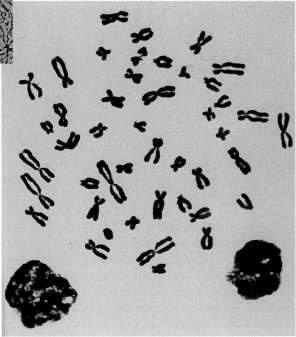

(b)

usually sugars and fats. Unlike power plants that generate electricity, mitochondria (plural) crank out ATP. Cells use ATP for many purposes: to synthesize chemical substances, to transport molecules across membranes, to divide, to contract, and to move about.

Although mitochondria vary considerably in form and number from cell to cell depending on the cell's energy requirement, they "share" several common characteristics. All mitochondria, for instance, are membrane-bound. In addition to the outer membrane, which holds the structure intact, all mitochondria contain an inner membrane, which is thrown into folds or **cristae** (CHRIS-tee) (**FIGURE 3-18B**). This membrane therefore divides the mitochondrion into two distinct compartments, the inner compartment and the outer compartment, each with its own function. They also increase the surface area on which chemical reactions take place.

Mitochondria play a key role in the liberation of energy in the body. During the breakdown of glucose, the principal source of energy, about one-third to two-thirds of the energy released during the chemical reactions is captured by the cell to produce ATP. The rest of the energy once housed in glucose molecules is given off as heat. This "waste" product creates body heat. It, in turn, is necessary for maintaining normal enzymatic function in warm-blooded animals like you and me. Without it, many processes would cease, as would we.

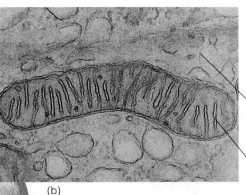

Cytoplasm

Matrix

(b)

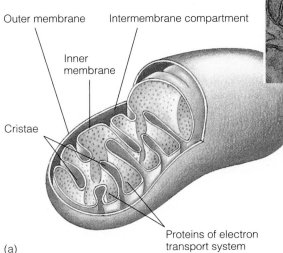

Outer membrane   Intermembrane compartment

Inner membrane

Cristae

Proteins of electron transport system

(a)

FIGURE 3-18 The Mitochondrion (a) Like the nucleus, the mitochondrion is delimited by a double membrane. The inner membrane, however, is infolded, forming cristae. (b) Electron micrograph of a mitochondrion. The infolding creates two distinct compartments: an inner compartment filled with a material called matrix, and an outer compartment lying between the two membranes. Compartmentalization is essential for ATP synthesis because it isolates function. Membranes that form compartments also increase the surface area on which reactions occur.

Three Organelles Are Involved in Manufacturing Protein and Other Cellular Products

Many cells are miniature factories synthesizing a variety of molecules vital for growth, reproduction, and day-to-day maintenance. Many of these molecules are used within the cell; others, such as hormones, are released from the cell into the bloodstream where they travel to distant cells, eliciting specific responses. This section examines three organelles principally involved in cellular synthesis: the endoplasmic reticulum, ribosomes, and the Golgi complex.

Endoplasmic Reticulum.

Coursing throughout the cytoplasm of many cells is a branched network of membranous channels, the **endoplasmic reticulum** (END-oh-PLAZ-mick rah-TICK-u-lum; meaning intracellular network) (**FIGURE 3-19A**). These membranous channels are derived chiefly from the nuclear envelope and may be coated with ribosomes (**FIGURE 3-19B**). Ribosome-studded endoplasmic reticulum is referred to as the **rough endoplasmic reticulum (RER)**. The RER produces a variety of proteins, among them digestive enzymes, protein hormones, plasma membrane proteins, and lysosomal enzymes (described later).

The proteins produced by the RER are made on the outside surface of the structure but are quickly transferred into its interior, possibly to protect them from damage. The interior of the RER is yet another specialized compartment. Not only does it protect the proteins, it contains 30–40 enzymes that convert some proteins into glycoproteins. Proteins and glycoproteins are then trans-ferred to yet another organelle, the Golgi complex, for final packaging.

Cells also contain endoplasmic reticulum lacking ribosomes (**FIGURE 3-19C**). Known as the **smooth endoplasmic reticulum (SER),** this organelle specializes in the production of phospholipids, which are needed to replenish the plasma membrane. The SER performs a variety of functions in different cells. In the human liver, for instance, it detoxifies certain drugs, such as barbiturates, one type of sedative. In the stomach of humans and other mammals, the SER of certain cells produces hydrochloric acid, necessary for protein digestion. In the adrenal glands, the SER produces steroid hormones.

Ribosomes.

As noted earlier, ribosomes are tiny particles made of protein and ribosomal RNA (rRNA). In the cytoplasm, ribosomes attach to one class of RNA molecules known as messenger RNA (mRNA). Messenger RNA molecules are produced in the nucleus of the cell and, as you shall see in Chapter 17, contain all of the information required to synthesize protein.

In the cytoplasm, several ribosomes attach to a single strand of mRNA, forming a structure known as a **polyribosome** or **polysome.** Together, mRNA and its attached ribosomes begin to produce protein. If it is a protein destined for inclusion in lysosomes or the plasma membrane or if it is intended for extracellular use (hormones and digestive enzymes), the polysome attaches to the RER, and the protein is transferred into the cisterna. If it is a protein for intracellular use, like those of the cytoskeleton or cytoplasmic enzymes, the polysome remains free within the cytoplasm.

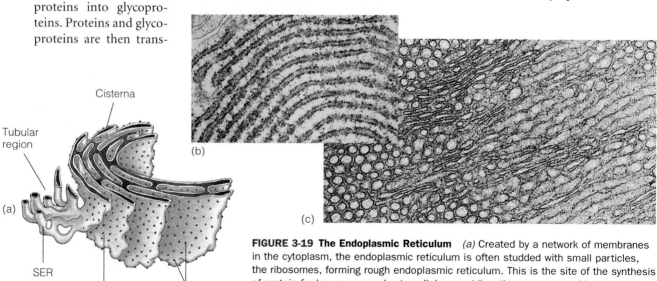

Cisterna

Tubular region

(a)

(b)

(c)

SER

RER   Ribosomes

FIGURE 3-19 The Endoplasmic Reticulum *(a)* Created by a network of membranes in the cytoplasm, the endoplasmic reticulum is often studded with small particles, the ribosomes, forming rough endoplasmic reticulum. This is the site of the synthesis of protein for lysosomes and extracellular use (digestive enzymes and hormones). *(b)* Electron micrographs of RER and *(c)* SER surrounding liquid droplets.

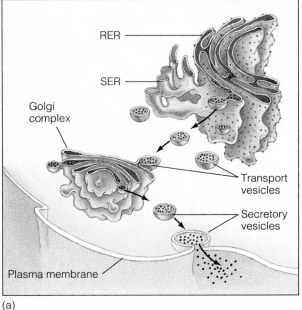

(a)

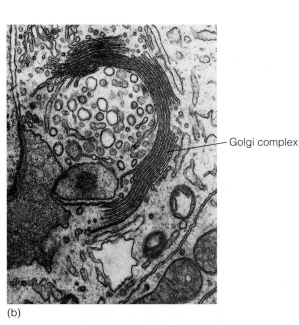

Golgi complex

(b)

FIGURE 3-20 Protein Synthesis and Secretion *(a)* Protein packed in lysosomes and secretory granules (for later export) is synthesized on the RER and then transferred in tiny transport vesicles to the Golgi complex. Protein is sometimes chemically modified in the cisternae of both the

RER and the Golgi complex. The Golgi complex separates protein by destination and repackages it into secretory granules, which remain in the cytoplasm until secreted by exocytosis. *(b)* Transmission electron micrograph of the Golgi complex.

The Golgi Complex.

RER transports the proteins and glycoproteins it manufactures to its terminal ends (**FIGURE 3-20A**). As these products build up inside the RER, the ends bulge and eventually pinch off, forming tiny membrane-bound **vesicles** (VESS-sick-kuls). The vesicles protect the proteins and glycoproteins as they are transported to the final packaging unit, the Golgi complex.

The **Golgi complex** (GOL-gee) consists of a series of membranous sacs (**FIGURE 3-20B**). It performs three functions. First, it sorts the molecules it receives. The Golgi complex separates lysosomal enzymes from hormones and plasma membrane proteins, ensuring that products end up in the right place. Second, like the RER, the Golgi complex chemically modifies many proteins, often adding carbohydrates or other small molecules. Finally, the Golgi complex packages proteins into lysosomes or secretory vesicles (or secretory granules) (**FIGURE 3-20A**).

Secretory vesicles are membranous organelles that reside in the cytoplasm. They act as temporary storage sites for hormones and digestive enzymes that will eventually be "shipped" to distant sites. When the signal for release comes, secretory vesicles migrate to the plasma membrane and fuse with it. The fused membranes soon dissolve, releasing the contents of the secretory vesicle into the extracellular space. This is an example of exocytosis.

Lysosomes Are Membrane-Bound Organelles That Contain Digestive Enzymes

Lysosomes (LIE-so-ZOMES; literally, digestive bodies) have been likened to tiny time bombs inside cells that, fortunately, rarely go off (**FIGURE 3-21**). Containing numerous digestive enzymes produced by the RER and packaged by the Golgi complex, these organelles—when functioning properly—break down materials that are phagocytized by cells. A better description of the lysosome might be to liken it to a wrecking crew that's kept under lock and key until the cell takes in something it needs to break apart.

As illustrated in the top of **FIGURE 3-22**, material engulfed by the cell (endocytosis) is enclosed in a membrane derived from the plasma membrane. The resultant structure is called a **food vacuole.** Lysosomes attach to food vacuoles and release their enzymes into them. The enzymes then break down the molecules inside the food vacuole into smaller molecules. These, in turn, diffuse into the cytoplasm, where they are often used by the cell. The undigested material left behind is expelled from the cell by exocytosis.

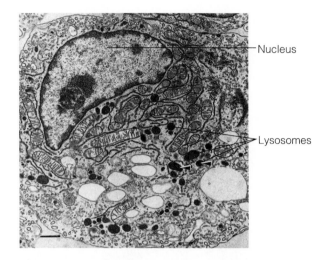

FIGURE 3-21 The Lysosome Electron micrograph of lysosomes within a macrophage. The dark-staining bodies are the lysosomes, filled with digestive enzymes that are used by these cells to break down ingested material.

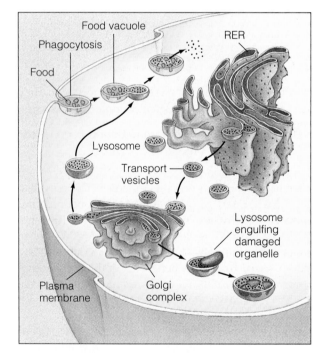

FIGURE 3-22 Functions of Lysosomes The digestive enzymes of the lysosome are produced on the RER and transported to the Golgi complex for repackaging. Lysosomes produced by the Golgi complex fuse with food vacuoles; this allows their enzymes to mix with the contents of the food vacuole. The enzymes digest the contents, which diffuse through the membrane into the cytoplasm, where they are used. The membrane surrounding the lysosome helps protect the cell from digestive enzymes.

Besides providing the cell with nutrients, lysosomes are cellular custodians that destroy aged or defective cellular organelles, such as mitochondria. Lysosomes may also digest protein and other molecules that have become defective. This process helps cells maintain their structure and function. How cells recognize defective molecules and organelles remains a mystery.

Lysosomes also rid the body of injured or aged cells. In these cells, lysosomes break open, releasing their enzymes and destroying the cell from the inside out.

Interestingly, lysosomal enzymes are released from damaged cells and, because of this, are the basis of a useful diagnostic tool. During a heart attack, for example, heart muscle cells that die release very specific enzymes into the blood, which can be measured to confirm a physician's diagnosis. Several other diseases can also be detected by blood enzyme levels.

Lysosomes play an important role in embryonic development. In humans, for example, the fingers of an early embryo are webbed (**FIGURE 3-23**). The cells of the webbing, however, are genetically programmed to self-destruct. At a certain point in development, the lysosomes of these cells release their enzymes, which destroy the cells and eliminate the webbing.

Most cells in the human body contain only a few lysosomes; these are used primarily to recycle outdated organelles or perhaps to put an end to the cell when its useful lifetime is over. However, certain cells contain hundreds of lysosomes. An example is the neutrophil (NEW-trowe-PHIL). Neutrophils are one type of blood cell that scavenges the blood and body tissues looking for bacteria and viruses, which they then phagocytize and digest internally with the aid of lysosomes. Thus,

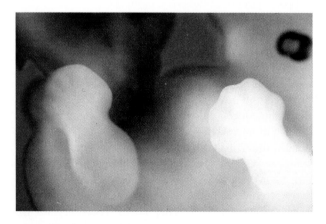

FIGURE 3-23 The Human Embryonic Hand The webs between the fingers present during the fetal stage are normally destroyed by lysosomes during development.

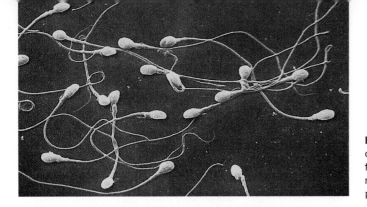

FIGURE 3-24 The Sperm Cell The sperm cell is a marvel of architecture, uniquely "designed" to streamline the cell for its long journey to fertilize the ovum. The nuclear material is compacted into the sperm head. A flagellum propels the sperm through the female reproductive tract.

FIGURE 3-25 Flagella *(a)* A sperm flagellum, showing the 9 + 2 arrangement and additional fibers thought to provide support and strength to the vigorously beating tail. *(b)* A basal body is found at the base of each cilium. It consists of nine sets of triplets that give rise to the microtubule doublets of the cilium.

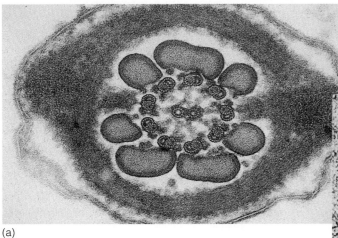

(a)

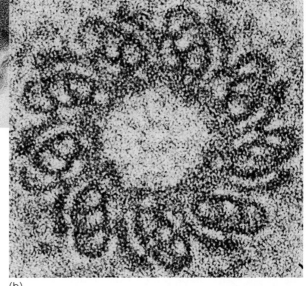

(b)

neutrophils play an important role in homeostasis by eliminating infectious agents.

Flagella Are Organelles That Permit Cellular Motility

Most cells in the human body are part of tissues and organs and remain in the same place throughout their life. One exception is the mammalian sperm cell, which must move out of the body to perform its specialized task: fertilization.

The human sperm cell moves with the aid of a flagellum (fla-GEL-um; plural, flagella; Latin for "whip"). As shown in **FIGURE 3-24**, a **flagellum** is a long, whiplike extension of the plasma membrane containing numerous small tubules, the **microtubules.** Microtubules produce the motive force.

As **FIGURE 3-25A** illustrates, the nine pairs of microtubules of the flagellum are arranged around a central pair. Biologists typically refer to this as the "9 + 2" arrangement. The flagella in sperm have additional fibers outside the central 9 + 2 array, which provide extra strength.

At the base of each flagellum is an anchoring structure, the **basal body** (BAZ-il) (**FIGURE 3-25B**). The basal body gives rise to the flagellum during development and contains nine sets of microtubules (with three microtubules each) but no central pair.

Movement Across the Surface of Cells Is Provided by Cilia

Several fixed cells in the body, such as those lining the trachea (TRAY-key-ah) or wind pipe, contain an organelle called the cilium (SILL-ee-um). **Cilia** (plural, pronounced SILL-ee-ah) are small extensions of the cell membrane (**FIGURE 3-26**). Like flagella, they have a central core of microtubules in a 9 + 2 arrangement.

Hundreds, even thousands, of cilia can be found on a single cell. In humans, for instance, thousands of cilia protrude from each cell lining the trachea (**FIGURE 3-26**). These cilia beat toward the mouth and transport mucus (MEW-kuss) that is produced by other cells in the lining to the oral cavity (mouth), where it can be either swallowed or expectorated (spit out). Since the mucus traps dust particles in the air we breathe, the cilia are part of an automatic cleansing mechanism, vital for protecting the lungs and maintaining our health.

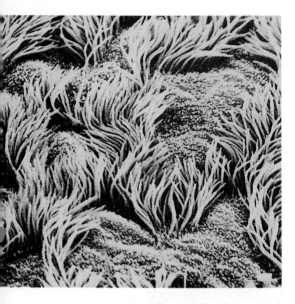

FIGURE 3-26 Cilia Cilia on cells lining the trachea of the human respiratory tract.

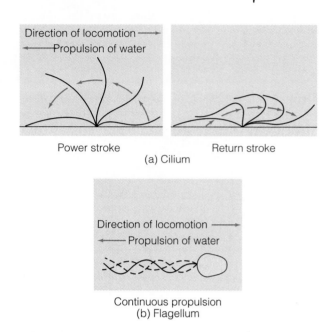

Power stroke   Return stroke
(a) Cilium

Direction of locomotion ⟶
⟵ Propulsion of water

Continuous propulsion
(b) Flagellum

FIGURE 3-27 Beating of Cilia and Flagella *(a)* Cilia beat with a stiff power stroke—something like a rowing motion. During the return stroke, the cilium is flexible. *(b)* The flagellum beats like a whip, over and over, propelling the sperm cell forward.

At the base of each cilium is a basal body, identical to those found in flagella. The basal body gives rise to the cilium during development and anchors the organelle in place.

Cilia and flagella are structurally similar, but flagella are much longer than cilia and far less numerous. Normal human sperm cells, for example, have one flagellum each, whereas each of the ciliated cells lining the respiratory system and parts of the female reproductive system (notably the Fallopian tubes) contains thousands of cilia. Flagella and cilia also beat differently. Cilia perform a stiff rowing motion with a flexible return stroke, whereas flagella beat more like whips, undulating in a continuous motion (**FIGURE 3-27**).

Because cilia and flagella often beat continuously, they require a more or less constant supply of energy, provided by ATP. As noted earlier, most ATP is produced by mitochondria, which are located nearby.

Some Cells Move by Amoeboid Motion

Another type of movement is called **amoeboid motion** (ah-ME-boyd). As **FIGURE 3-28** illustrates, during amoeboid movement, the cell sends out many small cytoplasmic projections, or **pseudopodia** (SUE-dough-POW-dee-ah; "false feet"). Pseudopodia attach to solid surfaces "in front" of the cell. Cytoplasm flows into the pseudopodia, moving the cell forward.

Phagocytic neutrophils, described earlier, are capable of amoeboid movement. These cells can escape the tiny capillaries (thin-walled blood vessels in body tissues) and migrate through tissues, gobbling up bacteria and damaged cells.

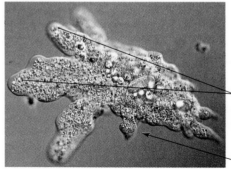

Pseudopodia

Direction of movement

FIGURE 3-28 Amoeboid Movement Some cells in the human body move by amoeboid movement. Studies of the amoeba (shown here) show that cells that move by amoeboid movement send out minute extensions called pseudopodia (false feet), which attach to solid surfaces. Cytoplasm then flows into the pseudopodia, moving the cell forward.

Section 3-6

Energy and Metabolism

Cells get the energy they need from the breakdown of carbohydrates (primarily glucose) and triglycerides. Under certain circumstances, they can even capture energy from proteins (Chapter 5). In this section, we'll look briefly at how cells in humans acquire energy from glucose, a type of sugar. Before we examine this process, however, a few words about glucose and other carbohydrates are in order.

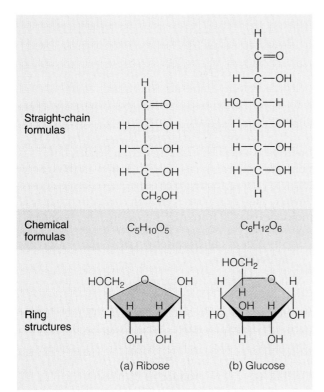

FIGURE 3-29 Two Monosaccharides

As noted in Chapter 2, carbohydrates are a general class of biological molecules. **Carbohydrates** are organic compounds that range in size from simple sugars such as glucose to very large molecules such as starch. Starch and other complex carbohydrates, as they are called, contain many molecules of glucose bonded to one another by covalent bonds (Chapter 2). The molecular structures of two simple carbohydrates—ribose, a five-carbon sugar, and glucose, a six-carbon sugar—are shown in **FIGURE 3-29**. These are also known as **monosaccharides.**

Simple sugars such as glucose and fructose (another six-carbon sugar that is found in fruits) can unite by covalent bonds to form **sucrose,** commonly known as table sugar (**FIGURE 3-30**). Sucrose is a **disaccharide,** because it consists of two monosaccharides. As noted earlier, monosaccharides can unite in long, branching chains (**FIGURE 3-31**). These are known as **polysaccharides,** and include such important molecules as starch, glycogen, and cellulose. These molecules will be discussed in more detail elsewhere in the book. For now, let us return to fascinating ways in which cells derive energy from glucose.

Glucose molecules, like other organic molecules, contain energy—a fact that you could demonstrate by burning a block of sugar. This energy is stored in the covalent bonds that hold the atoms of the molecule in place. Cells can tap into this energy by breaking glucose molecules apart in an orderly series of chemical reactions. This series begins in the cytoplasm and is completed in the mitochondrion. It is referred to as **cellular respiration** (because it requires oxygen and gives off carbon dioxide).

During cellular respiration, glucose, which contains six carbon atoms, is broken down into six molecules of carbon dioxide and six molecules of water. The overall reaction is

glucose + oxygen → carbon dioxide + water + energy

$$C_6H_{12}O_6 + 6 O_2 \rightarrow 6 CO_2 + 6 H_2O + 38 ATP$$

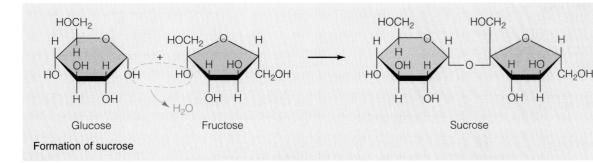

Formation of sucrose

FIGURE 3-30 Disaccharides Simple sugars such as glucose and fructose can combine to form the disaccharide known as sucrose.

FIGURE 3-31 Polysaccharides Simple sugars can also combine to form polysaccharides such as glycogen, which is a storage form of glucose in animals and is found in muscle and liver.

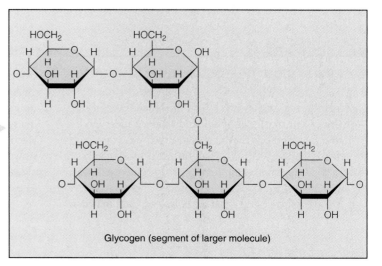

Portion of polysaccharide molecule (glycogen)

Glycogen (segment of larger molecule)

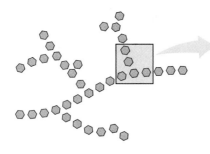

Glycolysis

Transition reaction

Citric acid cycle

ETS

FIGURE 3-32 Cellular Respiration This process involves four steps: glycolysis, the transition reaction, the citric acid cycle, and the electron transport system. Glycolysis occurs in the cytoplasm; the remaining steps take place in the mitochondrion. Most of the ATP is produced in the electron transport system (ETS) from the energy stripped from electrons given off by the previous three stages.

As shown, the complete breakdown of glucose requires oxygen. Without oxygen, glucose can be only partially broken down, and only a fraction of its energy can be captured. (You'll see why later.)

The breakdown of glucose releases lots of energy. A reaction that gives off energy is known as an **exergonic reaction** (energy-liberating). Cells can capture sizable portions of this energy and use it to drive the synthesis of ATP from inorganic phosphate (P_i) and ADP. The formation of ATP results in the storage of energy. ATP synthesis is an **endergonic reaction** (energy-requiring). Exergonic and endergonic reactions are often coupled—that is, linked so that energy from one is transferred to the other. The complete breakdown of glucose generates enough energy to produce 38 molecules of ATP from ADP and inorganic phosphate.

The Chemical Breakdown of Glucose Occurs in Four Steps

Cellular respiration in eukaryotes consists of four interconnected parts: (1) glycolysis, (2) the transition reaction, (3) the citric acid cycle, and (4) the electron transport system (**FIGURE 3-32**). Respiration involves numerous complex chemical reactions designed to break glucose apart and yield energy. The energy is captured by the cell to produce ATP, the only form of energy cells can use. **TABLE 3-4** summarizes the stages of

TABLE 3-4		
Overview of Cellular Energy Production		
Reaction	**Location**	**Description and Products**
Glycolysis	Cytoplasm	Breaks glucose into two three-carbon compounds, pyruvate; nets two ATP and two NADH molecules
Transition reaction	Mitochondrion	Removes one carbon dioxide from each pyruvate, producing two acetyl CoA molecules; produces two NADH molecules
Citric acid cycle	Inner compartment of mitochondrion	Completes the breakdown of acetyl cycle CoA; produces two ATPs per glucose; produces numerous NADH and FADH molecules
Electron transport	Inner membrane of mitochondrion	Accepts electrons from NADH and FADH, generated by previous system reactions; produces 34 ATPs

cellular respiration and lists their products. The following paragraphs provide a more detailed summary of cellular respiration for students required to have a more in-depth understanding of this process.

Glycolysis Is the Breakdown of Glucose into Two Pyruvate Molecules.

The first phase of cellular respiration in cells is known as glycolysis. **Glycolysis** (gly-COLL-eh-siss) is a metabolic pathway located in the cytoplasm of cells. As **FIGURE 3-33** shows, during glycolysis, glucose is split in half, forming two three-carbon molecules of pyruvic acid, or pyruvate. The energy released during this process nets the cell a modest two molecules of ATP. It also yields two molecules of NADH (nicotine adenine dinucleotide). NADH is a substance (technically a coenzyme) that picks up hydrogens and electrons given off by the reactions. Its importance will be clear shortly.

The Transition Reaction Is an Intermediate Step in Which One Carbon Atom Is Cleaved Off Each Pyruvate.

As **FIGURE 3-34** illustrates, pyruvate generated in glycolysis diffuses from the cytoplasm into the inner compartment of the mitochondrion and reacts there with a large molecule called Coenzyme A (CoA). During this

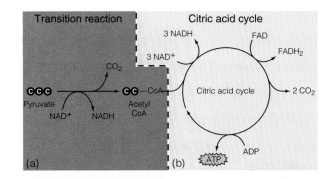

FIGURE 3-34 Overview of the Transition Reaction and the Citric Acid Cycle *(a)* The transition reaction cleaves off one carbon dioxide and binds the remaining two-carbon compound temporarily to a large molecule called Coenzyme A. The result is a molecule called acetyl-CoA. It then enters the citric acid cycle. *(b)* Occurring in the matrix of the mitochondrion, the citric acid cycle liberates two carbon dioxides and produces one ATP per pyruvate molecule. Its main products, however, are NADH and FADH$_2$, bearing high-energy electrons that are transferred to the electron transport system.

transition reaction, one carbon atom is cleaved off each pyruvate molecule, forming a two-carbon compound (acetyl group) that binds to Coenzyme A (**FIGURE 3-34**). The product of this reaction is **acetyl CoA** (ah-SEAT-ill). The transition reaction also yields two NADH molecules.

The Citric Acid Cycle Completes the Breakdown of Glucose.

Acetyl CoA enters the third stage of cellular respiration, a cyclic metabolic pathway also located inside the inner compartment of the mitochondrion. This elaborate pathway is called the **Krebs cycle,** after the scientist who worked out many of its details, or the **citric acid cycle,** after the very first product formed in the reaction sequence.

During the cycle, the two-carbon compound produced in the transition reaction binds to a four-carbon compound known as **oxaloacetate** (ox-AL-oh-ass-eh-tate). The result is a six-carbon product, citric acid. Citric acid proceeds through a series of chemical reactions, each catalyzed by its own enzyme. Along the way, the molecule is modified many times. During the complex molecular rearrangements, two molecules of carbon dioxide are removed from the six-carbon chain, and oxaloacetate is regenerated at the end, thus ensuring the continuation of the cycle.

As shown in **FIGURE 3-34**, one of the chemical reactions of the citric acid cycle yields an ATP molecule. Because two molecules of pyruvate enter the citric acid cycle for each glucose molecule broken down by

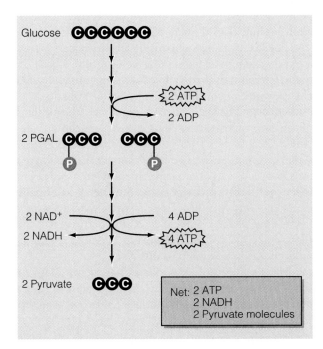

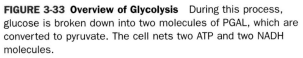

FIGURE 3-33 Overview of Glycolysis During this process, glucose is broken down into two molecules of PGAL, which are converted to pyruvate. The cell nets two ATP and two NADH molecules.

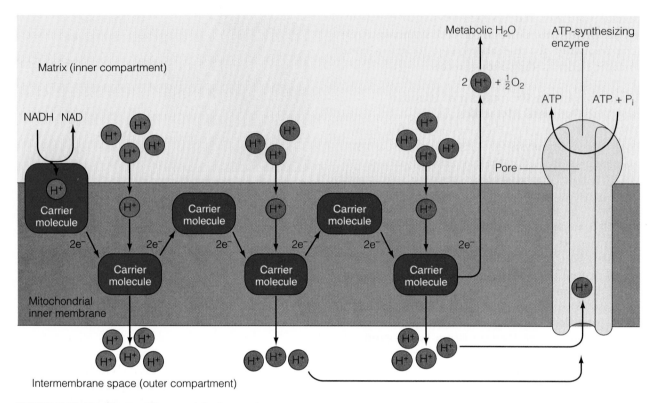

FIGURE 3-35 The Electron Transport System and Chemiosmosis Located on the inner surface of the mitochondrial inner membrane, the electron transport system produces ATP via chemiosmosis, described in the text.

glycolysis, the citric acid cycle nets the cell two ATP molecules. Several other reactions release electrons, which are funneled into the fourth, and final, part of cellular respiration, the electron transport system.

Electrons Given Off During the Citric Acid Cycle and Glycolysis Enter the Electron Transport System.

The electrons liberated during certain reactions of the citric acid cycle are passed to one of the two electron acceptors in the mitochondrial matrix: **nicotinamide adenine dinucleotide (NAD)** and **flavine adenine dinucleotide (FAD).** These molecules, in turn, transfer the electrons to a series of protein carrier molecules embedded on the inner surface of the inner membrane of the mitochondrion (**FIGURE 3-35**). The carrier proteins constitute the **electron transport system (ETS).**

The electron transport system accepts electrons from FAD and NAD. These electrons are then passed from one protein to another and are eventually given over to oxygen molecules inside the matrix. During

their rapid journey along this chain, the electrons lose energy, much like a hot potato passed along by a line of people. The energy lost along the way is used to make ATP. But how does the cell make ATP?

In 1961, British biochemist Peter Mitchell proposed a mechanism to explain how cells produce ATP in their electron transport systems. This hypothesis, called **chemiosmosis,** won Mitchell the Nobel Prize in 1978 (**FIGURE 3-35**). According to this hypothesis, the proteins of the electron transport chain are dual-purpose molecules. That is, they transport electrons, and they also serve as hydrogen-ion pumps, transporting hydrogen ions from the inner compartment of the mitochondrion to the outer compartment using the energy lost by the electrons (**FIGURE 3-35**). These hydrogens are released from some of the chemical reactions occurring in the citric acid cycle, and their export to the outer compartment results in a buildup in this place.

Hydrogen ions do not stay put, however. Many of them flow back into the inner compartment through pores in the inner mitochondrial membrane. It is this flow of hydrogen ions, Mitchell hypothesized, that powers ATP synthesis. How?

Just as electrons flowing in a wire provide energy to run motors or to power light bulbs, the flowing hydrogen ions are thought to drive the process of ATP synthesis.

All told, the electron transport system produces 34 ATP molecules per glucose molecule. Because 2 ATPs are generated by glycolysis and 2 are produced in the citric acid cycle, the total output for cellular respiration is 38 per glucose molecule.

As shown in **FIGURE 3-35**, the "de-energized" electrons from the electron transport system eventually combine with hydrogen ions and oxygen to produce water.

Enzymes Are Essential to Virtually All Chemical Reactions Occurring in the Cell

Before moving on, it is important to note that each of the chemical reactions of cellular respiration is regulated by an enzyme. Described earlier, enzymes are protein molecules that greatly accelerate chemical reactions. So important are they that a missing enzyme in a metabolic pathway will very likely shut the pathway down—in much the same way that a missing worker on an assembly line interrupts production.

Each enzyme is a large, globular protein with a small region known as the active site (**FIGURE 3-36A**). The **active site** is an indentation, or pocket, where the chemical reaction occurs. Its shape corresponds to that of the **substrate(s),** the molecule(s) undergoing reaction. Substrates fit into the active site in much the same way that a hand fits into a glove. Because each enzyme has its own uniquely shaped active site capable of binding to one or at most a few chemical substrates, enzymes are said to be specific. This feature allows the cells to regulate chemical reactions very precisely.

Interestingly, although enzymes play an active role in the metabolism of cells, they are unchanged by the reaction. As a result, they can be recycled—used over and over again.

Enzymes can also be controlled by cells, thus providing additional control over cellular metabolism. The most common regulators of enzyme activity are the end products of chemical reactions. These molecules often bind to specific control regions called **allosteric sites** (al-oh-STAIR-ick) on one of the enzymes in a metabolic pathway. Allosteric sites are the molecular equivalent of switches. When an end product binds to one, it turns the enzyme off, shutting down the entire metabolic pathway (**FIGURE 3-36B**). In some metabolic pathways, end products may bind directly to the active site of an

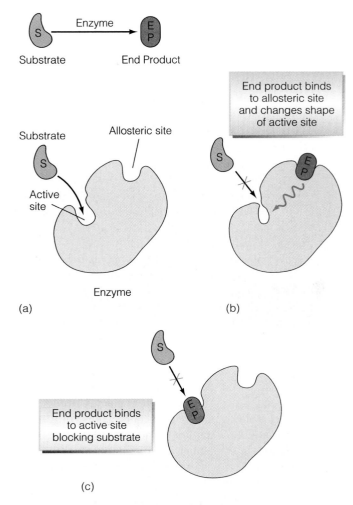

(a)

(b)

(c)

FIGURE 3-36 Enzyme Structure and Function *(a)* Two-dimensional drawing of an enzyme. The active site conforms to the shape of the reacting molecule(s). *(b)* The allosteric site is a molecular switch that regulates enzyme activity. In some metabolic pathways, the end products bind to the allosteric site. This may turn the active site on or off, depending on the enzyme. *(c)* In others, the end products of a reaction bind to the active site itself, blocking it entirely.

enzyme. This binding also blocks substrates from entering the site, thereby inhibiting the enzyme and shutting down the metabolic pathway (**FIGURE 3-36C**).

Cellular respiration can be controlled by the buildup of chemicals such as citrate and ATP, both of which shut down a specific enzyme in the glycolytic pathway.

Fermentation: Tired Muscles, Cheese, and Wine

What do tired muscles, cheese, and wine have in common? The answer is that they're all products of a process called fermentation. Fermentation occurs in human cells when oxygen levels run low. It also occurs in certain microorganisms. Basically, **fermentation** is a chemical reaction in which pyruvic acid produced from glycolysis is converted to one of several other end products (**FIGURE 3-37**). These reactions are important, in part, because they produce energy. Let's look briefly at this process in body cells.

During strenuous exercise, oxygen consumption in muscle cells may exceed replenishment. When this occurs, oxygen levels inside the cells fall. The cells are said to be anaerobic. With oxygen no longer available to accept electrons, the electron transport system (ETS) shuts down. When the ETS falters, so does the citric acid cycle. Fortunately, glycolysis still continues. But in the absence of oxygen, the end product of glycolysis, pyruvate, is converted to lactic acid as shown in **FIGURE 3-37A**.

Thus, rather than shutting energy production down completely, muscle cells continue to generate energy via glycolysis and fermentation. Unfortunately, fermentation is exceedingly inefficient, netting the cell only 2 ATPs for each glucose molecule, compared with 38 during the complete breakdown.

In humans, heavy exercise, such as weight lifting and running, not only depletes muscle oxygen, making cells become anaerobic, but also results in the buildup of lactic acid in muscle cells. This causes muscle fatigue. Lactic acid, however, diffuses out of the muscle within hours and is carried by the blood to the liver, where it is converted first to pyruvate and then to glucose in a series of chemical reactions that is essentially the reverse of glycolysis. Some of the glucose is used to synthesize glycogen, and the rest is released into the blood, where it is redistributed to body cells and used to generate energy.

Fermentation is also the chief source of energy for many bacteria and other single-celled organisms. Those that live in oxygen-free environments—for example, deep in muddy sediments of lakes and rivers—meet their energy needs by fermentation. Other microorganisms temporarily deprived of oxygen can also satisfy their energy needs through fermentation.

The most common type of fermentation is lactic acid fermentation, discussed above. Many popular foods, such as cheese and yogurt, are produced by lactic acid fermentation. Most cheeses, for instance, are pro-

duced from milk and selected lactic acid bacteria. To make cheese, lactic acid bacteria are added to milk. These bacteria break down milk sugar (lactose), releasing glucose molecules in the process. Glucose molecules are then broken down in the glycolytic and fermentation pathways, producing lactic acid. Proteolytic enzymes are then added to cause the liquid to curdle. The thin watery part (whey) is drawn off, leaving behind the unripened cheese (for example, cottage cheese). Additional bacteria and fungi may be added to break down milk fat and protein, yielding specific ripened cheeses, such as blue cheese.

Other microorganisms undergo **alcoholic fermentation,** which, as the name implies, produces ethanol and carbon dioxide from pyruvate (**FIGURE 3-37B**). Alcoholic fermentation is the basis of the production of beer, wine, and hard liquor. Alcoholic fermentation is also the basis of bread making. Yeast cells are added to bread dough. These microorganisms are anaerobic and produce carbon dioxide and ethanol. The carbon dioxide forms bubbles in the dough, causing it to rise, and the alcohol is driven off during baking, producing the pleasant smell of baking bread.

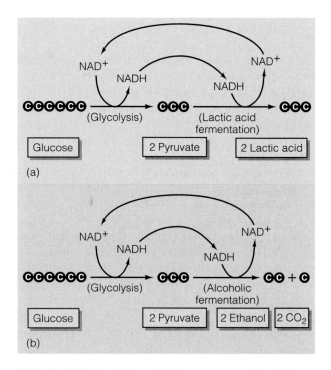

FIGURE 3-37 Fermentation (a) Pyruvate, or pyruvic acid, from glycolysis is converted into lactic acid in muscle cells when oxygen levels fall. (b) Fermentation can produce a variety of products in bacteria and yeast cells. In certain microorganisms, pyruvate is converted into ethanol and carbon dioxide.

Health, Homeostasis, and The Environment

Parkinson's and Pollution?

The cells of the body function independently, much like workers in a factory. Yet the entire body functions efficiently. Why? Each cell performs functions needed to survive, but each cell benefits from the specialized activities of others. The same is true in a factory.

But what does a cell gain from the specialized activities of its neighbors? Quite simply: excellent living and working conditions. That is to say, when the specialized cellular activities are combined, they function to maintain internal conditions essential to cellular life. In other words, they help maintain homeostasis.

The stability of the internal environment, however, is continuously threatened by disruptive forces, many of them of our own making. Chemical pollution is one of the many forces that can throw our cells and therefore our bodies out of balance. The following paragraphs describe a frightening example of this phenomenon.

In 1983, several relatively young men and women were admitted to a California hospital in a catatonic stupor. Dr. J. W. Langston, the neurologist who treated them, was baffled by their condition. His patients represented a novel medical puzzle unlike any he had ever seen. The patients lay immobile day after day. They could neither talk, nor feed themselves, nor move their limbs. It was as if they had been frozen.

Langston began an intensive study of the patients, all drug addicts. He found that each of them had taken a synthetic, heroinlike drug that had been contaminated with a paralyzing chemical known as MPTP. Made in the basement of one of California's small-time drug pushers, the drug was the latest in a long list of "designer" drugs then on the market. Sloppy chemistry resulted in the contamination.

MPTP attacks nerve cells in a part of the brain called the substantia nigra (sub-STAN-she-ah NIGH-gra). Degeneration of the brain cells in this region has long been thought to cause **Parkinson's disease,** a disorder that usually afflicts older individuals. Parkinson's disease is characterized by tremors, or involuntary rhythmic shaking, of the hands or head. If the disease worsens, patients have difficulty talking and writing. Walking becomes awkward and clumsy. In later stages, memory and thinking can be severely impaired.

Interestingly, individuals exposed to even minute amounts of MPTP develop symptoms similar to those seen in Parkinson's patients. In fact, a Swiss chemist who had worked briefly with the chemical was hospitalized because his body muscles, like those of the addicts whom Langston had encountered, "froze up" on him. Puzzled physicians put him in a mental hospital, assuming that he was suffering from a psychological disorder called catatonic schizophrenia.

MPTP is similar to several common industrial chemicals and pesticides. Some scientists speculated that MPTP-like chemicals in the air, water, and soil might be the underlying cause of the nerve cell damage in Parkinson's disease. However, several studies have failed to show a link between Parkinson's disease and industrial pollution. At a 1985 meeting at the National Institutes of Health, researchers agreed: Parkinson's disease had not been shown to be caused by environmental contamination.

No sooner had the researchers arrived home than a Canadian scientist announced the results of a study in Quebec showing a remarkable correlation between the use of MPTP-like pesticides (for example, paraquat) and the incidence of Parkinson's disease. He argued, however, that pesticides were probably not the only chemical villain. Many neurotoxins discharged from nearby chemical factories could have been present in the area.

In support of the hypothesis that Parkinson's disease is linked to industrial pollution and pesticide use, researchers note that Parkinson's disease was unheard of before the Industrial Revolution. As industrialization and environmental pollution spread, the incidence of the disease rose. It reached a plateau in the early 1900s. Although no increase has been seen since 1940, some researchers warn that an increase in the use of MPTP-like pesticides and an increase in industrial pollution could substantially raise the incidence of the disease in years to come.

This case study shows how the function of certain nerve cells may be affected by chemical pollution—altering their function and the function of the body. The delicate balance that keeps muscles in check is lost because of a chemical imbalance in the environment. Not only does this illustrate the connection between human health, homeostasis, and environmental conditions, it underscores the importance of a clean environment to a healthy future.

Parkinson's

www.jbpub.com/humanbiology

SUMMARY

1. The fundamental unit of all living organisms is the *cell*. All organisms are composed of one or more cells.
2. Two types of cells are found: prokaryotic and eukaryotic. *Prokaryotic cells* lack nuclei and are found in single-celled organisms such as bacteria. *Eukaryotic cells* are structurally more complex cells with membrane-bound nuclei. Some single-celled organisms and all multicellular organisms are eukaryotic.

CELLULAR EVOLUTION AND HOMEOSTASIS

3. During cellular evolution, prokaryotic organisms emerged first, about 3.5 billion years ago.
4. Eukaryotes evolved about 2.0 billion years later. The first eukaryotic cells were single-celled organisms.
5. As evolution proceeded, organisms became multicellular and began to acquire specialized cells. Over time, eukaryotes gave rise to three additional kingdoms: plants, fungi, and animals.
6. The increasing complexity of organisms was paralleled by the development of complex systems that ensure homeostasis.

MICROSCOPES: ILLUMINATING THE STRUCTURE OF CELLS

7. The *microscope* is one of many scientific instruments used to study cells. Microscopes enlarge objects and allow us to see detail.
8. Microscopes fall into two broad categories: (1) *light microscopes*, which use ordinary visible light to illuminate the specimen; and (2) *electron microscopes*, which use beams of electrons to visualize an image.

AN OVERVIEW OF CELL STRUCTURE

9. Cells of the human body differ widely, but all consist of two major compartments, the nuclear and the cytoplasmic.
10. The nuclear compartment, or *nucleus*, contains the genetic information that

regulates the structure and function of all eukaryotic cells.
11. The cytoplasmic compartment contains *cytoplasm*, which, in turn, contains numerous *cellular organelles*, structures that perform specific functions.
12. The cytoplasmic compartment is home to the *cytoskeleton*, a network of tubules and filaments that gives cells their three-dimensional shape and also binds to enzymes in metabolic pathways, increasing cellular efficiency.

THE STRUCTURE AND FUNCTION OF THE PLASMA MEMBRANE

13. The outermost boundary of the cell, the *plasma membrane*, consists of lipids, proteins, and small amounts of carbohydrate.
14. The plasma membrane ensures the structural integrity of the cell, regulates the flow of materials into and out of the cell, and participates in cellular communication and cellular identification.
15. *Lipids* are a diverse group of organic chemicals characterized by their lack of water solubility. The biologically important lipids serve many functions. Two main lipids were discussed in this chapter: (a) phospholipids and (b) steroids.
16. *Amino acids* are small organic molecules that join by *peptide bonds*, forming a variety of *peptides* (containing fewer than 20 amino acids) and *proteins* (containing more than 20 amino acids).
17. Thousands of different types of proteins are found in human cells; each type is structurally and functionally unique. Their uniqueness results from the sequence of amino acids in a protein.
18. Because it is mostly lipid, the membrane is a natural barrier to water-soluble molecules. Lipid-soluble materials pass through with ease.
19. Water-soluble molecules and ions are believed to pass through pores in the membrane by *diffusion*, the flow of

a substance from an area of higher concentration to an area of lower concentration.
20. *Carrier proteins* may also transport water-soluble molecules across the membrane down the concentration gradient. This process is called *facilitated diffusion*.
21. Some molecules are actively transported across the membrane. *Active transport* mechanisms use energy to pump materials into or out of the cell and require a special class of membrane proteins, the transport molecules. Active transport permits cells to transport substances from regions of low concentration to regions of high concentration.
22. Cells engulf liquids and their dissolved materials in a process called *pinocytosis*. Cells may also engulf large molecules and cells by *phagocytosis*. Pinocytosis and phagocytosis are collectively referred to as *endocytosis*.
23. During endocytosis, the plasma membrane indents and surrounds the material to be ingested. The material engulfed by the cell is incorporated in a membranous vesicle, which is taken into the cell.
24. Water is believed to move through tiny pores in plasma membranes. The diffusion of water molecules through a selectively permeable membrane from a region of higher water concentration to a region of lower water concentration is called *osmosis*.

CELLULAR COMPARTMENTALIZATION: ORGANELLES

25. The *nucleus* is one of the most conspicuous cellular organelles. It houses the DNA, which contains the genetic information that determines the structure and function of the cell.
26. The nucleus is bounded by a double membrane, the *nuclear envelope*, which contains numerous pores that allow many materials (but not the DNA) to pass freely between the nucleus and the cytoplasm.

27. Energy in the cell is produced principally by the breakdown of glucose. Glucose breakdown takes place, in part, in the cytoplasm but mostly in the *mitochondrion*. Energy liberated from glucose molecules is captured by ATP molecules.

28. All mitochondria have two membranes, an outer one and an inner one. The inner membrane is thrown into folds called *cristae*.

29. Cellular energy production captures 30–60% of the energy contained in glucose; the rest is liberated as heat.

30. The *endoplasmic reticulum* is a network of membranous channels in the cytoplasm of the cell and is the main location of cellular protein synthesis.

31. Endoplasmic reticulum may be smooth or rough. *Smooth endoplasmic reticulum* has no ribosomes. *Rough endoplasmic reticulum* contains many ribosomes on its outer surface.

32. Smooth endoplasmic reticulum produces phospholipids (needed to make more plasma membrane), and it has many additional functions in different cells. The rough endoplasmic reticulum is involved in synthesizing protein for extracellular use (for example, hormones) and in the production of enzymes that are contained in *lysosomes*.

33. Proteins produced on the surface of the RER enter the cavity, where they may be chemically modified. Proteins are transferred from the RER to the Golgi complex, where they are sorted, chemically modified, and repackaged into *secretory vesicles* or lysosomes.

34. Secretory vesicles accumulate in the cytoplasm and are released when needed. Lysosomes remain in the cell, where they bind to food vacuoles (material phagocytized or pinocytized by the cell) or destroy malfunctioning cellular organelles.

35. Most cells in the body are fixed. Still other cells can move about in tissues via amoeboid movement.

36. The human sperm cell utilizes another kind of motive force, the *flagellum*, a long, whiplike structure.

37. Some cells contain numerous smaller extensions, called *cilia*, which propel fluids and materials along their surfaces.

38. Both cilia and flagella contain nine pairs of microtubules arranged in a circle around a central pair.

ENERGY AND METABOLISM

39. Cells acquire energy from glucose, a carbohydrate.

40. Carbohydrates are a group of organic compounds that consist of simple sugars (monosaccharides) such as glucose, disaccharides such as sucrose, and long chains (polymers) of monosaccharides.

41. The complete breakdown of glucose is called *cellular respiration* and consists of four interdependent parts: glycolysis, the transition reaction, the citric acid cycle, and the electron transport system.

42. Glycolysis is a series of reactions occurring in the cytoplasm of cells. During glycolysis, glucose is split in two. Each molecule of glucose yields two molecules of pyruvic acid, or pyruvate.

43. Glycolysis nets the cell two molecules of ATP and two NADH molecules containing high-energy electrons.

44. Pyruvate produced during glycolysis next diffuses into the mitochondrion and undergoes the transition reaction. This step produces acetyl CoA and two NADH molecules.

45. Acetyl CoA enters the citric acid cycle, which produces two ATPs and numerous molecules of NADH and $FADH_2$. These electron carriers accept high-energy electrons from various reactions in the citric acid cycle and move them to the electron transport system.

46. The electron transport system consists of a series of protein molecules embedded in the inner surface of the inner membrane of the mitochondrion. Electrons from NADH and $FADH_2$ are passed down the chain and are eventually given to oxygen. During this process, numerous molecules of ATP are produced via *chemiosmosis*.

FERMENTATION: TIRED MUSCLES, CHEESE, AND WINE

47. Without oxygen, the electron transport system shuts down, as does the citric acid cycle. At this point, cells must rely on glycolysis and fermentation to generate energy.

48. Fermentation is a chemical reaction in which pyruvate is converted into lactic acid or some other product, such as ethyl alcohol.

49. Lactic acid fermentation is also an important source of food for humans, including cheeses and yogurts, which are made from milk and selected bacteria.

50. Alcoholic fermentation produces ethanol from pyruvate and occurs in certain microorganisms.

51. Wines, beer, and hard liquor are all produced by combining plant by-products (grains or fruits) with microorganisms that are alcoholic fermenters.

HEALTH, HOMEOSTASIS, AND THE ENVIRONMENT: PARKINSON'S AND POLLUTION?

52. MPTP is a chemical contaminant in a synthetic, heroinlike drug. MPTP attacks certain cells in the brain, the same cells, in fact, long thought to be associated with *Parkinson's disease,* a disorder that generally afflicts older individuals and that is characterized by involuntary rhythmic shaking of the hands or head.

53. Individuals exposed to even minute amounts of MPTP develop symptoms similar to those seen in Parkinson's patients.

54. MPTP is similar to several common industrial chemicals and pesticides. Some scientists now speculate that MPTP-like chemicals in the air, water, and soil may be the underlying cause of Parkinson's disease.

55. This example illustrates the connection between homeostasis, human health, and environmental quality. It shows how alterations in the level of cellular function can have dramatic effects on the whole body and, if the link between Parkinson's and industrial chemicals proves valid, underscores the importance of protecting human health by protecting the quality of our environment.

Critical Thinking

THINKING CRITICALLY— ANALYSIS

This analysis corresponds to the Thinking Critically scenario that was presented at the beginning of this chapter.

First, you might note that one study is never enough to test the validity of a hypothesis. Further work is needed before you can be certain that the results are valid. Some questions to ask are: Did the vitamin increase the life-span of all cells or just certain types of cells? Were there any other factors responsible for increased lifespan? Was the test done on cells from an infant, a toddler, a young-ster, a teenager, or an adult?

Second, you might also suggest that studies on ani-mals would be needed to test this hypothesis. Tissue cul-tures are very artificial environments. An organism is much more complicated than some of its cells in tissue culture. Many hormones and other substances are pres-ent that might alter the effect of the vitamin.

Finally, you might want your friends to consider the potential toxic effects of the vitamins they are thinking about taking. Would vitamin supplements cause more harm than good?

EXERCISING YOUR CRITICAL THINKING SKILLS

Reread the Health, Homeostasis, and the Environment section and list the evidence given in support of the hy-pothesis that environmental pollutants may be an un-derlying cause of Parkinson's disease. Can you make up your mind on the basis of this information? What other information might be needed? Design a study to provide the information you might need. Would you perform a study on animals or on human subjects? How would you design the study to get accurate, reproducible results?

TEST OF CONCEPTS

1. Describe the concepts of cellular and organismic homeostasis. How are they related? How does environ-mental homeostasis or environmen-tal quality affect them?

2. Cellular compartmentalization allows for greater efficiency. Why? Give an example to support your argument.

3. Draw a diagram of the plasma mem-brane, and describe the five routes by which molecules pass through the membrane.

4. In what ways does the plasma mem-brane participate in cellular com-munication? How is this function important for the maintenance of homeostasis?

5. Define the terms *osmosis, isotonic, hypotonic,* and *hypertonic.* In what way(s) is osmosis similar to diffusion, and in what way(s) is it different?

6. Describe the organelles involved in the synthesis, storage, and release of protein for extracellular use, starting

with the basic instructions needed to make the protein.

7. In what ways are cilia and flagella similar, and in what ways are they different?

8. What is the main energy currency of the cell? Where is it formed? Where does it get its energy?

9. Describe the four phases of cellular energy production.

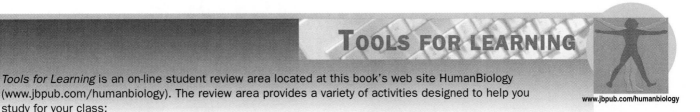

TOOLS FOR LEARNING

Tools for Learning is an on-line student review area located at this book's web site HumanBiology (www.jbpub.com/humanbiology). The review area provides a variety of activities designed to help you study for your class:

Chapter Outlines. We've pulled out the section titles and full sentence sub-headings from each chapter to form natural descriptive outlines you can use to study the chapters' material point by point.

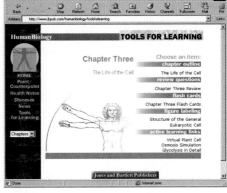

www.jbpub.com/humanbiology

Review Questions. The review questions test your knowledge of the important concepts and applications in each chapter. Written by the author of the text, the review provides feedback for each correct or incorrect answer. This is an excellent test preparation tool.

Flash Cards. Studying human biology requires learning new terms. Virtual flash cards help you master the new vocabulary for each chapter.

Figure Labeling. You can practice identifying and labeling anatomical features on the same art content that appears in the text.

Active Learning Links. Active Learning Links connect to external web sites that provide an opportunity to learn basic concepts through demonstrations, animations, and hands-on activities.

PRINCIPLES OF STRUCTURE AND FUNCTION

Thinking Critically

Health care costs are skyrocketing, and many critics are arguing that our health care dollars are being misused. In particular, they argue that expensive procedures, such as organ transplants for indigent people costing taxpayers $100,000 or more each, are draining dollars that could be invested in preventive medicine, such as prenatal care. Analyze this view using your critical thinking skills. ■

Human red blood cell trapped in a fibrin network of a blood clot.

GENETIC ENGINEERING IS HELPING SCIENTISTS alter the function of defective cells. Some geneticists, for instance, are experimenting with ways to introduce normal genes into defective cells such as those lining the respiratory tract of cystic fibrosis patients. These cells produce copious amounts of mucus that trap infectious organisms, resulting in frequent lung infections.

To correct the genetic defect, researchers are experimenting with certain cold viruses that infect the cells lining the respiratory tract. These cells inject their DNA into the lining cells, causing cold symptoms. To create a safe gene transplant tool, scientists cut out the viral genes that cause trouble. They then insert a human gene that corrects the defect in the genetic material of the cells lining the respiratory tract of cystic fibrosis patients. By inserting the normal genes into these viruses and exposing a cystic fibrosis patient to the virus, scientists hope to transport good genes into defective cells, creating a lasting cure for a disease that claims most of its victims' lives before they reach the age of 20.

This is just one form of "body work" scientists are now performing. Others are working at a level slightly higher—that is, at the level of tissues and organs (defined shortly). These scientists are trying to develop genetically engineered pigs whose cells contain human genes in the hope that pig organs could someday be transplanted into people to replace failed livers or kidneys. An organ, they speculate, could be specially made with a person's own genes to prevent tissue rejection. It would be available in a fraction of the time it now takes to find a genetically compatible organ from a suitable donor.

Health Note 4-1 describes other remarkable medical procedures used to repair or rebuild tissues and organs. This chapter presents some basic information you will need to understand the human body and many of the exciting research projects you hear about or read about in the news. This chapter also revisits homeostasis, elaborating on what you've already learned and setting the stage for your study of human organ systems, the subject of the next ten chapters.

Section 4-1

From Cells to Organ Systems

To begin our study of the structure and function of the human body, we will look at the beginning of life—that is, fertilization. The sperm and ovum combine to form a zygote (ZIE-goat). The zygote's genetic material (DNA) contains all of the information needed to develop a fully functional adult and to control complex life functions, such as growth, reproduction, and homeostasis.

Cells Unite to Form Tissues, and Tissues Combine to Form Organs

During embryonic development, the zygote divides many times, first producing a ball of cells called a **morula** (MORE-you-lah) (**FIGURE 4-1**). The cells continue to divide but soon start undergoing a process called **differentiation.** During this phase, the cells become structurally and functionally specialized. Cellular differentiation is akin to the kind of development students undergo as they proceed through an educational system. They start out in kindergarten pretty much interested in the same things: naps, finger painting, and playing in the mud. But as they proceed through primary and secondary school and college, they branch off. Some become physics majors. Others take up medicine. Others pursue the arts. Cells do pretty much the same thing. They start off pretty general, then gradually specialize.

The first sign of differentiation occurs in the early embryo with the emergence of three distinct cell types: ectoderm, mesoderm, and endoderm (**FIGURE 4-1D**). These cells can be told apart chiefly by their location in

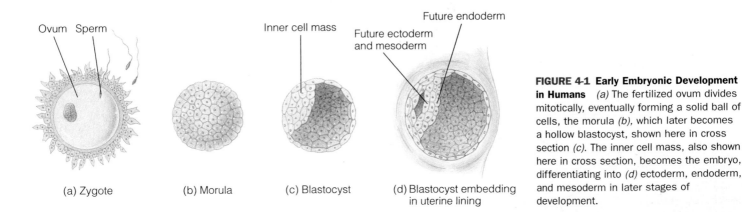

Ovum Sperm

Inner cell mass

Future endoderm

Future ectoderm and mesoderm

(a) Zygote

(b) Morula

(c) Blastocyst

(d) Blastocyst embedding in uterine lining

FIGURE 4-1 Early Embryonic Development in Humans *(a)* The fertilized ovum divides mitotically, eventually forming a solid ball of cells, the morula *(b)*, which later becomes a hollow blastocyst, shown here in cross section *(c)*. The inner cell mass, also shown here in cross section, becomes the embryo, differentiating into *(d)* ectoderm, endoderm, and mesoderm in later stages of development.

the early embryo. **Ectoderm,** for instance, lies on the outside of the embryo. **Mesoderm** lies in the middle. **Endoderm** is the innermost layer. During embryonic development in humans, ectodermal, mesodermal, and endodermal cells give rise to a variety of highly differentiated cell types, each of which carries out very specific functions. Ectoderm, for instance, gives rise to the skin, eyes, and nervous system. Mesodermal cells give rise to muscle, bone, and cartilage. Endoderm forms the lining of the digestive tract and several digestive glands. The specialized cells produced from the three embryonic cell layers are often bound together by extracellular fibers and other extracellular materials, forming **tissues** (from the Latin "to weave"). Extracellular materials may be liquid (as in blood), semisolid (as in cartilage), or solid (as in bone). Tissues, in turn, combine to form **organs,** discrete structures in the body that carry out specific functions. Organs, like their counterparts in cells, the organelles, provide for a division of labor.

Cells Combine to Form Four Primary Tissues

Four major tissue types are found in humans: (1) epithelial, (2) connective, (3) muscle, and (4) nervous. These are called **primary tissues. TABLE 4-1** lists the primary tissues. As you can see, each primary tissue

TABLE 4-1	
The Primary Tissues and Their Subtypes	
Epithelial tissue	Nervous tissue
Membranous	Conductive
Glandular	Supportive
Muscle tissue	Connective tissue
Cardiac	Connective tissue proper
Skeletal	Specialized connective tissue
Smooth	Blood
	Bone
	Cartilage

consists of two or three subtypes. Muscle tissue, for instance, comes in three varieties: cardiac, skeletal, and smooth. Cardiac muscle cells are found exclusively in the heart. Skeletal muscle cells are located chiefly in body muscles. Smooth muscle cells are found in the walls of the stomach, intestinal tract, and blood vessels. (More on these later.)

The primary tissues exist in all organs but in varying amounts. The lining of the stomach, for example, consists of a single layer of epithelial (ep-eh-THEEL-ee-ill) cells forming the surface epithelium (**FIGURE 4-2**). It's a protective coating. Just beneath the

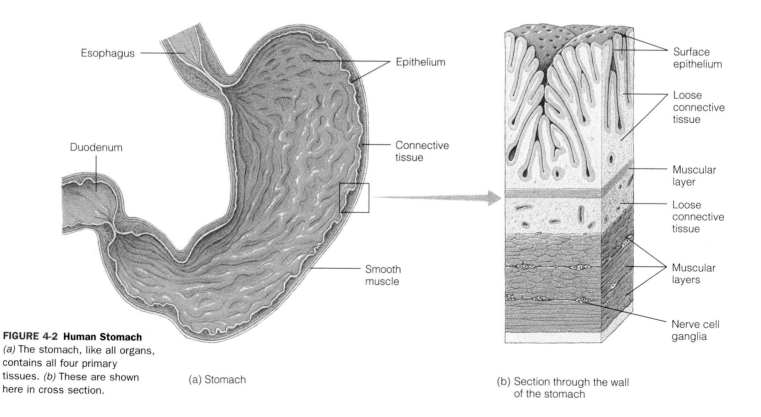

Esophagus

Duodenum

Epithelium

Connective tissue

Smooth muscle

Surface epithelium

Loose connective tissue

Muscular layer

Loose connective tissue

Muscular layers

Nerve cell ganglia

FIGURE 4-2 Human Stomach
(a) The stomach, like all organs, contains all four primary tissues. *(b)* These are shown here in cross section.

(a) Stomach

(b) Section through the wall of the stomach

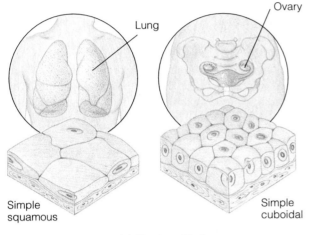

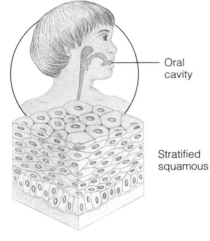

FIGURE 4-3 **Membranous Epithelia**   Single-celled (simple) epithelia *(a)* and stratified epithelia *(b)* exist in different parts of the body.

sheets of cells tightly packed together, forming the external coverings or linings of organs. Membranous epithelia come in a variety of forms, each specialized to protect underlying tissues. To simplify matters, histologists (scientists who study the microscopic structure of organisms) divide membranous epithelia into two broad categories: **simple epithelia,** consisting of a single layer of cells, and **stratified epithelia,** consisting of many layers of cells (**FIGURE 4-3**).

The **glandular epithelia,** the second major type of epithelium, consists of clumps of cells that form many of the glands of the body. Epithelial glands arise during embryonic development from tiny "ingrowths" of membranous epithelia, as illustrated in **FIGURE 4-4**. Some glands remain connected to the epithelium by hollow ducts and are called **exocrine glands** (EX-oh-crin; glands of external secretion); products of the

epithelial lining is a layer of connective tissue. It holds things together. Beneath that is a thick sheet of smooth muscle cells, which forms the bulk of the stomach wall. Smooth muscle cells are also found in blood vessels supplying the tissues of the stomach. Nerves enter with the blood vessels and control the flow of blood.

Epithelium Forms the Lining or External Covering of Organs and Also Forms Glands

Epithelial tissue exists in two basic forms: glandular and membranous. The **membranous epithelia** consist of

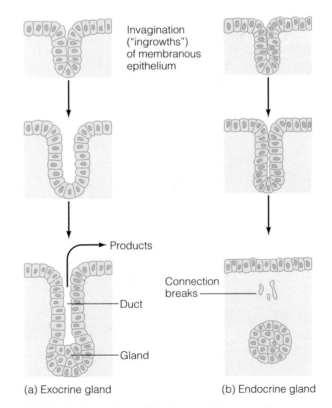

FIGURE 4-4 **Formation of Endocrine and Exocrine Glands** *(a)* Exocrine glands arise from invaginations of membranous epithelia that retain their connection. *(b)* Endocrine glands lose this connection and thus secrete their products into the bloodstream.

exocrine glands flow through ducts into some other body part (**FIGURE 4-4A**). In humans, sweat glands in the skin are exocrine glands. They produce a clear, watery fluid that is released onto the surface of the skin by small ducts. This fluid evaporates from the skin, and cools the body. Salivary glands are another exocrine gland. Located around the oral cavity, the salivary glands produce saliva, a fluid that is released into the mouth via small ducts.

Some glandular epithelial cells break off completely from their embryonic source, as shown in **FIGURE 4-4B**, to form **endocrine glands** (EN-doh-crin; glands of internal secretion). The endocrine glands produce hormones that are released from the cell and diffuse into the bloodstream, where they travel to other parts of the body (Chapter 14).

Epithelial Tissues Illustrate a Basic Biological Principle: That Structure Closely Correlates with Function.

One of the basic rules of architecture is that form (the structure of a building) often follows function—in other words, architectural design reflects underlying function. Animals exhibit a similar relationship. This rule is "obeyed" all the way down to the cellular level, so it is not surprising to find a remarkable degree of correlation between structure and function in tissues as well.

The membranous epithelia provide many examples of this relationship. A good example is the outer layer of the human skin, the **epidermis** (ep-eh-DERM-iss), which, among other things, protects us from excessive water loss. Shown in **FIGURE 4-5A**, the epidermis consists

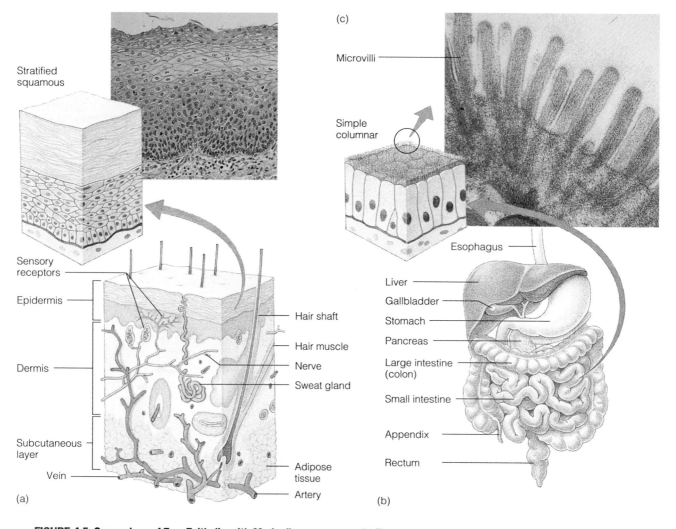

Stratified squamous

(c)

Microvilli

Simple columnar

Esophagus

Sensory receptors

Epidermis

Dermis

Subcutaneous layer

Vein

Hair shaft

Hair muscle

Nerve

Sweat gland

Adipose tissue

Artery

Liver

Gallbladder

Stomach

Pancreas

Large intestine (colon)

Small intestine

Appendix

Rectum

(a)

(b)

FIGURE 4-5 Comparison of Two Epithelia with Markedly Different Functions (a) A cross section of the skin showing the stratified squamous epithelium of the epidermis (above), which protects underlying skin from sunlight and desiccation.

(b) The simple columnar epithelium of the lining of the small intestine, which is specialized for absorption. (c) The plasma membranes of the cells lining the intestine are thrown into folds (microvilli) that greatly increase the surface area for absorption.

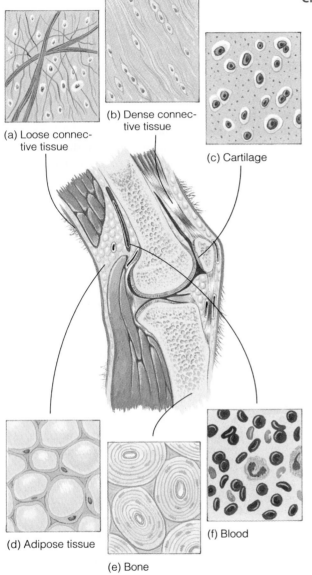

(a) Loose connective tissue

(b) Dense connective tissue

(c) Cartilage

(d) Adipose tissue

(e) Bone

(f) Blood

FIGURE 4-6 Connective Tissue Connective tissue consists of many diverse subtypes.

hance food absorption. As shown in **FIGURE 4-5C**, the surfaces of these cells have numerous tiny protrusions known as **microvilli** (MY-crow-VILL-eye), which markedly increase the surface area of the cell available for absorption. The larger the surface area, the more efficient the absorption of food.

Connective Tissue Binds the Cells and Organs of the Body Together

As the name implies, **connective tissue** is the body's glue. Connective tissue binds cells and other tissues together and is present in all organs in varying amounts.

The body contains several types of connective tissue, each with specific functions (**FIGURE 4-6**). Despite the differences, all connective tissues consist of two basic components: cells and varying amounts of extracellular material. Two types of connective tissue will be discussed here: connective tissue proper and the specialized connective tissues—bone, cartilage, and blood.

Connective Tissue Proper Consists of Two Subtypes, Determined by the Relative Proportion of Fibers and Cells.

Connective tissue proper is an important structural component of humans and consists of two types: dense connective tissue and loose connective tissue. As the name implies, **dense connective tissue** (**DCT**) consists primarily of densely packed fibers produced by cells interspersed between them. DCT is found in ligaments and tendons (**FIGURE 4-7A**). **Ligaments** join bones to bones at joints and provide support for joints. I use the mnemonic *LBJ*, which stands for one of our presidents, Lyndon Baines Johnson, if you're a history student, or *l*igaments connect *b*ones to bones at *j*oints if you're a biology student. **Tendons** join muscle to bone and aid in body movement.

DCT is also found at other sites. For instance, the layer of the skin underlying the epidermis, known as

of numerous cell layers, known as a stratified squamous epithelium (SQUAW-mus). The cells of this epithelium flatten toward the surface and are tightly joined by special connections. The outermost cells become isolated from the blood supply and die, forming a dry protective layer. Together, the thickness of the epidermis, the adhesion of one cell to another, and the dry protective layer of dead cells impede water loss. They also present a formidable barrier to microorganisms.

Another example of the relationship between structure and function is the epithelium of the small intestine. As shown in **FIGURE 4-5B**, the lining of the small intestine consists of a single layer of columnar cells, uniquely suited to absorb food molecules. These cells are also structurally modified in ways that further en-

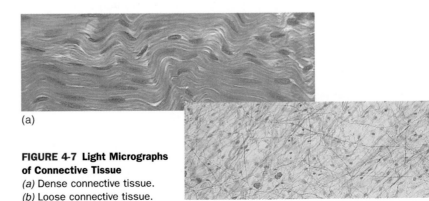

(a)

(b)

FIGURE 4-7 Light Micrographs of Connective Tissue
(a) Dense connective tissue.
(b) Loose connective tissue.

the **dermis,** is composed of dense connective tissue, although the fibers are less regularly arranged than those in ligaments and tendons.

If connective tissue were likened to packing materials in a factory, dense connective tissue would be akin to strapping tape that's wrapped around packages—it holds things together. **Loose connective tissue (LCT)** would be more like the styrofoam peanuts or excelsior used to fill the spaces between products jammed in a box. In other words, it's the body's packing material, but it also holds structures together.

As shown in **FIGURE 4-7B,** LCT contains cells in a loose network of collagen and elastic fibers. Both of these fibers are made of protein. Loose connective tissue forms around blood vessels in the body and in skeletal muscles, where it binds the muscle cells together. It also lies beneath epithelial linings of the intestines and trachea, filling the space and anchoring them to underlying structures.

The chief difference between dense and loose connective tissues lies in the ratio of cells to extracellular fibers. As you can see in **FIGURE 4-7,** dense connective tissue has far more fibers than loose connective tissue.

The extracellular fibers found in dense and loose connective tissue are produced by a connective tissue cell known as the **fibroblast** (FIE-bro-blast). In addition to producing the fibers that hold many tissues together, fibroblasts repair damage created by cuts or tears in body tissues. When the skin is cut, for example, fibroblasts in the dermis migrate into the injured area, where they begin producing large quantities of collagen. The collagen fibers fill the wound, closing it off, creating a temporary patch. The epidermis soon begins to grow over the damaged area, repairing the damage and restoring the integrity of the skin. When the cut is small, the epidermis covers the damaged area completely, leaving no scar, but in larger wounds, the epidermal cells are often unable to grow over the entire wound, leaving some of the underlying collagen exposed and producing a visible scar.

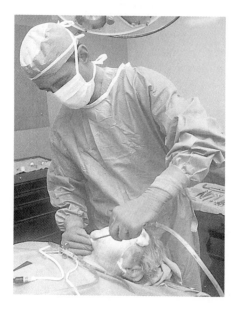

FIGURE 4-8 Liposuction During liposuction surgery, the physician aspirates fat from deposits lying beneath the skin, helping to reduce unsightly accumulations.

Besides helping to hold the body together, loose connective tissue gives refuge to several cell types that protect us against bacterial and viral infections. One of the most important of the protective cells is the **macrophage** (MACK-row-FAYGE; "big eater"). Containing numerous lysosomes for digesting foreign material, macrophages engulf microorganisms that penetrate the skin and underlying loose connective tissues as a result of injury. This prevents bacteria from spreading to other parts of the body and producing a systemic infection—one that affects the entire organism. Macrophages also play a role in immune protection, which will be discussed in Chapter 8.

Some LCT contains large, conspicuous fat cells. The **fat cell** is one of the most distinctive of all body cells. It contains huge fat globules that occupy virtually the entire cell, pressing the cytoplasm and the nucleus to the periphery. Fat cells occur singly or in groups of varying size. Large numbers of fat cells in a given region form a modified type of LCT known as **adipose tissue** (AD-eh-poze), or, less glamorously, fat. Adipose tissue is an important storage depot for lipids, particularly triglycerides, which are used as an energy source under certain conditions. Fatty deposits also provide insulation for humans and many other mammals and some birds, especially those that live in aquatic environments where heat loss can be substantial.

For humans, fatty deposits often become unsightly. To rid the body of these deposits, individuals can engage in exercise programs and can reduce their intake of food. The combination of the two can prove quite effective. Some individuals opt for a surgical measure called **liposuction** (LIE-poh-suck-shun) (**FIGURE 4-8**). In this procedure, a small incision is made in the skin through which surgeons insert a device to aspirate fat deposits under the skin in various locations, such as the thighs, buttocks, and abdomen. The fat cells extracted from one region can even be transferred to other regions, such as the breast, to resculpt the human body. Liposuction is a relatively safe technique but, like most surgery, not free from risk.

The Specialized Connective Tissues Are Structurally and Functionally Modified to Perform Specific Functions Essential to Homeostasis

The body contains three types of specialized connective tissue: cartilage, bone, and blood.

Cartilage.

Cartilage consists of cells embedded in an abundant and rather impervious extracellular material, the matrix

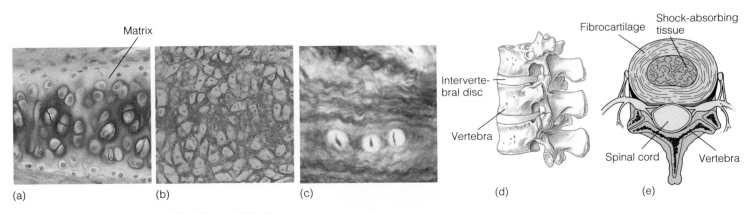

FIGURE 4-9 Light Micrographs of Cartilage   (a) Hyaline.
(b) Elastic. (c) Fibrocartilage. (d) Intervertebral disk showing location and (e) arrangement of fibrocartilage in a protective ring around the soft, spongy part of the disk that absorbs shock.

(FIGURE 4-9). Surrounding virtually all types of cartilage is a layer of dense, irregularly packed connective tissue, the **perichondrium** (PEAR-eh-CON-dree-um; "around the cartilage"). This layer contains the blood vessels that supply nutrients to cartilage cells through diffusion. No blood vessels penetrate the cartilage itself. Because cartilage cells are nourished by diffusion from perichondral capillaries, damaged cartilage heals very slowly. For this reason, joint injuries that involve the cartilage often take years to repair or may not heal at all.

Three types of cartilage are found in humans: hyaline, elastic, and fibrocartilage. Each serves a special purpose, which is reflected in its underlying structure. The most prevalent type is **hyaline cartilage** (HIGH-ah-lynn) (FIGURE 4-9A). Hyaline cartilage contains numerous collagen fibers, which appear white to the naked eye. Found on the ends of many bones, hyaline cartilage greatly reduces friction, so bones can move over one another with ease. Hyaline cartilage also makes up the bulk of the nose and is found in the larynx (voice box) and the rings of the trachea, which you can feel below the adam's apple. The ends of the ribs that join to the sternum (breastbone) are composed of hyaline cartilage. In embryonic development, the first skeleton is made of hyaline cartilage, which is later converted to bone.

Elastic cartilage is similar to hyaline cartilage but contains many wavy elastic fibers, which give it much greater flexibility (FIGURE 4-9B). Elastic cartilage is found in regions where support and flexibility are required—for example, in the external ears and eustachian tubes (ewe-STAY-shun), the cartilaginous ducts that help equalize pressure in the inner ear (Chapter 12).

Fibrocartilage is the rarest of all cartilage. Like hyaline cartilage, it consists of an extracellular matrix containing numerous bundles of collagen fibers. As shown in FIGURE 4-9C, however, fibrocartilage contains far fewer cells than either hyaline or elastic cartilage.

Fibrocartilage is found in the outer layer of **intervertebral disks** (in-ter-VER-tah-braul), the shock-absorbing tissue between the vertebrae (bones) of the spine (FIGURE 4-9D). An intervertebral disk consists of a soft, cushiony central region that absorbs shock. Fibrocartilage forms a ring around the central portion of the disk, holding it in place (FIGURE 4-9E).

As some of you will no doubt find out from first-hand experience, over time the fibrocartilage ring weakens and may tear, permitting the central part of the disk to bulge outward (or herniate). This condition is referred to as a slipped or herniated disk and usually occurs in the neck or lower back, resulting in a significant amount of pain in the neck, back, or one or both legs, depending on the location of the damaged disk. Pain is generated when the disk presses against nearby nerves.

Surgeons can correct the problem by removing the herniated portion of the disk. You can reduce your chances of "slipping" a disk in the first place by watching your weight, keeping in shape, sitting upright (not slouching) in a chair, and lifting heavy objects carefully (FIGURE 4-10).

Bone.

Bone is another form of specialized connective tissue. Contrary to what many might

FIGURE 4-10 Protecting Your Back
When lifting heavy objects, bend your legs and grasp the object, as the man in the top illustration is doing. With your back straight, stand up and lift with your legs.

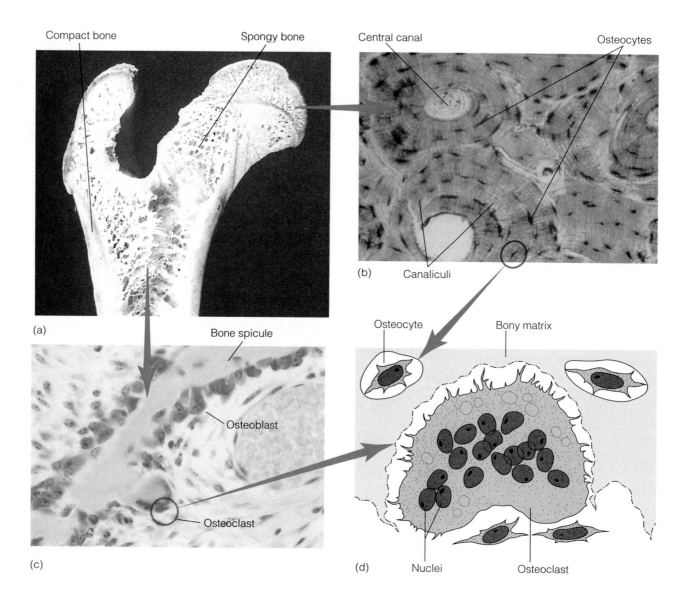

(a)

(b)

(c)

(d)

FIGURE 4-11 Bone *(a)* Compact and spongy bone, shown in a section of the humerus. *(b)* Light micrograph of compact bone. *(c)* Photomicrograph of spongy bone, showing osteoblasts and osteoclasts. *(d)* Osteoclast digesting surface of bony spicule.

think, bone is dynamic, living tissue. Besides providing internal support and protection to internal organs such as the brain, heart, and lungs, bone plays an important role in maintaining optimal blood calcium levels and is, therefore, a homeostatic organ. Calcium is required for many body functions such as muscle contraction, normal nerve functioning, and blood clotting.

Like all connective tissues, bone consists of cells embedded in an abundant extracellular matrix (**FIG-URE 4-11B**). Bone matrix consists primarily of collagen fibers, which give bone its strength and resiliency, interspersed with numerous needlelike salt crystals con-

taining calcium, phosphate, and hydroxide ions, which give bone its hardness. Calcium in bone can be dissolved by weak acids, leaving behind a collagen replica. It can be turned into a thick paste called **demineralized bone matter (DBM)**. DBM is rather remarkable stuff, and it is being used to repair severe bone damage, as described in our web page in the section on science in the news.

Two types of bone tissue are found in the body: compact bone and spongy bone (**FIGURE 4-11A**). **Compact bone** is, as its name implies, dense and hard. As illustrated in **FIGURE 4-11B**, the cells in compact bone (**osteocytes**) are located in concentric rings of calcified matrix. These surround a central canal through which the blood vessels and nerves pass. Each osteocyte has numerous processes that course through tiny canals in the bony matrix, known as **canaliculi** (CAN-al-ICK-u-LIE; literally "little canals"). The canaliculi provide a route for nutrients and wastes to flow to and from the osteocytes and the central canal.

Inside most bones of the body is a tissue known as spongy bone. **Spongy bone** consists of an irregular network of calcified collagen spicules (SPICK-yuls). As shown in **FIGURE 4-11C**, on the surface of the spicules are numerous **osteoblasts** (OSS-tee-oh-BLASTS), bone cells that produce collagen, which later becomes calcified. Once these cells are surrounded by calcified matrix, they are referred to as osteocytes.

Between the spicules are numerous cavities. In most of the bones of an adult, these cavities are filled with fat cells and form the **yellow marrow.** In other bones, the cavities are filled with blood cells and cells that give rise to new blood cells, thus forming **red marrow.**

Also on the surfaces of many bony spicules of spongy bone are large, multinucleated cells called **osteoclasts** (OSS-tee-oh-KLASTS; "bone breakers") (**FIGURE 4-11D**). These cells are part of a homeostatic system that ensures proper blood calcium levels. When calcium levels in the blood fall, osteoclasts are activated by the parathyroid hormone produced by the thyroid gland. Once activated, the osteoclasts digest small portions of the spongy bone. This, in turn, releases calcium into the bloodstream to restore proper levels.

Spongy bone is "remodeled" when bones are subjected to new stresses. For example, the leg bones of a desk-bound executive from Atlanta who goes on a skiing vacation in Jackson Hole, Wyoming, are remodeled as her legs are subjected to the rigors of skiing. This adjustment accommodates the new stresses and strains. During this process, osteoclasts tear down some of the spongy bone, while osteoblasts rebuild new bone in other areas to meet the new stresses. By the end of the 2-week ski trip, the executive's bones have been considerably refashioned and are much stronger than when she left home. When she is back at her desk, though, her bones revert to their previous, weaker state.

Blood.

Blood is also a specialized form of connective tissue and consists of two components: (1) the formed elements and (2) a large amount of extracellular material, a fluid called *plasma*. The **formed elements** of blood consist of the red and white blood cells and the platelets. They are responsible for 45% of the blood volume. Plasma makes up the remaining 55%. **Red blood cells,** or **erythrocytes,** transport oxygen and small amounts of carbon dioxide to and from the lungs and body tissues. **White blood cells,** or **leukocytes,** are involved in fighting infections. **Platelets** are fragments of large cells (**megakaryocytes**) located in the red bone marrow, the principal site of blood cell formation. Platelets play a key role in blood clotting.

Muscle Tissue Consists of Specialized Cells that Contract when Stimulated

Muscle is found in virtually every organ in the body. Muscle gets its name from the Latin word for "mouse" (mus). Early observers likened the contracting muscle of the biceps to a mouse moving under a carpet.

Muscle is an excitable tissue. When stimulated, it contracts, producing mechanical force. Muscle cells working in large numbers can create enormous forces. Muscles of the jaw, for instance, create a pressure of 200 pounds per square inch, forceful enough to snap off a finger. (Don't try this at home.) Muscle also moves body parts, propels food along the digestive tract, and expels the fetus from the uterus during birth. Heart muscle contracts and pumps blood through the 50,000 miles of blood vessels in the human body. Acting in smaller numbers, muscle cells are responsible for intricate movements, such as those required to play the piano or move the eyes.

As mentioned earlier, three types of muscle are found in humans: skeletal, cardiac, and smooth. The cells in each type of muscle contain two contractile protein filaments, actin and myosin. These very same fibers were first encountered in your study of the microfilamentous network lying beneath the plasma membrane of cells that is responsible for cytokinesis, the division of the cytoplasm. When stimulated, actin and myosin filaments of muscle cells slide over one another, shortening and causing contraction.

Skeletal muscle.

The majority of the body's muscle is called **skeletal muscle**—so named because it is frequently attached to the skeleton. When skeletal muscle contracts, it causes body parts (arms and legs, for instance) to move. Most skeletal muscle in the body is under voluntary, or conscious, control. In other words, signals from the brain cause the muscle to contract. A notable exception is the skeletal muscle of the upper esophagus, which contracts automatically during swallowing.

Skeletal muscle cells are long cylinders formed during embryonic development by the fusion of many embryonic muscle cells. Because of this, skeletal muscle cells are usually referred to as muscle fibers. Each muscle fiber contains many nuclei (**FIGURE 4-12A**). Because of the dense array of contractile fibers in the cytoplasm of the muscle fiber, the nuclei become pressed against the plasma membrane. As you might suspect, this highly specialized cell cannot divide, and damaged muscle cells cannot be replaced.

Skeletal muscle fibers appear banded, or striated, when viewed under the light microscope (**FIGURE 4-12A**). The striations result from the unique arrangement of actin and myosin filaments inside muscle cells, a topic discussed in Chapter 13.

Cardiac Muscle.

Like skeletal muscle, **cardiac muscle** is striated (**FIGURE 4-12B**). Unlike skeletal muscle, cardiac muscle is involuntary; that is, it contracts without conscious control. Found only in the walls of the heart, cardiac muscle cells contain a single nucleus. They also branch and interconnect freely, and individual cells are tightly connected to one another. This adaptation helps maintain the structural integrity of the heart, an organ subject to incredible strain as it pumps blood through the body day and night. The points of connection also provide pathways for electrical impulses to travel from cell to cell, allowing the heart muscle to contract uniformly when stimulated.

Smooth Muscle.

Smooth muscle, so named because it lacks visible striations, is involuntary. Actin and myosin filaments are present but are not organized like those found in other types (**FIGURE 4-12C**). Smooth muscle cells may occur singly or in small groups. Small rings of smooth muscle cells, for example, surround tiny blood vessels. When these cells contract, they shut off or reduce the supply of blood to tissues. Smooth muscle cells are most often arranged in sheets in the walls of organs, such as the stomach, uterus, and intestines (**FIGURE 4-2**). Smooth muscle cells in the wall of the stomach churn the food, mixing the stomach contents, and force tiny spurts of liquified food into the small intestine. Smooth muscle contractions also propel the food along the intestinal tract.

Nervous Tissue Contains Specialized Cells Characterized by Irritability and Conductivity

Last but not least of the primary tissues is nervous tissue, which consists of two types of cells: conducting cells and supportive cells. Many of the **conducting cells,** or **neurons,** are modified to respond to specific stimuli, such as pain or temperature. Stimulation results in bioelectric impulses, which the neuron transmits from one region of the body to another. As noted previously, the ability to respond to stimuli is a characteristic of all living things and is called *irritability.* The ability to transmit an impulse is called *conductivity.* The properties of irritability and con-

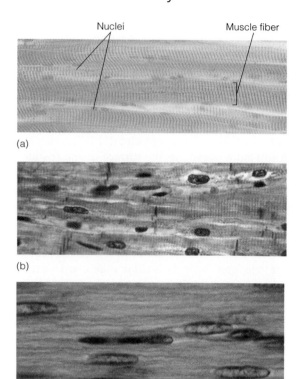

Nuclei   Muscle fiber

(a)

(b)

(c)

FIGURE 4-12 Light Micrograph of the Three Types of Muscle *(a)* Skeletal. *(b)* Cardiac. *(c)* Smooth.

ductivity in neurons allow us to be aware of our environment and to respond to internal and external stimuli. The evolution and refinement of the nervous system have been vitally important to our present success.

The supportive cells of the nervous system are a kind of nervous system connective tissue. These cells are incapable of conducting impulses, but they transport nutrients from blood vessels to neurons and also guard against toxins by creating a barrier to many potentially harmful substances. As you will see in Chapter 11, the supportive cells also increase the rate of conduction in neurons with which they are associated. Together, the neurons and their supporting cells form the brain, spinal cord, and nerves of the nervous system, described in more detail in Chapters 11 and 12.

Neurons Share Many Characteristics.

At least three distinct types of neurons are found in the body. Despite obvious anatomical differences, they share several common features. We will study these similarities by looking at one of the most common nerve cells, the multipolar neuron, shown in **FIGURE 4-13**. The **multipolar neuron** contains a prominent cell body to which are attached several short, highly branched processes, known as **dendrites** (DEN-drites). (It is the presence of these

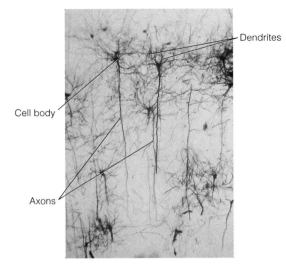

FIGURE 4-13 Multipolar Neurons Attached to the cell body of the multipolar neuron are many highly branched dendrites, which deliver impulses to the cell body. Multipolar neurons have one long, unbranching fiber called the axon, which transmits impulses away from the cell body.

processes that give this neuron its name, multipolar.) The dendrites receive impulses from receptors or other neurons and transmit them to the cell body. Also attached to the cell body of the multipolar neuron is a large, fairly thick process, the **axon** (AXE-on), which transports bioelectric impulses away from the nerve cell body.

Like muscle cells, nerve cells are highly differentiated and cannot divide. When a nerve cell is destroyed, it degenerates and cannot be replaced by cell division. A cut nerve axon, however, may partially regenerate. A new axon may grow from the damaged end, reestablishing previous connections and restoring some sensation or control over muscles. New research may someday provide ways to stimulate regeneration of nerve cells artificially, thus helping physicians to more fully restore nerve function to victims of accidents.

Tissues Combine to Form Organs; Organs Often Function in Groups Called *Organ Systems*

The cell contains organelles ("little organs") that carry out many of its functions in isolation from the biochemically active cytoplasm. As pointed out in Chapter 3, compartmentalization such

as this is an evolutionary adaptation, the importance of which is underscored by its ubiquity. Compartmentalization also occurs at the organismic level in organs.

Organs are discrete structures that have evolved to perform specific functions, such as digestion, enzyme production, and hormone synthesis. Most organs, however, do not function alone. Instead, they are part of a group of cooperative organs, called an **organ system.** The brain, spinal cord, and nerves are all organs that belong to an organ system known as the nervous system.

As you will see in upcoming chapters, components of an organ system are sometimes physically connected—as in the digestive system. In other cases, they are dispersed throughout the body—as in the endocrine system (**FIGURE 4-14**). Some organs belong to more than one system. For example, the pancreas pro-

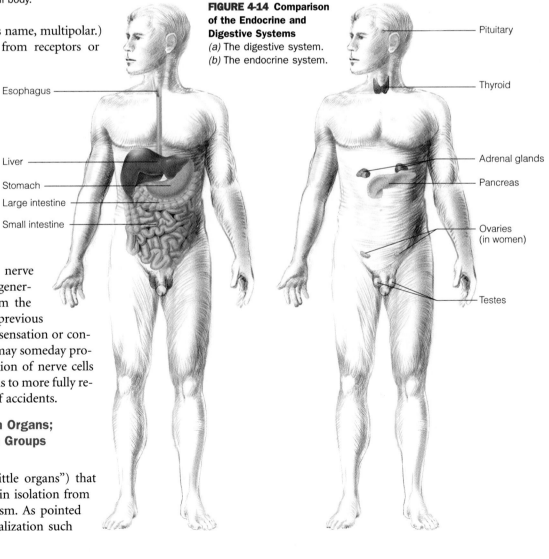

FIGURE 4-14 Comparison of the Endocrine and Digestive Systems
(a) The digestive system.
(b) The endocrine system.

FIGURE 4-15 Role of the Body Systems in Maintaining Homeostasis

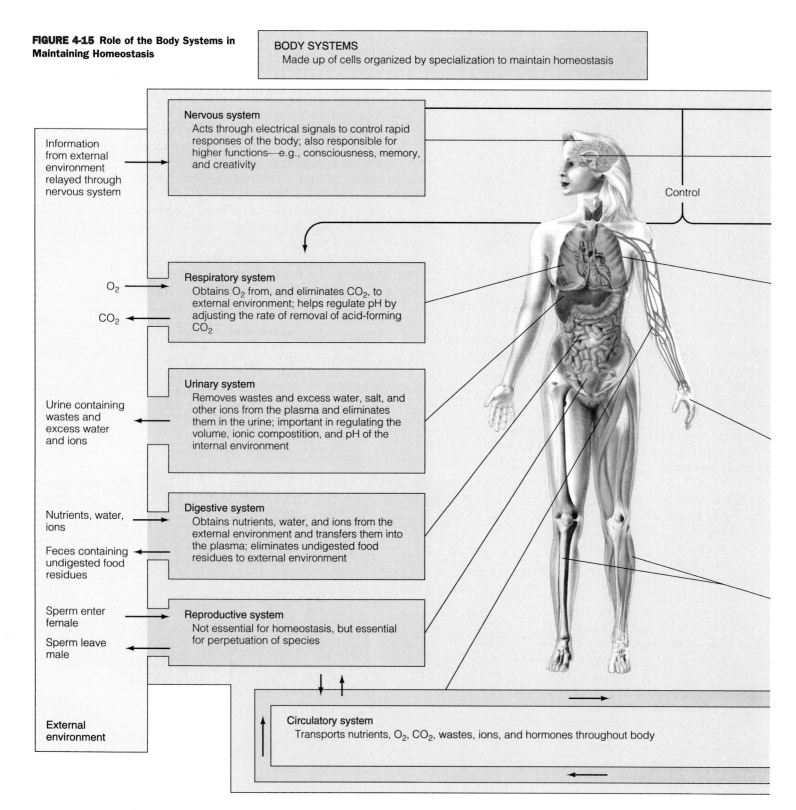

BODY SYSTEMS
Made up of cells organized by specialization to maintain homeostasis

Nervous system
Acts through electrical signals to control rapid responses of the body; also responsible for higher functions—e.g., consciousness, memory, and creativity

Information from external environment relayed through nervous system

Control

O_2

CO_2

Respiratory system
Obtains O_2 from, and eliminates CO_2, to external environment; helps regulate pH by adjusting the rate of removal of acid-forming CO_2

Urine containing wastes and excess water and ions

Urinary system
Removes wastes and excess water, salt, and other ions from the plasma and eliminates them in the urine; important in regulating the volume, ionic compostition, and pH of the internal environment

Nutrients, water, ions

Feces containing undigested food residues

Digestive system
Obtains nutrients, water, and ions from the external environment and transfers them into the plasma; eliminates undigested food residues to external environment

Sperm enter female

Sperm leave male

Reproductive system
Not essential for homeostasis, but essential for perpetuation of species

External environment

Circulatory system
Transports nutrients, O_2, CO_2, wastes, ions, and hormones throughout body

duces digestive enzymes that are secreted into the small intestine, where they break down food materials. The pancreas is therefore part of the digestive system. The pancreas also contains cells that produce insulin and glucagon, two hormones that help control blood glucose levels. Thus, the pancreas also belongs to the endocrine system.

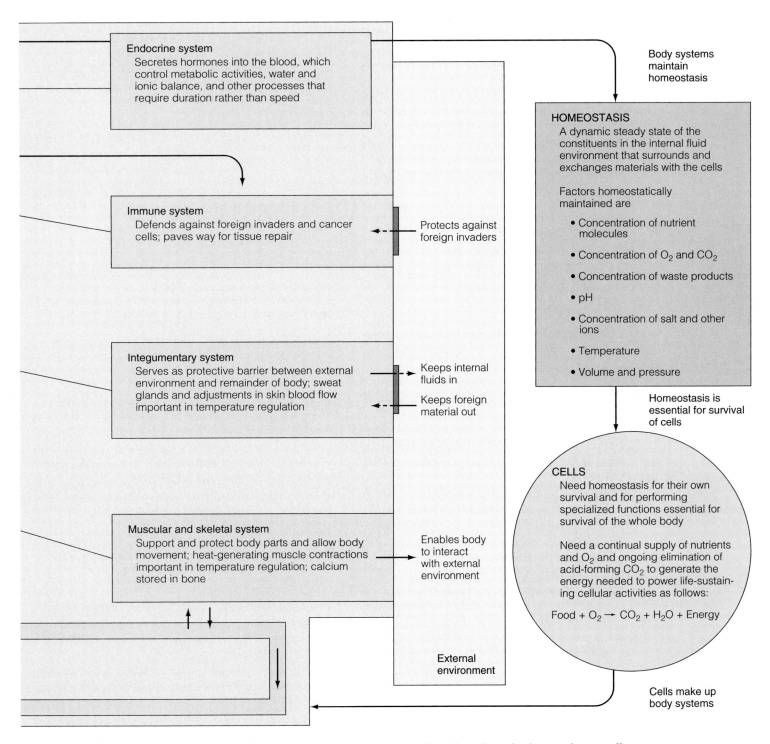

Endocrine system
Secretes hormones into the blood, which control metabolic activities, water and ionic balance, and other processes that require duration rather than speed

Body systems maintain homeostasis

HOMEOSTASIS
A dynamic steady state of the constituents in the internal fluid environment that surrounds and exchanges materials with the cells

Factors homeostatically maintained are

- Concentration of nutrient molecules
- Concentration of O_2 and CO_2
- Concentration of waste products
- pH
- Concentration of salt and other ions
- Temperature
- Volume and pressure

Immune system
Defends against foreign invaders and cancer cells; paves way for tissue repair

Protects against foreign invaders

Integumentary system
Serves as protective barrier between external environment and remainder of body; sweat glands and adjustments in skin blood flow important in temperature regulation

Keeps internal fluids in

Keeps foreign material out

Homeostasis is essential for survival of cells

CELLS
Need homeostasis for their own survival and for performing specialized functions essential for survival of the whole body

Need a continual supply of nutrients and O_2 and ongoing elimination of acid-forming CO_2 to generate the energy needed to power life-sustaining cellular activities as follows:

Food + O_2 → CO_2 + H_2O + Energy

Muscular and skeletal system
Support and protect body parts and allow body movement; heat-generating muscle contractions important in temperature regulation; calcium stored in bone

Enables body to interact with external environment

External environment

Cells make up body systems

The following 10 chapters describe the major organ systems and the functions they perform, paying special attention to their role in homeostasis. **FIGURE 4-15** summarizes the functions of each organ system and lists the role each plays in the overall economy of the human organism. You may want to take a moment to read the descriptions in the light blue boxes before proceeding.

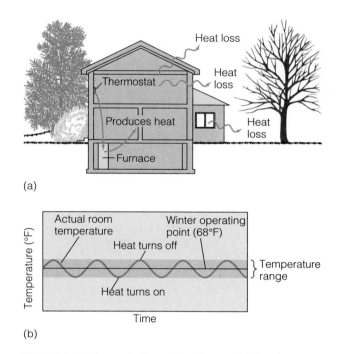

(a)

(b)

FIGURE 4-16 Homeostasis and the House *(a)* Heat is maintained in a house by a furnace, which compensates for heat loss. The thermostat monitors the internal temperature and switches the furnace on and off in response to temperature changes. *(b)* A hypothetical temperature graph showing temperature fluctuation around the operating point.

Principles of Homeostasis

FIGURE 4-15 shows that the body systems perform specific functions, many of which contribute to homeostasis. Homeostasis is essential for the survival of cells, which make up body systems.

Defined in Chapter 1 as a state of relative internal constancy, homeostasis occurs on a variety of levels—in cells, tissues, organs, organ systems, organisms, and even the environment. Homeostatic systems at all levels of biological organization have several common features essential for you to understand before you begin your study of the organ systems.

Homeostatic Systems Maintain Constancy Chiefly through Negative Feedback Mechanisms

Feedback mechanisms are control mechanisms. The most common type is called **negative feedback.** To understand how negative feedback systems work, let's consider a familiar example, the heating system in a typical house, which contains many of the features that are present in biological feedback mechanisms (**FIGURE 4-16A**).

In the winter, the heating system (the furnace and thermostat) of a house maintains a constant internal temperature, even though the outside temperature fluctuates sharply. Heat lost through ceilings, walls, windows, and tiny cracks is replaced by heat generated from the combustion of natural gas or oil in the furnace (**FIGURE 4-16A**).

The furnace is controlled by a thermostat, which detects changes in room temperature. When indoor temperatures fall below the desired setting, the thermostat sends a signal to the furnace, turning it on. Heat from the furnace is then distributed through the house, raising the room temperature. When the room temperature reaches the desired setting, the thermostat shuts the furnace off. Like all negative feedback mechanisms, the product of the system (heat) "feeds back" on the process, shutting it down.

A graph of a hypothetical house temperature is shown in **FIGURE 4-16B** and illustrates another important principle of homeostasis: Homeostatic systems do not maintain absolute constancy. Rather, they maintain conditions (such as body temperature) within a given range.

All homeostatic mechanisms in humans operate in a similar fashion, maintaining conditions within a narrow range around an operating point. The operating point is akin to the setting on a thermostat. As you will see in later chapters, our bodies maintain fairly constant levels of a great many chemical components, including hormones, nutrients, wastes, and ions. We also maintain physical parameters such as body temperature, blood pressure, blood flow, and others. Fortunately, all of this is done subconsciously and quite automatically. We'd never get anything done if we had to attend to all of these functions ourselves!

All Homeostatic Feedback Mechanisms Contain a Sensor (or Receptor) and an Effector

Biological homeostatic mechanisms contain **sensors** that detect change and **effectors** that correct conditions. In your home, the thermostat is the sensor, and the furnace is the effector.

The human body contains many sensors. In your body, for instance, specially modified nerve cell endings in the skin detect temperature changes in the environment. These sensors not only detect changes in the ambient (outside) temperature, but also send signals to the brain, alerting it to such changes. The brain then sends signals to the body to rectify matters—usually to reduce heat loss and increase heat output. The options for generating more heat are many. In other words, the body contains several different types of effectors—in the same way that heat can be generated and conserved in a house many ways (closing curtains, starting a fire in the woodstove, or cranking up the furnace).

One of the main sources of heat in the human body is the breakdown (catabolism) of glucose and other molecules. Chapter 3 noted that the cells of the body break down glucose to make ATP. During this process, heat is given off. Each cell, then, is a tiny furnace whose heat radiates outward and is distributed throughout the body by the blood. Unlike the furnace in your home, the cellular "furnaces" cannot be turned up very quickly. They respond much more slowly than a furnace and are part of a delayed response to low temperature. As winter progresses, energy catabolism increases, and the body produces more heat.

In order to respond to sudden changes in outside temperature, the body must rely on more rapid mechanisms. If you walked outdoors on a cold winter night dressed only in a light sweater and blue jeans, for example, receptors in your skin would sense the cold and send signals to the brain. The brain, in turn, would send signals to blood vessels in the skin, causing them to constrict, reducing the flow of blood in the skin. The restriction of blood flow through the skin reduces heat loss. If it is cold enough outside, the brain may also send signals to the muscles, causing them to undergo rhythmic contractions, known as shivering. Shivering burns additional glucose, releasing extra heat. (It's like adding additional logs to a fire. The more fuel burned, the more heat you get.) Many voluntary actions may also be "ordered" by the brain to reduce heat loss or generate more heat. For example, you might put on a hat or turn around and go back inside. In humans, conscious acts are often crucial components of homeostasis.

Homeostasis Is Maintained by Balancing Inputs and Outputs

The diagram in **FIGURE 4-17** illustrates the many ways the body regulates the level of various chemicals. Although it may look a bit complicated, it's pretty logical and easily understood if you take a look at one part at a time. Let's start with the input side first—that is, how chemicals enter our bodies.

As illustrated, ions (for example, calcium) and other chemical substances (for example, glucose) are ingested in food or water. Other essential chemicals such as oxygen enter our bodies in the air we breathe. Still other substances (such as certain amino acids) are actually produced in the body. Input from these sources tends to increase internal concentrations.

Now look at the output side. As illustrated, excretion and metabolism tend to lower concentrations. **Excretion** is the loss of materials through a number of specialized body systems. Carbon dioxide, for example, is excreted by the lungs, as is water. Water is also excreted by the kidneys. Metabolism (chemical reactions occurring in the body cells) also removes certain substances from the body. Blood glucose levels, for instance, are decreased by cellular catabolism.

Internal concentrations are kept in balance in large part by input and output. However, internal storage depots (regions where chemicals are stored) also participate in homeostatic balance. For example, glucose molecules in the blood can be stored in the liver. Thus, blood glucose levels are also determined by input and output in storage depots.

The relative importance of the various homeostatic pathways depends on the substance in question. Water, for example, enters the body primarily via the food and liquids we consume. Water is removed by the kidneys, lungs, and skin. For iron, the rate of absorption by the intestinal tract is a key determinant of blood levels. When iron levels decrease—say, because of a decrease in iron intake—the body reestablishes the balance by increasing the rate of absorption in the small intestine.

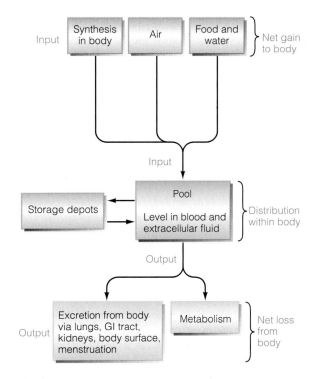

FIGURE 4-17 Generalized View of a Homeostatic System Inputs and outputs are balanced to maintain more or less constant levels of chemical and physical parameters.

Homeostasis Can Be Upset by Changes in the Input, Output, or Storage

Homeostasis can be thrown out of balance by changes in input, output, and storage. Consider output first. On a hot day, water escapes our bodies very rapidly through perspiration, an adaptation that helps cool us down. Although perspiration is beneficial, it can be quite damaging if water loss is severe. Severe water loss leads to dehydration, which may upset homeostasis—so much so, in fact, that death may occur.

Severe diarrhea—a drastic increase in water output—also results in a dangerous depletion of body fluids. It can be so severe that it too can result in death. In fact, dehydration resulting from severe diarrhea kills millions of children each year in the less developed nations.

Just as changes in the output of a substance drastically alter homeostasis, so do changes in input. For example, going without water for a prolonged period (approximately 3 days) can kill a human being. Excess input can also prove dangerous. Excess salt intake, for example, can result in hypertension (high blood pressure) in some people. Whatever the cause, imbalances in homeostasis can have dramatic impacts on human health.

Homeostatic Control Requires the Action of Nerves, Hormones, and Various Chemicals that Operate over Short Distances

Homeostatic mechanisms are **reflexes**—that is, automatic physiological responses triggered by various stimuli. Reflexes occur without conscious control. Two types of reflexes exist: nervous system and hormonal.

The Nervous System Reflex.

Homeostatic systems controlled by the nervous system are fairly straightforward. As shown in **FIGURE 4-18**, heat from a candle is detected by sensors, or receptors, in the skin. These receptors send nerve impulses to the brain or spinal cord via sensory nerves. The brain and spinal cord, in turn, direct an appropriate response to counterbalance the change. In this example, the spinal cord sends a signal to pull the hand away from the flame.

At any one time, the brain and spinal cord receive thousands of signals from receptors in the body. These signals alert the nervous system to a variety of internal and external conditions. The brain and spinal cord then send out special responses, after integrating (making sense of) the various inputs they have received. The brain and spinal cord control the activity of many effectors, generally muscles and glands.

Hormonal Control.

Hormones are chemical substances produced by the endocrine glands. Released into the blood, hormones travel

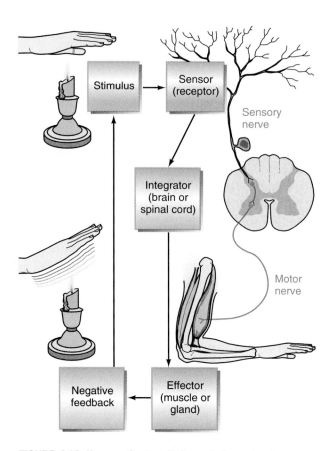

FIGURE 4-18 Nervous System Reflex   Reflexes involve some kind of stimulus, a sensor, an integrator, and an effector. The sensor detects a change and sends a signal to the integrator, the brain or spinal cord, which then elicits a response in the effector organs. Negative feedback from the effector eliminates the stimulus.

to distant sites, where they effect some sort of change. But hormones don't just pour out of endocrine glands in a constant stream. They are released only under certain circumstances and then usually as part of a chemical reflex essential for maintaining homeostasis. Consider an example.

The parathyroid glands (pair-ah-THIGH-roid) are four tiny nubbins of tissue embedded in the thyroid gland in the neck. They produce a hormone known as **parathyroid hormone,** or **PTH.** PTH is released from the cells of the parathyroid glands when calcium levels in the blood fall below a certain concentration. PTH then travels in the blood to the bone, where it stimulates osteoclasts, bone-destroying cells described earlier. These cells cause the release of calcium, which raises the calcium level of the blood, returning concentrations to normal.

In this reflex, the cells of the parathyroid gland are the sensors, which detect the drop in blood calcium. The bone is the effector, which produces the desired response (**FIGURE 4-19**). There's no intermediate command center such as the brain.

Some endocrine reflexes do operate through the nervous system, making the lines of cause and effect a bit more difficult to follow. A good example is the thy-

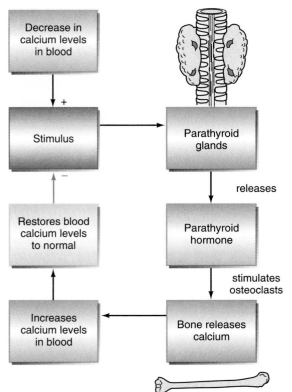

FIGURE 4-19 Endocrine Reflex Not Involving the Nervous System This reflex operates through the bloodstream. A decrease in calcium in the blood stimulates a series of reactions that restores normal blood calcium levels. A plus sign indicates that the stimulus increases parathyroid gland activity; a minus sign indicates the opposite effect.

roid gland, which produces a hormone called **thyroxine** (thigh-ROX-in) (**FIGURE 4-20**). Thyroxine increases the metabolic rate of cells and thus increases heat production in the same way that putting more logs on a fire increases its output of heat.

Thyroxine levels must be kept within a normal range and are monitored by cells in the brain (the receptors). When thyroxine levels fall, these brain cells release a chemical substance (thyroid-stimulating hormone-releasing factor) that travels from its site of production in the brain to another site, the pituitary gland, located just beneath the brain. The pituitary responds by releasing yet another hormone, thyroid-stimulating hormone, TSH, which travels in the blood to the thyroid gland. There, TSH steps up the production of thyroxine, correcting the deficiency.

This reflex has one sensor (cells in the brain), which detects levels of

thyroxine, but three effectors: the cells of the pituitary that release TSH, the cells of the thyroid that release thyroxine, and body cells that respond to thyroxine by increasing their rate of metabolism.

Local Chemical Control.

Nervous and endocrine feedback mechanisms generally occur over considerable distances. In the nervous system, for example, nerves carry impulses from the brain and spinal cord to distant parts of the body, often several feet away. In the endocrine system, the bloodstream carries the messages from endocrine glands to distant effectors. But not all systems require messages that are transported over long distances. In some cases, chemical control is exerted through the extracellular fluid only a cell or two away.

Chemicals that elicit local effects are called **paracrines** (PEAR-ah-crins). Produced by individual cells, paracrines diffuse to neighboring cells. Epidermal growth factor (EGF) produced by skin cells is one example. EGF stimulates cell division when skin cells are damaged or lost, providing a degree of local control over cell growth.

One of the best-known paracrines is a group of chemical substances known as **prostaglandins** (PROSS-tah-GLAND-ins). Prostaglandins comprise a rather large group of molecules with diverse functions. Some stimulate blood clotting; others stimulate smooth muscle contraction.

An interesting evolutionary adaptation akin to paracrines are the **autocrines,** molecules produced by a

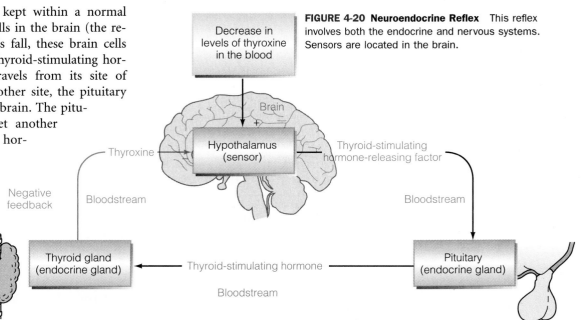

FIGURE 4-20 Neuroendocrine Reflex This reflex involves both the endocrine and nervous systems. Sensors are located in the brain.

Health Note 4-1

The Truth About Herbal Remedies

A male patient is treated by his physician for a painful infection of the prostate gland, one of the sex accessory glands that produces a large portion of the semen. She prescribes an antibiotic and advises the patient to take an herbal remedy, saw palmetto, to help prevent prostatic enlargement, a common problem in older men. A female patient suffering from restlessness and minor problems with sleep is advised by a friend to try some valerian root, another type of herbal remedy. A patient suffering from depression, who wants a "natural" remedy, is advised by her physician to try St. John's Wort.

Herbal remedies are derived from roots, bark, seeds, fruit, flowers, leaves, and branches of plants. In China, some herbal remedies even contain animal parts. An estimated 3000 different medicinal herbs are currently used throughout the world.

Herbal remedies are sold in the United States and in other countries by herbalists, and may be mixed on site. More commonly, people get their herbs in pill, capsule, or liquid form from health food stores, grocery stores, and even large discount stores such as Wal-Mart. In recent years,

more and more people have turned to alternative remedies for many conditions they can self-diagnose and treat—and even some that should be attended to by physicians.

Although herbs are the latest craze in the United States, the Chinese and other Asian peoples have been using them for thousands of years. In China, the history of herbal remedies goes back 5000 years. Unbeknownst to many doctors, certain herbs were used routinely in the United States and other developed nations before the advent of modern medicines. Echinacea (ek-eh-NAH-shah), for instance, was widely used to treat colds in the United States prior to the introduction of sulfa drugs (the first generation of synthetic antibiotics).

Herbal treatments not only have a long history of use, they have been the source of a surprisingly large proportion of the many prescription and over-the-counter drugs in use today. That is, many so-called "modern medicines" contain active ingredients originally extracted from plants. Aspirin, for instance, is derived from the inner bark of willow trees. Wild chimpanzees have been observed chewing wil-

low bark to combat tooth aches. But are herbal treatments effective?

There is no blanket answer to this question. While research shows that many claims about the effectiveness of numerous common herbal preparations appear to be unfounded, not all are. Echinacea, for instance, does appear to enhance the immune system function, and it can be used as a preventative and for treatment of colds and flus. Bilberry, dried blueberries, helps maintain eyesight and improves night vision. It was even used in World War II by the Royal Air Force on night bombing missions. Valerian root is an effective, nonaddicting sleep aid and mild relaxant. Saw palmetto appears to be effective in reducing enlargement of the prostate. Ginger is effective in the treatment of nausea and morning sickness, although it should not be used for an extended time during pregnancy. Garlic lowers serum cholesterol and may have antibacterial properties, a fact discovered by Louis Pasteur. Feverfew seems to help prevent migraine headaches. St. John's Wort is an antidepressant. Hawthorn fights hypertension. The list goes on. For those interested in

cell that affect the function of that cell. Autocrines are part of the simplest chemical reflexes in the body. Some prostaglandins function as autocrines. Like hormones, paracrines and autocrines are involved in maintaining homeostasis.

Human Health Depends on Homeostasis, Which, in Turn, Requires a Healthy Environment

This book emphasizes a theme that is important to all of us—notably, that human health is dependent on homeostasis. Homeostasis, in turn, requires a healthy, clean environment. The health of the environment, of course, is also dependent in part on properly functioning homeostatic systems in nature that maintain conditions conducive to life.

The relationship between homeostasis and health is shown in **FIGURE 4-21**. Take a moment to study this diagram. Notice that the body systems (right side of diagram) maintain overall homeostasis, which is essential for proper cell function. Cells make up body systems. They also contain homeostatic mechanisms that are vital to their own function as well as to the overall economy of the organism. Notice too that environmental factors, such as pollution, affect the function of body systems and cells in ways that can upset the internal balance, thus altering human mental and physical health.

Just as human health is affected by the condition of the environment, so too is the health of many other species. For a discussion of herbal remedies that help restore health, see Health Note 4-1.

finding out more, check out the web sites and books on the subject listed below.

The main advantage of herbs, say some proponents, is that they have no side effects and contain no chemicals; they are also purportedly safe and effective. Unfortunately, this claim is far from true. Herbal remedies contain naturally produced chemicals, and some can have very serious side effects. A number of herbs, for instance, damage the liver, including comfrey—an herb that reportedly promotes bone healing. When taken for long periods, however, its active chemical ingredient is toxic to the liver.

David Kroll, a pharmacologist from the University of Colorado Health Sciences Center, notes further problems. First and foremost, U.S. labeling information is inadequate. Herbs are sold as dietary supplements, and as such, no claims about their use are available. Individuals must often rely on advice from sales clerks. Very little is known about the long-term effects of the wide variety of herbal remedies or their potential interactions with conventional medicines, although our knowledge base is expanding rapidly. And others warn that without federal

regulations that require strict standards for drug content in various herbal remedies, a patient can't know for sure the exact dose he or she is receiving. In Europe, however, stricter laws require much better controls on product content. In Germany, in fact, herbal remedies are considered prescription medicines and are covered by health insurance.

Another bit of advice from Dr. Kroll. When given a choice between an over-the-counter remedy and an herbal one, he favors the over-the-counter treatment because of the standardization issue. He also notes that treatment with herbal remedies, as OTC preparations, may delay the diagnosis of more serious conditions. The rule: See a doctor first to be certain that you have what you think you have. And buy products from reputable companies.

Herbal remedies will invariably increase in popularity, and many measures should be taken to make them safer and more effective. Guidelines similar to those in Germany, for instance, could result in more consumer confidence in the quality and content of the product. Canada requires some testing before an herbal preparation can be sold. Although it is

not as rigorous as the testing required for pharmaceuticals in the United States, it does help ensure the effectiveness of the remedy. Canada has also established standards for the purity and content of herbal remedies.

The important lesson in all of this is *buyer beware*. Study an herb thoroughly before you take it, or better yet, see a qualified physician who practices both traditional and herbal medicine. More rigorous training of those who dispense herbs might also help raise consumer confidence. In China, herbalists graduate from a five-year program of study. Also needed is more research by herbal "manufacturers" similar to that required of drug companies—to fine tune our knowledge about these increasingly popular forms of treatment.

Visit *Human Biology's* Internet site, www.jbpub.com/ humanbiology, for links to web sites offering more information on this topic.

www.jbpub.com/humanbiology

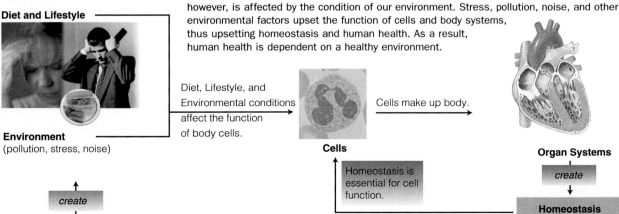

FIGURE 4-21 Health, Homeostasis, and the Environment Human health is dependent on maintaining homeostasis. Homeostasis, however, is affected by the condition of our environment. Stress, pollution, noise, and other environmental factors upset the function of cells and body systems, thus upsetting homeostasis and human health. As a result, human health is dependent on a healthy environment.

Diet and Lifestyle

Environment
(pollution, stress, noise)

create

Human Activities

Diet, Lifestyle, and Environmental conditions affect the function of body cells.

Cells

Cells make up body.

Homeostasis is essential for cell function.

Organ Systems

create

Homeostasis

Biological Rhythms

The previous discussion may have given you the impression that homeostasis establishes an unwavering condition of stability that remains more or less the same, day after day, year after year. In truth, many physiological processes undergo rhythmic change.

▌ Not All Physiological Processes Remain Constant over Time, But Those That Fluctuate Do So in Predictable Ways

Body temperature varies during a 24-hour period by as much as 0.5°C. Blood pressure may change by as much as 20%, and the number of white blood cells, which fight infection, can vary by 50% during the day. Alertness also varies considerably. About 1:00 P.M. each day, for instance, most people go through a slump. For most of us, activity and alertness peak early in the evening, making this an excellent time to study. Daily cycles, such as these, are called circadian rhythms (sir-KADE-ee-an; "about a day"). **Circadian rhythms** are natural body rhythms linked to the 24-hour day-night cycle.

Many hormones follow daily rhythmic cycles. The male sex hormone testosterone, for example, follows a 24-hour cycle. The highest levels occur in the night, particularly during dream sleep. Dream sleep occurs primarily in the early morning hours—the later the hour, the longer the periods of dream sleep (**FIGURE 4-22**).

Not all cycles occur over 24 hours, however. Some can be much longer. The menstrual cycle, for instance, is a recurring series of events in the reproductive functions of women that lasts, on average, 28 days. During the menstrual cycle, levels of the female sex hormone estrogen undergo dramatic shifts. Estrogen concentra-

tions in the blood are low at the beginning of each cycle and peak on day 14, when ovulation normally occurs. Throughout the remaining 14 days, estrogen levels are rather high. Then they drop off again when a new cycle begins. Estrogen levels follow this cycle month after month in women of reproductive age.

The important point here is that the body is not static. Although many chemical substances are held within a fairly narrow range by homeostatic mechanisms, others fluctuate widely in normal and quite essential cycles. Over the long run, these changes are quite predictable. They also occur within prescribed physiological limits; they do not run out of control. They are part of the body's dynamic balance, just as yearly weather changes are part of the dynamic balance of the planet's climate.

▌ In Humans, Internal Biological Rhythms Are Controlled by the Brain

Just how the body controls its many internal rhythms remains a mystery. Research suggests that the brain controls many biological cycles. One region in particular, the **suprachiasmatic nucleus** (SUE-pra-ki-as-MAT-tick), is thought to play a major role in coordinating several key rhythms.

The suprachiasmatic nucleus (SCN) is a clump of nerve cells in the base of the brain in a region called the hypothalamus. It may regulate other control centers. As a result, the suprachiasmatic nucleus is often referred to as the "master clock." Like a clock, the SCN ticks off the minutes, faithfully imposing its control on the body, turning body functions on and off like an automatic timer that controls lights in a home when the occupants are gone. Even in complete, prolonged darkness, the clock ticks on, directing many circadian rhythms.

In humans, research suggests, the master clock operates on a 25-hour cycle. That is, if isolated in a dark room, many of us would fall into a 25-hour sleep-wake cycle. But as with most other biological phenomena, there is considerable variability. Some individuals, for example, operate on 28-hour sleep-wake cycles. Other

FIGURE 4-22 Stages of Sleep   Numbers indicate the stages of sleep: the higher the number, the deeper the sleep. Note that around midnight sleep is deepest. As morning approaches, a person's sleep is lighter, and REM sleep, or dream sleep, occurs in longer increments.

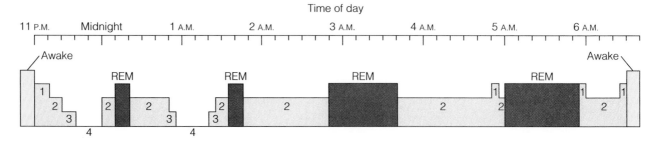

individuals' cycles are shorter than 24 hours. Sleep researchers believe that in the real world, the more common 25-hour clock is modulated by the 24-hour day-night cycle. In other words, the environment alters the clock and maintains many biological rhythms on a 24-hour cycle.

Ultimate control of the SCN is thought to reside in a gland in the brain known as the **pineal gland** (PIE-knee-al). It secretes a hormone thought to keep the suprachiasmatic nucleus in sync with the 24-hour day-night cycle.

The study of biological rhythms is a fascinating field that has yielded some important information and insights. One practical application is a better understanding of jet lag, that drowsy, uncomfortable feeling people get from the disruption of sleeping patterns caused by long-distance airplane travel. Studies suggest that jet lag occurs when the body's biological clock is thrown out of synchrony with the day-night cycle of a traveler's new surroundings. A businesswoman who travels from Los Angeles to New York, for instance, may be wide awake at 10:00 P.M. New York time because her body is still on Los Angeles time—3 hours earlier. When the alarm goes off at 6:00 A.M., our weary traveler crawls out of bed exhausted, because as far as her body is concerned, it is 3:00 A.M. Los Angeles time.

To avoid the discomfort associated with travel, sleep researchers suggest that you abide by your internal clock, following "home time" when in a new time zone, or, barring that, try to reset your biological clock before you get there. You can do this by going to bed an hour or two earlier when you are traveling from west to east. Do just the opposite for east-to-west travel.

Insomnia is another disruption of sleeping patterns. For more on this topic, see Health Note 4-2.

Research on body rhythms has also shown that people respond to drugs differently at different times of the day and night. By administering drugs at the body's most receptive times, physicians may be able to reduce the doses, lowering toxic side effects and fighting diseases more effectively.

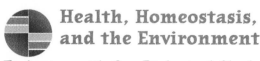

Health, Homeostasis, and the Environment

Tinkering with Our Biological Clocks

Human health is dependent on a healthy environment. Homeostasis is the intermediary. It is affected by our en-

vironment and its upset affects our health. Put another way, homeostasis is often the victim of unhealthy environments and the cause of an unhealthy body.

As I've pointed out previously in this book, the health of our environment depends on more than clean air and water. It requires a healthy social and psychological atmosphere with manageable levels of stress. As this section points out, though, many adults are thrown into fairly unhealthy workplaces, a situation caused in part by our society's obsession with economic growth. Let me elaborate further.

Modern life with its stress, noise, pollution, hectic pace, and weird work schedules can deal a blow to our internal biological rhythms. Dr. Richard Restak, a neurologist and author, notes that the "usual rhythms of wakefulness and sleep . . . seem to exert a stabilizing effect on our physical and psychological health." The greatest disrupter of our natural circadian rhythms, he says, is the variable work schedule, surprisingly common in the United States and other industrialized countries. Today, one out of every four working men and one out of every six working women is on a variable work schedule, shifting frequently between day and night work. In many industries, workers are at the job day and night to make optimal use of equipment and buildings. As a result, more restaurants and stores are open 24 hours a day, and more health care workers must be on duty at night to care for accident victims.

What business owners have forgotten is that for millions of years, humans have slept during the night and been awake during the day. Make a person work at night when he or she normally sleeps, say sleep experts, and you can expect more accidents and lower productivity. Consider an example.

At 4:00 o'clock in the morning in the control room of the Three Mile Island nuclear reactor in Pennsylvania, three operators failed to notice lights indicating that a valve in the system had remained open. When the morning-shift operators entered the control room, they quickly discovered the problem, but it was too late. Pipes in the system had burst, sending radioactive steam and water into the air and into two buildings. John Gofman and Arthur Tamplin, two radiation health experts, estimate that the radiation released from the accident, the worst nuclear accident in U.S. history, will result in at least 300 and possibly as many as 900 additional fatal cases of cancer in the residents living near the troubled reactor, although other experts (especially in the nuclear industry) contest these projections, saying that the accident will not have a noticeable effect. Whatever the outcome of this debate, one thing is

Health Note 4-2

Wide Awake at 3 A.M.: Causes and Cures of Insomnia

It's Thursday evening. The final exam in your human biology course is tomorrow. Forty percent of your grade depends on your performance. You've studied hard, but it's well past midnight and you can't get to sleep. It's happened before and you know your ability to think during the test, which is scheduled to start at 8 A.M., will be impaired. That puts more pressure on you, making it more difficult to sleep.

This frustrating problem, which affects almost everyone at some time in life, is just one of several sleep disorders that physicians call *insomnia* (in-SOM-knee-ah). Another common complaint of insomniacs is waking up in the middle of the night and not being able to fall asleep again for several hours, if at all. In other instances, individuals find that they wake up too early; still others complain of sleeping through the night, but waking up feeling unrefreshed. In summary, then, insomnia is best de-fined as a condition of inadequate or poor quality sleep. It's a problem that affects a great many people. In a recent study in Canada, for example, two of every five people surveyed said they had trouble sleeping at least once a week. Insomnia affects both men and women, but is more common in women and the elderly.

Insomnia may be transient (short term), intermittent (on and off), or chronic (persistent). Short-term and intermittent insomnia are generally of less concern than chronic insomnia, which can last for years. Chronic insomniacs complain of tiredness, a lack of energy, difficulty concentrating, and irritability. Insomnia can make people more susceptible to common illnesses, such as colds.

What causes insomnia? Perhaps the most common cause of insomnia is a change in one's daily routine. For example, travel, starting a new job or a new semester, going into a hospital, sleeping in a strange environment, moving into a new home, and anxiety over job interviews or tests are some possible causes. But these are generally short-lived stimuli that rob us of our sleep. Chronic insomnia is often related to more serious problems. Certain diseases that cause pain and nausea, for example, may affect sleep. Chronic anxiety, for example, caused by a project at work that may take five months to complete, with many deadlines along the way, can keep one from falling asleep.

Another very common cause of chronic insomnia is depression. Depression usually manifests itself in the wide-awake-at-3 A.M.-syndrome. A person who is depressed may fall asleep without trouble, but may wake up every evening at 2 or 3 A.M. and may stay awake for a few hours or through the entire night. Depression, which is discussed at length in Health Note 11-1, comes in a variety of shapes

certain: The 1979 accident at Three Mile Island will cost several billion dollars to clean up.

Late in April 1986, another nuclear power plant ran amok. This crisis, in the former Soviet Union, was far more severe. In the wee hours of the morning, two engineers were testing the reactor. They deactivated key safety systems in violation of standard operational protocol. Steam built up inside the reactor and blew the roof off the containment building. Workers battled for days to cover the molten radioactive core, which spewed radiation into the sky that spread worldwide.

Although no one will ever know for sure, the Chernobyl disaster, like the accident at Three Mile Island, may have been the result of workers operating at a time unsuitable for clear thinking. One has to wonder how many plane crashes, auto accidents, and acts of medical malpractice could be traced to judgment errors resulting from our insistence on working against the natural body rhythms.

Making matters worse, many companies that maintain shifts round the clock spread the burden evenly among employees. One week, workers are on the day shift; the next week, they are switched to the "graveyard shift" (midnight to 8:00 A.M.). The next week, they are put on the night shift (4:00 P.M. to midnight). Many workers subject to such disruptive changes report that they often feel run down and have trouble staying awake on the job. Their work performance suffers. When employees who have been on the graveyard shift arrive home, they are physically exhausted but cannot sleep because they are trying to sleep at a time when their body is trying to wake them up. What is more, weekly changes in schedule never permit workers' internal clocks to adjust. Studies show that most people require from 4 to 14 days to adjust to a new schedule.

Not surprisingly, workers on alternating shifts suffer more ulcers, insomnia, irritability, depression, and

and sizes, too—and can last for many years. Common symptoms include persistent sadness and anxiety. Feelings of emptiness, worthlessness, and pessimism are commonly reported. Loss of pleasure in activities that one once enjoyed and a loss of appetite or overeating are also common.

What can be done to treat chronic insomnia? If you have a medical problem, treatment of that problem can be helpful. If you are feeling anxious all of the time, stress-relieving techniques, including psychotherapy, all of which are described in Health Note 1-1, can be helpful. Depression can be treated by psychotherapy and drugs. Besides these important steps, experts recommend the following do's and don't's for dealing with insomnia.

- Go to bed and get up at the same time each day.
- Maintain a comfortable temperature in your bedroom—too hot or too cold can make it difficult to sleep.

- Sleep in a quiet, darkened room.
- Exercise regularly during the week—it helps to relieve stress.
- Don't exercise or engage in mentally stimulating activities just before going to bed.
- Exercise later in the afternoon or early in the evening.
- Engage in quiet tasks such as reading or listening to music in the hour or two before going to sleep.
- Try a light bedtime snack with protein just before going to bed.
- Avoid caffeine, which is found in chocolate, regular coffees and teas, and caffeinated sodas.
- Don't use alcohol to stimulate sleep.
- Don't fight to go sleep; if you can't fall asleep within a half an hour, get out of bed, engage in some relaxing activity, and then go to bed when you feel tired.

Sleeping pills can be used to treat insomnia, but they are mainly used

to treat short-term insomnia that results from occasional stressors. Sleeping pills are usually not taken for more than two weeks because they can actually worsen one's insomnia. Ironically, several common prescription drugs used to treat insomnia-causing depression—including Prozac and Paxil—actually cause sleeplessness in about one-third of the patients taking them. Sleeping pills, both over-the-counter remedies and prescription medicines, can cause a host of side effects, including confusion, heart problems, and decreased sexual desire.

Visit *Human Biology's* Internet site, www.jbpub.com/ humanbiology, for links to web sites offering more information on this topic.

www.jbpub.com/humanbiology

tension than workers on unchanging shifts. Many suffer from impaired judgment on the job and, in some circumstances, they may pose a threat to society.

Thanks to studies of biological rhythms, researchers are finding ways to reset the body's clock. These efforts could lessen the problems shift workers face and improve the performance of those on the graveyard shift. For instance, one simple measure is to place all shift workers on 3-week cycles instead of weekly cycles. This gives their biological clocks time to adjust. And instead of shifting workers from daytime to a graveyard shift, transfer them forward, rather than

backward (for example, from a daytime to an evening shift and from an evening shift to the graveyard shift). Studies suggest that moving forward is far easier for workers than shifting backward.

For reasons not well understood, special treatment with bright lights helps reset the biological clock. Patients who suffer from insomnia because their biological clocks are out of phase can receive daily doses of bright light, which somehow reset their internal clocks. Similar treatments are being tested for shift workers. They are a small price to pay for a healthy work force and a safer society.

SUMMARY

FROM CELLS TO ORGAN SYSTEMS

1. The basic structural unit of all organisms is the cell. In humans, cells and extracellular material form *tissues,* and tissues, in turn, combine to form *organs.*

2. Four *primary tissues* are found in the bodies of most multicellular animals such as humans: epithelial tissue, connective tissue, muscle tissue, and nervous tissue. Each primary tissue serves a unique function.

3. Organs contain all four primary tissues in varying proportions.

4. *Epithelial tissues* consist of two types: *membranous epithelia,* which form the coverings or linings of organs, and *glandular epithelia,* which form exocrine and endocrine glands.

5. *Connective tissues* bind other tissues together, provide protection, and support body structures. All connective tissues consist of two basic components: cells and extracellular fibers.

6. Two types of connective tissue are found in the body: *connective tissue proper* (dense and loose connective tissue) and *specialized connective tissue* (bone, cartilage, and blood).

7. *Cartilage* consists of specialized cells embedded in a matrix of extracellular fibers and other extracellular material.

8. *Bone* is a dynamic tissue that provides internal support, protects organs such as the brain, and helps regulate blood calcium levels.

9. Bone consists of bone cells (*osteocytes*) and a calcified cartilage matrix. Two types of bone tissue exist: spongy and compact.

10. The *osteoclast,* one type of bone cell, plays a major role in reshaping bone to meet the changing demands of the body and in releasing calcium to help maintain blood calcium levels.

11. *Blood* is another form of specialized connective tissue. It consists of numerous blood cells and *platelets* and an extracellular fluid, called *plasma.*

12. *Muscle* is an excitable tissue that contracts when stimulated. Three types of muscle tissue are found in the human body: skeletal, cardiac, and smooth muscle.

13. *Skeletal muscle* is under voluntary control, for the most part, and forms the muscles that attach to bones. *Cardiac muscle* is located in the heart and is involuntary. *Smooth muscle* is involuntary and forms sheets of varying thickness in the walls of organs and blood vessels.

14. *Nervous tissue* is the fourth primary tissue. It consists of two types of cells: conducting and nonconducting (supportive). The conducting cells, called *neurons,* transmit impulses from one region of the body to another. The *nonconducting cells* are a type of nervous system connective tissue.

15. Tissues combine to form organs in which specialized functions are carried out. Most organs are part of an organ system, a group of organs that cooperate to carry out some complex function.

PRINCIPLES OF HOMEOSTASIS

16. The nervous and endocrine systems coordinate the functions of organ systems, helping the body maintain homeostasis, or internal constancy. Homeostasis also occurs on the cellular level.

17. All homeostatic systems maintain constancy by balancing inputs and outputs as well as movement to and from storage depots. Homeostatic systems do not maintain absolute constancy and can be upset by alterations in input or output or by changes in movements in or out of storage depots. Imbalances can seriously affect an individual's health.

18. Homeostatic mechanisms are *reflexes,* occurring without conscious control. They operate by *negative feedback.*

19. Homeostatic reflexes occur at all levels of biological organization and involve two main mechanisms: nervous and hormonal.

20. Nervous and endocrine mechanisms generally occur over considerable distances. In some cases, control is exerted locally through the extracellular fluid.

BIOLOGICAL RHYTHMS

21. Many physiological processes undergo definite rhythmic changes. These natural rhythms may take place over a 24-hour period or over much longer or shorter periods.

22. The brain controls many biological cycles. A clump of nerve cells (the *suprachiasmatic nucleus*) located in the hypothalamus is thought to play a major role in coordinating several key functions and control centers. It is therefore often referred to as the "master clock."

HEALTH, HOMEOSTASIS, AND THE ENVIRONMENT: TINKERING WITH OUR BIOLOGICAL CLOCKS

23. The hectic pace of modern life, the shifting work schedules that many people follow, and the stressful environments in which we live can upset internal rhythms, with disastrous consequences.

24. The greatest disrupter of our natural circadian rhythms is the variable work schedule, which is surprisingly common among industrialized nations.

25. Workers on alternating shifts suffer from a higher incidence of ulcers, insomnia, irritability, depression, and tension than workers on regular shifts. Making matters worse, tired, irritable workers whose judgment is impaired by fatigue pose a threat not only to themselves, but also to society.

26. Researchers are finding ways to reset the biological clock, which could lessen the problems shift workers face and thereby increase work performance, particularly of those on the graveyard shift.

Critical Thinking

THINKING CRITICALLY— ANALYSIS

This Analysis corresponds to the Thinking Critically scenario that was presented at the beginning of this chapter.

With a little digging—a step vital to critical thinking—you will find that prenatal care consists of visits to doctors for checkups and advice on nutrition and other matters pertinent to pregnant women. Prenatal care is crucial to the development of a healthy fetus. If problems arise, they can be dealt with immediately, not when it's too late and too costly. You will also find that the cost of one organ transplant could supply medical care for 1000 pregnant women, assistance that itself could prevent ailments that necessitate costly transplant procedures.

Given this information, should Americans spend more on prevention and less on dramatic life-saving measures, such as organ transplants? Why?

After you have given your opinion on this matter, put yourself in the position of a parent whose child needs a kidney transplant to survive. How does this perspective affect your position? What lessons can be learned from this?

EXERCISING YOUR CRITICAL THINKING SKILLS

In the Health, Homeostasis, and the Environment section, I noted that workers on night shifts are more accident-prone and suffer from lower productivity. I then cited two possible examples: the Three Mile Island and Chernobyl nuclear reactor accidents, both of which occurred at night. One of the academic reviewers of this text wrote, "I am certain I can cite terrible worker error in daytime circumstances." The examples, she contended, are "biased and used for effect."

Using your critical thinking skills, how can you determine whether my statement that night-shift workers are more accident-prone is valid? Is there a way to determine whether the accidents at the nuclear power plants were the result of the late hour? If so, how? If not, why not? Is there a way to determine whether accidents at nuclear plants are more frequent at night or during the day? Is the reviewer's statement "I am certain I can cite terrible worker error in daytime circumstances" necessarily a valid refutation of the points I made? What is wrong with this line of reasoning?

TEST OF CONCEPTS

1. Define the following terms: tissues, extracellular material, organs, and organ systems.
2. List the four primary tissues and their subtypes.
3. Describe the similarities and differences in the embryonic origin of endocrine and exocrine glands. Explain why the two secrete differently.
4. Discuss some biological examples showing how structure reflects function.
5. Describe the two types of connective tissue and how they function.
6. Why do cartilage injuries repair so slowly or not at all? Bone is repaired much more easily than cartilage. Why is this true?

7. In what way is bone part of a homeostatic system?
8. Describe the chief differences among cardiac, skeletal, and smooth muscle.
9. What is an organ system? List some examples.
10. Define homeostasis, and describe the major principles of homeostasis presented in this chapter. Use an example to illustrate your points.
11. Homeostatic mechanisms are largely reflexes, involving chemical or nervous impulses. Describe each type of reflex, giving an example. You may find it helpful to include drawings of the systems.

12. Are biological rhythms an exception to the principle of homeostasis?
13. Describe the biological clock, and explain how it is synchronized with the 24-hour day-night cycle.
14. How does shift work upset the biological clock? How can these problems be mitigated?
15. The Health, Homeostasis, and the Environment section makes the point that human health and welfare is threatened by an environment of economic greed that compels people to work when they should be sleeping. Do you agree with this? Why or why not?

TOOLS FOR LEARNING

Tools for Learning is an on-line student review area located at this book's web site HumanBiology (www.jbpub.com/humanbiology). The review area provides a variety of activities designed to help you study for your class:

Chapter Outlines. We've pulled out the section titles and full sentence sub-headings from each chapter to form natural descriptive outlines you can use to study the chapters' material point by point.

Review Questions. The review questions test your knowledge of the important concepts and applications in each chapter. Written by the author of the text, the review provides feedback for each correct or incorrect answer. This is an excellent test preparation tool.

Flash Cards. Studying human biology requires learning new terms. Virtual flash cards help you master the new vocabulary for each chapter.

Figure Labeling. You can practice identifying and labeling anatomical features on the same art content that appears in the text.

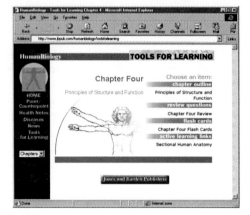

Active Learning Links. Active Learning Links connect to external web sites that provide an opportunity to learn basic concepts through demonstrations, animations, and hands-on activities.

CHAPTER 5

NUTRITION AND DIGESTION

Thinking Critically

Kidney stones begin as tiny crystals. These crystals, which are made of calcium and other ions, grow layer by layer and can eventually obstruct the flow of urine, causing considerable pain and discomfort. Because stones are made primarily of calcium, physicians often recommend a dietary solution to patients who have already had a stone: reducing their intake of calcium by cutting back on milk and milk products, such as cheese and yogurt. Sounds logical doesn't it?

Unfortunately, a new study of 45,000 men suggests that a low-calcium diet may actually increase the risk of developing kidney stones. This study showed that men who ate a diet rich in calcium actually had a 34% lower risk of developing kidney stones than men who ate a low-calcium diet. The researchers were perplexed by the results. Their study seemed counterintuitive.

Kidney specialist Gary Curhan offers a hypothesis to explain the results. He believes that a chemical called *oxalate*, which is present in many foods, may be the reason for the results. Oxalate binds to calcium and forms insoluble crystals that make up most kidney stones. Curhan believes that oxalate in the diets of people who ingest normal amounts of calcium binds to the calcium in their intestines. This prevents blood oxalate levels from rising and retards the formation of stones. Low-calcium diets permit an excess of oxalate to enter the body; this oxalate binds to calcium in the kidney and causes stones. How would you test this hypothesis? What lessons can be gleaned here? ■

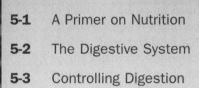

Electron micrograph of the villi of the small intestine.

T MUST BE A LAW OF HUMAN NATURE. ASK ALMOST any couple, and they will tell you: She lies shivering under the covers on a cold winter night while he bakes. Out on a hike in winter, he stays warm in a light jacket while she bundles up in stocking cap, down coat, and gloves. What causes this difference between many men and women?

Part of the answer may lie in iron—not pumping iron, but dietary iron. Quite simply, many American women do not consume enough iron to offset losses that occur during menstruation, the monthly discharge of blood and tissue from the lining of the uterus. Iron deficiencies in women may reduce internal heat production.[1]

John Beard, a researcher at Pennsylvania State University, recently published a study that supports this conclusion. Beard compared two groups of women, one with low levels of iron in the blood and another with normal levels. Beard found that body temperature dropped more quickly in iron-deficient women exposed to cold than in those with normal iron levels. He also found that iron-deficient women generated 13% less body heat.

Adding credence to the hypothesis that iron deficiency reduces body heat and makes women colder, Beard gave the iron-deficient women iron supplements for 12 weeks, after which they responded normally to cold. These findings illustrate how important proper nutrition is to normal physiological function, especially homeostasis.[2] This chapter provides further examples in discussions of nutrition and digestion, which offer background information essential to an understanding of the structure and function of the human body.

---

[1]Differences in body temperature between men and women may also result from other factors, such as body mass and surface area. Low blood pressure or poor circulation would also make the extremities feel cold.

[2]Because iron can be toxic if ingested in excess, people considering supplementing their diets with iron tablets should consult a physician.

A Primer On Nutrition

Some people eat to live, but many others seem to live to eat. No matter what your orientation, food probably occupies a central part of your life. If you're like many others, you may often plan your daytime activities around meals. In addition, you spend a good part of your life shopping for, preparing, and eating meals. Depending on your income, 10%–20% of the money you earn goes to buy food. Thus, 1.5–3 hours of each workday goes to providing money for food. (The rest goes to paying income tax!)

If you live to be 65, you will consume over 70,000 meals. Because foods affect your body in many ways, what you eat will determine how you feel in your later years. That's how important nutrition is to your health. Despite the increased emphasis on nutrition today, studies suggest that most Americans pay little attention to their diet. To perform and feel our very best, though, we must eat a balanced diet to acquire the energy and nutrients needed by our cells, tissues, and organs.

A **balanced diet** is attained by eating a variety of foods. In 1992, the U.S. Department of Agriculture released a helpful tool called a **food pyramid**, to help Americans eat better. It places foods in six major groups and prescribes allotments from each group necessary for proper nutrition (**FIGURE 5-1**). Take a few moments to study the pyramid and compare it to your diet. If you would like to assess your eating habits, you can take the test in **TABLE 5-1**.

The nutrients we need to survive and prosper physically and mentally can be divided into two broad cate-

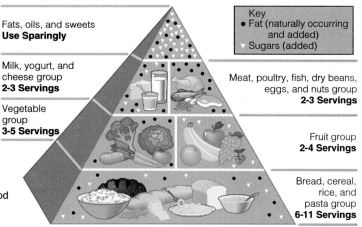

Fats, oils, and sweets
Use Sparingly

Milk, yogurt, and cheese group
2-3 Servings

Vegetable group
3-5 Servings

Key
● Fat (naturally occurring and added)
▼ Sugars (added)

Meat, poultry, fish, dry beans, eggs, and nuts group
2-3 Servings

Fruit group
2-4 Servings

Bread, cereal, rice, and pasta group
6-11 Servings

FIGURE 5-1 Food Guide Pyramid   Replacing the four food groups, the food pyramid divides food into six groups. Recommended daily intake is shown for each group.

TABLE 5-1

How Healthy Is Your Diet?

This quiz enables you to assess your eating habits. The more points you get, the better your nutritional health is likely to be.

PART I
1. I usually limit my meat, fish, poultry, or egg servings to once or twice a day. — Yes/No
2. I eat red meats (beef, ham, lamb, or pork) not more than about three times a week. — Yes/No
3. I remove fat or ask that fat be trimmed from meat before cooking. — Yes/No
4. I eat about three or four eggs per week, including those cooked with other foods. — Yes/No
5. I sometimes have meatless days and eat such protein-rich foods as legumes and nuts. — Yes/No
6. I usually broil, boil, bake, or roast meat, fish or poultry; I usually don't fry it. — Yes/No
Total "YES" answers: _____

PART II
7. I have two or more cups of milk or the equivalent in milk products every day. — Yes/No
8. I drink low-fat or nonfat milk (2% or less butterfat) rather than whole milk. — Yes/No
9. I eat ice cream or ice milk only twice a week or less. — Yes/No
10. I seldom have more than about 3 tsp of margarine or butter per day. — Yes/No
Total "YES" answers: _____

PART III
11. I usually have one serving ($\frac{1}{2}$ c) of citrus fruit or juice (oranges, grapefruit, etc.) each day. — Yes/No
12. I have at least one serving of dark green or deep orange vegetables each day. — Yes/No
13. I eat fresh fruits and vegetables when I can get them. — Yes/No
14. I cook vegetables without fat (if I use margarine, it's measured and added after cooking). — Yes/No
15. I eat fresh fruit for dessert more often than pastries. — Yes/No
Total "YES" answers: _____

PART IV
16. I generally eat whole-grain breads. — Yes/No
17. Most of the cereals I use are whole-grain and good sources of fiber. — Yes/No
18. The cereals I use have little or no sugar added. — Yes/No
19. I use brown rice in preference to white rice. — Yes/No
20. I generally have at least four servings of bread or cereal grain products each day. — Yes/No
Total "YES" answers: _____

PART V
21. I am usually within 5 to 10 lbs of the weight considered appropriate for my height. — Yes/No
22. I drink no more than $1\frac{1}{2}$ oz of alcohol (one to two drinks) per day. — Yes/No
23. I do not add salt to food after preparation and prefer foods salted lightly or not salted at all. — Yes/No
24. I try to avoid foods high in refined sugar and use sugar sparingly. — Yes/No
25. I always eat a breakfast of at least cereal and milk, egg and toast, or other protein-carbohydrate combination with fruit or fruit juice. — Yes/No
Total "YES" answers: _____

For each "yes" answer, give yourself 1 point. How are your points distributed among the various areas of nutrition?

		Excellent	Good	Fair	Poor	Your score
Part I	Meat and meat alternate choices	5–6	4	3	0–2	_____
Part II	Dairy choices	4	3	2	0–1	_____
Part III	Fruit and vegetable choices	5	4	3	0–2	_____
Part IV	Grain choices	5	4	3	0–2	_____
Part V	Weight control, other choices	5	4	3	0–2	_____

How is your overall nutrition score? — **Total points earned:** _____

(24–25 is excellent; 19–23 is good; 14–18 is fair; 13 or lower is poor.)

SOURCE: Adapted with permission from Roger Sargent, *Have a Good Life Series,* Greenville, S.C.: Liberty Life.

TABLE 5-2	
Macronutrients and Micronutrients	
Nutrients	Foods Containing Them
Macronutrients	
Water	All drinks and many foods
Amino acids and proteins	Milk and milk products, meats, eggs
Lipids	Milk and milk products, meats, eggs, nuts, oils
Carbohydrates	Breads, pastas, cereals, sweets
Micronutrients	
Vitamins	Many vegetables, meats, and fruits
Minerals	Many vegetables, meats, fruits, nuts, and seeds

gories: macronutrients and micronutrients. **TABLE 5-2** lists the basic nutrients and includes some of the foods and beverages that provide them. The following section describes the macronutrients and micronutrients more fully and explains why they are important to the proper functioning of the human body and the maintenance of health.

▮ Macronutrients Are Required in Relatively Large Quantities and Include Four Substances: Water, Carbohydrates, Lipids, and Proteins

Water.

Water is one of the most important of all the substances we ingest. Without it, a person can survive only about 3 days. Despite its importance, water rarely shows up on nutrition charts. In part, that is because it is supplied in so many different ways. For example, water constitutes the bulk of the liquids we drink and is present in virtually all of the solid foods we eat; water is even produced internally during cellular metabolism (Chapter 3).

Maintaining the proper level of water in the body is important for several reasons. First, water participates in many chemical reactions in the body. One of the most important reactions is known as **hydrolysis** (high-DROL-ah-siss). Hydrolysis means to break up (lyse) with water (hydro). Thus, hydrolytic reactions occur anytime a covalent bond in a molecule is broken by the addition of water (**FIGURE 5-2**). Many large molecules

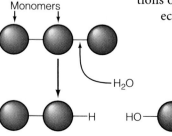

Monomers

H₂O

H   HO

FIGURE 5-2 Hydrolysis Hydrolysis is a reaction in which water is added across a covalent bond, causing the bond to split.

such as carbohydrates in the food we eat are broken down by hydrolytic enzymes in the small intestine. These enzymes cause bonds to split by adding water as shown in **FIGURE 5-2**.

Because of water's importance in metabolism, a decrease in its level can impair metabolism, including energy production. According to some studies, athletic performance may drop significantly when the body's water level falls even slightly.

Maintaining an adequate water volume also helps to stabilize body temperature. A decline in the amount of water in your body decreases your blood volume and your extracellular fluid. These declines cause your body temperature to rise, because the heat normally produced by body cells is being absorbed by a smaller volume of water. A rise in body temperature can impair cellular function and can, if high enough, lead to death. This is one reason physicians tell you to drink plenty of water when you are suffering from a fever.

Maintaining proper water levels also helps individuals maintain normal concentrations of nutrients and toxic waste products in the blood and extracellular fluid. If you don't drink enough liquid, your urine will become more concentrated; the rise in the concentration of chemicals in the blood and urine increases your chance of developing a kidney stone, a deposit of calcium, and other materials that can block the flow of urine, causing extensive damage to this organ.

Maintaining proper water levels is so important that animals have evolved a variety of homeostatic mechanisms to carry out this function. Several of these, which are present in humans, are featured in Chapter 10.

Carbohydrates Serve Many Functions in Humans; One of the Most Important Is as a Source of Energy.

Carbohydrates are a group of organic compounds. It includes such well-known examples as table sugar (sucrose), blood sugar (glucose), and starch. Carbohydrates consist primarily of carbon, oxygen, and hydrogen. Carbohydrates belong to three broad groups: monosaccharides, disaccharides, and polysaccharides.

Monosaccharides are the smallest carbohydrates and often combine to form polymers. **Monosaccharides** are often called simple sugars. These water-soluble molecules contain three to seven carbons, and many of them form ring structures when dissolved in water.

One of the most common and most important monosaccharides is **glucose** (GLUE-kose), a six-carbon sugar introduced in Chapter 3. Made by plants during photosynthesis, glucose molecules can be chemically united to form long, branching molecules such as starch. **Starch** molecules are stored in plant roots (as in potatoes) or seeds (as in wheat and rice) and are an important nutrient for plant-eating animals. In the digestive tract of humans and other animals, starch is broken down into glucose molecules, which enter the bloodstream. These molecules are then distributed to body cells, where they may either be stored for later use or broken down to release energy.

As the name implies, **disaccharides** (dye-SACK-are-ides) consist of two monosaccharides covalently bonded to each other. The disaccharide sucrose, for instance, consists of two monosaccharides, glucose and fructose.

In animals and plants, many monosaccharides react to form long polymers. The most common building block of polysaccharides is glucose. Plants synthesize two important polysaccharides from glucose: (1) **cellulose** (CELL-you-lose), one of the molecules found in the walls around plant cells, which gives wood its rigidity and protects the cells of leaves and other tissues; and (2) starch, a molecule that serves, as mentioned earlier, as a storage depot for glucose molecules.

In animals, glucose combines to form yet another polysaccharide, **glycogen** (GLYE-co-gen). It is often called "animal starch" because it serves a similar purpose and has a structure similar to starch found in plants. In humans, glycogen is composed of thousands of glucose molecules.

Glycogen is an important component in the homeostatic mechanism that controls blood sugar (glucose) levels. Glucose molecules derived from the food we eat may be immediately used by cells to generate energy. However, most glucose molecules absorbed from our meals enter muscle and liver cells and combine to form insoluble glycogen molecules.

In between meals, though, blood glucose levels tend to fall as cells use up circulating glucose molecules to make energy. To prevent glucose levels from falling below a certain level, the pancreas releases a hormone called *glucagon,* which stimulates glycogen breakdown in the liver. As glycogen is broken down, glucose is released into the bloodstream, thus maintaining normal blood sugar levels and keeping us from fainting. Incidentally, muscle glycogen does not participate in the maintenance of blood sugar levels; it's broken down to provide energy for active muscles.

Animals derive the glucose they need primarily from plant starch. A few animals such as cattle, sheep, and other grazers are able to acquire glucose from cellulose in the plants they consume.

Cellulose molecules consist of glucose, joined by a slightly different covalent bond. Although cellulose is prevalent in the diet of some animals, it cannot be digested by most animals because they lack the enzymes needed to break the covalent bonds joining its glucose subunits. (**Enzymes** are protein molecules that speed up the rate of many chemical reactions in cells.) Consequently, cellulose passes through the digestive system of most animals relatively unchanged.

Even though it is not digested, cellulose is important to normal body function. In the human diet, for example, nondigestible cellulose is a principal form of dietary fiber. For reasons explained shortly, cellulose facilitates the passage of feces through the large intestine and also reduces the incidence of colon cancer.

Those species that can digest cellulose rely on bacteria and other single-celled organisms within their digestive system to perform this function. The latter contain the enzymes required to break the bonds linking the glucose molecules in cellulose.

Humans need a continuous supply of energy, even the most ardent couch potatoes! Cells, for instance, require energy to carry out the thousands of functions required to grow, to divide, and to transport molecules across their membranes. Surprisingly, 70%–80% of the total energy required by sedentary humans (individuals who get little exercise) goes to perform basic functions—metabolism, food digestion, absorption, and so on. The remaining energy is used to power body movements such as walking, talking, and turning on the television via the remote control. For more active people, these percentages shift considerably. A bicycle commuter, for example, uses proportionately more energy for vigorous muscular activity than an office worker who rides a bus or drives a car to work.

At rest, the body relies on nearly equal amounts of carbohydrate (mostly glucose) and fat (triglycerides) to supply basic body needs. As you may recall from Chapter 3, glucose is broken down (catabolized) during cellular respiration.

The most common source of glucose is the polysaccharide starch. Starch is present in grains and their by-products (pasta and bread). It is also provided by many vegetables. Starch contributes on average about 22% of the total energy requirements of most Americans (**TABLE 5-3**).

In the digestive system of humans, starch is broken down into glucose molecules, which are absorbed into the bloodstream and distributed to body cells. The cells break down some glucose immediately to produce energy. Most of the rest is converted into glycogen in muscles and the liver and stored for later use (**FIGURE 5-3**). This ability to store glucose is vital to our survival. Without it, we'd need to eat constantly.

TABLE 5-3	
Energy Sources in the Average American Diet	
Source	Percentage
Fat	42
Sucrose	24
Starch	22
Protein	12

Between meals and during exercise, the liver breaks down stored glycogen to form glucose, which is released into the bloodstream. As noted earlier, glycogen in muscle is catabolized to provide energy for muscular contraction, but is not released into the bloodstream.

Another important carbohydrate, not involved in energy production, is fiber. **Dietary fiber** is nondigestible polysaccharide (such as cellulose) that is found in fruits, vegetables, and grains. As noted earlier, humans cannot digest cellulose because the body lacks the enzymes needed to break the covalent bonds joining the monosaccharide units (glucose) in the molecule. Consequently, fiber passes through the intestine largely unaffected by stomach acidity or digestive enzymes. So why is it so important to us?

Let me explain. Fiber exists in two basic forms: water-soluble and water-insoluble. **Water-soluble fibers** are gummy polysaccharides in fruits, vegetables, and some grains—including apples, bananas, carrots, barley, and oats. Several studies suggest that water-soluble fiber helps lower blood cholesterol by acting as a sponge that absorbs dietary cholesterol inside the digestive tract. This prevents cholesterol from being absorbed into the bloodstream.

In contrast, **water-insoluble fibers** are rigid cellulose molecules in foods such as celery, cereals, wheat products, and brown rice. Some foods, such as green beans and green peas, contain a mixture of both types.

Water-insoluble fiber such as that in whole wheat bread increases the water content of the **feces** (FEE-seas), the semisolid waste produced by the large intestine. This makes the feces softer and facilitates their transport through the large intestine. Increasing the water content also reduces constipation. To understand why, you must understand a little bit about the composition of a grain of wheat and the making of flour.

FIGURE 5-4 shows that wheat grain consists of three parts: (1) the **bran**, or shell, which contains most of the fiber; (2) the **endosperm**, the starchy portion that provides nutrients for the growing plant; and (3) the **germ**, the vitamin- and mineral-rich portion that becomes a new plant. When the wheat grain is ground, it produces whole-wheat flour, a brown powder containing the entire wheat grain. In contrast, white flour is produced from grain whose bran and germ have been removed. Removing the bran eliminates most, if not all, of the water-insoluble fiber (cellulose). Removing the germ eliminates most of the vitamins and minerals. White flour, therefore, consists mostly of starch. Research suggests that water-insoluble dietary fiber, such as that pres-

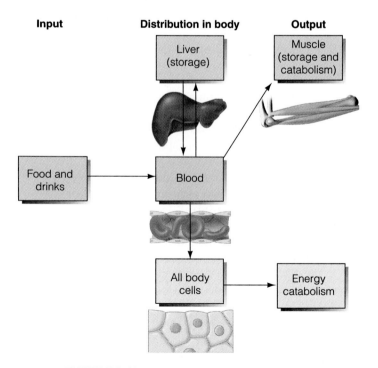

Input **Distribution in body** **Output**

FIGURE 5-3 **Glucose Balance** Glucose levels in the blood result from a balance of input and output.

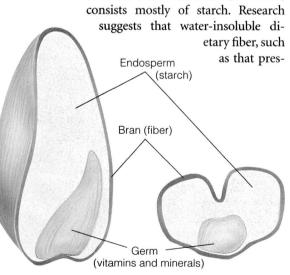

Endosperm (starch)

Bran (fiber)

Germ (vitamins and minerals)

FIGURE 5-4 **The Wheat Seed** Parts of a wheat grain and the nutrients and minerals they supply.

ent in the bran, also reduces the incidence of colon cancer, which afflicts about 3% of the U.S. population.

Lipids Are Fat-Soluble Molecules That Serve as a Source of Energy and Are Also Important Structural Components of Cells.

Lipids are a structurally diverse group of organic molecules. Lipids are insoluble in water and soluble in nonpolar solvents such as ethyl alcohol. Lipids are the waxy, greasy, or oily compounds found in plants and animals. Fats and oils in animals are energy-rich molecules that serve as an important storage depot. Still other lipids are structural components of cells. Two biologically important lipids will be discussed in this chapter: triglycerides and steroids. (Phospholipids were discussed in Chapter 3.)

Triglycerides.

Triglycerides are known to most of us as fats and oils. Cooking oil, for example, is a triglyceride, as is the fat in a steak or the butter on a piece of bread.

Triglycerides are composed of four subunits, each an organic molecule: one molecule of glycerol and three fatty acid molecules (**FIGURE 5-5**). Glycerol is a three-carbon compound; fatty acids are long molecules containing many carbons and hydrogens and a COOH, or **carboxyl group** (car-BOX-ul), on one end.

Triglycerides contain many covalent bonds, each of which stores a small amount of energy. When cells break down (catabolize) triglycerides, these bonds are broken

FIGURE 5-5 Triglycerides The triglycerides are the fats and oils. Triglycerides consist of glycerol and three fatty acids, covalently bonded as shown.

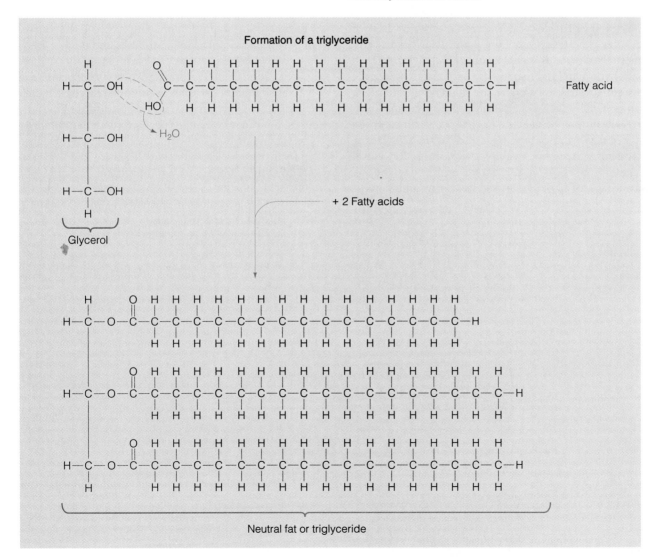

Fat cells

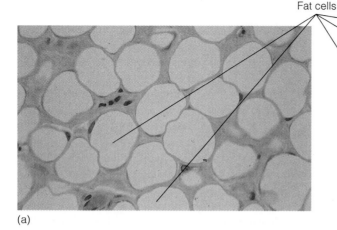

(a)

(b)

FIGURE 5-6 Fat Cells (a) A light micrograph of fat tissue. The clear areas are regions where the fat has dissolved during tissue preparation. Notice that the cytoplasm is reduced to a narrow region just beneath the plasma membrane. (b) A scanning electron micrograph of fat cells.

and energy is released. Energy liberated during the breakdown of triglycerides is used to drive a variety of cellular processes. Gram for gram, triglycerides yield more than twice as much energy as carbohydrates.

In humans and many other mammals, triglycerides are stored in fat cells under the skin and in other locations (**FIGURE 5-6**). In adult humans, triglycerides provide about half of the cellular energy consumed at rest, with carbohydrates providing most of the rest. During moderate (aerobic) exercise, triglycerides provide a larger proportion of the body's energy demand, explaining why aerobic exercise like jogging and bicycling helps people lose weight.

Fatty deposits around certain organs also cushion them from damage. Horseback riders, motorcyclists, and runners, for instance, can engage in their sports for hours at a time without damaging their internal organs in part because of nature's natural cushions.

With a few exceptions, fats (from most animals) are solid at room temperature, and oils (from plants and fish) are liquid. The reason for this difference lies in their chemical structures. In fats, the carbon atoms of the fatty acids are joined by single covalent bonds (**FIGURE 5-5**). The remaining bond sites on the carbon atoms are taken up by hydrogens. Thus, the fatty acids are said to be *saturated* with hydrogens. When the carbon backbone of a fatty acid consists solely of single covalent bonds, the structure zigzags but still remains fairly linear (**FIGURE 5-7A**). This allows the triglyceride molecules of a fat to pack together, forming a solid at room temperature.

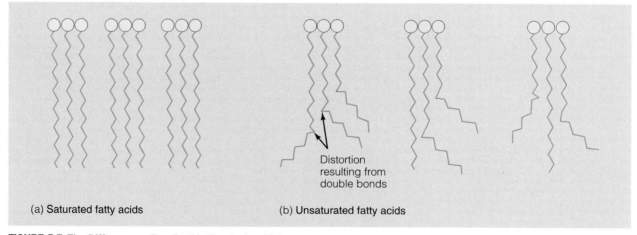

Distortion resulting from double bonds

(a) Saturated fatty acids   (b) Unsaturated fatty acids

FIGURE 5-7 The Difference a Few Double Bonds Can Make (a) Saturated fatty acids are principally derived from animal fats. The side chains are relatively straight and allow the molecules to pack tightly together, which explains why fats are solid at room temperature. (b) Double bonds in unsaturated fatty acids in oils cause the fatty acid chains to bend and thus prohibit tight packing. Oils, derived chiefly from plants, are therefore liquid at room temperature.

In contrast, the fatty acids in oils contain a number of double covalent bonds and these molecules are said to be *unsaturated* (**FIGURE 5-7B**). Double covalent bonds cause a bend in the fatty acid molecules that is believed to prevent tight packing. This somewhat "looser" arrangement of triglyceride molecules produces a liquid at room temperature.

When numerous double bonds exist in a fatty acid molecule, the molecule is said to be *polyunsaturated.* Studies show that polyunsaturated fats reduce one's risk of developing **atherosclerosis** (AH-ther-oh-skler-OH-siss), a disease that results from a buildup of cholesterol deposits, or **atherosclerotic plaque,** on the walls of arteries (**FIGURE 5-8**). Plaque restricts blood flow to the heart and brain, causing heart attacks and strokes.

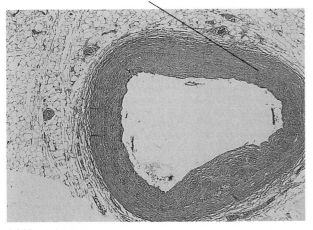

Wall of artery

(a) Normal artery

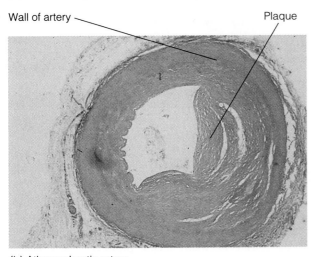

Wall of artery   Plaque

(b) Atherosclerotic artery

FIGURE 5-8 Atherosclerosis These cross sections of *(a)* a normal artery and *(b)* a diseased artery show how atherosclerotic plaque can obstruct blood flow.

For reasons not well understood, saturated fatty acids increase cholesterol production by the liver, so diets rich in saturated fat (animal fats) tend to increase the chances of developing atherosclerosis. Atherosclerosis is also more common in sedentary people and smokers. Some individuals are genetically predisposed to develop atherosclerosis. Their livers simply produce abnormally high levels of cholesterol.

Lowering the level of saturated fat in the diet can be accomplished by switching from whole milk to low-fat milk, reducing the consumption of red meat, trimming fat from chicken and other meats, and cooking with unsaturated vegetable oil instead of animal fat (lard). (For more on cholesterol, see Health Note 5-1.)

Vegetable oil can be converted to a solid (margarine) by chemically adding hydrogens to it. This process decreases the number of double bonds and tends to straighten the fatty acid chains, resulting in a solid. Even as such, polyunsaturated margarines, advertised for their health benefits, still contain a large number of double bonds.

Steroids.

Steroids are also present in the plasma membrane. As a group, the **steroids** (STEER-oids) are structurally quite different from the triglycerides and phospholipids. As shown in **FIGURE 3-12**, steroids consist of three six-carbon rings and one five-carbon ring all joined in one large structure resembling chicken wire. One of the best-known steroids is cholesterol (**FIGURE 3-12A**). **Cholesterol** is a component of the plasma membrane in animal cells (but not in plants). In humans, it is a raw material needed to synthesize other steroids, such as vitamin D, bile salts, and the sex hormones estrogen and testosterone. As noted earlier, cholesterol is also a major component of atherosclerotic plaque. In humans, cholesterol comes primarily from the liver. A lesser amount comes from the diet.

Most Americans consume lipids in excess of what is required. On average, fats provide about 42% of the dietary caloric intake (**TABLE 5-3**). To lower the risk of heart attack, fat intake should only be about 30%, perhaps even lower, and animal fat should be eaten in small quantities. (For more on lipids and their effects on heart disease, see Health Note 5-1.)

Amino Acids and Protein.

Amino acids and proteins are important nutrients. Their structure is discussed in Chapter 3. In healthy, well-nourished individuals, dietary protein is used to synthesize enzymes, hormones, and structural proteins such as collagen. Dietary proteins, however, cannot be

Health Note 5–1

Lowering Your Cholesterol

Let there be no doubt about it: Diseases of the heart and arteries are leading causes of death in the United States.

Atherosclerosis, the accumulation of plaque on artery walls, and the problems it creates are responsible for nearly two of every five deaths in the United States each year. Thanks to improvements in medical care and diet, the death rate from atherosclerosis has been falling steadily in recent years, but it is still a major concern. New research, in fact, shows that atherosclerotic plaque is present even in children.

Researchers believe that atherosclerotic plaques begin to form after minor injuries to the lining of blood vessels. High blood pressure, they think, may damage the lining, causing platelets (small cell-like structures in the blood) and cholesterol in the blood to adhere to the injured site. The blood vessel responds by producing cells that grow over the fatty deposit. This thickens the wall of the artery, reducing blood flow. Additional cholesterol is then deposited in the thickened wall, forming a larger and larger obstruction.

Cholesterol deposits impair the flow of blood in the heart and other organs, cutting off oxygen to vital tissues. Blood clots may form in the restricted sections of arteries, further reducing blood flow. When the oxygen supply to the heart is disrupted, cardiac muscle cells can die, resulting in heart attacks and death. Blood clots originating in other parts of the body may also lodge in diseased vessels, obstructing blood flow. Oxygen deprivation can weaken the heart, impairing its ability to pump blood. When the oxygen supply to the heart is restricted, the result is a type of heart attack known as a myocardial infarction (my-oh-CAR-dee-ill in-FARK-shun). Victims of a myocardial infarction feel pain in the center of the chest and down the left arm. If the oxygen-deprived area is extensive, the heart may cease functioning altogether.

Atherosclerotic plaque also impairs the flow of blood to the brain. Blood clots catch in the restricted areas and block the flow of blood to vital regions of the brain. Victims may lose the ability to speak or to move limbs. If the damage is severe enough, they may die. Thanks to ad-

vances in medical treatment, however, many victims can be saved. And over time, they can recover lost functions as other parts of the brain take over for the damaged regions.

Atherosclerosis and cardiovascular disease have been associated with nearly 40 factors. Several of these risk factors, such as old age and sex (being male), cannot be changed. Other factors are controllable. These include high blood pressure, high blood cholesterol, stress, smoking, inactivity, and excessive food intake. Of all the risk factors, three emerge as the primary contributors to cardiovascular disease: elevated blood cholesterol, smoking, and high blood pressure.

Consider cholesterol. Cholesterol is essential to normal body function. It is part of the plasma membrane and is needed to synthesize certain hormones. Interestingly, in most people, the majority of the cholesterol in the blood is produced by the liver. The liver synthesizes and releases about 700 milligrams of cholesterol per day. Only about 225 milligrams of cholesterol are derived from the food we eat each day. Interestingly, the

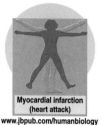

Myocardial infarction (heart attack)

www.jbpub.com/humanbiology

absorbed by the small intestine; they must first be broken down into amino acids by hydrolytic enzymes in the digestive tract.

Proteins in the human body contain 20 different amino acids—all of which can be provided from the diet. The body, however, is capable of synthesizing 11 of the amino acids it needs from nitrogen and smaller molecules derived from carbohydrates and fats. Thus, if they are not present in the diet, they can be made. The remaining nine amino acids cannot be synthesized by body cells under any circumstances—that is, they *must* be provided by the diet. These amino acids are called **essential amino acids**. A deficiency of even one of the essential amino acids can cause severe physiological

problems. As a result, nutritionists recommend a diet containing many different protein sources so individuals receive all of the amino acids they need.

Proteins are divided into two major groups by nutritionists: complete and incomplete. **Complete proteins** contain ample amounts of all of the essential amino acids. Complete proteins are found in milk, eggs, meat, fish, poultry, and cheese. **Incomplete proteins** lack one or more essential amino acids and include those found in many plant products: nuts, seeds, grains, most legumes (peas and beans), and vegetables. Vegetarians who avoid all animal products, including milk and eggs, must acquire the essential amino acids they need by combining incomplete protein sources

concentration of cholesterol in the blood tends to stay fairly constant. If dietary input falls, the liver increases its output. If the amount of cholesterol in the diet rises, the liver reduces its production. So what's all the fuss about cholesterol in a person's diet?

Even though the liver regulates cholesterol levels, it cannot work fast enough. That is, it may simply be unable to absorb, use, and dispose of cholesterol quickly enough. Consequently, excess cholesterol circulates in the blood after a meal and is deposited in the arteries.

Cholesterol is carried in the bloodstream bound to protein. These complexes of protein and lipid fall into two groups: high-density lipoproteins (HDLs) and low-density lipoproteins (LDLs). HDLs and LDLs function very differently. LDLs, for example, transport cholesterol from the liver to body tissues. In contrast, HDLs are scavengers, picking up excess cholesterol and transporting it to the liver, where it is removed from the blood and excreted in the bile. Research shows that the ratio of HDL to LDL is an accurate predictor of cardiovascular disease. The higher the ratio, the lower the risk of cardiovascular disease.

A high cholesterol level (or hypercholesterolemia) tends to run in families. Thus, if a parent died of a heart attack or suffers from this genetic disease, his or her offspring are more likely to have high cholesterol levels.

To reduce your chances of atherosclerosis, the American Heart Association recommends (1) limiting dietary fat to less than 30% of the total caloric intake; (2) limiting dietary cholesterol to 300 milligrams per day; and (3) acquiring 50% or more of one's calories from carbohydrates, especially polysaccharides (notably, starches found in potatoes, rice, vegetables and other foods). Reductions in saturated fats (animal fats) can also help lower cholesterol levels. You can cut back on saturated fat by reducing your consumption of red meat and trimming the fat off all meats before cooking. You can also increase your consumption of fruits, vegetables, and grains, letting these low-fat foods displace some of the fatty foods you might otherwise have eaten.

Blood cholesterol levels can also be lowered with drugs and exercise.

Research spanning several decades shows that a lower cholesterol level translates into a decline in cardiovascular disease.

High cholesterol is also surprisingly common in children, leading many health experts to believe that steps should be taken to prevent problems later in life. In children under the age of 2, however, diets should not be restricted. A diet that is too restrictive may actually impair physical growth and development. What all children need is a well-balanced diet, low in fats, especially animal fat, with sufficient calories from other sources. If nothing else, such a diet could encourage the eating habits necessary for good health throughout adult life.

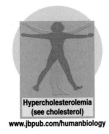

Hypercholesterolemia
(see cholesterol)
www.jbpub.com/humanbiology

Visit *Human Biology's* Internet site, www.jbpub.com/humanbiology, for links to web sites offering more information on this topic.

www.jbpub.com/humanbiology

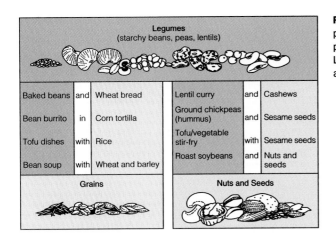

Legumes (starchy beans, peas, lentils)					
Baked beans	and	Wheat bread	Lentil curry	and	Cashews
Bean burrito	in	Corn tortilla	Ground chickpeas (hummus)	and	Sesame seeds
Tofu dishes	with	Rice	Tofu/vegetable stir-fry	with	Sesame seeds
Bean soup	with	Wheat and barley	Roast soybeans	and	Nuts and seeds
Grains			Nuts and Seeds		

FIGURE 5-9 Complementary Protein Sources By combining protein sources, a vegetarian who consumes no animal by-products can be assured of getting all of the amino acids needed. Legumes can be combined with foods made from grains or nuts and other seeds.

(**FIGURE 5-9**). Although many people think of proteins as a source of energy, proteins are only used for this purpose when dietary intake of carbohydrates and fats is severely restricted or when protein intake far exceeds demand.

Protein deficiencies occur in millions of children throughout the world. The arms and legs of these children are often thin and wiry because their muscle cells break down protein in an effort to provide energy (**FIGURE 5-10**).

On the other end of the spectrum are the overfed populations of the world. The average American, for example, consumes twice the daily requirement of protein. Amino acids derived from the surplus dietary protein are broken down in the body to produce energy and fat.

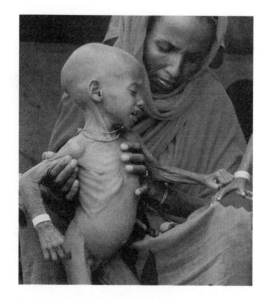

FIGURE 5-10 A Protein-Deficient Child   This child suffers from severe protein deficiency (kwashiorkor; kwash-EE-or-core). His arms and legs are emaciated because muscle protein has been broken down to supply energy. His belly is slightly swollen because of a buildup of fluid in the abdomen.

TABLE 5-4

Important Information on Vitamins

Vitamin	Major Dietary Sources	Major Functions	Signs of Severe, Prolonged Deficiency	Signs of Extreme Excess
Fat-soluble				
A	Fat-containing and fortified dairy products; liver; provitamin carotene in orange and deep green fruits and vegetables	Vitamin A is a component of rhodopsin; carotenoids can serve as antioxidants; retinoic acid affects gene expression; still under intense study	Night blindness; keratinization of epithelial tissues including the cornea of the eye (xerophthalmia) causing permanent blindness; dry, scaling skin; increased susceptibility to infection	*Preformed vitamin A:* damage to liver, bone; headache, irritability, vomiting, hair loss, blurred vision *13-cis retinoic acid:* some fetal defects *Carotenoids:* yellowed skin
D	Fortified and full-fat dairy products, egg yolk (diet often not as important as sunlight exposure)	Promotes absorption and use of calcium and phosphorus	Rickets (bone deformities) in children; osteomalacia (bone softening) in adults	Calcium deposition in tissues leading to cerebral, CV, and kidney damage
E	Vegetable oils and their products; nuts, seeds	Antioxidant to prevent cell membrane damage; still under intense study	Possible anemia and neurologic effects	Generally nontoxic, but at least one type of intravenous infusion led to some fatalities in premature infants; may worsen clotting defect in vitamin K deficiency
K	Green vegetables; tea	Aids in formation of certain proteins, especially those for blood clotting	Defective blood coagulation causing severe bleeding or injury	Liver damage and anemia from high doses of the synthetic form menadione
Water-soluble				
Thiamin (B-1)	Pork, legumes, peanuts, enriched or whole-grain products	Coenzyme used in energy metabolism	Nerve changes, sometimes edema, heart failure; beriberi	Generally nontoxic, but repeated injections may cause shock reaction

SOURCE: Adapted from *Nutrition for Living,* 4th ed., by J. L. Christian and L. L. Greger, Copyright © 1994 by The Benjamin/Cummings Publishing Company.

With this overview of the macronutrients, we now turn to the micronutrients.

Micronutrients Are Substances Needed in Small Quantities and Include Two Broad Groups: Vitamins and Minerals

Vitamins.

Vitamins are a diverse group of organic compounds present in very small amounts in many of the foods we eat. These molecules are absorbed by the lining of the digestive tract without being broken down.

The 13 known vitamins play an important role in many metabolic reactions. Because vitamins are recycled many times during metabolic reactions, they are needed only in very small amounts. In fact, 1 gram of vitamin B-12, about 1 teaspoon, would supply over 300,000 people for a day.

Most vitamins are not synthesized in the cells of the body or, if they can be made, are not produced in sufficient amounts to satisfy cellular demands, making dietary intake essential. Vitamin D, for instance, is manufactured by the skin when exposed to sunlight. However, most Americans spend so much time indoors that dietary input is essential to good health. **TABLE 5-4** lists the vitamins and their functions.

Vitamin	Major Dietary Sources	Major Functions	Signs of Severe, Prolonged Deficiency	Signs of Extreme Excess
Riboflavin (B-2)	Dairy products, meats, eggs, enriched grain products, green leafy vegetables	Coenzyme used in energy metabolism	Skin lesions	Generally nontoxic
Niacin	Nuts, meats; provitamin tryptophan in most proteins	Coenzyme used in energy metabolism	Pellagra (multiple vitamin deficiencies including niacin)	Flushing of face, neck, hands; potential liver damage
B-6	High-protein foods in general	Coenzyme used in amino acid metabolism	Nervous, skin, and muscular disorders; anemia	Unstable gait, numb feet, poor coordination
Folic acid	Green vegetables, orange juice, nuts, legumes, grain products	Coenzyme used in DNA and RNA metabolism; single carbon utilization	Megaloblastic anemia (large, immature red blood cells); GI disturbances	Masks vitamin B-12 deficiency, interferes with drugs to control epilepsy
B-12	Animal products	Coenzyme used in DNA and RNA metabolism; single carbon utilization	Megaloblastic anemia; pernicious anemia when due to inadequate intrinsic factor; nervous system damage	Thought to be nontoxic
Pantothenic acid	Animal products and whole grains; widely distributed in foods	Coenzyme used in energy metabolism	Fatigue, numbness, and tingling of hands and feet	Generally nontoxic; occasionally causes diarrhea
Biotin	Widely distributed in foods	Coenzyme used in energy metabolism	Scaly dermatitis	Thought to be nontoxic
C (ascorbic acid)	Fruits and vegetables, especially broccoli, cabbage, cantaloupe, cauliflower, citrus fruits, green pepper, kiwi fruit, strawberries	Functions in synthesis of collagen; is an antioxidant; aids in detoxification; improves iron absorption; still under intense study	Scurvy; petechiae (minute hemorrhages around hair follicles); weakness; delayed wound healing; impaired immune response	GI upsets, confounds certain lab tests

Because vitamins are needed in almost all cells of the body, a dietary deficiency in just one vitamin can cause wide-ranging effects. Vitamins also interact with other nutrients. Vitamin C, for example, increases the absorption of iron in the small intestine. Large doses of vitamin C, however, decrease copper utilization by the cells. Consequently, maintaining good health requires ingesting the proper balance of vitamins and other nutrients. Imbalances can result in upsets in homeostasis.

Vitamins fall into two broad categories: water-soluble and fat-soluble. **Water-soluble vitamins** include vitamin C and eight different forms of vitamin B. Water-soluble vitamins are transported in the blood plasma. Because they are water-soluble, they are readily eliminated by the kidneys and are not stored in the body in any appreciable amount.

Water-soluble vitamins generally work in conjunction with enzymes, promoting the cellular reactions that supply energy or synthesize cellular materials. Contrary to common myth, the vitamins themselves do not provide energy.

Many proponents of vitamin use believe that megadoses of water-soluble vitamins are harmless because these vitamins are excreted in the urine and do not accumulate in the body. New research, however, shows that this is not entirely true. Some water-soluble vitamins, such as vitamin C, when ingested in excess, can be toxic. (See **TABLE 5-4** for some examples.)

The **fat-soluble vitamins** are vitamins A, D, E, and K. They perform many different functions. Vitamin A, for example, is converted to light-sensitive pigments in receptor cells of the retina, the light-sensitive layer of the eye. These pigments play an important role in vision. Another member of the vitamin A group removes harmful chemicals (oxidants) from the body.

Unlike water-soluble vitamins, the fat-soluble vitamins are stored in body fat and accumulate in the fat reserves. The accumulation of fat-soluble vitamins can have many adverse effects (**TABLE 5-4**). An excess of vitamin D, for example, can cause weight loss, nausea, irritability, kidney stones, weakness, and other symptoms. Large doses of vitamin D taken during pregnancy can cause birth defects.

Vitamin excess is encountered largely in the developed countries and usually occurs only in people taking vitamin supplements. Each year, in fact, approximately 4000 Americans are treated for vitamin supplement poisoning. To avoid problems from excess vitamins, nutritionists recommend eating a balanced diet that provides all of the vitamins the body needs, rather than taking vitamin pills. Megadoses, they say, should be avoided.

Vitamin deficiencies, like dietary excesses, can lead to serious problems. A deficiency of vitamin D, for ex-

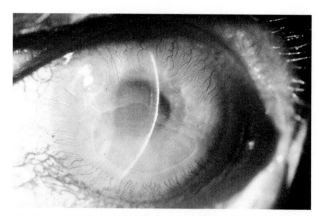

FIGURE 5-11 Vitamin A Deficiency   Vitamin A deficiency causes the cornea of the eye to dry and become irritated. If the deficiency is not corrected, corneal ulcers may form and rupture, resulting in permanent blindness.

ample, can produce **rickets** (RICK-its), a disease that results in bone deformities. Vitamin K deficiencies can result in severe bleeding on injury and internal hemorrhaging. Vitamin C deficiency can result in delayed wound healing and reduced immunity, making people more susceptible to infectious disease.

One of the most common dietary illnesses is caused by a deficiency of vitamin A. This deficiency afflicts over 100,000 children worldwide each year. If not corrected, vitamin A deficiency causes the eyes to dry. Ulcers may form on the eyeball and can rupture, causing blindness (**FIGURE 5-11**).

Most people afflicted by vitamin deficiencies are those who fail to get enough to eat, although even apparently well-fed individuals may suffer from a vitamin deficiency if they are not eating a well-rounded diet. Sadly, about one of every six people living in the nonindustrialized nations of the world, or about 800 million people, go to bed hungry; most of these people suffer from multiple vitamin deficiencies and exhibit many symptoms. People suffering from vitamin deficiency typically complain of weakness and fatigue. Children with insufficient vitamin intake fail to grow. All told, about 10 million children under the age of 5 suffer from extreme malnutrition in the less-developed nations of the world. Another 90 million under the age of 5 are moderately malnourished. In the United States, one of every five children is born into poverty and is a candidate for vitamin deficiency.

Minerals.

On average, an adult contains about 5 pounds of **minerals**, naturally occurring inorganic substances vital to many life processes. Humans require about two dozen minerals, such as calcium, sodium, iron, and potassium, to carry out normal body functions.

Minerals, like vitamins, are micronutrients and are derived from the food we eat and the beverages we drink. Minerals are divided into two groups: the major minerals and the trace minerals (**TABLE 5-5**). The **major minerals** are present in the body in amounts larger than 5 grams; the **trace minerals** are found in lesser quantities. Calcium and phosphorus, for example, are major minerals, and make up three-fourths of the total mineral content of the human body. These two minerals form part of the dense extracellular matrix of bone and are required in a much greater quantity than zinc or copper, two trace minerals that are components of some enzymes.

The distinction between major and trace minerals is not meant to imply that one group is more important than the other. In fact, a daily deficiency of a few micrograms (a microgram is $\frac{1}{1000}$ of a gram, and a gram is $\frac{1}{454}$ of a pound) of iodine is just as serious as a deficiency of several hundred milligrams of calcium.

TABLE 5-5 lists some of the minerals, their function, and problems that arise when they are ingested in excess or deficient amounts.

TABLE 5-5				
Important Information on Minerals				
Mineral	**Major Dietary Sources**	**Major Functions**	**Signs of Severe, Prolonged Deficiency**	**Signs of Extreme Excess**
Major minerals				
Calcium	Milk, cheese, dark green vegetables, legumes	Bone and tooth formation; blood clotting; nerve transmission	Stunted growth; perhaps less bone mass	Depressed absorption of some other minerals; perhaps kidney damage
Phosphorus	Milk, cheese, meat, poultry, whole grains	Bone and tooth formation; acid-base balance; component of coenzymes	Weakness; demineralization of bone	Depressed absorption of some minerals
Magnesium	Whole grains, green leafy vegetables	Component of enzymes	Neurologic disturbances	Neurologic disturbances
Sodium	Salt, soy sauce, cured meats, pickles, canned soups, processed cheese	Body water balance; nerve function	Muscle cramps; reduced appetite	High blood pressure in genetically predisposed individuals
Potassium	Meats, milk, many fruits and vegetables, whole grains	Body water balance; nerve function	Muscular weakness; paralysis	Muscular weakness; cardiac arrest
Chloride	Same as for sodium	Plays a role in acid-base balance; formation of gastric juice	Muscle cramps; reduced appetite; poor growth	High blood pressure in genetically predisposed individuals
Trace minerals				
Iron	Meats, eggs, legumes, whole grains, green leafy vegetables	Component of hemoglobin, myoglobin, and enzymes	Iron-deficiency anemia, weakness, impaired immune function	Acute: shock, death; Chronic: liver damage, cardiac failure
Iodine	Marine fish and shellfish; dairy products; iodized salt; some breads	Component of thyroid hormones	Goiter (enlarged thyroid)	Iodide goiter
Fluoride	Drinking water, tea, seafood	Maintenance of tooth (and maybe bone) structure	Higher frequency of tooth decay	Acute: GI distress; Chronic: mottling of teeth; skeletal deformation

Source: Adapted from *Nutrition for Living*, 4th ed., by J. L. Christian and L. L. Greger. Copyright © 1994 by The Benjamin/Cummings Publishing Company.

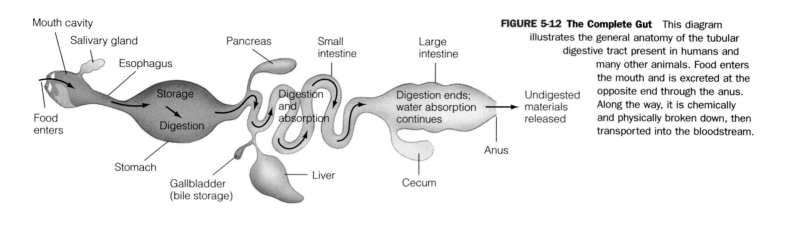

Mouth cavity
Salivary gland
Esophagus
Pancreas
Small intestine
Large intestine
Food enters
Storage
Digestion
Digestion and absorption
Digestion ends; water absorption continues
Undigested materials released
Anus
Stomach
Gallbladder (bile storage)
Liver
Cecum

FIGURE 5-12 The Complete Gut   This diagram illustrates the general anatomy of the tubular digestive tract present in humans and many other animals. Food enters the mouth and is excreted at the opposite end through the anus. Along the way, it is chemically and physically broken down, then transported into the bloodstream.

Section 5-2

The Digestive System

The human digestive system consists of a tube opened at both ends (**FIGURE 5-12**). Food enters the mouth and is chemically and physically altered as it travels along this tube, called the gut. Waste is excreted at the opposite end, the anus (EH-nuss). In the course of evolution, the various parts of the digestive tract have evolved to perform specific functions, as shown in the figure. Some regions grind and mix the food. Others store food. Still others play a role in digestion and absorption. In the mouth, for example, food is sliced, torn, and crushed by the teeth into smaller particles. This greatly increases the surface area presented to the digestive enzymes in the stomach and small intestine, which increases the efficiency of digestion. Enzymes participate in a chemical breakdown of food. In the small intestine, amino acids, monosaccharides, and other small molecules produced by enzymatic digestion are absorbed into the bloodstream for distribution to body cells.

■ The Physical Breakdown of Food Occurs in the Mouth

The mouth is a complex structure in which food is broken down mechanically and, to a much lesser degree, chemically. Food taken into the mouth is sliced into smaller pieces by the sharp teeth in front; it is then ground into a pulpy mass by the flatter teeth toward the back of the mouth. As the food is pulverized in the mouth, it is liquefied by **saliva** (sah-LIVE-ah), a watery secretion released by the salivary glands, three sets of exocrine glands located around the oral cavity (**FIGURE 5-13**). The release of saliva is triggered by the smell, feel, taste, and sometimes even by the thought of food.

Saliva performs at least five functions. It liquefies the food, making it easier to swallow. It kills or neutralizes some bacteria via the enzymes and antibodies it contains. It dissolves substances so they can be tasted. It begins to break down starch molecules with the aid of the enzyme **amylase** (AM-ah-lase). Saliva also cleanses the teeth, washing away bacteria and food particles. Because the release of saliva is greatly reduced during sleep, bacteria tend to accumulate on the surface of the teeth, where they break down microscopic food particles, producing some foul-smelling chemicals that give us "dragon breath," or "morning mouth."

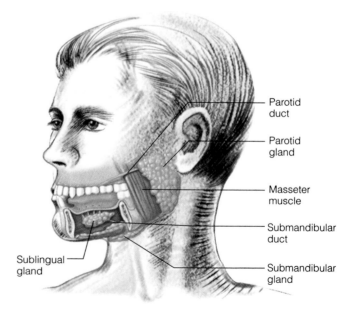

Parotid duct
Parotid gland
Masseter muscle
Submandibular duct
Submandibular gland
Sublingual gland

FIGURE 5-13 Salivary Glands   Three salivary glands (parotid, submandibular, and sublingual) are located in and around the oral cavity and empty into the mouth via small ducts.

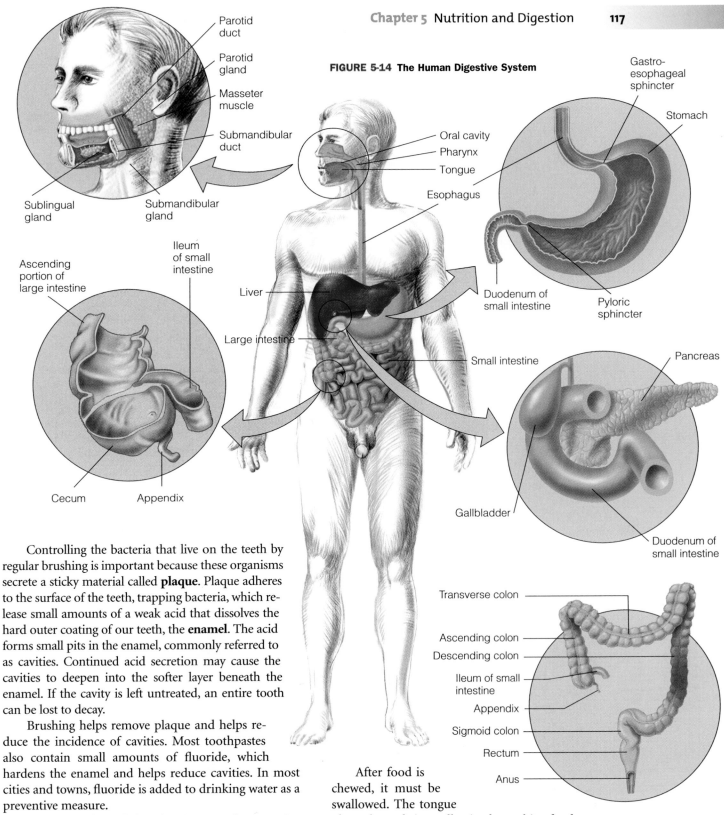

FIGURE 5-14 The Human Digestive System

Controlling the bacteria that live on the teeth by regular brushing is important because these organisms secrete a sticky material called **plaque**. Plaque adheres to the surface of the teeth, trapping bacteria, which release small amounts of a weak acid that dissolves the hard outer coating of our teeth, the **enamel**. The acid forms small pits in the enamel, commonly referred to as cavities. Continued acid secretion may cause the cavities to deepen into the softer layer beneath the enamel. If the cavity is left untreated, an entire tooth can be lost to decay.

Brushing helps remove plaque and helps reduce the incidence of cavities. Most toothpastes also contain small amounts of fluoride, which hardens the enamel and helps reduce cavities. In most cities and towns, fluoride is added to drinking water as a preventive measure.

One study showed that chewing sugarless gum increases the flow of saliva and cleanses the teeth. By chewing sugarless gum within 5 minutes after a meal, and for at least 15 minutes, the researchers showed that individuals can reduce the incidence of cavities.

After food is chewed, it must be swallowed. The tongue plays a key role in swallowing by pushing food to the back of the oral cavity into the **pharynx** (FAIR-inks), a funnel-shaped chamber that connects the oral cavity with the **esophagus** (ee-SOFF-ah-gus), a long muscular tube that leads to the stomach (**FIGURE 5-14**).

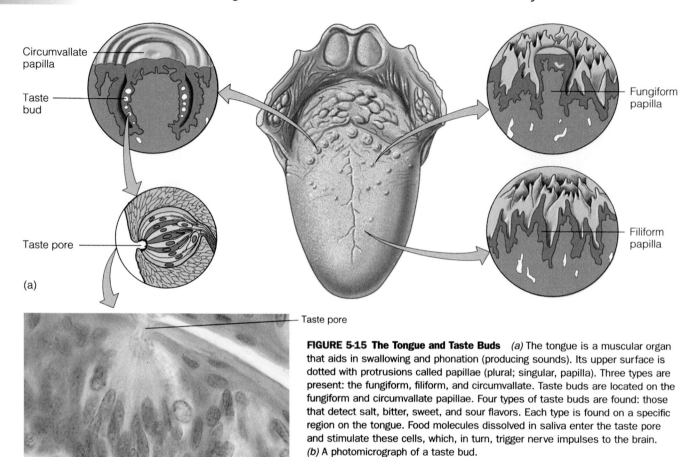

(a)

Circumvallate papilla

Taste bud

Taste pore

Fungiform papilla

Filiform papilla

Taste pore

(b)

FIGURE 5-15 The Tongue and Taste Buds   *(a)* The tongue is a muscular organ that aids in swallowing and phonation (producing sounds). Its upper surface is dotted with protrusions called papillae (plural; singular, papilla). Three types are present: the fungiform, filiform, and circumvallate. Taste buds are located on the fungiform and circumvallate papillae. Four types of taste buds are found: those that detect salt, bitter, sweet, and sour flavors. Each type is found on a specific region on the tongue. Food molecules dissolved in saliva enter the taste pore and stimulate these cells, which, in turn, trigger nerve impulses to the brain. *(b)* A photomicrograph of a taste bud.

FIGURE 5-16   The Epiglottis   This trapdoor prevents food from entering the trachea during swallowing. As illustrated, the trachea is lifted during swallowing, pushing against the epiglottis, which bends downward.

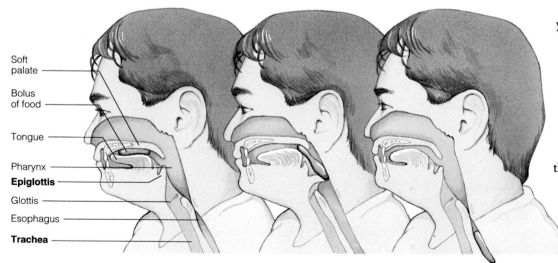

Soft palate

Bolus of food

Tongue

Pharynx

Epiglottis

Glottis

Esophagus

Trachea

Esophagus closed; glottis open; food in mouth.

Esophagus open; glottis closed; food in pharynx.

Esophagus closed; glottis open; food in esophagus.

The tongue, which also aids in speech, contains taste receptors, or **taste buds**, on its upper surface (**FIGURE 5-15**). Taste buds are stimulated by four basic flavors: sweet, sour, salty, and bitter. Various combinations of these flavors (combined with odors we smell) give us a rich assortment of tastes.

Food propelled from the pharynx into the esophagus is prevented from entering the **trachea** (TRAY-key-ah), or windpipe, which carries air to the lungs and lies in front of the esophagus, by the **epiglottis** (ep-ah-GLOT-tis) (**FIGURE 5-16**). The epiglottis is a flap of tissue that acts like a trapdoor, closing off the trachea during swallowing (Chapter 9).

Swallowing begins with a voluntary act—the tongue pushing food into the back of the oral cavity. Once food enters the pharynx, however, the process becomes automatic. Food in the pharynx stimulates stretch receptors in the wall of this organ, which, in turn, trigger the **swallowing reflex**, an involuntary contraction of the muscles in the wall of the pharynx. This forces the food into the esophagus.

The Esophagus Transports Food to the Stomach Via Peristalsis

Involuntary contractions of the muscular wall of the esophagus propel food to the stomach. As **FIGURE 5-17** shows, the muscles of the esophagus contract above the swallowed food mass, squeezing it along. This involuntary muscular action is called **peristalsis** (pear-eh-STALL-sis). It is so powerful that you can swallow when hanging upside down. Peristalsis also propels food (and waste) along the rest of the digestive tract.

Scientists once thought that esophageal peristalsis could proceed in the opposite direction under certain conditions and called this process reverse peristalsis, or, more commonly, vomiting.

Vomiting is a reflex action that occurs when irritants are present in the stomach. It is, therefore, believed to be a protective adaptation that allows the body to rid the stomach of bad food and harmful viruses and bacteria with obvious survival benefits to humans and other vertebrates (back-boned animals). Vomiting may also be caused by (1) emotional factors, such as stress, (2) rotation or acceleration of the head, as in motion sickness, (3) stimulation of the back of the throat, and (4) elevated pressure inside the brain caused by injury. Today we know that during vomiting food is expelled from the stomach as a result of contractions of muscles in the abdomen and diaphragm (DIE-ah-fram) (the diaphragm separates the thoracic and abdominal cavities and plays an important role in breathing).

The Stomach Stores Food, Releasing It Into the Small Intestine in Spurts

In humans, the stomach (**FIGURES 5-14** and **5-18**) lies on the left side of the abdominal cavity, partly under the protection of the rib cage. Food enters the stomach via the esophagus. The opening to the stomach, however, is closed off by the **gastroesophageal sphincter** (GAS-trow-eh-SOFF-a-gee-al SFINCK-ter), a thickened layer of smooth muscle at the juncture of the esophagus and stomach (**FIGURE 5-17B**). As food enters the lower esophagus, the gastroesophageal sphincter opens, allowing the food to enter the stomach. The sphincter then promptly closes, preventing food and stomach acid from percolating upward. If the sphincter fails to close, acid rising in the esophagus causes irritation, a condition known as "heartburn."

Inside the stomach, food is liquefied by acidic secretions of tiny glands in the wall of the stomach, the

gastric glands. These glands produce a watery secretion called **gastric juice**, which contains hydrochloric acid (HCl) and an enzyme precursor called pepsinogen (pep-SIN-oh-gen) (discussed below).

Inside the stomach, food is churned by peristalsis and mixed with gastric juice. The churning action of the stomach's muscular walls helps break down large pieces

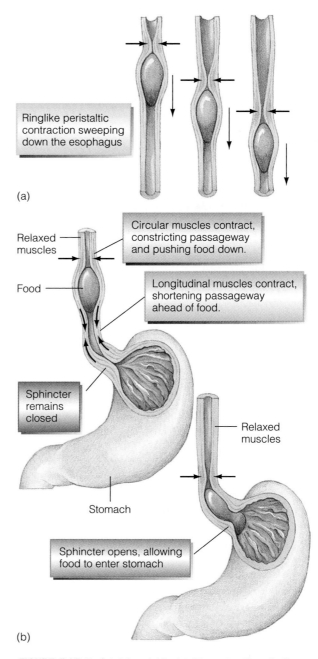

FIGURE 5-17 Peristalsis *(a)* Peristaltic contractions in the esophagus propel food into the stomach. *(b)* When food reaches the stomach, the gastroesophageal sphincter opens, allowing food to enter.

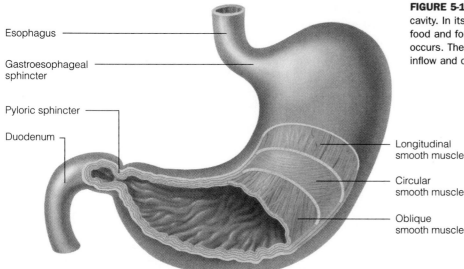

Esophagus

Gastroesophageal sphincter

Pyloric sphincter

Duodenum

Longitudinal smooth muscle

Circular smooth muscle

Oblique smooth muscle

FIGURE 5-18 The Stomach   The stomach lies in the abdominal cavity. In its wall are three layers of smooth muscle that mix the food and force it into the small intestine, where most digestion occurs. The gastroesophageal and pyloric sphincters control the inflow and outflow of food, respectively.

of food. Combined with the liquid from salivary glands and the gastric glands, the food becomes a rather thin, watery paste known as **chyme** (KIME). The stomach can hold 2–4 liters (2 quarts to 1 gallon) of chyme, which is gradually released into the small intestine at a rate suitable for efficient digestion and absorption.

Contrary to what many think, very little enzymatic digestion occurs in the stomach. The stomach's role is largely to prepare most food for the enzymatic digestion that will occur in the small intestine. There are some exceptions to this rule, however. Protein is one of them.

Proteins are denatured by HCl in the stomach. HCl also acts on **pepsinogen,** converting it to the active form pepsin. **Pepsin** is a proteolytic (pronounced pro-tee-oh-LIT-ic; protein-digesting) enzyme that catalyzes the breakdown of proteins into large fragments. These fragments are further broken down in the small intestine, the next stop in the digestive system.

Hydrochloric acid creates an acidic environment (pH 1.5–3.5) in the stomach. Besides activating pepsinogen and denaturing protein (rendering it digestible), HCl also breaks down connective tissue fibers in meat and kills most bacteria, thereby protecting the body from infection.

Also contrary to popular belief, the stomach does not absorb foodstuffs. Only a few substances, such as alcohol, actually penetrate the lining of the stomach to enter the bloodstream. Alcohol consumed on an empty stomach passes quickly into the blood, often producing an immediate dizzying effect. The presence of food in the stomach, however, slows alcohol absorption.

The stomach lining is protected from destruction by an alkaline secretion known as **mucus** (MEW-kuss), produced by certain cells in the lining. Mucus coats the epithelium, protecting it from acid and pepsin. The tissues beneath the epithelium are protected from acid leakage by the cells of the epithelium, which are tightly joined to one another, forming a leakproof barrier.

Until recently, scientists believed that the stomach's protective mechanisms could be hampered by certain factors such as stress; excess coffee (caffeine); excess aspirin, nicotine, and alcohol; or combinations of them. These factors were thought to increase the levels of acid in the stomach, overwhelming the mucous layer. When this happened, scientists believed that HCl and pepsin came in contact with the epithelial cells and digested parts of the wall of the stomach, forming painful **ulcers** (**FIGURE 5-19**)[3]. Accordingly,

---

[3]Ulcers also occur in the esophagus and, more commonly, in the small intestine. Excessive acid percolating into the esophagus and the release of excess acid into the small intestine can overwhelm the protective mucous layer found in both organs.

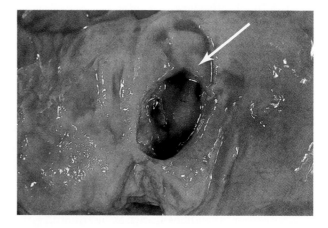

FIGURE 5-19 Ulcers   An ulcer in the lining of the stomach.

most ulcers that were detected early were treated by reducing stress and changing one's diet (eating smaller amounts of food and reducing coffee, aspirin, and alcohol) to reduce acid secretion in the stomach (see Health Note 1-1). When ulcers were not detected early and when damage was severe, parts of the stomach had to be removed surgically.

Now, once again, new research has thrown these beliefs into doubt. One researcher has discovered that most ulcers are caused by a bacterium. He has successfully cured many ulcers by treating patients with antibiotics. Although more work is needed to confirm this hypothesis, it could potentially result in a dramatic shift in medical treatment.

Chyme Leaves the Stomach and Enters the Small Intestine.

Chyme is ejected from the stomach into the small intestine by peristaltic muscle contractions. When the wave of contraction reaches the **pyloric** (pie-LORE-ick) **sphincter** (a ring of smooth muscle at the juncture of the small intestine and stomach), it opens, and chyme squirts into the small intestine.

The stomach contents are emptied in 2–6 hours, depending on the size of the meal and the type of food. The larger the meal, the longer it takes to empty. Solid foods (meat) empty slower than liquid foods (milk shakes). Peristaltic contractions continue after the stomach is empty and are felt as hunger pangs.

As chyme and protein leave the stomach, gastric gland secretion declines and the stomach slowly shuts down its production of HCl until the next meal.

▌ The Small Intestine Serves as a Site of Food Digestion and Absorption

The **small intestine** is a coiled tube in the abdominal cavity about 6 meters (20 feet) long in adults (**FIGURE 5-14**). So named because of its small diameter, the small intestine consists of three parts in the following order: the duodenum (DUE-oh-DEEN-um), jejunum (jeh-JEW-num), and ileum (ILL-ee-um).

Inside the small intestine, macromolecules are broken into smaller molecules (that is, digested) with the aid of enzymes. These molecules are then transported into the bloodstream and the lymphatic system. The **lymphatic system**, discussed more fully in Chapter 8, is a network of vessels that carries extracellular fluid from the tissues of the body to the circulatory system. In addition, tiny lymphatic vessels in the wall of the small intestine, known as **lacteals** (LACK-teels), absorb fats and transport them to the bloodstream.

Digestion in the Small Intestine Requires Enzymes From Two Major Sources.

The digestion of food molecules inside the small intestine requires enzymes produced from two distinctly different sources: the pancreas, an organ that lies beneath the stomach, and the lining of the small intestine itself.

The **pancreas** is nestled in a loop formed by the first portion of the small intestine, the duodenum (**FIGURE 5-20**). The pancreas is a dual-purpose organ—that is, it has endocrine and exocrine functions. As an exocrine gland, it produces enzymes and sodium bicarbonate essential for the digestion of foodstuffs in the small intestine. Its endocrine function is fulfilled by special cells that produce hormones (insulin and glucagon) that help regulate blood glucose levels and thus assist in maintaining homeostasis.

The digestive enzymes of the pancreas flow through the large pancreatic duct into the duodenum. Each day, approximately 1200–1500 milliliters (1.0–1.5 quarts) of pancreatic juice is produced and released into the small intestine. This liquid is composed of water, sodium bicarbonate, and several important digestive enzymes (**TABLE 5-6**).

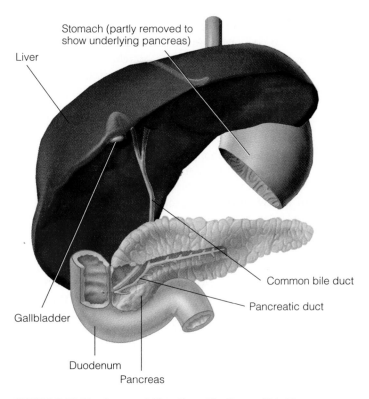

Stomach (partly removed to show underlying pancreas)

Liver

Common bile duct

Pancreatic duct

Gallbladder

Duodenum

Pancreas

FIGURE 5-20 The Organs of Digestion The liver, gallbladder, and pancreas all play key roles in digestion. All empty by the common bile duct into the small intestine, in which digestion takes place.

TABLE 5-6

Digestive Enzymes

Site of Production	Enzyme	Action
Salivary glands	Amylase	Digests polysaccharides in oral cavity
Stomach	Pepsin	Breaks proteins into peptides
Pancreas	Trypsin	Cleaves peptide bonds of polypeptides and proteins
	Chymotrypsin	Same as trypsin
	Carboxypeptidase	Cleaves peptide bonds on carboxyl end of polypeptides
	Amylase	Breaks starch molecules into smaller units (maltose)
	Phospholipase	Cleaves fatty acids from phosphoglycerides to form monoglycerides
	Lipase	Cleaves two fatty acids from triglycerides
	Ribonuclease	Breaks RNA into smaller nucleotide chains
	Deoxyribonuclease	Breaks DNA into smaller nucleotide chains
Epithelium of small intestine	Maltase	Breaks maltose into glucose subunits
	Sucrase	Breaks sucrose into glucose and fructose subunits
	Lactase	Breaks lactose into glucose and galactose subunits
	Aminopeptidase	Breaks peptides into amino acids

Sodium bicarbonate neutralizes the acidic chyme released by the stomach and thus protects the small intestine from stomach acid. It also gives the pancreatic juice a pH of about 8, creating an environment optimal for the function of the pancreatic enzymes.

The pancreatic enzymes released act on the large molecules in food (proteins, starches, and so on), as shown in **TABLE 5-6**. As a result of pancreatic enzymatic activity, fats are completely reduced to monoglycerides (glycerol attached to one fatty acid) and fatty acids, which can be absorbed without further action. Proteins are broken into small peptide fragments and some amino acids. Carbohydrates are broken down into disaccharides and some monosaccharides.

Like the stomach, the pancreas secretes its enzymes in an inactive form. This protects the gland from self-destruction. **Trypsinogen** (trip-SIN-oh-gin), for example, is the inactive form of the protein-digesting enzyme **trypsin** (TRIP-sin). Produced by the pancreas, trypsinogen is activated by a substance on the epithelial lining of the small intestine. Trypsin, in turn, activates other digestive enzymes.

The final stage of digestion occurs with the aid of enzymes produced by the epithelial cells of the small intestine. These enzymes are embedded in the membranes of the epithelial cells (**TABLE 5-6**). As a result, the final phase of digestion occurs just before the nutrient is absorbed into the cell.

The Liver Produces an Emulsifying Agent, Bile, Which Plays a Key Role in the Digestion of Fat.

The **liver** is one of the largest and most versatile organs in the body. By various estimates, the liver performs as many as 500 different functions. Situated on the right side of the abdomen under the protection of the rib cage, the liver is one of the body's storage depots for glucose (**FIGURE 5-3**). It also stores fats, iron, copper, and many vitamins. By storing glucose and lipids and releasing them as they are needed, the liver helps ensure a constant supply of energy-rich molecules to body cells. The liver also synthesizes some key blood proteins involved in clotting, and it is an efficient detoxifier of potentially harmful chemicals such as nicotine, barbiturates, and alcohol. These functions contribute significantly to homeostasis.

The liver also plays a key role in the digestion of fats through the production of a fluid called **bile**, which contains water, ions, and molecules such as cholesterol, fatty acids, and bile salts. **Bile salts** are steroids that emulsify fats, which means that they break fat globules into smaller ones. This process is essential for lipid digestion, because lipid-digesting enzymes in the small intestine do not work well on large fat globules. Unless fat globules can be broken into smaller ones, fat digestion is incomplete.

Produced by the cells of the liver, bile is first transported to the **gallbladder**, a sac attached to the underside of the liver (**FIGURE 5-20**). The gallbladder concentrates the bile by removing water from it. Bile is stored in the gallbladder until needed. When chyme is present in the small intestine, the gallbladder contracts and bile flows out through the duct system into the small intestine (**FIGURE 5-20**).

Bile flow to the small intestine may be blocked by **gallstones**, deposits of cholesterol and other materials that form in the gallbladder of some individuals. Gallstones may lodge in the ducts draining the organ, thus reducing—even completely blocking—the flow of bile to the small intestine. The lack of bile salts greatly reduces lipid digestion. Because lipid digestion is reduced, fat is not digested and absorbed.

Gallstones occur more frequently in older, overweight individuals, and the incidence in the elderly is about one in five. When they cause problems, gallstones are usually removed surgically. This procedure requires that the entire gallbladder be removed. Bile continues to be produced in these patients but is stored in the common bile duct, which becomes distended to accommodate the liquid bile. Scientists are now testing drugs that dissolve gallstones in many patients, hoping that they may someday eliminate or reduce the need for surgery.

The Intestinal Epithelium is Specially Modified for Absorption.

Virtually all food digestion occurs in the duodenum and jejunum. Once food molecules are digested, they must be absorbed—that is, transported across the epithelial lining of the small intestine into the bloodstream or lymphatic system, a network of vessels that drains excess fluid from tissues into the bloodstream. In these regions, absorption is facilitated by three structural modifications of the small intestine. **FIGURE 5-21A** shows, for

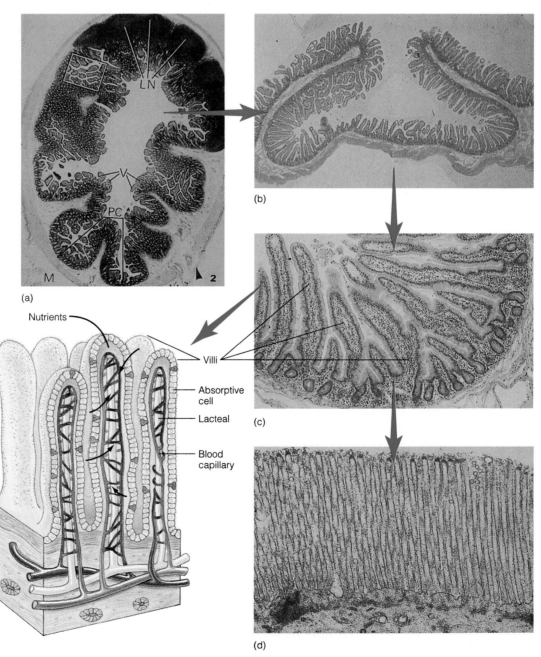

FIGURE 5-21 The Small Intestine The small intestine is uniquely "designed" to increase absorption. *(a)* A cross section showing the folds. LN means lymph nodules (aggregation of lymphocytes); V means villi; PC means plica circulares (circular fold). *(b)* A light micrograph of folds and villi. *(c)* Higher magnification of villi. *(d)* An electron micrograph of the surface of the absorptive cells showing the microvilli. *(e)* Each villus contains a loose core of connective tissue; a lacteal, or lymph, capillary; and a network of blood capillaries. Nutrients pass from the lumen of the small intestine through the epithelium and into the interior of the villi, where they are picked up by the lymph and blood capillaries.

Labels in figure: LN, V, PC, M, 2 (a); (b); Villi, (c); (d); Nutrients, Villi, Absorptive cell, Lacteal, Blood capillary, (e)

instance, that the lining of the small intestine is thrown into folds, which increase the overall surface area. On the surfaces of the folds are many fingerlike projections known as **villi** (VILL-eye), which also increase the surface area available for the absorption of food molecules.

FIGURE 5-22 Fat Digestion and Absorption

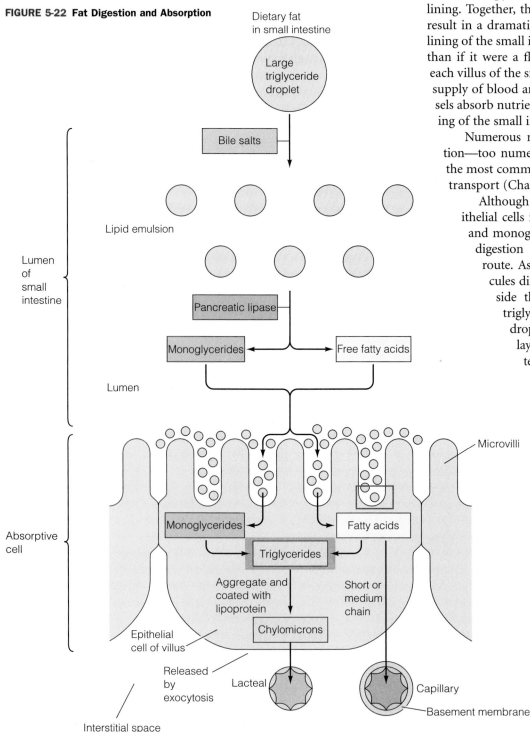

The intestinal surface area is further increased by **microvilli**, tiny protrusions of the plasma membranes of the epithelial cells lining the villi (**FIGURE 5-21D**). Each epithelial cell that functions in absorption contains an estimated 3000 microvilli. Two hundred million microvilli occupy a single square millimeter of intestinal lining. Together, the circular folds, villi, and microvilli result in a dramatic increase in the surface area of the lining of the small intestine, making it 600 times greater than if it were a flat layer. As shown in **FIGURE 5-21E**, each villus of the small intestine is endowed with a rich supply of blood and lymph capillaries. These tiny vessels absorb nutrients that have passed through the lining of the small intestine.

Numerous mechanisms are involved in absorption—too numerous to be discussed here. Three of the most common are diffusion, osmosis, and active transport (Chapter 3).

Although most nutrients diffuse from the epithelial cells into the blood capillaries, fatty acids and monoglycerides produced by the enzymatic digestion of triglycerides follow a different route. As shown in **FIGURE 5-22**, these molecules diffuse into the cells lining the villi; inside these cells they combine to reform triglycerides. The triglycerides form fat droplets, which are then coated with a layer of lipoprotein (lipid bound to protein), which is produced by the endoplasmic reticulum. This treatment renders the fat droplets, or **chylomicrons** (KIE-low-MIE-krons), water-soluble. They are released by the epithelial cells into the interstitial fluid by exocytosis (**FIGURE 5-22**). Because blood capillaries are relatively impermeable to chylomicrons, most of the lipid globules enter the more porous lymph capillaries in the villi. As shown in **FIGURE 5-22**, small- or medium-chain fatty acids not incorporated in triglycerides pass directly into blood capillaries in the villi.

The Large Intestine Is the Site of Water Resorption

The large intestine is about 1.5 meters (5 feet) long and consists of four regions, the cecum, appendix, colon, and rectum (**FIGURE 5-23**). The **cecum** (SEA-come) is a pouch that forms below the juncture of the large and small intestine. A small, worm-like structure, the **appendix**, attaches to the bottom of the cecum. Most of the large intestine consists of the **colon** (COE-lun). Unlike the small intestine, which is coiled and packed in a small volume, the colon consists of three relatively straight portions, the ascending colon, the transverse colon, and the descending colon. The colon empties into the **rectum** (WRECK-tum).

So named because of its large diameter, the large intestine receives materials from the small intestine. The material entering the large intestine consists of a mixture of water, undigested or unabsorbed food molecules, and undigestible food residues, such as cellulose. It also contains sodium and potassium ions.

The colon absorbs approximately 90% of the water and sodium and potassium ions that enter it. The undigested or unabsorbed nutrients feed a rather large population of *E. coli* bacteria. These bacteria synthesize several key vitamins: B-12, thiamine, riboflavin, and, most importantly, vitamin K, which is often deficient in the human diet. These vitamins are absorbed by the large intestine.

The contents of the large intestine (after water and salt have been removed) are known as the feces. The feces consist primarily of undigested food, indigestible materials, and bacteria. Bacteria, in fact, account for about one-third of the dry weight of the feces.[4]

The feces are propelled by peristaltic contractions along the colon until they reach the rectum (**FIGURE 5-23**). As fecal matter accumulates in the rectum, it distends the organ. This action, in turn, stimulates stretch receptors in the wall of the rectum. These receptors stimulate the **defecation reflex** (DEAF-eh-KA-shun). In this reflex, nerve impulses from the stretch receptors in the rectum travel via nerves to the spinal cord. In the spinal cord, the nerve impulses stimulate nerve cells that supply the smooth muscle in the wall of the rec-

tum, causing them to contract and expel the feces. Other nerve impulses from the spinal cord travel to the **internal anal sphincter**, a ring of smooth muscle, which normally keeps fecal matter from entering the anal canal. In the defecation reflex, however, the sphincter relaxes and the fecal matter can escape.

Defecation does not occur until the **external anal sphincter** relaxes. This sphincter is composed of skeletal muscle and is under conscious control. If the time and place are appropriate, the external anal sphincter is relaxed, and defecation can occur. Defecation is usually assisted by voluntary contractions of the abdominal muscles and a forcible exhalation, both of which increase intraabdominal pressure.

If circumstances are inappropriate, voluntary tightening of the external anal sphincter prevents defecation, despite prior activation of the reflex. When defecation is

FIGURE 5-23 The Large Intestine This organ consists of four basic parts, the cecum, appendix, colon, and rectum.

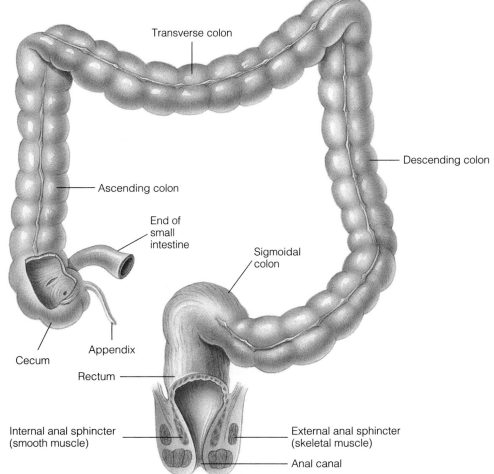

Transverse colon

Descending colon

Ascending colon

End of small intestine

Sigmoidal colon

Appendix

Cecum

Rectum

Internal anal sphincter (smooth muscle)

External anal sphincter (skeletal muscle)

Anal canal

[4]The feces are about 30% dead bacteria, 10%–20% fat, 10%–20% inorganic matter, 2%–3% protein, and about 30% undigested roughage (cellulose).

delayed, the muscle in the wall of the rectum relaxes, and the urge to defecate subsides—at least until the next movement of feces into the rectum occurs. If defecation is delayed too long, a person may become constipated. **Constipation** results from excess water absorption, which makes the feces hard and dry. Besides being uncomfortable, constipation may result in a dull headache, loss of appetite, and depression. Constipation is generally caused by (1) ignoring the urge to defecate; (2) decreases in colonic contractions, which occur with age, emotional stress, or low-fiber diets; (3) tumors or colonic spasms that obstruct the movement of feces; and (4) nerve injury that impairs the defecation reflex.

Constipation is not only uncomfortable, it can result in serious problems. Hardened fecal material, for instance, may become lodged in the appendix. This, in turn, may lead to inflammation of the organ, a condition called **appendicitis** (ah-PEND-eh-SITE-iss). When this occurs, the appendix becomes swollen and filled with pus and must be removed surgically to prevent the organ from bursting and spilling its contents into the abdominal cavity. Fecal matter leaking into the abdominal cavity introduces billions of bacteria and can result in a deadly infection.

Controlling Digestion

Digestion is a complex and varied process that is controlled largely by nerves and hormones. This section discusses some of the key events involved in the control of digestion.

Salivation Is Stimulated by a Nervous Reflex

Digestion begins in the oral cavity. As noted earlier, the sight, smell, taste, and sometimes even the thought of a hot fudge sundae stimulates the release of saliva. Chewing has a similar effect. The secretion of saliva via these routes is controlled by the nervous system and is largely a reflex response. Take a moment to review the reflex shown in **FIGURE 5-24**. For a discussion of the discovery of this phenomenon, see Scientific Discoveries 5-1.

Gastric Gland Secretion Is Controlled by a Variety of Different Mechanisms

Besides activating salivary production, the stimuli listed above also cause the brain to send nerve impulses along the **vagus nerve** (VAG-gus) to the stomach (**FIGURE 5-24**). These nerve impulses initiate the secretion of hydrochloric acid (HCl) and pepsinogen from cells of the gastric glands. Nerve impulses also stimulate the secretion of the hormone gastrin. As shown in **FIGURE 5-24**, gastrin is released into the blood and acts on the gastric glands, stimulating additional HCl and pepsinogen secretion.

The most potent stimulus for gastric secretion is the presence of protein in the stomach. Protein stimulates chemical receptors inside the stomach that activate networks of nerves in the stomach wall. These nerves, in turn, stimulate the gastric glands to secrete HCl and pepsinogen.

The concentration of HCl in the stomach is also regulated by a negative feedback mechanism, illustrated in **FIGURE 5-24**. If the acid content rises too high, HCl inhibits gastrin secretion, thus shutting off production by the gastric glands.

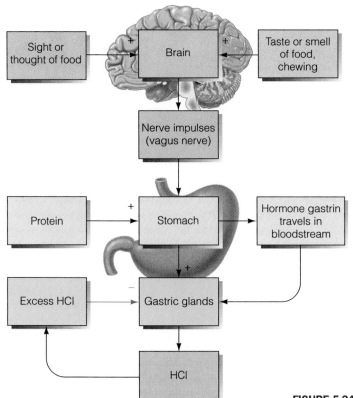

FIGURE 5-24 Pathway Leading to the Release of HCl by the Stomach

Pancreatic Secretions Released Into the Small Intestine Are Stimulated by Two Intestinal Hormones

Acidic chyme leaves the stomach and enters the small intestine, where it stimulates the release of the hormone **secretin** (se-CRETE-in), as shown in **FIGURE 5-25**. Secretin is produced by the cells of the duodenum and, like all hormones, travels in the bloodstream. In the pancreas, secretin stimulates the release of sodium bicarbonate. Sodium bicarbonate, in turn, is excreted into the small intestine, where it neutralizes the acidic chyme and creates an environment optimal for pancreatic enzymes.

The release of pancreatic enzymes is triggered by yet another intestinal hormone with the tongue-twisting name **cholecystokinin** (COAL-ee-sis-toe-KIE-nin) or **CCK.** This hormone is produced by cells of the duodenum in the presence of chyme (**FIGURE 5-25**). CCK also stimulates the gallbladder to contract, releasing bile into the small intestine.

Interestingly, recent evidence links abnormally low secretion of CCK to **bulimia** (boo-LEEM-ee-ah), a disorder characterized by recurrent binge eating followed by intentional vomiting. Some researchers believe that CCK may be involved in a range of other behaviors, too. Approximately 4% of America's young women, and a far smaller fraction of men, suffer from bulimia. This disorder is thought to have both biological and psychological roots, but researchers had failed to identify a biochemical cause until recently. Although no single chemical is likely to control a complex behavior such as appetite, it appears that CCK plays an important role. CCK has been found in the hypothalamus (a region of the brain) with other hormones.

Health, Homeostasis, and the Environment

Eating Right/Living Right

There are few places where the relationship between homeostasis, human health, and the environment is as evident as in nutrition. Because cells and organs involved in homeostasis require an adequate supply of nutrients, nutritional deficiencies can have noticeable impacts on body functions and health. Supporting this idea, many studies show that an unhealthy diet can increase the risk of contracting certain diseases, including cancer, heart disease, and hypertension.

Consider an example. Magnesium (a major mineral) is routinely ingested in insufficient amounts. New research suggests that such deficiencies may underlie a number of medical conditions, including diabetes, high blood pressure, problems during pregnancy, and cardiovascular disease.

Research shows that adding magnesium to the drinking water of rats with hypertension can eliminate the disease. Studies in rabbits show that magnesium reduces lipid levels in the blood and also reduces plaque formation in blood vessels. Rabbits on a high-cholesterol,

Bulimia

www.jbpub.com/humanbiology

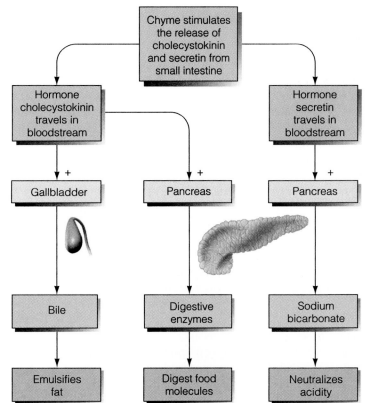

FIGURE 5-25 Control of the Pancreas and Gallbladder

Scientific Discoveries That Changed The World 5-1

Discovering the Nature of Digestion

Featuring the Work of van Helmont, Beaumont, and Pavlov

The foundation for the modern understanding of digestion was laid by a number of pioneering scientists. Two noteworthy contributors were a seventeenth-century Flemish physician, Jan Baptista van Helmont, and a nineteenth-century American, William Beaumont.

In van Helmont's time, most people thought that the digestion of food in the human was akin to cooking—that is, that food inside the stomach was broken down by body heat. Van Helmont, however, dismissed this theory with the simple observation that digestion occurs in cold-blooded animals such as fish. The lack of heat in such animals suggested to him that an alternative process was at work. But what was it?

The road to discovery began in a rather bizarre way. One day, a tame sparrow that often visited the scientist and would sit on his shoulder attempted to bite van Helmont's tongue. (We hope this was a playful gesture.) The physician detected a slight acidity in the bird's throat and hypothesized that acid must digest food.

To test this hypothesis, van Helmont tried to digest meat with a vinegar solution, but he could not. Consequently, he hypothesized that the body must contain "ferments," chemical substances in the stomach and small intestine that are specific for different types of food. Today, we know these "ferments" to be digestive enzymes.

Little insight into digestion was forthcoming in the period between van Helmont's work and the studies of the American scientist William Beaumont. Beaumont was the adventurous son of a Connecticut farmer who left his home in 1806. Soon thereafter, he became a schoolmaster in New York, where he studied medicine and associated sciences in his spare time. Several years later, he became an apprentice to a physician in Vermont, and 2 years later, he received a license to practice medicine. (Medical education was much less involved back then!)

Soon after becoming a licensed physician, Beaumont joined the army, where he would make his important discoveries about digestion. He owed his great work to an accident. In 1822, a French-Canadian porter and servant in the army was wounded in the abdomen by a musket that had accidentally discharged at close range. The servant, Alexis St. Martin, who was only 18, was brought to Beaumont for treatment. The doctor found that the bullet had ripped a hole in St. Martin's stomach, out of which poured food the man had eaten.

---

low-magnesium diet, for example, have 80%–90% more atherosclerotic (ah-ther-oh-skler-AH-tick) plaque than rabbits on a high-cholesterol, high-magnesium diet.

Studies have shown that magnesium deficiencies during pregnancy result in migraine headaches, high blood pressure, miscarriages, stillbirths, and babies with low birth weight. Research suggests that magnesium deficiency causes spasms in the blood vessels of the placenta, which reduce blood flow to the fetus. This can retard fetal growth and may even kill a fetus in utero. Magnesium supplements greatly reduce the incidence of these problems.

Researchers believe that 80%–90% of the American public may be magnesium deficient. One reason the American diet may be deficient in magnesium is that phosphates in many carbonated soft drinks bind to magnesium in the intestine, preventing it from being absorbed into the blood. Magnesium deficiencies can be reversed by eating more green leafy vegetables, seafood, and whole grain cereals and reducing consumption of carbonated soft drinks with high phosphate levels. Mineral supplements could help as well, but they should be used with extreme caution, because excess magnesium can cause neurological disorders.

Over the years, numerous dietary recommendations have been issued to help Americans live healthier lives and reduce the risk of cancer and heart attack. Nutritionists recommend that we daily consume (1) fruits and vegetables, especially cabbage and greens, (2) high-fiber foods, such as whole-wheat bread and celery, and (3) foods high in vitamins A and C. A healthy diet also minimizes the consumption of animal fat; red meat; and salt—cured, nitrate-cured, smoked, or pickled foods, including bacon and lunch meat.

A healthy diet results largely from habit and circumstance—that is, our social environment. How does our environment affect our nutrition and health?

In the hustle and bustle of modern society, many of us ignore proper nutrition, grabbing fat-rich foods because we haven't the time to sit down to a nutritionally balanced meal. Our fast-paced world places a high premium on saving time, often ignoring the importance of eating and living right. Fast food may allow us to hurry on to our next appointment, but unless it is nutritionally sound, it probably decreases the quality of our lives in the long run.

As was the custom of the time, Beaumont bled the victim. Despite this treatment, the patient survived, but the wound was slow to heal. For over 8 months, Beaumont tried in vain to close the hole in the stomach. During this time, it dawned on the young physician that he had stumbled across a golden opportunity to study digestion.

Beaumont's experiments consisted of feeding the patient various types of food, then studying what took place inside the man's stomach. In the next 9 years, Beaumont studied St. Martin's stomach contents with a wide assortment of foods, looking at the rate and temperature of digestion and the chemical conditions that favored different stages of the process.

Beaumont's studies helped settle a controversy over the nature of digestion. Some of his contemporaries contended that gastric juice was a kind of chemical solvent. Others believed that it merely liquefied food and that digestion occurred as a result of a vital force present in living organisms. By showing that some digestion could take place outside the stomach in the presence of gastric juice, Beaumont demonstrated that gastric secretions did more than moisten food and that no "vital force" was at work. Digestion, Beaumont asserted, was a chemical phenomenon.

Beaumont did not address one question that could have been answered with existing techniques—that is, what caused gastric juice to be secreted? Was it the presence of food in the stomach or something else?

The answer came in 1889 from the work of Ivan Pavlov, a notable scientist whose experiments on behavior are legendary. In studies on dogs, Pavlov showed that the secretion of gastric juice was stimulated by the nervous system. How did he make this determination?

In one of many experiments, the Russian scientist surgically connected a fold of a dog's stomach to an opening in the animal's side; this permitted him to examine the production of gastric secretions. He next cut and tied off the esophagus of the dog, so that food the animal swallowed could not enter the stomach. Following the surgery, Pavlov gave the dog food and found that as soon as food entered the dog's mouth, gastric juice began to be secreted. These secretions continued as long as food was present, thus suggesting nervous system involvement.

Since that time, a great deal has been learned about digestion and its control. For example, as you learned in this chapter, hormones are also involved in the control of many digestive processes. But don't close the book on this subject; much more will inevitably be discovered.

SUMMARY

1. Studies suggest that iron levels in the body may affect heat production, a relationship that, if substantiated by additional research, illustrates the importance of diet to physiological processes.

A PRIMER ON NUTRITION

2. Humans acquire energy and nutrients from the food they eat. These nutrients fall into two categories: *macronutrients*, substances needed in large quantity, and *micronutrients*, substances required in much lower quantities.

3. The four major macronutrients are water, carbohydrates, lipids, and proteins.

4. Water is contained in the liquids we drink and the foods we eat. Maintaining adequate water intake is important, because water is involved in many chemical reactions in the body. It also helps maintain body temperature and a constant level of nutrients and wastes in body fluids, both vital to homeostasis.

5. Carbohydrates and lipids are major sources of energy; 70%–80% of all energy required by the body is for basic functions.

6. Glucose is a six-carbon sugar that is broken down in the body to produce energy. In the human diet, glucose comes primarily from the polymer, starch.

7. Glucose molecules are stored in the liver and muscle as glycogen.

8. Another important dietary carbohydrate is cellulose. Although not digested, cellulose facilitates the passage of fecal matter through the intestinal tract and reduces the incidence of colon cancer.

9. Dietary protein is chiefly a source of amino acids for building proteins, enzymes, and protein hormones.

10. Contrary to popular myth, protein is not a source of energy, except when lipid and carbohydrate intake is low or when protein intake exceeds daily requirements.

11. Protein is broken down into amino acids and absorbed into the bloodstream. However, the body can synthesize some amino acids. These are known as *nonessential amino acids*. Others cannot be synthesized in the body and must be supplied in the food

we eat. These are known as *essential amino acids.*

12. To ensure an adequate supply of all amino acids, individuals should eat *complete proteins,* such as those found in milk or eggs, or combine lower quality protein sources.

13. Lipids are a diverse group of organic chemicals characterized by their lack of water solubility. The biologically important lipids serve many functions.

14. Some lipids (triglycerides) provide energy. They also form layers of heat-conserving insulation. Other lipids (phospholipids and steroids) are part of the plasma membranes of cells.

15. Triglycerides are fats and oils.

16. Triglycerides with many double bonds in their fatty acid side chains (the polyunsaturated fatty acids) lower one's risk of developing atherosclerosis, a build-up of placque on arterial walls. Triglycerides low in saturated fatty acids tend to increase this disease; they are commonly found in animal fats.

17. Micronutrients are needed in much smaller quantities and include two groups: vitamins and minerals.

18. *Vitamins* are a diverse group of organic compounds that are required in relatively small quantities for normal metabolism. A deficiency or surplus of one or more vitamins may alter homeostasis, with serious effects on human health.

19. Human vitamins fit into two categories: water-soluble and fat-soluble. The *water-soluble vitamins* include vitamin C and the B-complex vitamins. The *fat-soluble vitamins* include vitamins A, D, E, and K.

20. *Minerals* fit into one of two groups: *trace minerals,* those required in very small quantity, and *major minerals,* those required in greater quantity. Deficiencies and excesses of both types of minerals can lead to serious health problems.

THE DIGESTIVE SYSTEM

21. Food is physically and chemically broken down in the digestive system. Small molecules produced during di-

gestion are absorbed by the intestinal tract into the bloodstream and circulated throughout the body for use by the cells.

22. Food digestion begins in the mouth. The teeth mechanically break down the food. *Saliva* liquefies it, making it easier to swallow. *Salivary amylase* begins to digest starch molecules.

23. Food is pushed by the tongue to the *pharynx,* where it triggers the swallowing reflex. *Peristaltic contractions* propel the food down the *esophagus* to the stomach.

24. The *stomach* is an expandable organ that stores and further liquefies the food. The churning action of the stomach, brought about by peristaltic contractions, mixes the food, turning it into a paste referred to as *chyme.*

25. The stomach releases food into the small intestine in timed pulses, ensuring efficient digestion and absorption. Very limited chemical digestion and absorption occur in the stomach.

26. The stomach produces *hydrochloric acid,* which denatures protein, allowing it to be acted on by enzymes. The stomach also produces a proteolytic enzyme called *pepsin,* which breaks proteins into peptides. The lining of the stomach is protected from acid by mucus.

27. The functions of the stomach are regulated by neural and hormonal mechanisms.

28. The *small intestine* is a long, coiled tubule in which most of the enzymatic digestion and absorption of food occur. Digestive enzymes come from the lining of the intestine and the pancreas.

29. Pancreatic enzymes break macromolecules into smaller fragments. The intestinal enzymes break these molecules into even smaller fragments that can be absorbed by the epithelial lining of the small intestine.

30. The pancreas also releases *sodium bicarbonate,* which neutralizes HCl entering the small intestine with the chyme and creates an environment suitable for pancreatic enzyme function.

31. The *liver* plays an important role in digestion. It produces a liquid called *bile* that contains, among other substances, bile salts. Bile is stored in the *gallbladder* and released into the small intestine when food is present. Bile salts emulsify fats, breaking them into small globules that can be acted on by enzymes.

32. Undigested food molecules pass from the small intestine into the *large intestine,* which absorbs water, sodium, and potassium, as well as vitamins produced by intestinal bacteria. It also transports the waste, or feces, to the outside of the body.

CONTROLLING DIGESTION

33. Digestive processes are largely controlled by the nervous and endocrine systems. The release of saliva is stimulated by the sight, smell, taste, and the thought of food. These stimuli also cause the brain to send nerve impulses to the gastric glands of the stomach, initiating the secretion of HCl and gastrin, a hormone that also stimulates HCl secretion.

HEALTH, HOMEOSTASIS, AND THE ENVIRONMENT: EATING RIGHT/LIVING RIGHT

34. Human health is dependent on good nutrition. Numerous studies suggest that a healthy, balanced diet can decrease the risk of cancer, heart disease, hypertension, and other diseases.

35. Nutritionists recommend the daily consumption of (a) fruits and vegetables, especially cabbage and greens; (b) high-fiber foods, such as whole-wheat bread and celery; and (c) foods high in vitamins A and C.

36. In addition, nutritionists recommend reducing the consumption of animal fat; red meat; and salt-cured, nitrate-cured, smoked, or pickled foods, including bacon and lunch meat.

37. Studies suggest, however, that Americans have not taken these recommendations to heart. Many of us ignore proper nutrition, because we don't take the time to sit down to a nutritionally balanced meal. Living fast-paced lives, we often ignore the importance of eating right.

Critical Thinking

THINKING CRITICALLY— ANALYSIS

This Analysis corresponds to the Thinking Critically scenario that was presented at the beginning of this chapter.

To test this hypothesis, one might simply measure oxalate in the urine and blood of people eating both high- and low-calcium diets. If levels were highest in the people eating low-calcium diets, the hypothesis might very well be valid. Knowing that one study is not enough to validate a hypothesis, it might be wise to perform others.

This exercise illustrates the importance of questioning basic assumptions or popularly held beliefs. It shows how dead wrong popular wisdom can be and is a startling testimony to the need for scientific testing of beliefs. One has to wonder how many other assumptions are in error.

EXERCISING YOUR CRITICAL THINKING SKILLS

Science News recently reported the results of a study on premenstrual syndrome, a condition that results in irritability, tension, bloating, and discomfort, among other symptoms, prior to menstruation. This study showed that high doses of calcium successfully reduce mood swings and physical discomfort in women before and during menstruation.

Researchers at the U.S. Agricultural Research Service in Grand Forks, North Dakota, studied 10 healthy women who experienced mild behavioral and physical symptoms before and during menstruation. The women were assigned to one of two groups. The first group received a high daily dose of calcium (1300 milligrams); the second got a lower dose (600 milligrams) in liquid form added to their food. Halfway through the 6-month study, the women switched doses.

According to the report, 9 of the 10 women reported a reduction in crying, irritability, and depression while on the high dose. The high dose also seemed to reduce physical discomfort. In fact, 7 out of the 10 women reported a reduction in cramps and backaches while on the high-calcium diet. (The daily requirement for women 25 and older is 800 milligrams per day.)

Unfortunately, the *Science News* report failed to mention whether the low-dose group benefited from the treatment. It did, however, note that the project leader suggested that women boost their intake of calcium by eating more calcium-rich foods, such as skim milk and nonfat yogurt. Knowing what you do about proper experimentation, how would you critique this study? What are its flaws? What further work is needed to be certain of these results? Why?

TEST OF CONCEPTS

1. The body requires proper nutrient input to maintain homeostasis. Give an example, and explain how the nutrient affects homeostasis.

2. Describe the conditions during which protein provides cellular energy.

3. If you were considering becoming a strict vegetarian, eating no animal products, even milk and eggs, how would you be assured of getting all of the amino acids your body needs?

4. Describe how the different types of dietary fiber help protect human health.

5. What are vitamins, and why are they needed in such small quantities?

6. A dietary deficiency of one vitamin can cause wide-ranging effects. Why?

7. What organs physically break food down, and what organs participate in the chemical breakdown of food?

8. Describe the process of swallowing.

9. Describe the function of hydrochloric acid and pepsin in the stomach. How does the stomach protect itself from these substances?

10. How do ulcers form, and how can they be treated?

11. Describe the endocrine and nervous system control of the stomach function.

12. The small intestine is the chief site of digestion and absorption. Where do the enzymes needed for this process come from, and how is the release of these enzymes stimulated? What other molecules are needed for proper digestion?

13. Describe the functions of the large intestine.

TOOLS FOR LEARNING

www.jbpub.com/humanbiology

Tools for Learning is an on-line student review area located at this book's web site HumanBiology (www.jbpub.com/humanbiology). The review area provides a variety of activities designed to help you study for your class:

Chapter Outlines. We've pulled out the section titles and full sentence sub-headings from each chapter to form natural descriptive outlines you can use to study the chapters' material point by point.

Review Questions. The review questions test your knowledge of the important concepts and applications in each chapter. Written by the author of the text, the review provides feedback for each correct or incorrect answer. This is an excellent test preparation tool.

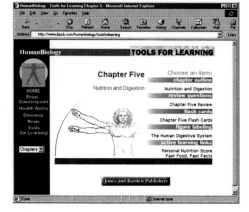

Flash Cards. Studying human biology requires learning new terms. Virtual flash cards help you master the new vocabulary for each chapter.

Figure Labeling. You can practice identifying and labeling anatomical features on the same art content that appears in the text.

Active Learning Links. Active Learning Links connect to external web sites that provide an opportunity to learn basic concepts through demonstrations, animations, and hands-on activities.

THE CIRCULATORY SYSTEM

Thinking Critically

High blood pressure, or hypertension, is a disease that afflicts many people the world over. African Americans in the United States, however, are more prone to this disease than most people. In fact, they are twice as likely to contract the disease as Caucasians.

Some scientists believe that this anomaly may be the result of genetic differences in the population and that these differences arose when many blacks were captured in Africa and transported to plantations where they were held as slaves. According to this hypothesis, many African Americans are hypertensive today because of a rare genetic trait that helped a small subset of their ancestors survive the inhumane conditions of slavery. That trait is an inherited tendency to conserve salt within the body.

According to the hypothesis, blacks that could conserve water were more likely to survive the grueling trip to the United States and the arduous conditions on the plantations. Conserving water meant they were less likely to die of dehydration. This trait, which proved to be an asset for slaves, may now be a deadly attribute among slavery's modern descendants. As a scientist how could you test this hypothesis? ■

Red blood cells in a small vein.

DAVID MCMAHON, A 48-YEAR-OLD NEW YORK attorney, collapsed in his office one morning and was rushed to a nearby hospital. There a team of cardiologists discovered a blood clot lodged in a narrowed section of his right coronary artery, which supplies the heart. The physicians began immediate action to prevent further damage to the oxygen-starved heart muscle. Through a small incision in McMahon's groin, they inserted a tiny plastic catheter into the artery that carries blood to the leg. They then threaded the

catheter up through the arterial system to his heart (**FIGURE 6-1**). Once they reached the heart, physicians guided the catheter into the coronary artery, where they injected an enzyme, called *streptokinase* (strep-toe-KI-nace), through the catheter. Streptokinase dissolved the blood clot, restoring blood flow to McMahon's heart muscle.

Although David McMahon survived, many others who suffer similar attacks are not so lucky. They either arrive at the hospital too late or, if they do make it to the hospital, do not receive blood-clot-dissolving agents or

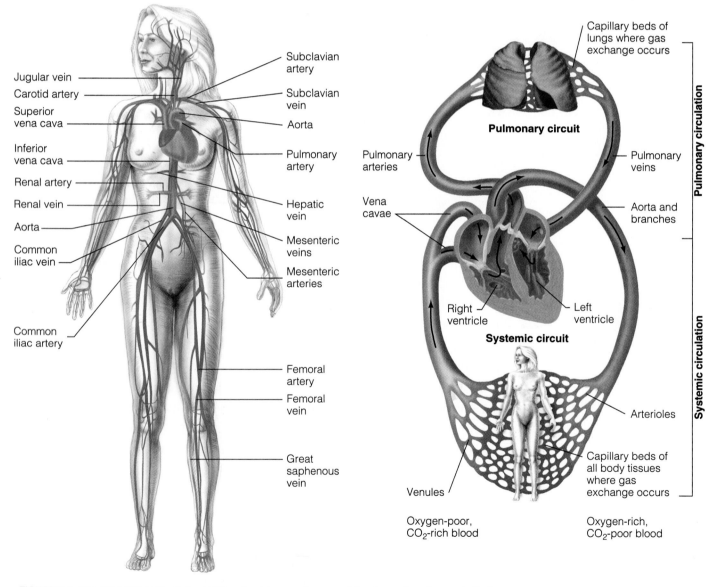

FIGURE 6-1 The Circulatory System   *(a)* The circulatory system consists of a series of vessels that transport blood to and from the heart, the pump. *(b)* The circulatory system has two major circuits, the pulmonary circuit, which transports blood to and from the lungs, and the systemic circuit, which transports blood to and from the body (excluding the lungs).

other treatments in time to avoid extensive damage to their heart muscle.

Heart attacks strike thousands of Americans each year. They are one of a handful of diseases of the circulatory system caused by the stressful conditions of modern life, smoking, poor eating habits, and a host of other factors. This chapter examines the circulatory system—in sickness and in health. It also discusses the lymphatic system, which functions in both circulation and immune protection.

Section 6-1
The Circulatory System's Function: An Overview

The human circulatory system consists of a muscular pump, the heart, and a network of vessels that transport blood throughout the body (**FIGURE 6-1**). The transport of blood, in turn, ensures the transport of oxygen and nutrients throughout the body. Oxygen enters the body in the respiratory system; nutrients enter via the digestive system. The circulation of blood also ensures the distribution of body heat and nutrients. Wastes pro-

TABLE 6-1		
Functions of the Circulatory System		
Transports oxygen to body cells		
Transports nutrients from the digestive system to body cells		
Transports hormones to body cells		
Transports wastes from body cells to excretory organs		
Distributes body heat		

duced by body cells are picked up by the blood and transported to organs of excretion such as the kidneys. Each of these functions serves a higher purpose: to maintain the relatively constant internal conditions (homeostasis) necessary for cellular function. For a review of these functions, see **TABLE 6-1**.

Section 6-2
The Heart

The **heart** is a muscular pump in the thoracic (chest) cavity. Often referred to as the workhorse of the cardiovascular system, the heart propels blood through the 50,000 miles of blood vessels in the body. Each day, this tireless organ beats approximately 100,000 times, adjusting its rate to meet the changing needs of the body. If you had a dollar for every heartbeat, you would be a millionaire in 10 days. Over a 70-year lifetime, you would collect $2.5 billion for your heart's work!

The heart, shown in **FIGURE 6-2**, is a fist-sized organ whose walls are composed of three layers, the pericardium, the myocardium, and the endocardium. The **pericardium** (pear-ah-CARD-ee-um) forms a thin, closed sac that surrounds the heart and the bases of large vessels that enter and leave the heart. The pericardial sac is filled with a clear, slippery aqueous fluid that reduces the friction produced by the heart's repeated contraction. The middle layer, the **myocardium** (my-oh-CARD-ee-um), is the thickest part of the wall and is composed chiefly of cardiac muscle cells (Chapter 4). The inner layer, the **endocardium** (end-oh-CARD-ee-um), is the endothelial layer, which lines the heart chambers.

The Circulatory System Has Two Distinct Circuits Through Which Blood Flows

To understand how the heart's anatomy relates to its function, it is important to first know a bit about circulation. As shown in **FIGURE 6-1B**, the circulatory system

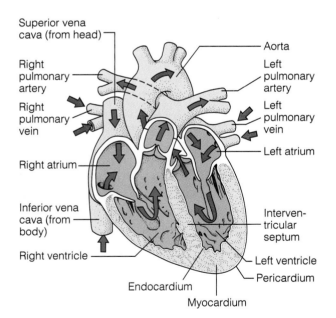

Superior vena cava (from head)
Aorta
Right pulmonary artery
Left pulmonary artery
Right pulmonary vein
Left pulmonary vein
Left atrium
Right atrium
Inferior vena cava (from body)
Interventricular septum
Right ventricle
Left ventricle
Pericardium
Endocardium
Myocardium

FIGURE 6-2 Blood Flow Through the Heart Deoxygenated (carbon-dioxide-enriched) blood (blue arrows) flows into the right atrium from the systemic circulation and is pumped into the right ventricle. The blood is then pumped from the right ventricle into the pulmonary artery, which delivers it to the lungs. In the lungs, the blood releases its carbon dioxide and absorbs oxygen. Reoxygenated blood (red arrows) is returned to the left atrium, then flows into the left ventricle, which pumps it to the rest of the body through the systemic circuit.

consists of two distinct circuits, the **pulmonary circuit,** which carries blood to and from the lungs, and the **systemic circuit,** which transports blood to and from the rest of the body.

Because these circuits deliver blood to different "customers," their functions are also quite different. The pulmonary circuit is involved in supplying oxygen to the systemic circuit and getting rid of carbon dioxide gathered by the blood as it flows throughout the body. The systemic circuit distributes oxygen and nutrients to body cells and picks up wastes, especially carbon dioxide.

As shown in **FIGURE 6-1B,** the heart consists of four hollow chambers—two on the right side of the heart and two on the left. Blood is pumped through the pulmonary circuit by the right side of the heart—the right atrium and right ventricle. Blood is pumped through the systemic circuit by the left side of the heart—the left atrium and left ventricle.

FIGURE 6-2 illustrates the course that blood takes through the heart. Drawn in blue, blood low in oxygen (and rich in carbon dioxide) enters the right side of the heart from the **superior** and **inferior vena cavae** (VEEN-ah CAVE-ee), which are part of the systemic circulation. These veins, which deliver blood that has circulated through the body, empty directly into the **right atrium** (A-tree-um), the uppermost chamber on the right side of the heart. The blood is pumped from here into the **right ventricle** (VEN-trick-el), the lower chamber on the right side. When the right ventricle is full, the muscles in its wall contract, forcing blood into the pulmonary arteries, which lead to the lungs.

In the lungs, this blood is oxygenated, then returned to the heart via the pulmonary veins. The pulmonary veins, in turn, empty directly into the **left atrium,** the upper chamber on the left side of the heart. It's the first part of the systemic circuit.

Next, the oxygen-rich blood is pumped to the left ventricle. When it's full, the left ventricle's thick, muscular walls contract and propel the blood into the aorta (a-OR-tah). The **aorta** is the largest artery in the body. It carries the oxygenated blood away from the heart, delivering it to the cells and tissues of the body.

The flow of blood just described presents a slightly misleading view of the way the heart really works. As shown in **FIGURE 6-3,** both atria actually fill and contract simultaneously, delivering blood to their respective ventricles. The right and left ventricles also fill simultaneously, and when both ventricles are full, they too contract in unison, pumping the blood into the systemic and pulmonary circulations. The coordinated contraction of heart muscle is brought about by an internal timing device, or pacemaker (described later).

Heart Valves Are Located Between the Atria and Ventricles and Between the Ventricles and the Large Vessels into Which They Empty

The human heart contains four valves that control the direction of blood flow, ensuring a steady flow from the atria to the ventricles and from the ventricles to the large vessels that lead away from the heart (**FIGURE 6-4A**).

The valves between the atria and ventricles are known as **atrioventricular valves** (AH-tree-oh-ven-TRICK-u-ler). Each valve consists of two or three flaps of tissue anchored to the inner walls of the ventricles by slender tendinous cords, the **chordae tendineae**

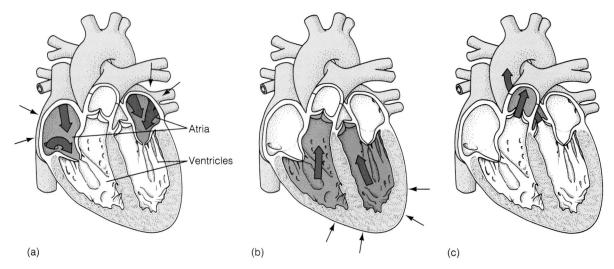

(a) (b) (c)

FIGURE 6-3 Blood Flow Through the Heart (a) Blood enters both atria simultaneously from the systemic and pulmonary circuits. When full, the atria pump their blood into the ventricles. (b) When the ventricles are full, they contract simultaneously, (c) delivering the blood to the pulmonary and systemic circuits.

Aorta

Superior vena cava

Right pulmonary veins

Right atrium

Right atrio-ventricular (tricuspid) valve

Inferior vena cava

Chordae tendineae   Right ventricle   Septum

Left pulmonary artery

Left pulmonary veins

Left atrium

Left atrio-ventricular (bicuspid) valve

Semilunar valves

Left ventricle

(a)

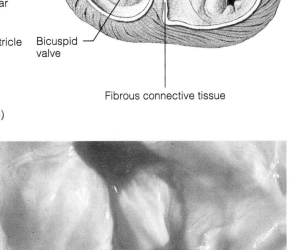

Aortic semilunar valve

Pulmonary semilunar valve

Tricuspid valve

Bicuspid valve

Fibrous connective tissue

(c)

(b)

(d)

FIGURE 6-4 Heart Valves *(a)* A cross section of the heart showing the four chambers and the location of the major vessels and valves. *(b)* Photograph of chordae tendineae.

(c) A view of the heart from above, with the major vessels removed to show the valves. *(d)* Pulmonary semilunar valve photographed from the ventricle.

(CORD-ee ten-DON-ee-ee), resembling the strings of a parachute (**FIGURES 6-4A** and **6-4B**). The right atrioventricular valve, between the right atrium and right ventricle, is called the **tricuspid valve** (try-CUSS-pid), because it contains three flaps. The left atrioventricular valve is the **bicuspid valve** (bye-CUSS-pid).[1] To remember the valves, imagine you are wearing a jersey with the number 32 on the front. This reminds you that the *tri*cuspid valve is on the right side and the *bi*cuspid valve is on the left.

Between the right and left ventricles and the arteries into which they pump blood (pulmonary artery and aorta, respectively) are the **semilunar valves** (SEM-eye-LUNE-ir) (**FIGURES 6-4A** and **6-4C**). The semilunar valves (literally, "half moon") consist of three semicircular flaps of tissue (**FIGURES 6-4C** and **6-4D**).

The atrioventricular and semilunar valves are one-way valves that open when blood pressure builds on one side and close when it increases on the other, much like the purge valves in scuba diving masks, which allow divers to force water out of their masks, or the ball valves in snorkels, which operate similarly. When the ventricles contract, blood forces the semilunar valves open. Blood flows out of the ventricles into the large

---

[1]The bicuspid valve is also called the *mitral valve,* because it resembles a miter, a hat worn by the Pope and Catholic bishops.

arteries. The backflow of blood causes the valve to close, preventing blood from draining back into the ventricles. The atrioventricular valves function in similar fashion.

Heart Sounds Result from the Closing of Various Heart Valves

When physicians listen to your heart, they are actually listening to sounds of the heart valves closing. The noises they hear are called the heart sounds and are often described as "LUB-dupp." The first heart sound (LUB) results from the closure of the atrioventricular valves. It is longer and louder than the second heart sound (dupp), produced when the semilunar valves shut.

Interestingly, the right and left atrioventricular valves do not close at precisely the same time. Nor do the semilunar valves. Thus, by careful placement of the stethoscope, a physician can listen to each valve individually to determine whether it is functioning properly.

For most of us, our heart valves function flawlessly throughout life. However, in some individuals, diseases alter the function of the valves. This, in turn, may dramatically decrease the efficiency of the heart and the circulation of blood. Rheumatic (RUE-mat-tick) fever, for example, is caused by a bacterial infection and affects many parts of the body, including the heart. Although it is relatively rare in developed countries, rheumatic fever is still a significant problem in the Third World. Rheumatic fever begins as a sore throat caused by certain types of **streptococcus** (STREP-toe-COCK-iss) **bacteria.** The sore throat—known as **strep throat**—is usually followed by general illness. During this infection, the body forms **antibodies** (proteins made by cells of the immune system) to the bacteria. These antibodies circulate in the blood and can damage the heart valves, preventing them from closing completely. This causes blood to leak back into the atria and ventricles after contraction and results in a distinct "sloshing" sound, called a **heart murmur.** This condition reduces the efficiency of the heart and causes the organ to work harder to make up for the inefficient pumping. Increased activity, in turn, causes the walls of the heart to enlarge and in severe cases, can result in heart failure. To prevent heart failure, damaged valves can be replaced by artificial implants.

Tumors (benign and malignant) and scar tissue have an opposite effect—that is, they reduce blood flow through the heart valves. This condition is known as **valvular stenosis** (sten-OH-siss; from the Greek word steno, meaning "narrow"). Valvular stenosis prevents the ventricles from filling completely. As in valvular incompetence, the heart must beat faster to ensure an adequate supply of blood to the body's tissues. This acceleration also puts additional stress on the organ.

Strep throat

www.jbpub.com/humanbiology

Heart Rate Is Largely Controlled by an Internal Pacemaker

The human heart functions at different rates under different conditions. At rest, it generally beats slowly. When one is excited or working hard, it beats much faster. This variation in heart rate helps the body adjust for differences in oxygen requirements by cells and tissues. Heart rate is controlled by a number of mechanisms.

One of the most important mechanisms is an internal pacemaker, the **sinoatrial (SA) node** (SIGN-oh-A-tree-ill noad) (**FIGURE 6-5**). Located in the wall of the right atrium, the SA node is composed of a clump of specialized cardiac muscle cells. These cells contract spontaneously and rhythmically. Each contraction produces a bioelectric impulse, akin to those produced by nerve cells. This impulse spreads rapidly from the SA node to the cardiac muscle, and then from muscle cell to muscle cell in both atria. Because cardiac muscle cells are tightly joined, and because the impulse travels quickly, the two atria contract simultaneously and uniformly.

Left to their own devices, cardiac muscle cells would contract independently and in a disorderly way, creating an ineffective pumping action. The SA node, however, imposes a single rhythm on all of the atrial heart muscle cells. The SA node is therefore like the conductor of an orchestra.

The electrical impulse generated by the SA node and transmitted throughout the atria next passes to the ventricles. However, its passage is briefly slowed by a barrier of unexcitable tissue that separates the atria from the ventricles. The impulse is delayed approximately 1/10 second. This delay gives the blood-filled atria time to contract and empty their contents into the ventricles. It also provides the ventricles plenty of time to fill before they are stimulated to contract.

After this brief delay, the impulse is channeled through a second mass of specialized muscle cells, the **atrioventricular (AV) node,** shown in **FIGURE 6-5.** From the AV node, the impulse travels along a tract of specialized cardiac muscle cells, known as the **atrioventricular bundle.**

As **FIGURE 6-5** shows, the atrioventricular bundle divides into two branches (called the *bundle branches*) that travel on either side of the wall separating the ventricles. The bundle branches give off smaller branches that terminate on specialized muscle cells, **Purkinje fibers** (per-KIN-gee), named after the scientist who discovered them. These fibers terminate on the cardiac muscle cells in the walls of the ventricles, stimulating them to contract.

Unlike the muscle cells of the atria, the cardiac muscle cells of the ventricles do not contract in unison, in large part because the impulse is not transmitted as

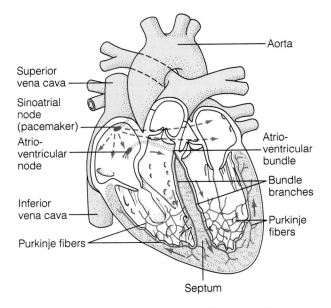

Aorta

Superior vena cava

Sinoatrial node (pacemaker)

Atrio-ventricular node

Inferior vena cava

Purkinje fibers

Atrio-ventricular bundle

Bundle branches

Purkinje fibers

Septum

FIGURE 6-5 Conduction of Impulses in the Heart The sinoatrial node is the heart's pacemaker. Located in the right atrium, it sends timed impulses into the atrial heart muscles, coordinating muscle contraction. The impulse travels from cell to cell in the atria, then passes to the atrioventricular node and into the ventricles via the atrioventricular bundle and its two branches, which terminate on the Purkinje fibers.

quickly and as uniformly through the ventricles as it is through the atria. Instead, contraction begins at the bottom of the heart and proceeds upward, squeezing the blood out of the ventricles into the aorta and pulmonary arteries.

The SA node of the human heart produces a steady rhythm of about 100 beats per minute when isolated from outside influences, but this is much too fast for most human activities. To bring the heart rate in line with body demand, the SA node must therefore be dampened. The SA node is curbed by impulses transmitted by nerves that connect the heart with a control center in the brain. At rest or during nonstrenuous activity, these impulses slow the heart to about 70 beats per minute, thus aligning heart rate with body demands. During exercise or stress, when the heart rate must increase to meet body demands, the decelerating impulses from the brain are reduced.

Other nerves also influence heart rate. These nerves carry impulses that accelerate the heart rate even further, allowing the heart to attain rates of 180 beats or more when the cells' demand for oxygen is great.

Several hormones also play a role in controlling heart rate. One of these is **epinephrine** (EP-eh-NEFF-rin), also known as **adrenalin.** This hormone is secreted during stress or exercise by the adrenal glands located on top of the kidneys. Epinephrine increases the heart rate, increasing the flow of blood through the body.

The nervous and endocrine system mechanisms described above are important evolutionary adaptations that help us and other vertebrates cope with the changing demands of our lives.

Electrical Activity in the Heart Can Be Measured on the Surface of the Chest

When the electrical impulse that stimulates muscle contraction in the heart reaches a cardiac muscle cell, it causes the cell to contract. Normally, the outside surface of the cardiac muscle cell is slightly positive. The inside surface is slightly negative. When the impulse arrives, it causes a rapid change in the permeability of the cardiac muscle cell's plasma membrane to sodium ions. Sodium ions flow inward, changing the polarity of the membrane and temporarily making the inside of the cell more positive than the outside. This change in polarity causes the release of calcium from internal storage depots, which, in turn, causes the cell to contract.

The shift in cardiac muscle cell polarity, or **depolarization,** can be detected by surface electrodes, small metal plates connected to wires and a voltage meter (**FIGURE 6-6A**). The electrodes are placed on a person's chest.

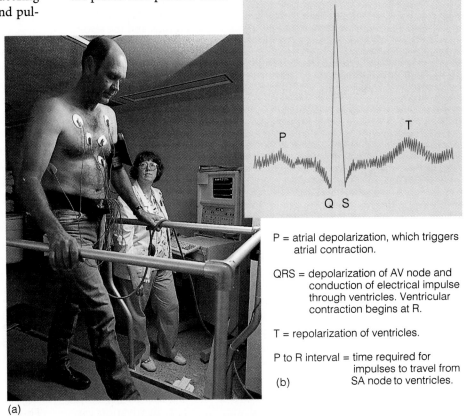

P = atrial depolarization, which triggers atrial contraction.

QRS = depolarization of AV node and conduction of electrical impulse through ventricles. Ventricular contraction begins at R.

T = repolarization of ventricles.

P to R interval = time required for impulses to travel from SA node to ventricles.

FIGURE 6-6 The Electrocardiogram (a) This patient taking a treadmill test to check his heart's performance is wired to a meter that detects electrical activity produced by the heart. (b) An electrocardiogram.

The resulting reading on a voltage meter is called an **electrocardiogram** (**ECG** or, sometimes, **EKG**) (**FIGURE 6-6B**).

For a normal person, the tracing produced on the voltage meter has three distinct waves (**FIGURE 6-6B**). The first wave, the P wave, represents the electrical changes occurring in the atria of the heart. The second wave, the QRS wave, is a record of the electrical activity taking place during ventricular contraction, and the third wave, the T wave, is a recording of electrical activity occurring as the ventricles relax.

Diseases of the heart may disrupt one or more waves of the ECG. As a result, an ECG is often a valuable diagnostic tool for cardiologists. Bear in mind, though, that the ECG detects only those diseases that alter the heart's electrical activity.

Cardiac Output Varies from One Person to the Next, Depending on Activity and Conditioning

The total amount of blood pumped by the ventricles each minute is called the **cardiac output.** Cardiac output is a function of two factors: **heart rate,** the number of contractions the heart undergoes per minute, and **stroke volume,** the amount of blood pumped by each ventricle during each contraction. At rest, the heart beats approximately 70 times per minute, and the stroke volume is about 70 milliliters. This produces a cardiac output of 5000 milliliters, or 5 liters per minute.

Cardiac output varies among individuals, depending on their physical condition and their level of activity. The heart of a trained athlete, for example, can pump 35 liters of blood per minute (seven times the cardiac output at rest). Most nonathletes, however, can increase the cardiac output to only about 20 liters per minute.

Myocardial infarction (heart attack)

www.jbpub.com/humanbiology

Section 6-3

Heart Attacks: Causes, Cures, and Treatments

Heart attacks come in several varieties. The most common heart attack is a **myocardial infarction** (my-oh-CARD-ee-al in-FARK-shun). Myocardial infarctions are caused by **thromboses** (throm-BOW-seas), blood clots that block one or more of the coronary arteries, usually those narrowed by atherosclerotic plaque (discussed next). A blood clot lodged in a coronary artery restricts the flow of blood to the heart muscle, cutting off the supply of oxygen and nutrients. This deprivation can damage and even kill the heart muscle cells. The damaged region is called an **infarct** (in-FARKT)—hence, the name myocardial infarction.

Myocardial Infarctions Usually Occur When Blood Clots Lodge in Arteries Narrowed by Atherosclerosis

As noted in Health Note 5-1, the formation of atherosclerotic plaque results from a combination of factors: stress, poor diet, lack of exercise, smoking, heredity, and several others. Narrowing of a coronary artery by plaque does not usually block the flow of blood enough to cause a heart attack, however, unless it is quite severe. Nonetheless, less severe narrowing does make the vessel more susceptible to blood clots. That is, clots often form in the vessel at the site of narrowing. Also, clots that form in other parts of the body can lodge in the narrowed vessels.

The outcome of heart attacks varies. If the size of the damaged area is small and if the change in electrical activity of the heart is minor and transient, a heart attack is usually not fatal. If the damage is great or electrical activity is severely disrupted, myocardial infarctions can prove fatal.

Heart attacks can occur quite suddenly, without warning, or may be preceded by several weeks of **angina** (an-GINE-ah), pain that is felt when the supply of oxygen to the myocardium is reduced. Anginal pain appears in the center of the chest and can spread to a person's throat, upper jaw, back, and arms (usually just the left one). Angina is a dull, heavy, constricting pain that appears when an individual is active, then disappears when he or she ceases the activity.

Angina may also be caused by stress and exposure to carbon monoxide, a pollutant that reduces the oxygen-carrying capacity of the blood. Angina begins to show up in men at age 30 and is nearly always caused by coronary artery disease. In women, angina tends to occur at a much later age. Interestingly, about 90% of all "chest pain" patients report to physicians turns out to be unrelated to the heart. What causes it? Many people who are stressed feel pain in the wall of the chest, typically in the muscles between the ribs. The muscles become sore because they're constantly contracted. When one is tense, deep breathing, relaxation, and stress reduction are effective means of eliminating this pain.

Heart Muscle Cells Unleashed from Their Control Beat Independently, Greatly Reducing the Heart's Effectiveness

Another type of heart attack results from a kind of cardiac anarchy known as **fibrillation** (FIB-ril-LAY-shun). This occurs when the SA node loses control of the heart. With the SA node no longer in charge, the cardiac muscle cells beat independently. The lack of coordination converts the heart into an ineffective, quivering mass that pumps little, if any, blood. If the heart stops beating altogether, the condition is known as **cardiac arrest.**

Physicians treat fibrillation by applying a strong electrical current to the chest, a procedure known as **defibrillation.** The electrical current passes through the wall of the chest and is often sufficient to restore normal electrical activity and heartbeat. A normal heartbeat can also be restored by **cardiopulmonary resuscitation (CPR),** in which the heart is "massaged" externally by applying pressure to the sternum (breastbone).

Prevention Is the Best Cure, But in Cases Where Damage Has Already Occurred, Medical Science Has a Great Deal to Offer

Proper diet, exercise, and stress management can reduce the risk of heart problems, as noted in Health Notes 1-1 and 5-1. Research also shows that a daily dose of aspirin (half a tablet a day is enough) taken over long periods can substantially reduce an individual's chances of a heart attack. Studies suggest that aspirin reduces heart attacks by reducing the formation of blood clots.[2]

Prevention should be the first line of attack against heart disease. It could save Americans hundreds of millions of dollars each year in medical bills, lost work time, and decreased productivity. But given human nature, the fast pace of modern life, and our inattentiveness to exercise and proper diet, heart disease will probably be around for a long time. To reduce the death rate, physicians therefore also look for ways to treat patients after they have had a heart attack.

One promising development is the use of blood-clot-dissolving agents such as streptokinase, mentioned at the beginning of this chapter. When administered within a few hours of the onset of a heart attack, streptokinase can greatly reduce the damage to heart muscle and accelerate a patient's recovery.

Ironically, streptokinase is an enzyme derived from the bacterium that causes rheumatic fever. Because streptokinase is a foreign substance, it evokes an immune reaction. In some people, the reaction is quite severe and may

even cause death. The immune reaction to this drug has inspired a search for similar chemicals without the dangerous side effects. One promising clot-busting enzyme is **urokinase** (YOUR-oh-kine-ace), which is produced by human cells.

Scientists are also testing another naturally occurring clot dissolver, called **TPA (tissue plasminogen activator).** Tests in humans suggest that TPA may also be free of the dangerous side effects of streptokinase. TPA has been approved for use in humans in the United States since 1987. Nevertheless, its use is not without problems. Two of the most significant are (1) the recurrence of blood clots in many patients and (2) the high cost of the drug.

In cases where the coronary arteries are completely blocked by atherosclerotic plaque, it is necessary to reestablish full blood flow to the heart muscle (**FIGURE 6-7A**). To restore blood flow, physicians often perform **coronary bypass surgery** in which they transplant segments of veins from the leg into the heart (**FIGURE 6-7B**). These venous bypasses transport blood around the clogged coronary arteries, restoring blood flow to the heart muscle. Once hoped to be a long-term answer to a widespread problem, it now appears that coronary bypass surgeries are only a temporary solution. Studies show that bypass patients have a significantly higher rate of survival in the

Cardiac Arrest
(see heart attack)
www.jbpub.com/humanbiology

---

[2]You should consult your physician if you are thinking about taking aspirin as a preventive measure.

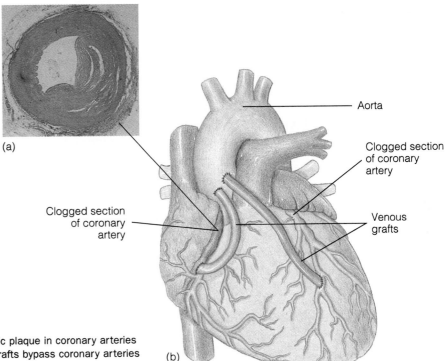

(a)

Aorta

Clogged section of coronary artery

Clogged section of coronary artery

Venous grafts

(b)

FIGURE 6-7 Coronary Bypass Surgery *(a)* Atherosclerotic plaque in coronary arteries can block the flow of blood to heart muscle. *(b)* Venous grafts bypass coronary arteries occluded by atherosclerotic plaque.

5 years following surgery than patients who just receive drugs. In the next 7 years, however, studies show that long-term survival from coronary bypass surgery is about the same as that of patients treated with diet and medications. In the long run, bypass surgery is only slightly more effective than nonsurgical medical treatments. Why?

Venous grafts often fill fairly quickly with plaque. The recurrence of plaque in grafts has led researchers to turn to marginally important arteries, such as the internal mammary artery, for this procedure. Many researchers believe that arteries will prove more resistant to plaque buildup than veins.

Physicians can clean clogged blood vessels by inserting a small catheter with a tiny balloon attached to its tip. After chemical clot dissolvers are administered to a patient, the balloon is inflated, forcing the artery open and loosening the plaque from the wall. This procedure is called **balloon angioplasty** (AN-gee-oh-PLAST-ee). Scientists are experimenting with lasers that burn away plaque in artery walls. Unfortunately, as in other techniques, cholesterol builds up again in the walls of arteries within a few months.

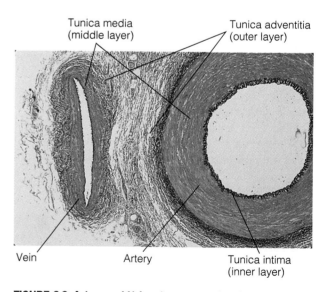

FIGURE 6-8 Capillary Network A network of capillaries between the arteriole and the venule delivers blood to the cells of body tissues (not shown).

Labels: Arteriole, Venule, Capillary network

Section 6-4
The Blood Vessels

The circulatory system can be divided into four functional parts. The first is the heart, which pumps blood throughout the body. The second is the arteries, which form a delivery system that transports blood from the heart to the body tissues. The third is an exchange system, consisting of networks of tiny vessels known as capillaries, found in body tissues. The fourth is the return system, consisting of veins that carry oxygen-depleted and waste-enriched blood back to the heart from the body tissues. (For a discussion of some of the discoveries that led to our understanding of circulation, see Scientific Discoveries 6-1.)

Arteries, which transport blood *away* from the heart, branch many times, forming smaller and smaller vessels. The smallest of all arteries is the **arteriole** (are-TEAR-ee-ol). As shown in **FIGURE 6-8**, arterioles empty into **capillaries** (CAP-ill-air-ees), tiny, thin-walled vessels that permit wastes and nutrients to pass through with relative ease. Capillaries form extensive, branching networks in body tissues, referred to as **capillary beds.** Capillaries have very thin walls that permit water and various molecules to pass through with ease.

Blood flows out of the capillaries into the smallest of all veins, the **venules.** Venules, in turn, converge to

form small veins, which unite with other small veins, in much the same way that small streams unite to form a river. Blood in veins flows toward the heart.

FIGURE 6-9 shows a cross section of an artery and a vein. As illustrated, these two vessels are structurally different. Veins, for example, tend to be smaller and to have thinner walls. Despite their obvious differences, arteries and veins have a common architecture. Both consist of three layers: (1) an external layer of connective tissue, which binds the vessel to surrounding tissues; (2) a mid-

Labels: Tunica media (middle layer), Tunica adventitia (outer layer), Vein, Artery, Tunica intima (inner layer)

FIGURE 6-9 Artery and Vein A cross section through a vein shows that the muscular layer, the tunica media, is much thinner than it is in an artery. Veins typically lie alongside arteries and, in histological sections such as these, usually have irregular lumens (cavities).

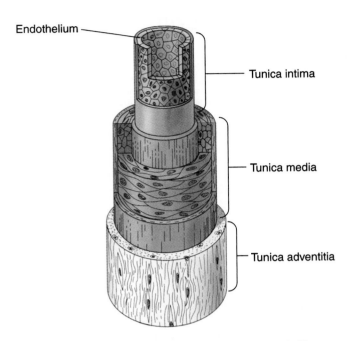

Endothelium

Tunica intima

Tunica media

Tunica adventitia

FIGURE 6-10 General Structure of the Blood Vessel The artery shown here consists of three major layers, the tunica intima, tunica media, and tunica adventitia.

dle layer, which is primarily made of smooth muscle; and (3) an internal layer, which is composed of a layer of flattened cells, the endothelium, and a thin, nearly indiscernible layer of connective tissue, which binds the endothelium to the middle layer (**FIGURE 6-10**).

Arteries and Arterioles Deliver Oxygen-Rich Blood to Tissues and Organs

The largest of all arteries is the aorta, a massive vessel that carries oxygenated blood from the left ventricle of the heart to the rest of the body. The aorta loops over the back of the heart, then descends through the chest and abdomen, giving off large branches along its way. These branches carry blood to the head, the extremities (arms and legs), and major organs, such as the stomach, the intestines, and the kidneys (**FIGURE 6-1**). The very first branches of the aorta are the coronary arteries.

The aorta and many of its chief branches contain numerous wavy elastic fibers interspersed among the smooth muscle cells of the wall. As blood pulses out of the heart, these arteries expand to accommodate it. Like a stretched rubber band, the elastic fibers cause the arterial walls to recoil. This helps push the blood along the arterial tree, maintaining an even flow of blood through the capillaries.

The elastic arteries branch to form smaller vessels, the **muscular arteries.** Muscular arteries contain fewer elastic fibers, but still expand and contract with the flow of blood. You can feel this expansion and contraction in the arteries lying near the skin's surface in your wrist and neck. It's the pulse that health care workers use to measure heart rate.

The smooth muscle of the muscular arteries responds to a number of stimuli, including nerve impulses, hormones, carbon dioxide, and lactic acid. These stimuli cause the blood vessels to open or close to varying degrees. This allows the body to adjust blood flow through its tissues to meet increased demands for nutrients and oxygen. Arterioles in muscles, for instance, dilate (DIE-late) when a person is threatened by danger. This increases blood flow to the muscle, allowing the person to flee or to meet the danger head on. At the same time, vessels in the digestive system constrict, reducing the digestive process and increasing the amount of blood available to the muscles. Regulating the flow of blood to body tissues is also required to control body temperature.

Blood Pressure.

The force that blood applies to the walls of a blood vessel is known as the **blood pressure.** Like many other physical conditions in the human body, blood pressure varies from time to time. For example, it changes in relation to one's activity and stress levels. When someone makes you angry, they really are raising your blood pressure! In a given artery, blood pressure rises and falls with each heart beat. Blood pressure also varies throughout the cardiovascular system, being the highest in the aorta and the lowest in the veins. Blood pressure is also rather low in the capillaries, a feature that enhances the rate of exchange between the blood and the tissues.

Blood pressure is measured by using an inflatable device with the tongue-twisting name of **sphygmomanometer** (SFIG-mo-ma-NOM-a-ter), or, more commonly, **blood pressure cuff** (**FIGURE 6-11A**, page 146). The blood pressure cuff is first wrapped around the upper arm. A stethoscope is positioned over the artery just below the cuff. Air is pumped into the cuff until the pressure stops the flow of blood through the artery (**FIGURE 6-11B**). The pressure in the cuff is then gradually reduced as air is released. When the blood pressure in the artery exceeds the external pressure of the cuff, the blood starts flowing through the vessel once again. This point represents the **systolic pressure** (sis-STOL-ick), the peak pressure at the moment the ventricles contract. Systolic pressure is the higher of the two numbers in a blood pressure reading (120/70,

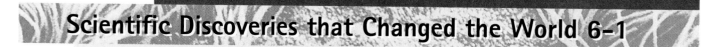

Scientific Discoveries that Changed the World 6-1

The Circulation of Blood in Animals

Featuring the Work of Harvey and Hales

The seventeenth-century British physician William Harvey is generally credited with the discovery of the circulation of blood in animals (**FIGURE 1**). Harvey is known as a scientist with a short temper who wore a dagger in the fashion of the day, which he reportedly brandished at the slightest provocation. He was probably not the kind of professor you might "argue" with over grades.

Temperament aside, Harvey is generally honored for his work on the role of the heart and the flow of blood in animals and is often praised as a pioneer of scientific methodology. His application of quantitative procedures to biology, some say, ushered in the modern age of this science.

In Harvey's medical school days, anatomists thought the intestines produced a substance called *chyle*. Chyle, they thought, was a fluid derived from the food people ate. It was passed from the intestines to the liver. The liver, in turn, converted the chyle to venous blood, then distributed the blood through arteries and veins.

As a medical student, Harvey was told that blood oozing through arteries and veins supplied organs and tissues with nourishment. He was also told that the blood merely ebbed back to the heart and lungs, where impurities were removed. In other words, there was no form of circulation, just an ebb and flow similar to the tides. These ideas had been proposed by the Greek physician Galen 14 centuries earlier and persisted nearly without challenge until Harvey's time.

As a teacher in the Royal College of Physicians in 1616, Harvey began to describe the circulation of blood, based on the results of his experiments and observations on animals. Apparently rebuked for his ideas by some of his colleagues, Harvey engaged in many years of research to provide supporting evidence. He described the muscular character of the heart and the origin of the heartbeat. He also demonstrated that the pulse felt in arteries resulted from the impact of blood pumped by the heart. Furthermore, he described the pulmonary and systemic circuits and proposed that blood flowed to the tissues and organs of the body via the arteries and returned via the veins.

A brief examination of one of his experiments illustrates that even though Harvey is a key figure in the

FIGURE 1 William Harvey This flamboyant scientist greatly advanced our knowledge on circulation in animals.

for example).[3] The pressure at the moment the heart relaxes to let the ventricles fill again is the **diastolic pressure** (DIE-ah-STOL-ick) and is the lower of the two readings. It is determined by continuing to release air from the cuff until no arterial pulsation is audible. At this point, blood is flowing continuously through the artery.

A typical reading for a young, healthy adult is about 120/70, although readings vary considerably from one person to the next. Thus, what is normal for one person may be abnormal for another. As a person ages, blood pressure tends to rise. Thus, a healthy 65-year-old might have a blood pressure reading of 140/90.

Hypertension is a prolonged elevation in blood pressure. Like other cardiovascular diseases, it has many causes, including kidney disease, high salt intake, obesity, and genetic predisposition. Nearly always a symptomless disease early on, hypertension is often characterized by a gradual increase in blood pressure over time. A person may feel fine and display no physical problems whatsoever for years. Symptoms, such as headaches, palpitations (rapid, forceful beating of the heart), and a general feeling of ill health, usually occur only when blood pressure is dangerously high. Consequently, early detection and

---

[3]Blood pressure is measured in millimeters of mercury (mm Hg; see **FIGURE 6-11B**).

history of biological science and played a key role in promoting quantitative study, some of his work was less than exceptional. It was sometimes based on poor assumptions and inaccurate observations. As an example, consider the work he used to rebut Galen's hypothesis that the blood was produced by the food people ate.

Harvey first approximated the amount of blood the heart ejected with each heartbeat (stroke volume), then determined the pulse rate. He called on earlier observations of a heart from a human cadaver to determine stroke volume. At that time, he had noted that the left ventricle contained more than 2 ounces of blood, and then, for reasons not entirely clear to historians of science, he hypothesized that the ventricle ejected "a fourth, a fifth, a sixth or only an eighth" of its contents. (Today, studies indicate that the heart ejects nearly all of its contents.) Based on this assumption, Harvey estimated that the stroke volume was about 3.9 grams of blood per beat. Modern estimates put it at 89 grams per beat.

Harvey also made a grave error in determining pulse. His value of 33 beats per minute is about half of the actual rate in humans. No one knows how he could have been so wrong. Armed with two erroneous measurements, Harvey derived a figure for the amount of blood that circulated through the body that was $\frac{1}{36}$ of the lowest value accepted today.

Regardless, Harvey "proved" his point—that each half-hour the blood pumped by the heart far exceeds the total weight of blood in the body. From this he concluded that blood must be circulated. It is not, as Galen proposed, produced by the food we eat. The amount of food one eats could not produce blood in such volume.

Harvey debunked another falsehood perpetrated through the centuries—the Galenic myth that blood flowed into the extremities in both arteries and veins. Harvey first wrapped a bandage around an extremity. This obstructed the flow of blood through the veins but not the arteries. He noted that the veins swelled because, as he conjectured, blood was

being pumped into them via underlying arteries and there was nowhere for the blood to go. Tightening the bandage further cut the blood flow in the arteries as well and thus prevented the veins from swelling. From these observations, Harvey correctly surmised that the arteries deliver blood to the extremities and the veins return it to the heart.

Harvey's work laid the foundation for modern cardiovascular physiology but left many questions unanswered. Many of these were addressed by the highly industrious English biologist Stephan Hales, who was born a century after Harvey. In a long series of scientifically rigorous experiments on horses, dogs, and frogs, Hales explored many aspects of the cardiovascular system. Benefiting from more modern methods of study, he charted blood pathways and examined blood flow and blood pressure in different parts of the circulatory system. After settling many of the unanswered questions left by Harvey, Hales went on to study plant physiology and is perhaps best known for his work on the circulation of sap in plants.

treatment are essential to prevent serious problems, including heart attacks.

Capillaries Permit the Exchange of Nutrients and Wastes Between Blood and Body Cells

As described above, the heart, arteries, and veins form an elaborate system that propels and transports blood to and from the capillaries. Capillaries form branching networks, the capillary beds, among the cells of body tissues. It is in these extensive networks of vessels that wastes and nutrients are exchanged between the cells of the body and the blood.

As shown in **FIGURE 6-12**, the walls of the capillaries consist of flattened endothelial cells. These cells permit dissolved substances to pass through them with ease—

and provide another illustration of the correlation between structure and function in the body.

If you could remove all of the capillaries from the body and line them up end to end, they would extend over 80,000 kilometers (50,000 miles)—enough to circle the globe at the equator two times. The extensive branching of capillaries not only brings them in close proximity to body cells, but also slows the rate of blood flow through capillary networks and decreases pressure, both of which increase the efficiency of capillary exchange.

As blood flows into a capillary bed, nutrients, gases, water, and hormones carried in the blood immediately begin to diffuse *out of* the tiny vessels. Meanwhile, water-dissolved wastes in the tissues, such as carbon dioxide, begin to diffuse inward.

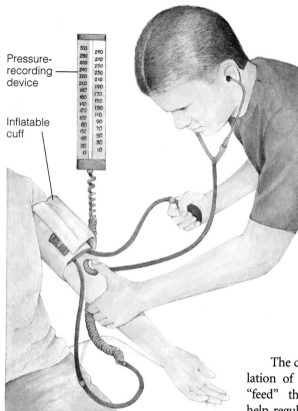

Pressure-recording device

Inflatable cuff

(a)

FIGURE 6-11 Blood Pressure Reading *(a)* A sphygmomanometer (blood pressure cuff) is used to determine blood pressure. *(b)* As shown, the blood pressure (indicated by the red line) rises and falls with each contraction of the heart. When the pressure in the cuff exceeds the arterial peak pressure, blood flow stops ①. No sound is heard. Cuff pressure is gradually released. When pressure in the cuff falls below the arterial pressure, blood starts flowing through the artery once again. This is the systolic pressure ②. The first sound will be heard. Cuff pressure continues to drop. When cuff pressure is equal to the lowest pressure in the artery, the artery is fully open and no sound is heard ③. This is the diastolic pressure.

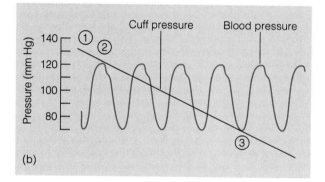

(b)

The constriction and dilation of the arterioles that "feed" the capillaries also help regulate body temperature. On a cold winter day, for example, the arterioles close down, restricting blood flow through the capillaries and conserving body heat. Just the reverse happens on a warm day. The flow of blood through the skin increases, releasing body heat and often creating a pink flush. Capillaries are therefore a part of the body's system of homeostasis.

Veins and Venules Transport the Oxygen-Poor and Waste-Laden Blood Back to the Heart

Blood leaves the capillary beds stripped of its nutrients and loaded with cellular wastes. As it drains from the capillaries, the blood enters the smallest of all veins, the venules. Venules converge to form small veins, which join with others to form larger and larger vessels. Unlike the arteries, then, the veins start off small and converge with other veins, forming larger and larger vessels. Eventually, all blood returning to the heart in the systemic circuit enters the superior or inferior vena cavae, the two main veins that drain into the right atrium of the heart (**FIGURE 6-2**). These vessels drain the upper and lower parts of the body, respectively.

Veins and arteries generally run side by side throughout the body. The arteries take blood away from the heart and toward body tissues, and the veins return blood to the heart.

As noted earlier, blood pressure in the veins is low, and veins have relatively thin walls with fewer smooth muscle cells than arteries (**FIGURE 6-9**). Because the veins' walls are so thin, obstructions can cause blood to pool in them, in much the same way that a tree down across a small stream can cause water to pool upstream. Blood pools in the obstructed veins, forming rather unsightly bluish bulges called **varicose veins** (VEAR-uh-cose) (**FIGURE 6-13**).

Some people inherit a tendency to develop varicose veins, but most cases can be attributed to various factors that reduce the flow of blood back to the heart: abdominal tumors, pregnancy, obesity, and even sedentary life-styles.

Varicose veins are not only unsightly, they can result in considerable discomfort. The restriction of blood flow, for example, may result in muscle cramps and the buildup of fluid, **edema** (uh-DEEM-ah; swelling), in the ankles and legs.

Varicose veins may also form in the wall of the anal canal. The veins in this region are known as the **internal hemorrhoidal veins** (hem-eh-ROID-il). A swelling of the internal hemorrhoidal veins results in a condition known as **hemorrhoids** (hem-eh-ROIDS). Because the internal hemorrhoidal veins are supplied by numerous pain fibers, this condition can be quite painful.

How Do the Veins Work?

FIGURE 6-14 shows that blood pressure is lowest in the veins. With such low pressure, how do the veins return blood to the heart?

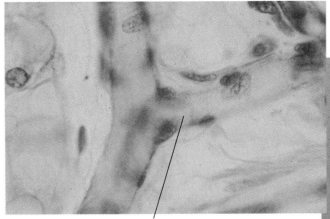

(a)   Capillary

FIGURE 6-12 The Capillary *(a)* A light micrograph of a capillary showing the endothelial cells that make up the wall of this vessel. *(b)* A cross section of a capillary showing the nucleus of an endothelial cell and capillary lumen.

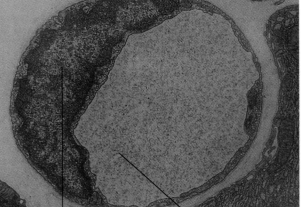
(b) Endothelial cell nucleus   Capillary lumen

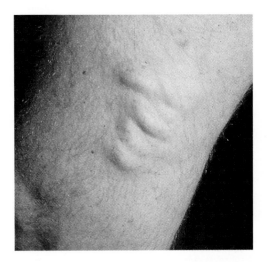

FIGURE 6-13 Any restriction of venous blood flow to the heart causes veins to balloon out, creating bulges commonly known as varicose veins.

For blood in veins above the heart, gravity is the chief means of propulsion. But for veins below the heart, return flow depends on the movement of body parts, which "squeezes" the blood upward. As you walk to class, for example, the contraction of muscles in your legs pumps the blood in the veins. This forces the blood upward, slowly and surely causing it to move against the force of gravity. Even the nervous muscle contractions that occur when you're studying help move the blood back to the heart.

Valves also facilitate the return of blood. Valves are flaps of tissue that span the veins and prevent the backflow of blood. The structure of the valves is shown in **FIGURE 6-15**. As illustrated, the semilunar flaps of the veins resemble those found in the heart. Just as in the valves of the heart, blood pressure, however slight,

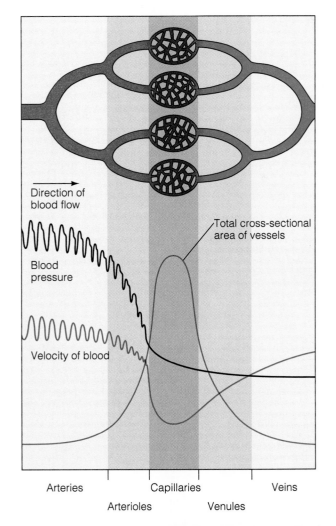

Direction of blood flow

Blood pressure

Velocity of blood

Total cross-sectional area of vessels

Arteries   Capillaries   Veins
     Arterioles   Venules

FIGURE 6-14 Blood Pressure in the Circulatory System Blood pressure declines in the circulatory system as the vessels branch. Arterial pressure pulses because of the heartbeat, but pulsation is lost by the time the blood reaches the capillary networks, creating an even flow through body tissues. Blood pressure continues to decline in the venous side of the circulatory system.

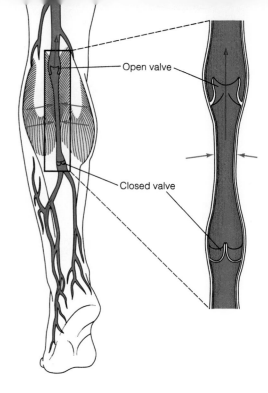

FIGURE 6-15 Valves in Veins The slight hydrostatic pressure in the veins and the contraction of skeletal muscles propel the blood along the veins back toward the heart. The one-way valves stop the blood from flowing backward.

pushes the flaps open (**FIGURE 6-15**). This allows the blood to move forward. As the blood fills the segment of the vein in front of the valve, it pushes back on the valve flaps and forces them shut.

To locate a valve, hold your arm out in front of you and make a fist. The veins should stick out or at least be apparent beneath the skin of your forearm. To locate a valve in the superficial veins on your forearm, press gently on a vein, then run your finger toward your wrist. You will note that the vein collapses behind your finger until it crosses a valve.

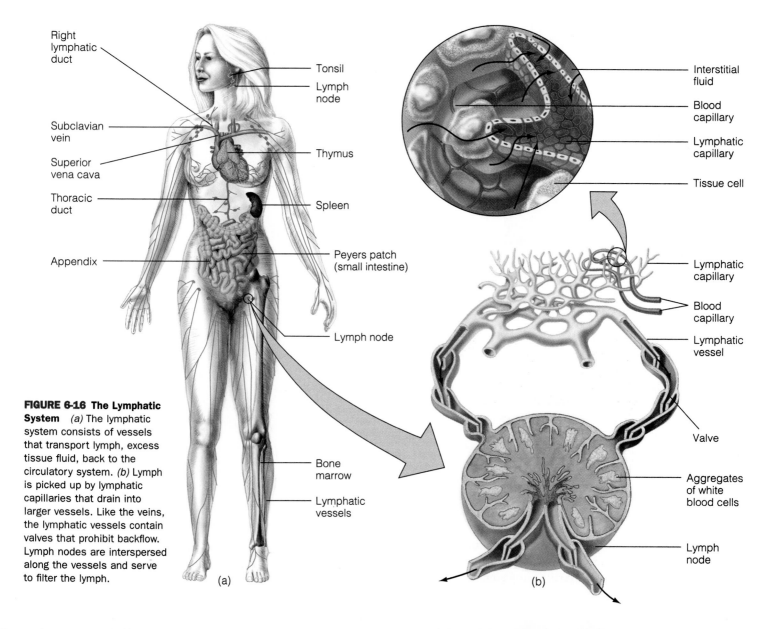

FIGURE 6-16 The Lymphatic System (a) The lymphatic system consists of vessels that transport lymph, excess tissue fluid, back to the circulatory system. (b) Lymph is picked up by lymphatic capillaries that drain into larger vessels. Like the veins, the lymphatic vessels contain valves that prohibit backflow. Lymph nodes are interspersed along the vessels and serve to filter the lymph.

Section 6-5

The Lymphatic System

The lymphatic system is an extensive network of vessels and glands (**FIGURE 6-16**). It is functionally related to two systems: the circulatory system and the immune system. This section examines the circulatory role of the lymphatic system.

You may recall from Chapter 4 that the cells of the body are bathed in a liquid called **interstitial fluid** (inter-STISH-il). Interstitial fluid provides a medium through which nutrients, gases, and wastes can diffuse between the capillaries and the cells.

Tissue fluid is replenished by water that diffuses out of the capillaries. The flow of water out of the capillaries, however, normally exceeds the return flow by about 3 liters per day. The "excess" water is picked up by small **lymph capillaries** in tissues. Like the capillaries of the circulatory system, these vessels have thin, highly permeable walls through which water and other substances pass with ease.

Lymph drains from the capillaries into larger ducts. These vessels, in turn, merge with others, creating larger and larger ducts, which empty into the large veins at the base of the neck.

Lymph moves through the vessels of the lymphatic system in much the same way that blood is transported in veins. In the upper parts of the body, it flows by gravity. In regions below the heart, it is propelled largely by muscle contraction. Breathing pumps lymph out of the chest and walking pumps the lymph out of the extremities. Lymphatic flow is also assisted by valves similar to those in the veins.

The lymphatic system also consists of several lymphatic organs: the lymph nodes, the spleen, the thymus, and the tonsils. The lymphatic organs function primarily in immune protection and are discussed in Chapter 8. Lymph nodes, however, play a role worth considering here.

Varying in size and shape, lymph nodes are found in association with lymphatic vessels in small clusters in the armpits, groin, neck, and other locations (**FIGURE 6-16**). A lymph node consists of a network of fibers and irregular channels that slow down the flow of lymph and filter out bacteria, viruses, cellular debris, and other particulate matter transported in the lymph. Lining the channels are numerous cells (macrophages) that phagocytize microorganisms and other materials.

Normally, lymph is removed from tissues at a rate equal to its production. In some instances, however, lymph production exceeds the capacity of the system. A burn, for example, may cause extensive damage to blood

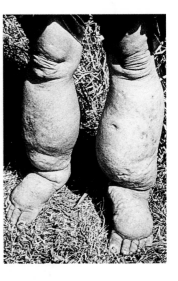

FIGURE 6-17 Elephantiasis A parasitic worm that invades the body stimulates the production of scar tissue, which blocks the flow of lymph through the nodes, causing tissue fluid to build up. This condition is known as elephantiasis.

capillaries, increasing their leakiness and overwhelming the lymphatic capillaries. This "flood" results in a buildup of fluid in tissues called edema.

Lymphatic vessels may also become blocked. One of the most common causes of blockage is an infection by tiny parasitic worms that are transmitted to humans by mosquitoes in the tropics. The worm larvae (an immature form) enter lymph vessels and take up residence in the lymph nodes. An inflammatory reaction causes the buildup of scar tissue in the nodes, which blocks the flow of lymph. After several years, the lymphatic drainage of certain parts of the body may be almost completely obstructed. A leg may swell so much, in fact, that it weighs as much as the rest of the body (**FIGURE 6-17**). This condition is known as **elephantiasis** (el-uh-FUN-TIE-ah-sis).

Health, Homeostasis, and the Environment

Cadmium and Hypertension

In recent years, scientists have discovered numerous chemical pollutants that are released into the environment and ingested in our food, air, and water, affecting our health by altering homeostatic mechanisms. Cadmium is one of those pollutants.

Cadmium is a heavy metal. In earlier times, cadmium was once used to treat syphilis and malaria, a remedy soon abandoned when physicians learned of its toxic effects. Today, cadmium is used for many commercial purposes. One of the most common uses is as a chemical stabilizer in polyvinyl chloride, a compound used to make vinyl (found in children's toys such as beach balls). Cadmium is combined with gold and other metals to make jewelry and also used to make camera batteries, engine parts, and radio and television sets.

Cadmium is released into the air and water during the manufacture of many products. Incinerators that burn municipal garbage also release small amounts of cadmium once contained in rubber tires, plastic bottles, furniture, and other items. Recycling facilities that melt down radiators and scrap steel also emit cadmium.

Elephantiasis

www.jbpub.com/humanbiology

Health Note 6-1

Hypertension, Atherosclerosis, and Aneurysms: Causes and Cures

Hypertension, or high blood pressure, is one of the most common health problems of our time. Its causes vary widely. Hypertension can be hereditary—that is, passed from parent to offspring. It is also thought to result from stress and diet, especially high salt intake in some individuals. People who are overweight when they are young are more likely to suffer from hypertension when they reach adulthood. In others, it is caused by disorders such as kidney failure or hormonal imbalances. Pregnancy can lead to hypertension, as can the use of oral contraceptives. Cadmium in food, air, and water may also contribute to this disease, as noted in this chapter.

If hypertension is untreated, blood pressure rises steadily over the years. Unfortunately, hypertension is a nearly symptomless disease for many people. Individuals feel fine for many years, despite their gradually rising blood pressure. Many exhibit no signs of the disease until it has progressed to the dangerous stage. Because of this, it is important for people over the age of 40 to have their blood pressure checked each year.

Hypertension is more common in men than in women and is twice as prevalent in African Americans than Caucasians. The disease is dangerous because the increased pressure in the circulatory system forces the heart to work harder. Elevated blood pressure may also damage the lining of arteries, creating a site at which atherosclerotic plaque forms. Atherosclerosis increases the risk of heart attack: A hypertensive person is six times more likely to have a heart attack than an individual with normal blood pressure. Hypertension also increases the chances for an occlusion in the arteries supplying the brain, which can result in strokes.

Another common problem of modern times is atherosclerosis, the buildup of cholesterol plaque in the walls of arteries, discussed in previous chapters. Arteries clogged with cholesterol force the heart to work harder, putting strain on this organ. Perhaps the most significant problem arises from blood clots that lodge in narrowed coronary arteries, reducing the flow of blood to the heart muscle.

Arteries can not only clog, they can also rupture. Certain infectious diseases (such as syphilis), atherosclerotic plaque, and hypertension can all result in a weakening of the wall of arteries. This weakening causes ar-

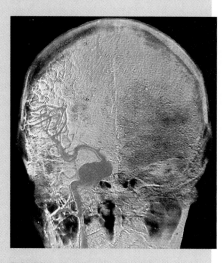

FIGURE 1 Aneurysm This X-ray shows a ballooning of one of the arteries in the brain. If untreated, an aneurysm can break, causing a stroke.

teries to balloon outward, a condition known as an **aneurysm** (FIGURE 1). Like a worn spot on a tire, an aneurysm can rupture when pressure builds inside or when the wall becomes too thin.

When an aneurysm breaks, blood pours out of the circulatory system. Because it happens so quickly, most aneurysms lead to death. An estimated 30,000 Americans die each

Together, these facilities emit over 4 million pounds of cadmium into the air each year in the United States alone.

Cadmium, which ends up in food, air, and water, is toxic to practically all the body systems of humans and other animals. It is absorbed into the body and retained, so its levels increase as we get older. The average American has 30 milligrams of cadmium in his or her body at death.

What are the effects of low-level cadmium exposure over a lifetime? One probable result is hypertension, a subject described more fully in Health Note 6-1. Cadmium acts on the smooth muscle of blood vessels, causing hypertension. Laboratory animals that consume even small amounts of cadmium develop hypertension. Humans with hypertension secrete 40 times more cadmium in their urine than individuals with normal blood pressure.

No one knows how many people are suffering from hypertension caused by cadmium, but the number is probably fairly large. Given the rise in cadmium use and the increases in recycling and incineration of trash, exposure can only get worse unless concerted efforts are made to tighten controls on facilities that release this toxic element and, perhaps more importantly, to reduce the use of cadmium in the first place.

year from ruptured aneurysms in the brain, and nearly 3000 die from ruptured aortic aneurysms.

As in most diseases, the first line of defense against aneurysms is prevention. By reducing or eliminating the two main causes—atherosclerosis and high blood pressure—individuals can greatly lower their risk.

Physicians recommend a number of steps to reduce your chances of developing atherosclerosis and hypertension. If you smoke, stop. If you are overweight, exercise and lose weight. If you're fond of salt, cut back on your intake. If you suffer from stress at work and at home, find ways to reduce your stress levels. If you drink alcohol, consume it in moderation, or quit altogether. If you are fond of foods that are rich in fats and cholesterol, cut down on them and consume foods containing water-soluble fiber such as apples, bananas, citrus fruits, carrots, barley, and oats. These reduce cholesterol uptake by the intestine, as explained in Chapter 5.

The second line of defense against cardiovascular disease is early detection and treatment. As noted earlier, hypertension can be detected by regular blood pressure readings. Athero-

sclerosis can be discovered by blood tests that measure cholesterol. Aneurysms can be detected by X-rays. Pain also alerts patients and physicians that something is wrong. Once any of these diseases is detected, physicians have many options.

In the event of an aneurysm, surgeons can remove the weakened section of the artery and replace it with a section of a vein. In cases where venous grafts are not practical (in the brain, for example), surgeons can clamp or tie off the artery just before the bulge, preventing blood flow through the damaged section. This works only when other arteries provide adequate blood flow to the area served by the damaged artery. In larger arteries, pieces of Dacron or other synthetic materials can be sewn into the wall of the artery, protecting it from breaking.

Researchers are also experimenting with an alloy of nickel and titanium called *nitinol*. Nitinol is a "metal with a memory." When a fine nitinol wire is heated and wrapped around a cylinder, it forms a tightly coiled spring. When the spring is cooled, it reverts to a straight wire. When reheated, the metal returns to a coil.

In experiments with dogs, scientists have created a nitinol wire coil that corresponds to the internal diameter of an artery. Next they cool the wire, causing it to revert to the straight form. The wire is then pushed through a catheter inserted in the artery. As the wire emerges from the cooled catheter inside the artery, body heat causes it to coil again. When in place, the coil adds strength to the wall of the artery, preventing rupture.

Experiments with dogs show that the endothelial cells of the innermost layer soon grow over the implant, making it a permanent part of the artery's wall. If successful in humans, this procedure could help save hundreds, perhaps thousands, of lives each year. It is, however, no substitute for a healthy diet and a healthy environment.

Visit *Human Biology's* Internet site, www.jbpub.com/humanbiology, for links to web sites offering more information on this topic.

www.jbpub.com/humanbiology

SUMMARY

THE CIRCULATORY SYSTEM'S FUNCTION: AN OVERVIEW

1. The *circulatory system* is one of the body's chief homeostatic systems. It helps maintain constant levels of nutrients and wastes, helps regulate body temperature, distributes body heat, protects against microorganisms, and, through clotting, protects against blood loss.

THE HEART

2. The circulatory system consists of a pump, the *heart,* and two circuits, the

pulmonary circuit, which transports blood to and from the lungs, and the *systemic circuit,* which delivers blood to the body and returns it to the heart.

3. The human heart consists of four chambers: two atria and two ventricles.

4. The right atrium and right ventricle service the pulmonary circuit. The left atrium and left ventricle pump blood into the aorta and are part of the systemic circuit.

5. The right atrium receives blood from the superior and inferior vena cavae. This blood, returning from body tissues, is low in oxygen and rich in carbon dioxide.

6. From the right atrium, blood is pumped into the right ventricle, then to the lungs, where it is resupplied with oxygen and stripped of most of its carbon dioxide. Blood returns to the heart via the pulmonary veins, which empty into the left atrium.

7. Blood is pumped from the left atrium to the left ventricle, then out the aorta to the body, where it supplies cells of tissues and organs with oxygen and picks up cellular wastes.

8. *Heart valves* help control the direction of blood flow. The *atrioventricular valves* permit blood to flow from the atria to the ventricles and then prevent it from flowing in the reverse direction when the ventricles contract. The *semilunar valves* in the base of the aorta and pulmonary artery prevent blood from flowing back into the ventricles.

9. The closing of the valves produces distinct *heart sounds*, which can be detected through the chest using a stethoscope. Irregularities in heart sounds indicate the presence of diseased or damaged valves.

10. Cardiac muscle cells contract rhythmically and independently; contraction is coordinated by the heart's pacemaker, the *sinoatrial node* located in the upper wall of the right atrium.

11. The cells of the SA node discharge periodically, sending impulses to all atrial muscle cells, which cause them to contract in unison.

12. The impulse next travels to the ventricles, but its passage is delayed, providing time for the ventricles to fill before contracting.

13. The impulse travels from the *atrioventricular node* down the *atrioventricular bundle* into the myocardium of the ventricles.

14. Contraction in the ventricles begins at the tip of the heart and proceeds upward.

15. Left on its own, the SA node would produce 100 contractions per minute. Nerve impulses, however, reduce the heart rate to about 70 beats per minute when a person is inactive. During exercise or stress, the heart rate increases to meet body demands.

16. Depolarization of the heart muscle produces weak electrical currents that can be detected by surface electrodes. The change in electrical activity is detected by a voltage meter.

17. The tracing produced on the voltage meter is called an *electrocardiogram (ECG)* and has three distinct waves. Diseases of the heart may disrupt one or more of the waves, making the ECG a valuable diagnostic tool.

HEART ATTACKS: CAUSES, CURES, AND TREATMENTS

18. The most common type of heart attack is a *myocardial infarction,* caused by a blood clot that either forms in an already narrowed *coronary artery* or arises elsewhere and breaks loose, only to become lodged in a coronary artery. The obstruction decreases the flow of blood and oxygen to heart muscle, sometimes killing the cells.

19. Heart attacks can occur quite suddenly without warning, or they may be preceded by several weeks of *angina*, pain felt when blood flow to heart muscle is reduced. Angina appears when an individual is active, then disappears when he or she rests.

20. The risk of heart attack can be reduced by proper diet, exercise, and daily doses of aspirin. Numerous treatments are available for patients who have had a heart attack.

THE BLOOD VESSELS

21. The heart pumps blood into the *arteries*, which distribute the blood to *capillary beds* in body tissues where nutrient and waste exchange occurs. Blood is returned to the heart in the *veins*.

22. The largest of all arteries is the *aorta*. It carries oxygenated blood from the left ventricle of the muscles, glands, and organs of the head and the rest of the body (except the lungs).

23. The aorta and many of its chief branches are *elastic arteries*. As blood pulses out of the heart, the elastic arteries expand to accommodate the blood, then contract, helping pump the blood and ensuring a steady flow through the capillaries.

24. As they course through the body, the elastic arteries branch to form *muscular arteries*, which also expand and contract with the flow of blood.

25. The smooth muscle in the walls of muscular arteries responds to a variety of stimuli. These stimuli cause the blood vessels to open or close to varying degrees, controlling the flow of blood through body tissues.

26. Blood pressure and flow rate are highest in the aorta and drop considerably as the arteries branch. By the time the blood reaches the capillaries, its flow and pressure are greatly reduced. This dramatic decline enhances the rate of exchange between the blood and the tissues.

27. *Capillaries* are thin-walled vessels that form branching networks, or capillary beds, among the cells of body tissues. Cellular wastes and nutrients are exchanged between the cells of body tissues and the blood in capillary networks.

28. Blood flow through a capillary network is regulated by constriction and relaxation of the arterioles that empty into them.

29. Blood draining from capillary beds enters *venules*, which join to form veins. Veins return blood to the heart and generally run alongside the arteries.

30. Because the walls of veins have very little smooth muscle, they are easily affected by obstructions, which cause the walls to balloon, forming bluish bulges called *varicose veins*.

31. In the veins above the heart, blood drains by gravity. Veins below the heart, however, rely on the movement of body parts to squeeze the blood upward and on *valves,* flaps of tissue that span the veins and prevent the backflow of blood.

THE LYMPHATIC SYSTEM

32. The *lymphatic system* is a network of vessels that drains excess interstitial fluid from body tissues and transports it to the blood.

33. The lymphatic system also consists of several lymphatic organs, such as the lymph nodes, the spleen, the thymus, and the tonsils, which function primarily in immune protection.

34. Along the system of lymphatic vessels are small nodular organs called *lymph nodes,* which filter the lymph.

35. Under normal circumstances, lymph is removed from tissues at a rate equal to its production, keeping tissues from swelling. In some cases, however, lymph production exceeds the capacity of the lymphatic capillaries, and swelling (edema) results.

HEALTH, HOMEOSTASIS, AND THE ENVIRONMENT: CADMIUM AND HYPERTENSION

36. Cadmium is a toxic heavy metal that is released into the air and water during the manufacture of numerous products, the incineration of municipal garbage, and the recycling of steel and other metals.

37. Cadmium ends up in food, air, and water and increases in humans and other organisms over time. Cadmium acts on the smooth muscle of blood vessels, causing hypertension.

Critical Thinking

THINKING CRITICALLY— ANALYSIS

This Analysis corresponds to the Thinking Critically scenario that was presented at the beginning of this chapter.

This hypothesis suggests that a selective survival process was in operation—that is, that salt-retaining (dehydration-resistant) slaves were more likely to survive the rigors of slavery than those who lacked this trait. To test the validity of this idea, one might begin by looking at hypertensive African Americans to see whether they actually have the capacity to retain excess salt and, if so, whether this capacity leads to hypertension.

Another research study that might lead to useful results would require one to locate populations of blacks consisting of individuals of African and non-African heritage. By studying differences in blood pressure and the ability to retain salt, one might be able to draw some conclusions as to the validity of the salt-retention hypothesis. One such example is the Island of Barbados in the Caribbean. Its population contains some African blacks who were imported as slaves to work on the island's sugar plantations.

One set of researchers actually performed this analysis and found that blacks of African descent had a slightly higher incidence of high blood pressure than blacks of non-African descent. What does this suggest? The researchers thought that it helped support the salt-retention hypothesis.

Yet another study might involve tests of blacks who reside in the regions of Africa from which slaves were drawn. Finding a low rate of hypertension in this group would suggest that a selective survival mechanism was in effect. One such study showed that, in rural parts of Nigeria, blacks did not suffer from hypertension despite the fact that they consumed large amounts of salt. The researchers concluded that if these people are descendants of the ancestors of slaves, as are African American blacks, then the hypothesis supporting selective survival seems valid.

Critics of this hypothesis argue that this kind of evolutionary change in such a short period is unlikely. Critically analyze this argument. (You may want to read the chapter on evolution to form an argument.) Would it affect your view if you found out that 70% of African slaves died within 4 years of their capture?

EXERCISING YOUR CRITICAL THINKING SKILLS

In June 1988, an Environmental Protection Agency scientist, Joel Schwartz, published data showing a strong relationship between levels of lead in the blood of men and hypertension. In March 1989, Schwartz published another study showing that even relatively low lead levels in men's blood increased their risk of developing heart disease. That same month, a professor at the University of Rochester, in New York, published the results of studies suggesting that elderly men are extremely sensitive to elevated levels of lead in the blood.

Schwartz and two other scientists have also found that bone loss (osteoporosis) in women who have passed menopause releases large quantities of lead stored in bone. In the blood, this lead may harm sensitive organs such as the liver and kidneys. Lead is also a neurotoxin, which, even in low doses, has been shown to affect mental development and coordination.

In 1989, in response to these and other studies, the EPA announced new guidelines for lead levels in drinking water. The rules allow lead levels no higher than 5 parts per million, compared with the old standard of 50 parts per million. Some newspapers praised the action as a major accomplishment. Would it affect your view of this announcement if you knew that the measurements for the old standard were taken at people's faucets and that the measurements for the new standard are to be taken at water treatment plants, before the water is released into the pipes that distribute it to customers? What do you need to know before you can answer this question?

Would it affect your view of the announcement if you knew that most of the lead in a water system comes from lead pipes in old homes and lead solder used in copper pipes? What questions need to be answered before you can determine the importance of the EPA's announcement? What does this exercise suggest about analyzing news reports?

TEST OF CONCEPTS

1. List and describe several ways in which the circulatory system functions in homeostasis.
2. Describe how the pacemaker, the sinoatrial node, coordinates muscular contractions of the heart.
3. Define the pulmonary and systemic circuits.
4. Based on what you know about the heart, what would happen if the septum separating the two atria failed to form completely during embryonic development?

5. Describe how the valves control the direction of blood flow in the heart. What are valvular incompetence and valvular stenosis? What causes these conditions, and how do they affect the heart?
6. How does atherosclerosis of the coronary arteries affect the heart?
7. Describe the general structure of the arteries and veins. How are they different? How are they similar?
8. How do the elastic fibers in the major arteries help ensure a continuous blood flow through the capillaries?

9. Capillaries illustrate the remarkable correlation between structure and function. Do you agree with this statement? Why or why not?
10. Describe how the movement of blood through capillary beds is controlled. Why is this homeostatic mechanism so important? Give some examples.
11. Explain how the blood is returned to the heart via the veins.
12. Explain the role of the lymphatic system in circulation.

TOOLS FOR LEARNING

www.jbpub.com/humanbiology

Tools for Learning is an on-line student review area located at this book's web site HumanBiology (www.jbpub.com/humanbiology). The review area provides a variety of activities designed to help you study for your class:

Chapter Outlines. We've pulled out the section titles and full sentence sub-headings from each chapter to form natural descriptive outlines you can use to study the chapters' material point by point.

Review Questions. The review questions test your knowledge of the important concepts and applications in each chapter. Written by the author of the text, the review provides feedback for each correct or incorrect answer. This is an excellent test preparation tool.

Flash Cards. Studying human biology requires learning new terms. Virtual flash cards help you master the new vocabulary for each chapter.

Figure Labeling. You can practice identifying and labeling anatomical features on the same art content that appears in the text.

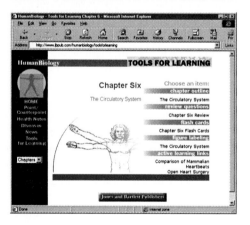

Active Learning Links. Active Learning Links connect to external web sites that provide an opportunity to learn basic concepts through demonstrations, animations, and hands-on activities.

THE BLOOD

Scanning electron micrograph of human red blood cells.

Thinking Critically

High concentrations of a bloodborne chemical, lipoprotein(a) (LIE-po-PRO-teen), are a known risk factor of atherosclerosis. This molecule transports cholesterol in the blood and deposits it in arterial walls, causing a thickening of the walls. Studying muscle cells isolated from a normal human artery and grown in tissue culture, two researchers also found that lipoprotein(a) stimulates smooth muscle cells to proliferate. The researchers concluded that high concentrations of lipoprotein(a) have two effects—they deposit cholesterol and cause muscle cells to divide—which may account for the elevated risk of cardiovascular disease they present.

Analyze this assertion based on what little you know about the experiment. What critical thinking rules helped you analyze this matter? ■

N 1966, DR. LELAND C. CLARK OF THE UNIVERSITY of Cincinnati's College of Medicine immersed a live laboratory mouse in a clear fluorocarbon solution saturated with oxygen. To the amazement of the audience, the mouse did just fine, "breathing" the oxygen-rich liquid as if it were air. Some time later, Clark extracted the mouse from the solution. After a moment, the rodent began to move about, apparently unharmed by the ordeal.

The point of this demonstration was not to show that an animal could "breathe" this fluid into its lungs and survive but, rather, to show that the solution held enormous amounts of oxygen, so much so that it could possibly be used as a substitute for blood. Clark and his colleagues hoped that their "artificial blood" would someday be a boon to medical science, helping paramedics keep accident victims who have lost substantial amounts of blood alive while they are being transported to hospitals.[1] In rural America, where a trip to a hospital takes considerable time, artificial blood could save thousands of lives a year.

---

[1]"Artificial blood" is, of course, a misnomer. The fluorocarbon solution takes over the role of the plasma and the red blood cells, which transport oxygen and carbon dioxide, but not the role of the white blood cells, which protect the body against infection, among other things.

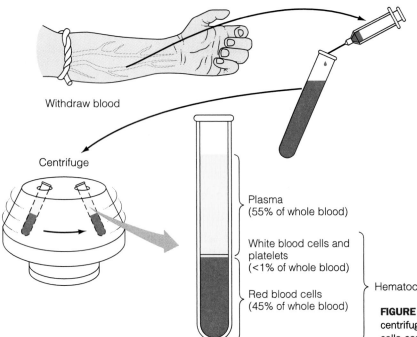

Withdraw blood

Centrifuge

Plasma
(55% of whole blood)

White blood cells and platelets
(<1% of whole blood)

Red blood cells
(45% of whole blood)

Hematocrit

Section 7-1

Blood: Its Composition and Functions

Human blood is a far cry from Clark's artificial substitute. The blood in our circulatory system, for example, is a water-based fluid, not a fluorocarbon, and consists of two basic components: plasma and formed elements. **Plasma** is the liquid portion of the blood and is about 90% water. It contains many dissolved substances. Three types of **formed elements** are suspended in the plasma: (1) white blood cells (or leuko-cytes; LUKE-oh-sites), (2) red blood cells (or erythrocytes; eh-RITH-row-sites), and (3) platelets (or thrombocytes; THROM-bow-sites).

Blood accounts for about 8% of our total body weight. A man weighing 70 kilograms (150 pounds), has about 5.6 kilograms (12 pounds)—or about 5 to 6 liters (1.3 to 1.5 gallons) of blood. On average, women have about a liter less.

FIGURE 7-1 shows that the plasma makes up about 55% of the blood volume and formed elements about 45%. The volume of the blood occupied by blood cells is referred to as the **hematocrit** (he-MAT-o-krit).

The hematocrit varies in individuals living at different altitudes. In people living in Denver, a mile above sea level, for example, hematocrits are typically about 5% higher than in people living at sea level. The slight increase in the hematocrit (red blood cell [RBC] concentration) in Denver residents is the result of a physiological mechanism that compensates for the slightly lower level of oxygen in the

White blood cells

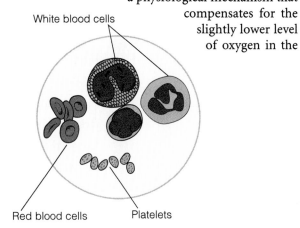

Red blood cells   Platelets

FIGURE 7-1 Blood Composition   Blood removed from a person can be centrifuged to separate plasma from the cellular component. Red blood cells constitute about 45% of the blood volume, except at higher altitudes where they make up about 50% of the volume to compensate for the lower oxygen levels.

atmosphere at the higher altitude. This homeostatic mechanism helps the body deliver a steady supply of oxygen to body cells.

In this day of high-tech athletic competition, where trainers and athletes look for any competitive edge they can find, many athletes train at high altitudes. Their rationale is that the higher hematocrit improves their performance at lower altitudes because their blood contains more oxygen-carrying hemoglobin. In fact, it is partly for this reason that the U.S. Summer Olympic team is headquartered in Colorado Springs, Colorado.

The Blood Plasma Is a Watery Transport Medium

Plasma is a light yellow (straw-colored) fluid. Dissolved in the plasma are (1) gases such as nitrogen, carbon dioxide, and oxygen; (2) ions such as sodium, chloride, and calcium; (3) nutrients such as glucose and amino acids; (4) hormones; (5) proteins; (6) various wastes; and (7) lipid molecules. Lipids are either suspended in tiny globules or bound to certain plasma proteins, which transport them through the bloodstream.

Plasma proteins are the most abundant of all dissolved substances in the plasma. Blood proteins contribute to osmotic pressure (described briefly in Chapter 3). Osmotic pressure helps regulate the flow of materials in and out of capillaries. It is therefore essential to the proper distribution of wastes and nutrients in the body, so vital to homeostasis.[2]

Blood proteins also help to regulate the pH of the blood. They do so by binding to hydrogen ions, thus preventing the H^+ concentration in the blood from rising, which is also essential to homeostatic balance.

Some plasma proteins serve as carrier proteins. Human blood contains three types of protein: (1) albumins, (2) globulins, and (3) fibrinogen (**TABLE 7-1**). Albumins (al-BEW-mins) and two types of globulins (GLOB-u-lins), for example, bind to hormones, ions, and fatty acids, and transport these molecules through the bloodstream. Carrier proteins are large, water-soluble molecules. Their binding to much smaller lipid molecules renders the latter water-soluble, and facilitates their transport through the largely aqueous bloodstream. Carrier proteins also protect smaller molecules from destruction by the liver. These proteins are there-

TABLE 7-1	
Summary of Plasma Proteins	
Protein	**Function**
Albumins	Maintain osmotic pressure and transport smaller molecules, such as hormones and ions
Globulins	Alpha and beta globulins transport hormones and fat-soluble vitamins; gamma globulins (antibodies) bind to foreign substances.
Fibrinogen	Converted into fibrin network that form blood clots

fore essential to maintaining chemical balance in the blood and tissue fluids.

Another group of globulins, known as the **gamma globulins,** are **antibodies,** proteins that "neutralize" viruses and bacteria or target them for destruction by phagocytic cells in body tissues known as macrophages (Chapter 8).

Still another important blood protein is **fibrinogen** (fie-BRIN-oh-gin). This unique protein is converted into **fibrin** (FIE-brin), which forms blood clots in the walls of injured blood vessels. Clots prevent blood loss and thus also help maintain homeostasis.

Red Blood Cells Are Flexible, Highly Specialized Cells That Transport Oxygen and Small Amounts of Carbon Dioxide in the Blood

The **red blood cell** is the most abundant cell in human blood. In fact, 20 drops of blood (equal to 1 milliliter) contain approximately 5 billion RBCs.[3] If RBCs were people, a single milliliter of blood would contain nearly the entire world population!

Human RBCs are highly flexible, biconcave disks that transport oxygen and, to a lesser degree, carbon dioxide in the blood (**FIGURE 7-2; TABLE 7-2**). The unique shape of the human RBC, shown in **FIGURE 7-2**, increases its surface area. The greater the surface area, the more readily gases move in and out of RBCs. Like so many other structures encountered in biology, the human RBC is a remarkable example of the marriage of form

---

[2]Osmotic pressure is the force responsible for the movement of water across a selectively permeable membrane; it is created by the difference in solute concentrations on either side of the membrane. The greater the concentration difference, the greater the osmotic pressure.

[3]Many texts state the concentration of RBCs per cubic millimeter, which is about 5 million RBCs. There are 1000 cubic millimeters in 1 cubic centimeter, or 1 milliliter.

FIGURE 7-2 Red Blood Cells *(a)* Transmission electron micrograph of human RBCs showing their flexibility. *(b)* Scanning electron micrograph of human RBCs.

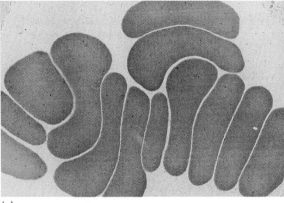

(b)

(a)

and function forged during the evolution of life on Earth.

Swept along in the bloodstream, RBCs travel many times through the circulatory system each day. As they pass through the capillaries, the RBCs bend and twist.

TABLE 7-2

Summary of Blood Cells

Name	Light Micrograph	Description	Concentration (Number of Cells/mm³)	Life Span	Function
Red blood cells (RBCs)		Biconcave disk; no nucleus	4–6 million	120 days	Transports oxygen and carbon dioxide
White blood cells Neutrophil		Approximately twice the size of RBCs; multi-lobed nucleus; clear-staining cytoplasm	3000 to 7000	6 hours to a few days	Phagocytizes bacteria
Eosinophil		Approximately same size as neutrophil; large pink-staining granules; bilobed nucleus	100 to 400	8–12 days	Phagocytizes antigen-antibody complex; attacks parasites
Basophil		Slightly smaller than neutrophil; contains large, purple cytoplasmic granules; bilobed nucleus	20 to 50	Few hours to a few days	Releases histamine during inflammation
Monocyte		Larger than neutrophil; cytoplasm grayish-blue; no cytoplasmic granules; U- or kidney-shaped nucleus	100 to 700	Lasts many months	Phagocytizes bacteria, dead cells, and cellular debris
Lymphocyte		Slightly smaller than neutrophil; large, relatively round nucleus that fills the cell	1500 to 3000	Can persist many years	Involved in immune protection, either attacking cells directly or producing antibodies
Platelets		Fragments of megakaryocytes; appear as small dark-staining granules	250,000	5–10 days	Play several key roles in blood clotting

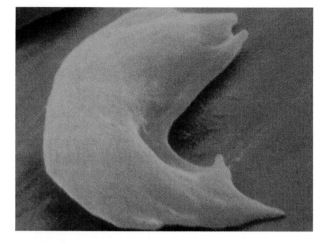

FIGURE 7-3 Sickle-Cell Anemia Scanning electron micrograph of a sickle cell.

This permits them to pass through the many miles of tiny capillaries whose internal diameters are often smaller than the RBCs.

The flexibility of RBCs is an adaptation that serves us well. However, not all people are so fortunate. According to various estimates, approximately one of every 500–1000 African Americans suffers from a disease called **sickle-cell anemia** (ah-NEEM-ee-ah). This disease results in a marked decrease in the flexibility of RBCs and is caused by a genetic mutation—a slight alteration in the hereditary material (DNA) (Chapter 16). This mutation, in turn, results in the production of hemoglobin that contains one incorrect amino acid. This small defect dramatically alters the three-dimensional structure of the hemoglobin molecule, causing RBCs to transform from biconcave disks into sickle-shaped cells when they encounter low levels of oxygen in capillaries (**FIGURE 7-3**). Sickle-shaped cells are considerably less flexible than the biconcave discs and are unable to bend and twist. As a result, the sickle cells collect at branching points in capillary beds like logs in a logjam. This blocks blood flow, disrupting the supply of nutrients and oxygen to tissues and organs. This, in turn, results in a condition known as **anoxia** (ah-NOCKS-ee-ah). Anoxia causes considerable pain. Because the lack of oxygen can kill body cells, blockages in the lungs, heart, and brain can be life-threatening. Blockages in the heart and brain often lead to heart attacks and serious brain damage. Many people who have sickle-cell anemia die in their late twenties and thirties; some die even earlier.

On average, RBCs live about 120 days. At the end of their life span, the liver and spleen remove the aged RBCs from circulation. The iron contained in the hemoglobin, however, is recycled by these organs and used to produce new RBCs in the red bone marrow. The recycling of iron is not 100% efficient, however, so small amounts of iron must be ingested each day in the diet. Loss of blood from an injury or, in women, during menstruation increases the body's demand for iron. Without adequate iron intake, oxygen transport may become impaired.

Human RBCs are highly specialized "cells" that lose their nuclei and organelles during cellular differentiation (**FIGURE 7-2A**). Because of this, RBCs cannot divide to replace themselves as they age. In humans, new RBCs are produced in the bone marrow. In the bone marrow, RBCs are produced by **stem cells,** undifferentiated cells that trace back to embryonic development. These cells give rise to 2 million RBCs per second!

In infants and children, almost all of the bone marrow is involved in the production of RBCs. As growth slows, however, the red marrow of many bones becomes inactive and gradually fills with fat cells and becomes **yellow marrow,** a fat storage depot. By the time an individual reaches adulthood, only a few bones, such as the hip bones, sternum (breastbone), ribs, and vertebrae are still engaged in RBC production. In severe, prolonged anemia, however, yellow marrow can be converted back into active red marrow.

The number of RBCs in the blood remains more or less constant over long periods. Maintaining a constant concentration of RBCs is essential to homeostasis. This process is controlled by a hormone with a rather forbidding name, **erythropoietin** (eh-RITH-row-po-EAT-in).

Erythropoietin is produced by cells in the kidney when blood oxygen levels decline—for example, when a person moves to high altitudes or loses a significant amount of blood in an accident. In the red bone marrow, this hormone stimulates the stem cells to divide and multiply, thus increasing RBC production. As the RBC concentration increases, oxygen supplies increase. When oxygen levels return to normal, erythropoietin levels fall, reducing the rate of RBC formation in a classical negative feedback mechanism so common among the body's homeostatic mechanisms.

Hemoglobin Is an Oxygen-Transporting Protein Found in RBCs.

Hemoglobin (HEME-oh-GLOBE-in) is a large protein molecule composed of four subunits. It is found exclusively in the RBCs. As shown in **FIGURE 7-4**, each hemoglobin subunit contains a **heme group** consisting of a large, organic ring structure, called a **porphyrin ring** (POUR-for-in). In the center of the ring is an iron ion.[4]

When blood from the pulmonary arteries flows through the capillary beds of the lungs, oxygen diffuses into the blood, then into the RBCs. Inside the RBC,

---

[4]The iron is Fe^{++}, the ferrous ion. To be effective, iron supplements should contain ferrous iron.

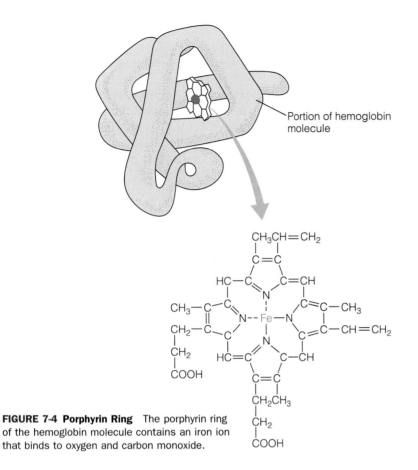

FIGURE 7-4 Porphyrin Ring   The porphyrin ring of the hemoglobin molecule contains an iron ion that binds to oxygen and carbon monoxide.

oxygen binds to the iron in hemoglobin molecules for transport through the circulatory system. In fact, 98% of the oxygen in the blood is transported bound to iron in hemoglobin. Only 2% is dissolved in the blood plasma.

Carbon dioxide, a waste product of cellular respiration, also binds to hemoglobin, but to a much lesser degree. Most carbon dioxide molecules react with water in RBCs to form bicarbonate ions (HCO_3^-). This reaction is catalyzed by the enzyme **carbonic anhydrase** (car-BON-ick an-HIGH-drace) inside the RBCs. Most of the bicarbonate ions then diffuse out of the RBCs and are transported in the plasma.

Anemia Is a Condition Resulting in a Decrease in the Ability of Blood to Transport Oxygen.

Homeostasis depends on the normal operation of the heart and blood vessels. It also requires that the blood be able to absorb a sufficient amount of oxygen as it passes through the lungs. Unfortunately, the oxygen-carrying capacity of the blood can be impaired by a number of conditions. A reduction in the oxygen-carrying capacity of the blood is known as **anemia** and may result from (1) a decrease in the number of circulating RBCs, (2) a reduction in the RBC hemoglobin content, or (3) the presence of abnormal hemoglobin in RBCs.

The causes of anemia are many. We will consider only a few. First, the number of RBCs in the blood may decline because of excessive bleeding or because of the presence of a tumor in the red marrow that reduces RBC production. Several infectious diseases (such as malaria) also decrease the RBC concentration in the blood. A reduction in the amount of hemoglobin in RBCs may be caused by iron deficiency or a deficiency in vitamin B-12, protein, and copper in the diet. Abnormal hemoglobin is produced in sickle-cell anemia and other genetic disorders.

Anemia generally results in weakness and fatigue. Individuals are often pale and tend to faint or become short of breath easily. People suffering from anemia often have an increased heart rate, because the heart beats faster to offset the reduction in the oxygen-carrying capacity of the blood. As a rule, anemia is not a life-threatening condition. However, it does weaken one's resistance to other diseases or injuries and also limits a person's productivity and energy level. Therefore, no matter what the cause, anemia should be treated quickly.

White Blood Cells Are a Diverse Group That Protects the Body from Infections

White blood cells (WBCs), or **leukocytes,** are nucleated cells that are part of the body's protective mechanism, a homeostatic system that combats harmful microorganisms such as bacteria and viruses. White blood cells are produced in the bone marrow and circulate in the bloodstream, but constitute less than 1% of the blood volume.

Interestingly, although WBCs are officially blood cells, they do most of their work *outside* of the bloodstream, in the tissues. The blood and circulatory system merely serve to transport the WBCs to sites of infection. When WBCs arrive at the "scene," they escape through the walls of the capillaries by "squeezing" between the endothelial cells (**FIGURE 7-5**).

TABLE 7-2 lists and describes the five types of WBCs found in the blood. The three most numerous, which are discussed here, are neutrophils, monocytes, and lymphocytes.

Neutrophils (NEW-trow-fills) are the most abundant of the WBCs. Approximately twice the size of the RBC, these cells are distinguished by their multi-lobed nuclei. So named because their cytoplasm has a low affinity for stains, neutrophils circulate in the blood like a cellular police force awaiting microbial invasion. At-

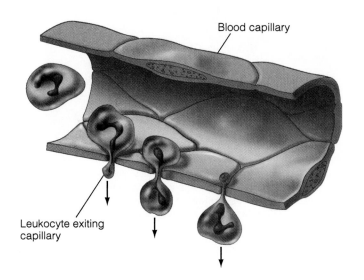

FIGURE 7-5 **White Blood Cell Exit** White blood cells (leukocytes) escape from capillaries by squeezing between endothelial cells.

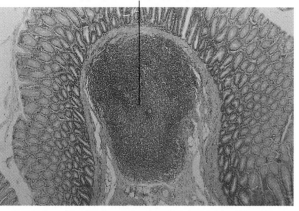

Aggregation of lymphocytes

FIGURE 7-6 **Lymphoid Tissue** The loose connective tissue beneath the lining of the large intestine and other sites is often packed with lymphocytes that have proliferated in response to invading bacteria.

tracted by chemicals released by infected tissue, neutrophils escape from the bloodstream, then migrate to the site of infection by amoeboid movement. Neutrophils are usually the first WBCs to arrive on the scene. When they arrive, they immediately begin to engulf (phagocytize) microorganisms, thereby preventing the spread of bacteria and other organisms from the site of invasion. When a neutrophil's lysosomes are used up, the cell dies and becomes part of the yellowish liquid, or **pus,** that exudes from wounds. Pus is a mixture of dead neutrophils, cellular debris, and bacteria, both living and dead.

Monocytes (MON-oh-sites) are also phagocytic cells. Slightly larger than neutrophils, monocytes contain distinctive U-shaped or kidney-shaped nuclei. Like neutrophils, monocytes leave the bloodstream to do their "work" and migrate through body tissues via amoeboid motion. Once at the site of an infection, they begin phagocytizing microorganisms, dead cells, and dead neutrophils. Thus, while neutrophils are the "first-line" troops, monocytes constitute something of a mop-up crew.

Monocytes also take up residence in connective tissues of the body, where they are referred to as macrophages (MACK-row-FAY-ges). These cells remain more or less stationary, like watchful soldiers ever ready to attack and phagocytize invaders.

The second most abundant WBC is the **lymphocyte.** Although they are found in the blood, most lymphocytes exist outside the circulatory system in lymphoid organs (LIM-foid). This includes organs such as

the spleen, thymus, and lymph nodes, and lymphoid tissue. Aggregations of lymphocytes also exist beneath the lining of the intestinal and respiratory tracts. It is in these locations that lymphocytes attack microbial intruders (**FIGURE 7-6**). Unlike neutrophils and monocytes, these cells are not phagocytic.

Two types of lymphocytes are found in the body. The first type is the **T lymphocyte,** or **T cell.** T cells attack foreign cells such as fungi, parasites, and tumor cells directly.[5] They are thus said to provide cellular immunity.

The second type is called the **B lymphocyte,** or **B cell.** When activated, B cells transform into another kind of cell, known as the **plasma cells.** Plasma cells, in turn, synthesize and release antibodies, proteins that circulate in the blood and bind to foreign substances. The binding of antibodies to "foreigners" coats them and neutralizes them. Or, it marks microorganisms and tumor cells for destruction by macrophages (Chapter 8).

Like the other formed elements of blood, the WBCs are involved in homeostasis. Their numbers can increase greatly during a microbial infection and other diseases. In fighting off a disease, they help return the body to normal function.

An increase in the number of WBCs, called **leukocytosis** (lew-co-sigh-TOE-siss), is a normal homeostatic

---

[5]The T lymphocytes usually attack large eukaryotic cells such as fungi and parasites. Most bacteria are controlled by antibodies. Only the few bacteria that are intracellular parasites, such as *M. tuberculosis,* are attacked by T cells, but even then, the lymphocyte attacks the host cell, not the bacterium directly.

response to intruders. It ends when the microbial invaders have been destroyed. Increases and decreases in various types of WBCs can be used to diagnose many medical disorders. For example, a dramatic increase in lymphocytes and lower abdominal pain are usually signs of appendicitis (Chapter 5). Because variations in the WBC count accompany many diseases, a blood test is a standard procedure for patients undergoing diagnostic testing.

Diseases Involving WBCs Also Affect the Body's Internal Balance.

Leukemia
www.jbpub.com/humanbiology

Like many other components of homeostasis, WBCs can malfunction. For example, some WBCs can become cancerous, dividing uncontrollably in the bone marrow, then entering the bloodstream. A cancer of WBCs is called **leukemia** (lew-KEEM-ee-ah; literally, "white blood"). The most serious type of leukemia is acute leukemia, so named because it kills victims quickly. Children are the primary victims of this disease.

In acute leukemia, WBCs fill the bone marrow, crowding out the cells that produce RBCs and platelets. This results in a decline in the production of RBCs, which leads to anemia. It also results in a reduction in platelet production, which reduces clotting and increases internal bleeding. Making matters worse, the cancerous WBCs produced in leukemia are often incapable of fighting infection. Leukemia patients typically succumb to infections and internal bleeding.

Infectious
mononucleosis
www.jbpub.com/humanbiology

FIGURE 7-7 Rosy Periwinkle   Many tropical plants contain chemicals that provide extraordinary medical benefits. One substance from the rosy periwinkle has helped physicians treat leukemia. Unfortunately, the tropical rain forests are being cut at an alarming rate, reducing our chances of finding other cures.

Leukemia can be treated by irradiating the bone marrow and by administering a drug called vincristine (VIN-chris-teen) that stops mitosis. Twenty years ago, only one of every four children with leukemia survived. Today, thanks to vincristine, three of every four children with the disease survive! Vincristine was discovered in a plant known as the rosy periwinkle found in tropical rain forests (**FIGURE 7-7**). Thousands of other drugs have come from other tropical plants, underscoring the importance of protecting the rain forests (**FIGURE 7-8**).

Another common disorder of the WBCs is **infectious mononucleosis** (MON-oh-NUKE-clee-OH-siss), commonly called "mono" or "kissing disease." Mono is

(a)

FIGURE 7-8 Forest Destruction   (a) Rampant deforestation not only leaves the tropics denuded, but also robs humankind of many potential cures for disease. (b) The loss of forests and other habitats in the industrial countries is also of grave importance. The bark of this tree, the Pacific yew, contains a chemical that may prove effective in fighting breast cancer. However, each year thousands of acres of old-growth forest, where the tree lives, are cut down.

(b)

caused by a virus transmitted through saliva and may be spread by kissing; by sharing silverware, plates, and drinking glasses; and possibly even through drinking fountains. The virus spreads through the body. In the blood, the virus infects only lymphocytes in the bloodstream; however, the number of monocytes and lymphocytes both increase rapidly during an infection. Individuals suffering from mono complain of fatigue, aches, sore throats, and low-grade fever. Physicians recommend that victims get plenty of rest and drink lots of liquids while the immune system eliminates the virus. Within a few weeks, symptoms generally disappear, although weakness may persist for two more weeks.

Platelets Are a Vital Component of the Blood-Clotting Mechanism

The capillaries are rather delicate structures. Even minor bumps and scrapes can cause them to leak. Leakage is normally prevented by blood clotting, surely one of the most intricate homeostatic systems to have evolved. A simplified version is discussed here.

Among the most important agents of blood clotting are the **platelets,** tiny formed elements produced in the bone marrow by fragmentation of a huge cell known as the megakaryocyte (MEG-ah-CARE-ee-oh-site) (**FIGURE 7-9**). Like RBCs, platelets lack nuclei and organelles and therefore are not true cells. Also like RBCs, platelets are unable to divide. Carried passively in the bloodstream, platelets are coated by a layer of a sticky material, which causes them to adhere to irregular surfaces such as tears in blood vessels or atherosclerotic plaque.

Clotting is a chain reaction stimulated by the release of a chemical called **thromboplastin** (THROM-

bow-PLASS-tin) from injured cells lining damaged blood vessels (**FIGURE 7-10A**). Thromboplastin is a lipoprotein that acts on an inactive plasma enzyme in the blood known as **prothrombin** (produced by the liver). Thromboplastin causes prothrombin to be converted into its active form, **thrombin.** Thrombin, in turn, acts on another blood protein, **fibrinogen,** also produced by the liver. When activated, fibrinogen is converted into **fibrin,** long, branching fibers that produce a weblike network in the wall of the damaged blood vessel (**FIGURE 7-10B**). The fibrin web traps RBCs and platelets, forming a plug that stops the flow of blood to the tissue. Platelets captured by the fibrin web release additional thromboplastin, known as **platelet thromboplastin,** which causes more fibrin to be laid down, thus reinforcing the fibrin network.

Blood clotting occurs fairly quickly. In most cases, a damaged blood vessel is sealed by a clot within 3–6 minutes of an injury; 30–60 minutes later, platelets in the clot begin to draw the clot inward, stitching the wound together. How do they perform this remarkable task?

Platelets contain contractile proteins like those in muscle cells. Contraction of the protein fibers draws the fibrin network inward, pulling the edges of the cut or damaged blood vessel together. This closes the wound like a surgical stitch.

Blood clots do not stay in place indefinitely. If they did, the circulatory system would eventually become clogged, and blood flow would come to a halt. Instead, clots are dissolved by a blood-borne enzyme known as **plasmin** (PLAZ-min), which is produced from an inactive form, **plasminogen** (plaz-MIN-oh-gin). Plasminogen is incorporated in the clot as it forms. It is then gradually converted to plasmin by an activating factor secreted by the endothelial cells of the blood vessel. This ensures that the plasmin dissolves the clot after the blood vessel damage has been repaired.

As important as blood clots are in protecting the body, clotting can also cause problems. For example, blood clots can break loose from their site of formation, circulate in the blood, and become lodged in arteries, especially those narrowed by plaque. In such cases, the blood clots restrict blood flow, causing considerable damage to tissues served by the vessel. Blood clots typically lodge in the narrow vessels serving the heart muscle, brain, and other vital organs.

Clotting Disorders Not Only Upset the Homeostatic Balance, They Can Be Life-Threatening.

In some individuals, blood clotting is impaired because of: (1) an insufficient number of platelets; (2) liver

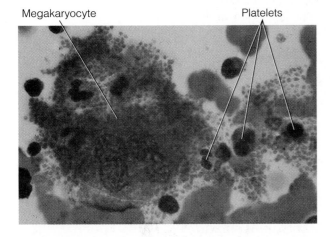

FIGURE 7-9 Megakaryocyte A light micrograph of a megakaryocyte, a large, multinucleated cell found in bone marrow; the megakaryocyte fragments, giving rise to platelets.

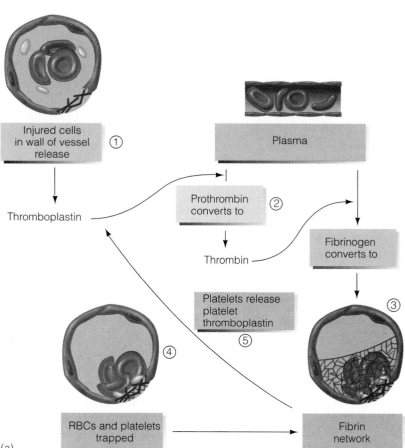

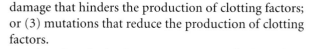

(a)

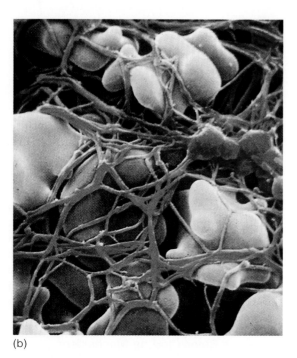

(b)

FIGURE 7-10 Blood Clotting Simplified *(a)* Injured cells in the walls of blood vessels release the chemical thromboplastin (1). Thromboplastin stimulates the conversion of prothrombin, found in the plasma, into thrombin (2). Thrombin, in turn, stimulates the conversion of the plasma protein fibrinogen into fibrin (3). The fibrin network captures RBCs and platelets (4). Platelets in the blood clot release platelet thromboplastin (5), which converts additional plasma prothrombin into thrombin. Thrombin, in turn, stimulates the production of additional fibrin. *(b)* A scanning electron micrograph of a fibrin clot that has already trapped platelets and RBCs, plugging a leak in a vessel. The RBCs are yellow, and the fibrin network is red.

damage that hinders the production of clotting factors; or (3) mutations that reduce the production of clotting factors.

A reduced platelet count may result from leukemia (noted earlier). It may also result from an exposure to excess radiation, which damages bone marrow where the megakaryocytes reside. Liver damage, which impairs the production of blood-clotting factors, can be caused by hepatitis, liver cancer, or excessive alcohol consumption. The most common genetic defect is **hemophilia** (he-moe-FEAL-ee-ah), a disease in which the liver fails to produce the necessary clotting factors. Problems begin early in life, and even tiny cuts or bruises can bleed uncontrollably, threatening one's life. Because of repeated bleeding into the joints, victims suffer great pain and often become disabled; they often die at a young age.

Hemophiliacs can be treated by transfusions of blood-clotting factors. This therapy, however, is expensive and is required every few days. It has also put hemophiliacs at risk for AIDS, acquired immunodeficiency syndrome, which is caused by a virus transmitted primarily by sexual contact (Chapter 8). The AIDS virus invades certain white blood cells, known as **helper T cells,** resulting in a gradual deterioration of the immune function. For years, AIDS has been considered 100% fatal—that is, everyone who gets it eventually dies. Although it is too early to tell, new treatments may change the prognosis.

Unfortunately, testing for the AIDS virus began late, and many transfusions of whole blood, blood plasma, and clotting factors were contaminated by the virus. Clotting factors produced by genetic engineering, however, have eliminated the need for transfusions of clotting agents taken from whole blood, reducing the risk to hemophiliacs of contracting AIDS.

Hemophilia
www.jbpub.com/humanbiology

Health, Homeostasis, and the Environment

Carbon Monoxide

The blood is vitally important to homeostasis. It transports materials, especially oxygen, to and from the cells. It buffers changes in pH. It transports excess heat to the body's surface, where it is eliminated. It plays a key role in the body's defense system, and it seals injuries in blood vessels through clots.

It is no surprise then to learn that cells depend mightily on the proper functioning of blood. Anything that upsets this function can upset cellular function. A good example is carbon monoxide gas, a pollutant produced from many sources in our society.

Carbon monoxide (CO) is a colorless, odorless gas produced by the incomplete combustion of organic fuels. It emanates from stoves and furnaces in our homes and spews out of the tailpipes of our automobiles, causing increased levels along streets and highways, in parking garages, and in tunnels. It pours out of power plants and factory smokestacks, polluting the air in our cities. It is even a major pollutant in tobacco smoke. According to estimates by the Environmental Protection Agency, over 40 million Americans (one of every six people) are exposed to levels of CO harmful to their health.

What makes CO so dangerous? Carbon monoxide, like oxygen, binds to hemoglobin. However, hemoglobin has a much greater affinity for CO than for oxygen (about 200 times greater). Consequently, CO "outcompetes" oxygen for the binding sites on the hemoglobin molecules and reduces the blood's ability to carry oxygen, which is essential for energy production.

For healthy people, low levels of CO in their blood do not create much of a problem. Their bodies simply produce more RBCs or increase the heart rate to augment the flow of oxygen to tissues. CO does create a problem, however, when levels are so high that the body cannot compensate, and at very high levels, CO becomes deadly.

New research suggests that CO levels currently deemed acceptable by federal standards can trigger chest pain (angina) in active adults with coronary artery disease. Chest pains result from a lack of oxygen in the heart muscle. Levels such as those once thought safe in many workplaces or levels in the blood encountered after 1 hour in heavy traffic cause angina and abnormal ECGs in these people.

In levels commonly found in and around cities, CO is also troublesome for the elderly. Carbon monoxide places additional strain on their heart, making the already weakened organ work harder. Thus, the elderly and people suffering from cardiovascular disease are advised to stay inside on high-pollution days.

Carbon monoxide is just one of many pollutants we breathe. No one knows what, if any, long-term effects it might have on our health. In the short term, however, it is clear that CO upsets oxygen delivery, which is essential to homeostasis in all of us, and especially in the elderly and the infirm. It has become another risk factor in an increasingly risky society, but it is not an insolvable problem. Much more efficient automobiles, widespread use of mass transit, overall reductions in driving, and alternative automobile fuels (such as hydrogen) can all reduce the level of CO in and around the cities. More efficient factories, home furnaces, and power plants can also reduce CO levels in emissions. Passive-solar homes and improvements in insulation can also cut our demand for fossil fuel energy. Bans on smoking indoors could go a long way, too, in ridding our air of this transparent killer.

SUMMARY

BLOOD: ITS COMPOSITION AND FUNCTIONS

1. *Blood* is a watery tissue consisting of two basic components: the *plasma*, a fluid that contains dissolved nutrients, proteins, gases, and wastes, and the *formed elements*, red blood cells, white blood cells, and platelets, which are suspended in the plasma.

2. The functions of the formed elements are listed in **TABLE 7-2**.

3. The plasma constitutes about 55% of the volume of a person's blood, and the formed elements make up the remainder. The volume occupied by the blood cells and platelets is called the *hematocrit*.

4. *Red blood cells (RBCs)* are highly specialized cells that lack nuclei and organelles. Produced by the red bone marrow, RBCs transport oxygen in the blood.

5. The concentration of RBCs in the blood is maintained by the hormone *erythropoietin*, produced by the kidneys in response to falling oxygen levels.

6. *White blood cells (WBCs)* are nucleated cells and are part of the body's protective mechanism to combat microorganisms. WBCs are produced in the bone marrow and circulate in the bloodstream but do most of their work outside it, in the body tissues.

7. The most abundant WBCs are the *neutrophils*, which are attracted by chemicals released from infected tissue. Neutrophils leave the bloodstream and migrate to the site of infection by amoeboid movement.

8. Neutrophils are the first WBCs to arrive at the site of an infection, where they phagocytize microorganisms, preventing the spread of bacteria and other organisms.

9. The second group of cells to arrive are the *monocytes*, which phagocytize microorganisms, dead cells, cellular debris, and dead neutrophils.

10. *Lymphocytes* are the second most numerous WBCs and play a vital role in immune protection.
11. *Platelets* are fragments of large bone marrow cells and are involved in blood clotting.
12. Platelets are coated by a layer of a sticky material, which causes them to adhere to irregular surfaces such as tears in blood vessels. Blood clotting is summarized in **FIGURE 7-10**.

HEALTH, HOMEOSTASIS, AND THE ENVIRONMENT: CARBON MONOXIDE

13. Carbon monoxide is produced by the incomplete combustion of organic fuels in our homes, automobiles, factories, and power plants.
14. Carbon monoxide binds to hemoglobin and reduces the oxygen-carrying capacity of the blood, which impairs cellular energy production, upsetting homeostasis.
15. At high concentrations, carbon monoxide can be lethal. At lower concentrations, it is harmful to people with cardiovascular disease and lung disease. For individuals with heart disease, it puts additional strain on the heart.

Critical Thinking

THINKING CRITICALLY— ANALYSIS

This Analysis corresponds to the Thinking Critically scenario that was presented at the beginning of this chapter.

The researchers themselves freely admit that this study only looks at the effect of lipoprotein(a) on smooth muscle cells grown in laboratory culture dishes. They have not yet determined whether the effect occurs in the body itself.

Caution is advised when examining studies of cells and tissues in culture. Although studies in tissue culture are essential to medical science, it's important to remember that a culture dish is an artificial environment. Because of this, such studies may lead to erroneous conclusions. This exercise requires one to scrutinize the experimental method.

EXERCISING YOUR CRITICAL THINKING SKILLS

The stem cell in bone marrow is the subject of intense research. Scientists believe that by manipulating the cell they can find ways to better treat—perhaps even cure—a number of diseases such as cancer, sickle-cell anemia, and AIDS. In order to manipulate the cell, however, scientists recognize that they need to know more about cell replication and differentiation. Two schools of thought exist on the subject. The first says that the stem cell is subject to external influences such as hormones, which direct its differentiation. The second school of thought contends that replication and differentiation occur randomly—that is, that there are no outside controls. At this point, can you see any problems in the way the debate is framed? You may want to review the critical thinking rules in Chapter 1.

If after reviewing the debate and the critical thinking rules, you said the problem might be that researchers may be oversimplifying matters, you have pinpointed the problem. Researchers seem to have fallen into the dualistic thinking trap and may have set up a false dichotomy. It's quite possible that outside influences and random replication and differentiation both occur.

In an effort to settle this debate, Makio Ogawa of the Medical University of South Carolina devised a method for growing human stem cells in a gel. He removed single cells from the stem cell colonies and used them to start new colonies. He found that the cells in the secondary colonies developed into a variety of cell types. That is, they appeared to differentiate randomly along several lines. This finding is taken to be support of the second hypothesis cited above.

Think about the experiment for a few moments. Can you find any weaknesses in its design?

If you said that the researcher isolated the cells from outside influences, such as hormones, that might be present in bone marrow and that might influence differentiation, you're right. Cell culture studies—that is, studies of cells in culture dishes—although extremely useful, are often criticized because they do not expose cells to potentially important influences such as hormones. The fact that stem cells differentiate into a variety of types in cell culture does not mean that the endocrine system is not influential in some way.

Can you think of any ways to solve this problem—to study cell differentiation while preserving potential hormonal influences?

TEST OF CONCEPTS

1. Describe the structure and function of each of the following: red blood cells, platelets, lymphocytes, monocytes, and neutrophils.

2. Define each of the following terms: leukemia, anemia, and infectious mononucleosis.

3. Explain how a blood clot forms and how it helps prevent bleeding.

4. Describe the many ways blood participates in homeostasis. How can these homeostatic mechanisms be upset?

TOOLS FOR LEARNING

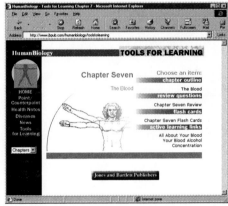

www.jbpub.com/humanbiology

Tools for Learning is an on-line student review area located at this book's web site HumanBiology (www.jbpub.com/humanbiology). The review area provides a variety of activities designed to help you study for your class:

Chapter Outlines. We've pulled out the section titles and full sentence sub-headings from each chapter to form natural descriptive outlines you can use to study the chapters' material point by point.

Review Questions. The review questions test your knowledge of the important concepts and applications in each chapter. Written by the author of the text, the review provides feedback for each correct or incorrect answer. This is an excellent test preparation tool.

Flash Cards. Studying human biology requires learning new terms. Virtual flash cards help you master the new vocabulary for each chapter.

Figure Labeling. You can practice identifying and labeling anatomical features on the same art content that appears in the text.

Active Learning Links. Active Learning Links connect to external web sites that provide an opportunity to learn basic concepts through demonstrations, animations, and hands-on activities.

THE IMMUNE SYSTEM

Electron micrograph of a macrophage
binding to three tumor cells.

Thinking Critically

You're the parent of a young
child and are debating whether to have your child vaccinated. Some friends tell
you that it's dangerous—that in the past some children have been paralyzed by
vaccinations. They also tell you that vaccination damages the immune system
and that there's no need to subject a child to this treatment because the truly
dangerous infectious diseases have been eliminated.

After reading this chapter, how would you respond? Is there merit to
the claim that a vaccination damages the immune system? Have the diseases
been eliminated? Do you have enough information to make an informed
decision? ■

EACH DAY AS YOU VENTURE OUTSIDE YOUR home, you are entering a war zone. Floating in the air and circulating in the water are billions upon billions of microorganisms, viruses, and bacteria. Although many are innocuous, some are not. Fortunately, humans and other multicellular organisms, which have evolved in a world teeming with viruses and single-celled organisms such as bacteria, have also evolved mechanisms to protect themselves against potentially harmful adversaries. This chapter examines three lines of defense against disease-causing (pathogenic) microorganisms. It also looks at defenses against cancer and discusses AIDS. To set the stage for understanding the body's protective mechanisms, we begin with a brief overview of infectious agents.

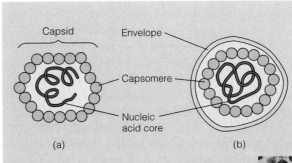
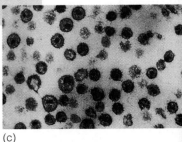
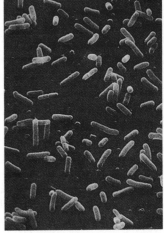

FIGURE 8-1 General Structure of a Virus *(a)* The virus consists of a nucleic acid core of either RNA or DNA. Surrounding the viral core is a layer of protein known as the capsid. Each protein molecule in the capsid is known as a capsomere. *(b)* Some viruses have an additional protective coat known as the envelope. *(c)* Electron micrograph of the Human Immunodeficiency Virus (HIV).

Capsid	Envelope
Capsomere	
Nucleic acid core	
(a)	(b)
	(c)

Section 8-1

Viruses and Bacteria: An Introduction

Two of the most important infectious agents are viruses and bacteria. **Viruses** consist of a nucleic acid core, consisting of either DNA or RNA, and an outer protein coat, the **capsid** (**FIGURE 8-1**). Some viruses have an outer **envelope** that lies outside the capsid and is structurally similar to the plasma membranes of eukaryotic cells.

Viruses can be likened to pirates because, after they invade cells, they commandeer their metabolic machinery in much the same way that pirates take over ships. In the process, viruses convert host cells into miniature virus factories. However, in the process of taking over a cell's metabolic machinery, they often kill their "hosts."

Viruses most often enter the body through the respiratory and digestive systems and spread from cell to cell in the bloodstream and lymphatic system. However, other avenues of entry are also possible—for example, sexual contact.

The immune system (discussed more fully later) kills many viruses. However, some viruses evade it and take refuge in body cells, reemerging under stress or some other influence. One example is the virus responsible for genital herpes (HER-pees), tiny sores that periodically emerge on the genitals, thighs, and buttocks of infected men and women.

Viruses are not considered living organisms, because they cannot reproduce on their own, and unless they've invaded a host cell, they cannot metabolize. In contrast, **bacteria** (singular, bacterium) are living organisms, for they are capable of reproducing without taking over host cells. As pointed out in Chapter 3, bacteria are prokaryotes. These simple cells consist of a circular strand of DNA, not bound by any kind of membrane, and cytoplasm, which is enveloped by a plasma membrane (**FIGURE 8-2**). Outside the plasma membrane is a thick, rigid cell wall. Many bacteria also contain tiny circular pieces of extrachromosomal DNA, called plasmids.

Although best known for their role in causing sickness and death, most bacteria perform useful functions. Soil bacteria, for example, help recycle nutrients in rotting plants and animals. Although this may not seem like a glamorous occupation, it's absolutely essential to the continuation of life on Earth. Without these bacteria, biological systems would quickly become impoverished. Some of the biologically useful bacteria will be discussed in Chapter 23.

This chapter concerns itself primarily with harmful bacteria that invade the

FIGURE 8-2 General Structure of a Bacterium *(a)* Bacteria come in many shapes and sizes, but all have a circular strand of DNA, cytoplasm, and a plasma membrane. Surrounding the membrane of many bacteria is a cell wall. *(b)* Electon micrograph of samonella bacteria.

- Cell wall
- Cytoplasm
- Plasma membrane
- Circular DNA

(a)

(b)

human body and the body's defensive mechanisms. The mechanisms that evolved to protect multicellular organisms such as humans from the hordes of potentially harmful microorganisms fall into three categories—or three lines of defense.

The First and Second Lines of Defense

In Humans, the First Line of Defense Consists of the Skin; Epithelial Linings of the Respiratory, Digestive, and Urinary Systems; and Body Secretions That Destroy Harmful Microorganisms

The human body is like a fort with an outer barrier that wards off potential invaders. One component of this outer wall is the skin, a protective layer that repels many potentially harmful microorganisms.

As noted in Chapter 4, the skin consists of a relatively thick and impermeable layer of epidermal cells overlying the rich vascular layer, the dermis. Epidermal cells are produced by cell division in the base of the epidermis. As the basal cells proliferate, they move outward, become flattened, and die. The dead cells contain a protein called **keratin** (CARE-ah-tin). Keratin forms a fairly waterproof protective layer that not only reduces moisture loss, but also protects underlying tissues from microorganisms. The cells of the epidermis are joined by special structures known as **tight junctions**, which impede water loss and microbial penetration.

The skin protects the outer surfaces of the human body, but the human body contains three passageways that penetrate into its interior, making protection difficult. The passages into the interior are the respiratory, digestive, and urinary tracts. All three of them are protected by epithelial linings, which usually keep potentially harmful microorganisms from invading the underlying tissues. A break in these linings, however, permits microorganisms to enter.

The body's first line of defense consists of more than passive physical barriers. It also actively engages in "chemical warfare." For instance, the skin produces slightly acidic secretions that impair bacterial growth. The stomach lining produces hydrochloric acid, which destroys many ingested bacteria. Tears and saliva contain an enzyme called **lysozyme** (LIE-so-zime) that dissolves the cell wall of bacteria, killing them. Cells in the lining of the respiratory tract produce mucus, which has antimicrobial properties. These protective mechanisms

are all nonspecific. Like a castle wall, they operate indiscriminately against all invaders.

The Body's Second Line of Defense Combats Infectious Agents That Penetrate the First Line and Consists of Cellular and Chemical Responses

The first line of defense is not impenetrable. Even tiny breaks in the skin or in the lining of the respiratory, digestive, or urinary tracts permit viruses, bacteria, and other microorganisms to enter the body. Fortunately, a second line of defense exists. It involves a host of chemicals and cellular agents that work together to combat invaders. Like the first line, these mechanisms are nonspecific. This section discusses four components of the second line of defense: the inflammatory response, pyrogens, interferons, and complement.

The Inflammatory Response Is a Major Part of the Second Line of Defense and Involves Chemical and Cellular Responses.
Damage to body tissues triggers a series of reactions collectively referred to as an **inflammatory response.** Not a string of expletives released in anger, the inflammatory response is a protective mechanism. The word *inflammatory* refers to the heat given off by a wound. The inflammatory response is also characterized by redness, swelling, and pain.

The inflammatory response is a kind of chemical and biological warfare waged against bacteria, viruses, and other microorganisms. It begins with the release of a variety of chemical substances by the injured tissue (**FIGURE 8-3**). Some chemicals attract macrophages that reside in body tissues and neutrophils in the blood (Chapter 7). These cells phagocytize bacteria that have entered the wound. Soon after these cells begin to work, a yellowish fluid begins to exude from the wound. Called **pus,** it contains dead white blood cells (mostly neutrophils), microorganisms, and cellular debris, which accumulate at the site of inflammation.

Other chemical substances released by injured tissues cause blood vessels to dilate (expand) and leak (**FIGURE 8-3**). One such substance is **histamine** (HISS-tah-mean). Histamine stimulates the arterioles in the injured tissue to dilate, causing the capillary networks to swell with blood.[1] The increase in the flow of blood

---

[1]Histamine is produced by mast cells, platelets, and basophils, a type of white blood cell. Mast cells are a connective tissue cell described later in the chapter.

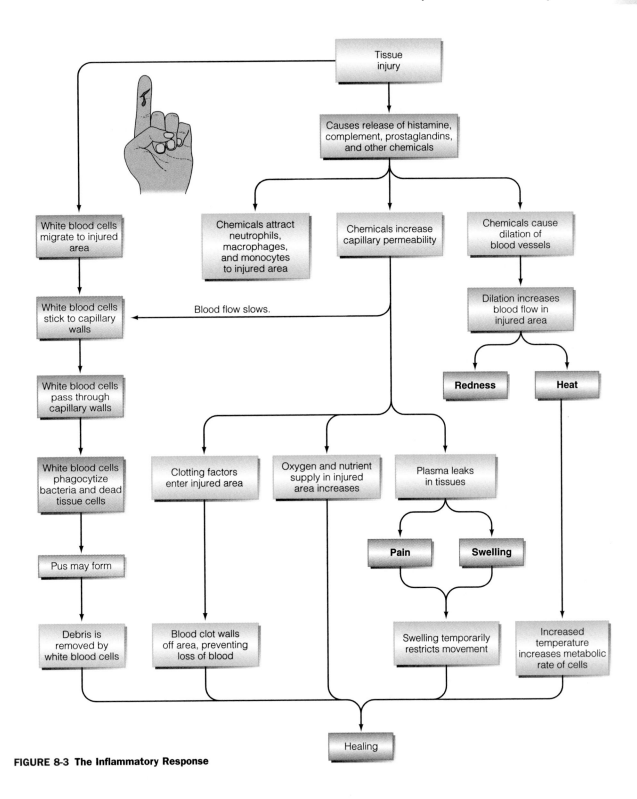

FIGURE 8-3 The Inflammatory Response

through an injured tissue is responsible for the heat and redness around a cut or abrasion. Heat, in turn, increases the metabolic rate of cells in the injured area and accelerates healing.

Still other substances released by injured tissues increase the permeability of capillaries, augmenting the flow of plasma into a wounded region. Plasma carries with it oxygen and nutrients that facilitate healing. It also carries the molecules necessary for blood clotting. As noted in Chapter 7, the clotting mechanism walls off injured vessels and reduces blood loss.

As illustrated in **FIGURE 8-3**, plasma leaking into injured tissues causes swelling, which stimulates pain receptors in the area. Pain receptors send nerve impulses to the brain. Pain also results from chemical toxins released by bacteria and from chemicals released by injured cells themselves. One important pain-causing chemical is **prostaglandin** (PROSS-tah-GLAN-din). Aspirin and other mild painkillers work by inhibiting the synthesis and release of prostaglandins.

Although it evokes pain, the flow of fluid into body tissues is helpful. Injury to joints, for example, results in local swelling that helps immobilize joints. In effect, swelling is nature's way of protecting joints and allowing tissues to mend. Inflammation even comes equipped with its own cleanup crew in the form of late-arriving monocytes that phagocytize dead cells, cell fragments, dead bacteria, and viruses. Take a moment to study **FIGURE 8-3** to review the inflammatory response.

The Second Line of Defense Consists of Three Additional Chemicals.

In a nuclear power plant, safety is ensured in part by redundant systems—that is, backup systems that kick in if the main systems fail. In the human body, redundant systems also exist, but not as a backup. In the body's protective mechanism, the redundant systems function as supplementary systems—that is, they operate in addition to the inflammatory response. Three such systems are described in this section: pyrogens, interferons, and complement.

Pyrogens (PIE-rah-gins) are molecules released primarily by macrophages that have been exposed to bacteria and other foreign substances. Pyrogens travel to a region of the brain called the **hypothalamus** (high-po-THAL-ah-mus). In the hypothalamus is a group of nerve cells that controls the body's temperature, in much the same way that a thermostat regulates the temperature of a room. Pyrogens turn the thermostat up, increasing body temperature and producing a fever.

Fever is actually an adaptation that helps combat bacterial infections. How? Mild fevers cause the spleen and liver to remove additional iron from the blood. Interestingly, many pathogenic bacteria require iron to reproduce. Fever therefore reduces the replication of bacteria. Fever also increases metabolism, which facilitates healing and accelerates cellular defense mechanisms such as phagocytosis. Important as it is, fever can also be debilitating, and a severe fever (over 105°F) is potentially life-threatening because it begins to dena-

ture vital body proteins, especially enzymes needed for biochemical reactions in body cells.

Another chemical safeguard is a group of small proteins known as the **interferons** (in-ter-FEAR-ons). Interferons are released from cells infected by viruses and bind to receptors on the plasma membranes of noninfected body cells (**FIGURE 8-4**). This, in turn, triggers the synthesis of cellular enzymes that inhibit viral replication, thus protecting the cell.

Interferons do not protect cells already infected by a virus; they simply stop the spread of viruses from one cell to another. In essence, the production and release of interferon are the dying cell's last act to protect other cells of the body. Interferons are a remarkable adaptation that stops the spread of viruses while the immune response attacks and destroys the viruses outside the cells.

Another group of chemical agents that fight infection are the complement proteins. These blood proteins form the **complement system,** so named because it complements the action of antibodies, briefly described in Chapter 7.

The details of the complement system are very complex. A few points will demonstrate how this remarkable system works. Complement proteins circulate in the blood in an inactive state. When foreign cells such as bacteria invade the body, the complement protein is

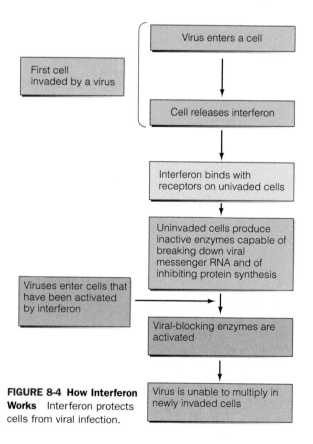

FIGURE 8-4 How Interferon Works Interferon protects cells from viral infection.

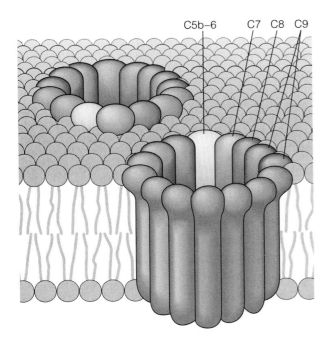

C5b–6   C7   C8   C9

FIGURE 8-5 The Membrane-Attack Complex Five complement proteins combine and embed in a cell's membrane, causing it to leak, swell, and burst. (Source: By Dana Burns from John Ding-E Young and Zanvil A. Kohn, "How Killer Cells Kill." Copyright © January 1988 by *Scientific American, Inc.* All rights reserved.)

activated. This triggers a chain reaction in which one complement protein activates the next.

Five proteins in the complement system join to form a large protein complex, known as the **membrane-attack complex** (**FIGURE 8-5**). The membrane-attack complex embeds in the plasma membrane of bacteria, creating an opening into which water flows. The influx of water causes bacterial cells to swell, burst, and die.

Several of the activated complement proteins also function on their own and are part of the inflammatory response. Some of them, for example, stimulate the dilation of blood vessels in an infected area, described earlier. Others increase the permeability of the blood vessels, allowing white blood cells and nutrient-rich plasma to pass more readily into an infected zone. Certain complement proteins may also act as chemical attractants, drawing macrophages, monocytes, and neutrophils to the site of infection, where they phagocytize foreign cells. Yet another complement protein (C3b) binds to microorganisms, forming a rough coat on the intruders that facilitates their phagocytosis.

Section 8-3

The Third Line of Defense: The Immune System

The immune system is the third line of defense. Unlike the respiratory or digestive systems, the immune sys-

tem is rather diffuse—spread out and indistinct. Lymphocytes, for example, circulate in the blood and lymph and also take up residence in the **lymphoid organs** such as the spleen, thymus, lymph nodes, and tonsils, as well as other body tissues. The cells of the immune system selectively target foreign substances and foreign organisms. As a result, the immune system is said to be specific.

The immune system, like the first and second lines of defense, is an important homeostatic mechanism that eliminates foreign organisms—including bacteria, viruses, single-celled fungi, and many parasites. It comes into play when foreign organisms penetrate the outer defenses of the body. The immune system also helps prevent the emergence of cancer cells. Thus, in a world filled with infectious agents and natural mutagens (agents that cause mutation, some of which might lead to cancer), the immune system struggles to provide the internal constancy needed by body cells to carry out their functions and, as such, is an important evolutionary advance.

Lymphocytes Detect Foreign Substances in the Body and Mount an Attack on Them

One of the chief functions of the immune system is to identify what belongs in the body and what does not. Once a foreign substance has been detected, the immune system mounts an attack to eliminate it. Therefore, like all homeostatic systems, the immune system requires receptors to detect a change and effectors to bring about a response. In the immune system, the lymphocytes serve both functions.

Foreign Substances That Trigger an Immune Response Are Proteins and Polysaccharides with Large Molecular Weights

The immune response is triggered by large foreign molecules, notably proteins and polysaccharides. These molecules are called **antigens** (AN-tah-gins), which is an abbreviation for *anti*body-*gen*erating substances. The larger the molecule is, the greater its antigenicity (AN-tah-gen-ISS-eh-tee).

As a rule, small molecules generally do not elicit an immune reaction. In some individuals, however, small, nonantigenic molecules such as formaldehyde, penicillin, and the poison ivy toxin bind to naturally occurring proteins in the body, forming complexes. These large complexes are unique compounds that are foreign to the body and are therefore capable of eliciting an immune response.

The immune system reacts to viruses, bacteria, and single-celled fungi in the body. It also responds to parasites such as the protozoan that causes malaria. Viruses, bacteria, fungi, and parasites elicit a response because they are enclosed by membranes, or coats, that contain large-molecular-weight proteins or polysaccharides—that is, antigens.

Cells transplanted from one person to another also elicit an immune response, because each individual's cells contain a unique "cellular fingerprint," resulting from the unique array of plasma membrane glycoproteins (Chapter 3). The immune system is activated by these antigens on the foreign cells. Cancer cells also present a slightly different chemical fingerprint, making them essentially foreign cells within our own bodies to which the immune system responds. Although cancer cells evoke an immune response, it is often not sufficient to stop the disease.

Antigens stimulate two types of lymphocytes: **T lymphocytes**, commonly called **T cells**, and **B lymphocytes**, also called **B cells** (Chapter 7). As you will see in later sections, B and T cells react differently and respond to different types of antigens. As a rule of thumb, B cells recognize and react to microorganisms such as bacteria and bacterial toxins, chemical substances released by bacteria. They also respond to a few viruses. When activated, B cells produce antibodies to these antigens.

In contrast, T cells recognize and respond to our own body cells that have gone awry. This includes cancer cells as well as body cells that have been invaded by viruses. T cells also respond to transplanted tissue cells and larger disease-causing agents, such as single-celled fungi and parasites. Unlike B cells, T cells attack their targets directly.

Immature B and T Cells Are Incapable of Responding to Antigens, but Soon Gain This Ability

Lymphocytes are produced in the red bone marrow and released into the bloodstream. These immature cells circulate in the blood and lymph, but are not able to function until they become immunologically competent. This process occurs in specific organs in the body. Consider the T cell.

T Cells.

Some immature, undifferentiated lymphocytes take up (temporary) residence in the **thymus** (THIGH-muss), a lymphoid organ located above the heart. In a few days, these lymphocytes mature and become functional T cells—they are said to develop **immunocompetence** (IM-you-know-COM-pah-tense)—the capacity to respond to specific antigens. T cells are so named because they become immunocompetent in the thymus.

Each cell produces a unique type of membrane receptor that will bind to one—and only one—type of antigen. Over an individual's lifetime, thousands upon thousands of antigens will be encountered. Thanks to the immunocompetence developed during fetal development, each of us is equipped with millions of uniquely programmed T cells that respond to the onslaught of antigens.

B Cells.

B cells mature and differentiate in the bone marrow.[2] Afterwards, immunologically competent B cells circulate in the blood and take up residence in connective and lymphoid tissues. They therefore become part of the body's vast cellular reserve, stationed at distant outposts, awaiting the arrival of the microbial invaders.

By various estimates, several million immunologically distinct B and T cells are produced in the body early in life. Over a lifetime, only a relatively small fraction of these cells will be called into duty.

B Cells Provide Humoral Immunity Through the Production of Antibodies

The immune response consists of two separate but related reactions: humoral immunity, provided by the B cells, and cell-mediated immunity, involving T cells (**TABLE 8-1**). Let's consider humoral immunity and the B cells first.

When an antigen first enters the body, it binds to B cells programmed during their residence in the bone marrow to respond to that particular antigen

---

[2]B cells were so named because they develop immunocompetence in a part of the chicken's digestive system known as the bursa. Humans lack this organ.

TABLE 8-1

Comparison of Humoral and Cell-Mediated Immunity

Humoral	Cell-Mediated
Principal cellular agent is the B cell.	Principal cellular agent is the T cell.
B cell responds to bacteria, bacterial toxins, and some viruses.	T cell responds to cancer cells, virus-infected cells, single-cell fungi, parasites, and foreign cells in an organ transplant.
When activated, B cells form memory cells and plasma cells, which produce antibodies to these antigens.	When activated, T cells differentiate into memory cells, cytotoxic cells, suppressor cells, and helper cells; cytotoxic T cells attack the antigen directly.

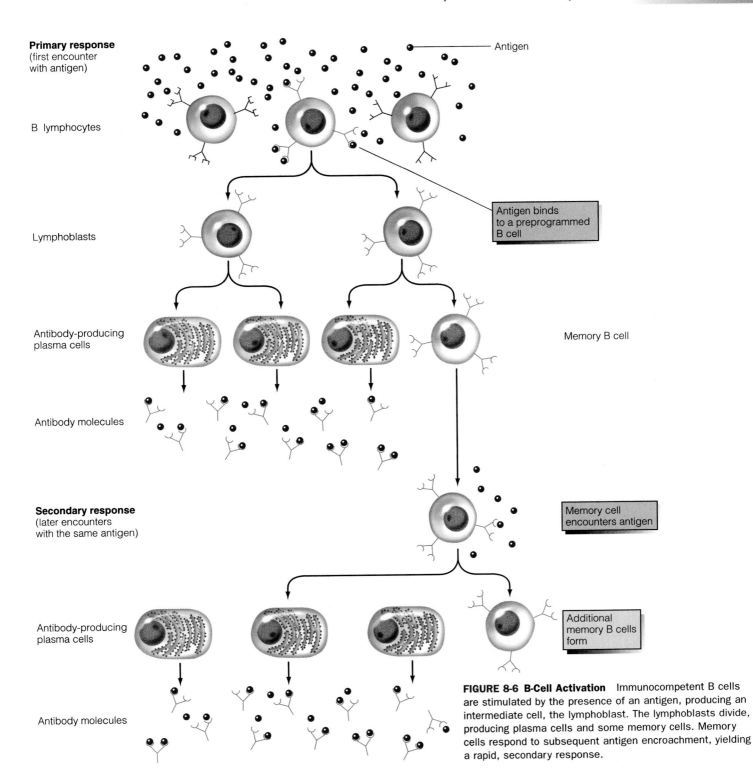

Primary response
(first encounter
with antigen)

Antigen

B lymphocytes

Antigen binds
to a preprogrammed
B cell

Lymphoblasts

Antibody-producing
plasma cells

Memory B cell

Antibody molecules

Secondary response
(later encounters
with the same antigen)

Memory cell
encounters antigen

Antibody-producing
plasma cells

Additional
memory B cells
form

Antibody molecules

FIGURE 8-6 B-Cell Activation Immunocompetent B cells
are stimulated by the presence of an antigen, producing an
intermediate cell, the lymphoblast. The lymphoblasts divide,
producing plasma cells and some memory cells. Memory
cells respond to subsequent antigen encroachment, yielding
a rapid, secondary response.

(**FIGURE 8-6**).[3] These cells soon begin to divide, producing a population of immunologically similar cells. Some
of the B cells differentiate and form another kind of cell,
the **plasma cell.** Plasma cells produce copious amounts
of antibody.

Antibodies released from plasma cells circulate in
the blood and lymph, where they bind to the antigens

[3]As you will soon see, this process is a bit more complex and
involves the macrophage.

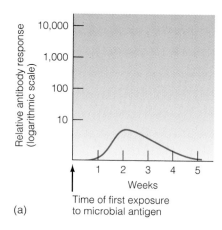

(a)

Time of first exposure to microbial antigen

Relative antibody response (logarithmic scale)

Weeks

(b)

Time of subsequent exposure to microbial antigen

Relative antibody response (logarithmic scale)

Weeks

FIGURE 8-7 Primary and Secondary Responses *(a)* The primary (initial) immune response is slow. It takes about 10 days for antibody levels to peak. Almost no antibody is produced during the first week as plasma cells are being formed. *(b)* The secondary response is much more rapid. Antibody levels rise almost immediately after the antigen invades. T cells show a similar response pattern.

that triggered the response. Because the blood and lymph were once referred to as body "humors," this arm of the protective immune response is called **humoral immunity**.

The Initial Reaction to an Antigen Is Slower and Weaker Than Subsequent Responses.

The first time an antigen enters the body, it elicits an immune response, but the initial reaction—or **primary response**—is relatively slow (**FIGURE 8-7**). During the primary response, antibody levels in the blood do not begin to rise until approximately the

beginning of the second week after the intruder was detected, partly explaining why it takes most people about 7–10 days to combat a cold or the flu. This delay occurs because it takes time for B cells to multiply and form a sufficient number of plasma cells. Antibody levels usually peak about the end of the second week, then decline over the next three weeks.

If the same antigen enters the body at a later date, however, the immune system acts much more quickly and more forcefully (**FIGURE 8-7**). This stronger reaction constitutes the **secondary response.** As **FIGURE 8-7** illustrates, during a secondary response, antibody levels increase rather quickly—a few days after the antigen has entered the body. The amount of antibody produced also greatly exceeds quantities generated during the primary response. Consequently, the antigen is quickly destroyed, and a recurrence of the illness is prevented.

The Rapidity of the Secondary Response Is the Result of the Production of Memory Cells During the Primary Response.

Why is the secondary response so different? **FIGURE 8-6** shows that during the primary response, some lymphocytes divide to produce **memory cells.** Memory cells are immunologically competent B cells. Although they do not transform into plasma cells, they remain in the body awaiting the antigen's reentry. These cells are produced in large quantity and thus create a relatively large reserve of antigen-specific B-cells. When the antigen reappears, memory cells proliferate rapidly, producing numerous plasma cells that quickly crank out antibodies to combat the foreign invaders. Memory cells also generate additional memory cells during the secondary response. These remain in the body in case the antigen should enter the body at some later date. Immune protection afforded by memory cells

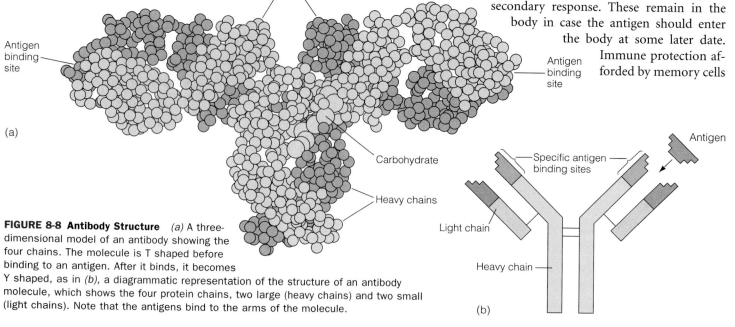

(a)

Light chains

Antigen binding site

Antigen binding site

Carbohydrate

Heavy chains

Specific antigen binding sites

Antigen

Light chain

Heavy chain

(b)

FIGURE 8-8 Antibody Structure *(a)* A three-dimensional model of an antibody showing the four chains. The molecule is T shaped before binding to an antigen. After it binds, it becomes Y shaped, as in *(b)*, a diagrammatic representation of the structure of an antibody molecule, which shows the four protein chains, two large (heavy chains) and two small (light chains). Note that the antigens bind to the arms of the molecule.

can last 20 years or longer, which explains why a person who has had a childhood disease such as the mumps or chicken pox is unlikely to contract it again.

Antibodies Destroy Antigens in One of Four Ways.

Antibodies belong to a class of blood proteins called the **globulins** (GLOB-you-lynns), introduced in Chapter 7. Antibodies are specifically called **immunoglobulins** (im-YOU-know-GLOB-you-lynns).

Each antibody is a T-shaped molecule consisting of four peptide chains (**FIGURE 8-8**). The arms of the T bind to antigens. Like active sites on enzymes, these binding regions confer antibody specificity.

Antibodies destroy foreign organisms and antigens via one of four mechanisms: (1) neutralization, (2) agglutination, (3) precipitation, and (4) complement activation. Let's consider each one briefly.

Neutralization. Some antibodies bind to viruses and bacterial toxins and form a complete coating around them. This prevents viruses from binding to plasma membrane receptors of body cells. If a virus cannot bind to a plasma membrane receptor, it cannot get inside most cells (**FIGURE 8-9**, far right), and it is effectively neutralized. Toxins and viruses neutralized by their antibody coating are eventually engulfed by macrophages and other phagocytic cells (**FIGURE 8-9**).

Agglutination. Antibodies deactivate foreign cells (bacteria and red blood cells transfused into another person) by agglutination (ah-GLUTE-tin-A-shun). During agglutination, a single antibody may bind to several antigens. This causes them to clump together (**FIGURE 8-9**). The antigen-antibody complexes are then phagocytized by macrophages and other phagocytic cells.

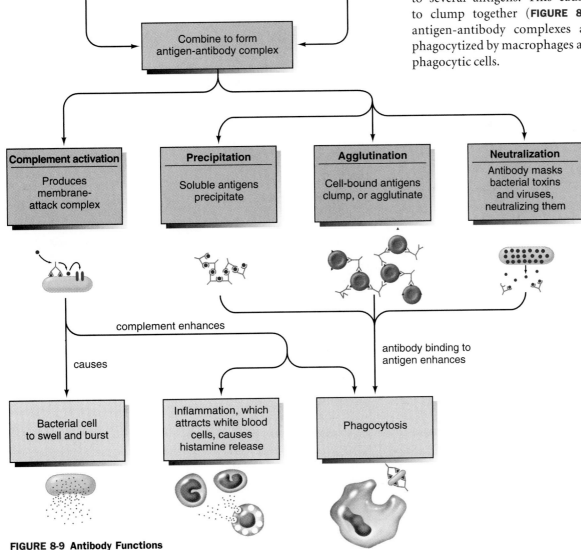

FIGURE 8-9 Antibody Functions

Precipitation. Antibodies bind to soluble antigens (for example, a protein), forming much larger, water-insoluble complexes that precipitate out of solution, where they are engulfed by phagocytic cells.

Activation of the Complement System. Antibodies also help to rid the body of bacteria by activating the complement system. As noted earlier, the complement system is a family of blood-borne proteins that is part of the nonspecific immune response to antigens. The complement system is activated by the presence of antigen-antibody complexes (antibodies bound to antigens).

Macrophages in the Body's Tissues Play a Key Role in Activating B Cells.

Macrophages are phagocytic cells found in connective tissue, lymphoid tissue, and the organs of the lymphatic system (for example, lymph nodes). As noted in Chapter 7, macrophages arise from monocytes that escape from the bloodstream and set up residence in body tissues.

Macrophages play several important roles in the immune response. First, they phagocytize bacteria and other antigens at the site of infection, thereby lessening the initial assault. Second, they play a mop-up role by phagocytizing antigen-antibody complexes, dead cells, and dead microorganisms—cleaning up the debris. Third, macrophages are initiators that activate T- and B-cell differentiation. B cells, in fact, cannot differentiate into plasma cells and produce antibodies without macrophages.

FIGURE 8-10 is a simplified illustration showing the role of the macrophage in B-cell activation. As illustrated,

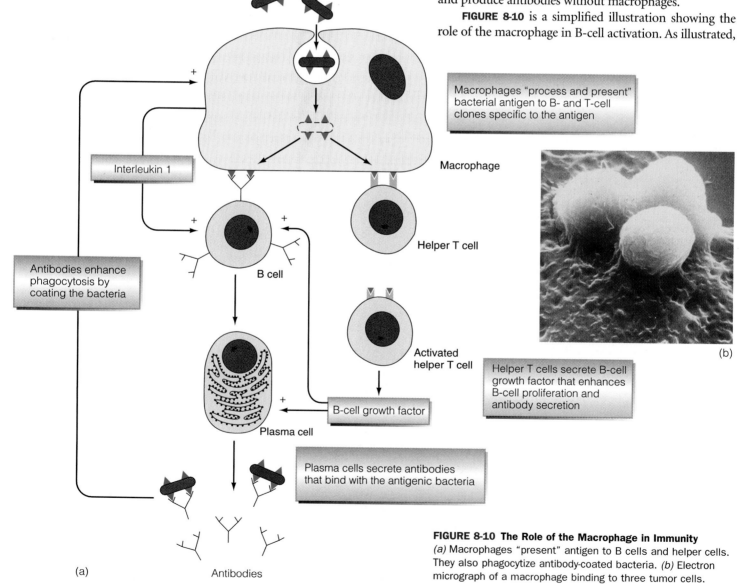

FIGURE 8-10 The Role of the Macrophage in Immunity
(a) Macrophages "present" antigen to B cells and helper cells. They also phagocytize antibody-coated bacteria. *(b)* Electron micrograph of a macrophage binding to three tumor cells.

macrophages first engulf invading bacteria. The macrophages then transfer antigens from the surface of the bacteria to their own plasma membrane. Next, macrophages cluster around B cells, "presenting" the bacterial antigen to them. B cells programmed to respond to the bacterial antigen concentrate on the surface of macrophages and are activated. The B cells begin to divide and differentiate, forming antibody-producing plasma cells and memory cells. Macrophages also secrete **interleukin 1** (in-ter-LEW-kin), a chemical that enhances the proliferation and differentiation of activated B cells.

Macrophages also present antigens to certain T cells, called helper T cells (described in detail shortly). As shown in **FIGURE 8-10**, the helper T cells are activated by this contact and begin producing a chemical substance known as **B-cell growth factor.** It enhances the proliferation and differentiation of B cells such as interleukin 1. The B-cell growth factor also enhances antibody production by the plasma cells (**FIGURE 8-10**).

◼ T Cells Differentiate into at Least Four Cell Types, Each with a Separate Function in Cell-Mediated Immunity

T cells provide a much more complex form of protection than B cells. Like B cells, they respond to the presence of antigens by undergoing rapid proliferation. T cells, however, differentiate into at least four cell types: (1) memory T cells, (2) cytotoxic T cells, (3) helper T cells, and (4) suppressor T cells (**TABLE 8-2**).

Memory T cells form a cellular reserve force vital to mounting a rapid secondary response. **Cytotoxic T cells** perform two roles (**TABLE 8-2**). First, some attack and kill body cells that have been infected by viruses. When a virus infects a cell, antigenic proteins in the virus's envelope become incorporated in the plasma membrane of the host cell. Cytotoxic T cells bind to that antigen and destroy the host cell. Other cytotoxic T cells attack and kill bacteria, parasites, single-celled fungi, cancer cells, and foreign cells introduced during blood transfusions or tissue or organ transplants.

Cytotoxic T cells bind to antigenic molecules in the membranes of cells they attack and release a chemical known as **perforin-1** (purr-FOR-in). As shown in **FIGURE 8-11**, perforin-1 molecules embed in the plasma membrane of the target cell and then join to form pores, similar to those produced by the membrane-attack

FIGURE 8-11 How Cytotoxic T Cells Work   Cytotoxic T cells, containing perforin-1 granules, bind to their target and release perforin-1, then detach in search of other invaders. Perforin-1 molecules congregate in the target plasma membrane, forming a pore that disrupts the plasma membrane, causing the cell to die. (SOURCE: By Dana Burns from John Ding-E Young and Zanvil A. Kohn, "How Killer Cells Kill." Copyright © January 1988 by *Scientific American, Inc.* All rights reserved.)

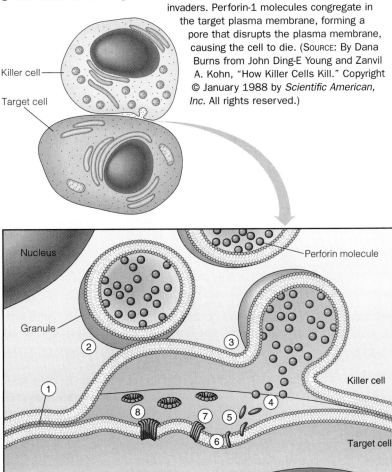

TABLE 8-2	
Summary of T Cells	
Cell Type	**Action**
Cytotoxic T cells	Destroy body cells infected by viruses, and attack and kill bacteria, fungi, parasites, and cancer cells
Helper T cells	Produce a growth factor that stimulates B-cell proliferation and differentiation and also stimulates antibody production by plasma cells; enhance activity of cytotoxic T cells
Suppressor T cells	May inhibit immune reaction by decreasing B- and T-cell activity and B- and T-cell division
Memory T cells	Remain in body awaiting reintroduction of antigen, at which time they proliferate and differentiate into cytotoxic T cells, helper T cells, suppressor T cells, and additional memory cells

complex of the complement system. These pores cause the plasma membrane to leak, destroying the target cell within a few hours. After it has delivered its lethal payload, the cytotoxic cell detaches and is free to hunt down other antigens.

Helper T cells enhance the immune response and are activated by the presence of certain antigens presented by macrophages, as noted earlier. Helper T cells stimulate B cells, but also activate T cells via the release of a chemical known as **interleukin 2.** It increases the activity of cytotoxic T as well as that of suppressor T and other helper T cells. Its name *helper T cell* is therefore somewhat inaccurate. The cell "helps" just about all lymphocytes!

Helper T cells are the most abundant of all the T cells (comprising 60%-70% of the circulating T cells). Some immunologists liken helper T cells to the immune system's master switch. Without them, antibody production and T-cell activity is greatly reduced. In fact, without helper T cells, antigens would stimulate a few B and T cells and the process would come to a halt. We would have no immune protection. Interestingly, the AIDS virus (discussed later) preferentially infects helper T cells, disabling patients' immune responses. Patients with AIDS therefore contract several infectious diseases to which they are unable to mount an effective immune response. Unless a cure can be found, all people infected with the AIDS virus will eventually die from these diseases.

The role of **suppressor T cells** is less understood. Research suggests that they "turn off" the immune reaction as the antigen begins to disappear—that is, as the antigen is phagocytized. The activity of suppressor T cells, therefore, increases as the immune system finishes its job. Suppressor cells release chemicals that reduce B- and T-cell division.

Two Types of Immunity Are Possible: Active and Passive

One of the major medical advances of the last century was the discovery of **vaccines** (vac-SEENS), used to prevent bacterial and viral infections. Vaccines contain inactivated or weakened viruses, bacteria, or bacterial toxins. When injected into the body, the "disabled" antigens in vaccines elicit an immune response. Many vaccines provide immunity or protection from microorganisms for long periods, sometimes for life. Others give only short-term protection.

Vaccines stimulate the immune reaction because the weakened or deactivated organisms (or toxins) they contain still possess the antigenic proteins or carbohydrates that trigger B- and T-cell activation. Because they have been seriously weakened or deactivated, however,

FIGURE 8-12 Poison and Antidote
Poisonous snakes like this rattler inject venom into their victims. Venom can be milked from the snake and used to produce antivenom, a serum containing immunoglobulins that neutralize the venom.

viruses, bacteria, and bacterial toxins in vaccines usually do not cause disease.

Vaccination provides a form of protection that immunologists call **active immunity**—so named because the body actively produces memory T and B cells that protect a person against future infections. Viral or bacterial infections also produce active immunity.

Vaccinations are vital in controlling deadly diseases such as polio, typhus, and smallpox—diseases that can kill people before their immune system mounts an effective response. In fact, in the wealthier nations of the world, such as the United States, vaccines have nearly eliminated many infectious diseases such as smallpox.

The second type of immunity, called **passive immunity,** is a temporary form of protection, resulting from the injection of immunoglobulins (antibodies to specific antigens). These antibodies are produced by injecting antigens in other animals such as sheep. The antibodies are then extracted from the blood for use in humans.

Passive immunity is so named because the cells of the immune system are not activated. Immunoglobulins remain in the blood for a few weeks, protecting an individual from infection. Because the liver slowly removes these molecules from the blood, a person gradually loses protection.

Immunoglobulins are administered to prevent or counteract certain infections already under way. Travelers to developing nations, for instance, are often given immunoglobulins to viral hepatitis (liver infection) as a preventive measure. Immunoglobulins are also used to treat individuals who have been bitten by poisonous snakes (**FIGURE 8-12**). The venom in poisonous snakes is a mixture of proteins, enzymes, and polypeptides that attack body cells, especially nerve cells and heart muscle cells. In the United States, the most common poisonous snakebites come from rattlesnakes.[4] Bites of poisonous snakes can be treated by antivenom, immunoglobulins that quickly destroy or deactivate the immunogenic molecules in snake venom before they can have adverse effects.

Passive immunity can also be conferred naturally—for instance, from a mother to her fetus (FEE-tuss). A fetus receives a dose of antibodies via the **placenta**

---

[4]In the United States, 15% of untreated rattlesnake bites and only about 1% of treated bites are fatal.

(plah-SEN-tah), the organ that transfers nutrients from the mother's bloodstream to the fetal blood. Maternal antibodies transferred via the placenta remain in the blood of a newborn infant for several months, protecting the youngster from bacteria and viruses while its immune system is developing. Mothers also transfer antibodies to their babies in breast milk. The maternal antibodies in milk attack bacteria and viruses in the intestine, protecting the infant from infection. (For more on this topic, see Health Note 8-1.)

Vaccination Fears in the United States.

Vaccines have lowered the incidence of many infectious diseases in the United States and other relatively affluent industrialized nations by 99% or more. Vaccines for diphtheria, tetanus, whooping cough, polio, measles, mumps, and congenital rubella (German measles) have all but eliminated these deadly or crippling disease organisms.

Despite the tremendous successes of vaccines, publicity concerning their rare side effects has created something of a medical dilemma in the United States, Japan, and Great Britain. In 1976 and 1977, for example, a mass-immunization program in the United States for the swine flu, one type of influenza, resulted in the paralysis of a number of people. As a result of public concern over this and other incidents, many parents have chosen *not* to have their children vaccinated. Proponents of vaccinations argue that the excessive media attention given to the rare but serious complications has harmed efforts to promote vaccination.

Another cause for the decline in vaccination stems from the success of previous immunization programs, which, as noted above, have greatly reduced the incidence of most infectious diseases. Parents who were reared in an environment free from such diseases are often unaware of the dangers of infectious disease.

Having their children immunized seems unimportant. Some people, notably Christian Scientists, refuse to vaccinate children on religious grounds. Public health officials are quick to point out that pathogenic organisms that once took a huge toll on humans have not been eradicated. They fear that without widespread protection, epidemics could occur again.

Harmful side effects from conventional vaccines are often caused by reactions to certain "nonessential" antigens on the injected microorganism. These antigens frequently play little or no role in immunity. By eliminating them from vaccines, researchers hope to develop safer alternatives to the vaccines in use today.

Section 8-4

Practical Applications: Blood Transfusions and Tissue Transplantation

Although the immune system is important in protecting us from microorganisms, it presents something of a challenge to physicians during blood transfusions and tissue transplants—biological interventions unwitnessed in evolution. Let's consider blood transfusions first.

Blood Transfusions Require Careful Cross-Matching of Donors and Recipients

The surface of the red blood cell (RBC) membrane, like that of other cells, contains many glycoproteins that form a unique cellular fingerprint (Chapter 3). These glycoproteins are the basis of blood-typing. As noted in Chapter 7, four blood types exist: A, B, AB, and O. The letters refer to one type of glycoprotein (antigen) present on the plasma membrane of RBCs of an individual. As illustrated in **TABLE 8-3**, individuals with type A blood

TABLE 8-3				
Summary of Blood Types				
			Safe to Transfuse	
Blood Type	**Antigens on Plasma Membranes of RBCs**	**Antibodies in Blood**	*To*	*From*
A	A	b*	A, AB	A, O
B	B	a	B, AB	B, O
AB	A + B	—	AB	A, B, AB, O
O	—	a + b	A, B, AB, O	O

*Lowercase *b* indicates antibody to B antigen.

Health Note 8-1

Bringing Baby Up Right: The Immunological and Nutritional Benefits of Breast Milk

A baby is born into a dangerous world in which bacteria and viruses abound. Complicating matters, the immune system of a newborn child is poorly developed. Fortunately, the newborn is protected by passive immunity—antibodies that have traveled from its mother's blood. Antibodies also travel to the infant in breast milk (**FIGURE 1**).

FIGURE 1 Breast-Feeding Mother breast-feeding her newborn infant.

Several immunoglobulins are present in breast milk. One of these is called secretory IgA. It is present in very high quantities in colostrum, a thick fluid produced by the breast immediately after delivery—before the breast begins full-scale milk production. Colostrum is so important, in fact, that some hospitals give "colostrum cocktails" to newborns who are not going to be breast-fed by their mothers. Nurses remove the colostrum from the mother's breast with a breast pump and feed it to the baby in a bottle.

Colostrum, says Sarah McCamman, a nutritionist at the University of Kansas Medical School, coats the lining of the intestines. The IgA antibodies in colostrum prevent bacteria ingested by the infant from adhering to the epithelium and gaining entrance. Breast milk also contains lysozyme, an enzyme that breaks down the cell walls of bacteria, destroying them.

Unfortunately, not all medical personnel agree on the benefits of breast milk. One problem is that breast milk has unusually low levels of iron. This fact has led many physi-

cians to recommend iron supplements for newborns. A more careful analysis, however, shows that breast-fed infants generally do not suffer from iron deficiency because the percentage of iron absorbed from breast milk is extraordinarily high. Thus, low levels of iron in breast milk are offset by the high absorption.

The wisdom of iron supplements for newborns has also been questioned on other grounds. Researchers have found that iron supplements increase the incidence of harmful bacterial infections in newborns. As noted earlier in the chapter, iron is a limiting factor in many pathogenic bacteria. Low levels of iron in breast milk, therefore, may reduce bacterial replication in an infant's intestinal tract.

In general, breast-fed babies are healthier than bottle-fed babies. The incidence of gastroenteritis (inflammation of the intestine), otitis (ear infections), and upper respiratory infections is lower in breast-fed babies. Studies also show that children breast-fed for at least 6 months contract fewer childhood cancers than

have RBCs whose plasma membranes contain the A antigen (glycoprotein). The RBCs in individuals with type B blood contain type B antigens. People with AB blood have both A and B antigens, and people with type O have neither antigen.

Physicians learned a long time ago that blood could be successfully transfused from one person to another, but only if their blood matched. In other words, an individual with type A blood could only receive type A blood, and individuals with type B blood can only receive type B blood.

Cross-matching blood is essential to prevent life-threatening immune reactions. These reactions result from antibodies found in the blood of most people. As shown in **TABLE 8-3**, each blood type carries a specific

type of antibody. People with type A blood, for example, naturally contain antibodies to the B antigen. (This is why individuals with type A blood cannot receive type B blood.) People with type B blood contain antibodies to the A antigen. For reasons not well understood, these antibodies appear in the blood during the first year of life.

Serious problems arise when incompatible blood types are mixed. For example, consider what happens if an individual with type A blood is accidentally given type B blood (**FIGURE 8-13**). The antibodies to type B blood found in the recipient bind to the transfused RBCs (which contain type B antigens). This causes the type B RBCs to agglutinate (clump) and hemolyze (HEEM-oh-lize) (burst). Hemolysis and agglutination constitute the **transfusion reaction**.

their bottle-fed counterparts. The incidence of childhood lymphoma, a cancer of the lymph glands, in bottle-fed babies is nearly double the rate in breast-fed children for reasons not yet understood.

Research also suggests that certain proteins in breast milk may stimulate the development of a newborn's immune system. In laboratory experiments, the proteins speed up the maturation of B cells and prime them for antibody production. These soluble proteins may also activate macrophages, which play a key role in the immune system.

Breast milk is also more digestible and more easily absorbed by infants than formula. Formula is a mixture of cow's milk, proteins, vegetable oils, and carbohydrates. It is only an approximation of mother's milk and is not broken down and absorbed as completely as breast milk.

Because of a growing awareness of the benefits of breast-feeding, virtually every national and international organization involved with maternal and child health supports breast-feeding, says McCamman. Breast-feeding can have a major health impact in this country. Unfortunately,

the benefits are not as widely known as many people would like, even among health care professionals. Fortunately, more and more health care workers are beginning to understand the benefits of breast-feeding and are promoting this option. As a result, many middle-class American women are now choosing to breast-feed.

Unfortunately, says McCamman, there is "a huge population of low-income, poorly educated . . . women who choose not to nurse." The federal government may be playing an unwitting role in their decision. A national program aimed at improving child nutrition provides free formula to needy women, perhaps discouraging mothers from breast-feeding.

Another reason is economic. Low-income women often work at jobs that do not provide maternity leave. Thus, these women must return to work soon after giving birth, and breast-feeding is difficult to do under these circumstances.

Still another reason for the low rate of breast-feeding among low-income women is that women need a lot of support to nurse. "People think nursing is innate, natural, and easy," says McCamman, "but this is not al-

ways the case." Getting started often requires guidance and education. Without such support, breast-feeding can be a difficult and painful experience. Breast-feeding among all women, rich and poor, may also be discouraged by cultural attitudes and fear of embarrassment. In fact, many otherwise open-minded people find breast-feeding in public or even semi-public settings embarrassing.

Given the many benefits of breast-feeding, McCamman recommends it to all mothers who can. Health care workers can help by educating their patients on the benefits of breast-feeding. "Doctors should present the information on breast- and bottle-feeding," says McCamman, "outlining the pros and cons of both methods. Then, let the woman choose. Too few doctors do that today, so women aren't making informed decisions."

Visit *Human Biology's* Internet site, www.jbpub.com/humanbiology, for links to web sites offering more information on this topic.

www.jbpub.com/humanbiology

RBC clumping restricts blood flow through capillaries, reducing oxygen and nutrient flow to cells and tissues. Massive hemolysis (heme-OL-eh-siss) results in the release of large amounts of hemoglobin into the blood plasma. Hemoglobin precipitates in the kidney, blocking the tiny tubules that produce urine. This can result in acute kidney failure, which can be life-threatening.

Because of the possibility of this potentially life-threatening reaction, successful transfusions require careful matching of the blood types of the donor and recipient. As **TABLE 8-3** shows, RBCs from individuals with type O blood have neither A nor B antigens. Therefore, type O blood can be transfused into individuals with all four blood types. Type O individuals are said to be **universal donors.** However, type O blood, while free of

antigens, contains antibodies to both A and B antigens. Therefore, individuals with type O blood can receive only type O blood.

As shown in **TABLE 8-3**, individuals with type AB blood contain RBCs with both A and B antigens, but no antibodies related to the ABO system. These people can therefore *receive* blood from all others and are consequently referred to as **universal recipients.** Note, however, that AB blood can be safely *transfused* only into individuals with AB blood.

The terms *universal donor* and *universal recipient* are somewhat misleading, however, because RBCs also contain other antigens that can cause transfusion reactions. The most important of these is the **Rh factor.** This antigen was first identified in rhesus (REE-suss)

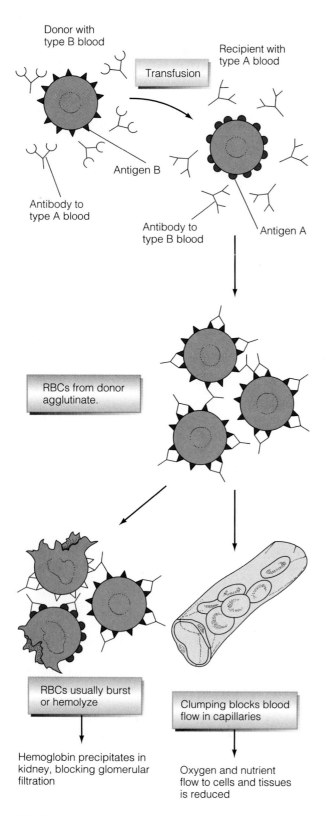

Donor with type B blood

Transfusion

Recipient with type A blood

Antigen B

Antibody to type A blood

Antibody to type B blood

Antigen A

RBCs from donor agglutinate.

RBCs usually burst or hemolyze

Clumping blocks blood flow in capillaries

Hemoglobin precipitates in kidney, blocking glomerular filtration

Oxygen and nutrient flow to cells and tissues is reduced

FIGURE 8-13 Transfusion Reaction Type B blood transfused into an individual with type A blood results in a transfusion reaction, characterized by agglutination and hemolysis.

monkeys; hence the designation. People whose cells contain the Rh antigen, or Rh factor, are said to be Rh-positive. Those without it are Rh-negative.

Unlike the ABO system, in the Rh system, antibodies are produced only when Rh-positive blood is transfused into the bloodstream of a person with Rh-negative blood. The first transfusion of Rh-positive blood into an Rh-negative person generally does not result in a transfusion reaction, but a second transfusion does. To reduce the likelihood of a transfusion reaction, Rh-negative people should receive only Rh-negative blood, and Rh-positive people should receive only Rh-positive blood.

The Rh factor becomes particularly important during pregnancy. Problems can arise if an Rh-negative mother has an Rh-positive baby (**FIGURE 8-14**). Even though the maternal and fetal bloodstreams are separate, small amounts of fetal blood usually enter the maternal bloodstream at birth. Rh antibodies form in the maternal bloodstream, and the woman becomes sensitized to the Rh factor.

To prevent antibody production in Rh-negative women who give birth to Rh-positive babies, physicians routinely inject antibodies to fetal Rh-positive RBCs into the mother soon after she has given birth. (The antibody-containing serum is called *RhoGAM.*) These antibodies bind to Rh-positive RBCs from the fetus before a woman's immune system responds to them. This, in turn, prevents a woman from being sensitized. To be effective, however, the treatment must be given within 72 hours after the baby is born.

If the woman is not treated at the time and becomes pregnant again with an Rh-positive baby, maternal antibodies to the Rh factor will cross the placenta. In the fetal circulation, these antibodies cause fetal RBCs to agglutinate, then break down, resulting in anemia and hypoxia (lack of oxygen to tissues). Unless the baby receives a blood transfusion (of Rh-negative blood) before birth and several after birth, it is likely to have brain damage and may even die.

Tissue Transplantation Often Evokes Cell-Mediated Immunity, Which Can Be Blocked by Certain Drugs

Organ and tissue transplantation is a much more complex matter. Only three conditions exist in which a person can receive a transplant and not reject it. One is if the tissue comes from an individual's own body. For burn victims, surgeons might use healthy skin from one part of the body to cover a badly damaged region elsewhere. The second instance is when a tissue is transplanted between identical twins—individuals derived

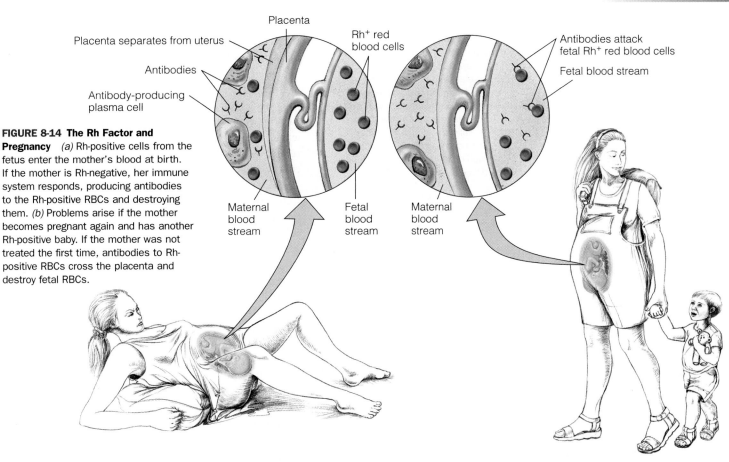

FIGURE 8-14 The Rh Factor and Pregnancy *(a)* Rh-positive cells from the fetus enter the mother's blood at birth. If the mother is Rh-negative, her immune system responds, producing antibodies to the Rh-positive RBCs and destroying them. *(b)* Problems arise if the mother becomes pregnant again and has another Rh-positive baby. If the mother was not treated the first time, antibodies to Rh-positive RBCs cross the placenta and destroy fetal RBCs.

(a) First child. Rh⁺ red blood cells enter the maternal bloodstream during birth, evoking an immune reaction.

(b) Second child. Rh⁺ antibodies cross the placenta, destroying fetal red blood cells.

from a single fertilized ovum that splits early in embryonic development to form two embryos. These individuals are genetically identical and have identical cellular antigens.

A third instance occurs when tissue rejection is inhibited by specific drugs. For example, heart, liver, and kidney transplants are successful only when recipients are treated with drugs that suppress the immune system. This treatment must be continued throughout the life of the patient. Unfortunately, most immune suppressants have numerous side effects and often leave the patient vulnerable to bacterial and viral infections. Without them, however, transplants from individuals not genetically identical to the recipient are quickly rejected.

In the 1980s, a new drug known as **cyclosporin** (SIGH-clow-SPORE-in) was introduced. This drug suppresses the formation of interleukin 2 by helper T cells, thus greatly reducing cell-mediated immunity without affecting B cells. Patients who receive the drug are therefore able to combat many bacterial infections with antibodies.

Diseases of the Immune System

The immune system, like all other body systems, can malfunction. This section looks at two disorders: allergies and autoimmune diseases.

The Most Common Malfunctions of the Immune System Are Allergies

An **allergy** is an overreaction to some environmental substance, or antigen, such as pollen or a food (**FIGURE 8-15**). Antigens that stimulate allergic reactions are called **allergens** (AL-er-gens). Allergens cause the production of one class of immunoglobulins, the IgE antibodies, from plasma cells.[5] As **FIGURE 8-15** shows, these antibodies bind to mast cells, which are found in many tissues, but especially in the connective tissue

---

[5]Some allergies involve IgG or IgM, and some apparently do not involve antibodies at all.

FIGURE 8-15 Allergic Reaction   Antigen stimulates the production of massive amounts of IgE, a type of antibody produced by plasma cells. IgE attaches to mast cells. This is the sensitization stage. When the antigen enters again, it binds to the IgE antibodies on the mast cells, triggering a massive release of histamine and other chemicals. Histamine, in turn, causes blood vessels to dilate and become leaky. This triggers the production of mucus in the respiratory tract. In some people, the chemicals released by the mast cells also cause the small air-carrying ducts in the lungs to constrict, making breathing difficult.

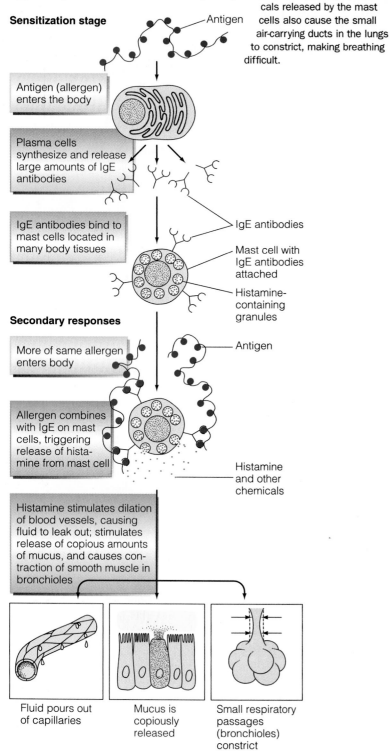

surrounding blood vessels. Mast cells have large cytoplasmic vesicles containing the chemical histamine.

Allergens then bind to the IgE antibodies attached to the mast cells, triggering the release of histamine from the vesicles via exocytosis (**FIGURE 8-15**). Histamine, in turn, causes nearby arterioles to dilate. Histamine released in the lungs causes the bronchioles to constrict, reducing airflow and making breathing difficult. This condition is called **asthma** (AS-ma).

Allergic reactions usually occur in specific body tissues, where they create local symptoms that, while irritating, are not life-threatening. For example, an allergic response may occur in the eyes, causing redness and itching. Or, it may occur in the nasal passageway, causing stuffiness. However, the allergic response can also occur in the bloodstream, where it can be fatal if not treated quickly. For example, the presence of penicillin or bee venom in the bloodstream of certain people causes massive release of histamine and other chemicals. This, in turn, causes extensive dilation of blood vessels in the skin and other tissues. The blood pressure then falls precipitously, shutting down the circulatory system. Histamine released by mast cells also causes severe constriction of the bronchioles (the ducts in the lungs that open onto the alveoli), making breathing difficult. The decline in blood pressure and constriction of the bronchioles result in **anaphylactic shock** (ANN-ah-fah-LACK-tic). Death may follow if measures are not taken to reverse this catastrophic event. One such measure is an injection of the hormone epinephrine (commonly known as adrenalin), which rapidly reverses the constriction of the bronchioles.

Allergies are treated in three ways. First, patients are advised to avoid allergens—for instance, to avoid milk and milk products or stay clear of dogs and cats. Second, patients may also be given **antihistamines** (an-tea-HISS-tah-means), drugs that counteract the effects of histamine. Third, patients may also be given allergy shots, injections of increasing quantities of the offending allergen. In many cases, this treatment makes an individual less and less sensitive to the allergen. Desensitization results from the production of another class of antibodies, the **IgG antibodies,** which bind to allergens. This, in turn, blocks the antigen from binding to the mast cells, thus preventing the release of histamine and other chemical substances responsible for the allergic reaction.

Autoimmune Diseases Result from an Immune Attack on the Body's Own Cells

Occasionally, the immune system mounts an attack on the body's own cells. This unfortunate state of affairs is known as an **autoimmune disease.** Autoimmune diseases result from many causes. For example, in some instances, normal body proteins can be modified by environmental pollutants, viruses, or genetic mutations so that they are no longer recognizable by the body as self. In other cases, normal body proteins usually isolated from the immune system enter the bloodstream and evoke an immune response. For example, a protein called *thyroglobulin* is produced by the thyroid gland in the neck. Thyroglobulin is stored inside the gland and not exposed to cells of the immune system. If the gland is injured, however, thyroglobulin may enter the bloodstream. Lymphocytes encountering this protein may then mount an immune response to it.

Yet another cause of autoimmune reaction is exposure to antigens that are nearly identical to body proteins. The bacterium that causes strep throat, for example, contains an antigen structurally similar to one of the proteins found in the plasma membranes of the cells lining the heart valves of some individuals. The body mounts an attack on the bacterium, but antibodies may also bind to the lining of the heart valve, causing a local inflammation and scar tissue to develop. This can damage the valve, resulting in valvular incompetence, discussed in Chapter 6.

Section 8-6

AIDS: The Deadly Virus

In 1985, Damion Knight, a bright, young cabinetmaker, began to lose weight and experience bouts of unexplained fever. His lymph nodes became swollen, and he felt weak and drowsy. A doctor found that Damion had **acquired immunodeficiency syndrome,** commonly known as **AIDS.** Like thousands of others, Knight died several years after his diagnosis. His doctor could not help him.

AIDS is caused by a virus that attacks and weakens the immune system. This virus is known as the **human immunodeficiency virus** or **HIV** for short (**FIGURE 8-16**).

HIV is an RNA virus that attacks helper T cells, severely impairing the immune system. AIDS patients grow progressively weaker and generally fall victim to other infectious agents. For example, many die from an otherwise rare form of pneumonia. AIDS rose to infamy in the early 1980s. The first documented case was that of a young Missouri boy who died at age 15 in 1969. Studies of tissue samples of a British sailor revealed that he probably died of AIDS ten years earlier. In 1998, scientists announced the presence of HIV in a blood sample of a Bantu man from Africa's Democratic Republic of the Congo who died in 1959.

The incidence of AIDS (number of cases) and the number of deaths from AIDS in the United States have

FIGURE 8-16 HIV and Kaposi's Sarcoma *(a)* AIDS viruses. *(b)* Kaposi's sarcoma on foot.

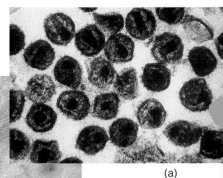

(a)

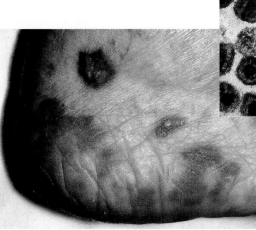

(b)

increased dramatically since the early 1980s. The sharp increase in incidence and deaths prompted one researcher to liken AIDS to the bubonic plague that spread through Europe in the fourteenth and fifteenth centuries. In the 1300s, the plague killed one quarter of the adult population and numerous children. In some African villages today, AIDS runs rampant with as many as 50% of the residents testing HIV-positive. This fact suggests that, at least in some locations, predictions of a plaguelike scourge are not that far-fetched.

HIV infection is a global epidemic. To date, approximately 28 million people have been infected with the virus. Of these, nearly 6 million have died. By the year 2000, researchers estimate that the number will reach 40 million. In the United States, 548,000 people have been reported with AIDS; 343,000 have died. The vast majority of the AIDS victims are men. Some good news emerged in 1998, however. The incidence of AIDS has stabilized in the West, although it continues to grow in the developing nations. The incidence of AIDS in the United States has decreased among white males and children, but continues to increase in women, Blacks, and Hispanics. In 1996, Blacks, Hispanics, and women accounted for 42%, 19%, and 20%, respectively, of all new cases of AIDS—a fact that suggests the need for better education in these sectors.

AIDS Is a Progressive Disease That Exhibits Three Distinct Phases

AIDS progresses through three distinct phases: asymptomatic, AIDS-Related Complex (ARC), and full-blown AIDS. **FIGURE 8-17** shows several key physiological parameters and symptoms during each phase. The top panel, for instance, plots the number of helper T cells, also called **T4 cells.** Symptoms are listed in the middle panel. The bottom panel shows the concentration of HIV and the HIV antibody levels.

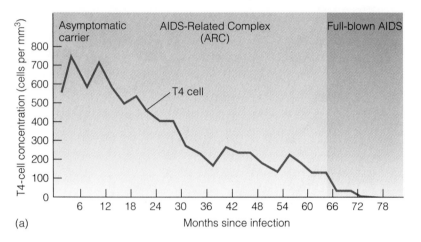

(a)

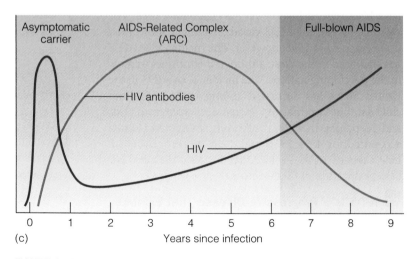

(b)

(c)

FIGURE 8-17 **Tracking a Killer** (a) T4 cell concentration, (b) symptoms, and (c) HIV and antibody levels (labeled immune system). (Source: Data for parts a and c from R. R. Redfield and D. S. Burke, "HIV Infection: The Clinical Picture" in *Scientific American,* October 1988.)

As illustrated in **FIGURE 8-17**, most patients show no symptoms during the first few months after infection. A few, perhaps 1%–2%, may exhibit symptoms similar to those of infectious mononucleosis—that is, fever, chills, aches, and swollen lymph nodes. However, these symptoms vanish shortly thereafter, and individuals go on about their business, unaware that they have contracted a deadly disease.

During the first phase, the T4-cell count is high (**FIGURE 8-17**). As the bottom panel in **FIGURE 8-17** shows, the number of viruses (blue line) rises rapidly during the first phase of AIDS. Antibodies to HIV begin to increase as a result.

During the second phase of this disease, T4-cell count begins to fall in response to the rise in HIV. Antibody levels rise initially, then fall as the immune system begins to falter. During the second phase, patients begin to show outward signs of the disease caused by the decline in the number of T4 cells. Severe fatigue and unexplained, persistent fever are two common symptoms. Some patients complain of a persistent cough and loss of memory, difficulty thinking, and depression. When recurring infections set in, the third and final phase of AIDS is about to begin.

The final phase, known as full-blown AIDS, is characterized by persistent infections, extreme loss of weight, and weakness. Most patients succumb to one of a handful of infectious organisms—microbes not ordinarily capable of producing infections. When a patient is in a state of extreme immune compromise, though, these organisms take hold and become life-threatening.

HIV Also Causes Cancer and Produces a Substance That May Cause Deterioration of Brain Function

HIV affects more than a person's immune system. AIDS patients, for example, often contract a rare form of skin cancer called **Kaposi's sarcoma** (kah-PO-sees sar-KOME-ah) (**FIGURE 8-16B**).

Studies show that HIV carries a gene (a segment of DNA) that's incorporated into the genetic material of body cells. This gene may cause certain cells to proliferate uncontrollably, forming a cancerous tumor or it may stimulate the production of a chemical substance that causes rapid cell growth (cancer) in neighboring cells.

AIDS patients also experience a number of neurological disorders, beginning in the second phase. These include memory loss and progressive mental deterioration. A recent laboratory study may explain the cause of these symptoms.

Researchers have found that inside helper T cells, HIV produces several proteins that are incorporated into the viral capsid. One of those proteins is known as gp120. This protein kills fetal brain cells in culture. The researchers believe that gp120 travels in the blood to the brains of some patients, where it kills neurons, thus producing neurological defects.

Loss of mental function may also be caused by a single-celled parasite that is normally found in cats, but sets up residence in people whose immune systems are compromised by HIV. This parasite causes a brain infection (encephalitis) that leads to a loss of brain cells, seizures, and weakness.

HIV Is Transmitted in Many Ways, but Not by Casual Contact

Research has shown that HIV is passed primarily via three routes: (1) sexual contact, (2) blood transfusions, and (3) contaminated needles shared by intravenous drug abusers. Homosexual men, hemophiliacs, and drug addicts are the primary victims in the United States. But HIV is also transmitted among the heterosexual population and can even be transmitted from an infected mother to her baby through the placenta. A recent study showed that a man infected with AIDS is many times more likely to transmit the disease to a female partner than vice versa.

Individuals with the genetic disorder hemophilia were once at risk for AIDS. Hemophiliacs are given clotting factors from pooled human plasma. Before 1984, blood donors were not screened for HIV. Consequently, many of the preparations were contaminated with HIV. As a result, a majority of the estimated 15,000 hemophiliacs in the United States who received clotting factors between 1975 and 1984 have HIV antibodies in their blood.

To prevent the spread of AIDS through blood transfusions, blood is now routinely tested for HIV. Tissues and organs for transplantation are also tested. Despite improvements in screening, blood transfusion is not a fail-safe proposition. Individuals who will need blood for an operation are therefore encouraged to donate blood ahead of time.

HIV, although lethal, does not spread as readily as the flu virus or cold viruses; individuals can protect themselves by practicing sexual abstinence before marriage, by engaging in safe sex (using condoms, for example), and by avoiding multiple sexual partners. To prevent the spread of HIV among intravenous drug users, some countries and some U.S. cities distribute clean hypodermic needles to addicts.

The Battle Against AIDS Has Been Facilitated by New Screening Tests and by Drugs That Slow Down the Development of the Disease

Health care workers determine the presence of HIV via an immunologic test, which detects antibodies to HIV in the blood. Recently, scientists announced the development of a new, more sensitive genetic test that could improve the screening of blood and tissue.

Although no cure has been discovered, a drug called AZT (zidovudine) may prolong the lives of people who have tested HIV-positive.[6] AZT inhibits viral replication and is effective in people with full-blown AIDS, or people who show early signs of HIV infection. It is now also used in people in who are infected, but who show no signs other than impaired immunity, and children. It is even used to prevent the transmission of HIV from mothers to fetuses and newborns (through breast milk). Unfortunately, AZT is costly and may be carcinogenic.

To date, 33 drugs are available to combat HIV and to treat infections, Kaposi's sarcoma, and other complications including weight loss. One of the newest treatments that has proved useful in prolonging the lifespan of people infected with HIV is a combination of AZT, another similar drug, and a more recent development known as **protease inhibitors,** which are drugs that block certain stages that are vital in the replication of HIV. The use of these and other drugs have resulted in a substantial decrease in deaths due to AIDS in the United States and other developed nations.

Because of an outcry among the homosexual community, the U.S. Food and Drug Administration, which regulates all drug testing on humans, has relaxed its standards in hopes of bringing potential AIDS drugs to the public more quickly. New drugs could help physicians and their patients hold the disease in abeyance, greatly prolonging the lives of those who are infected with this deadly virus.

AIDS will undoubtedly remain a significant public health threat in the world for many years. To bring this disease under control, more intensive efforts are needed to educate all people, especially individuals in high-risk groups, on ways to prevent the disease. (For a discussion of some of the social and political implications of efforts to control the spread of AIDS, see the Point/Counterpoint in this chapter.) Many scientists believe that a vaccine is the only way to ultimately bring this disease under control.

Although Some Researchers Are Optimistic About Finding a Vaccine for HIV, Not All Share Their View

HIV is notorious for its ability to mutate, a feature that is making the task of developing a vaccine extremely

---

[6]This drug was first called *azidothymidine,* hence the name AZT.

Point/Counterpoint

Tracking People with AIDS

ANONYMOUS TESTING IS THE ANSWER

Earl F. Thomas

Earl F. Thomas has served on the Governor's AIDS Council and is on the boards of directors for the Colorado AIDS Project and the National Association of People with AIDS. Thomas was diagnosed as having AIDS in 1986.

Confidential testing for HIV (in which the names and addresses of individuals who test positive are reported to health officials) versus anonymous testing continues to be a highly controversial topic. In today's society, the word *AIDS* still breeds fear in the general populace, which, for AIDS patients, translates into fear of discrimination. These fears, quite simply, are keeping people from being tested.

Health departments and AIDS organizations stress that persons who have reason to believe that they may have been infected with the HIV virus should be tested, for several reasons. First, the earlier a person knows he or she is HIV-positive, the earlier treatment can be started to slow the progression of the disease or simply to buy time while researchers explore better treatments. Second, knowledge of one's HIV status is crucial in determining what behavioral changes need to be made.

If the testing system discourages individuals from obtaining knowledge about their HIV status, however, no one benefits. Unfortunately, name reporting tends to create an atmosphere of distrust between health officials and those who wish to be tested. They believe that information concerning their HIV status goes beyond the health department to others who may have a need to know, and they fear that the information will eventually find its way to persons who have no need to know. For instance, numerous links in the information chain (nurses, laboratory workers, therapists, and secretaries) all have access to a person's medical information. Confidential medical information is not so confidential after all. As a result of these concerns, many persons, even in high-risk groups, refuse to be tested at all.

In a survey conducted by the Educational Department of the Colorado AIDS Project in 1989, 32% of 1,112 respondents (homosexual and bisexual men) cited confidentiality concerns as their reason for not being tested. Despite the health department's insistence that fears of information leaks are poorly founded, they are very real to many individuals. More importantly, these fears prevent individuals who have engaged in high-risk behaviors from being tested.

Further information that supports anonymous testing comes from the state of Oregon. When anonymous HIV testing was offered along with confidential testing, there was a 50% increase in the demand to be tested during the first 4 months of the program. People from all segments of society sought out the anonymous test sites. Confidential testing sites reported no increase during the same period.

Fear of testing is justified. There is documented evidence of discrimination against people who are HIV-positive. Individuals have been denied housing and, in certain cases, have been evicted from their homes, lost their jobs, been denied access to public education, and have been shunned by families, friends, and co-workers.

Partner notification (contact tracing) may have a place when trying to control the spread of AIDS, but at what cost? Some feel that contact tracing is not a viable option due to the monetary cost. In 6 months in 1988, the state of Colorado spent $450,000 on partner notification. The result of this enormous expenditure was that 52 people were found.

The experiences of other health departments clearly show that partner notification can be carried out with reasonable success when testing is anonymous. If lists are being maintained, people shy away from testing and forfeit the possibility of early intervention treatment and partner notification information. When health officials act as contact tracers, they are perceived as police. In our society, police-state tactics will never work, and no one benefits.

Anonymous testing would reduce a serious impediment to a powerful, collective anti-AIDS effort. And most important, we could get an answer as to whether people are truly avoiding testing because of reportability.

Addendum: Anonymous testing began in Denver in September of 1990. It is a huge success with other Denver sites offering anonymous testing on request, along with the cities of Boulder, Longmont, and Lafayette.

NOTIFICATION WORKS
John Potterat

The process of reporting people with the AIDS virus (HIV) by name to public health officers and in turn tracing their contacts should not be controversial. Such procedures have been standard public health practice for serious communicable diseases for nearly a century. Formal notification allows society to define the disease burden (surveillance) and to counterattack (control). You cannot control a communicable disease if you do not know who has it and who might be next to have it; moreover, you need to find those directly affected.

Notifying partners of people infected with sexually transmitted disease has been an effective control tool for 50 years. The fundamental reason that health officers are involved in this notification process is that STD patients are not good at referring their own sexual partners. Such "self-referral" fails more often than it succeeds: less than a third of STD partners are successfully referred for medical evaluation. Partner referral by HIV patients is even less successful (despite frequent assurances by patients that they "will take care of it!"). Part of this failure is due to the reluctance of HIV patients to face their partners (fear of anger or reprisal); part is due to selective notification (denial that "nice" partners can be infected); and part to failure to convince partners (partner denial). Trusting the notification process to infected people alone is a luxury that society can ill afford.

Those exposed to HIV have a right to know. Important sexual (safer practices) and reproductive (postponing pregnancy) decisions depend on knowledge of exposure and its outcome. Many persons are unaware that their partners have histories of needle exposures or of bisexuality. The duty to warn people has compelling moral, legal, and historical foundations. In free societies, notification is a straightforward, confidential process. Medical workers who detect HIV infection report the case by name and address to the local health officer who then discreetly contacts the patient to counsel him or her and to obtain identifying information on sexual and needle partners. People are persuaded, not coerced, into voluntary cooperation. Although counseling is "mandatory," blood testing is optional.

Even if "treatment" for partners were to consist solely of personal counseling to discourage behaviors that facilitate transmission or accelerate disease progression, partner notification would be worthwhile.

A disease control procedure should be acceptable to people. Partner notification by health officers has been well received by the affected populations. The majority (70%–80%) of notified partners accept blood testing, and almost all who decline testing accept counseling. Although organized gay advocacy groups have generally opposed both HIV reporting by name and partner notification by health officers, when approached individually and sympathetically, gay men have generally cooperated.

Health officers are responsible for maintaining the physical security of HIV records; such records are also immune from any discovery process. They cannot be subpoenaed or released to potentially adversarial agents like insurance, police, or employer investigators. Whatever discrimination is suffered by infected people, none of it stems from disease notification to, or by, public health officers.

Notification initiatives are affordable, acceptable to patients, and effective in reaching high-risk people. It is well known to health officers that those at highest risk are least inclined to appear for counseling and least likely to use safer practices. While notification is not a panacea, it is one of the most useful measures for containing this tragic epidemic.

John Potterat *is an authority on AIDS and sexually transmitted disease (STD) control. He has published numerous articles in medical journals dealing with STD and AIDS control and is currently director of the STD and AIDS programs in Colorado Springs.*

▌SHARPENING YOUR CRITICAL ▌ THINKING SKILLS

1. Summarize Thomas's reasons for keeping AIDS testing anonymous.
2. Summarize Potterat's views in support of name reporting.
3. Do you agree or disagree with the following statement? Both writers believe that their approach will provide the greatest protection to the public health, but they differ in their approach. Explain.
4. Of the two basic approaches, which do you think would be most effective in reducing the spread of AIDS?

Visit *Human Biology's* Internet site, www.jbpub.com/humanbiology, to research opposing web sites and respond to questions that will help you clarify your own opinion. (See Point/Counterpoint: Furthering the Debate.)

FIGURE 8-18 Duping the AIDS Virus   The AIDS virus has a protein, gp120, in its capsid. When it infects cells, it produces more gp120 to make new capsids. However, some gp120 ends up in the infected cell's plasma membrane, thereby marking it. By genetically engineering a bacterium that can locate the infected cells through the gp120 marker, medical researchers may be able to hunt down and kill infected cells, stopping the spread of the virus.

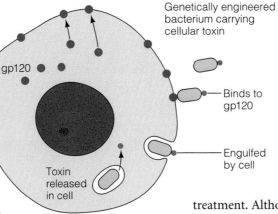
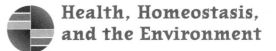

AIDS-infected cell

difficult, if not impossible.[7] Even within the body, HIV mutates fairly freely. In one study, for example, researchers analyzed viruses isolated from two infected patients. Over a 16-month period, they found 9–17 different varieties, all thought to have been formed from the original virus.

Making matters worse, HIV may be able to hide in the body. In 1988, a research team announced that out of 100 homosexual men studied, 4 initially showed antibodies to the AIDS virus, but slowly lost them. This process usually occurs only in the late stages of AIDS, when the immune system is too weak to produce antibodies. These men, however, had no overt symptoms of AIDS. Why?

Research suggests that HIV may take up residence in bone marrow stem cells that give rise to lymphocytes. If this is true, the virus can then be transmitted to new white blood cells by cell division. Thus, once the virus is in the body, it may be there forever. Eliminating the virus from the body may be virtually impossible. One of the upshots of these troubling findings is that if HIV can go into hiding, AIDS-infected blood donors may escape detection, even with the new genetic tests. AIDS-infected blood cells could unknowingly be passed to thousands of patients over the coming years. Despite these discouraging findings, researchers remain determined to find a cure for AIDS and a way to prevent it. In 1998, U.S. researchers launched the first global test of an HIV vaccine. This vaccine was the first of 25 approved for widespread testing on humans.

On another front, researchers have developed a genetically engineered weapon that could kill cells infected with HIV, possibly eliminating the disease after it has developed. As noted earlier, cells infected with the AIDS virus produce a protein known as gp120, which is part of the capsid (**FIGURE 8-18**). This protein also ends up in the plasma membrane of infected cells.

Researchers have genetically engineered a bacterium that binds to the gp120 protein. The bacterium carries with it a toxin that kills the HIV-infected cells. Preliminary studies indicate that noninfected cells are unharmed by this treatment. Although initial studies were disappointing, researchers hope that the technique can be improved or modified, making it possible to kill enough infected cells in AIDS patients to halt the disease. One question that must be answered before this procedure can be tried in people is whether AIDS-infected cells killed by this technique will degenerate and release active AIDS viruses that spread to other body cells.

Health, Homeostasis, and the Environment

Multiple Chemical Sensitivity

Human beings would not survive past early infancy without some means of protection against potentially harmful viruses and microorganisms that abound in our world. But what does protection have to do with homeostasis?

The answer is, plenty. Without the protective mechanisms you've studied in this chapter, homeostasis could not be maintained. Bacteria and viruses would take over. Therefore, the protective mechanisms of our bodies play a vital role in maintaining internal constancy. By regulating the concentration of viruses, bacteria, and other infectious agents, they help keep the cells of our tissues and organs alive so they can perform their specialized functions. Many of these cells, of course, help maintain homeostasis.

Despite its prowess, the immune system is not impenetrable. A growing body of evidence suggests that the chemically polluted world we live in may be having a tremendous impact on the immune systems of many people. That is, chemical imbalance in our world is creating physiological imbalances in people. Put yet another way, internal homeostasis is being compromised by alterations in environmental homeostasis. Consider the case of Richard Sharp.

Richard Sharp was a physicist for a major aviation company in California. Today, he is confined to two

---

[7]The AIDS virus is a retrovirus, an RNA virus whose RNA is used to produce DNA after invading a cell. This process, called *reverse transcription*, is fairly inaccurate and results in many mutant forms of the virus.

stripped-down rooms equipped with special filters that remove all air contaminants. Why?

Sharp is one of many Americans who has developed a disease known as **multiple chemical sensitivity (MCS).** Much to his regret, Sharp has become a prisoner in his own home, unable to venture forth into modern society without suffering extreme discomfort, even debilitation.

MCS is a truly modern disease thought to be caused by exposure to a number of common household and industrial chemicals, including formaldehyde, solvents, acrylic resins, mercury compounds, and pesticides. How do these chemicals create such debilitating symptoms?

The answer is, by acting through the immune system. Consider formaldehyde, which is commonly found in carpeting, plywood, furniture, and many other household products. Formaldehyde binds to naturally occurring proteins in the body. This process produces foreign substances that the immune system attacks. In other words, common household and industrial substances can turn the immune system against the body. Individuals become sensitive to low levels of chemicals over long periods.[8] As a result, this condition is typically referred to as **hypersensitivity.** Patients that have become sensitized to one chemical often react to other chemically similar substances.

The symptoms of MCS vary, ranging from life-threatening to mild. The most common symptoms are tension, memory loss, fatigue, sleepiness, headaches, confusion, and depression. Many victims of MCS experience gastrointestinal problems such as nausea, indigestion, and cramps. Some exhibit respiratory symptoms as well, including frequent colds, bronchitis, and shortness of breath. Skin rashes are not uncom-

mon. Many people report allergylike symptoms such as nasal stuffiness and sinus infections.

MCS is puzzling to victims, their families, and physicians. Individuals often experience a sudden deterioration in their health and are often labeled "psychiatric cases." People suffering from MCS must often get rid of all cleaning agents, pesticides, perfumes, deodorants, and other household chemicals.

The National Research Council estimates that 15% of the U.S. population experiences some degree of chemical hypersensitivity. Studies show that 5% of the workers exposed to an agent used in the manufacture of plastics, TDI (toluene diisocyanate), develop asthmalike symptoms. TDI apparently binds to proteins in the respiratory tract, creating foreign substances that stimulate a hypersensitivity reaction. Individuals who have been hypersensitized have difficulty breathing when exposed to TDI, tobacco smoke, and air pollutants. In Japan, 15% of all cases of asthma in men have been attributed to industrial exposure to chemicals.

Other chemicals bind to proteins in the skin, creating foreign substances to which the immune system reacts. Formaldehyde, for example, results in a condition called contact dermatitis, characterized by a skin rash. T cells attack and destroy the cells of the skin. Even low levels of formaldehyde in newsprint dyes, some cosmetics, and photographic papers are sufficient to induce rashes.

Other chemicals apparently act by suppressing immune function, making individuals more susceptible to infectious agents. Dioxins, PCBs, ozone, certain pesticides, and a variety of other chemical pollutants suppress the immune response in laboratory animals and humans. However, the overall significance of immune suppression and hypersensitivity in human populations remains unknown. Nonetheless, there are subtle and potentially far-reaching effects of toxic chemicals, again underscoring how homeostasis and health are dependent on a healthy environment.

---

[8]Massive exposures to certain chemicals may also elicit a hypersensitivity reaction.

SUMMARY

VIRUSES AND BACTERIA: AN INTRODUCTION
1. Two of the most important common agents are viruses and bacteria. *Viruses* are submicroscopic structures that consist of a nucleic acid core, consisting of either DNA or RNA, and an outer protein coat, the *capsid.*

2. Viruses are not true living organisms and must invade other organisms to reproduce. Viruses most often enter the body through the respiratory and digestive systems and spread from cell to cell in the bloodstream and lymphatic system. However, other avenues of en-

try are also possible—for example, sexual contact.

3. *Bacteria* (singular, bacterium) are single-celled microorganisms that consist of a circular strand of DNA and cytoplasm which is enclosed by a plasma membrane.

4. Although they are best known for their role in causing sickness and death, most bacteria perform useful functions.

THE FIRST AND SECOND LINES OF DEFENSE

5. The first line of defense against viruses, bacteria, and other infectious agents is the skin and the epithelia of the respiratory, digestive, and urinary systems. Some epithelia also produce protective chemical substances that kill microorganisms.

6. The second line of defense consists of cells and chemicals that the body produces to combat infectious agents that penetrate the epithelia.

7. One of the chief combatants in the second line of defense is the *macrophage*, a cell derived from the monocyte. Macrophages are found in connective tissue beneath epithelia, where they phagocytize infectious agents, preventing their spread. Neutrophils and monocytes also invade infected areas from the bloodstream and destroy bacteria and viruses.

8. Another combatant in the second line of defense consists of the chemicals released by damaged tissue, which stimulate arterioles in the infected tissue to dilate. The increase in blood flow raises the temperature of the wound. Heat stimulates macrophage metabolism, accelerating the rate of the destruction of infectious agents. Heat also speeds up the healing process.

9. Still other chemicals increase the permeability of the capillaries, causing plasma to flow into the wound and increasing the supply of nutrients for macrophages and other protective cells.

10. The increase in blood flow, the release of chemical attractants, and the flow of plasma into the wound constitute the *inflammatory response*.

11. Another part of the secondary line of defense are the *pyrogens*, chemicals released primarily by macrophages exposed to bacteria, which raise body temperature and lower iron availability, thus decreasing bacterial replication.

12. *Interferons*, a group of proteins released by cells infected by viruses, are also part of the second line of defense. Interferons travel to other virus-infected cells, where they inhibit viral replication.

13. The blood also contains the *complement proteins*, which circulate in an inactive state, becoming activated only when the body is invaded by bacteria. They too are part of the second line of defense.

14. Some of the complement proteins stimulate the inflammatory response. Others embed in the plasma membrane of bacteria. There they combine to form *membrane-attack complexes*, which create holes in the bacterial plasma membranes, killing these pathogens. Another complement protein binds to the invader, making it more easily phagocytized by macrophages.

THE THIRD LINE OF DEFENSE: THE IMMUNE SYSTEM

15. The *immune system* consists of billions of *lymphocytes* that circulate in the blood and lymph and take up residence in the *lymphoid organs* and *lymphoid tissues*, which are also part of the immune system.

16. The lymphocytes recognize *antigens*—foreign cells and foreign molecules, mostly proteins and large-molecular-weight polysaccharides.

17. *T and B cells* are two types of lymphocytes produced in red bone marrow. The T cells become immunocompetent—able to respond to a particular antigen in the thymus. B cells gain this ability in the bone marrow. During this process, the T and B cells produce membrane receptors that bind to specific antigens.

18. Immunocompetent B cells encounter antigens (often presented to them by macrophages) to which they are programmed to respond. They then begin to divide, forming *plasma cells* and *memory cells*. Plasma cells produce *antibodies*. The memory cells enable the body to respond more quickly to future invasions by the same antigen.

19. Antibodies are small protein molecules that bind to specific antigens, destroying them either by precipitation, agglutination, or neutralization, or by activation of the complement system.

20. When T and B cells first encounter an antigen, they react slowly. The initial response is called the *primary response*. Because the body responds slowly at first, there is often a period of illness before the pathogen is removed by the immune system.

21. Numerous memory cells produced during the first assault ensure that a reappearance of the antigen will elicit a much faster and more powerful reaction, the *secondary response*. Consequently, the pathogen is usually vanquished before symptoms of illness occur. The resistance created by a response to an antigen is called *immunity*.

22. When activated by an antigen, T cells multiply and differentiate, forming *memory T cells, cytotoxic T cells, helper T cells*, and *suppressor T cells* whose functions are summarized in **TABLE 8-2**.

23. A solution containing a dead or weakened virus, bacterium, or bacterial toxin that is injected into people to create *active immunity* is called a *vaccine*.

24. *Passive immunity* can be achieved by injecting antibodies into a patient or by the transfer of antibodies from a mother to her baby through the bloodstream or breast milk. Passive immunity is short-lived, lasting at most only a few months, compared to active immunity which lasts for years.

PRACTICAL APPLICATIONS: BLOOD TRANSFUSIONS AND TISSUE TRANSPLANTATION

25. Blood transfusions require careful matching of donor and recipient blood types. Tissue transplantation requires similar matching. In such instances, only cells from the same individual or an identical twin will be accepted. All others are rejected by the T cells, unless the system is suppressed with drugs.

DISEASES OF THE IMMUNE SYSTEM

26. The most common malfunctions of the immune system are *allergies*, extreme reactions to some antigens.

27. Allergies are caused by *IgE antibodies*, produced by plasma cells. IgE antibodies bind to mast cells, which causes them to release histamine and other chemical substances that induce the symptoms of an allergy—production of mucus, sneezing, and itching.

28. *Autoimmune diseases*, another immune system disorder, result from an immune attack on the body's own cells. Autoimmune diseases may occur when normal proteins are modified by chemicals or genetic mutations so that they are no longer recognizable as self. Other possible causes are the sudden presence of proteins that are normally isolated from the immune system and exposure to antigens that are nearly identical to body proteins.

AIDS: THE DEADLY VIRUS

29. *AIDS* is a disease of the immune system caused by *HIV*, a virus that attacks helper T cells (T4 cells), severely impairing a person's immune system.

30. AIDS progresses through three stages: asymptomatic carrier, AIDS-Related Complex, and full-blown AIDS.

31. During the first phase, no symptoms appear, although an individual is highly infectious—able to transmit the disease to others.

32. During the second phase, patients grow progressively weaker as their immune system falters. Lymph nodes swell and patients report persistent or recurrent fevers and a persistent cough. Mental deterioration may also occur.

33. During the last phase, patients suffer from severe weight loss and weakness. Many develop cancer and bacterial infections because of their diminished immune response.

34. AIDS is spread through body fluids during sexual contact and blood transfusions or through needles shared by drug users.

35. Stopping the virus has proved difficult, in large part because symptoms of AIDS do not appear until several months to several years after the initial HIV infection.

36. Fortunately, several drugs have been developed that slow down the progression of the disease. Numerous researchers are developing vaccines that they hope will protect people and eventually eradicate the virus.

HEALTH, HOMEOSTASIS, AND THE ENVIRONMENT: MULTIPLE CHEMICAL SENSITIVITY

37. The immune system and protective mechanisms that constitute the first and second lines of defense protect the body from harmful viruses and microbes. In a sense, they help maintain a constant internal state either directly, by warding off infectious agents, or secondarily, by protecting other body cells that are essential to homeostasis.

38. Chemicals can damage the immune system, upsetting homeostasis.

39. Many individuals suffer from *multiple chemical sensitivity*. Chronic exposure to low levels of pollutants or short-term exposure to high levels may alter the immune system, causing a wide range of symptoms.

40. Some toxic chemicals cause *hypersensitivity*, evoking allergylike symptoms. Others stimulate autoimmune responses, and still others cause immune suppression.

Critical Thinking

THINKING CRITICALLY— ANALYSIS

This Analysis corresponds to the Thinking Critically scenario that was presented at the beginning of this chapter.

This chapter offers a brief overview of the topic of vaccination. The first conclusion that you might draw is that the text does not provide enough information to analyze the claims made by your friends about vaccination. You'd be advised to study an immunology book as well as the scientific literature. Once you've read some more, you will be better able to discern whether vaccination is worth the relatively small risk.

Despite the brief coverage, this chapter does point out that infectious diseases have not been eliminated. Vaccination has only kept them under control by reducing the potential population of hosts. Thus, if large numbers of people are unvaccinated against infectious disease, outbreaks could occur. The results could be quite dramatic because we're living on a crowded planet, and infectious organisms spread quickly in crowded environments.

EXERCISING YOUR CRITICAL THINKING SKILLS

You have been selected as a juror for a trial. The case you are about to hear involves a physician who refused to perform surgery on an AIDS patient. The patient was in an automobile accident that ruptured his spleen, causing internal bleeding. The physician refused to perform an operation that could have saved the patient's life because she was afraid of contracting AIDS.

Consider the following facts presented by the attorneys for the plaintiffs (the AIDS patient's family). The plaintiffs argue that the physician violated her code of ethics, which obligates her to treat all patients. They also argue that she knowingly allowed a patient to die and that she should be punished by having to pay damages as compensation for the lost life. Several years earlier, the AIDS patient received a transfusion of HIV-contaminated blood, and had at the time of the accident, only 3–6 months to live. Nevertheless, these months were valuable to him and to his family.

The defense attorneys admit that the physician refused treatment, thereby contributing to the premature death of her patient. They say, however, that she acted rightfully. During surgery, sharp instruments frequently pierce the protective gloves of the surgical team, exposing them to the patient's blood, which, in this case, was contaminated with HIV. Refusing to operate protected not only the physician, but also her entire surgical team, a group of people who will, over the course of the ensuing years, save hundreds of lives. The physician was thus considering the greater good—the benefit to the public of her services and those of her team. The surgeon also had a husband and two children whose needs she was also taking into account.

How would you decide such a case? Should the physician be forced to pay damages? Why? Or was she correct in her decision? Why? What would you have done if you were the physician?

As a side note, hospitals often assemble teams of doctors and other medical personnel who are willing to treat AIDS patients to avoid problems such as the one described above.

TEST OF CONCEPTS

1. Describe the structure of a typical virus.
2. Based on your previous studies (Chapter 3), in what ways are bacteria similar to human cells, and in what ways are they different?
3. The human body consists of three lines of defense. Describe what they are and how they operate.
4. Describe the inflammatory response, and explain how it protects the body.
5. Define each of the following terms, and explain how they help protect the body: pyrogen, interferon, and complement proteins.
6. The first and second lines of defense differ substantially from the third line of defense. Describe the major differences.
7. How does the immune system detect foreign substances?
8. Describe how the B cell operates. Be sure to include the following terms in your discussion: bone marrow, immunocompetence, plasma membrane receptors, primary response, plasma cell, antigen, antibody, and secondary response.
9. Describe the four mechanisms by which antibodies "destroy" antigens.
10. Describe the events that occur after a T cell encounters its antigen.
11. What is the difference between active and passive immunity?
12. A child is stung by a bee, swells up, and collapses, having great difficulty breathing. What has happened? What can be done to save the child's life?
13. What is an autoimmune disease? Explain the reasons it forms.
14. What is AIDS? What are the symptoms? What causes it?

TOOLS FOR LEARNING

Tools for Learning is an on-line student review area located at this book's web site HumanBiology (www.jbpub.com/humanbiology). The review area provides a variety of activities designed to help you study for your class:

Chapter Outlines. We've pulled out the section titles and full sentence sub-headings from each chapter to form natural descriptive outlines you can use to study the chapters' material point by point.

Review Questions. The review questions test your knowledge of the important concepts and applications in each chapter. Written by the author of the text, the review provides feedback for each correct or incorrect answer. This is an excellent test preparation tool.

Flash Cards. Studying human biology requires learning new terms. Virtual flash cards help you master the new vocabulary for each chapter.

Figure Labeling. You can practice identifying and labeling anatomical features on the same art content that appears in the text.

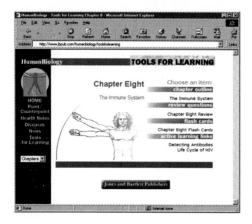

Active Learning Links. Active Learning Links connect to external web sites that provide an opportunity to learn basic concepts through demonstrations, animations, and hands-on activities.

THE VITAL EXCHANGE

RESPIRATION

Air-conducting portion of the human lung.

Thinking Critically

Representatives of the tobacco industry still claim that there is no proof that cigarette smoking causes cancer, despite the fact that over 40 epidemiological studies show a strong correlation between smoking and lung cancer. Most of these studies correct for other factors that might cause lung cancer such as exposure to carcinogenic chemical pollutants in the workplace.

Why do representatives from the tobacco industry persist in claiming that there is no proof that smoking causes lung cancer? How would you respond to their assertion? Can a study be devised to prove the connection? What critical thinking rules does this exercise rely on? ■

GEORGE F. EATON WAS A ROBUST AND HAND-some Irishman who grew up in eastern Massachusetts, married, and raised three children. To support his family, Eaton worked as a fire fighter for 30 years. Fire fighting is dangerous work, in part because it exposes men and women to smoke containing numerous potentially harmful air pollutants.

Upon retirement, Eaton moved to Cape Cod, but his retirement years were cut short by **emphysema**—a debilitating respiratory disease resulting from the breakdown of the air sacs, or **alveoli,** in the lungs, where oxygen and carbon dioxide are exchanged between the air and the blood. As the walls of the alveoli degenerate, the surface area for the diffusion of oxygen and carbon dioxide gradually decreases. Emphysema is irreversible and incurable. Patients suffer from shortness of breath; eventually, even mild exertion becomes trying, prompting one patient to describe the disease as a kind of "living hell."

The degeneration of the lungs in patients with emphysema creates a domino effect. One of the dominoes is the heart. Because oxygen absorption in the lungs declines quite substantially over time, the heart of an emphysemic patient must work harder and harder. This effort puts additional strain on this already hardworking organ and can lead to heart failure.

George Eaton died a slow and painful death as his lungs grew increasingly more inefficient. Relatives mourning his death blamed it on the pollution to which he was exposed while working for the fire department. Fire fighting, however, was probably only part of the cause, for Eaton had smoked most of his adult life. Cigarette smoking is the leading cause of emphysema.

This chapter describes the respiratory system—its structure and function and the diseases that affect it, such as emphysema.

Structure of the Human Respiratory System

The human respiratory system functions automatically, drawing air into the lungs, then letting it out, in a cycle that repeats itself about 16 times per minute at rest—or about 23,000 times per day. If you got a dollar for each time you took a breath, you'd be a millionaire in a month and a half.

The respiratory system supplies oxygen to the body and gets rid of carbon dioxide. Oxygen, of course, is needed for cellular respiration; carbon dioxide is a waste product of this process. In its role as a provider of oxygen and a disposer of carbon dioxide waste, the respiratory system helps to maintain the constant internal environment necessary for normal cellular metabolism. Thus, like many other systems, it plays an important role in homeostasis.

The respiratory system consists of two basic parts: an air-conducting portion and a gas-exchange portion (**TABLE 9-1**). The air-conducting portion is an elaborate set of passageways that transports air to and from the **lungs,** two large, saclike organs in the thoracic cavity (**FIGURE 9-1A**). Like the arteries of the body, these passageways start out large, then become progressively smaller and more numerous, branching profusely in the lungs, where the exchange of oxygen and carbon dioxide between the air and the blood takes place.

The lungs are the gas-exchange portion of the respiratory system. Each lung has millions of tiny, thin-walled alveoli (singular, alveolus) (**FIGURE 9-1B**). The walls of the alveoli contain numerous capillaries that absorb oxygen from the inhaled air and release carbon dioxide (**FIGURE 9-1B**).

The Conducting Portion of the Respiratory System Moves Air In and Out of the Body and Also Filters and Moistens Incoming Air

Air enters the respiratory system through the nose and mouth, then is drawn backward into the **pharynx**

TABLE 9-1

Summary of the Respiratory System

Organ	Function
Air conducting	
Nasal cavity	Filters, warms, and moistens air; also transports air to pharynx
Oral cavity	Transports air to pharynx; warms and moistens air; helps produce sounds
Pharynx	Transports air to larynx
Epiglottis	Covers the opening to the trachea during swallowing
Larynx	Produces sounds; transports air to trachea; helps filter incoming air; warms and moistens incoming air
Trachea and bronchi	Warm and moisten air; transport air to lungs; filter incoming air
Bronchioles	Control air flow in the lungs; transport air to alveoli
Gas exchange	
Alveoli	Provide area for exchange of oxygen and carbon dioxide

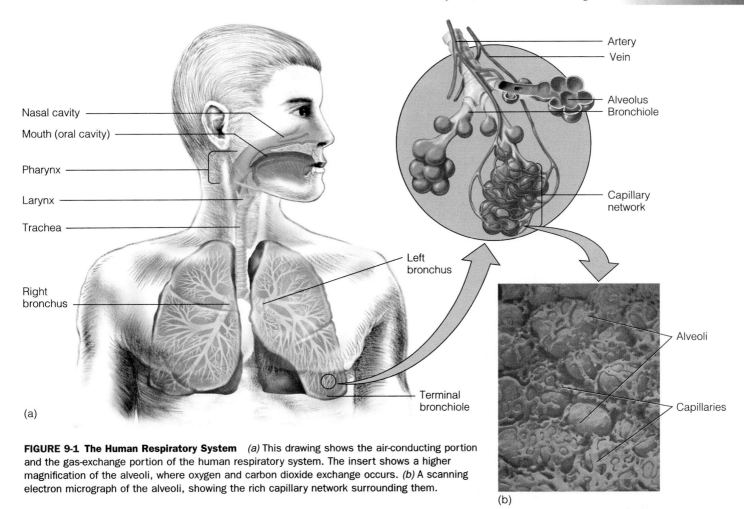

Nasal cavity

Mouth (oral cavity)

Pharynx

Larynx

Trachea

Right bronchus

(a)

Artery

Vein

Alveolus
Bronchiole

Capillary network

Left bronchus

Terminal bronchiole

Alveoli

Capillaries

(b)

FIGURE 9-1 The Human Respiratory System *(a)* This drawing shows the air-conducting portion and the gas-exchange portion of the human respiratory system. The insert shows a higher magnification of the alveoli, where oxygen and carbon dioxide exchange occurs. *(b)* A scanning electron micrograph of the alveoli, showing the rich capillary network surrounding them.

(FAIR-inks) (**FIGURE 9-2**). The pharynx opens into the nose and mouth in the front and joins below with the larynx (LAIR-inks). The **larynx,** or voice box, is a **rigid**, hollow structure that houses the **vocal cords** (**FIGURE 9-2**). To feel it, gently put your fingers alongside your throat, then swallow. The structure that moves up and down is the larynx.

The larynx opens into the **trachea** (TRAY-kee-ah) or windpipe below. You can feel the trachea beneath your Adam's apple, the protrusion of the laryngeal cartilage on your neck.

As explained in Chapter 5, food is prevented from entering the larynx by the **epiglottis** (ep-eh-GLOT-tis), a flap of tissue that closes off the opening to the larynx during swallowing. Occasionally, however, food goes the wrong way, accidentally entering the larynx and trachea.

This unfortunate event leads to violent coughing, a reflex that helps eject the food from the trachea. If the food cannot be dislodged by coughing, steps must be taken to remove it—and fast—or the person will suffocate. Health Note 9-1 explains what to do when a person chokes.

The trachea is a short, wide duct. Starting in the neck

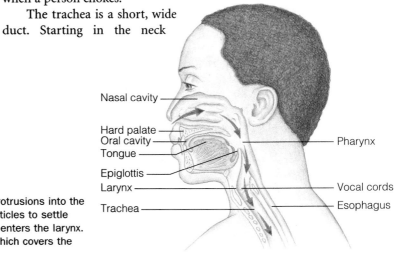

Nasal cavity

Hard palate
Oral cavity
Tongue

Epiglottis
Larynx

Trachea

Pharynx

Vocal cords

Esophagus

FIGURE 9-2 Uppermost Portion of the Respiratory System Bony protrusions into the nasal cavity (not shown here) create turbulence that causes dust particles to settle out on the mucous coating. Notice that air passing from the pharynx enters the larynx. Food is kept from entering the respiratory system by the epiglottis, which covers the laryngeal opening during swallowing.

Health Note 9-1

First Aid for Choking That May Save Someone's Life

What do you do if you encounter a person who is choking? If the person can cough, speak, or breathe, do not interfere. Encourage the individual to continue coughing. If he or she cannot cough, speak, or breathe, **CALL 911 OR YOUR LOCAL EMERGENCY NUMBER IMMEDIATELY.**

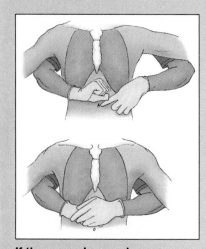

If the person is conscious:
- Stand behind him and wrap your arms around his waist.
- Make a fist with one hand (thumb outside the fist), and place the fist just above the navel in the

middle of the abdomen. Be sure that your fist is well below the lower tip of the sternum.
- Wrap the other hand around the fist and perform cycles of five quick inward and upward thrusts. After every five thrusts, check the person and repeat the cycles until the person begins to cough, speak, or breathe on his own.

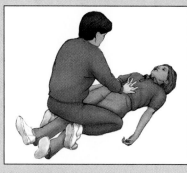

If the person is unconscious:
- Carefully position her on her back.
- Open the airway by placing one hand on the forehead and one hand on the chin and gently tilt the head backward (head-tilt/chin-lift). Place your ear next to

the person's mouth, listening for air exchange; feel for air exchange with your cheek; and look to see the chest rise and fall during air exchange.
- If you do not see, hear, or feel air exchange, pinch the person's nose closed and seal your mouth around the victim's mouth and give two slow breaths. The chest should rise and fall gently.
- If the first breath did not go in, reposition the victim's head and attempt another breath.
- If the breaths did not go in, straddle the victim's thighs and place the heel of one hand just above the navel and place the other hand on top, interlocking your fingers.
- Give five quick inward and upward thrusts.
- Return to the victim's head; perform a finger sweep of her mouth, searching for a dislodged object; perform a head-tilt/chin-lift; and give one slow breath.
- If the breath will not go in, give five more abdominal thrusts, followed by a finger sweep, and one breath.

---

below the larynx, it enters the thoracic cavity, where it divides into two large branches, the right and left bronchi (BRON-kee). The **bronchi** (singular, bronchus) enter the lungs alongside the arteries and veins. Inside the lungs, the bronchi branch extensively, forming progressively smaller tubes that carry air to the alveoli.

The trachea and bronchi are reinforced by hyaline cartilage, which prevents the organs from collapsing during breathing, thus ensuring a steady flow of air in and out of the lungs.

The smallest bronchi in the lungs branch to form **bronchioles** (BRON-kee-ols), which lead to the alveoli. Like the arterioles of the circulatory system, the walls of the bronchioles consist largely of smooth muscle. This

permits them to open and close, and thus provides a means of controlling air flow in the lungs. During exercise or times of stress, the bronchioles open and the flow of air into the lungs increases. This homeostatic mechanism helps meet the body's need for more oxygen, in much the same way that the arterioles of capillary beds dilate to let more blood into body tissues in times of need.

The respiratory system is in direct contact with the external environment and is therefore quite vulnerable to infectious organisms and pollutants present in the atmosphere. Not surprisingly, the respiratory system has evolved protective mechanisms to maintain homeostasis. The conducting portion, for example, filters many impurities from the air we breathe, especially airborne

- Continue this cycle until the victim begins to cough, speak, or breathe on his or her own.

Perform the same technique for children as you would for an adult. Children are considered those who are over one year of age and under the age of eight. The size of the child must also be a consideration.

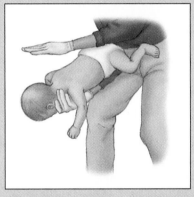

For infants and very small children who cannot cough, speak, or breathe:

- Pick up the infant and place her face down on your forearm.
- Give the infant five back blows with the heel of your hand between the shoulder blades.

- Reposition the infant face up on the opposite forearm and give five chest thrusts using the pads of two or three fingers in the lower half of the sternum.
- Continue this cycle until the infant can breathe on his or her own or until the infant goes unconscious.

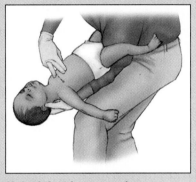

If the infant is unconscious:

- Perform a head-tilt/chin-lift and look, listen, and feel for air exchange.
- If no exchange is present, seal your mouth around the mouth and nose of the infant and give two slow breaths. The chest should rise and fall gently.

- If the first breath did not go in, reposition the infant's head and attempt another breath.
- If the breaths did not go in, position the infant for five back blows, five chest thrusts; then, check his mouth for an object and one give breath.
- Continue this cycle until the infant begins to breathe on his or her own.

In order to be skilled in these and other lifesaving skills, you should take a first aid and CPR class. Classes are offered in most communities through the National Safety Council. Call 1-800-621-7618 for information on classes in your area.

SOURCE: Adapted by Jennifer Belcher, *EMT-B* from the guidelines of the American Red Cross, National Safety Council and the American Heart Association.

Visit *Human Biology's* Internet site, www.jbpub.com/humanbiology, for links to web sites offering more information on this topic.

www.jbpub.com/humanbiology

particles such as dust and bacteria. Some of these particles are small and capable of penetrating deeply into the lung. Some particles contain toxic metals such as mercury, which can cause lung cancer.

As a rule, larger particles are deposited as the inhaled air travels through the nose and the passageways leading to the lungs. Particles removed from the air in the nose, trachea, and bronchi are trapped in a layer of **mucus** (MEW-kuss), a thick, slimy secretion deposited on the inside of these structures (**FIGURE 9-3**). Mucus is produced by cells in the epithelium, the **mucous cells.**[1]

The epithelium of the respiratory tract also contains numerous ciliated cells. The cilia of these cells beat upward toward the mouth, transporting mucus containing bacteria and dust particles. Operating day and night, they sweep the mucus toward the oral cavity, where it can be swallowed or expectorated (spit out). This protects the respiratory tract and lungs from bacteria and potentially harmful particulates. It is another of the body's homeostatic mechanisms.

Like all homeostatic mechanisms, the respiratory mucous trap is not invincible. Bacteria and viruses do occasionally penetrate the lining, causing respiratory infections. Making matters worse, sulfur dioxide, a pollutant in cigarette smoke and urban air pollution, temporarily paralyzes, and may even destroy, cilia. Sulfur

[1]You will notice that the adjective form is spelled "mucous" (as in mucous cell) and the noun form is spelled "mucus."

FIGURE 9-3 Mucous Trap
(a) Drawing of the lining of the trachea. Mucus produced by the mucous cells of the lining of much of the respiratory system traps bacteria, viruses, and other particulates in the air. The cilia transport the mucus toward the mouth.
(b) Higher magnification of the lining showing a mucous cell and ciliated epithelial cells.

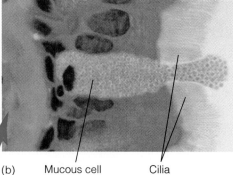

(b)   Mucous cell   Cilia

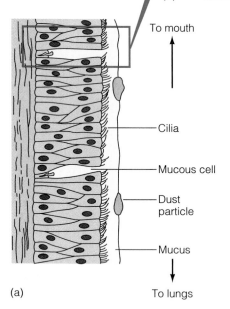

To mouth

Cilia

Mucous cell

Dust
particle

Mucus

To lungs

(a)

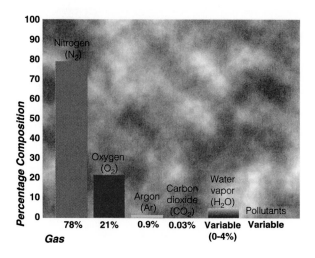

FIGURE 9-4 Composition of Air

dioxide gas in the smoke of a single cigarette, for instance, will paralyze the cilia for an hour or more, permitting bacteria and toxic particulates to be deposited on the lining of the respiratory tract, even enter the lungs. Ironically, the cilia of a smoker are paralyzed when they are needed the most!

Because smoking impairs a natural protective mechanism, it should come as no surprise that smokers suffer more frequent respiratory infections than non-smokers. Research shows that alcohol also paralyzes the respiratory system cilia, explaining why alcoholics are prone to respiratory infections.

The conducting portion of the respiratory system also moistens and warms the incoming air. Beneath the epithelium of the respiratory tract is a rich network of capillaries that releases moisture and heat. Moisture protects the lungs from drying out, and heat protects them from cold temperatures. Except in extremely cold weather, by the time inhaled air reaches the lungs, it is nearly saturated with water and is warmed to body temperature.

On the way out of the respiratory system, much of the water that was added to the air condenses on the lining of the nasal cavity, which was cooled by the evaporation of water as the air was drawn in. This mechanism is an adaptation that conserves water and also accounts for the reason our noses tend to drip in cold weather.

The Alveoli Are the Site of Gaseous Exchange

The air we breathe consists principally of nitrogen and oxygen, with small amounts of carbon dioxide and

other gases (**FIGURE 9-4**). Oxygen in the atmosphere is generated by the photosynthetic activity of plants and photosynthetic single-celled organisms. Oxygen is vital to humans and virtually all other living organisms. A constant supply must be delivered to body cells in order for us to maintain cellular energy production.

Oxygen is delivered to the lungs by the conducting portion of the respiratory system. Oxygen is transported via the bronchioles to the alveoli. Each lung contains an estimated 150 million alveoli, giving the lung the appearance of an angel food cake (**FIGURES 9-5A** and **9-5B**). The alveoli are the site of oxygen absorption. If the alveoli could be flattened, they'd produce a membrane with a surface area of 60–80 square meters—approximately the size of a tennis court.

As shown in **FIGURE 9-6**, the alveoli are lined by a single layer of flattened cells, called **Type I alveolar cells,** and are surrounded by an extensive capillary bed. These cells permit gases to move into and out of the alveoli with great ease. Thus, the large surface area of the lungs created by the alveoli and the relatively thin barrier between the blood and the alveolar air result in a rather rapid diffusion of gases across the alveolar wall. The alveoli therefore provide another example of the marriage of form and function produced by evolution.

Another important cell in the alveoli is the **alveolar macrophage,** sometimes known as the **dust cell.** Alveolar macrophages remove dust and other particulates that reach the lungs (**FIGURE 9-6**). Dust cells wander freely around and through the alveoli, an ever-vigilant police force, engulfing foreign material that has escaped filtration. Once filled with particulates, the macrophages accumulate in the connective tissue surrounding the alveoli. A smoker's lungs or the lungs of

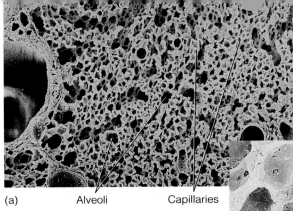

(a) Alveoli Capillaries

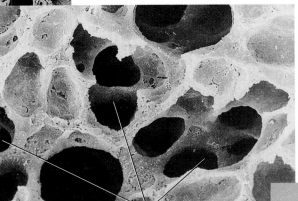

(b) Alveoli

FIGURE 9-5 Alveoli (a) A scanning electron micrograph of the lung showing many alveoli. The smallest openings are capillaries surrounding the alveoli. (b) A higher magnification scanning electron micrograph of lung tissue showing alveoli. (c) A transmission electron micrograph showing several alveoli and the close relationship of the capillaries containing RBCs (dark structures inside the capillaries). This close relationship between capillaries and alveoli ensures the rapid transport of oxygen and carbon dioxide.

an urban resident are therefore often blackened by the accumulation of smoke and dust particles.

Another important cell is the **Type II alveolar cell** (**FIGURE 9-6**). Type II alveolar cells are large, round cells that produce a chemical substance called **surfactant** (sir-FACK-tant), a detergentlike substance that dissolves in the thin layer of water lining the alveoli. Surfactant is nature's tension remover. Let me explain.

The water covering the alveolar lining produces **surface tension,** which results from hydrogen bonds that form between water molecules (Chapter 2). In water, hydrogen bonds draw water molecules together. At the surface of a watery fluid, the hydrogen bonds draw water molecules together more tightly than elsewhere, creating a slightly denser region referred to as surface tension. Surface tension on a pond permits some insects such as water striders to walk on water and is the reason a drop of water beads up on your car windshield (**FIGURE 9-7**).

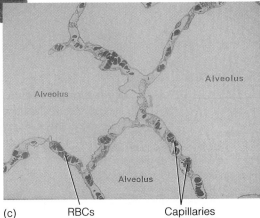

(c) RBCs Capillaries

In the alveoli, surface tension tends to draw the walls of the alveoli inward. Surfactant, however, reduces surface tension in the alveoli, decreasing forces that might otherwise cause the alveoli to collapse.

Some premature babies lack sufficient surfactant. High surface tension in the alveoli causes the larger alveoli to collapse, resulting in a dramatic, life-threatening condition known as **respiratory distress syndrome** or **RDS.** Children can be treated with an artificial surfactant until they produce enough of their own.

Respiratory distress syndrome
www.jbpub.com/humanbiology

FIGURE 9-6 The Alveolar Macrophage Drawing of the alveolus showing Type I and Type II alveolar cells and macrophages or dust cells.

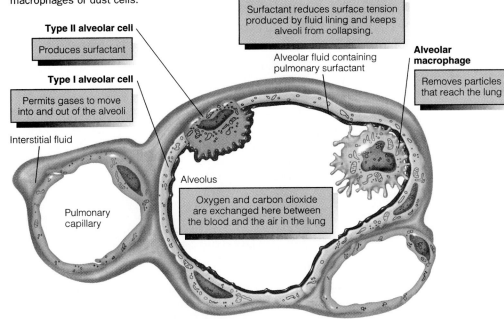

Type II alveolar cell
Produces surfactant

Type I alveolar cell
Permits gases to move into and out of the alveoli

Interstitial fluid

Alveolus

Pulmonary capillary

Oxygen and carbon dioxide are exchanged here between the blood and the air in the lung

Surfactant reduces surface tension produced by fluid lining and keeps alveoli from collapsing.

Alveolar fluid containing pulmonary surfactant

Alveolar macrophage
Removes particles that reach the lung

FIGURE 9-7 Surface Tension Some insects can walk on water because of surface tension, the tight packing of water molecules along the surface of a pond.

Functions of the Respiratory System

The chief functions of the respiratory system are to (1) replenish the blood's oxygen supply and (2) rid the blood of excess carbon dioxide, but the respiratory system serves other functions as well. The vocal cords, located in the larynx, produce sounds that allow people to communicate. The respiratory system houses the **olfactory membrane** (ol-FAC-tore-ee), a specialized patch of epithelium in the roof of the nasal cavity that allows humans to perceive odors. The respiratory system also helps maintain pH balance by its influence on carbon dioxide levels.

In Humans, Sound Is Produced by the Vocal Cords and Is Influenced by the Tongue and Oral Cavity

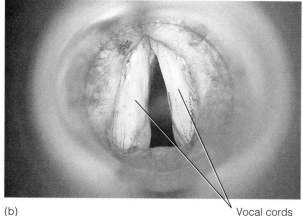

FIGURE 9-8 A Coyote Howls The vocal cords enable coyotes and many other species to create a wide assortment of sounds essential to communication.

Phonation, the production of sounds, is critical to many members of the animal kingdom. The eerie cry of the coyote, for example, signals to the pack a member's whereabouts and helps the members of the pack stay in contact (**FIGURE 9-8**). The coyote's growl can signal to an intruder its intention to defend itself.

Except perhaps at sporting events, humans exhibit a much wider range of sounds for communication than do other animals. These sounds are produced by two elastic ligaments inside the larynx, the **vocal cords,** which vibrate as air is expelled from the lungs (**FIGURE 9-9**). The sounds generated by the vocal cords are modified by changing the position of the tongue and the shape of the oral cavity.

The vocal cords vary in length and thickness from one person to the next. They also vary between men and women. Most men, for example, have longer, thicker vocal cords than women and therefore have deeper voices as a result of testosterone, the male sex hormone produced by the testes.

The vocal cords are like the strings of a guitar or violin: they can be tightened or loosened, producing sounds of different pitch. The tighter the string on a guitar, the higher the note. In humans, muscles in the larynx that attach to the vocal cords make this adjustment possible. Relaxing the muscles lowers the tension on the cords, dropping the tone. Tightening the vocal cords has the opposite effect.

Bacterial and viral infections of the larynx result in a condition known as laryngitis (lair-in-JITE-iss). **Laryngitis** is an inflammation of the lining of the larynx and the vocal cords. This thickens the cords, causing a person's voice to lower. Laryngitis may also be caused by tobacco smoke, alcohol, excessive talking, shouting, coughing, or singing, all of which irritate the vocal cords. In young children, inflammation results in a swelling of the lining that may impede the flow of air and impair breathing, resulting in a condition called the **croup** (crewp).

Oxygen and Carbon Dioxide Diffuse Rapidly across the Alveolar and Capillary Walls

Deoxygenated blood entering the lungs arrives via the pulmonary arteries. This blood, as noted in Chapter 6, is laden with carbon dioxide picked up as it circulates through body tissues. In the capillary beds of the lungs,

Epiglottis

Thyroid cartilage

Ventricular fold (false vocal cord)

True vocal cord

Tracheal cartilages

(a)

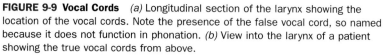

(b)

Vocal cords

FIGURE 9-9 Vocal Cords *(a)* Longitudinal section of the larynx showing the location of the vocal cords. Note the presence of the false vocal cord, so named because it does not function in phonation. *(b)* View into the larynx of a patient showing the true vocal cords from above.

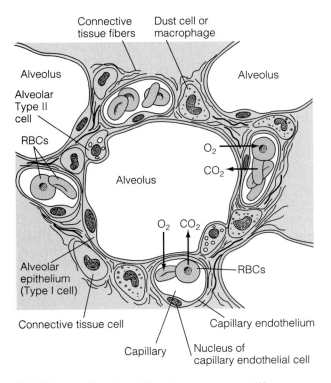

FIGURE 9-10 Close-Up of the Alveolus Oxygen diffuses out of the alveolus into the capillary. Carbon dioxide diffuses in the opposite direction, entering the alveolar air that is expelled during exhalation.

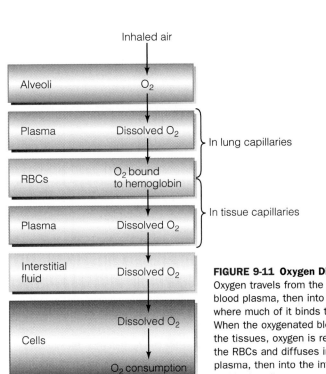

FIGURE 9-11 Oxygen Diffusion Oxygen travels from the alveoli into the blood plasma, then into the RBCs, where much of it binds to hemoglobin. When the oxygenated blood reaches the tissues, oxygen is released from the RBCs and diffuses into the plasma, then into the interstitial fluid and body cells.

carbon dioxide is released, and oxygen is added, replenishing supplies depleted as blood flows through body tissues.

Carbon dioxide and oxygen readily diffuse across the capillary and alveolar walls, a process driven by the concentration difference between the alveoli and capillaries.[2] **FIGURE 9-10** illustrates the direction in which these gases flow. Oxygen in the alveolar air first diffuses through the alveolar epithelium, then into the extracellular fluid surrounding the capillaries. It then diffuses through the capillary wall and into the blood plasma (**FIGURE 9-11**). From here, oxygen molecules cross the plasma membrane of the red blood cells (RBCs) and bind to hemoglobin molecules in their cytoplasm. About 98% of the oxygen in the blood is carried in the RBCs bound to hemoglobin; the rest is dissolved in the plasma and cytoplasm of the RBCs.

In order to understand the details of carbon dioxide diffusion in the lung, we must go

back to the body cells, where CO_2 is formed. In body tissues, carbon dioxide diffuses out of the cells and enters the blood plasma. Much of it then diffuses into the RBCs, where it is converted to carbonic acid, H_2CO_3 (**FIGURE 9-12**). This reaction is catalyzed by the enzyme carbonic anhydrase, found inside RBCs.

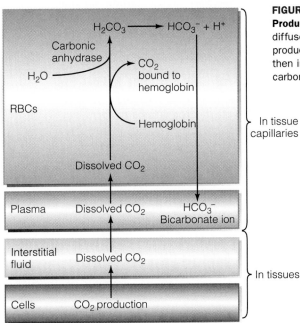

FIGURE 9-12 Bicarbonate Ion Production Carbon dioxide (CO_2) diffuses out of body cells where it is produced and into the tissue fluid, then into the plasma. Although some carbon dioxide binds to hemoglobin and some is dissolved in the plasma, most is converted to carbonic acid (H_2CO_3) in the RBCs. Carbonic acid dissociates and forms hydrogen ions and bicarbonate ions. Hydrogen ions remain inside the RBCs, but most bicarbonate diffuses into the plasma where it is transported.

---

[2]Physiologists actually speak of differences in partial pressure. The partial pressure of a gas is caused when gas molecules collide with a surface. The partial pressure of oxygen is proportional to the impact of all the oxygen molecules striking the alveolar wall. Thus, the partial pressure is proportional to the concentration of the gas molecules.

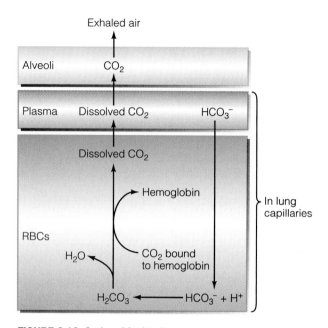

FIGURE 9-13 Carbon Dioxide Production from Bicarbonate
When the carbon dioxide-laden blood reaches the lungs, bicarbonate ions diffuse back into the RBCs and combine with hydrogen ions in RBCs forming carbonic acid, which dissociates, forming carbon dioxide gas. CO_2 diffuses out of the RBCs into the plasma, then into the alveoli.

As shown at the top of **FIGURE 9-12**, carbonic acid molecules readily dissociate to form bicarbonate ions and hydrogen ions. Many of the bicarbonate ions then diffuse out of the RBCs into the plasma, where they are carried with the blood. Hydrogen ions stay behind. A small percentage (15%–25%) of the carbon dioxide given off by body cells binds to hemoglobin (but not at the oxygen binding site), and an even smaller percentage (7%) dissolves in the plasma (neither is shown in **FIGURE 9-12**).

When blood rich in carbon dioxide reaches the lungs, bicarbonate ions in the plasma reenter the RBCs, where they combine with hydrogen ions to form carbonic acid (**FIGURE 9-13**). Carbonic acid, in turn, reforms carbon dioxide. The CO_2 then diffuses out of the RBCs into the blood, and then into the alveoli down a concentration gradient. Carbon dioxide is eventually expelled from the lungs during exhalation.

The uptake of oxygen and the discharge of carbon dioxide in the lungs "replenish" the blood in the alveolar capillaries. The oxygenated blood then flows back to the left atrium of the heart via the pulmonary veins. From here it empties into the left ventricle and is pumped to the body tissues via the aorta and its multitude of branches.

Section 9-3

Breathing and the Control of Respiration

Air moves in and out of the lungs in much the same way that it moves in and out of the bellows that blacksmiths use to fan their fires. Breathing, however, is largely an involuntary action, controlled by the nervous system.

Air Is Moved In and Out of the Lungs by Changes in the Intrapulmonary Pressure

During breathing, air must first be drawn into the lungs. This process is known as **inspiration,** or **inhalation** (**TABLE 9-2**). Following inspiration, air must be expelled. This is known as **expiration,** or **exhalation.**

Inhalation is an active process controlled by the brain. Nerve impulses traveling from the brain stimulate the **diaphragm,** a dome-shaped muscle (unique to mammals) that separates the abdominal and thoracic cavities (**FIGURE 9-14A**). These impulses cause the diaphragm to contract. When it contracts, the diaphragm flattens and lowers. Much in the same way that pulling the plunger of a syringe draws in air, the contraction of the diaphragm draws air into the lungs.

Inhalation also involves the intercostal muscles, the short, powerful muscles that lie between the ribs. (They

TABLE 9-2

Summary of Inhalation and Exhalation

Inhalation

Nerve impulses from the breathing center stimulate the muscles of inspiration—the diaphragm and intercostal muscles.

Contraction of the intercostal muscles causes the rib cage to move up and out.

Contraction of the diaphragm causes it to flatten.

Volume of the thoracic cavity increases.

Intrapulmonary pressure decreases.

Air flows into the lungs through the nose and mouth.

Exhalation

Nerve impulses from the breathing center feed back on it, shutting off stimuli to muscles of inspiration.

The intercostal muscles relax and the rib cage falls.

The diaphragm relaxes and rises.

The lungs recoil.

Air is pushed out of the lungs.

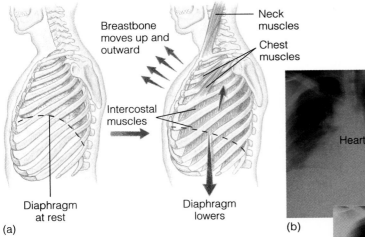

Breastbone moves up and outward

Neck muscles

Chest muscles

Intercostal muscles

Diaphragm at rest

Diaphragm lowers

(a)

(b)

FIGURE 9-14 The Bellows Effect *(a)* The rising and falling of the chest wall through the contraction of the intercostal muscles (muscles between the ribs) is shown in the diagram, illustrating the bellows effect. Inspiration is assisted by the diaphragm, which lowers. Like pulling a plunger out on a syringe, the rising of the chest wall and the lowering of the diaphragm draw air into the lungs. *(b)* X-rays showing the size of the lungs in full exhalation *(top)* and full inspiration *(bottom)*.

Heart

Diaphragm

Heart

Diaphragm

are the meat on barbecued ribs.) Nerve impulses traveling to these muscles cause them to contract as the diaphragm is lowered. When the intercostal muscles contract, the rib cage lifts up and out. Together, the contractions of the intercostal muscles and the diaphragm increase the volume of the thoracic cavity (**FIGURE 9-14B**, bottom). This, in turn, decreases the **intrapulmonary pressure,** the pressure in the alveoli. The decrease in pressure draws air in through the mouth or nose into the trachea, bronchi, and lungs.

At rest, each breath delivers about 500 milliliters of air to the lung. This is known as the **tidal volume,** the amount of air inhaled or exhaled with each breath when a person is at rest.

In contrast to inhalation, exhalation is a passive process—that is, one that does not require muscle contraction, at least in a person at rest. Exhalation begins after the lungs have filled. At this point in the cycle, the diaphragm and intercostal muscles relax. The relaxed diaphragm rises and resumes its domed shape, and the chest wall falls slightly inward. These changes reduce the volume of the thoracic cavity, raising the pressure and forcing the air out in much the same way that squeezing an inflated beach ball forces air out of the opening.

The lungs also contribute to passive exhalation. Containing numerous elastic connective tissue fibers, they fill like balloons during inspiration. When inhalation ceases, the lungs simply recoil (shrink), forcing air out.

Although exhalation is a passive process in an individual at rest, it can be made active by contracting the muscles of the wall of the chest and abdomen. The forceful expulsion of air is called **forced exhalation.**

Inhalation can also be consciously augmented by a forceful contraction of the muscles of inspiration. (You can test this by taking a deep breath.) Forced inhalation increases the amount of air entering your lungs. Athletes often actively inhale and exhale just before an

event to increase oxygen levels in their blood. A competitive swimmer, for example, may take several deep breaths before diving into the pool for a race. Deep breathing, while effective, can be dangerous, for reasons explained shortly.

The Health of a Person's Lungs Can Be Assessed by Measuring Air Flow In and Out of Them under Various Conditions

Children who are exposed to tobacco smoke at home experience a decrease in lung capacity—that is, a decrease in their ability to move air in and out of their lungs. Several measurements of lung capacity are routinely used to determine the health of a person's lungs.

Measurements of lung function are taken under controlled conditions. As shown in **FIGURE 9-15A**, pa-

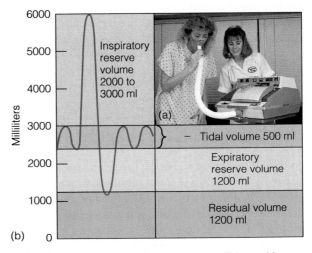

FIGURE 9-15 Measuring Lung Capacity *(a)* This machine allows health-care workers to determine tidal volume, inspiratory reserve volume, and other lung-capacity measurements to determine the health of an individual's lung. *(b)* This graph shows several common measurements.

tients breathe into a machine that measures the amount of air moving in and out of the lung at various times. **FIGURE 9-15B** shows a graph of some of the common measurements. The first is the tidal volume, which, as noted earlier, is the amount of air that moves in and out during passive breathing. After exhalation, under resting conditions, the lungs still contain a considerable amount of air—about 2400 milliliters. Forced exhalation will expel about half of that air. The amount that can be exhaled after a normal exhalation is called the **expiratory reserve volume.** The remaining 1200 milliliters is known as the **residual volume.**

Another important measurement is the **inspiratory reserve volume**—the amount of air that can be drawn into the lungs during active inspiration—deep inhalation. Deep inhalation draws in four to six times more air than the tidal volume, or 2000–3000 milliliters, depending on the size of the individual.

Lung diseases often result in changes in the amount of air that can be moved in and out of the lung or changes in the residual volume. Asthma, for example, reduces the inspiratory reserve volume—the amount of air that can be inhaled during forced inspiration—because the constricted bronchioles limit air flow.

Breathing Is Controlled Principally by the Breathing Center in the Brain

The **breathing center** is located in a region of the brain called the *brain stem* (or medulla, pronounced meh-DEW-lah). Somewhat similar to the sinoatrial node of the heart, the breathing center contains nerve cells that generate periodic impulses that stimulate contraction of the intercostal muscles and the diaphragm, resulting in inhalation. When the lungs fill, the nerve impulses cease and the muscles relax, forcing air out of the lungs.

Several mechanisms are responsible for the termination of the impulses to the muscles involved in inspiration. The first is a negative feedback loop, shown on the right side of **FIGURE 9-16.** Here's how it works. When the breathing center sends nerve impulses to the diaphragm and intercostal muscles, it also sends impulses to a nearby region of the brain stem, a kind of "relay center" that transmits nerve impulses back to the breathing center. When these impulses arrive, they inhibit the neurons in the breathing center, shutting off the signals to the muscles of inspiration, terminating the inspiration. The second control mechanism consists of sensory nerve fibers known as **stretch receptors,** which are found in the lungs. When the lung is full, nerve impulses from the stretch receptors are transmitted to the breathing center. These impulses turn off the breathing center. Stretch receptors probably function only during

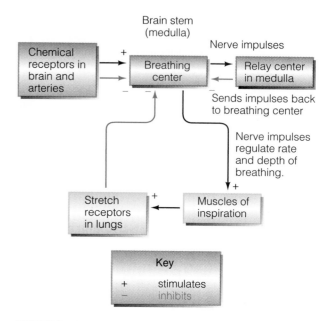

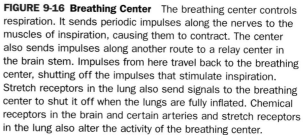

FIGURE 9-16 Breathing Center   The breathing center controls respiration. It sends periodic impulses along the nerves to the muscles of inspiration, causing them to contract. The center also sends impulses along another route to a relay center in the brain stem. Impulses from here travel back to the breathing center, shutting off the impulses that stimulate inspiration. Stretch receptors in the lung also send signals to the breathing center to shut it off when the lungs are fully inflated. Chemical receptors in the brain and certain arteries and stretch receptors in the lung also alter the activity of the breathing center.

exercise, when large volumes of air are moved in and out of the lungs.

Changes in the depth and rate (frequency) of breathing are thought to result from nerve impulses from chemical receptors in the brain and certain arteries (**FIGURE 9-16**, left). These receptors detect the concentration of carbon dioxide, hydrogen ions, and oxygen in the body. Since these chemicals are key components of cellular metabolism, it is therefore no surprise that they are involved in the control of breathing.

The most important chemical involved in controlling respiration is carbon dioxide. As shown in **FIGURE 9-17**, carbon dioxide levels are monitored by receptors in the aorta (the large artery that carries oxygenated blood out of the heart) and the carotid arteries (kah-RAW-tid) (which carry oxygenated blood to the brain from the aorta). When carbon dioxide levels rise, these receptors transmit impulses to the breathing center. This increases the rate of breathing. A decline in carbon dioxide levels has the opposite effect.

As **FIGURE 9-17** shows, carbon dioxide also diffuses into the **cerebrospinal fluid (CSF),** a clear liquid found in cavities in the brain, the **ventricles.** In the CSF, carbon dioxide is converted into carbonic acid, which then dissociates to form bicarbonate and hydrogen ions. A rise in carbon dioxide in the blood, therefore, results in an increase in the H^+ concentration of the CSF. The increase in H^+, in turn, is detected by chemical receptors, or **chemoreceptors** (KEY-moe-ree-CEP-ters), in the brain. These receptors send impulses to the breathing center, triggering an increase in the rate and depth of breathing (**FIGURE 9-17**).

The chemoreceptors in the brain and arteries allow the body to align respiration with cellular demands. During exercise, for example, cellular respiration increases to meet body demands for energy. As cellular respiration increases, oxygen demand increases. Carbon dioxide production also climbs. The carbon dioxide produced during exercise increases the depth and rate of breathing with two effects. First, it lowers the concentration of carbon dioxide in the blood. (Breathing slows when levels return to normal.) Second, increased ventilation also makes more oxygen available for energy production.

The body also contains a set of oxygen receptors. These receptors are not as sensitive as the H^+ receptors, so oxygen levels must fall considerably before the oxygen receptors begin generating impulses. This fact can have profound consequences for divers and swimmers. Repeated deep and rapid breathing, or **hyperventilation,** for example, makes it possible for divers to hold their breath under water longer. Hyperventilation decreases carbon dioxide levels in the blood and H^+ concentrations in the CSF, reducing the urge to breathe. When the diver enters the water, oxygen levels in the blood may fall so low that the brain is deprived of oxygen, causing the individual to lose consciousness. Ironically, the decrease in blood oxygen levels is not enough to stimulate breathing; the diver blacks out well before the H^+ concentration in the CSF reaches the level needed to stimulate breathing.

Section 9-4

Diseases of the Respiratory System

Bacterial and viral infections of the respiratory tract can cause considerable discomfort, and some can be fatal. Infections may settle in many different locations in the respiratory system and are named by their site of residence. An infection in the bronchi is therefore known as **bronchitis** (bron-KITE-iss). An infection of the sinuses is known as **sinusitis** (sigh-nu-SITE-iss). (Bacterial and

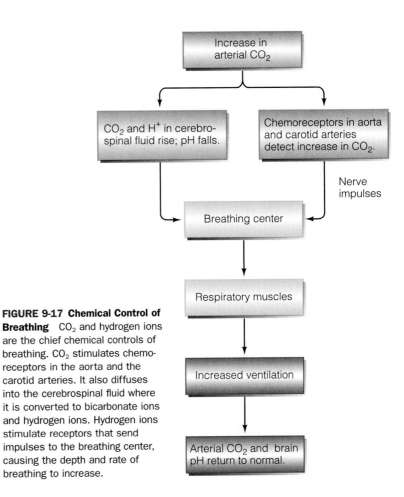

FIGURE 9-17 Chemical Control of Breathing CO_2 and hydrogen ions are the chief chemical controls of breathing. CO_2 stimulates chemoreceptors in the aorta and the carotid arteries. It also diffuses into the cerebrospinal fluid where it is converted to bicarbonate ions and hydrogen ions. Hydrogen ions stimulate receptors that send impulses to the breathing center, causing the depth and rate of breathing to increase.

viral infections were discussed in the previous chapter; a few of the more common ones are listed in **TABLE 9-3.**)

Once inside the respiratory tract, bacteria, viruses, and other microorganisms can spread to other organ systems. For example, **meningitis** (MEN-in-JITE-iss) is a bacterial or viral infection of the **meninges** (meh-NIN-jees), the fibrous layers surrounding the brain and spinal cord. This potentially fatal disease usually starts out as an infection of the sinuses or the lungs.

The lungs are also susceptible to airborne materials, among them asbestos fibers, which can cause two types of lung cancer and a debilitating disease known as asbestosis (ass-bes-TOE-sis). **Asbestosis** is a buildup of scar tissue that reduces the lung capacity. Because asbestos is believed to be dangerous, many of its uses have been banned in the United States, and asbestos used for insulation and decoration is being removed from buildings.

Another common disease of the respiratory system is **asthma,** which is characterized by periodic episodes of wheezing and difficult breathing. Unlike sinusitis, colds, and other respiratory diseases, asthma is a chronic disorder—a disease that persists for many years. Asthma

TABLE 9-3

Common Respiratory Diseases

Disease	Symptoms	Cause	Treatment
Emphysema	Breakdown of alveoli; shortness of breath	Smoking and air pollution	Administer oxygen to relieve symptoms; quit smoking; avoid polluted air. No known cure.
Chronic bronchitis	Coughing, shortness of breath	Smoking and air pollution	Quit smoking; move out of polluted area; if possible, move to warmer, drier climate.
Acute bronchitis	Inflammation of the bronchi; yellowy mucus coughed up; shortness of breath	Many viruses and bacteria	If bacterial, take antibiotics, cough medicine; use vaporizer.
Sinusitis	Inflammation of the sinuses; mucus discharge; blockage of nasal passageways; headache	Many viruses and bacteria	If bacterial, take antibiotics and decongestant tablets; use vaporizer.
Laryngitis	Inflammation of larynx and vocal cords; sore throat; hoarseness; mucus buildup and cough	Many viruses and bacteria	If bacterial, take antibiotics, cough medicine; avoid irritants, like smoke; avoid talking.
Pneumonia	Inflammation of the lungs ranging from mild to severe; cough and fever; shortness of breath at rest; chills; sweating; chest pains; blood in mucus	Bacteria, viruses, or inhalation of irritating gases	Consult physician immediately; go to bed; take antibiotics, cough medicine; stay warm.
Asthma	Constriction of bronchioles; mucus buildup in bronchioles; periodic wheezing; difficulty breathing	Allergy to pollen, some foods, food additives; dandruff from dogs and cats; exercise	Use inhalants to open passageways; avoid irritants.

is not an infectious disease. Most cases of asthma are caused by allergic reactions (abnormal immune reactions) to common stimulants such as dust, pollen, and skin cells (dander) from pets. In some individuals, certain foods such as eggs, milk, chocolate, and food preservatives trigger asthma attacks. Still other cases are triggered by drugs, vigorous exercise, and physiological stress.

In asthmatics, irritants such as pollen and dander cause a rapid increase in the production of mucus by the bronchi and bronchioles. Irritants also stimulate the constriction of the bronchioles. Mucus production and constriction of the bronchioles make it difficult for asthmatics to breathe. Asthmatics also suffer a chronic inflammation of the lining of the respiratory tract.

Although asthma is fairly common in school children, it often disappears as they grow older. As a result, only about 2% of the adult population suffers from asthma. Nevertheless, asthma is a serious disease. Periodic attacks can be quite disabling; some even lead to death. By one estimate, several thousand Americans die each year from severe asthma attacks. Victims are generally elderly individuals who are suffering from other diseases.

The severity of asthma attacks can be greatly lessened by proper medical treatment. One of the most common treatments is an oral spray (inhalant) containing the hormone epinephrine, which stimulates the bronchioles to open. Anti-inflammatory drugs (steroids) can be administered to treat chronic inflammation. Screening tests can help a patient find out what substances trigger an asthmatic attack so they can be avoided. One of the most prevalent respiratory diseases is lung cancer caused by smoking, a topic discussed in Health Note 9-2.

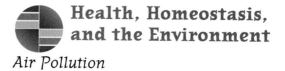

Health, Homeostasis, and the Environment

Air Pollution

The respiratory system plays a vital role in homeostasis. It helps regulate oxygen and carbon dioxide levels in the blood and in body tissues. It helps maintain the pH (acidity) of extracellular fluid by controlling the rate at which acid-forming carbon dioxide is removed. And it helps protect us from infectious disease. However, its function can be gravely thrown out of kilter by chemical contaminants in the environment; this, in turn, alters homeostasis.

The air of the industrialized world contains a multitude of potentially harmful chemical pollutants that damage human health by upsetting homeostasis. In fact, air pollutants generated in our homes, factories, and cities now claim the lives of thousands of Americans each year.

Air pollution upsets the homeostatic balance in our bodies and affects millions of us on a daily basis. Unfortunately, most people are unaware of the dangers of air pollution because the line between cause and effect is not always clear. Consider, for instance, the headache you experienced in traffic going home from school or work. Was it caused by tension, or could it have been caused by carbon monoxide emissions from cars, buses, and trucks? And what about the runny nose and sinus condition you experienced last winter? Were they caused by a virus or bacterium or by pollution?

One classic study of air pollution on the East Coast illustrates the relationship between air pollution and upper respiratory problems. Researchers found that the level of sulfur dioxide, a pollutant produced by automobiles, power plants, and factories, increased during the winter months in New York City. Certain weather conditions trapped the pollutants, substantially raising ground-level sulfur dioxide concentrations. During one episode, upper respiratory illness in New York residents skyrocketed (**FIGURE 9-18**). Colds, coughs, nasal irritation, and other symptoms increased fivefold in a few days. Soon after the pollution levels returned to normal, the symptoms subsided. Few people knew they'd been poisoned by the air they were breathing and not stricken, as they had supposed, by some infectious agent.

Air pollution is partly responsible for other long-term diseases. One of these is **chronic bronchitis,** a persistent irritation of the bronchi. Characterized by excess mucus production, coughing, and difficulty in breathing, chronic bronchitis afflicts one out of every five American men between the ages of 40 and 60.

Although the leading cause of chronic bronchitis is cigarette smoking, urban air pollution also contributes to this disease. Three air pollutants have been identified as causative agents: sulfur dioxide, nitrogen oxides, and ozone. Each of these irritates the lung and bronchial passages and arises from the combustion of fossil fuel by cars, buses, power plants, factories, and homes.

A far more troublesome disease is emphysema. Emphysema, discussed earlier in the chapter, is one of the fastest growing causes of death in the United States. Resulting principally from smoking and air pollution, emphysema afflicts over 1.5 million Americans. This condition, which is more common among men, is a progressive and incurable disease. As it worsens, lung function deteriorates, and victims eventually require supplemental oxygen to perform even routine functions, such as walking or speaking.

The leading cause of emphysema is smoking, a habit of 47 million American adults. Emphysema is also caused by urban air pollution. Not surprisingly, smokers who live in polluted urban settings have the highest incidence of the disease. Like chronic bronchitis, emphysema is caused by ozone, sulfur dioxide, and nitrogen oxides.

No one knows the exact role of urban air pollution. It is not something that can be determined easily, if at all, because people are exposed to many different pollutants over their life-times. However, a recent report issued by the federal government estimates that approximately 51,000 Americans die each year from lung disease caused by urban air pollution. The number of victims could climb to 60,000 per year around the year 2000, illustrating once again that human health is clearly dependent on a clean environment. This statistic also illustrates that the respiratory system of humans is not well adapted to the physical environment we have made for ourselves. Cultural evolution has progressed in ways that are overwhelming biological evolution.

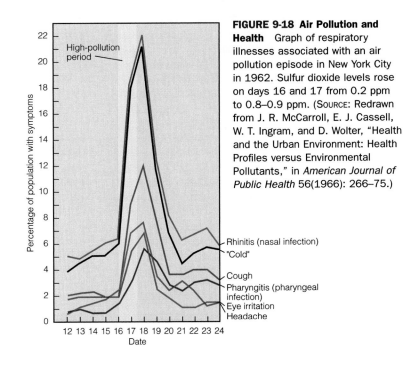

FIGURE 9-18 Air Pollution and Health Graph of respiratory illnesses associated with an air pollution episode in New York City in 1962. Sulfur dioxide levels rose on days 16 and 17 from 0.2 ppm to 0.8–0.9 ppm. (SOURCE: Redrawn from J. R. McCarroll, E. J. Cassell, W. T. Ingram, and D. Wolter, "Health and the Urban Environment: Health Profiles versus Environmental Pollutants," in *American Journal of Public Health* 56(1966): 266–75.)

Health Note 9-2

Smoking and Health: The Deadly Connection

Smoking - related diseases
www.jbpub.com/humanbiology

Urban air pollution worries many Americans, and with good reason. However, city air that many of us breathe is benign compared with the "air" that 47 million Americans over the age of 18 voluntarily inhale from cigarettes. Loaded with dangerous pollutants in concentrations far greater than those of our cities, cigarette smoke takes a huge toll on citizens of the world. In the United States, for example, an estimated 419,000 people—about 1150 every day—die from the many adverse health effects of tobacco smoke, including heart attacks, lung cancer, and emphysema. Making matters worse, smoking is addictive. Nicotine, one of the many components of tobacco smoke, hooks many people in a dangerous activity.

Smoking costs society a great deal in medical bills and lost productivity. According to the Worldwatch Institute, every pack of cigarettes sold in the United States costs our society about $1.25–3.17 in medical costs, lost wages, and reduced productivity—that's about $125–$400 billion a year!

Smoking is a principal cause of lung cancer, claiming the lives of an estimated 130,000 men and women in the United States each year. Smokers are 11–25 times more likely to develop lung cancer than nonsmokers. The more one smokes, the more risk one suffers.

Unfortunately, nonsmokers are also affected by the smoke of others. Nonsmokers inhale tobacco smoke in meetings, in restaurants, at work, and at home. Research has shown that nonsmokers—sometimes referred to as "passive smokers" because they "smoke" involuntarily—are more likely to develop lung cancer than those nonsmokers who manage to steer clear of smokers. In a study of Japanese women married to men who smoked, researchers found that the wives were as likely to develop lung cancer as people who smoked half a pack of cigarettes a day! A recent report by the U.S. Environmental Protection Agency estimates that passive smoking causes 500–5000 cases of lung cancer a year in the United States. Passive smokers who are exposed to tobacco smoke for long periods also suffer from impaired lung function equal to that seen in light smokers (people who smoke under a pack a day).

Cigarette smoke in closed quarters can cause angina (chest pains) in smokers and nonsmokers afflicted by atherosclerosis of the coronary arteries. Carbon monoxide in cigarette smoke is responsible for angina attacks.

Smokers are also more susceptible to colds and other respiratory infections. And smoking affects children. One study showed that children from families in which both parents smoked suffered twice as many upper respiratory infections as children from nonsmoking families. Recently, researchers from the Harvard Medical School reported finding a 7% decrease in lung capacity in children raised by mothers who smoked. The researchers believe that this change may lead to other pulmonary problems later in life.

Numerous studies also show that smoking is a major contributor to impotence in men. In fact, smokers are 50% more likely to be impotent than nonsmokers. Smoking also causes arterial changes that restrict blood flow to the penis. Changes begin to occur early in life. Teenage smokers could have problems in their thirties if they continue to smoke.

For years, passive smokers have had little to say about their exposure to other people's smoke. Today, however, as a result of a growing awareness of the dangers, new regulations are banning or restricting smoking in many public places and in the workplace.

The effects of smoking extend way beyond respiratory disease. Recent studies, for instance, show that smoking even decreases fertility in women. For example, women who smoke more than a pack of cigarettes a day are half as fertile as nonsmokers. Smoking may also affect the outcome of pregnancy. According to the 1985 U.S. Surgeon General's Report, women who smoke several packs a day during pregnancy are much more likely to miscarry and are also more likely to give birth to smaller children. On average, the children of these women are 200 grams (nearly 0.5 pounds) lighter than children born to nonsmoking mothers. Finally, children of women who smoke heavily during pregnancy generally score lower on mental aptitude tests

during early childhood than children whose mothers do not smoke.

Tobacco smoke contains numerous hazardous substances that damage the lining of the respiratory system. Nicotine and sulfur dioxide, for example, paralyze the cilia lining the respiratory tract.

Tobacco smoke is also laden with microscopic carbon particles. These carbon particles penetrate deeply into the lungs, where they accumulate in the alveoli and alveolar walls, turning healthy tissue into a blackened mass that often becomes cancerous (**FIGURE 1**). Tobacco smoke may also paralyze the alveolar macrophages, making a bad situation even worse.

Toxic chemicals in cigarette smoke, many of which are known carcinogens, attach to carbon particles. Toxin-carrying particles adhere to the lungs, larynx, trachea, and bronchi. Virtually any place they stick, they can cause cancer, explaining why smokers are five times more likely than nonsmokers to develop laryngeal cancer and four times more likely to develop cancer of the oral cavity.

Nitrogen dioxide and sulfur dioxide in tobacco smoke penetrate deep into the lungs, where they dissolve in the watery layer inside the alveoli. Nitrogen dioxide is converted to nitric acid; sulfur dioxide is converted to sulfuric acid. Both acids erode the alveolar walls, leading to emphysema.

If tobacco smoke is so dangerous, why don't we ban smoking or discontinue generous government subsidies to tobacco growers? Surely, an air pollutant or food contaminant that killed hundreds of thousands of Americans each year would be prohibited immediately.

Part of the answer lies in the fact that tobacco use has a long history in the United States. Furthermore, many people think that because smoking is a voluntary act, individuals should have the right to make their own decision. Government, they say, has no right to regulate their pleasure. Furthermore, smoking supports a $30-billion-a-year tobacco industry that employs about 2 million people, including tobacco farmers, advertisers, and retailers. The tobacco industry lobbies diligently to protect the rights of smokers.

The dangers of smoking are becoming well known. As a result of widespread publicity and public pressure, smoking has dropped substantially in the United States. In 1996, for example, only 25% of American adults smoked, down from 34% in 1985 and down substantially from the 1950s and 1960s, when well over half of all men and over one-third of all women engaged in this potentially lethal habit.

Despite the downturn in smoking, 48 million American adults (over the age of 18) still smoke. The number of adult smokers has remained constant since 1990 and the number of adolescent smokers has remained constant over the past decade, indicating no further progress in reducing the number of smokers. An estimated six million teenagers and 100,000 children under the age of 13 smoke. Smoking occurs in greatest proportion in certain minority groups—particu-

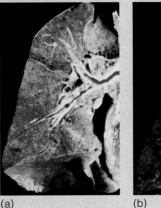
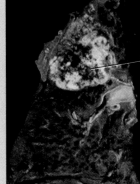

(a)   (b)

FIGURE 1 The Normal and Cancerous Lung *(a)* The normal lung appears spongy. *(b)* The cancerous lung from a smoker is filled with particulates and a large tumor.

larly black men, Native Americans, and Native Alaskans.

On a global level, smoking is skyrocketing—increasing about 3% a year. A 3% growth rate means that the number of smokers will double in about 23 years. By one estimate, the number of deaths related to smoking could increase from 2.5 million per year today to about 12 million per year by 2050.

Visit *Human Biology's* Internet site, www.jbpub.com/humanbiology, for links to web sites offering more information on this topic.

SUMMARY

STRUCTURE OF THE HUMAN RESPIRATORY SYSTEM

1. The respiratory system consists of an air-conducting portion and a gas-exchange portion.

2. The air-conducting portion transports air from outside the body to the alveoli in the *lungs*, the site of gaseous exchange.

3. The *alveoli* are tiny, thin-walled sacs formed by a single layer of flattened epithelial cells that facilitate diffusion. Surrounding the alveoli are capillary beds that pick up oxygen and expel carbon dioxide.

4. The lining of the alveoli is kept moist by water. *Surfactant*, a chemical substance produced in the lung, reduces the surface tension inside the alveoli and prevents their collapse.

FUNCTIONS OF THE RESPIRATORY SYSTEM

5. The respiratory system conducts air to and from the lungs, exchanges gases, and helps produce sounds. Sound is generated as air rushes past the *vocal cords*, causing them to vibrate. The sounds are modified by movements of the tongue and changes in the shape of the oral cavity.

6. Oxygen and carbon dioxide diffuse across the alveolar wall, driven by concentration differences between the blood and alveolar air. Oxygen diffuses into the blood plasma, then into the RBCs, where most of it binds to hemoglobin. Carbon dioxide diffuses in the opposite direction.

7. Carbon dioxide, a waste product of cellular respiration, is picked up by the blood flowing through capillaries. Some carbon dioxide is dissolved in the blood. Most of it, however, enters the RBCs in the bloodstream, where it is converted into carbonic acid. Carbonic acid dissociates, forming hydrogen and bicarbonate ions. The latter diffuse out of the RBCs and are transported in the plasma.

BREATHING AND THE CONTROL OF RESPIRATION

8. Breathing is an involuntary action with a conscious override. It is controlled by the *breathing center* in the brain stem. Nerve cells in the breathing center send impulses to the *diaphragm* and *intercostal muscles*, which contract. This increases the volume of the thoracic cavity, which draws air into the lungs through the nose or mouth.

9. When the impulses stop, inspiration ends. Air is then expelled passively as the chest wall returns to the normal position, and the diaphragm rises. The recoil of the lungs also assists in expelling the air.

10. The breathing center is regulated by a negative feedback loop that it generates itself. It is also regulated by outside influences—notably, levels of carbon dioxide in the blood.

11. Expiration can be augmented by enlisting the aid of abdominal and chest muscles, as can inspiration.

12. The rate of respiration can be increased by rising blood carbon dioxide levels or falling oxygen levels and by an increase in physical exercise.

DISEASES OF THE RESPIRATORY SYSTEM

13. Bacterial and viral infections of the respiratory tract can cause considerable discomfort, and some can be fatal.

14. Once inside the respiratory tract, bacteria, viruses, and other microorganisms can spread to other organ systems.

15. The lungs are also susceptible to airborne materials, among them asbestos fibers, which can cause two types of lung cancer and a debilitating disease called *asbestosis*.

16. Another common disease of the respiratory system is asthma, an allergy (abnormal immune reaction) to dust, pollen, and other common substances. It is characterized by periodic episodes of wheezing and difficult breathing.

HEALTH, HOMEOSTASIS, AND THE ENVIRONMENT: AIR POLLUTION

17. The respiratory system contributes mightily to homeostasis by regulating concentrations of oxygen and carbon dioxide in the blood and body tissues. It also helps control the pH of the blood and tissues. Proper functioning of the respiratory system is therefore essential for health.

18. Respiratory function can be dramatically upset by microorganisms as well as by pollution from factories, automobiles, power plants, and even our own homes.

19. *Chronic bronchitis*, a persistent irritation of the bronchi, is caused by sulfur dioxide, ozone, and nitrogen oxides, lung irritants sometimes found in dangerous levels in urban air.

20. *Emphysema* is a breakdown of the alveoli that gradually destroys the lung's ability to absorb oxygen. It is also caused by lung irritants. Despite the role of air pollution in causing emphysema and chronic bronchitis, smoking remains the number one cause of these diseases.

Critical Thinking

THINKING CRITICALLY— ANALYSIS

This Analysis corresponds to the Thinking Critically scenario that was presented at the beginning of this chapter.

Representatives from the tobacco industry persist in saying that there is no link between smoking and lung cancer because there is no proof positive. Semantically, they're quite correct. Science does not *prove* anything. Sure, scientists have performed many studies on people that show *correlations* between lung cancer and smoking. In other words, they've shown that smokers are much more likely to develop lung cancer than nonsmokers and that the more one smokes, the higher his or her chances are of developing this disease. Technically, they haven't proved anything, they've shown a correlation.

The tobacco industry is being a bit deceptive, though, for when a correlation like that between smoking and lung cancer shows up again and again or when the correlation is high, you can have a fair amount of confidence in the cause-and-effect relationship under study.

Can a study be devised to prove the connection? It's unlikely. As I mentioned earlier, science is in the business of supporting hypothesis and generating theory, not providing proof positive.

This exercise requires one to look at definitions (what constitutes proof and the difference between proving something and showing a correlation). It also shows that one usually needs to dig deeper to understand controversies. In this instance, the tobacco industry is hanging its argument on semantics. This exercise also suggests the importance of looking at who's talking—and what their hidden biases are. The tobacco industry clearly has a vested interest in assuring people that smoking is not harmful.

EXERCISING YOUR CRITICAL THINKING SKILLS

An inventor has devised a pollution-control device that removes particulates from the smokestacks of factories. He claims that the device removes 80% of all particulates and will bring factories into compliance with federal law, thereby protecting nearby residents from harmful pollutants.

What is your reaction? What questions might you ask? Would you approve installing the device, given this information? Would your decision be affected by data showing that, although the device does reduce total particulates by 80%, it does not capture finer particulates and that these particles are typically inhaled deeply into the lungs? Would your decision be affected if you found that the small (respirable) particles in the smokestack contained toxic metals such as mercury and cadmium? What rule(s) of critical thinking does this example illustrate?

TEST OF CONCEPTS

1. Trace the flow of air from the mouth and nose to the alveoli, and describe what happens to the air as it travels along the various passageways.
2. Draw an alveolus, including all cell types found there. Be sure to show the relationship of the surrounding capillaries. Show the path that oxygen and carbon dioxide take.
3. Trace the movement of oxygen from alveolar air to the blood in alveolar capillaries. Describe the forces that cause oxygen to move in this direction. Do the same for the reverse flow of carbon dioxide.
4. Why would a breakdown of alveoli in emphysemic patients make it difficult for them to receive enough oxygen?
5. Describe how sounds are generated and refined.
6. A baby is born prematurely and is having difficulty breathing. As the attending physician, explain to the parents what the problem is and how it could be corrected.
7. Smoking irritates the trachea and bronchi, causing mucus to build up and paralyzing cilia. How do these changes affect the lung?
8. Describe inspiration and expiration, being sure to include discussions of what triggers them and the role of muscles in effecting these actions.
9. How does the breathing center regulate itself to control the frequency of breathing?
10. Exercise increases the rate of breathing. How?
11. Turn to **FIGURE 9–15** on page 207. Explain each of the terms.
12. What is asthma? What are its symptoms? How does it affect one's inspiratory reserve volume?
13. Debate this statement: Urban air pollution has very little overall impact on human health.
14. Do you agree or disagree that the single most effective way of reducing deaths in the United States would be to ban smoking? Explain.

www.jbpub.com/humanbiology

TOOLS FOR LEARNING

Tools for Learning is an on-line student review area located at this book's web site HumanBiology (www.jbpub.com/humanbiology). The review area provides a variety of activities designed to help you study for your class:

Chapter Outlines. We've pulled out the section titles and full sentence sub-headings from each chapter to form natural descriptive outlines you can use to study the chapters' material point by point.

Review Questions. The review questions test your knowledge of the important concepts and applications in each chapter. Written by the author of the text, the review provides feedback for each correct or incorrect answer. This is an excellent test preparation tool.

Flash Cards. Studying human biology requires learning new terms. Virtual flash cards help you master the new vocabulary for each chapter.

Figure Labeling. You can practice identifying and labeling anatomical features on the same art content that appears in the text.

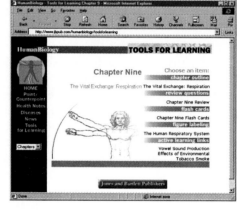

Active Learning Links. Active Learning Links connect to external web sites that provide an opportunity to learn basic concepts through demonstrations, animations, and hands-on activities.

THE URINARY SYSTEM

RIDDING THE BODY OF WASTES AND MAINTAINING HOMEOSTASIS

Color-enhanced X ray of human urinary
system showing ureters and kidneys.

Thinking Critically

Each year, 24,000 new cases of kidney cancer are diagnosed in the United States. About 10,000 people a year die from this disease. Chemotherapy is virtually the only treatment available for patients whose cancers have spread to other locations, but it helps fewer than 10% of the patients. Patients receiving chemotherapy survive about 9 months after initial diagnosis.

Recently, a new treatment was introduced that holds some promise. The treatment, known as autolymphocyte therapy (ALT), is marketed by a company in Massachusetts. Costing about $22,000 per patient, ALT is designed to strengthen the immune system of patients. Here's how it works: Blood is withdrawn from patients with kidney cancer. Lymphocytes are extracted, then treated with monoclonal antibodies (antibodies to the kidney tumor). This process activates the lymphocytes, causing them to produce a chemical substance known as cytokines, natural compounds that boost the activity of lymphocytes. The cytokines are divided into six batches and stored for later use. Once a month, for the next six months, patients donate more of their own lymphocytes. The cells are treated with cytokines, then reinjected, where they apparently go to work on the tumors.

Results of one study on the effectiveness of ALT were published in April 1990 in the British medical journal *Lancet*. This study of 90 individuals showed that patients who had undergone the procedure survived 22 months after diagnosis, two and a half times longer than patients treated with chemotherapeutic agents. Less than a year after the publication of this clinical trial, a treatment center opened in Boston.

Assume that you are considering investing in the company that markets ALT. Using your critical thinking skills and your knowledge of scientific method, what concerns do you have, and what questions would you ask before investing in this company?

LEON MARKOWITZ IS LOWERED INTO A LARGE pool of warm water in a special room in the hospital (**FIGURE 10-1**). Physicians position a large, cylindrical device in the water in front of one of his kidneys, the organs that filter the blood, helping to maintain normal blood concentrations of nutrients and wastes, and helping to rid the body of wastes. Over the next few hours, as Markowitz listens to tapes of Paul Simon, ultrasound waves, undetectable by the human ear, will smash the kidney stones that are obstructing the flow of urine and causing excruciating pain.

A decade earlier, surgeons would have had to cut an incision 15–20 centimeters (6–8 inches) long in Markowitz's side to remove the stone. Leon would have spent 7–10 days recovering in the hospital and another 8 weeks at home recuperating before returning to work. With this new technique, known as **ultrasound lithotripsy** (LITH-oh-TRIP-see) ("stone crushing"), patients usually return home within a day.

In this chapter we will look at the ways in which humans rid themselves of potentially harmful wastes derived from metabolism. We will focus principally on one of the body's most important excretory systems, the urinary system. We will also discuss how the urinary system contributes to homeostasis and then look at some common diseases such as kidney stones that disturb the function of this important organ system.

Organs of Excretion: a Biological Imperative

Ralph Waldo Emerson once said that as soon as there is life, there is risk. A biologist might say, as soon as there is life, there is waste. All living things produce waste and humans are no exception. Fortunately for life on Earth, elaborate systems of recycling have evolved in the biosphere. Thus, the waste of one organism is often a resource for another. The best example is carbon dioxide, a waste product of energy production in plants and animals. CO_2 produced in this process is a vital component of photosynthesis.

Even though wastes are often recycled, they cannot accumulate in organisms without causing harm. Thus, all organisms face a common challenge—ridding themselves of internally produced wastes such as carbon dioxide, ammonia, and urea.

TABLE 10-1 lists the major metabolic wastes and other chemicals excreted from the body. The first three entries are nitrogen-containing wastes—ammonia (ah-MOAN-ee-ah), urea (yur-EE-ah), and uric acid (YUR-

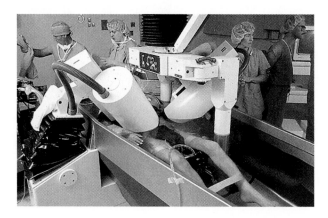

FIGURE 10-1 Lithotripsy Ultrasound waves, undetectable to the human ear, bombard the kidney stones in this man, smashing them into sandlike particles that can be passed relatively painlessly in the urine.

ick). **Ammonia** (NH_3), is a product of the metabolic pathway in which amino acids are broken down, occurring primarily in the liver. Amino acid breakdown generally occurs when there's excess protein in the diet or a

TABLE 10-1		
Important Metabolic Wastes and Substances Excreted from the Body		
Chemical	**Source**	**Organ of Excretion**
Ammonia	Deamination (removal of amine group) of amino acids in liver	Kidneys
Urea	Derived from ammonia	Kidneys, skin
Uric acid	Nucleotide breakdown	Kidneys
Bile pigments	Hemoglobin breakdown in liver	Liver (into small intestine)
Urochrome	Hemoglobin breakdown in liver	Kidneys
Carbon dioxide	Breakdown of glucose in cells	Lungs
Water	Food and water; breakdown of glucose	Kidneys, skin, and lungs
Inorganic ions*	Food and water	Kidneys and sweat glands

*Ions are not a metabolic waste product like the other substances shown in this table. Nonetheless, ions are excreted to maintain constant levels in the body.

shortage of carbohydrate, which causes the body to break down protein to acquire amino acids for energy production. The amino groups (NH_2) removed from the amino acids in a reaction called **deamination** (dee-AM-in-A-shun) are converted into ammonia.

Ammonia is a highly toxic chemical. A small amount of the ammonia produced by the liver is excreted in the urine, but the liver converts most ammonia to **urea** (**FIGURE 10-2**).

Another by-product of metabolism in humans is **uric acid,** which is produced in the liver during the breakdown of nucleotides, the building blocks of DNA and RNA. Uric acid is excreted in the urine. In adults, excess production may result in the deposition of uric acid crystals in the bloodstream and in joints, where they cause considerable pain. This condition is known as **gout** (gowt). Uric acid may also appear in the urine of some babies as orange crystals. Although alarming to a parent, their presence is generally not a problem.

The **bile pigments** are derived from the breakdown of hemoglobin in RBCs in the liver. Bile pigments are transferred from the liver to the gallbladder. Bile pigments are released with the bile salts required for fat digestion and are passed along the digestive tract. Some are reabsorbed, and the rest are eliminated with the feces.

The liver also produces another water-soluble pigment during the breakdown of hemoglobin. Known as **urochrome** (YUR-oh-chrome), this yellow pigment is dissolved in the blood and passes to the kidneys, where it is excreted with the urine. Urochrome gives urine its yellowish color.

TABLE 10-1 also lists inorganic ions. Even though they are not end-products of metabolism, inorganic ions are excreted from the body by various organs. This is essential to maintain constant levels in the blood, the tissue fluid, and the cytoplasm of cells.

Section 10-2

The Urinary System

Evolution has "provided" several avenues by which animals get rid of, or excrete, wastes. In humans, excretion of wastes occurs in the lungs, the skin, the liver, the kidneys, and even the intestines.

Of all the organs that participate in removing waste, however, the kidneys rank as one of the most important, for they rid the body of the greatest variety of dissolved wastes. In so doing, the kidneys also play a key role in regulating the chemical constancy of the blood.

FIGURE 10-2 Structure of Urea, Uric Acid, and Ammonia These three waste products are released by cells.

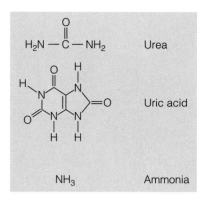

As **FIGURE 10-3A** shows, humans come equipped with two kidneys, which are a part of the **urinary system.** The urinary system also includes the (1) ureters (YUR-eh-ters), (2) the urinary bladder, and (3) the urethra (you-REETH-rah). The functions of these organs are described in more detail below and are listed in **TABLE 10-2**.

The Urinary System Consists of the Kidneys, Ureters, Bladder, and Urethra

The kidneys lie on either side of the vertebral column. About the size of a person's fist, the kidneys are surrounded by a layer of fat and are located high in the posterior (back) abdominal wall beneath the diaphragm. The human kidneys are oval structures, slightly indented on one side, much like kidney beans (**FIGURE 10-3B**). Arterial blood flows into the kidneys through the **renal arteries,** which enter at each indented region. The renal arteries are major branches of the abdominal aorta, a large blood vessel that delivers blood to the abdominal organs and lower limbs.

Much of the blood-borne wastes are removed by the kidneys. After the blood has been filtered, it leaves the kidneys via the **renal veins,** which drain into the inferior vena cava, a vessel that transports venous blood to the heart.

TABLE 10-2	
Components of the Urinary System and Their Functions	
Component	**Function**
Kidneys	Eliminate wastes from the blood; help regulate body water concentration; help regulate blood pressure; help maintain a constant blood pH
Ureters	Transport urine to the urinary bladder
Urinary bladder	Stores urine; contracts to eliminate stored urine
Urethra	Transports urine to the outside of the body

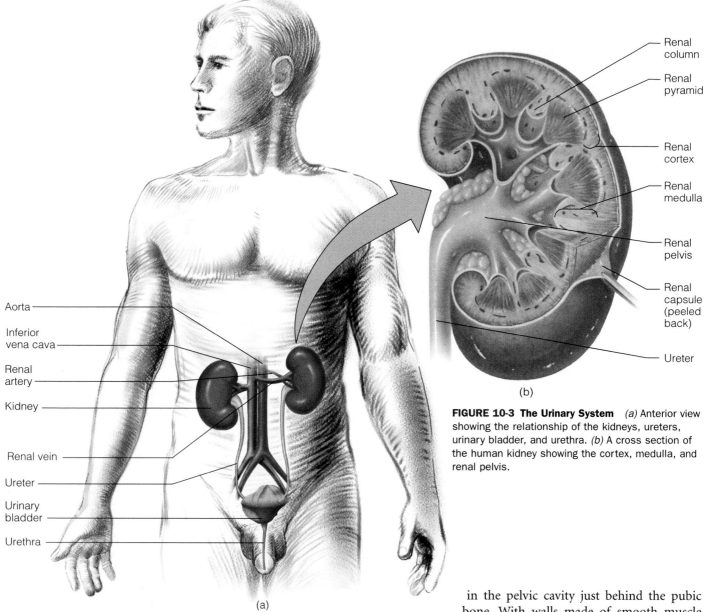

Aorta

Inferior
vena cava

Renal
artery

Kidney

Renal vein

Ureter

Urinary
bladder

Urethra

(a)

Renal
column

Renal
pyramid

Renal
cortex

Renal
medulla

Renal
pelvis

Renal
capsule
(peeled
back)

Ureter

(b)

FIGURE 10-3 The Urinary System   *(a)* Anterior view showing the relationship of the kidneys, ureters, urinary bladder, and urethra. *(b)* A cross section of the human kidney showing the cortex, medulla, and renal pelvis.

Wastes are eliminated in the **urine** (YUR-in), a yellowish fluid containing water, nitrogenous wastes (such as urea), small amounts of hormones, ions, and other substances. Dissolved wastes are removed from the blood by numerous microscopic filtering units in the kidney, the **nephrons** (NEFF-rons).

The urine produced by the nephrons is drained from the kidney by the **ureters.** These muscular tubes transport urine to the urinary bladder with the aid of peristaltic contractions, smooth muscle contractions like those in the digestive tract. The **urinary bladder** lies

in the pelvic cavity just behind the pubic bone. With walls made of smooth muscle that stretch as the bladder fills, the bladder serves as a temporary receptacle for urine. When the bladder is full, its walls contract, forcing the urine out through the urethra.

The **urethra** is a narrow tube, measuring approximately 4 centimeters (1.5 inches) in women and 15–20 centimeters (6–8 inches) in men. The additional length in men largely results from the fact that the urethra travels through the penis (**FIGURE 10-4**).

The difference in the length of the urethra between men and women has important medical implications. The shorter urethra in women, for example, makes women more susceptible to bacterial infections of the

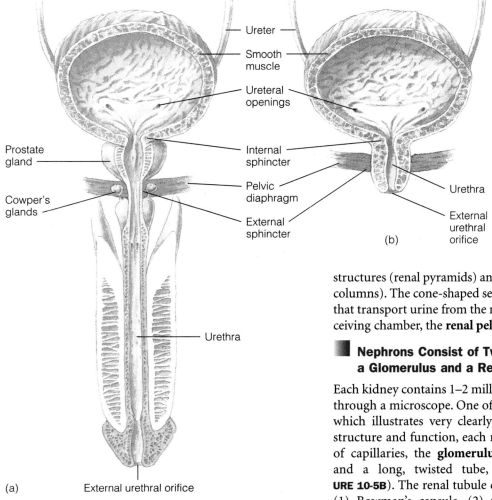

Ureter
Smooth muscle
Ureteral openings
Prostate gland
Cowper's glands
Internal sphincter
Pelvic diaphragm
External sphincter
Urethra
Urethra
External urethral orifice
(b)
External urethral orifice
(a)

FIGURE 10-4 The Urinary Bladder and Urethra These drawings show the differences in the urethras of men *(a)* and women *(b)*. The smooth muscle at the juncture of the urinary bladder and urethra forms the internal sphincter. The pelvic diaphragm is a flat sheet of muscle covering the lower boundary of the pelvic cavity. It forms the external sphincter and is under voluntary control.

structures (renal pyramids) and intervening tissue (renal columns). The cone-shaped sections contain small ducts that transport urine from the nephrons into a central receiving chamber, the **renal pelvis** (**FIGURE 10-3B**).

Nephrons Consist of Two Parts, a Glomerulus and a Renal Tubule

Each kidney contains 1–2 million nephrons, visible only through a microscope. One of the marvels of evolution, which illustrates very clearly the correlation between structure and function, each nephron consists of a tuft of capillaries, the **glomerulus** (glom-ERR-you-luss),[2] and a long, twisted tube, the **renal tubule** (**FIGURE 10-5B**). The renal tubule consists of four segments: (1) Bowman's capsule, (2) the proximal convoluted tubule, (3) the loop of Henle, and (4) the distal convoluted tubule (**TABLE 10-3**). As illustrated in **FIGURE 10-5B**, the glomerulus is surrounded by a saclike portion of the renal tube, called **Bowman's capsule** (after the scientist who first described it). It is a double-walled structure; the inner wall fits closely over the glomerular capillaries and is separated from the outer wall by a small space, **Bowman's space.**

To understand the relationship between the glomerulus and Bowman's capsule, imagine that your fist is a glomerulus. Then imagine that you are holding a balloon in your other hand. If you push your fist (glomerulus) into the balloon, the layer immediately surrounding your fist would be the inner layer of Bowman's capsule. It is separated from the outer layer of the capsule by Bowman's space.

The outer wall of Bowman's capsule is continuous with the **proximal convoluted tubule (PCT),** a

urinary bladder.[1] Bladder infections may result in an itching or burning sensation and an increase in the frequency of urination. They may also cause blood to appear in the urine. Urinary tract infections can be treated with antibiotics, but untreated infections may spread up the ureters to the kidneys, where they can seriously damage the nephrons and impair kidney function.

The Human Kidney Consists of Two Zones, an Outer Cortex and Inner Medulla

Each kidney is surrounded by a connective tissue capsule, appropriately called the **renal capsule.** Immediately beneath the capsule is a region known as the **renal cortex.** This region contains many nephrons. The inner zone is the **renal medulla.** It consists of cone-shaped

---

[1]Sexual intercourse increases the frequency of urinary bladder infections in many women. To reduce chances of developing an infection, women are advised to empty their bladders soon after intercourse.

---

[2]Glomerulus is the Latin word for "ball of yarn."

FIGURE 10-5 The Nephron   *(a)* A cross section of the kidney showing the location of the nephrons. *(b)* A drawing of a nephron.

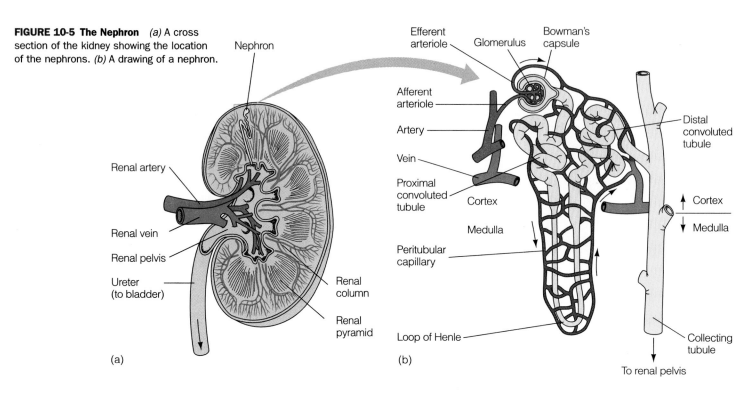

(a)   (b)

contorted section of the renal tubule. The PCT soon straightens, then descends a bit, makes a sharp hairpin turn, and ascends. In the process, it forms a thin, U-shaped structure known as the **loop of Henle** (HEN-lee). The loops of Henle of some nephrons extend into the medulla.

The loop of Henle drains into the fourth and final portion of the renal tubule, the **distal convoluted tubule (DCT),** another contorted segment. Each DCT drains into a straight duct called a **collecting tubule.** The collecting tubules merge to form larger ducts that course through the cone-shaped renal pyramids and eventually empty their contents into the renal pelvis.

The nephrons filter enormous amounts of blood and produce from 1 to 3 liters of urine per day, depending on how much fluid a person ingests.

Section 10-3

Function of the Urinary System

With the basic anatomy of the kidney and urinary system in mind, let us turn our attention to the function of the urinary system, focusing first on the nephron.

Blood Filtration in Nephrons Involves Three Processes: Glomerular Filtration, Tubular Reabsorption, and Tubular Secretion

Glomerular Filtration.
The first step in the purification of the blood is **glomerular filtration** (**FIGURE 10-6**). As blood flows through the glomerular capillaries, water and dissolved materials are forced through the cellular lining of the capillaries by blood pressure. This liquid, called the **glomerular filtrate,** travels through the inner layer of Bowman's capsule.

TABLE 10-3

Components of the Nephron and Their Function

Component	Function
Glomerulus	Mechanically filters the blood
Bowman's capsule	Mechanically filters the blood
Proximal convoluted tubule	Reabsorbs 75% of the water, salts, glucose, and amino acids
Loop of Henle	Participates in countercurrent exchange, which maintains the concentration gradient
Distal convoluted tubule	Site of tubular secretion of H^+, potassium, and certain drugs

Glomerular Filtration Is Controlled by Regulating the Flow of Blood and Blood Pressure in the Glomerulus.

Blood pressure in the glomerulus can be increased by dilation (opening) of the **afferent arterioles,** the small vessels leading into the glomerulus. This permits more blood to flow in and raises the blood pressure.

The **efferent arterioles** drain blood from the glomerular capillaries and raise glomerular blood pressure because they are slightly smaller in diameter than the afferent arterioles. Slight constrictions of the efferent arterioles further increase blood pressure within the glomerular capillaries. Any increase in blood pressure inside the glomerular capillaries increases the rate of filtration.

Tubular Reabsorption Helps Conserve Valuable Nutrients and Ions.

Each day, approximately 180 liters (45 gallons) of filtrate is produced in the glomeruli. However, the kidneys produce only about 1–3 liters of urine each day. Thus, only about 1% of the filtrate actually leaves the kidneys as urine. What happens to the rest of the fluid filtered by the glomerulus?

Most of the fluid filtered by the glomeruli is reabsorbed—it passes from the renal tubule *back* into the bloodstream. The movement of water, ions, and molecules from the renal tubule to the bloodstream is referred to as **tubular reabsorption.** Water containing valuable nutrients and ions leaves the renal tubule and enters the networks of capillaries that surround the nephrons. As shown in **FIGURE 10-5**, these capillaries are branches of the efferent arterioles. **TABLE 10-4** shows the reabsorption rates of various molecules.

TABLE 10-4		
Fate of Various Substances Filtered by Kidneys		
	Filtered Substance (Average %)	
Substance	**Reabsorbed**	**Excreted**
Water	99	1
Sodium	99.5	0.5
Glucose	100	0
Waste products		
Urea	50	50
Phenol	0	100

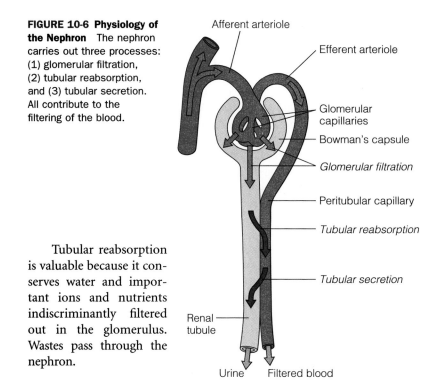

FIGURE 10-6 Physiology of the Nephron The nephron carries out three processes: (1) glomerular filtration, (2) tubular reabsorption, and (3) tubular secretion. All contribute to the filtering of the blood.

Afferent arteriole

Efferent arteriole

Glomerular capillaries

Bowman's capsule

Glomerular filtration

Peritubular capillary

Tubular reabsorption

Tubular secretion

Renal tubule

Urine   Filtered blood

Tubular reabsorption is valuable because it conserves water and important ions and nutrients indiscriminantly filtered out in the glomerulus. Wastes pass through the nephron.

Tubular Secretion Is the Transport of Waste Products from the Peritubular Capillaries into the Renal Tubule.

Waste disposal is supplemented by a third process, **tubular secretion.** Here is how it operates. Wastes not filtered from the blood as it passes through the glomerular capillaries remain in the blood that flows through the capillaries surrounding the nephron. Some of these wastes are then transported *into* the renal tubule *from* these peritubular capillaries (pear-ee-TUBE-you-ler). Tubular secretion helps rid the blood of wastes and also helps regulate the H^+ concentration of the blood.

Blood draining from the capillaries surrounding the nephrons is purged of wastes. Blood draining from the peritubular capillaries empties into small veins. These veins converge to form the **renal vein,** which transports the filtered blood out of the kidney and into the inferior vena cava and then on to the heart.

The urine leaving the nephron consists mostly of water and a variety of dissolved waste products. Urine leaves the distal convoluted tubule and enters the collecting tubules. These tubules descend through the medulla and converge to form larger ducts that empty their contents into a central chamber known as the **renal pelvis.** As the collecting tubules descend through the medulla, much of the remaining water escapes by osmosis, further concentrating the urine and conserving body water.

Urination: Controlling a Reflex

Urine is produced continuously by the kidneys and flows down the ureters to the urinary bladder. Leakage out the bladder and into the urethra is prevented by two sphincters—muscular "valves" similar to those found in the stomach (**FIGURE 10-4**).

The first sphincter, the **internal sphincter,** is formed by a smooth muscle in the neck of the urinary bladder at its junction with the urethra. The second valve, the **external sphincter,** is a flat band of skeletal muscle that forms the floor of the pelvic cavity. When both sphincters are relaxed, urine is propelled into the urethra and out of the body. How is urination stimulated?

When 200-300 milliliters of urine accumulate in the bladder, stretch receptors in the wall of the organ begin sending impulses to the spinal cord via sensory nerves (**FIGURE 10-7**). In the spinal cord, incoming nerve impulses stimulate special nerve cells. Nerve impulses generated in these cells leave the spinal cord and travel along nerves that terminate on the muscle cells in the wall of the bladder. These impulses stimulate the smooth muscle cells to contract. Muscular contraction forces the internal sphincter to open, letting urine enter the urethra.

In babies and very young children, nerve impulses arriving in the spinal cord from the stretch fibers also inhibit the nerve cells that supply the external sphincter. These nerves send impulses to the sphincter, keeping it closed. Nerve impulses from the stretch receptors, however, inhibit these impulses, causing the sphincter to relax. In babies and very young children, then, urination is a reflex—that is, there is no conscious override. Not until children grow older (2–3 years) can they begin to control urination.

In older children and adults, the external sphincter is under *conscious* control. It will not relax until you consciously permit it to do so.

Adults sometimes lose control over urination, resulting in a condition referred to as **urinary incontinence** (in-KAN-teh-nance). Urinary incontinence may be caused by a traumatic injury to the spinal cord, which disrupts descending nerve fibers that carry the reflex-overriding impulses from the brain. In such cases, the urinary

Superior wall of *distended* bladder

Superior wall of *empty* bladder

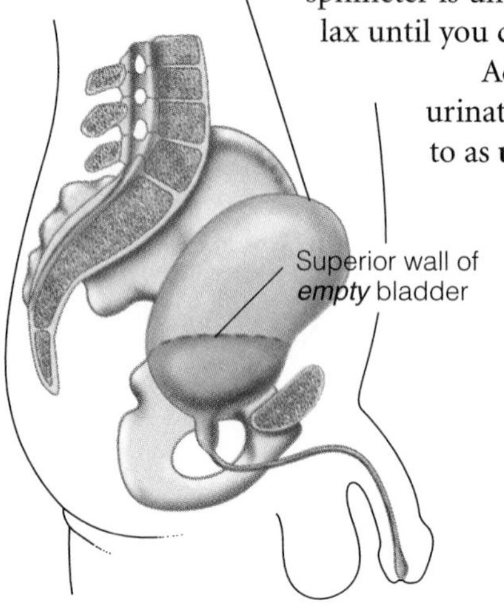

FIGURE 10-7 Urination in Babies
The bladder before and after it fills, showing how much this organ can expand to accommodate urine.

reflex remains intact, and the bladder empties as soon as it reaches a certain size, much as it does in a baby or young child.

Mild urinary incontinence is much more common. It is characterized by the escape of urine when a person sneezes or coughs. This is most common in women and usually results from damage to the external sphincter during childbirth. Childbirth stretches the skeletal muscles of the external sphincter, reducing their effectiveness and making such accidents embarrassingly common. To avoid this, many women undertake exercise programs to strengthen these muscles before and after childbirth. Urinary incontinence may also occur in men whose external sphincters have been injured in surgery on the prostate gland, which surrounds the neck of the urinary bladder.

Controlling Kidney Function and Maintaining Homeostasis

The kidneys help the body control the chemical composition of the blood and maintain homeostasis. This section briefly describes the kidney's role in maintaining water balance (that is, proper levels of water).

Water Balance Is Maintained by Conserving Water When Intake Falls or by Ridding the Body of It When Intake Is Excessive

As noted earlier, much of the water filtered by the glomeruli is passively reabsorbed by the renal tubules and returned to the bloodstream. The rate of water reabsorption can be increased or decreased to alter urine production. The ability to adjust water reabsorption and urine output allows the body to rid itself of excess water or to conserve water when an individual is becoming dehydrated. Two hormones play a major role in this important homeostatic process.

The Hormone ADH Increases Water Reabsorption and Conserves Body Water.
Water reabsorption is controlled in part by **antidiuretic hormone** (ANN-tie-DIE-yur-eh-tick) or **ADH.** ADH is released by an endocrine gland at the base of the brain known as the **pituitary** (peh-TWO-eh-TARE-ee) (Chapter 14).[3] ADH secretion is regulated by two receptors (**FIGURE 10-8**). The first is a group of nerve cells in a region of the brain called the **hypothalamus** (HIGH-poe-THAL-ah-muss), which is located just above the

---

[3] ADH is manufactured by the hypothalamus and transported to the posterior lobe of the pituitary gland via modified nerve cells, called *neurosecretory cells,* which are described in Chapter 14.

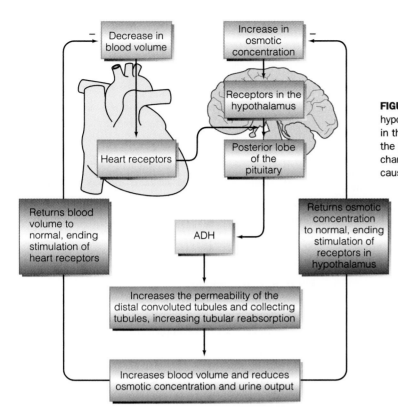

FIGURE 10-8 ADH Secretion ADH secretion is under the control of the hypothalamus. When the osmotic concentration of the blood rises, receptors in the hypothalamus detect the change and trigger the release of ADH from the posterior lobe of the pituitary. Detectors in the heart also respond to changes in blood volume. When it drops, they send signals to the brain, causing the release of ADH.

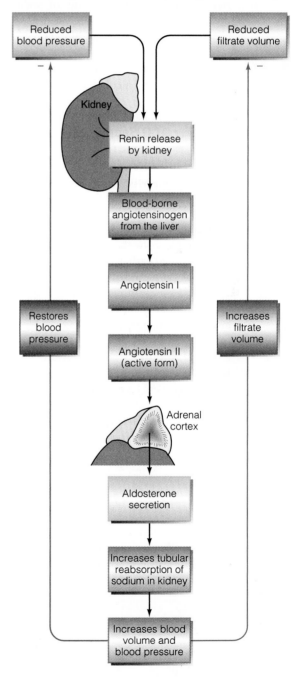

FIGURE 10-9 Aldosterone Secretion Aldosterone is released by the adrenal cortex. Its release, however, is stimulated by a chain of events that begins in the kidney.

pituitary gland. These cells monitor the osmotic concentration of the blood. The production and release of ADH are also controlled by receptors in the heart, which detect changes in blood volume (which reflect water levels).

When the body loses water, blood volume decreases and the osmotic concentration increases (**FIGURE 10-8**). The decrease in blood volume and the rise in osmotic concentration trigger the release of ADH. ADH circulates in the blood to the kidney, where it increases the rate of tubular reabsorption of water. This reduces urinary output and restores the volume and osmotic concentration of the blood.

Excess water intake has just the opposite effect. That is, it increases the blood volume and decreases its osmotic concentration. These changes reduce ADH secretion. As ADH levels in the blood fall, tubular reabsorption decreases and more water is lost in the urine. This decreases blood volume and also restores the proper osmotic concentration of the blood.

The Hormone Aldosterone Also Stimulates Water Reabsorption in the Kidney.

Water balance is also regulated by the hormone aldosterone (al-DOS-ter-own). **Aldosterone** is a steroid hormone produced by the adrenal glands, which sit atop the kidneys like loose-fitting stocking caps (**FIGURE 14-1A**). Aldosterone levels in the blood are controlled by three factors: (1) blood pressure, (2) blood volume, and (3) osmotic concentration (I use the mnemonic PVC—pressure, volume, and concentration to remember them).

FIGURE 10-9 shows that aldosterone is controlled by a complex sequence of events involving several chemical intermediaries. To understand this process, begin at the top of **FIGURE 10-9**. As illustrated, a reduction in blood pressure or a reduction in the volume of filtrate in the renal tubule causes certain cells in the kidney to produce an enzyme called **renin** (REE-nin). In the blood, renin cleaves a segment off a large plasma protein called **angiotensinogen** (AN-gee-oh-TEN-SIN-oh-gin). The result is a small peptide molecule called **angiotensin I.** Angiotensin I is converted into the active form, **angiotensin II,** by further enzymatic action in the lungs. Angiotensin II stimulates aldosterone secretion. Aldosterone increases the amount of sodium reabsorbed by the nephrons. Water follows the sodium and this increases blood volume and blood pressure.

Water balance is also affected by several chemicals in many people's daily diet, two of the most influential being caffeine and alcohol.

Diabetes (sugar)
www.jbpub.com/humanbiology

Caffeine Increases Urine Output without Affecting ADH Secretion.

Caffeine is a **diuretic** (DIE-yur-ET-ick), a chemical that increases urination. Found in coffee, (most nonherbal) teas, and many soft drinks, caffeine increases urine production in two ways.[4] First, it increases glomerular blood pressure, which, in turn, increases glomerular filtration. (As a result, more filtrate is formed.) Second, caffeine decreases the tubular reabsorption of sodium ions. As noted earlier, water follows sodium ions out of the renal tubule during tubular reabsorption. Thus, a decrease in sodium reabsorption results in a decline in the amount of water leaving the renal tubule and an increase in urine output.

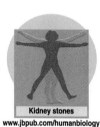

Kidney stones
www.jbpub.com/humanbiology

Ethanol Inhibits the Secretion of ADH by the Pituitary.

Alcoholic beverages contain ethanol, which is a diuretic. It inhibits the secretion of ADH by the pituitary, which reduces water reabsorption and increases water loss, giving credence to the quip that you don't buy wine or beer, you rent it!

Section 10-6

Diseases of the Urinary System

Like other body systems, the urinary system can malfunction, creating a homeostatic nightmare in the body, not unlike the situation that might exist in a factory if its

wastes could not be released or shipped elsewhere. This section discusses three disorders: kidney stones, kidney failure, and diabetes insipidus.

Kidney Stones Can Block the Outflow of Urine and Cause Severe Kidney Damage

Urine contains numerous ions and dissolved wastes. Ninety percent of the dissolved waste consists of three substances: urea, sodium ions, and chloride ions. (Use the mnemonic USC to remember them.) Varying amounts of other chemical substances are also present, but in minute amounts.

In many ways, the urine provides a window to the chemical workings of the body. Thus, increases or decreases in the level of substances in the urine may signal underlying problems—upsets in homeostasis and possible health problems. Consequently, physicians routinely analyze the urine of their patients to test for metabolic disorders. One of the most common is **diabetes mellitus** (DIE-ah-BEE-tees mell-EYE-tus), or sugar diabetes. This disease results in a defect in glucose uptake by cells in the body. Because body cells cannot absorb glucose, blood levels and urine levels are elevated.

Excess chemicals and ions in the urine not only signal problems elsewhere, they can be damaging to the kidney itself. For instance, higher than normal concentrations of calcium, magnesium, and uric acid in the urine, often caused by inadequate fluid intake, may crystallize inside the kidney, most often in the renal pelvis. Small deposits enlarge by accretion (ah-CREE-shun)—the deposition of materials on the outside of the stone, which causes them to grow in much the same way that a pearl grows in an oyster (**FIGURE 10-10**). These small crystals, or **kidney stones,** eventually grow into fairly large deposits. Unlike pearls, however, kidney stones are a health threat and an economic liability.

Fortunately, many kidney stones are flushed out of the kidney on their own. They enter the ureter and are passed to the urinary bladder. Small stones that enter the urinary bladder are often excreted in the urine. The sharp edges of the stone often dig into the walls of the ureters and urethra, causing considerable pain.

Problems arise when stones become lodged inside the kidney or the ureters and obstruct the flow of urine. This causes internal pressure to increase, resulting in considerable damage to the nephrons if left untreated.

For years, kidney stones were removed surgically. Today, however, the relatively new medical technique of **ultrasound lithotripsy** is generally used. In this technique, mentioned in the opening paragraph of this chapter, physicians bombard kidneys with ultrasound

---

[4]Part of the increase is due to the fluid contained in these drinks.

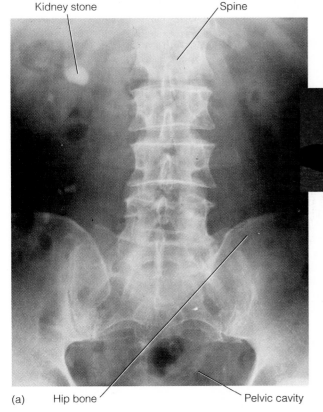

Kidney stone Spine

(a) Hip bone Pelvic cavity

FIGURE 10-10 Kidney Stones (a) An X-ray of a kidney stone. (b) Kidney stones removed by surgery. (See nail for size.)

(b)

Kidney failure
www.jbpub.com/humanbiology

Renal Failure May Occur Suddenly or Gradually and Can be Treated by Dialysis, a Mechanical Filtering of the Blood

The importance of the kidneys is most obvious when they stop working, a condition known as **renal failure.** Renal failure generally results from one of four causes. In some instances, it is caused by the presence of certain toxic chemicals in the blood. In others, it results from immune reactions to certain antibiotics. In still other cases, it is caused by severe kidney infections or sudden decreases in blood flow—for example, after an injury.

Renal failure may occur suddenly, over a period of a few hours or a few days. This is referred to as **acute renal failure.** Kidney function may also deteriorate slowly over many years, resulting in **chronic renal impairment** (so named because the kidney function declines over time). Chronic renal impairment may lead to a complete or nearly complete shutdown, known as **end-stage failure.**

waves, which shatter the stones, producing fine, sandlike grains that are passed in the urine without incident. The procedure is nearly painless and much safer than surgery. **TABLE 10-5** lists additional urinary system disorders.

TABLE 10-5

Common Urinary Disorders

Disease	Symptoms	Cause
Bladder infections	Especially prevalent in women; pain in lower abdomen; frequent urge to urinate; blood in urine; strong smell to urine.	Nearly always bacteria
Kidney stones	Large stones lodged in the kidney often create no symptoms at all; pain occurs if stones are being passed to the bladder; pains come in waves a few minutes apart.	Deposition of calcium, phosphate, magnesium, and uric acid crystals in the kidney, possibly resulting from inadequate water intake
Kidney failure	Symptoms often occur gradually; more frequent urination, lethargy, and fatigue; should the kidney fail completely, patient may develop nausea, headaches, vomiting, diarrhea, water buildup, especially in the lungs and skin, and pain in the chest and bones.	Immune reaction to some drugs, especially antibiotics; toxic chemicals; kidney infections; sudden decreases in blood flow to the kidney resulting, for example, from trauma
Pyelonephritis	Infection of the kidney's nephrons; sudden, intense pain in the lower back immediately above the waist, high temperature, and chills	Bacterial infection

Point/Counterpoint

Prioritizing Medical Expenditures

WE NEED TO LEARN TO PRIORITIZE MEDICAL EXPENDITURES

Richard D. Lamm

Richard D. Lamm *is the director of the Center for Public Policy and Contemporary Issues at the University of Denver, where his primary research and teaching interests have been in the area of health policy. Before assuming his present position, he served three terms as governor of Colorado, from 1975 until 1987.*

Heath care in the United States finds itself in (to paraphrase Dickens) "the best of times and the worst of times." It is the epoch of medical miracles, yet at the same time millions go without even basic health care.

The basic dilemma of American medicine is that we have invented more health care than we can afford to deliver. Health care costs are rising at three times the rate of inflation. They are absorbing funds desperately needed elsewhere in our system to educate our kids, rebuild our infrastructure, and revitalize our industries.

Health care costs are rising for many reasons. One major reason is that no one prioritizes the myriad of procedures that health care can deliver. No one asks, "How do we buy the most health for the most people with our limited funds?" Recent policy decisions in Oregon and California, relating to the public funding of transplant operations, illustrate how two states attempted to deal with a crisis in health care.

Oregon recently received adverse publicity for its decision not to publicly fund soft tissue transplant operations. As is often the case, much of the focus has been on the handful of individuals who have been adversely af-

fected by the policy rather than the numerous, but anonymous, people who will benefit. The Oregon policy sparked a public outcry over society's lack of compassion for individuals who desperately need transplants. Yet, when Oregon policymakers weighed the needs of a few transplant patients against the basic health care needs of the medically indigent, they decided against the former in favor of the latter. They have now set up a process that sets priorities throughout the health care system.

California took a contrary approach. Policymakers in California decided to publicly fund transplant operations. Then, one week later, they removed 270,000 low-income people from their state's medical assistance program.

Which was the wiser decision? The answer seems clear. Oregon bought more health for more of its citizens for its limited funds. Some say Oregon should have raised its taxes and funded transplants. That is very easy for nonpoliticians to say. Polls show people believe all Americans should have access to quality health care. But the polls also show that most are unwilling to accept even modest tax increases to provide the care they say they favor.

Illinois recently passed legislation giving "universal access to major organ transplants," but appropriated less than one-third of the funds needed. Politically, of course, Illinois played it far safer than Oregon but, in doing so, failed to confront one of the most pressing social issues of our time. Avoidance may be a politically expedient tactic, but sooner or

later our society will be forced to allocate scarce health resources.

Clearly, medical science is inventing faster than the public is willing or able to pay. Realistically, no system of health care can avoid rationing medicine. In fact, rationing already exists. But instead of rationing medicine according to need or a patient's prospects for recovery, we ration by seniority and ability to pay. Those who pretend that we do not ration medicine forget that 31 million Americans do not have full access to health care in America. Most states have staggering numbers of medically indigent, yet they provide a small percentage with a full program of coverage. Oregon is the first state to cover 100% of the people living under the federal poverty line with basic health care.

I suggest this yardstick: Which state best asked itself, "how do we buy the most health for the most people?" Which state tried to maximize limited resources? Which state benefited the greatest number of people and which state harmed the greatest number of people?

Oregon attempted to weigh the basic health needs of the medically indigent against other programs for which the state paid. It recognized that the money the state paid for transplants and other high-cost procedures for a few people could buy more health care elsewhere, providing basic low-technology services to people not covered at that time.

Prioritizing health care does not abandon the poor; it seeks to serve the largest number with the more effective procedures.

Kidney failure (whether acute or end stage) is life-threatening, for when the kidneys stop working, water and toxic wastes begin to accumulate in the body. This disrupts homeostasis. If untreated, a patient will die in 2–3 days. Patients usually die from an increase in the concentration of potassium ions in the blood and tissue fluids. Why? Although normal levels are essential for the function of heart muscle, excess potassium destroys the

MEDICAL PRIORITIZATION IS A BAD IDEA

Arthur L. Caplan

The American health care system, the experts say, is going bust at a rapid rate. Efforts to contain our burgeoning $600-billion-plus tab have been a total failure.

The dilemma of how to pay for health care is forcing some public officials to think the unthinkable. The state of Oregon has plans to institute rationing policies for health care.

But, before you applaud the realism, consider that the Oregon plan will ration access to health care only for the poor. Those eligible for Medicaid in Oregon will be required by law to forego life-saving medical care.

Officials in Oregon note that the poor have always had less access to health care than the rich. This is true. But, our society's failure to meet the health care needs of the poor hardly justifies a public policy that asks the poor to bear the burden of rationing as a matter of law.

Who concocted this blatantly unethical scheme? Incredibly, the inspiration for the Oregon plan for pocketbook triage comes in part from those in my line of work—medical ethicists.

A California bioethics consulting firm was paid by Oregon state officials to provide moral rationales for dropping the poor out of the health care lifeboat. The consultants appeared to approach their task with gusto.

"You have to draw the line somewhere," one moralist-for-hire said in a recent newspaper article about rationing for the poor. "We'll provide all services to a diminishing segment of the population, and literally we'll throw the rest of the people overboard."

No hint is given of the theoretical position that would justify aiming all rationing efforts at the poor. But it is hard to think of an ethic that holds that when a nation cannot pay its doctor bills, it is the poor and only the poor who should be denied the right to see a doctor.

It is hard to understand how any ethicist could become involved in a scheme so blatantly unfair as that of rationing necessary health care only for the poor. What is worse is that the same ethicists and the officials taking their advice are not asking whether it is really necessary to institute the rationing of necessary medical care for anyone.

Before saying goodbye to the indigent, why aren't public officials in Oregon thinking about cracking down on practices that add tens of millions of dollars to state-financed health care costs each year? Before saying no to a bone marrow transplant for a three-year-old whose mother is on Medicaid, couldn't county and state legislators insist that every licensed hospital and physician be required by law to provide a fixed percentage of care for those who cannot pay?

Before creating laws that would send some of the poor to a premature demise, county and state officials ought to require private health insurers to charge subscribers an additional premium that could be used to supplement the pitifully small budgets of Medicaid and public hospitals. And would it not make some sense to insist on a luxury tax, which could be used to help meet the crucial health care needs of the poor, from the rich who avail themselves of psychotherapy, vitamins, cosmetic surgery, diet clinics, and stress-management seminars?

It is wrong to make the poor and only the poor bear the burden of rationing. It is unethical to institute rationing of necessary health services for any group of Americans unless we have made every effort to be as efficient as we can be in spending our health care dollars.

At a time when some can indulge their wants by buying a face-lift, it seems extraordinarily hard for ethicists or legislators to convincingly argue that they have no other option but to condemn the poor to die for want of money.

Arthur L. Caplan *is director of the Center for Biomedical Ethics at the University of Minnesota.*

SHARPENING YOUR CRITICAL THINKING SKILLS

1. Summarize Lamm's and Caplan's key points. List supporting information given for each position or main point.
2. Do you agree with the views of Lamm or Caplan? Explain why.

Visit *Human Biology*'s Internet site, www.jbpub.com/humanbiology, to research opposing web sites and respond to questions that will help you clarify your own opinion. (See Point/Counterpoint: Furthering the Debate.)

www.jbpub.com/humanbiology

---

rhythmic contraction of the heart, causing fibrillation (Chapter 6). Patients most often die of heart attack.

Treating renal failure depends on the underlying cause. If the problem is caused by an acute loss of blood, transfusions may be required. Patients whose kidneys have shut down, even temporarily, may require **renal dialysis** (DIE-AL-eh-siss). In this procedure, blood is drawn out of a vein and passed through a piece of tubing that transports the blood to a filter that removes wastes. After filtration, the blood is pumped back into the patient's bloodstream. Dialysis requires several hours and must be repeated every 2 or 3 days. Some patients have dialysis units at home and simply hook themselves up each night before they go to bed.

Another more recent method is **continuous ambulatory peritoneal dialysis** or **CAPD.** In this procedure, 2 liters of dialysis fluid are injected into a person's abdomen through a permanently implanted tube or catheter (CATH-eh-ter). Waste products diffuse out of the blood vessels into the abdominal cavity across a thin membrane that lines the organs and wall of abdominal cavity, the **peritoneum** (PEAR-eh-tah-KNEE-um). The fluid, containing waste products, is drained from the abdominal cavity a couple of times a day.

This form of dialysis is much simpler. Patients can take care of it themselves. In addition, it allows for more frequent filtering of the blood and permits patients to continue their daily activities without having to strap themselves to a machine.

Complete kidney failure can be treated by kidney transplants. Transplants are generally most successful when they come from closely related family members, for reasons explained in Chapter 8.

For years, the complete or nearly complete destruction of kidney function was almost always fatal. Thanks to renal dialysis and kidney transplantation, many patients today can live normal, healthy lives. These procedures, especially transplants, are costly, however. The public currently picks up the tab for individuals without insurance—or they go without needed treatments and die.[5] For a debate on prioritizing medical expenditures, see this chapter's Point/Counterpoint.

Diabetes insipidus

Diabetes Insipidus Is a Rare Disease Caused by a Lack of ADH

Hormonal imbalances can also lead to disruptions in the urinary system. Severe head injuries, for example, may halt the production of ADH, leading to a disease known as **diabetes insipidus** (DIE-ah-BE-teas in-SIP-eh-duss). This condition is not to be confused with diabetes mellitus (commonly called *sugar diabetes*), a disorder involving the hormone insulin.

Diabetes insipidus is characterized by frequent urination and excessive liquid intake. Diabetes insipidus results from insufficient ADH output. Remarkably, patients with this disease produce up to 20 liters (5 gal-

lons) of colorless, dilute urine per day—fluid that must be replaced by frequent consumption of liquids. Diabetes insipidus gets its name from the fact that the urine is dilute and tasteless (insipid).[6] Sleeping through the night is impossible, for patients are continually awakened by thirst or the urge to urinate.

Diabetes insipidus can be treated in a variety of ways, depending on the severity of the disorder and the cause. In patients whose urine output is only slightly elevated, dietary salt restrictions and antidiuretics (drugs that reduce urine output) work. In severe cases, patients must receive synthetic ADH, which is administered by injections or nose drops. If the disease is caused by a head injury, treatment may be required only for a year or so. If the damage is permanent, however, treatment will be required for the rest of the person's life.

Health, Homeostasis, and the Environment

Mercury Poisoning

The kidneys are elaborate biological filters that help maintain the proper levels of nutrients and help eliminate various cellular wastes and potentially harmful substances such as food additives, pesticides, and toxic chemicals ingested in the food we eat and the water we drink.

Unfortunately, the kidneys can be seriously damaged by a variety of environmental contaminants. As the previous section showed, damage to the kidney has severe physiological repercussions, primarily resulting from homeostatic imbalance.

One of the most toxic by-products of the modern world are heavy metals, such as mercury and cadmium. Heavy metals are used in large quantities in many manufacturing processes and in many products. As a group, the heavy metals are potent **nephrotoxins** (NEFF-row-TOX-ins)—toxic chemicals that affect the nephrons. Even relatively low doses of heavy metals can damage the kidneys.

Fortunately, the nephrons possess several protective mechanisms to reduce the impact of heavy metals. Lysosomes inside the cells of the renal tubule, for example, engulf heavy metals, decreasing their cytoplasmic concentration. As blood levels of these toxins increase, however, the protective mechanisms are overwhelmed, and the cells of the kidney begin to die.

---

[5]If national health care is successful, this problem may no longer exist.

[6]Diabetes mellitus is so named because the urine is sweet. The Latin word for "honey" is *mellifer.*

Exposure to low levels of various heavy metals increases urinary output and increases the concentrations of amino acids and glucose in the urine. Both of these symptoms are good indicators that tubular reabsorption is not working properly. At higher levels, heavy metals can cause renal failure and death. The proximal convoluted tubule is the part of the nephron most sensitive to heavy metals.

One of the most common and hence more troublesome heavy metals is mercury. Found in the water we drink and occasionally even in the food we eat, mercury in high concentrations produces acute renal failure. This condition results from vasoconstriction of the afferent arterioles, which reduces the flow of blood into the glomeruli. Mercury also produces cellular deterioration.

In the 1950s, Japan announced an outbreak of mercury poisoning in residents who had consumed fish and shellfish taken from Minimata Bay, which had been contaminated by a nearby plastics factory. Over 100 people developed numbness of the limbs, lips, and tongue and lost muscular control, becoming clumsy. Most suffered from blurred vision, deafness, and mental derangement. All told, 17 people died and 23 were permanently disabled, in large part due to the inability of their kidneys to cope with the mercury and subsequent nervous system effects. The tragedy was deepened by the discovery of birth defects in 19 babies born to women who had eaten contaminated seafood. Many of their mothers showed no signs of mercury poisoning, illustrating a general principle of toxicology: the younger an organism, the more sensitive it is to toxins.

The Minimata Bay incident was not an isolated event. A similar tragedy occurred on the Japanese island of Honshu. In Sweden in the early 1960s, mercury poisoning killed large numbers of birds that had been feeding on seeds treated with a mercury fungicide, a coating that retards mildew. Swedes who ate pheasant and other birds were also poisoned.

Fortunately, incidents of overt mercury poisoning are rare. However, mercury is one of the more common water pollutants in the industrialized world, and people are continually exposed to low levels from a variety of sources. For example, mercury is a by-product of the production of the plastic polyvinyl chloride (commonly called PVC plastic). PVC is used to manufacture children's toys, beach balls, car seats, and other products. Chances are you have some PVC nearby. Mercury is emitted into waterways by a variety of chemical manufacturers and also released by coal-fired power plants and garbage incinerators that burn batteries thrown out in our trash.

Many sources of mercury production are on the rise. Without tighter controls on emissions, low-level mercury poisoning may become more and more common in the years to come. If so, we can expect a rise in kidney disease and other disorders caused by impairment of the kidney's homeostatic functions.

SUMMARY

ORGANS OF EXCRETION: A BIOLOGICAL IMPERATIVE

1. All organisms produce waste and all organisms face the same challenge: getting rid of internally generated wastes.
2. Fortunately, humans have excretory organs that remove waste and help regulate internal concentrations of ions and water vital for homeostasis.
3. **TABLE 10-1** lists the major metabolic wastes and other molecules excreted from the body.
4. These wastes are removed by the skin, lungs, liver, kidneys, and intestines.

THE URINARY SYSTEM

5. One of the most important organs of excretion in humans (and other vertebrates) is the *kidney.* Kidneys remove impurities from the blood and also help regulate the water levels and ionic concentrations of the blood.
6. Blood enters the kidneys in the *renal arteries* whose branches deliver it to the millions of nephrons located in the kidney.
7. The nephrons produce urine, which drains from the kidneys into the *ureters,* slender muscular tubes that lead to the urinary bladder. Urine is stored in the *urinary bladder,* then voided through the *urethra.*
8. Each *nephron* consists of a *glomerulus,* a tuft of highly porous capillaries, and a *renal tubule,* where urine is produced.
9. The renal tubule consists of four parts: (a) *Bowman's capsule,* (b) the *proximal convoluted tubule,* (c) the *loop of Henle,* and (d) the *distal convoluted tubule.* The distal convoluted tubules of nephrons drain into *collecting tubules,* which converge and empty urine into the *renal pelvis.*

FUNCTION OF THE URINARY SYSTEM

10. Blood filtration is accomplished by: glomerular filtration, tubular reabsorption, and tubular secretion.
11. *Glomerular filtration* occurs in the glomerulus, producing a liquid called the *filtrate.*
12. The filtrate is processed as it flows along the renal tubule. Water, ions, and nutrients are largely reabsorbed as they travel along the tubule. This process is

called *tubular reabsorption.* Water and reabsorbed nutrients and ions pass into a network of capillaries, the *peritubular capillaries,* surrounding each nephron. What is left is a concentrated liquid, the *urine.*

13. Not all wastes are filtered from the blood in the glomerulus. Those that remain pass into the peritubular capillaries with the blood. These substances may be transported out of the peritubular capillaries into the renal tubule in a process known as *tubular secretion.* Hydrogen and potassium ions, for example, are secreted into the renal tubule.

URINATION: CONTROLLING A REFLEX

14. *Urination* is a reflex in babies and very young children. In older children and adults, the urination reflex still operates, but it is overridden by a conscious control mechanism.

CONTROLLING KIDNEY FUNCTION AND MAINTAINING HOMEOSTASIS

15. Each drop of blood in your body flows through the kidneys many times in a single day. This flow allows for a thorough filtering of the blood and also helps the body control the chemical composition of the blood and extracellular fluid.

16. The concentration of water and dissolved substances in the blood is controlled by two hormones: ADH and aldosterone.

17. *ADH* or *antidiuretic hormone* is secreted by the posterior lobe of the pituitary gland. It is released when the osmotic concentration of the blood increases or when blood volume decreases.

18. ADH increases the permeability of the distal convoluted tubules and the collecting tubules to water. When ADH is present, water reabsorption increases.

19. *Aldosterone* is produced by the adrenal cortex. Aldosterone stimulates the reabsorption of sodium ions by the nephron. Water follows the sodium out of the renal tubule, increasing blood pressure and blood volume.

DISEASES OF THE URINARY SYSTEM

20. Calcium, magnesium, and other materials can precipitate out of the urine in the renal pelvis, forming *kidney stones.*

21. Smaller stones may be passed along the ureters to the bladder and are often eliminated during urination. Larger stones that remain in the kidney must be removed surgically or via *ultrasound lithotripsy.*

22. Renal failure is a disease in which the kidneys stop working. Renal failure may occur suddenly or gradually and is a life-threatening disorder. It can be treated by dialysis, a mechanical filtering of the blood.

23. Diabetes insipidus is a rare but troublesome disease resulting from a lack of ADH. The disease is characterized by frequent urination and excessive fluid intake.

HEALTH, HOMEOSTASIS, AND THE ENVIRONMENT: MERCURY POISONING

24. The kidneys help regulate the levels of nutrients and toxins produced by the cells of the body. They also help eliminate toxins taken into the body from air, water, and food.

25. Unfortunately, the kidneys are not immune to many harmful substances. Heavy metals, for example, can destroy cells of the renal tubule and can restrict blood flow to the glomeruli. As a result, heavy metals can damage the kidney, impairing the function of this important homeostatic organ.

Critical Thinking

THINKING CRITICALLY— ANALYSIS

This Analysis corresponds to the Thinking Critically scenario that was presented at the beginning of this chapter.

After preparing your list, compare your concerns and questions with mine to see how we match.

- Concern: A clinical trial on 90 patients is extremely small. With so few patients, the reliability of the results is in question.
- Question 1: Are more clinical tests under way?
- Question 2: Does the treatment have any adverse impacts?
- Question 3: Is the marketing company unbiased?

- Question 4: Is it promoting a product that may turn out to be ineffective, opening the company to lawsuits for fraud? In other words, is it letting financial concerns outweigh the need for good scientific research and carefully controlled experiments?

Now suppose that you have kidney cancer and have the $22,000 to pay for ALT. Would you do it? Are any of the issues above still relevant? How would you go about determining whether you should try the procedure?

To be fair to all parties concerned, let me point out that approximately two-thirds of the insurance companies in the United States pay for ALT treatment. As a rule, most companies do not pay for experimental procedures. In other words, they must be satisfied that ALT is an effective treatment before reimbursing clients.

EXERCISING YOUR CRITICAL THINKING SKILLS

After reading the following hypothetical scenario, you will be asked to make a decision. Then you will be asked some questions about your decision that may help you begin to clarify your values. Clarifying your values is essential to critical thinking because it helps you understand your own biases.

Here's the scenario. You are a state legislator considering legislation on prioritizing medical expenditures. Your subcommittee will make a recommendation to the legislature to adopt or reject a plan that would shift state funding from an organ transplant program to a prenatal care program. The state currently pays for organ transplants for needy families, spending over $5 million per year. A group of legislators, however, is proposing that this money be used to fund free checkups for pregnant women as well as advice on drug and alcohol use during pregnancy. The program would also offer information on maternal and infant nutrition to expectant mothers.

Proponents of medical care prioritization say that the money now required for one organ transplant could fund prenatal care for about 1000 mothers. By spending the money on prenatal care, the state could reach thousands of pregnant women who are too poor to see a doctor. Proponents also estimate that 10% of all newborns in your state are born addicted to cocaine, which affects mental and physical development.

Opponents of the bill point out that if needy families are denied money for organ transplants, dozens of children will die. They present the case of Jason Lowry to illustrate what will happen. Jason is 12 years old and needs a liver transplant that will cost $100,000. Unfortunately, his family lives on welfare. Without the transplant, Jason is certain to die.

If the state chooses to fund prenatal care instead of organ transplants, dozens of children needing liver transplants will die each year. You have a choice. Would you recommend the bill that transfers funding to prenatal care or continue funding organ transplants? Why?

Make a list of reasons why you supported or opposed the bill. Now take a moment to ponder them. Was your decision based on economics? Was it based on relative benefits—that is, the benefits for a few versus the benefits for many? Was your decision based on benefits for future generations? Or were you mostly concerned with immediate effects—for example, saving a few lives now?

Imagine, if you will, that Jason was your son. How does that affect your decision? Does the issue take on a different meaning? What general observations can you make about your objectivity? Did it change as the issue came "closer to home"? To what extent do personal interests affect your decisions about other questions, such as environmental issues? Give some examples.

TEST OF CONCEPTS

1. Draw the various parts of the urinary system, and describe what each one does.
2. Draw a nephron, then label its parts. Explain what happens to the filtrate in each section of the nephron.
3. Trace the flow of blood into and out of the kidney. Be sure to include details of the pathway once it reaches the afferent arteriole.
4. Describe the three ways in which the kidney filters the blood.
5. A drug inhibits the uptake, or reabsorption, of water by the distal convoluted tubules and collecting tubules. What effect would this drug have on urine output, urine concentration, blood pressure, blood volume, and the concentration of the blood?
6. Describe how ADH controls blood pressure and the water content of the body. Describe the hormonal and physiological changes in the body that take place when excess liquid is ingested. Do the same for dehydration.
7. How does urination differ between newborns and adults? Explain what is meant by this statement: In older children and adults, urination is a reflex with a conscious override.
8. Aldosterone helps regulate blood pressure and water content. In what ways is this hormone different from ADH?
9. You have just finished your residency in family medicine. A patient comes to your office complaining that he drinks water all day long and spends much of the rest of the day in the bathroom urinating. What tests would you order? What diagnosis would you suspect?

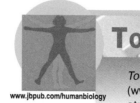

TOOLS FOR LEARNING

Tools for Learning is an on-line student review area located at this book's web site HumanBiology (www.jbpub.com/humanbiology). The review area provides a variety of activities designed to help you study for your class:

Chapter Outlines. We've pulled out the section titles and full sentence sub-headings from each chapter to form natural descriptive outlines you can use to study the chapters' material point by point.

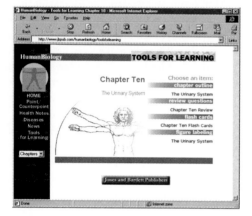

Review Questions. The review questions test your knowledge of the important concepts and applications in each chapter. Written by the author of the text, the review provides feedback for each correct or incorrect answer. This is an excellent test preparation tool.

Flash Cards. Studying human biology requires learning new terms. Virtual flash cards help you master the new vocabulary for each chapter.

Figure Labeling. You can practice identifying and labeling anatomical features on the same art content that appears in the text.

Active Learning Links. Active Learning Links connect to external web sites that provide an opportunity to learn basic concepts through demonstrations, animations, and hands-on activities.

THE NERVOUS SYSTEM

INTEGRATION, COORDINATION, AND CONTROL

Light micrograph of the cerebellum of the human brain.

Thinking Critically

Researchers at the Washington University School of Medicine in St. Louis recently published a report in the *Journal of Neuroscience* on brain activity in people with severe depression. They compared brain activity in normal individuals with that in people suffering from severe depression, which is characterized by dreary moods, apathy, and hopelessness.

The scientists injected the subjects with radioactive oxygen, which is taken up by metabolically active brain cells. Inside the brain, radioactive oxygen molecules emit gamma rays; these are picked up by a special device that sends signals to a computer. The computer then maps brain cell metabolic activity.

In the study under discussion, researchers compared 6 patients with depression to 18 control subjects. To compare the results, they used a computer program that combines the "brain maps" of each group to produce a composite—sort of an average of the entire group.

The scientists found that a region of the brain known as the left prefrontal cortex was extremely active in patients suffering from depression. None of the controls showed a similar response. How would you interpret this data? What further studies would you perform?

ERNEST HEMINGWAY'S NOVELS AND SHORT stories won him the Nobel Prize in literature. Despite success and widespread popularity, Hemingway was a troubled man who eventually committed suicide. What ultimately caused him to take his life no one can know, but some believe that one contributing factor was a rare and painful nervous system disorder known as **trigeminal neuralgia** (try-GEM-in-al ner-AL-gee-ah). People suffering from this disease complain of periodic, unexplained flashes of pain along the course of the **trigeminal nerve,** which supplies the face. A slight breeze or the pressure of a razor can set off this intense pain, which lasts a minute or more. In some patients, the pain reappears for no apparent reason every few minutes for weeks on end. Some people believe that the pain Hemingway felt may have become unbearable. Combined with personal conflict, it may have caused the writer to take his life.

Today, physicians treat mild cases of trigeminal neuralgia with drugs. In extreme cases, however, they may elect to sever the nerve as it leaves the brain. This procedure ends the pain, but because it cuts off the inflow of other sensory information, it leaves half of the victim's face, tongue, and oral cavity numb—a little like the feeling you get when a dentist injects novocaine into your gums.

This chapter describes the anatomy and physiology of the human nervous system. You will briefly study the nerve that may have caused Hemingway so much trouble and will also examine ways in which the nervous system contributes to homeostasis.

Section 11-1

An Overview of the Nervous System

The human nervous system governs the functions of the body, exerting control over muscles, glands, and organs. It also controls heartbeat, breathing, digestion, and urination. It helps regulate blood flow as well as the osmotic concentration of the blood. As such, it plays a major role in maintaining homeostasis.

The nervous system receives input from a large number of sources in the body. This input helps the nervous system "manage" body functions in much the same way that letters from citizens help elected officials govern society.

The human nervous system provides functions not seen in other animal species. For example, the brain is the site of ideation—the formation of ideas. Our brain allows us to think about and plan for the future. It enables us to reason—that is, to judge right from wrong, logical from illogical. Although some species display a rudimentary ability to reason, this function is best expressed in humankind.

The nervous system also allows us to manipulate our environment to our own liking. Much more than any other species alive today, we humans are reshaping the planet. We level tropical forests to make room for cattle, drain and fill swamps to build homes and factories, split atoms to generate energy, and catapult men and women into space. Joan McIntyre, an author and critic, once wrote, "The ability of our minds to imagine, coupled with the ability of our hands to devise our images, brings us a power almost beyond control."

Today, many of the advances we humans have made, in our unrelenting march toward "progress," now threaten our very existence (Chapter 24). Pollution, resource depletion, and other troubles stand in the way of our own success. The human brain gives us the power to create yet also an incredible power to destroy. Numerous examples in this book show that the by-products of many technologies are overwhelming environmental and human physiological systems involved in homeostasis and are consequently threatening our health and well-being. Fortunately, our hope of survival itself also depends on our brain, an evolutionary by-product unrivaled in the biological world.

◼ The Nervous System Consists of Two Main Anatomical Subdivisions, the Central and Peripheral Nervous Systems

The human nervous system consists of three components: the brain, the spinal cord, and nerves (**FIGURE 11-1**). The brain and spinal cord constitute the **central nervous system (CNS)** and are housed in the skull and vertebral canal, respectively. Three layers of connective tissue, known as the **meninges** (men-IN-gees), surround the brain and spinal cord (**FIGURE 11-2**). The outer layer consists of fibrous connective tissue and is known as the **dura mater** (DURE-ah MAH-ter; meaning "hard mother"). The middle covering is the **arachnoid layer** (ah-RACK-noid), so named for its spider-weblike appearance. The innermost layer is the **pia mater** (PEE-ah MAH-ter; literally, "tender mother"). The pia mater is a delicate, vascular layer that adheres closely to the brain and spinal cord. The space between the arachnoid layer and pia mater is filled with a liquid called **cerebrospinal fluid (CSF)** (sir-REE-bro-SPIE-nal).

Nerves consist of bundles of nerve fibers. Each fiber is part of a nerve cell, or **neuron** (NER-on). Nerves transport messages to and from the CNS. These nerves end in various receptors that respond to a variety of internal and external stimuli. Together the nerves and sensors constitute the **peripheral nervous system (PNS).**

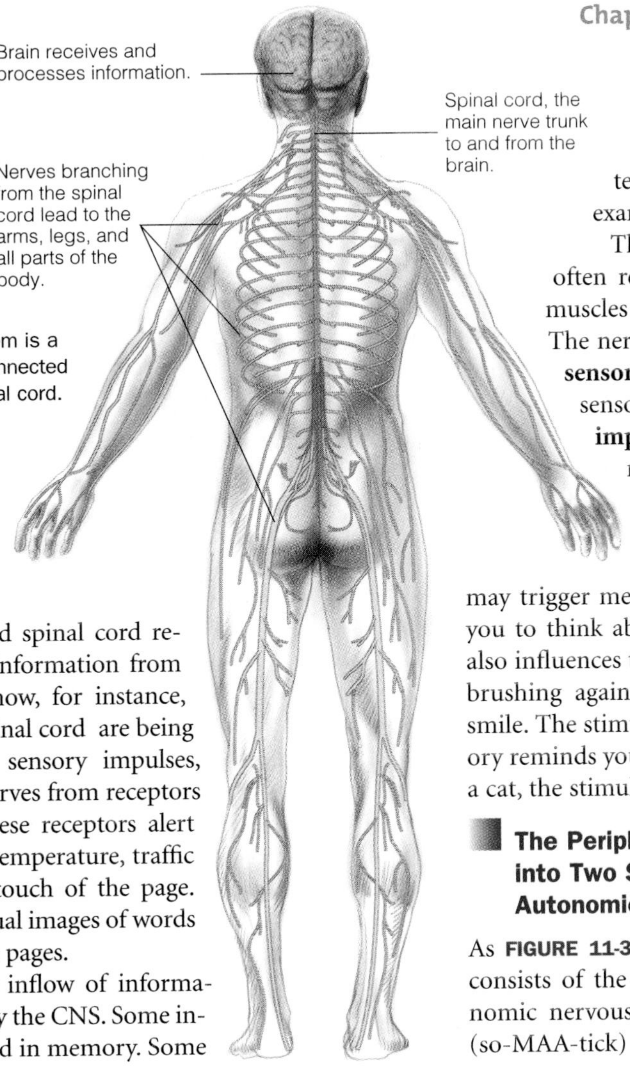

Brain receives and processes information.

Spinal cord, the main nerve trunk to and from the brain.

Nerves branching from the spinal cord lead to the arms, legs, and all parts of the body.

FIGURE 11-1 The Nervous System The human nervous system is a network of nerves connected to the brain and spinal cord. Nerves comprise the peripheral nervous system. The spinal cord and brain make up the central nervous system.

The brain and spinal cord receive all sensory information from the body. Right now, for instance, your brain and spinal cord are being bombarded with sensory impulses, which travel by nerves from receptors in your body. These receptors alert you to the room temperature, traffic sounds, and the touch of the page. They transmit visual images of words and figures on the pages.

This massive inflow of information is managed by the CNS. Some information is stored in memory. Some

incoming information is ignored, and some elicits responses. A particularly interesting and exciting section you read, for example, might accelerate your heart rate.

The brain processes incoming stimuli and often responds by sending nerve impulses to muscles and glands (effectors) via the nerves. The nerves carry two types of information: (1) **sensory impulses** traveling to the CNS from sensory receptors in the body, and (2) **motor impulses** traveling away from the CNS to muscles and glands.

Sensory information entering the CNS is integrated with information stored in memory. Thus, a new fact may trigger memories of previous knowledge, causing you to think about a problem in a new way. Memory also influences the way we respond to stimuli. A pet cat brushing against your leg, for example, may elicit a smile. The stimulus is not startling, because your memory reminds you of the cat's presence. If you didn't own a cat, the stimulus might send you through the roof!

The Peripheral Nervous System Is Divided into Two Subdivisions: The Somatic and Autonomic

As **FIGURE 11-3** shows, the peripheral nervous system consists of the somatic nervous system and the autonomic nervous system. The **somatic nervous system** (so-MAA-tick) is that portion of the PNS that controls

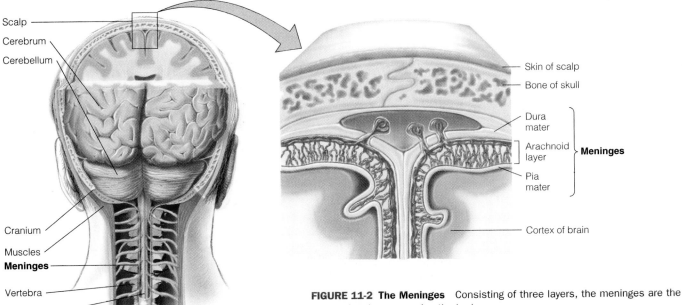

Scalp

Cerebrum

Cerebellum

Cranium

Muscles

Meninges

Vertebra

Spinal cord

Skin of scalp

Bone of skull

Dura mater

Arachnoid layer } **Meninges**

Pia mater

Cortex of brain

FIGURE 11-2 The Meninges Consisting of three layers, the meninges are the connective tissue covering the brain.

voluntary functions, such as muscle contractions that cause the limbs to move. It also controls certain involuntary reflex actions, such as the knee-jerk response.

That part of the PNS that controls the rest of the involuntary functions such as heart rate and breathing is the **autonomic nervous system** (au-toe-NOM-ick) or **ANS** (**FIGURE 11-3**). Many other functions are under the control of the autonomic nervous system, including digestion and body temperature regulation. These are controlled by negative feedback loops.

The ANS is absolutely essential to survival. To understand why, imagine how much more difficult our lives would be if we had to consciously control our breathing and other automatic functions.[1] We'd have very little time for anything else.

FIGURE 11-3 shows that each branch of the peripheral nervous system consists of two types of neurons. The first are **sensory,** or **afferent, neurons** (A-fair-ent),

---

[1]Note that breathing can be controlled voluntarily, but for the most part, it is an involuntary action.

which transmit information to the CNS. The second are **motor,** or **efferent, neurons** (EE-fair-ent), which transmit information to the effectors.

Structure and Function of the Neuron

Before we examine the structure and function of the nervous system in more detail, let us look at the neuron, or nerve cell. The **neuron** is the fundamental structural unit of the nervous system. This highly specialized cell generates bioelectric impulses and transmits them from one part of the body to another. Such signals alert us to a variety of internal and external stimuli and permit us to respond to them.

All Neurons Consist of a Cell Body and Two Types of Processes That Transmit Impulses

Neurons come in several shapes and sizes. Despite these differences, nerve cells share several characteristics. All neurons, for example, consist of a more or less spherical central portion, the **cell body** (**FIGURE 11-4**). It houses the nucleus, most of the cell's cytoplasm, and numerous organelles. Metabolic activities in the cell body sustain the entire neuron, providing energy and synthesizing materials necessary for proper cell function. Two organelles of particular interest are microtubules and microfilaments. These structures form the cytoskeleton that gives rise to the neuron's characteristic shape.

All nerve cells contain two types of processes that transmit bioelectric impulses. Those that transmit impulses *to* the cell body are **dendrites** (DEN-drights). Those that transmit impulses away from the cell body are **axons** (AXE-ons).

FIGURE 11-3 Subdivisions of the Nervous System The nervous system is divided into two parts, the central nervous system (CNS) and the peripheral nervous system (PNS). The PNS consists of autonomic and somatic divisions. The activities of the autonomic and somatic divisions often overlap.

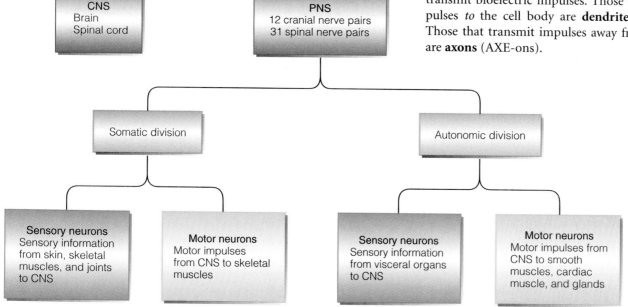

Neurobiologists classify neurons by their structure—notably, the type of processes they contain. Accordingly, three distinct types of neuron are found in the human nervous system. Shown in **FIGURE 11-5**, they are the unipolar, the bipolar, and the multipolar.

The **unipolar neuron** (you-nee-POLE-er) has a single cellular process that splits into an axon and a dendrite soon after it leaves the cell body. The **bipolar neuron** has two cellular processes, one axon and one dendrite, on opposite sides of the cell body. The **multipolar neuron** contains a single long axon and numerous short, branching dendrites that are attached to the cell body. Multipolar neurons are the most abundant and will be the focus of our discussion.[2] In multipolar neurons, the dendrites transmit impulses to the cell body. The cell bodies of multipolar neurons are located in the spinal cord and brain. After reaching the cell body, impulses travel down the long, unbranched axon. Some axons connect one part of the CNS to another and thus remain inside the CNS. Other axons exit from the CNS to form nerves. Axons occasionally also give off side branches, **axon collaterals** (**FIGURE 11-4B**). When an axon reaches its destination, it often branches profusely, giving off many small fibers. These fibers terminate in tiny swellings called **terminal boutons** (boo-TAWNS; *boutons* is the French word for buttons). Terminal bou-

tons serve as a communication link with other neurons, muscle fibers, or glands.

Axons in the CNS and PNS Contain an Insulative Layer, the Myelin Sheath, Which Greatly Increases the Rate of Transmission of Nerve Impulses.

The axons of many multipolar neurons in both the central and peripheral nervous systems are coated with a protective layer called the **myelin sheath** (MY-eh-lin)

---

[2]Multipolar neurons are involved in the efferent pathways of the PNS and carry motor information to effector organs such as glands and muscles.

FIGURE 11-4 A Neuron *(a)* A scanning electron micrograph of the cell body and dendrites of a multipolar neuron. The multipolar neuron resides within the central nervous system. Its multiangular cell body has several highly branched dendrites and one long axon. *(b)* Collateral branches may occur along the length of the axon. When the axon terminates, it branches many times, ending on individual muscle fibers.

Axon   Cell body   Dendrites

(a)

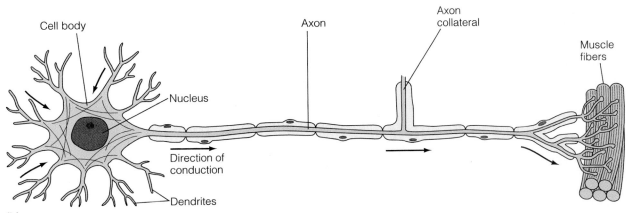

Cell body   Axon   Axon collateral   Muscle fibers

Nucleus

Direction of conduction

Dendrites

(b)

(**FIGURE 11-6A**). The myelin sheath is formed by non-conducting cells of the nervous system known as **glial cells** (GLEE-al). In the PNS, the glial cells that form the myelin sheath are the **Schwann cells,** named after the German cytologist Theodore Schwann. In the CNS, the myelin sheath is laid down by glial cells known as **oligodendrocytes** (oh-LEE-go-DEN-drow-SITES).

During embryonic development, Schwann cells in the PNS attach to growing axons, then begin to encircle them (**FIGURE 11-6D**). As they do, they leave behind a trail of plasma membrane that wraps around the axon, forming many concentric layers—a little like an elastic bandage wrapped around your wrist (**FIGURES 11-6B** and **11-6C**).

The entire myelin sheath of an axon is formed by numerous Schwann cells that align themselves along the length of the axon. Each lays down a separate patch of myelin. Because the plasma membrane of the Schwann cell is about 80% lipid, the myelin sheath is mostly lipid (mostly triglyceride) and appears glistening white when viewed with the naked eye. As shown in **FIGURE 11-6A**, each segment of myelin is separated by a small unmyelinated segment known as a **node of Ranvier** (RON-vee-A). Although the term *node* implies a bump, the nodes are actually regions of axon that contain no myelin.

For reasons explained later, the myelin sheath permits nerve impulses to travel with great speed down the axons, "jumping" from node to node like a stone skipping along the surface of the water (**FIGURE 11-6A**). In the CNS, a single oligodendrocyte produces myelin for several axons (**FIGURE 11-6E**).

Destruction of the myelin sheath of nerve cells in the central nervous system results in a condition known

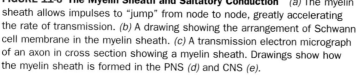

(a) Unipolar   (b) Bipolar   (c) Multipolar

FIGURE 11-5 Three Types of Neurons

FIGURE 11-6 The Myelin Sheath and Saltatory Conduction   *(a)* The myelin sheath allows impulses to "jump" from node to node, greatly accelerating the rate of transmission. *(b)* A drawing showing the arrangement of Schwann cell membrane in the myelin sheath. *(c)* A transmission electron micrograph of an axon in cross section showing a myelin sheath. Drawings show how the myelin sheath is formed in the PNS *(d)* and CNS *(e)*.

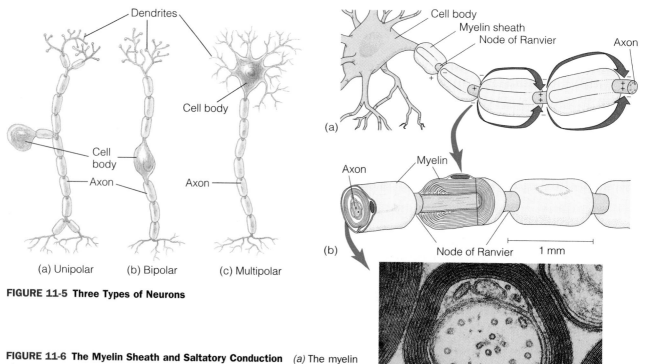

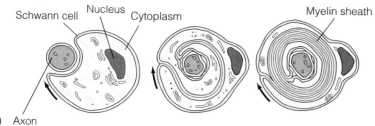

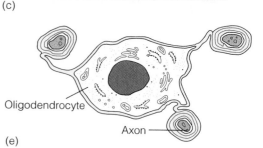

as **multiple sclerosis** (skler-OH-siss). The destroyed myelin is replaced by plaque that disrupts the transmission of impulses. Thought to be an autoimmune disease, multiple sclerosis can affect any part of the CNS. Early symptoms are generally mild weakness or a tingling or numb feeling in one part of the body. Temporary weakness may cause a person to stumble and fall. Some people report blurred vision, slurred speech, and difficulty controlling urination. In many cases, these symptoms disappear, never to return. Other individuals suffer repeated attacks. Because recovery after each attack is incomplete, patients gradually deteriorate, losing vision and becoming progressively weaker. Fortunately, many treatments are available, and only a small number of multiple sclerosis patients are crippled by the disease.

Although most axons are covered with myelin, some are unmyelinated. Found in both the central and peripheral nervous systems, unmyelinated axons conduct impulses much more slowly than their myelinated counterparts. The reduced rate of transmission in unmyelinated axons results from the fact that the impulse must travel along the entire membrane of the axon—more like a wave moving along the surface of a pond than a stone skipping across it. As a rule, the most urgent types of information are transmitted via myelinated fibers; less urgent information is transmitted via unmyelinated fibers.

Neurons Lose the Ability to Divide.

During development, nerve cells lose the ability to divide. That's bad news for people, because it means that nerve cells that die as a result of injury or old age cannot be replaced by cell division. But not all damage leads to nerve cell death and not all nerve cell death leads to irreversible damage. Why?

Consider what happens in a stroke. Strokes or cerebrovascular accidents result from one of three causes: (1) cerebral hemorrhages, breaks in arteries of the brain; (2) cerebral thromboses (throm-BOW-sees), blood clots that form in brain arteries narrowed by atherosclerosis; and (3) cerebral embolisms (EM-bow-LIZ-ims), blood clots from other sources that lodge in arteries of the brain (**FIGURE 11-7**).

All three problems have the same end result: They terminate blood flow to brain cells, often killing them. But this doesn't mean a person is always left permanently impaired. If a person survives a stroke, undamaged neurons in the brain can take over the function of damaged brain cells, thus permitting partial recovery. Recovery also occurs when cells that were injured but not killed by the injury regain their function. This generally takes a long time and explains why stroke patients require long-term rehabilitation.

FIGURE 11-7 Brain Damage Caused by Stroke PET (positron emission tomography) scan of brain revealing damaged region of the cerebral cortex following a stroke. This color image was generated by a computer that converts readings of the rate of emission from radioactive glucose molecules injected into the patient. Damaged (dark region) areas show the lowest glucose uptake. Highest glucose uptake and highest emissions are in red.

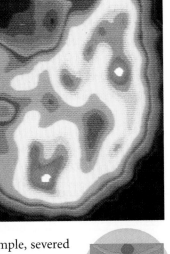

Nerves can be severed in accidents or other injuries. The fate of a severed axon depends in large part on its location in the nervous system. In the brain and spinal cord, for example, severed axons *cannot* be repaired. In the peripheral nervous system, nerve regeneration is possible because severed axons can regrow. As **FIGURE 11-8** illustrates, a severed axon in the PNS generally degenerates from the point of injury to the muscle or gland it supplied. The segment

Multiple sclerosis
www.jbpub.com/humanbiology

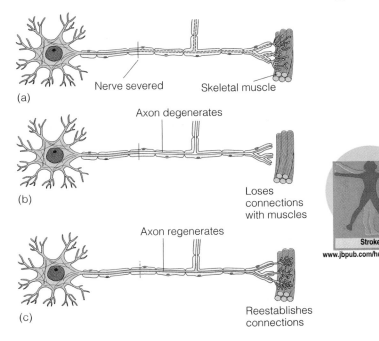

(a) Nerve severed   Skeletal muscle

Axon degenerates

(b) Loses connections with muscles

Axon regenerates

(c) Reestablishes connections

Stroke
www.jbpub.com/humanbiology

FIGURE 11-8 Axonal Regeneration *(a)* A severed axon can regenerate in the peripheral nervous system. *(b)* The segment from the cut to the effector organ degenerates (white section). The myelin sheath remains, providing a tunnel *(c)* through which the axonal stub can regrow, often reestablishing previous contacts and restoring motor function.

of the axon still attached to the cell body may elongate and replace the degenerated section. During regeneration, the axon extends along the hollow tunnel left in the myelin sheath by the degenerated part. Eventually, the axon reestablishes connections with the muscles or glands it once supplied, making partial or nearly complete recovery of control possible.

Neurosurgeons can facilitate axonal regeneration in the PNS by **microsurgery**—surgery performed under a dissecting microscope. With painstaking care, surgeons sew the severed ends of nerves together after they've lined up the empty myelin sheaths with those of the regenerating nerve fibers.

Neurons Have a High Metabolic Demand, Making Them Highly Susceptible to Loss of Oxygen and Glucose.

Besides being unable to divide, nerve cells have an extraordinarily high metabolic rate. Consequently, they require a constant supply of oxygen for energy production. Making matters worse, neurons cannot generate ATP in the absence of oxygen via fermentation (glucose breakdown to lactic acid) the way most other cells can. Thus, if the amount of oxygen flowing to the brain is drastically reduced, neurons begin to die within minutes.

To prevent brain damage from occurring in someone who has collapsed with a severe heart attack or has drowned or suffered an electric shock, rescuers must start resuscitating the victim within 4–5 minutes. Although victims may be revived after this crucial period, the lack of oxygen in the brain often results in brain damage. Generally, the longer the deprivation, the greater the damage.

If one's "lucky" enough to drown in cold water, though, resuscitation may be successful if begun within an hour. In most cases, victims recover without any detectable brain damage. Recovery is possible in such instances because cold water greatly slows brain metabolism. This, in turn, dramatically reduces oxygen demand. As a result, brain cells are preserved, and brain damage is minimized or avoided. In one exceptional case, a young boy from Fargo, North Dakota, was allegedly underwater for 5 hours before he was resuscitated. Much to his parents' delight, the boy not only survived the incident but also has no apparent ill effects.

Nerve cells are also highly dependent on glucose for energy production. Unlike most other body cells, they cannot use fatty acids to generate energy. Neither can they store glucose as liver and muscle cells do. As a result, when blood glucose levels fall, nerve cells are first to "feel" the ill effects.

A decline in blood glucose is normally prevented by homeostatic mechanisms described in Chapter 4. Serious problems can result in people whose glucose homeostasis is not working properly. In diabetics, for instance, blood glucose levels can fall dangerously low if too much insulin is taken or if not enough glucose is ingested. Deprived of glucose, brain cells begin to falter. Individuals can become dizzy and weak. Vision may blur. Speech may become awkward. Diabetics are sometimes mistakenly considered drunk. Chronic low blood glucose often results in severe headaches. Some diabetics become aggressive when blood sugar levels fall. In extreme cases, low blood sugar triggers convulsions and death. Other body cells are not adversely affected by a decline in blood glucose because they switch to alternative fuels, fats and proteins.

Bioelectric Impulses in Nerve Cells Result from the Flow of Ions Across Their Plasma Membranes

Nerve impulses are not like the electric current that powers computers or light bulbs, which is formed by the flow of electrons. Rather, nerve impulses are small ionic changes in the membrane of the neuron that move along the plasma membrane of a nerve cell.

To understand the nerve cell impulse, or **bioelectric impulse,** we begin by examining the plasma membrane of a neuron. If you placed tiny electrodes on the outside and inside of the plasma membrane of a neuron and hooked them up to a voltmeter, you would measure a small voltage, much like that produced in a battery. Voltage is a measure of the tendency of charged particles to flow from one pole of the battery to the other. The higher the voltage, the greater the tendency for electrons to flow through a wire connected to the poles. In the nerve cell, however, electrons do not flow from one side of the membrane to another; sodium and potassium ions do.

In neurons, the potential difference, or voltage, is a measure of the force that can drive sodium ions from one side of the membrane to the other. For now, it is important just to remember that a small voltage exists across the plasma membrane of the neuron. It is so small, in fact, that it is measured in millivolts (MILL-ee-volts). A millivolt is 1/1000 of a volt.

As **FIGURE 11-9** shows, the potential difference in a typical neuron is about −70 millivolts. This is known as the **membrane potential,** or **resting potential**—so named because it is the membrane potential of a nerve cell at rest. The minus sign is added because the plasma membrane is positively charged on the outside and negatively charged on the inside, for reasons beyond the scope of this book.

To understand the bioelectric impulse, it is important to know that sodium ions are found in greater con-

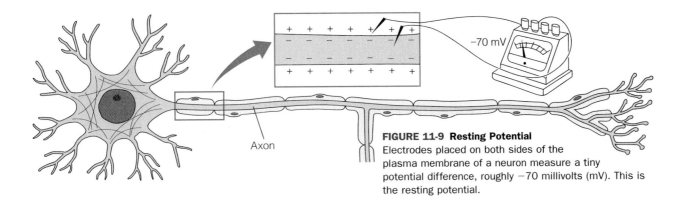

Axon

FIGURE 11-9 Resting Potential
Electrodes placed on both sides of the
plasma membrane of a neuron measure a tiny
potential difference, roughly −70 millivolts (mV). This is
the resting potential.

centration *outside* the neuron. Potassium ions are found
in greater concentration *inside* the cell. In fact, neurons
expend a great deal of energy to maintain this concentration imbalance, so essential to the production of
nerve impulses. This energy fuels the active transport
pumps (proteins) in the plasma membrane of neurons,
which pump sodium ions that have leaked into the cytoplasm of the neuron back out into the surrounding
fluid and transport potassium ions that have leaked into
the extracellular fluid back into the cell.

After a Nerve Cell Is Stimulated, the Membrane Undergoes Dramatic Changes in Permeability to Various Ions, Resulting in Sudden Shifts in Membrane Potential.

The plasma membrane of a nerve cell has a built-up
charge. The charge consists of sodium ions concentrated on the outside of the cell. When the neuron is
stimulated, the membrane undergoes a rapid change,
"discharging" the load. The first change occurring in the
membrane is a rapid increase in its permeability to
sodium ions. Neurophysiologists believe that stimulating the nerve cell causes protein pores in the plasma
membrane to open. Sodium ions flow *into* the cell
through these pores.

Electrodes implanted in a nerve cell like those used
to measure resting potential detect the sudden influx of
positively charged sodium ions when the neuron is activated. The electrodes then register a shift in the resting
potential from −70 millivolts to +30 millivolts. The
change in voltage occurs at the site of stimulation and is
called **depolarization.**[3] The membrane is said to be depolarized. Immediately after depolarization, the membrane returns to its previous state, which is referred to
as **repolarization.**

The depolarization/repolarization of the membrane is shown graphically in **FIGURE 11-10A**. This tracing is an **action potential.** The action potential consists
of (1) a brief upswing, depolarization, as the voltage
goes from −70 millivolts to +30 millivolts, and (2) a
rapid downswing, repolarization, the return to the resting potential.

The depolarization occurs so rapidly and the membrane returns to the resting state so quickly (in about
3/1000 of a second) that neurons can be stimulated in
rapid succession. Such brisk recovery allows us to respond swiftly and forcefully to danger and to perform
rapid muscle movements. Nerve cells can also transmit
many impulses in sequence—one after another—
because only a small number of sodium ions are exchanged with each impulse.

As noted above, depolarization results in the rapid
inflow of sodium ions. But what causes repolarization,
the shift from +30 millivolts back to −70 millivolts? Repolarization results from two factors. The first is a sudden decrease in the membrane's permeability to sodium
ions. This stops the influx of sodium ions. Repolarization also results from a rapid efflux of positively charged
potassium ions (**FIGURE 11-10D**). Potassium ions flow out
of the axon down a concentration gradient. Because the
outside of the membrane becomes negatively charged
during depolarization, an electrical gradient also facilitates repolarization. The net outward movement of
potassium ions reestablishes the resting potential.

When a neuron is no longer stimulated, it quickly
reestablishes the sodium- and potassium-ion concentrations inside and outside the cell. It does this by
pumping sodium ions that flowed into the cell during
activation out of the axon and by pumping potassium
ions that flowed out of the neuron during repolarization back in. As noted earlier, the reestablishment of the
chemical disequilibrium results from the action of numerous sodium-potassium active transport pumps in

---

[3]Depolarization means that the membrane loses its previous
polarization.

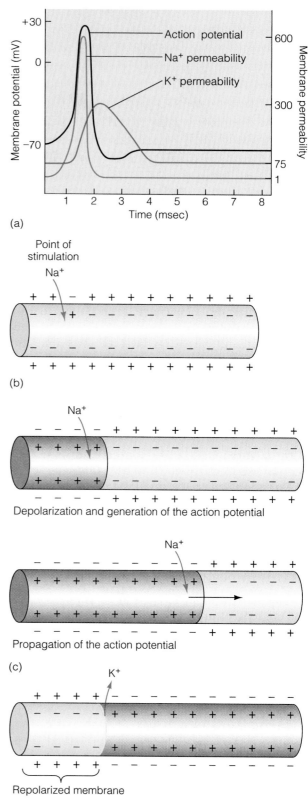

FIGURE 11-10 Action Potential   *(a)* Stimulating the neuron creates a bioelectric impulse, which is recorded as an action potential. The resting potential shifts from −70 millivolts to +30 millivolts. The membrane is said to be depolarized. This graph shows the shift in potential and the change in the permeability of sodium (Na⁺) and potassium (K⁺) ions, which is largely responsible for the action potential. *(b)* The influx of sodium ions and the depolarization that occur at the point of stimulation. *(c)* The impulse travels along the membrane as a wave of depolarization (gray area). *(d)* The efflux of potassium ions restores the resting potential, allowing the neuron to transmit additional impulses almost immediately.

the plasma membranes of nerve cells. (You may recall from Chapter 3 that active transport pumps require energy in the form of ATP.) The sodium-potassium pumps, in fact, consume about 30% of the energy your body needs.[4]

Depolarization in One Region Stimulates Depolarization in Adjacent Regions and Allows Nerve Impulses to Move Along the Membrane of the Neuron.

Researchers who have explored the neuron's mode of impulse conduction have found that a change in membrane permeability in the stimulated region, which results in depolarization, causes a change in the sodium permeability in adjacent regions. Thus, depolarization in one location on the axon stimulates depolarization in adjacent regions. This process continues along the length of the axon.

In unmyelinated fibers in the human nervous system, nerve impulses travel like waves in water from one region to the next. In myelinated fibers, however, the depolarization "jumps" from one node of Ranvier (the section between adjacent Schwann cells) to another. Shown in **FIGURE 11-6A**, impulse jumping greatly increases the rate of transmission. In fact, a nerve impulse travels along an unmyelinated fiber at a rate of about 0.5 meter per second (1.5 feet per second). In a myelinated neuron, the impulse travels 400 times faster or about 200 meters per second. That's about 400 miles per hour! The difference in the rate of transmission of these two modes is largely due to a difference in the total amount of membrane that must be depolarized and repolarized.

Nerve Impulses Travel from One Neuron to Another Across Synapses

Nerve impulses travel from one neuron to another across a small space that separates them (**FIGURE 11-11**). This juncture is called a **synapse** (SIN-apse). A synapse consists of (1) a terminal bouton (or some other kind of

axon terminus), (2) a gap between the adjoining neurons, the **synaptic cleft,** and (3) the membrane of the dendrite or postsynaptic cell (**FIGURE 11-11B**). The neuron that transmits the impulse is the **presynaptic neu-**

---

[4]The sodium-potassium pump is found in all cells, not just neurons.

ron; the one that receives the impulse is the **postsynaptic neuron.**

How does a nerve impulse get across the synapse? Quite frankly, it doesn't. It ends in the terminal bouton. However, there it triggers a series of chemical changes that stimulate an impulse in the postsynaptic neuron. Let me explain.

When an impulse reaches a terminal bouton, depolarization of the plasma membrane of the bouton stimulates a rapid influx of calcium ions into the bouton from the extracellular fluid. Calcium ions, in turn, stimulate the release (by exocytosis) of a chemical substance stored in small vesicles in the terminal bouton. These chemicals are known as **neurotransmitters.**

At least 30 chemicals serve as neurotransmitters. Produced and packaged in vesicles in the cell body of the neuron, neurotransmitters move down the axon along the microtubules to the terminal bouton, where they are stored until needed. When the bioelectric impulse arrives, calcium flows in, and the vesicles bind to the presynaptic membrane and release the neurotransmitter into the synaptic cleft.

Neurotransmitters are chemical messengers that travel from nerve cell to nerve cell—or from nerve cell to effector (such as muscle). Neurotransmitters diffuse across the synaptic cleft between adjoining nerve cells and bind to receptors in the plasma membrane of the postsynaptic neuron. The binding of most neurotransmitters to the postsynaptic membrane stimulates a rapid increase in the permeability of the membrane of the postsynaptic cell to sodium ions. Do they stimulate another nerve impulse? Not necessarily.

Neurotransmitters May Excite or Inhibit the Postsynaptic Membrane.

In the brain, a single nerve cell may have as many as 50,000 synapses. In some of these synapses, neurotransmitters stimulate the uptake of sodium ions, which slightly depolarizes the postsynaptic neuron. Synapses that depolarize the postsynaptic neuron are known as **excitatory synapses.** In other synapses, however, neurotransmitters hyperpolarize the membrane—that is, increase the voltage difference across the membrane, making it less excitable. These neurotransmitters cause chloride channels to open in the postsynaptic membrane. Because

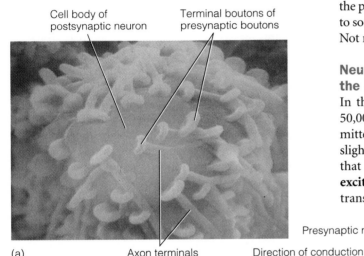

Cell body of postsynaptic neuron — Terminal boutons of presynaptic boutons

(a) Axon terminals

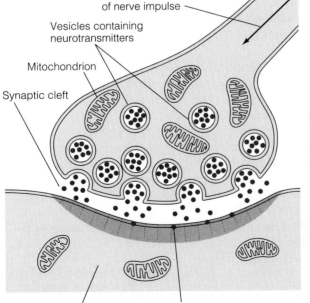

Presynaptic neuron
Direction of conduction of nerve impulse
Vesicles containing neurotransmitters
Mitochondrion
Synaptic cleft
Postsynaptic neuron
Receptors on postsynaptic membrane bound to neurotransmitter
(b)

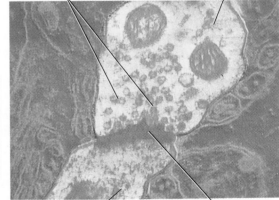

Synaptic vesicles — Presynaptic neuron
(c) Postsynaptic neuron — Synaptic cleft

FIGURE 11-11 The Terminal Bouton and Synaptic Transmission *(a)* A scanning electron micrograph showing the terminal boutons of an axon ending on the cell body of another neuron. *(b)* The arrival of the impulse stimulates the release of neurotransmitters held in membrane-bound vesicles in the axon terminals. Neurotransmitter diffuses across the synaptic cleft and binds to the postsynaptic membrane, where it elicits another action potential that travels down the dendrite to the cell body. *(c)* A transmission electron micrograph showing the details of the synapse.

the chloride-ion concentration outside the cell is higher than it is inside, the opening of the chloride channels results in an influx of chloride ions, making the interior of the postsynaptic cell and hence the resting potential more negative, which renders the neuron less excitable. Such synapses are called **inhibitory synapses.**[5]

Whether a neuron fires (generates an action potential) depends on the summation of excitatory and inhibitory impulses. If the number of excitatory impulses exceeds the number of inhibitory impulses, a nerve impulse will be generated. If not, the neuron will not fire. This phenomenon provides the nervous system with a way of integrating incoming information—determining a response by the kinds of input it receives.

Because only terminal boutons contain neurotransmitters and because receptors for these substances are found only in the postsynaptic membrane, transmission across a synapse occurs only in one direction.

Neurotransmitters Are Quickly Removed from the Synaptic Cleft.

Transmission across the synapse is remarkably fast, requiring only about 1/1000 of a second, or 1 millisecond. Synaptic transmission is also a transitory event. A short burst of neurotransmitter is released each time an impulse reaches the terminal bouton. It binds to the postsynaptic membrane, elicits a change, and is then removed from the synaptic cleft by three routes: (1) enzymatic destruction, (2) reabsorption by the terminal bouton or absorption by glial cells in the brain, and (3) diffusion away from the synapse.

Many Chemical Substances, Including Drugs and Insecticides, Exert Their Effect by Altering Synaptic Transmission.

The synapses are vital links in the body's system of internal communication. As such, they are vulnerable to outside influences. Certain drugs and common environmental chemicals (for example, insecticides) impair the removal of neurotransmitters from the synaptic cleft. As long as neurotransmitters remain bound to the receptors on the postsynaptic membrane, the postsynaptic neuron remains activated.

Many insecticides inhibit the removal of the neurotransmitter **acetylcholine** (a-SEA-tol-KO-leen), the main excitatory neurotransmitter in the neurons that supply skeletal muscle cells. Insecticides kill insects by disrupting nerve transmission—creating a nervous

system overload. Unfortunately, insecticides have the same effect on people such as farm workers and pesticide applicators exposed to high levels of the pesticides at work. Low doses can cause blurred vision, headaches, rapid pulse, and profuse sweating; higher doses can be fatal.

Each year, an estimated 100,000–300,000 Americans (mostly farm workers) are poisoned by pesticides. By various estimates, 200–1000 of them die. Worldwide, an estimated 500,000 people are poisoned, and 5000–14,000 people die each year from pesticides.

These problems and the widespread contamination of the environment have led some farmers to reduce pesticide use and to rely on other, nonpolluting methods of pest control. Crop rotation, insect-resistant crops, natural predators (ladybugs and praying mantises), and a variety of alternatives are available. Most are quite practical and cost-effective. Such strategies also benefit the soil and surrounding environment.

Anesthetics (an-es-THET-icks) are chemicals used to deaden pain or to put people to sleep for surgery. Some anesthetics may alter synaptic transmission, decreasing the transmission of pain impulses. Others seem to operate on the nerve cell itself. They apparently alter protein pores in the plasma membrane of neurons that regulate the flow of sodium ions into and out of the nerve cells. By blocking the flow of sodium, these anesthetics "paralyze" sensory nerves carrying pain messages to the brain.

Caffeine and cocaine also affect synaptic transmission. Caffeine increases synaptic transmission, thus increasing overall neural activity. It is no wonder that coffee makes some people so jittery. Cocaine, on the other hand, blocks the uptake of neurotransmitters by terminal boutons in the brain. Because the neurotransmitter remains in the synaptic cleft for a longer time, neural activity is greatly increased. This increase results in a heightened state of alertness and euphoria, commonly known as a "high," which lasts for 20–40 minutes. Euphoria, however, is followed by a period of depression and anxiety, which causes the user to seek another high. Excessive cocaine use can result in serious mental derangement—in particular, paranoid delusions that others are out to get the user. In this state, heavy users may become violent.

For a discussion of the effect of addictive chemicals see Health Note 11-1.

Nerve Cells Can Be Grouped Into Three Functional Categories

As noted earlier, nerve cells can be categorized by structure. For our purposes, though, a functional classifica-

---

[5]Inhibitory neurotransmitters may also stimulate the opening of potassium channels. Potassium ions flow out of the postsynaptic cell, making the interior more negative.

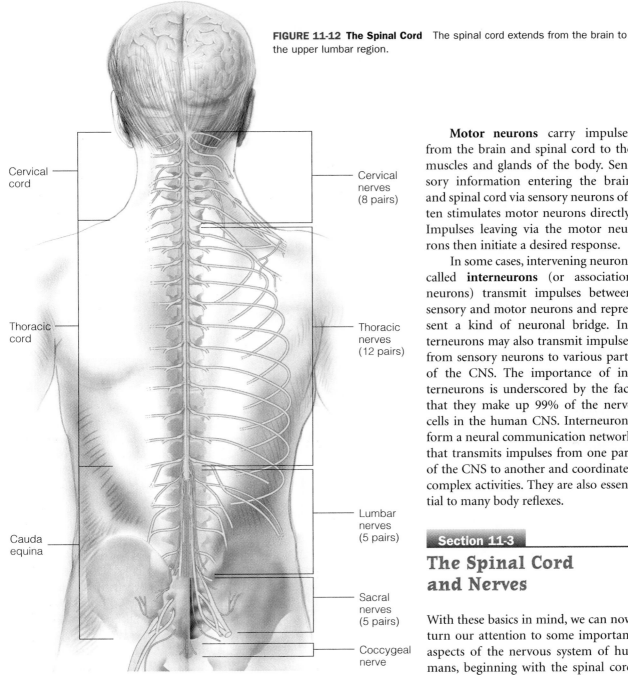

FIGURE 11-12 The Spinal Cord The spinal cord extends from the brain to the upper lumbar region.

Cervical cord

Thoracic cord

Cauda equina

Cervical nerves (8 pairs)

Thoracic nerves (12 pairs)

Lumbar nerves (5 pairs)

Sacral nerves (5 pairs)

Coccygeal nerve

Motor neurons carry impulses from the brain and spinal cord to the muscles and glands of the body. Sensory information entering the brain and spinal cord via sensory neurons often stimulates motor neurons directly. Impulses leaving via the motor neurons then initiate a desired response.

In some cases, intervening neurons called **interneurons** (or association neurons) transmit impulses between sensory and motor neurons and represent a kind of neuronal bridge. Interneurons may also transmit impulses from sensory neurons to various parts of the CNS. The importance of interneurons is underscored by the fact that they make up 99% of the nerve cells in the human CNS. Interneurons form a neural communication network that transmits impulses from one part of the CNS to another and coordinates complex activities. They are also essential to many body reflexes.

Section 11-3

The Spinal Cord and Nerves

With these basics in mind, we can now turn our attention to some important aspects of the nervous system of humans, beginning with the spinal cord and nerves.

The Spinal Cord Transmits Information to and from the Brain and Also Houses Many Reflexes

The **spinal cord** is a long, ropelike structure about the diameter of a person's little finger. It connects to the brain and courses downward through the **vertebral canal** (VER-teh-brill) formed by the bones of the spine, the **vertebrae** (ver-tah-BREE) (**FIGURE 11-12**). The spinal cord gives off nerves along its course that supply the skin,

tion is more useful. According to this system, nerve cells fall into three groups: (1) sensory neurons, (2) motor neurons, and (3) interneurons.

Sensory neurons carry impulses from sensory receptors in the body to the CNS. Sensory receptors come in many shapes and sizes and respond to a variety of stimuli such as pressure, pain, heat, and movement (Chapter 12).

Health Note 11-1

Exploring the Root Causes of Addiction

Marleen Whitehead lives a lie. Each morning, she kisses her husband good-bye, then drives off to work. On her way to the office, she pulls off the highway for a moment, opens a flask she keeps under the seat, and takes a drink. Throughout the day, she laces her coffee with scotch, so no one will suspect that she is drinking.

Like millions of other Americans, Whitehead is on the road to certain

disaster. Her boss is aware of her drinking, and her job is now at risk. What makes her and millions of others like her, who cannot get through a day without a drink or a fix from a needle, so dependent on a drug?

Many biologists think that genetics plays a key role in alcoholism, while many behavioralists argue that a person's upbringing and psychology create alcohol dependency. Recent research suggests that there may be no single cause of alcohol addiction. Biological (genetic) and behavioral factors may both be involved in alcoholism and other addictive behavior.

Let's review some of the facts about addiction. A recent study in mice showed that addictive drugs such as alcohol and cocaine stimulate the brain's pleasure center. Researchers believe that this stimulation may underlie all forms of drug addiction (**FIGURE 1**).

Experiments in which researchers implanted electrodes in the pleasure centers of rats showed that the rodents could be trained to press a bar to stimulate the center. Some rats, in

fact, pressed the bar hundreds of times per hour, ignoring food, water, and sex. After 15–20 hours of continual pressing, the rats collapsed from exhaustion. But when they awoke, they began pressing again.

Experiments with rats and mice indicate that the pleasure center is activated by cocaine and amphetamines. These drugs stimulate the production of large amounts of a neurotransmitter known as dopamine. In a recent study, researchers implanted small tubes into rats' pleasure centers and areas involved in muscle movement. Researchers then administered various doses of addictive and nonaddictive drugs while the rats moved freely in their cages. Researchers then extracted fluid from the tubes to measure dopamine levels.

Drugs that are addictive to humans and rewarding to rats (amphetamine, cocaine, morphine, methadone, ethanol, and nicotine) increased dopamine concentrations in both areas, but levels were much higher in the pleasure center. Although these results cannot be extrapolated to humans, some re-

FIGURE 1 An addict injects heroin. Heroin is a highly addictive derivative of morphine that induces euphoria in users.

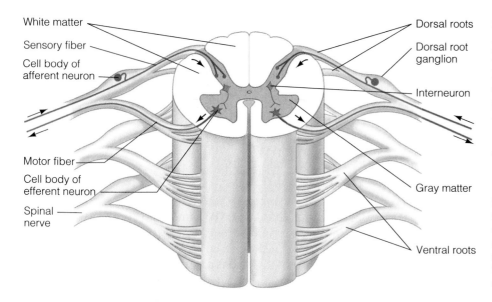

White matter
Sensory fiber
Cell body of afferent neuron
Motor fiber
Cell body of efferent neuron
Spinal nerve

Dorsal roots
Dorsal root ganglion
Interneuron
Gray matter
Ventral roots

muscles, bones, and organs of the body. The spinal cord ends at the lower back (at about the level of the second lumbar vertebra), at which point it gives off a series of nerves forming the **cauda equina** (CAW-dah eh-QUINE-ah; meaning "horse's tail"), which supply the lower sections of the body.

As **FIGURE 11-13** shows, the central portion of the spinal cord is an H-shaped zone of gray

FIGURE 11-13 The Spinal Cord and Dorsal Root Ganglia Spinal nerves are attached to the spinal cord by two roots, the dorsal and ventral roots. The dorsal root carries sensory information into the spinal cord. The ventral root carries motor information out of the spinal cord. The spinal nerve often contains both sensory and motor fibers.

searchers believe that they contribute to the case for the theory that dopamine release in the pleasure center is a common denominator in all forms of drug addiction. Other researchers think that other neurotransmitters may also be involved.

This research begins to explain the neural and biochemical basis of addiction, but what about the underlying cause? Why do some people become addicted to drugs while others don't?

Research into the genetics of alcoholism has built a case for the assertion that alcoholism is largely the result of defective genes. Donald W. Goodwind of the University of Kansas Medical Center in Kansas City found that children of alcoholics had an increased risk of becoming alcoholic even when reared by adoptive (nonalcoholic) parents. This and a number of other similar findings suggest that the environment is less important than a person's genetic heritage. If one or both of your biological parents is an alcoholic, say researchers, you are much more likely to be one yourself.

Nevertheless, some researchers think that the scientific community

has accepted the genetic case too readily and uncritically. They say that the genetic theory of alcoholism may be a simplified view of the causes of the disease. Alcoholism probably results from a complex interaction of environmental factors and genetics. New research shows that alcoholism results from physical, personal, and social characteristics that predispose a person to drink excessively.

Herbert Fingarette, a professor at the University of California at Santa Barbara contends that alcoholism is psychological, not biological. Fingarette's arguments reflect the findings of many psychological studies. This line of research indicates that alcoholism and other addictions are more habits than diseases. Addictive behavior, the studies suggest, typically revolves around immediate gratification.

Alcohol enhances social and physical pleasure, increases sexual responsiveness and assertiveness, and reduces tension up to a point. Unfortunately, the initial physical stimulation, brought on by low doses of alcohol, can lead some people into

an addictive cycle. The expectation of improved feelings drives people to drink. But higher doses of alcohol dampen arousal, sap energy, and cause hangovers. This, in turn, say psychologists, leads to a craving for alcohol's stimulating effects—that is, a craving to feel good again. The repetitive cycle of pleasure and displeasure is addiction.

The controversy over the roots of alcoholism and other addictive behaviors will undoubtedly continue for years, pitting biologists against psychologists. Although it is impossible to predict the outcome of future research, it seems likely that the intermediate position will hold sway: that addiction may be explained by both genetics and psychology.

Visit *Human Biology's* Internet site, www.jbpub.com/humanbiology, for links to web sites offering more information on this topic.

matter. **Gray matter** consists of nerve cell bodies of interneurons and motor neurons and is so named because it appears gray to the naked eye. Surrounding the gray matter are fiber tracts consisting of axons and a much smaller number of dendrites that travel up and down the spinal cord, carrying information to and from the brain. The fiber tracts form the white matter, whose characteristic color comes from the myelin sheaths of the axons coursing through it.

▌ The Nerves of the PNS Contain Motor and Sensory Fibers

As noted earlier, nerves are part of the peripheral nervous system. They carry sensory information to the spinal cord and brain and motor information out. Some nerves are strictly motor and some are strictly sensory, but many are mixed, having both motor and sensory fibers.

Nerves arising from the brain are **cranial nerves** (CRAY-knee-al) (**FIGURE 11-14**). The cranial nerves supply structures of the head and several key body parts such as the heart and diaphragm. The trigeminal nerves mentioned in the introduction are one of the 12 pairs of cranial nerves.

Nerves associated with the spinal cord are known as **spinal nerves.** Each spinal nerve has two roots, the dorsal (back side) and ventral (belly side), which attach to the spinal cord (**FIGURE 11-13**). The **dorsal root** is the inlet for sensory information traveling to the spinal cord. On each dorsal root is a small aggregation of nerve cell bodies, the **dorsal root ganglion** (GANG-lee-on). The dorsal root ganglia house the cell bodies of sensory neurons. As **FIGURE 11-13** shows, the dendrites of these bipolar nerve cells conduct impulses from receptors in the body to the dorsal root ganglia; the axons, in turn,

FIGURE 11-14 Cranial Nerves The 12 pairs of cranial nerves arise from the underside of the brain and brain stem.

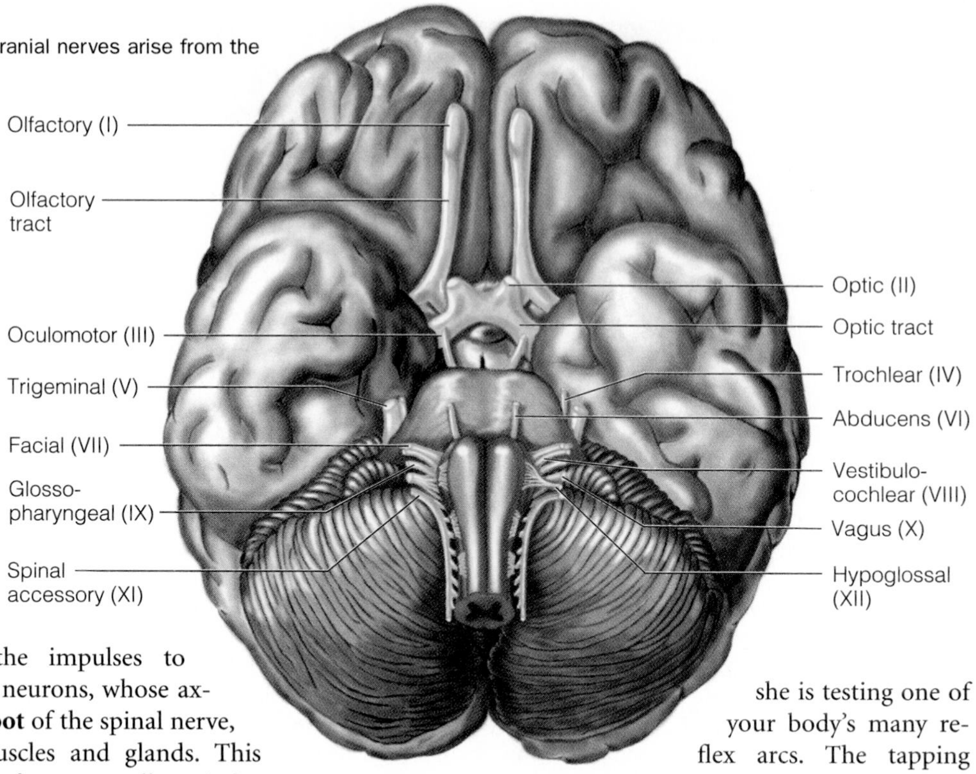

Olfactory (I)

Olfactory tract

Oculomotor (III)

Trigeminal (V)

Facial (VII)

Glosso-pharyngeal (IX)

Spinal accessory (XI)

Optic (II)

Optic tract

Trochlear (IV)

Abducens (VI)

Vestibulo-cochlear (VIII)

Vagus (X)

Hypoglossal (XII)

carry the impulse into the spinal cord along the dorsal root.

As noted earlier, sensory fibers entering the spinal cord often end on interneurons (**FIGURE 11-13**). Interneurons receive input from many sensory neurons and process this information, acting like a receptionist in a busy corporate office. Interneurons transmit the impulses to nearby multipolar motor neurons, whose axons leave in the **ventral root** of the spinal nerve, carrying impulses to muscles and glands. This anatomical arrangement of neurons allows information to enter and leave the cord quickly and forms the basis of the **reflex arc,** a neuronal pathway by which sensory impulses from receptors reach effectors without traveling to the brain (**FIGURE 11-15**). Some reflex arcs contain interneurons, and some do not.

When a physician taps a rubber hammer on the tendon just below your kneecap (patellar tendon), he or she is testing one of your body's many reflex arcs. The tapping stimulates stretch receptors in the tendon. These receptors generate nerve impulses that travel to the spinal cord via sensory neurons. In this reflex, each sensory neuron ends directly on a motor neuron, which supplies the muscles of the thigh. Thus, a quick tap on the tendon results in a motor impulse sent to the anterior thigh muscles (quadriceps), causing them to contract and the knee to jerk.

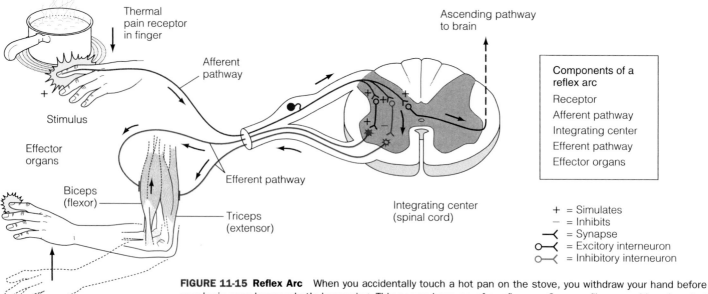

Thermal pain receptor in finger

Afferent pathway

Ascending pathway to brain

Stimulus

Effector organs

Biceps (flexor)

Triceps (extensor)

Efferent pathway

Integrating center (spinal cord)

Response

Components of a reflex arc

Receptor
Afferent pathway
Integrating center
Efferent pathway
Effector organs

+ = Simulates
− = Inhibits
= Synapse
= Excitory interneuron
= Inhibitory interneuron

FIGURE 11-15 Reflex Arc When you accidentally touch a hot pan on the stove, you withdraw your hand before your brain even knows what's happening. This occurs because of a reflex arc. Sensory fibers send impulses to the spinal cord. The sensory impulses stimulate motor neurons in the spinal cord. This causes muscle contraction in the flexor muscles (+) and inhibits muscle contraction in the extensor muscles (−), allowing you to withdraw your hand. Nerve impulses also ascend to the brain to let it know what is happening.

Reflexes are mechanisms that protect the body from harm. Touching a hot stove, for example, elicits the withdrawal reflex. Sensory impulses stimulate the muscles of your arm to contract so that you pull your hand away from the stove before your brain is aware of what is happening.

Babies come equipped with a number of important reflexes. Rub your finger on the cheek of a newborn, and it immediately turns its head toward your finger. This reflex helps babies find the mother's nipple. Crying is also a reflex. When a baby is hungry, thirsty, wet, or uncomfortable, it cries, a reflex sure to get attention.

The spinal cord also transmits sensory information to the brain. Sensory impulses travel to the brain along special tracts lying outside the gray matter of the spinal cord. Although the incoming sensory information may elicit a reflex, the brain is still informed of the problem, allowing for appropriate follow-up action.

Damage to the Spinal Cord Can Cause Permanent Damage

The Severity of the Injury Depends on Its Location and the Extent of the Damage.
An automobile crash, a bullet wound, or even a bad fall can damage or sever the fiber tracts of the spinal cord. Injury to the cord usually results in permanent damage because, as noted previously, axons of the CNS do not usually regenerate.

The amount of damage depends on where the cord is injured and the severity of the injury. Sensory and motor fibers traveling to and from the brain, respectively, run in separate tracts in the white matter in the spinal cord. If both tracts are severed—for example, by a severe vertebral fracture—all of the sensory and motor functions below the level of the injury are lost. That is, muscles supplied by nerves below the injury become paralyzed and unable to contract voluntarily and the portion of the body below the injury loses all sensation. If the spinal cord injury occurs high in the neck (above the fifth cervical vertebra), it cuts the nerve fibers traveling to the muscles that control breathing. These muscles become paralyzed, and the person dies quickly. This is how a hangman's noose kills its victim.

Damage to the cord just below the fifth cervical vertebra does not affect breathing, but it does paralyze the legs and arms. This condition is known as **quadriplegia** (QUAD-reh-PLEE-gee-ah). If the spinal cord injury occurs below the nerves that supply the arms, the result is **paraplegia** (PEAR-ah-PLEE-gee-ah), paralysis of the legs.

New studies show that the administration of large doses of an anti-inflammatory steroid drug shortly after a head or neck injury stops swelling of the spinal cord. Swelling is thought to greatly increase the amount of damage resulting from such injuries. This treatment may therefore greatly reduce paralysis in such cases. Research is also under way to find ways to stimulate the regeneration of axons in the CNS. Some early results are promising, suggesting that physicians may one day be able to "repair" spinal cord damage.

Section 11-4

The Brain

About the size of a cantaloupe, the human brain is an extraordinary organ, the product of many millions of years of evolutionary trial and error. Our brains have given us art and music, remarkable feats of engineering, and abstract reasoning, not to mention the ability to ponder right and wrong and remember a vast amount of information.

The Cerebral Hemispheres Function in Integration, Sensory Reception, and Motor Action

The brain is housed in the skull, a bony shell that protects it from injury. **FIGURE 11-16** shows some of the externally visible parts of the human brain. The largest and most conspicuous is the **cerebrum** (sir-EE-brum), a convoluted mass of nervous tissue that constitutes about 80% of the total brain mass.

The cerebrum is divided into two halves, the right and left **cerebral hemispheres.** Each hemisphere has a thin outer layer of gray matter, the **cerebral cortex**

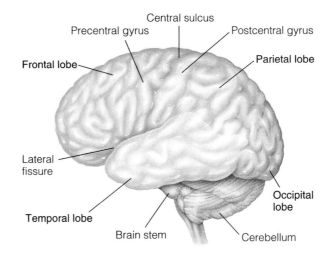

FIGURE 11-16 The Brain The cerebral cortex consists of the lobes shown here. The lobes, in turn, can be divided into sensory, association, and motor areas (not shown).

FIGURE 11-17 Structures of the Brain A cross section through the brain showing the gray and white matter of the cortex and deeper structures, notably the thalamus and hypothalamus.

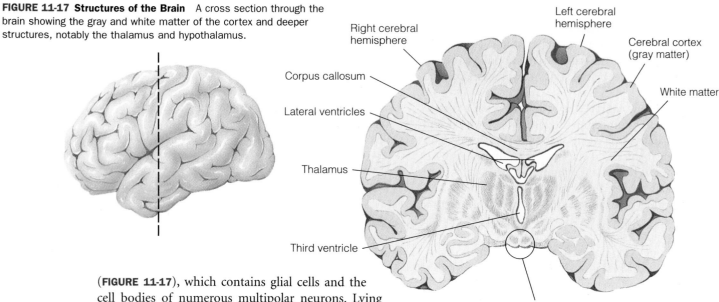

(**FIGURE 11-17**), which contains glial cells and the cell bodies of numerous multipolar neurons. Lying beneath the cerebral cortex is a thick central core of white matter containing bundles of myelinated axons that give it a white appearance. The axons of the white matter carry impulses from the cerebral cortex to the spinal cord, enabling information to flow from the brain to motor neurons that control many of the body muscles. Some axons also transmit impulses from one part of the cerebral cortex to another, permitting the integration of the activities of various parts of the cortex.

As shown in **FIGURE 11-17**, the cerebral cortex is thrown into numerous folds, called **gyri** (JI-reye; singular, gyrus, JI-russ). The gyri are separated by numerous valleys, known as **sulci** (SUL-see; singular, sulcus, SUL-kuss).

Deep within the white matter of the cerebral hemispheres are masses of gray matter, the **basal ganglia** (BAY-zill GANG-lee-ah). The nerve cells of the basal ganglia help to fine-tune muscular control.

The Cerebrum Is Divided into Four Major Lobes, and Each Lobe Contains Areas That House Specific Functions.

The cerebral hemispheres are divided into four major lobes, shown in **FIGURE 11-16**. They are the frontal lobe, the parietal lobe, the occipital lobe, and the temporal lobe. Within each of these lobes are areas that house specific functions. These functional areas (**FIGURE 11-18**) can be broadly classified into three types: motor cortex, sensory cortex, and association cortex. The **motor cortex** stimulates muscle activity. The **sensory cortex** receives sensory stimuli, and the **association cortex** integrates information, bringing about coordinated responses. Let's examine the key regions, beginning with the primary motor cortex.

The Primary Motor Cortex Controls Voluntary Movement.

The **primary motor cortex** occupies a single gyrus (ridge) on each hemisphere, just in front of the central sulcus (**FIGURE 11-18**). This area controls voluntary motor activity—for example, the muscles in your hand that are turning the pages of your book.

The neurons in the primary motor cortex are arranged in a very specific way, so that each region of the motor cortex controls a specific body part. As **FIGURE 11-19A** illustrates, the neurons that control the muscles of the knee are located in the uppermost region of the primary motor cortex. Hip muscle control occurs below that. Muscles of the hand are controlled by neurons located even lower.

To bring about a voluntary movement, a conscious thought stimulates the neurons of the primary motor cortex to generate an impulse. It travels from the brain down the spinal cord to the motor neurons in the spinal cord. The axons of these neurons exit the spinal cord and terminate on the muscles.

In front of the primary motor area is the **premotor cortex.** The premotor cortex is also involved in controlling muscle contraction. However, the movements that it controls are less voluntary—for example, the fingering required to play a musical instrument.

The Primary Sensory Cortex Receives Sensory Information from the Body.

Just behind the central sulcus is a ridge of tissue running parallel to the primary motor area. Known as the **pri-**

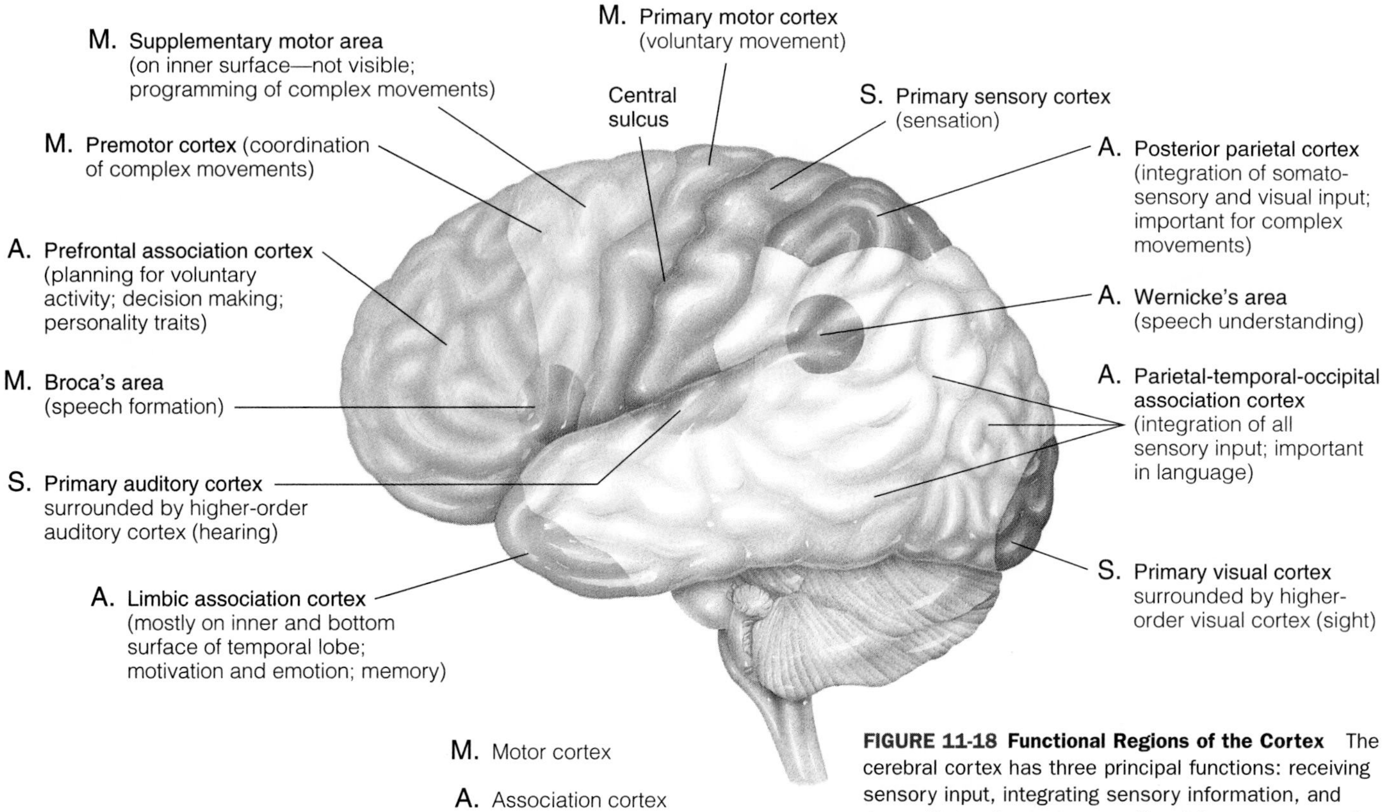

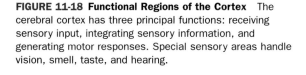

M. Supplementary motor area (on inner surface—not visible; programming of complex movements)

M. Premotor cortex (coordination of complex movements)

A. Prefrontal association cortex (planning for voluntary activity; decision making; personality traits)

M. Broca's area (speech formation)

S. Primary auditory cortex surrounded by higher-order auditory cortex (hearing)

A. Limbic association cortex (mostly on inner and bottom surface of temporal lobe; motivation and emotion; memory)

M. Primary motor cortex (voluntary movement)

Central sulcus

S. Primary sensory cortex (sensation)

A. Posterior parietal cortex (integration of somato-sensory and visual input; important for complex movements)

A. Wernicke's area (speech understanding)

A. Parietal-temporal-occipital association cortex (integration of all sensory input; important in language)

S. Primary visual cortex surrounded by higher-order visual cortex (sight)

M. Motor cortex

A. Association cortex

S. Sensory cortex

FIGURE 11-18 Functional Regions of the Cortex The cerebral cortex has three principal functions: receiving sensory input, integrating sensory information, and generating motor responses. Special sensory areas handle vision, smell, taste, and hearing.

mary sensory cortex, it is the destination of many sensory impulses traveling to the brain. As in the primary motor area, different parts of this ridge correspond to different parts of the body (**FIGURE 11-19B**).

The Association Cortex Is the Site of Integration and Complex Intellectual Activities.

The **association cortex** consists of large expanses of cerebral cortex where integration occurs. In the frontal lobe is a region of the association cortex (prefrontal association cortex) that houses complex intellectual activities such as planning and ideation. This region also modifies behavior, conforming human actions with social norms.

Another important association area lies posterior to the primary sensory cortex (posterior parietal cortex). It interprets sensory information and stores memories of past sensations. Other association areas interpret language in written and spoken form.

Hearing, Vision, Taste, and Smell Are Also Housed in Specific Cortical Regions.

FIGURE 11-18 shows patches of sensory cortex for hearing (primary auditory cortex) and vision (primary vi-

sual cortex). Separate regions (not shown) exist for taste and smell.

Unconscious Functions Are Housed in the Cerebellum, Hypothalamus, and Brain Stem

Consciousness resides in the cerebral cortex. However, many body functions occur at an unconscious level, among them heartbeat, breathing, and many homeostatic functions.[6] One region of the brain that controls unconscious actions is the **cerebellum** (sar-ah-BELL-um), the second largest structure of the brain. As **FIGURE 11-16** shows, it sits below the cerebrum on the brain stem.

The Cerebellum Controls Muscle Synergy and Helps Maintain Posture.

The cerebellum plays several key roles. One of them is synergy. Neurophysiologists define **synergy** as the coordination of skeletal muscle contraction and the movement of body parts to create smooth, efficient motion.

---

[6]Conscious control is possible for many automatic functions, but for the most part, breathing and swallowing are controlled automatically in lower brain centers.

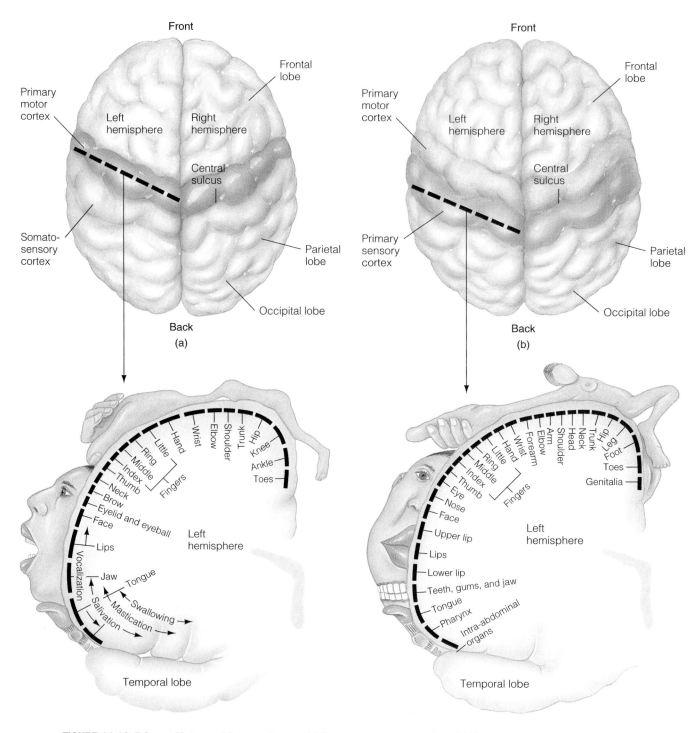

FIGURE 11-19 Primary Motor and Sensory Areas *(a)* Top view of the brain showing the primary motor cortex. Below is a map of the location of motor functions within the primary motor cortex. *(b)* The primary sensory cortex (somatosensory cortex). Below is a map of the location of regions within the primary sensory cortex.

To understand this concept, hold your arm straight out, then bring your hand to your chest. To perform this simple action, your biceps (muscles in the front of the upper arm) contracted while your triceps (muscles in the back of the upper arm) relaxed. For smooth, coordinated movement, some muscles must contract while others relax. The cerebellum ensures that opposing sets of muscles work together to create a smooth motion.

Like most body functions, synergy is not something we think about unless it goes wrong. Commonly, dysfunction arises when the cerebellum is damaged during childbirth—for example, if the baby's blood supply is interrupted, which sometimes happens when the umbilical cord is wrapped around the baby's neck. The ensuing lack of oxygen to the brain can permanently damage the cerebellum, resulting in a loss of synergistic control of the skeletal muscles. Mild damage generally causes a slight rigidness (called *spasticity*) and moderately jerky motions. More severe damage causes serious impairment, with body motions becoming extremely jerky and simple tasks requiring several attempts. This condition is known as **cerebral palsy** (PALL-zee).[7]

The cerebellum also maintains posture. It receives impulses from sense organs in the ear that detect body position. It then sends impulses to the muscles to maintain or correct posture.

The Thalamus Is a Relay Center.

Just beneath the cerebrum is a region of the brain called the **thalamus** (**FIGURE 11-17**). The thalamus is a relay center. It receives all sensory input, except for smell, then relays it to the sensory and association cortex.

---

[7]Cerebral palsy is also caused by abnormal brain development, possibly due to exposure to harmful chemicals (e.g., alcohol) during embryonic development.

The Hypothalamus Controls Many Autonomic Functions Involved in Homeostasis.

Beneath the thalamus is the **hypothalamus** (high-poe-THAL-ah-muss; hypo = under). It consists of many aggregations of nerve cells known as the *nuclei*, which should not to be mistaken for the nuclei in cells (**FIGURE 11-20**). Hypothalamic nuclei control a variety of autonomic functions, all of which are aimed at maintaining homeostasis. Appetite and body temperature, for example, are controlled by hypothalamic nuclei, as are water balance, blood pressure, and sexual activity.

The hypothalamus is a primitive brain center. One of its best-known functions is the control of the pituitary gland, an endocrine gland that regulates many body functions through the hormones it releases (Chapter 14).

Cerebral palsy

www.jbpub.com/humanbiology

The Limbic System Is the Site of Instinctive Behavior and Emotion.

Instincts, among other functions, reside in a complex array of structures called the **limbic system** (LIM-bick), shown in **FIGURE 11-21**. The limbic system operates in conjunction with the hypothalamus.

Instincts are among the most fundamental responses of organisms. They include the protective urge of a mother, the territorial assertions of a male, and the fight-or-flight response an animal experiences in the face of adversity.

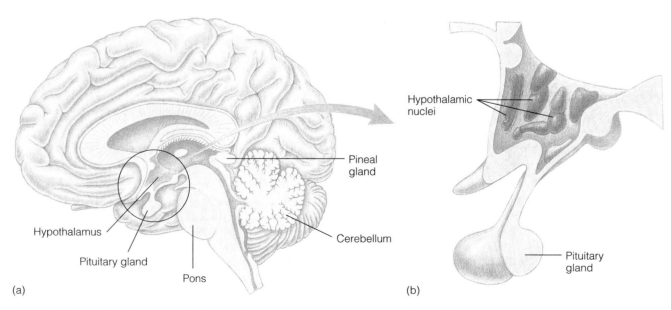

(a)   (b)

FIGURE 11-20 The Hypothalamus *(a)* The hypothalamus is clearly visible in a cross section of the brain. *(b)* The hypothalamus is at the base of the brain, just above the

pituitary gland. It regulates many autonomic functions and plays a particularly important role in controlling the release of hormones by the pituitary.

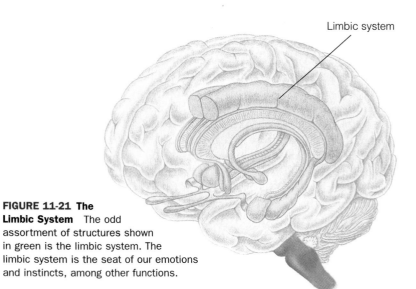

Limbic system

FIGURE 11-21 The Limbic System The odd assortment of structures shown in green is the limbic system. The limbic system is the seat of our emotions and instincts, among other functions.

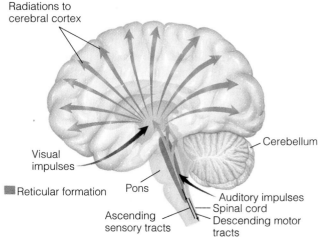

Radiations to cerebral cortex

Cerebellum

Visual impulses

■ Reticular formation   Pons

Ascending sensory tracts

Auditory impulses
Spinal cord
Descending motor tracts

FIGURE 11-22 The Reticular Activating System The reticular formation resides in the brain stem, where it receives input from incoming and outgoing neurons. Fibers projecting from the reticular formation to the cortex constitute the reticular activating system.

The limbic system also plays a role in emotions—fear, anger, and so on. Electrodes placed in some areas of the limbic system elicit primitive rage and in other areas, placidity. Stimulation of specific regions within the limbic system of humans may elicit sensations patients describe as joy, pleasure, fear, or anxiety, depending on the site of stimulation. For a discussion of depression and its effects on emotions, see Health Note 11-2.

Many Basic Body Functions Are Controlled by the Brain Stem.

The brain stem connects the brain to the spinal cord and gives rise to most of the cranial nerves. As shown in **FIGURE 11-21**, the brain stem consists of three parts. Like the hypothalamus, the brain stem contains aggregations of nerve cells that control many basic body functions such as heart rate, blood pressure, and breathing. It also regulates swallowing, coughing, vomiting, and many digestive functions. The hypothalamus and brain stem therefore often work in concert.

Besides containing control regions (nuclei), the brain stem is a main thoroughfare for information traveling to and from the brain. Nerve fibers passing through the brain stem, however, frequently give off branches that terminate in a special region of the brain stem known as the reticular formation (reh-TICK-you-ler) (**FIGURE 11-22**).

The **reticular formation** runs through the entire brain stem into the thalamus. It receives all incoming and outgoing information, monitoring activity much like a security guard at a doorway. Nerve impulses are then projected to the cerebral cortex via special nerve fibers. These impulses activate cortical neurons. These nerve fibers comprise the **reticular activating system (RAS)** and are like the alarm clock of the central nervous system, for they help maintain wakefulness or alertness. Consider what happens when you're asleep.

During sleep, the flow of information through the RAS is greatly reduced, and the cortex "sleeps." However, this sleep may be disturbed by a loud noise. Noise is picked up by the ear, which sends sensory impulses to the auditory cortex of the brain. But the cortex doesn't perceive the sound because it's asleep. To rouse the brain, nerve impulses branch off into the RAS. These fibers then send signals that awaken the cortex and the sleeper.

The RAS not only awakens us and keeps us alert, but it also prevents us from falling asleep from time to time. Pain from a bad sunburn, for example, prevents many people from falling asleep. Why? Pain impulses traveling to the brain from sensors in the skin enter the RAS. It, in turn, generates impulses that travel to the cortex and keep the cells active despite one's fatigue.

Cerebrospinal Fluid Cushions the CNS

The brain and spinal cord are housed in the skull and spinal column, respectively. These bony fortresses protect those vital and delicate organs. But as anyone who's ever packed a fragile object in a box knows, to provide true protection, you must supply some added cushion.

Evolution "solved" this dilemma by surrounding the brain with a watery liquid, called **cerebrospinal fluid (CSF).** Similar in composition to blood plasma

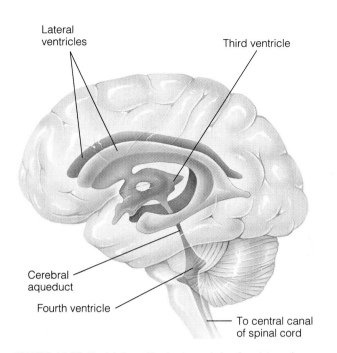

Lateral ventricles

Third ventricle

Cerebral aqueduct

Fourth ventricle

To central canal of spinal cord

FIGURE 11-23 Ventricles The brain contains four internal cavities, called *ventricles,* which are filled with a plasmalike fluid known as cerebrospinal fluid.

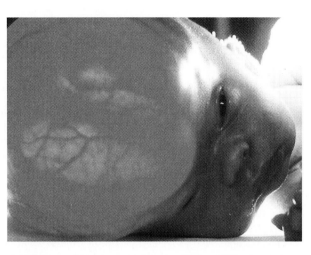

FIGURE 11-24 Hydrocephalus This birth defect results from a blockage in the ventricles, which causes CSF to build up, thinning the cortex and causing severe brain damage.

and interstitial fluid, CSF is found in the arachnoid space, which lies between the inner and outer layers of the meninges. CSF also fills internal cavities (ventricles) in the brain and spinal cord (**FIGURE 11-23**). CSF cushions these organs and protects them from injury.

Cerebrospinal fluid is produced by the lining of two large cavities, the **lateral ventricles,** in the cerebral hemispheres, (**FIGURE 11-23**). CSF is continuously generated and slowly drains out of the brain and into the bloodstream. As a rule, the amount of CSF produced by the ventricles equals the amount removed.

If the circulation of CSF is blocked in a newborn, the accumulation of fluid causes the brain and head to enlarge and produces a condition known as **hydrocephalus** (HIGH-drow-CEFF-eh-luss; literally, "water on the brain") (**FIGURE 11-24**). As the ventricles fill with fluid, the cortex thins, which damages brain cells. If detected early, surgeons can insert a plastic tube in the brain to drain the fluid from the ventricles (upstream from the blockage) into the thoracic cavity. This permits the brain to develop normally. If not detected early and the damage is severe, the child will be permanently impaired and may die.

Because CSF bathes the CNS, examining fluid provides a means of detecting infections in the brain, spinal cord, and meninges. Samples of CSF are obtained by inserting a needle between the third and fourth lumbar vertebrae. The needle projects into the space between the dura mater and pia mater. CSF is then withdrawn.

One dangerous condition detected by this method is **meningitis** (men-in-JI-tiss), an infection of the meninges caused by certain viruses or bacteria. As noted in Chapter 9, meningitis usually begins with an infection in the respiratory system. In adults, meningitis is characterized by fever, headache, and nausea. Patients often vomit and complain of a stiff neck and an inability to tolerate bright light (photophobia). These symptoms develop over the course of a few hours. In babies and children, the symptoms are often less obvious, causing parents to overlook a potentially fatal condition until it is too late. In both cases, meningitis requires immediate attention because it can be fatal.

Hydrocephalus
www.jbpub.com/humanbiology

Electrical Activity in the Brain Varies Depending on Activity Level or Level of Sleep

Electrodes applied to different parts of the scalp detect underlying electrical activity in the brain and produce a tracing known as an **electroencephalogram** (ee-LECK-trow-en-CEFF-eh-low-GRAM) or **EEG**. Some sample tracings of the electrical activity at different times are shown in **FIGURE 11-25** on page 260; note that the type of brain wave recorded depends on the level of cortical activity. The waves often appear irregular, but distinct patterns can sometimes be observed, especially during the different phases of sleep.

Because diseases or injuries of the cerebral cortex often give rise to altered EEG patterns, EEGs are used to diagnose brain dysfunction. Perhaps the most common disorder is epilepsy (EP-eh-LEP-see). **Epilepsy** is characterized by periodic seizures that occur when a large number of neurons fire spontaneously. The seizures can be accompanied by involuntary spasms of skeletal muscles and alterations in behavior. Research has shown that in about two-thirds of the cases, patients have no

Meningitis
www.jbpub.com/humanbiology

Epilepsy
www.jbpub.com/humanbiology

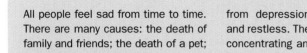

Health Note 11-2

Feeling Blue: Understanding and Treating Depression

All people feel sad from time to time. There are many causes: the death of family and friends; the death of a pet; the loss of a job; divorce; the loss of a valuable friendship; or poor performance on an exam. All of these adversities—and a great many more—can bring on feelings commonly called *depression*. But to a psychologist, occasional and transient feelings of sadness are a normal part of life.

Psychologists reserve the term *depression* for a condition characterized by one of several symptoms, including feelings of intense sadness that are evoked by even minor stimuli, things that people who are not depressed seem to cope with quite well. People suffering from depression may not feel sad, but rather experience persistent numbness or emptiness. And, they often complain of feeling little or no pleasure in things that once brought them happiness. Depression can be manifested in anxiety, feelings of hopelessness, pessimism, and a general sense of worthlessness. People suffering

from depression may feel irritable and restless. They may have difficulty concentrating and making decisions. Depressed individuals may feel helpless and often suffer from sleep disorders—waking up in the middle of the night or oversleeping (see Health Note 4-2 for a discussion of sleep disorders). Some people can't eat. Others overeat. They may complain of physical symptoms such as chronic pain and headaches that do not respond to treatment. Thoughts of suicide become prevalent.

Depression varies in intensity from mild to severe, and in duration from short term to long term. In its most extreme forms, depression can be life-threatening because severe depression can lead to suicide. Depression also comes in many forms. In this Health Note, we examine two of the many clinically recognizable forms: major depression and a less severe type called dysthymia (dis-THIGH-me-ah).

Major depression is so severe that it often incapacitates a person. Indi-

viduals suffering from depression are unable to work, sleep, eat, or enjoy anything. Dysthymia is less serious because it tends only to impair work, sleep, eating, and other functions. In other words, it does not "cripple" a person; it just keeps him or her from functioning at full capacity.

The causes of depression are something of a mystery to medical science. Researchers know that it results from a change in brain chemistry—involving the level of certain neurotransmitters. The same changes occur during periodic bouts of sadness. But in patients suffering from depression, the changes are not reversed, even after the triggering event or stimulus is eliminated. In other words, the changes become self-sustaining. New stresses then emerge. A person may become stressed about feeling unhappy all of the time. Lack of sleep may put emotional stress on an individual. Depression can tax relationships, creating an additional stimulus for depression. Feeling down often kills the impulse to get out-

identifiable structural abnormality in the brain. In the remaining cases, the seizures can usually be traced to brain damage at birth, severe head injury, or inflammation of the brain. In some instances, epileptic seizures are caused by brain tumors.

Headaches
www.jbpub.com/humanbiology

▮ Headaches Have Many Causes, But Are Rarely the Result of Life-Threatening Anomalies

No discussion of the brain would be complete without considering something we've all had experience with, headaches. **Headaches** are the most common form of pain. Many headaches are caused by tension—sustained tightening of the muscles of the head and neck when a person is nervous, stressed, or tired. Headaches also commonly result from swelling of the membranes lining the sinuses (cavities inside the bones of the skull surrounding the nasal cavity). Such swelling usually oc-

curs because of sinus infections or allergies. Eyestrain is another common cause of headaches, as is the dilation of cerebral blood vessels, which may be associated with high blood pressure or excessive alcohol consumption.

Not all headaches are created equal though. In fact, very serious headaches may result from increased intracranial pressure. This, in turn, may result from a brain tumor or intracranial bleeding. Inflammation caused by an infection of the meninges (meningitis) or the brain itself (encephalitis), although rare, also produce intense headaches that require immediate attention.

Section 11-5

The Autonomic Nervous System

As noted earlier, the nervous system consists of two parts: the peripheral nervous system and the central

side for fresh air, sunshine, or exercise—all of which can counteract depression.

When individuals become aware of their condition, often upon the advice of friends, family, or family physicians, most can pinpoint an event or several events that are responsible for their depression. Most stimuli seem to have one of three features in common: the loss of self-determination (the ability to make choices); the loss of empowerment (feeling you have power to make choices that make a difference); and the loss of self-confidence. In some instances, however, depression appears to have no link to a person's situation. The supposedly spontaneous changes in brain chemistry apparently occur without stimulus—or at least none that can be currently identified.

Although depression can be treated very effectively, many people slip into this state without even realizing it, and sink deeper and deeper in depression unaware that they are spiraling downward and are in need of help. Friends, relatives, and family physicians can help individuals realize that they are depressed. Interestingly, in most instances, the chemical changes in the brain seem to self-correct—but usually after one to three years without medical treatment. For those who don't want to wait that long to get over depression, there's great hope. Psychological therapy, which addresses the root causes, can be helpful for long-term recovery. Prescription medicines (antidepressants) can also help immensely. Many antidepressants are currently available, and they are very effective. In fact, two-thirds of those people treated by any one medication respond favorably. Those who fail to respond to one medication have a great chance of responding to another one.

Antidepressants work by elevating levels of neurotransmitters in the brain, which, for reasons unknown, seems to relieve depression. Antidepressants are not "happy" pills. They don't make one euphoric. They simply lift the feeling of depression.

Antidepressants take a while to work—up to two months—and have some annoying side-effects. Two of the most commonly used antidepressants, Prozac and Paxil, for instance, decrease sexual interest or cause difficulty in achieving an orgasm in about 30% of the patients. Both cause nausea in about 20% of the cases and insomnia in about 15% of the patients.

There are no safeguards against depression. But living well—eating right, getting plenty of exercise, balancing work with pleasure, and having a healthy psychological framework—can all help us reduce the likelihood of this sometimes trying condition. For those who contract it, there is hope, and the sooner the problem is recognized, the sooner the cure will take place.

Visit *Human Biology's* Internet site, www.jbpub.com/humanbiology, for links to web sites offering more information on this topic.

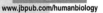
www.jbpub.com/humanbiology

nervous system. The PNS consists of a somatic division and an autonomic division. Each division contains motor and sensory neurons (**FIGURE 11-3**).

The **somatic division** receives sensory information from the skin, muscles, and joints, which it transmits to the CNS. Motor impulses from the CNS, in turn, are sent to skeletal muscles via motor neurons. In contrast, the **autonomic division** transmits sensory information from organs to the CNS. It then delivers motor impulses to smooth muscle, cardiac muscle, and glands. The latter constitutes the **autonomic nervous system (ANS).**

The ANS functions automatically, usually at a subconscious level. It innervates all internal organs, supplying smooth muscle, cardiac muscle, and glands. The autonomic nervous system has two subdivisions: the sympathetic and the parasympathetic.

The **sympathetic division** of the ANS functions in emergencies and is largely responsible for the **fight-or-flight response.** This response occurs when an individual is startled or faced with danger. A sudden scare results in sympathetic nerve impulses that bring about a whole host of responses, preparing us to fight or flee the scene. One result is a rapid increase in heart rate. The fight-or-flight response also results in pupil dilation and an increase in breathing rate. Dilation of the pupils lets more light in, and presumably helps us see better. Increased breathing delivers more oxygen to the blood. Finally, the fight-or-flight response results in a decrease in blood flow to the intestines, diverting blood from digestion, and an increase in blood flow to the skeletal muscles needed for a rapid escape.

In contrast, the **parasympathetic division** of the ANS brings about internal responses associated with the relaxed state. It therefore reduces heart rate, contracts the pupils, and promotes digestion.

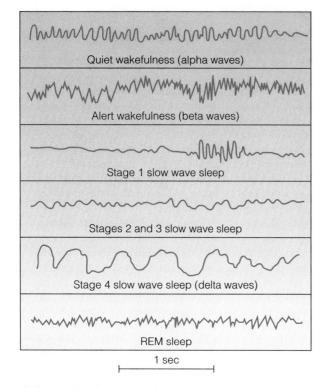

FIGURE 11-25 Electroencephalogram Several sample tracings of brain waves during different activities and stages of sleep.

Most organs are supplied with both parasympathetic and sympathetic fibers. As a general rule, these fibers have antagonistic (or opposite) effects. One stimulates activity, and the other reduces activity. This provides the body with a means of fine-tuning organ function.

Section 11-6

Learning and Memory

Few questions intrigue humans more than how the brain functions in learning and memory. **Learning** is a process in which an individual acquires and retains knowledge and skills. Learning usually results from experience, instruction, or some combination of the two, and it depends on one's ability to store information in the brain, a phenomenon called **memory.**

Newly acquired knowledge is first stored in **short-term memory**. Short-term memory holds information for periods of seconds to hours. The phone number you just looked up or the name of a new acquaintance are stashed in short-term memory. If you don't do something to "fix" these memories, they soon vanish. Cramming for tests places a lot of information into short-

term memory. Unfortunately, soon after the test is over, the information fades into oblivion. The reason people often cannot recall the details of a traumatic accident is because their short-term memory has been wiped out.

Long-term memory holds information for much longer periods—from days to years—and has a much greater storage capacity than short-term memory "banks."

Transferring information from short-term to long-term memory is called **consolidation.** In the Study Skills section at the front of this book, I described several ways to consolidate information, including repetition, mnemonics, and rhymes.

The process of recalling information stored in either short- or long-term memory is called **remembering.** Short-term memory recall is generally faster than long-term memory recall unless the long-term memory you are retrieving is one that is thoroughly ingrained (for example, your place of birth). By the same token, information lost from short-term memory is generally gone forever. Information you cannot recall from long-term memory is often still there; it just takes time or some special stimulus to extract it.

Memory Is Stored in Multiple Regions of the Brain

Neurons involved in storing memories are widely distributed in the cerebral cortex, especially the temporal lobe. Other regions also hold memories, including the cerebellum and the limbic system. The temporal lobes and a part of the limbic system known as the **hippocampus** (hip-poe-CAMP-us) appear essential for transferring short-term memories into long-term memory.

Short-Term and Long-Term Memory Appear to Involve Structural and Functional Changes of the Neurons

How are memories stored? Although no one knows the precise mechanism for this, evidence points toward certain anatomical and physiological changes in the neuronal circuitry. However, short- and long-term memories probably involve different mechanisms. Studies on slugs and snails suggest that short-term memory may involve transient modifications in the function of existing synapses. Whether this mechanism holds true for humans and other animals remains to be seen.

In contrast, long-term memory appears to involve relatively permanent structural and functional changes in the neurons of the brain. Studies that compared the brains of experimental animals reared in a sensory-deprived environment with those reared in a sensory-rich environment revealed marked differences in the microscopic structures of their brains. The brains of animals exposed to a sensory-rich environment displayed greater branching and elongation of dendrites in regions of the brain thought to serve as memory storage.

Some studies suggest that protein synthesis may be involved in long-term memory because drugs that block protein synthesis interfere with long-term memory. What proteins are involved remains unknown.

Health, Homeostasis, and the Environment

The Violent Brain

The nervous system is one of the body's two control systems (the other being the endocrine system). Receptors perceive changes in the internal and external environment and send signals to the brain and spinal cord, which then respond with corrective actions. This helps maintain physical and chemical conditions essential to life.

Like other organ systems, the nervous system can be damaged by a wide variety of factors. The following case study illustrates an extreme example of what happens when the nervous system comes under attack from poisonous chemicals in the work environment.

Perhaps one of the most disturbing stories of the impacts of hazardous chemicals on the brain is that of David Garabedian, a man described by his family and friends as mild-mannered, passive, even docile. Garabedian worked for a lawn-care company in Massachusetts. On March 29, 1983, he arrived at the home of Eileen Muldoon to estimate the cost of treating her lawn with insecticides. He knocked on the door but found no one at home. Garabedian decided to make an estimate and leave it for her.

After he had completed his work, he experienced a sudden urge to urinate. He went to the backyard and urinated near the house. Just then, Mrs. Muldoon appeared, yelling angrily at him. Garabedian became confused, apologized, and tried to explain his plight. According to his testimony, she turned away from him, refusing to listen. Garabedian tapped her on the shoulder to say he was sorry, but the woman turned on him and clawed his face with her fingernails. Garabedian exploded, grabbing the woman by the neck and strangling her. As she lay motionless on the ground, he hurled large rocks at her head, smashing her skull.

A month before the murder, Garabedian had undergone a remarkable personality change. Usually amicable, he angered easily and was abusive toward his family. He complained of tension, nervousness, impatience, and terrible nightmares. He also complained of numerous physical symptoms such as nausea, diarrhea, headaches, and frequent urination. These symptoms suggested to some that the young man was suffering from pesticide poisoning.

Dr. Peter Spencer, a toxicologist at the Albert Einstein College of Medicine in New York, testified at the trial that "David Garabedian was involuntarily intoxicated with a chemical in the lawn products that he was exposed to on a daily basis." He not only sprayed chemicals on lawns and trees, but also mixed them, pouring large amounts into trucks each night in preparation for the next day's spraying.

One chemical, in particular, has been singled out as the possible culprit. It is carbaryl, a powerful inhibitor of acetylcholinesterase. Physicians believe that the carbaryl, and possibly other pesticides that Garabedian routinely handled, inhibited acetylcholinesterase in his brain. Acetylcholine then accumulated in the synaptic clefts in neurons in his limbic system, triggering uncontrollable rage.

A psychiatrist who testified at the trial said, "In my view, David did not have the capacity to control his behavior because his brain was poisoned." Despite this testimony, Garabedian was found guilty of first-degree murder and is currently serving a life sentence.

Drugs, alcohol, pesticides, and industrial chemicals routinely enter our bodies. In small amounts, they may be harmless, but in higher concentrations or in combination, they can disrupt chemical balance and cellular structure, upsetting homeostasis with a wide range of consequences.

Damage in crucial areas of the brain, no matter how slight, can upset its normal operation, resulting in erratic behavior or violent rages. Dr. Richard Restak, a neurologist and author of numerous popular books on the brain, notes that the ingestion of alcohol and other drugs can dismantle (at least temporarily) the brain's intricate system of checks and balances and elicit unusual, and even violent, behavior. "It is humbling and frightening," says Restak, "to consider that our rationality is dependent on the normal function of tissue within our skulls." It is even more frightening to realize that the balance can be so easily upset. "We have the capacity, if everything is operating correctly within our brains, of composing a Bill of Rights or the Constitution," says Restak. "But in the presence of a barely measurable electrical impulse within the limbic system, our much vaunted rationality can be replaced by savage attacks and seemingly inexplicable violence."

SUMMARY

AN OVERVIEW OF THE NERVOUS SYSTEM

1. The *nervous system* controls a wide range of functions and plays a key role in ensuring homeostasis. The human nervous system also performs many other functions, such as ideation, planning for the future, learning, and remembering.

2. The nervous system consists of two anatomical subdivisions: the *central nervous system (CNS)*, made up of the *brain* and *spinal cord*, and the *peripheral nervous system (PNS)*, which consists of the *spinal* and *cranial nerves*.

3. Receptors in the skin, skeletal muscles, joints, and organs, transmit sensory input to the CNS via *sensory neurons*. The CNS integrates all sensory input and generates appropriate responses. Motor output leaves the CNS in *motor neurons*.

4. The PNS has two functional divisions. The *autonomic division* controls involuntary actions such as heart rate. The *somatic division* largely controls voluntary actions such as skeletal muscle contractions. It also provides the neural connections needed for many reflex arcs.

STRUCTURE AND FUNCTION OF THE NEURON

5. The fundamental unit of the nervous system is the *neuron*. This highly specialized cell generates and transmits bioelectric impulses from one part of the body to another.

6. Three types of neurons are found in the body: *sensory neurons, interneurons,* and *motor neurons*.

7. All neurons have more or less spherical *cell bodies*. Extending from the cell body are two types of processes: *dendrites*, which conduct impulses to the cell body, and *axons*, which conduct impulses away from the cell body.

8. Many axons are covered by a layer of myelin, which increases the rate of impulse transmission. In the PNS, myelin is laid down by *Schwann cells* during embryonic development. In the CNS, myelin is produced by *oligodendrocytes*.

9. The terminal ends of axons branch profusely, forming numerous fibers that end in small knobs called *terminal boutons*.

10. During cellular differentiation, nerve cells lose their ability to divide. Because of this, neurons that die cannot be replaced by existing cells. Axons often regenerate in the PNS, but only rarely in the CNS.

11. Nerve cells have exceptionally high metabolic rates and require a constant supply of oxygen. Because nerve cells rely exclusively on glucose for energy, decreases in blood glucose levels can have extremely deleterious effects on the nervous system.

12. The small electrical potential across the membrane of nerve cells is known as the *membrane potential* or *resting potential*.

13. When a nerve cell is stimulated, its plasma membrane increases its permeability to sodium ions. Sodium ions rush in, causing *depolarization*, which spreads down the membrane.

14. Depolarization is followed by *repolarization*, a recovery of the resting potential stemming largely from the efflux of potassium ions. The depolarization and repolarization of the neuron's plasma membrane constitute a *bioelectric impulse* or *action potential*.

15. Nerve impulses travel along the plasma membranes of dendrites and unmyelinated axons, but in myelinated axons, impulses "jump" from node to node.

16. When a bioelectric impulse reaches the terminal bouton, it stimulates the release of *neurotransmitters* contained in membrane-bound vesicles.

17. Released into *synaptic clefts*, neurotransmitters diffuse across the cleft, where they bind to receptors in the *postsynaptic membrane*. Neurotransmitters may excite or inhibit the postsynaptic membrane. Whether a neuron fires depends on the sum of excitatory and inhibitory impulses it receives.

18. After stimulating the postsynaptic membrane, neurotransmitters may diffuse out of the synaptic cleft, be reabsorbed by the terminal bouton, or be removed by enzymes.

19. Some insecticides inhibit the activity of the enzymes that deactivate neurotransmitters, creating a wide range of nervous system effects.

THE SPINAL CORD AND NERVES

20. The *spinal cord* descends from the brain through the *vertebral canal* to the lower back. It carries information to and from the brain, and its neurons participate in many reflexes.

21. Two types of nerves emanate from the CNS: spinal and cranial. *Spinal nerves* arise from the spinal cord and may be sensory, motor, or mixed. *Cranial nerves* attach to the brain and supply the structures of the head and several key body parts.

22. Spinal nerves are attached to the spinal cord via two roots, a *dorsal root*, which brings sensory information into the cord, and a *ventral root*, which carries motor information out.

23. Sensory fibers entering the cord often end on *interneurons*, which frequently end on motor neurons in the spinal cord, forming *reflex arcs*. Interneurons can also send axons to the brain.

THE BRAIN

24. The brain is housed in the skull. The *cerebrum* with its two *cerebral hemispheres* is the largest part of the brain.

25. The outer layer of each hemisphere is the *cortex*. It contains *gray matter*, which houses nerve cell bodies, and underlying *white matter*, which contains myelinated nerve fibers that transmit nerve impulses to and from the gray matter.

26. The cerebral cortex consists of many discrete functional regions, including motor, sensory, and association areas.

27. Consciousness resides in the cerebral cortex, but a great many functions occur at the unconscious level in parts of the brain beneath the cortex.

28. The *cerebellum* coordinates muscle movement and controls posture. The *hypothalamus* regulates many homeostatic functions. The *limbic system* houses instincts and emotions. The *brain stem*, like the hypothalamus, regulates basic body functions.

29. A watery fluid surrounds the brain and spinal cord. Known as *cerebrospinal fluid (CSF)*, it serves as a cushion that protects the brain and spinal cord from traumatic injury. It also fills the central canal of the spinal cord and cavities within the brain, the *ventricles*.

30. Electrodes applied to different parts of the scalp detect electrical activity in the brain and produce a tracing known as an *electroencephalogram (EEG)*. The type of brain wave recorded depends on one's level of cortical activity. EEGs are used to diagnose some brain dysfunctions.

31. *Headaches* are the most common form of pain and generally result from tension, swelling of the membranes lining the sinuses, eyestrain, or dilation of the cerebral blood vessels.

THE AUTONOMIC NERVOUS SYSTEM

32. The *autonomic nervous system* or *ANS* is a division of the PNS and transmits motor impulses to smooth muscle, cardiac muscle, and glands.

33. The ANS innervates all internal organs and has two subdivisions: the sympathetic and the parasympathetic.

34. The *sympathetic division* of the ANS functions in emergencies and is responsible in large part for the *fight-or-flight response*, accelerating heart rate and breathing, dilating the pupils, and shunting blood to the skeletal muscles, which provides additional oxygen and nutrients needed to combat an adversary or to escape.

35. The *parasympathetic division* of the ANS reduces heart rate, contracts the pupils, and promotes digestion—responses associated with the relaxed state.

LEARNING AND MEMORY

36. *Learning* is a process in which an individual acquires knowledge and skills from experience, instruction, or some combination of the two. Learning depends on one's ability to store information in the brain—that is, *memory*.

37. Newly acquired knowledge is first stored in *short-term memory*, which retains information for periods of seconds to hours. Short-term memories may be transferred to *long-term memory*, which holds information for periods of days to years.

38. Memory is stored in multiple regions of the brain.

39. Short- and long-term memory appear to involve structural and functional changes of the neurons.

HEALTH, HOMEOSTASIS, AND THE ENVIRONMENT: THE VIOLENT BRAIN

40. Abnormal brain activity can result in bizarre behavior in humans. Brain damage may result from harmful chemicals such as pesticides and drugs.

Critical Thinking

THINKING CRITICALLY— ANALYSIS

This Analysis corresponds to the Thinking Critically scenario that was presented at the beginning of this chapter.

The researchers hypothesized that the left prefrontal cortex of the brain may process negative thoughts. Since they found that this cortex is hyperactive in severely depressed subjects, they hypothesized that the hyperactivity results in symptoms of depression. The scientists also conjectured that chemical messengers in the brain that normally dampen the activity of this region are released in abnormally high amounts, causing the brain to malfunction.

To confirm their findings, the researchers compared another group of seven individuals with severe depression to a control group—and found similar results. Although this is encouraging, it is important to note that the sample sizes are rather small. Before one can accept these findings, it's essential that they be repeated by other researchers studying larger sample sizes.

EXERCISING YOUR CRITICAL THINKING SKILLS

Elevated blood glucose levels in diabetics damage their kidneys and nervous system, in some instances causing renal failure and blindness. Diabetics also suffer from intestinal disorders, which researchers think may be caused by damage to the nerves that innervate the intestines.

One researcher, intent on pinpointing the cause of the intestinal problems in diabetics, decided to examine the microscopic anatomy of the paravertebral ganglia. These ganglia are clusters of nerve cell bodies along the spine that are involved in the autonomic nervous system, which, as you learned in this chapter, controls the function of many organs in the body, including the intestines and urinary bladder.

The researcher first examined ganglia in older diabetic rats. In his studies, he found that nerve endings (terminal boutons) in the ganglia were quite swollen, some of them 30 times their normal size. This observation led him to hypothesize that diabetes was causing the swelling, which was due to the accumulation of vesicles containing neurotransmitters.

To test this hypothesis, the researcher examined control groups composed of nondiabetic rats of the same age that had been kept under similar conditions. To his surprise, many of the nerve terminals in the nondiabetic rats' ganglia were also swollen. (This example illustrates how important a control group is to good science.) The researcher concluded that some other factor was responsible for the swelling. From the small amount of detail given so far, can you come up with any ideas? (Read back over the material for a hint . . . it's there.)

If you said age, you are right. After thinking about the commonalities of the two groups, the researcher realized that all of his animals were old. He then hypothesized that age might have been responsible for the swelling. If you were involved in the study, how would you test this new hypothesis?

If you said that you would compare the ganglia of young and old nondiabetic rats, you're right again. That's exactly what he did. When the researcher made the comparison, he found no evidence of nerve terminal swelling in the young rats.

Like other good researchers, this person, intent on solving one problem, found a partial answer to another: aging. As the body ages, the autonomic nervous system can malfunction, causing a wide range of problems in elderly people, including loss of bladder control and fainting when standing too quickly.

The researcher in this example decided to see if this age-related swelling occurred in humans as well. He collected ganglia from 56 people aged 15 to 93 who had died from a variety of causes. In people under 60, there were few swollen nerve terminals, but after age 60 the number of swollen nerve terminals increased dramatically.

Although it is true that the researcher was unsuccessful in finding answers to the original problem and that much research is still needed to determine why nerve terminals swell in older animals, this case study is a good example of how using critical thinking skills and good experimental methods can lead to important results.

TEST OF CONCEPTS

1. The nervous system performs a great many functions. Describe them. What functions do you think are unique to humans?
2. Describe how the resting and action potentials are generated. Describe how the plasma membrane of a neuron is repolarized.
3. Draw a typical multipolar neuron, and label its parts. Show the direction in which an action potential travels.
4. Draw a typical synapse, label the parts, and explain how a nerve impulse is transmitted from one nerve cell to another.
5. Describe how each of the following substances affects synaptic transmission: caffeine, cocaine, and malathion (a pesticide).
6. In general, how do anesthetics function?
7. Name and describe the various divisions of the nervous system.
8. Draw a cross section through the spinal cord showing the spinal nerves. Label the parts, and explain a reflex arc and how nerve impulses entering a spinal nerve also travel to the brain.
9. A physician can stimulate various parts of the brain and get different responses. What effects would you expect if the electrodes were placed in the premotor area? The primary motor cortex? The sensory motor cortex?
10. The brain is a delicate organ, and slight shifts in electrical activity can create bizarre behavior. Do you agree with this statement? Give examples.

TOOLS FOR LEARNING

www.jbpub.com/humanbiology

Tools for Learning is an on-line student review area located at this book's web site HumanBiology (www.jbpub.com/humanbiology). The review area provides a variety of activities designed to help you study for your class:

Chapter Outlines. We've pulled out the section titles and full sentence sub-headings from each chapter to form natural descriptive outlines you can use to study the chapters' material point by point.

Review Questions. The review questions test your knowledge of the important concepts and applications in each chapter. Written by the author of the text, the review provides feedback for each correct or incorrect answer. This is an excellent test preparation tool.

Flash Cards. Studying human biology requires learning new terms. Virtual flash cards help you master the new vocabulary for each chapter.

Figure Labeling. You can practice identifying and labeling anatomical features on the same art content that appears in the text.

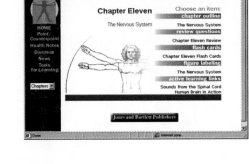

Active Learning Links. Active Learning Links connect to external web sites that provide an opportunity to learn basic concepts through demonstrations, animations, and hands-on activities.

CHAPTER 12

THE SENSES

Light micrograph of a pressure receptor found in human skin.

Thinking Critically

A television news program trumpeted the glories of an operation called radial keratotomy, which is designed to improve eyesight. In this procedure, surgeons make tiny incisions in the cornea (the clear portion of the eye), which flattens it and improves eyesight. In many cases, individuals are able to abandon eyeglasses altogether. In others, a patient's eyesight improves so much that he can reduce his prescription strength and get rid of the thick lenses required for extreme nearsightedness (myopia).

The television program featured the story of one woman whose eyesight improved so much that she was able to see without corrective lenses. All in all, it was a pretty compelling story.

Are there any dangers in this type of reporting? If so, what are they? If you were considering this operation, how would you go about researching its effectiveness? What questions would you ask? Whom would you ask? ■

DEBRA CARTWRIGHT NOTICED SOMETHING strange one day while she was eating dinner. For no apparent reason, she had lost her senses of smell and taste. Doctors were puzzled at first by the finding, for the young lady seemed to be in fine health. She was not suffering from a cold, sinus infection, or even any allergies that might have blocked her nasal passages and impaired her senses of smell or taste. A blood test, however, revealed the cause of her problem: low blood levels of the micronutrient zinc. Her physician prescribed a zinc supplement, and in a few days her senses of smell and taste returned.

This story illustrates the importance of a nutritionally balanced diet and the impact of a dietary deficiency on two important body functions—the senses of taste and smell[1] It is just one of many examples presented in this book that illustrates how body functions can be altered by external factors such as stress, pollution, and diet.

This chapter examines the senses, dividing the discussion into two broad categories: the general senses and the special senses.

Section 12-1

The General and Special Senses

Staying aware of one's environment and responding to changes in that environment require a system of surveillance not unlike that found at high-security government facilities, where both inside and outside activities are routinely monitored to detect intruders. The surveillance system of the human body, like that of many other animals, consists of numerous receptors, which detect internal and external conditions. In humans, receptors are strategically located in the skin, internal organs, bones, joints, and muscles. They detect stimuli that give rise to the **general senses**: pain, temperature, light touch, pressure, and a sense of body and limb position (**TABLE 12-1**).

The human body is also endowed with five additional senses, known as the **special senses**. They are taste, smell, vision, hearing, and balance. They are generally made possible by highly sophisticated sensory organs—for example, the eye. These marvelously "engineered" detectors increase our ability to perceive stimuli

---

[1]Zinc deficiencies are quite rare, and dietary supplements containing zinc are generally not needed if you are eating a well-balanced diet. Furthermore, dietary supplements can be dangerous. Just a few times the recommended daily allowance of zinc can cause serious problems.

TABLE 12-1

Summary of General and Special Senses

Sense	Stimulus	Receptor
General senses	Pain	Naked nerve endings
	Light touch	Merkel's discs; naked nerve endings around hair follicles; Meissner's corpuscles; Ruffini's corpuscles; Krause's end-bulbs
	Pressure	Pacinian corpuscles
	Temperature	Naked nerve endings
	Proprioception	Golgi tendon organs; muscle spindles; receptors similar to Meissner's corpuscles in joints
Special senses	Taste	Taste buds
	Smell	Olfactory epithelium
	Sight	Retina
	Hearing	Organ of Corti
	Balance	Crista ampularis in the semicircular canals; maculae in utricle and saccule

in the environment and are one of the most remarkable developments in the evolution of complex multicellular organisms.

Receptors in humans involved in the general and special senses fall into five functional categories: (1) mechanoreceptors, (2) chemoreceptors, (3) thermoreceptors, (4) photoreceptors, and (4) nociceptors (pain receptors). **Mechanoreceptors** are those activated by mechanical stimulation—for example, touch or pressure. **Chemoreceptors** are activated by chemicals in the food we eat, the air we breathe, or in our blood. **Thermoreceptors** are activated by heat and cold, and **photoreceptors** are sensitive to light. **Nociceptors** (no-see-SEP-tors) are stimulated by tissue damage such as that caused by pinching, tearing, or burning.

Section 12-2

The General Senses

Sit back in your chair for a moment, close your eyes, and concentrate on what you feel. You may detect the pressure of the chair on your buttocks and warmth emanating from a nearby reading lamp. You may feel your cat brushing against the hairs on your arm. You may detect a slight pain from a bruise or gas pressure in your intestines from the burritos you ate for dinner last night. Now move your arm. Even though your eyes are closed, you can feel it moving.

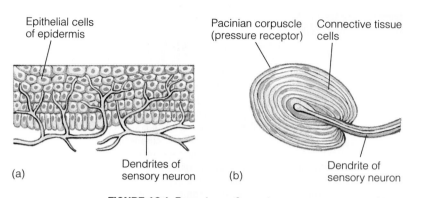

FIGURE 12-1 Receptors   General sense receptors are either (a) naked nerve endings or (b) encapsulated nerve endings.

The various sensations you have just experienced fall into the group of general senses. Receptors for the general senses detect internal and external stimuli and relay messages to the spinal cord and brain via sensory nerves (Chapter 11). Sensory input to the central nervous system may elicit a conscious response—for example, the touch of a cat may cause you to reach down and pat your furry friend. Some stimuli will cause unconscious responses—for example, heat may cause you to perspire, or cold may cause you to shiver. Others may simply be registered in the cerebral cortex, making you aware of the stimulus. Still other stimuli may be blocked so they elicit no response at all.

Receptors for the general body senses come in many shapes and sizes, but they generally fit into two groups based on structure: naked nerve endings and encapsulated receptors (**FIGURE 12-1**). As you shall soon see, the naked nerve endings and encapsulated receptors in the body serve very specific roles.

Naked Nerve Endings in Body Tissues Detect Pain, Temperature, and Light Touch

Located in the skin, bones, and internal organs and in and around joints, **naked nerve endings** are the terminal ends of the dendrites of sensory neurons. They are responsible for at least three sensations: pain, light touch, and temperature.

Pain.

Physiologists recognize two basic types of pain: somatic and visceral. **Somatic pain** results from injuries in the skin, joints, muscles, and tendons, which stimulates receptors in these areas. Pain receptors are part of a protective mechanism of considerable adaptive value in the evolution of complex multicellular organisms. They signal that something is awry and alert the brain so that appropriate action can be taken.

Somatic pain receptors respond to several types of stimuli. Some respond to cutting, crushing, and pinching. Others respond to temperature extremes—hot or cold. Still others respond to irritating chemicals released from injured tissues.

Visceral pain results from the stimulation of naked nerve endings in body organs (the viscera). In some body organs, pain receptors are stimulated by distension. In others, they're stimulated by **anoxia** (ah-NOCKS-see-ah), a lack of oxygen. For example, the intestinal pain you feel when you have gas results from the stretching of naked nerve fibers in the wall of the intestine. These nerve endings are mechanoreceptors. The pain felt during a heart attack results from a lack of oxygen in the heart muscle.

Visceral pain and somatic pain result from quite different stimuli and are perceived very differently as well. Somatic pain is easily pinpointed. Visceral pain, however, is vague and difficult to localize. Moreover, it is generally felt on the body surface at a site some distance from its origin. For example, pain caused by a lack of oxygen to the heart muscle appears in the chest and along the inside of the left arm (**FIGURE 12-2**).

Visceral pain that appears on the body surface away from the location of the pain is called **referred pain.** Physiologists do not know the cause of this phenomenon, but most think that it results from the fact that pain fibers from internal organs enter the spinal cord at the same location that the sensory fibers from the skin enter. The brain, they hypothesize, interprets the impulses from pain fibers supplying the organs as pain from a somatic source. (Health Note 12-1 describes techniques that relieve pain.)

Light Touch.

Light touch is perceived by two anatomically distinct mechanoreceptors. The first receptor is located at the base of the hairs in our skin. As shown in **FIGURE 12-3**, naked nerve endings (dendrites) wrap around the base of the hair follicles. When a hair is moved—for example, by a gentle touch—these nerve fibers are stimulated.

The second light-touch mechanoreceptor is the **Merkel's disc** (MUR-killz). Shown in **FIGURE 12-3**, Merkel's discs consist of small cup-shaped cells on which naked nerve endings terminate. These receptors, located in the outer layer of the epidermis of the skin, are activated by gentle pressure applied to the skin.

Temperature.

Naked nerve endings in the skin detect heat and cold. Heat receptors respond to temperatures from 25°C (77°F) to 45°C (113°F). If the temperature of the skin

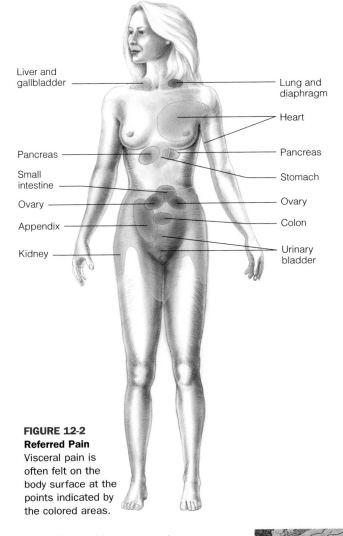

FIGURE 12-2
Referred Pain
Visceral pain is often felt on the body surface at the points indicated by the colored areas.

Labels on figure (left side, top to bottom): Liver and gallbladder; Pancreas; Small intestine; Ovary; Appendix; Kidney

Labels on figure (right side, top to bottom): Lung and diaphragm; Heart; Pancreas; Stomach; Ovary; Colon; Urinary bladder

puscles are smaller than Pacinian corpuscles. These oval receptors contain two or three spiraling dendritic ends surrounded by a thin cellular capsule (**FIGURE 12-3C**). Like the naked nerve endings surrounding the base of the hair follicle and Merkel's discs, Meissner's corpuscles are thought to respond to light touch. Located just beneath the epithelium in the outermost layer of the dermis, Meissner's corpuscles are most abundant in the sensitive parts of the body, such as the lips and the tips of the fingers.

Two additional encapsulated receptors are Krause's end-bulbs and Ruffini's corpuscles (**FIGURE 12-3A**). Most

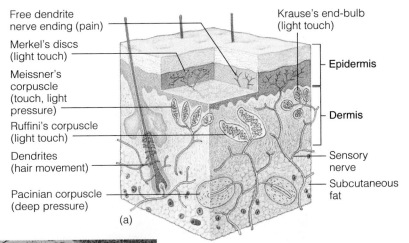

Labels on figure: Free dendrite nerve ending (pain); Merkel's discs (light touch); Meissner's corpuscle (touch, light pressure); Ruffini's corpuscle (light touch); Dendrites (hair movement); Pacinian corpuscle (deep pressure); Krause's end-bulb (light touch); Epidermis; Dermis; Sensory nerve; Subcutaneous fat

(a)

rises above this range, pain receptors are activated, creating a burning sensation. In contrast, cold receptors respond to temperatures from 10°C (50°F) to 20°C (68°F). If the temperature drops below 10°C, pain receptors respond.

Encapsulated Receptors Consist of Naked Nerve Endings Surrounded by One or More Layers of Cells

Shown in **FIGURE 12-3B**, the largest encapsulated nerve ending is the **Pacinian corpuscle** (pah-SIN-ee-an CORE-puss-'l). It consists of a naked nerve ending surrounded by numerous concentric cell layers. The Pacinian corpuscle resembles a small onion pierced by a thin wire (**FIGURE 12-3B**). Located in the deeper layers of the skin, in the loose connective tissue of the body, and elsewhere, Pacinian corpuscles are stimulated by pressure such as the pressure you feel sitting in your chair.

Another common encapsulated sensory receptor is the Meissner's corpuscle (MEIZ-ners). **Meissner's cor-**

(b)

FIGURE 12-3 General Sense Receptors
(a) The skin houses many of the receptors for general senses. Receptors fall into two categories: naked nerve endings and encapsulated receptors. *(b)* The Pacinian corpuscle, often located in the dermis of the skin, detects pressure. *(c)* Meissner's corpuscle, found just beneath the epidermis, detects light touch.

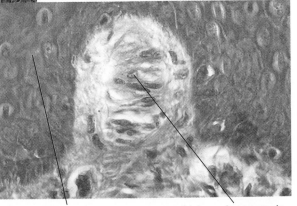

(c)   Epidermis   Meissner's corpuscle

Health Note 12-1

Old and New Treatments for Pain

Millions of people suffer from chronic or persistent pain. In the United States, for example, an estimated 70 million people are tormented by back pain. Another 20 million suffer from migraine headaches.

Despite its prevalence, pain is one of the least understood medical problems in the world. The study of pain, in fact, is so poorly funded and so widely ignored by the medical community that some have called it an orphan science. Making matters worse, physicians are often poorly trained in dealing with pain.

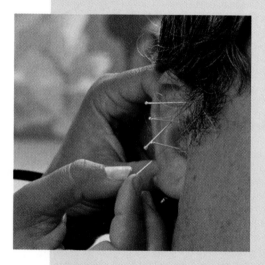

FIGURE 1 Patient Undergoing Acupuncture Acupuncture is generally used to alleviate pain, but it has other applications as well.

For many decades, pain was treated with painkillers such as morphine and codeine. Their addictive nature led researchers to look for other techniques. One promising option is acupuncture, a technique used by the Chinese for thousands of years. Acupuncture relies on thread-thin needles inserted in the skin near nerves (**FIGURE 1**).

No one knows how acupuncture works. The Chinese throught that it interrupted the flow of energy through various pathways in the body, called *meridians.* These meridians, the Chinese contend, are rivers of energy that flow through the body to irrigate and nourish the tissues. They contend that blood and nervous impulses also follow these meridians. Any obstruction in the movement of these energy rivers results in deficiencies of energy, blood, and nerve impulses that may lead to disease. The acupuncture needles, they contend, unblock obstructions and reestablish normal flow.

Neurologists are not certain how acupuncture works, but many think that it blocks pain by overloading the neuronal circuitry. They note that two types of nerve fibers transmit sensory information from the body to the central nervous system: small- and large-diameter fibers. Small-diameter fibers carry pain messages. Large-diameter fibers carry many other forms of sensory information from receptors in the skin—for example, pressure and light

touch. The dendrites of both small- and large-diameter sensory nerve cells often terminate in the same location in the spinal cord. From here they send impulses to the brain, signaling pain or some other sense.

Neurobiologists believe that acupuncture needles stimulate the large-diameter nerve fibers. This stimulation blocks nerve impulses carried by the smaller nerve fibers, thus blocking pain messages to the brain. Research on the mechanism of action of acupuncture points to several other possibilities as well. Acupuncture needles, for instance, may stimulate the release of endorphins, the body's natural pain killers. Many U.S. physicians trained to use drugs and surgery to solve most pain remain skeptical about the usefulness of acupuncture. Over the past 20 years, however, a small but steady stream of research has confirmed the pain-killing effect of this treatment.

Joseph Helms, a physician with the American Academy of Acupuncture in Berkeley, California, performed acupuncture on 40 women with menstrual pain. Some women received real acupuncture treatment. Others received placebo treatments (shallow needle treatments that did not reach the acupuncture points). In the group of women receiving acupuncture, 10 out of 11 showed a marked decrease in pain. Patients reported an approximately 50% de-

scientists believe that these receptors are structural variations of the Meissner's corpuscle and are stimulated by light touch.

Proprioception (PRO-pree-oh-CEP-shun; the sense of position) is provided by special encapsulated receptors located in the joints of the body. Resembling Meissner's corpuscles, these receptors inform us of the position of our limbs and alert us to movements of the body.

Proprioception is also served by the muscle spindle and the Golgi tendon organ. **Muscle spindles,** or **neuromuscular spindles,** are located in skeletal muscles. Muscle spindles consist of several modified muscle fibers with sensory nerve endings wrapped around them.

Spindle fibers are stimulated when muscles are stretched. Nerve impulses generated by the spindle are then transmitted to the spinal cord, where they may ascend to the cerebral cortex, helping us remain aware of

crease in pain. In the placebo group, only 4 out of 11 reported a lessening of pain. Only 1 of 10 people given no treatment showed improvement. Acupuncture also reduced the need for painkilling drugs during treatment by over half. Remember, however, that the Critical Thinking section in Chapter 1 suggested that experiments using small numbers of subjects are themselves subject to question. Further studies are needed to confirm these results.

While acupuncture is slowly earning a respected place in the treatment of pain, some health care practitioners are using a procedure called *transcutaneous electrical nerve stimulation,* or TENS. Patients are fitted with electrodes attached to the skin above the nerves that transmit pain signals to the central nervous system. A small battery supplies energy that stimulates the electrodes. When the pain begins, patients press a button on the battery pack that sends a tiny current to the electrode. The current is conducted through the skin and blocks the pain impulses. How it blocks pain is currently subject to debate. It may block pain by inhibiting abnormally firing or damaged nerves or blocking pain impulses at the spinal cord. Or, it may stimulate a region of the brain that in turn blocks pain. Finally, it may stimulate the release of endorphins (en-DOOR-fins), the body's own pain-killing chemicals.

TENS can be used to reduce pain after surgery. One study showed that this technique reduced the amount of painkillers that doctors had to administer by two-thirds and cut hospital stays by 1 or 2 days. Someday, TENS may be used to reduce the pain of childbirth or pain that athletes often endure.

TENS works like acupuncture, although part of its success may be psychological. In a study of 93 patients with chronic pain, researchers found that over one-third (36%) of them reported no pain or greatly reduced pain when they thought they were being stimulated but really were not. How effective was it when the battery really worked? About half of the patients reported no pain or greatly reduced pain. The slight difference suggests to some physicians that TENS may be overrated. To those suffering from chronic pain, it can be a godsend.

Severe pain can also be treated by surgery. Doctors may cut nerves or destroy small parts of the brain to get rid of chronic pain. Unfortunately, pain recurs in 9 of 10 patients who have undergone pain-relieving surgery, usually within a year or so. Recurring pain results from the partial regrowth of axons. Even after another operation, the pain frequently returns, often with much greater intensity. Consequently, physicians are now looking for alternative measures to eliminate chronic pain.

One promising measure is deep brain stimulation. Electrodes can be implanted in parts of the brain and stimulated to block pain impulses before they reach the sensory cortex, where pain is perceived. The electrodes are connected to a portable battery worn on the belt or implanted under the skin. When the pain begins, the patient turns on the current, blocking the pain impulses.

Research shows that deep brain stimulation is an effective blocker of even the most powerful pain stimuli. Yet it does not upset other brain functions. Unfortunately, this technique requires surgery, and implanting electrodes may cause hemorrhaging in some patients, which can result in permanent paralysis or a loss of feeling in parts of the body. Infections may also develop where the electrodes enter the skull. The 75% success rate and the reduced suffering make most patients more than willing to accept the risks.

Visit *Human Biology's* Internet site, www.jbpub.com/humanbiology, for links to web sites offering more information on this topic.

www.jbpub.com/humanbiology

---

body position. Others may stimulate motor neurons in the spinal cord, eliciting a reflex contraction of the muscle being stretched. This is what happens when you fall asleep sitting upright in a chair when your head falls forward and quickly snaps back.

Golgi tendon organs (GOAL-jee) are mechanoreceptors named after the Italian cytologist who discovered them. Functionally similar to the muscle spindle, these receptors are located in **tendons**—the structures that connect muscles to bones. Golgi tendon organs are composed of connective tissue fibers surrounded by dendrites and encased in a capsule. When a muscle contracts, the tendon stretches and stimulates the receptor. Like the muscle spindle, this receptor alerts the brain to movement and body position. Impulses from the Golgi tendon organ can also stimulate reflex contraction of muscles. The knee-jerk reflex described in Chapter 11 is a good example.

Many Receptors Stop Generating Impulses After Exposure to a Stimulus for Some Length of Time

Pain, temperature, and pressure receptors are all subject to a phenomenon known as **adaptation.**[2] This occurs when sensory receptors stop generating impulses even though the stimulus is still present. You have probably experienced the phenomenon many times. Recall, for example, the first time you wore a ring or contact lenses. At first, the sensation may have nearly driven you mad, but after a few days or perhaps a few hours, the stimulus seemed to have disappeared. What happened was that pressure receptors that were originally alerting the brain stopped generating impulses, relieving you of what would have otherwise been unrelenting discomfort.

Not all receptors adapt. Muscle stretch receptors and joint proprioceptors are two examples. Because the central nervous system must be continuously apprised of muscle length and joint position to maintain posture, adaptation in these receptors would be counterproductive.

Receptors Play an Important Role in Homeostasis

Many general sense receptors play an important role in homeostasis. Mechanoreceptors that detect changes in blood pressure and chemoreceptors that respond to the ionic concentration of the blood, for example, help

---

[2]Not to be confused with the adaptation occurring in evolution.

maintain blood pressure and proper water balance through mechanisms involving the kidney. Chemoreceptors that detect levels of carbon dioxide and hydrogen ions in the blood and cerebrospinal fluid help to regulate respiration. Chapter 14 describes additional chemoreceptors that detect levels of various hormones, nutrients, and ions in the blood and body fluids. These detectors may stimulate hormonal responses that correct potentially disruptive chemical imbalances.

Taste and Smell: The Chemical Senses

The special senses are taste, smell, vision, hearing, and balance.

Taste Buds Are the Receptors for Taste and Respond to Chemicals Dissolved in Food

In humans and other mammals, the tongue contains receptors for taste. Known as the **taste buds,** these microscopic, onion-shaped structures are located in the surface epithelium of the tongue and on small protrusions, **papillae** (pah-PILL-ee), on the upper surface of the tongue (**FIGURE 12-4A**). Taste buds are also found on the roof of the oral cavity, the pharynx, and the larynx, but in smaller numbers.

Taste buds are chemoreceptors and are stimulated by chemicals in the food we eat. These substances dis-

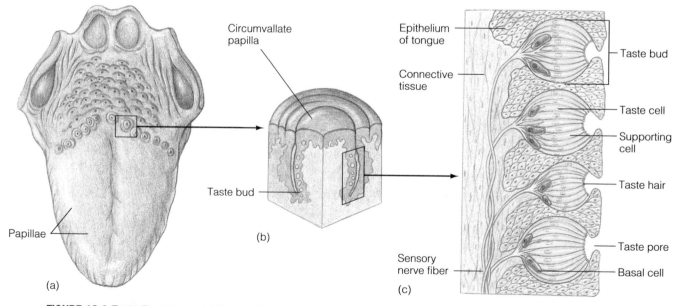

FIGURE 12-4 Taste Receptors *(a)* Taste buds are located on the upper surface of the tongue and are concentrated on the *(b)* papillae. *(c)* The structure of the taste bud.

solve in the saliva and enter the **taste pores,** small openings that lead to the interior of the taste bud (**FIGURE 12-4C**). Taste buds contain receptor cells, the ends of which possess large microvilli, known as **taste hairs.** Taste hairs project into the taste pore. The taste hairs contain receptors that bind to food molecules dissolved in water. This stimulates the receptor cells, which then stimulate the dendrites of the sensory nerves that are wrapped around the receptor cells.[3] Impulses from the taste buds are then transmitted to the taste centers in the cerebral cortex.

In the opening paragraph of this chapter, you were introduced to an unfortunate student who lost her sense of taste because of a dietary zinc deficiency. Research has shown that zinc, which is normally found in low concentrations in the saliva, stimulates division of the cells in taste buds. Because cells of the taste buds are lost from normal wear and tear, a zinc deficiency reduces cellular replacement, and the taste buds eventually cease operation.

The Taste Buds Respond to All Four Primary Flavors But Are Generally Preferentially Responsive to One.

Humans can discriminate among thousands of taste sensations. The taste sensations are a combination of four basic flavors: sweet, sour, bitter, and salty. Sweet flavors result from sugars and some amino acids. Sour flavors result from acidic substances. Salty tastes result from metal ions (like sodium in table salt). Bitter flavors result from chemical substances belonging to a group called **alkaloids,** among them caffeine, but also some nonalkaloid substances such as aspirin. All taste buds respond, in varying degree, to all four taste sensations, but they respond preferentially to one taste.

Taste buds are distributed unevenly on the surface of the tongue. A simple experiment in which drops of various substances are placed on the tongue shows that the tip of the tongue is most sensitive to sweet flavors for it contains a higher proportion of taste buds that respond preferentially to sweet flavors. The sides of the tongue are most sensitive to sour flavors, and the back of the tongue is most sensitive to bitter flavors. Salty taste is more evenly distributed, with slightly increased sensitivity on the sides of the tongue near the front. Food contains many different flavors. What we taste, therefore, depends on the relative proportion of the four basic flavors.

---

[3]Unlike the receptors in the previous section, the receptor cells of taste buds are not part of nerve cells but are independent cells.

The Olfactory Epithelium Is a Patch of Receptor Cells That Detects Odors

Smell, like taste, is a chemical sense. The receptors for smell are located in the roof of each nasal cavity in a patch of cells called the **olfactory epithelium,** or **membrane** (ole-FACK-tore-ee) (**FIGURE 12-5**). Odors are perceived by **receptor cells,** neurons whose dendrites extend to the surface of the olfactory membrane. These cells terminate in six to eight long projections, **olfactory hairs,** or **olfactory cilia** (**FIGURE 12-5**). The olfactory hairs contain receptors for molecules. When airborne molecules are trapped on the thin, watery layer on the surface of the cell, they bind to the receptors, activating the neurons. This causes the neurons to generate impulses that are sent to the **olfactory bulb,** a complex neural structure whose neurons synapse with the dendrites of the receptor cells. Axons of the neurons in the olfactory bulb cells then travel to the brain via the **olfactory nerve.**

Humans can distinguish tens of thousands of odors, many at very low levels. The olfactory receptors are so sensitive, in fact, that even a single molecule binding to the olfactory hairs can produce a bioelectric impulse. But the sense of smell in humans is not as sensitive as that of other animals such as dogs, wolves, and coyotes. A dog's keen sense of smell is due to an olfactory membrane that's about 20 times larger than ours.

Like taste, odor discrimination is thought to depend on combinations of primary odors. Unfortunately, neuroscientists do not agree on what the primary odors are. One system of classification recognizes seven primary odors ranging from pepperminty to floral to putrid. It is thought that molecules that produce a similar odor are similarly shaped and that specific receptor sites on the olfactory hairs bind to molecules with a common shape. Various combinations of the primary odors give rise to the many odors we perceive. Some researchers hypothesize that there may be thousands of kinds of smell receptors providing odor discrimination.

Even Though the Olfactory Receptors Are Sensitive and Able to Discriminate Many Odors, They Quickly Become Adapted.

Receptors for smell adapt within a short period—about a minute. If you have ever visited a dairy farm or lived near one, you understand (and are grateful for) olfactory adaptation.

Smell Influences Our Sense of Taste, and Vice Versa.

Hold a piece of hot apple pie to your nose and take a deep breath. It smells so good you can almost taste it. In fact, you *are* tasting it. Molecules given off by the pie

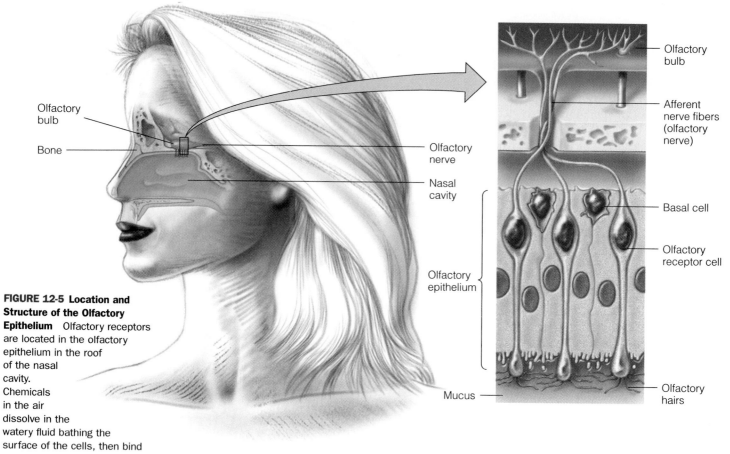

FIGURE 12-5 Location and Structure of the Olfactory Epithelium Olfactory receptors are located in the olfactory epithelium in the roof of the nasal cavity. Chemicals in the air dissolve in the watery fluid bathing the surface of the cells, then bind to receptors on the plasma membranes of the olfactory hairs. The olfactory receptors terminate in the olfactory bulb. From here, nerve fibers travel to the brain.

enter the nose, reach the mouth through the pharynx, and dissolve in the saliva, where they stimulate taste receptors. Just as odors stimulate taste receptors, food in our mouths also stimulates olfactory receptors. Molecules released by food enter the nasal cavities, dissolving in the water on the surface of the olfactory membrane and stimulating the receptor cells.

The complementary nature of taste and smell is abundantly evident when a person suffers from nasal congestion. Cold sufferers often complain that they cannot taste their food. This phenomenon results from the buildup of mucus in the nasal cavities, which blocks the flow of air. Mucus may also block the olfactory hairs. Therefore, food loses its "taste" when you have a stuffy nose because your sense of smell is impaired. (You can test this by holding your nose while you eat.)

Section 12-4

The Visual Sense: The Eye

The human eye is one of the most extraordinary products of evolution. It contains a patch of photoreceptors that permits us to perceive the remarkably diverse and colorful environment we live in.

The Human Eye Consists of Three Distinct Layers

Human eyes are roughly spherical organs located in the eye sockets, or **orbits,** cavities formed by the bones of the skull. The eye is attached to the orbit by six muscles, the **extrinsic eye muscles,** which control eye movement. Small tendons connect these muscles to the outermost layer of the eye.

The Sclera and Cornea.
As **FIGURE 12-6** shows, the wall of the human eye consists of three layers (**TABLE 12-2**). The outermost is a durable fibrous layer, which consists of the **sclera,** the white of the eye, and the **cornea,** the clear part in front, which lets light into the interior of the eye (**FIGURE 12-6**). Tendons of the extrinsic eye muscles attach to the sclera.

The Choroid, Ciliary Body, and Iris.
As **FIGURE 12-6** shows, the middle layer consists of three parts: the choroid, the ciliary body, and the iris. The **choroid** is the largest portion of the middle layer. It contains a large amount of **melanin** (MELL-ah-nin), a pigment that absorbs stray light the way the black interior

of a camera does. The blood vessels of the choroid supply nutrients to the eye. Anteriorly, the choroid forms the ciliary body. The **ciliary body** contains smooth muscle fibers, which control the shape of the lens, permitting us to focus incoming light. The **iris** is the colored portion of the eye visible through the cornea. Looking in a mirror, you can see a dark opening in the iris called the **pupil.** The pupil allows light to penetrate the eye. The blackness you see through the pupil is the choroid layer and the pigmented section of the retina, discussed below. Like

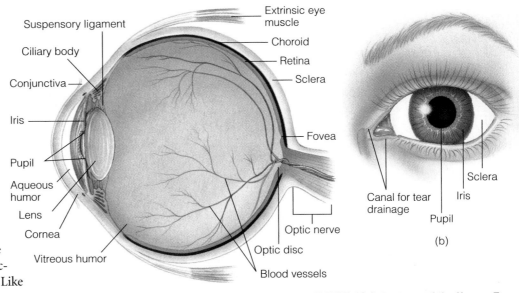

FIGURE 12-6 Anatomy of the Human Eye

(a)

(b)

the ciliary body, the iris contains smooth muscle cells. The smooth muscle of the iris regulates the diameter of the pupil. Opening the pupil lets more light in, and narrowing it reduces the amount of light that can enter. The pupils open and close reflexively in response to light intensity. This reflex is an adaptation that protects the light-sensitive inner layer, the retina.

The Retina.

The innermost layer of the eye is the **retina.** The retina consists of an outer, pigmented layer and an inner layer consisting of photoreceptors and associated nerve cells. The retina is weakly attached to the choroid and can become separated from it as a result of trauma to the head. A detached retina can lead to blindness if not repaired by surgery. Today, eye surgeons usually repair detached retinas with lasers.

The **photoreceptors** of the retina are modified nerve cells. Two types of photoreceptors are present in the retina: rods and cones (**TABLE 12-3**). The **rods,** so named because of their shape, are sensitive to low light (**FIGURES 12-7B** and **12-7C**). Rods function at night and produce grayish, somewhat vague black-and-white images. The **cones,** also named because of their shape, operate only in brighter light. They are responsible for visual acuity—sharp vision—and color vision.

TABLE 12-2

Structures and Functions of the Eye

Structure		Function
Wall		
Outer layer	Sclera	Provides insertion for extrinsic eye muscles
	Cornea	Allows light to enter; bends incoming light
Middle layer	Choroid	Absorbs stray light; provides nutrients to eye structures
	Ciliary body	Regulates lens, allowing it to focus images
	Iris	Regulates amount of light entering the eye
Inner layer	Retina	Responds to light, converting light to nerve impulses
Accessory structures and components		
	Lens	Focuses images on the retina
	Vitreous humor	Holds retina and lens in place
	Aqueous humor	Supplies nutrients to structures in contact with the anterior cavity of the eye
	Optic nerve	Transmits impulses from the retina to the brain

TABLE 12-3

Summary of Rods and Cones

Photoreceptor	Day or Night	Color Vision	Location
Rods	Night vision	No	Highest concentration in the periphery of the retina
Cones	Day vision	Yes	Highest concentration in the macula and fovea

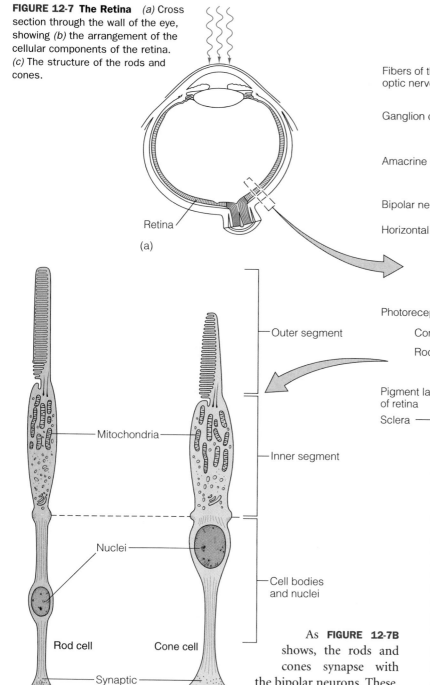

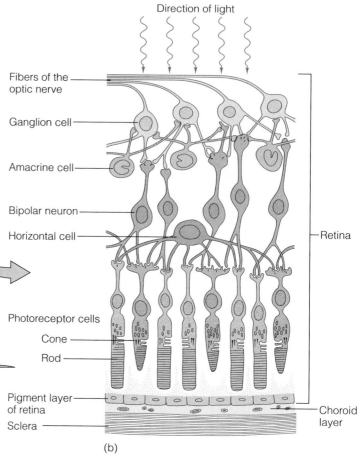

FIGURE 12-7 The Retina *(a)* Cross section through the wall of the eye, showing *(b)* the arrangement of the cellular components of the retina. *(c)* The structure of the rods and cones.

Direction of light

Fibers of the optic nerve

Ganglion cell

Amacrine cell

Bipolar neuron

Horizontal cell

Retina

Photoreceptor cells

Cone

Rod

Pigment layer of retina

Sclera

Choroid layer

(b)

Retina

(a)

Outer segment

Inner segment

Mitochondria

Nuclei

Cell bodies and nuclei

Rod cell

Cone cell

Synaptic endings

(c)

As **FIGURE 12-7B** shows, the rods and cones synapse with the bipolar neurons. These, in turn, synapse with ganglion cells. The axons of the ganglion cells unite at the back of the eye to form the **optic nerve.** The optic nerve leaves at the **optic disc,** or **blind spot,** so named because it contains no photoreceptors and is therefore insensitive to light. The blood vessels that enter and leave the eye do so with the optic nerve. These blood vessels and their branches can readily be seen by an ophthalmologist by shining a light

through the pupil onto the posterior wall of the eye (**FIGURE 12-8**).

Rods and cones are found throughout the retina, but the cones are most abundant in a tiny region of each eye lateral to the optic disc. This spot is called the **macula lutea** (MACK-you-lah LEW-tee-ah), literally "yellow spot." In the center of the macula is a minute depression, about the size of the head of a pin, known as the **fovea centralis** (FOE-vee-ah sen-TRAL-iss; meaning central depression). The fovea contains only cones. The sharpest vision occurs at the fovea because it contains the highest concentration of cones and because the bipolar neurons and ganglion cells do not cover the cones in this region as they do throughout the rest of the retina. The number of cones in the retina decreases progressively from the fovea outward, whereas the number of rods increases. Thus, the greatest concentration of rods is found in the periphery of the retina.

Images from our visual field are cast onto the retina, and impulses are transmitted to the visual cortex of the brain's occipital lobe, where they are interpreted. When we focus on an object, the image is projected upside down onto the fovea. Even though the image is up-

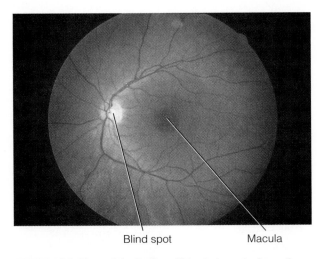

Blind spot   Macula

FIGURE 12-8 View of the Retina This photograph shows the inside back wall of the eye as seen through an ophthalmoscope. The optic disc (blind spot) and macula (site of the fovea centralis) are both indicated.

side down, the brain processes the information and gives us a right-side-up image; that is, it allows us to perceive objects in their true position.

The Lens.

Light is focused on the retina by the lens. The **lens** is a transparent, flexible structure that lies behind the iris (**FIGURE 12-9**). The lens is attached to the ciliary body by thin fibers of the **suspensory ligament.** This connection allows the smooth muscle of the ciliary body to alter the shape of the lens, an action necessary for focusing the eye.

In older individuals, the lens sometimes develops cloudy spots, or **cataracts.** The loss of transparency is especially prevalent in people who have been exposed to excessive sunlight or excessive ultraviolet light at work or elsewhere. Patients with this disease complain of cloudy vision. Looking out on the world to them is a little like looking through frosted glass. Interestingly, cataract risk may increase as the Earth's ozone layer, which blocks dangerous ultraviolet radiation from the sun, is eroded by chlorofluorocarbons from refrigerators, air-conditioning units, and other sources, as discussed in Chapter 24.

Research suggests that the color of one's eyes affects one's risk of developing cataracts. Dark-eyed people run the highest risk of cataracts. Brown- and hazel-eyed subjects had more cataracts than did blue-, gray-, and green-eyed patients. Researchers suggest that melanin in the irises of dark-eyed people may absorb solar radiation, causing more damage to the nearby lens. To lower your risk of cataracts, many eye doctors suggest wearing sunglasses with a coating that reduces ultraviolet penetration. But beware, there is evidence that some manufacturers' claims about ultraviolet protection may be inaccurate.

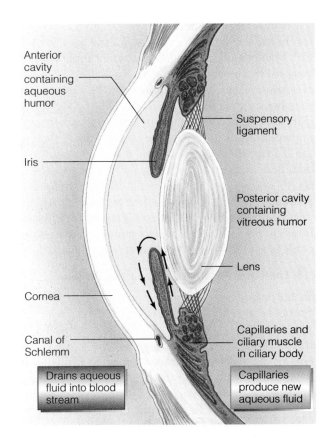

FIGURE 12-9 Detailed Cross Section of the Anterior Cavity Arrows show the flow of aqueous humor.

Eye doctors once treated cataracts surgically by removing the afflicted lens. Patients were fitted with a thick pair of glasses or a pair of contact lenses, which compensated for the missing lens. Today, however, the afflicted lenses are routinely replaced by artificial plastic lenses that provide nearly normal vision.

The lens separates the interior of the eye into two cavities of unequal sizes. Everything in front of the lens is the **anterior cavity;** everything behind it is the **posterior cavity.** The posterior cavity is filled with a clear, gelatinous material, the **vitreous humor** (VEH-tree-ous HUE-mur; "glassy liquid"). Formed during embryonic development, the vitreous humor remains throughout life, holding the lens and retina in place.

The anterior cavity contains a thin liquid, chemically similar to blood plasma, called the **aqueous humor** ("watery fluid").

Unlike the vitreous humor, the aqueous humor is constantly replaced. New fluid is produced by capillaries in the ciliary body and drains into a cavity located at the junction of the sclera and cornea. From here, the plasmalike fluid flows into the bloodstream.

The aqueous humor provides nutrients to the cornea and lens and carries away cellular wastes. In normal, healthy individuals, aqueous humor production is balanced by absorption. If the outflow is blocked, however, the aqueous humor builds up inside the anterior chamber, creating internal pressure. This disease, called **glaucoma** (glaw-COE-mah), progresses gradually and imperceptibly. If untreated, the pressure inside the eye can damage the retina and optic nerve, causing blindness. Because the incidence of glaucoma increases after age 40, doctors recommend an annual eye examination for people over 40. If diagnosed early, glaucoma can be treated with eye drops that increase the rate of drainage. In severe cases, surgery may be required to increase the outflow.

The Lens Focuses Light on the Retina

To understand how the lens operates, you must first understand a little bit about light.

Refraction.
First of all, visible light travels in waves. Light waves travel at a constant rate in any given medium such as air or water. When light passes from one medium to another, however, its velocity changes. For example, when light passes from air to a denser medium such as the cornea of the eye, it slows down. Anytime light changes speed in passing from one substance to another, it bends. The bending of light is called **refraction** (ree-FRACK-shun) (**FIGURE 12-10**).

Focusing the Image.
The lens of the eye bends incoming light rays, focusing images on the photoreceptors of the retina. Lying in front of the lens is the cornea; it also bends incoming light rays (**FIGURE 12-10**). Although we usually think of the lens as the structure that allows us to focus, most of the bending of incoming rays takes place in the cornea. However, the cornea is a fixed structure. Like the lens on a fixed-focus camera, the cornea cannot be adjusted to focus on nearby objects. Without the adjustable lens, the eye would be unable to focus on objects close at hand.

The lens is a flexible structure whose shape is controlled by the muscles in the cil-

FIGURE 12-10 Refraction The pencil in the glass of water appears to bend. Actually, the light rays coming to our eyes are bent as they pass through the water and glass, making the pencil appear bent.

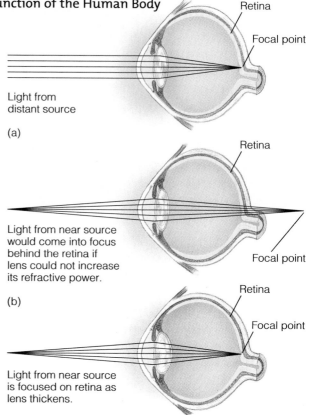

FIGURE 12-11 Refraction of Light by the Cornea and Lens
(a) Light rays from distant objects are parallel when they strike the eye. The refractive power of the cornea and resting lens are sufficient to bring them into focus on the retina. *(b)* Light rays from nearby objects are divergent. The cornea has a fixed refractive power and cannot change. The rays would focus behind the retina if the lens could not alter its refractive power. *(c)* To focus the image, the ciliary muscles contract. This lessens pressure on the suspensory ligaments, which allows the lens to thicken and shorten, becoming more curved and more refractive.

iary body that are attached via the suspensory ligament. When the muscles of the ciliary body are relaxed, the suspensory ligament is taut, and the lens is somewhat flattened. As **FIGURE 12-11A** shows, light from distant objects comes to the eye as nearly parallel rays. The fixed refractive power of the cornea and the refractive power of the lens in its relaxed state are sufficient to bend these beams to bring them into focus on the retina.

As **FIGURE 12-11B** shows, light rays from nearby objects are divergent. To focus on nearby objects, the lens must become more curved. When the eye focuses on nearby objects, the ciliary smooth muscle contracts, causing the suspensory ligament to relax (**FIGURE 12-12B**). The lens thickens and shortens, becoming more curved, in much the same way that a rubber ball flattened between your hands will return to normal when you reduce the pressure on it. This automatic adjustment in the curvature of the lens as it focuses on a nearby object is called **accommodation** (**FIGURE 12-12**). Accommodation increases the refractive (bending) power of the lens.

Accommodation Is Enhanced by Pupillary Constriction.

As the eyes focus on a nearby object, the pupils constrict. Pupillary constriction is a reflex that eliminates divergent rays of light that would otherwise strike the periphery of the lens. Without pupillary constriction, images of nearby objects would be quite blurred, for the lens would be unable to bend these rays sufficiently to bring them into focus on the retina.

Synchronized Eye Movement and Convergence.

Six muscles located outside the eye, the extrinsic eye muscles, are responsible for eye movement (**FIGURE 12-13**). As noted earlier, these muscles attach to the orbit (the bony eye socket) and to the sclera (the white of the eye).

The eyes generally move in unison like a pair of synchronized swimmers. Synchronized movement is an evolutionary adaptation that ensures that images are focused on the foveas of both eyes at the same time. To test the synchronized movement, close one of your eyes. Place an index finger gently over the closed lid. Then hold the other hand in front of your face and move it back and forth, then up and down, following it with your opened eye. You should feel your closed eye moving in sync.

When a nearby object is viewed, the eyes turn inward—that is, they converge. **Convergence** ensures that near images are focused on each fovea. Because convergence occurs during all near-point work—reading, writing, sewing, knitting—it tends to strain the extrinsic eye muscles, creating eyestrain.

▮ Alterations in the Shape of the Lens and Eyeball Cause the Most Common Visual Problems

In a relaxed eye with perfect vision, objects farther than 6 meters (20 feet) away fall into perfect focus on the retina (**FIGURE 12-14A**). Many individuals, however, have imperfectly shaped eyeballs or defective lenses. These imperfections result in three visual problems: nearsightedness, farsightedness, and astigmatism.

Myopia.

Nearsightedness or myopia (my-OH-pee-ah), results when the eyeball is slightly elongated (**FIGURE 12-14B**). Without corrective lenses, parallel light rays arising from distant images will fall into focus in front of the retina,

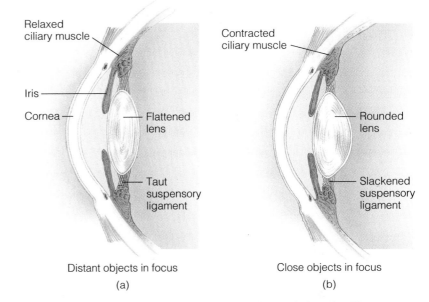

FIGURE 12-12 Accommodation *(a)* The lens is flattened when the ciliary muscles are relaxed. *(b)* When the ciliary muscles contract, tension on the suspensory ligaments is reduced and the lens shortens and thickens.

FIGURE 12-13 Extrinsic Eye Muscles These muscles move the eye in all directions. They attach to the bony orbit and the sclera.

creating a rather fuzzy image. In contrast, nearby images with much more divergent light rays tend to be in focus in the uncorrected eye. People with myopia, therefore, can see near objects without corrective lenses—hence the name nearsightedness. Myopia may also result when the lens is too strong—that is, too concave. This lens bends the light too much, causing the image to come into focus in front of the retina.

Myopia is quite common. Approximately one of every five Americans needs glasses to correct for it. Myopia is caused by many factors and tends to run in families; it generally appears around age 12, often worsening until a person reaches 20.

Myopia can be corrected by contact lenses or prescription glasses that cause incoming light rays to diverge (**FIGURE 12-14B**). Contact lenses fit on the surface of the cornea and bend the incoming light rays outward, thereby compensating for the shape of the eye or a defective lens.

Nearsightedness (myopia)

www.jbpub.com/humanbiology

FIGURE 12-14 Common Visual Problems

(a) Normal eye

Correction

Focal plane

None required

Parallel light rays from distant object

(b) Myopic eye (nearsighted)

Concave lens

Divergent light rays from near object

(c) Hyperopic eye (farsighted)

Convex lens

Hyperopia.

Hyperopia (HIGH-per-OPE-ee-ah), or **farsightedness,** is the opposite of myopia. It results when the eyeballs are too short or the lens is too weak (too convex). In the eyes of farsighted individuals, parallel light rays from distant objects usually fall into focus on the retina, but divergent rays from nearby objects cannot be focused sufficiently. Without corrective lenses, farsighted individuals see distant objects well, but nearby objects are fuzzy. Glasses or contact lenses that bend the light inward bring near objects into sharp focus on the retina (**FIGURE 12-14C**).

Hyperopia is generally present from birth and is usually diagnosed during childhood. Like myopia, it tends to run in families.

Astigmatism.

The cornea and lens have uniformly curved surfaces, but either of these surfaces can be slightly disfigured. The surface of the cornea, for example, can have a slightly different curvature in the vertical plane than it does in the horizontal plane. This unequal curvature is called an **astigmatism** (a-STIG-mah-tiz-em). It creates fuzzy images because light rays are bent differently by the different parts of the cornea. Astigmatism is usually present from birth and does not grow worse with age. It can also be corrected with glasses or contact lenses.

Radial Keratotomy and Laser Surgery.

Surgeons have also developed a method to correct nearsightedness, called **radial keratotomy (RK)** (RAY-dee-al CARE-ah-TOT-oh-mee). To correct myopia, numerous small, superficial incisions are made in the cornea, radiating from the center like the spokes of a bicycle wheel. This procedure flattens the cornea and reduces its refractive power, causing the rays to diverge and come into focus on the back of the retina.

To correct astigmatism, surgeons make curved or straight incisions in the cornea that help create a round cornea. A round cornea eliminates the blurring caused by distortions in the cornea, as described earlier. Radial keratotomy patients generally recover quickly and suf-

Although they're useful in correcting vision, contact lenses can result in serious problems. The largest study of their use in the United States suggests that extended-wear contact lenses, those worn overnight and for up to a week at a time, carry a far greater risk of serious complications than daily-wear lenses. This study of over 22,000 lens wearers showed that sight-threatening complications, such as corneal abrasion, growth of blood vessels into the cornea, corneal ulcers, and severe corneal scarring, were two to four times more prevalent in users of extended-wear contacts than in those using daily-wear lenses. In fact, researchers found that 1 in 2000 daily-wear lens users developed corneal ulcers; and 1 in 300 suffered from other serious complications. In extended-wear users, however, 1 in 500 suffered from corneal ulcers, and 1 in 100 had other serious reactions, suggesting that you should think carefully before choosing extended-wear contacts.

Farsightedness (hyperopia)

fer minimal discomfort. Ninety percent of all RK patients achieve a visual acuity of 20/40 or better. However, the deep incisions required for RK weaken the structural integrity of the eye and make the eye susceptible to traumatic injury. Other problems have also been reported.

Surgeons are now also using lasers to correct myopia—a procedure only recently approved for use in the United States. Lasers reshape the surface of the cornea and reduce the amount of refractive error. Worldwide, hundreds of thousands of patients have been successfully treated with this procedure. Unlike radial keratotomy, in which incisions are made deep into the cornea, excimer laser refractive keratectomy (PRK), as it is called, only shaves off microscopic amounts of the cornea, ensuring better structural integrity of the eye. The computer ensures better control of the procedure than RK. In addition, the procedure appears to be more successful than radial keratotomy as well, with 95% of patients achieving 20/40 visual acuity or better. It seems to take longer than RK for the eye to heal, however.

Eyestrain.

For most of human evolution, people have used their eyes principally for viewing objects at a distance—for example, watching their children play or wild animals roam. Near-point vision probably occupied little of their time. It should come as no surprise, then, to learn that the human eye is best suited for distance vision. Today, however, near-point work is becoming commonplace. This strains the eyes and can cause a progressive deterioration of eyesight. Those who read a lot or spend long periods staring at computer monitors (like textbook authors) often become more nearsighted as they become older. No one knows why, but research suggests that the eye may elongate as a result of constant near-point use.

To reduce eyestrain and the deterioration of eyesight, eye doctors advise that computer operators look away from their screens and that readers look up from their materials regularly, letting their eyes focus on distant objects. This action relaxes the ciliary muscles, reducing eyestrain. (Try it and see if you can feel a difference.) Some ophthalmologists suggest that computer users "blink at every period and look up after every paragraph."

Presbyopia.

The aging process brings with it many joys, among them hair loss, hearing loss, and arthritis. Aging also results in a decline in the resiliency of the lens. Thus, when the ciliary muscles contract to allow one to focus on a nearby object,

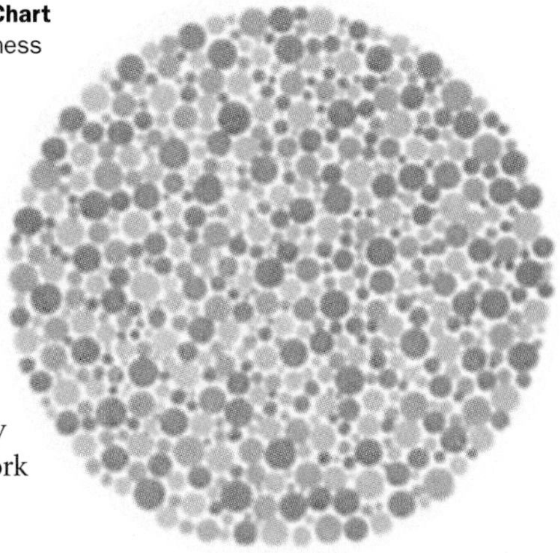

FIGURE 12-15 Color-Blindness Chart
People with red-green color blindness cannot detect the number 29 in this chart.

the lens responds slowly or only partially, making it difficult to focus. This condition, known as **presbyopia** (PREZ-bee-OPE-ee-a), usually begins around the age of 40 and can be corrected by glasses worn for near-point work such as reading.[4]

Color Blindness.

About 5% of the human population suffers from color blindness. **Color blindness** is a hereditary disorder, more prevalent in men than women. Characterized by a deficiency in color perception, color blindness ranges from an inability to distinguish certain shades to a complete inability to perceive color. The most common form of this disorder is red-green color blindness.

In individuals with red-green color blindness, the red or green cones may either be missing altogether or be present in reduced number. If the red cones are missing, red objects appear green. If the green cones are missing, however, green objects appear red. Color blindness can be detected by simple tests (**FIGURE 12-15**).

Many color-blind people are unaware of their condition or untroubled by it. They rely on a variety of visual cues such as differences in intensity to distinguish red and green objects. They also rely on position cues. In vertical traffic lights, for example, the red light is always at the top of the signal; green is on the bottom. Although the colors may appear more or less the same, the position of the light helps color-blind drivers determine whether to hit the brakes or step on the gas.

Overlapping Visual Fields Give Us Depth Perception

Human eyes are located in the front of the skull, looking forward, much like those of other predatory animals. Each eye has a visual field of about 170 degrees, but the visual fields overlap considerably (**FIGURE 12-16**). This overlapping gives us the ability to judge the relative position of objects in our visual field—that is, it gives us **depth perception,** or **stereoscopic vision.**

---

[4]You may have seen your parents or grandparents holding a phone book at arm's length to read it. They probably argued that their eyes weren't going bad, their arms were just too short!

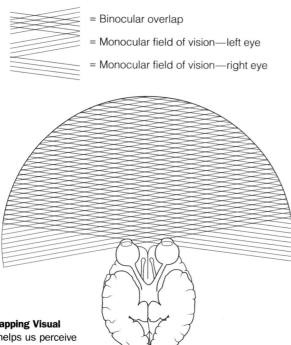

= Binocular overlap

= Monocular field of vision—left eye

= Monocular field of vision—right eye

FIGURE 12-16 Overlapping Visual Fields The overlap helps us perceive depth—three-dimensional relationships.

TABLE 12-4		
Structures and Functions of the Ear		
Part	**Structure**	**Function**
Outer ear	Auricle	Funnels sound waves into external auditory canal
	Ear lobe	
	External auditory canal	Directs sound waves to the eardrum
Middle ear	Tympanic membrane, or eardrum	Vibrates when struck by sound waves
	Ossicles	Transmit sound to the cochlea in the inner ear
Inner ear	Cochlea	Converts fluid waves to nerve impulses
	Semicircular canals	Detect head movement
	Saccule and utricle	Detect head movement and linear acceleration

Section 12-5

Hearing and Balance: The Cochlea and Middle Ear

The human ear is an organ of special sense. It serves two functions: it detects sound, and it detects body position, enabling us to maintain balance (**TABLE 12-4**).

The Ear Consists of Three Anatomically Separate Portions: The Outer, Middle, and Inner Ears

The Outer Ear.
The **outer ear** consists of an irregularly shaped piece of cartilage covered by skin, the **auricle** (OR-eh-kul), and the earlobe, a flap of skin that hangs down from the auricle (**FIGURE 12-17**). The outer ear also consists of a short tube, the **external auditory canal,** which transmits airborne sound waves to the middle ear (**FIGURE 12-17A**). The external auditory canal is lined by skin containing modified sweat glands that produce earwax. Earwax traps foreign particles, such as bacteria, and contains a natural antibiotic substance that may reduce ear infections.

The Middle Ear.
The **middle ear** lies entirely within the temporal bone of the skull (**FIGURE 12-17B**). The eardrum, or **tympanic membrane** (tim-PAN-ick), separates the middle ear

cavity from the external auditory canal. The tympanic membrane vibrates when struck by sound waves in much the same way that a guitar string vibrates when a note is sounded by another nearby instrument.

Inside the middle ear are three minuscule bones, the **ossicles** (OSS-eh-kuls). Starting from the outside, they are the **malleus** (pronounced MAL-ee-us; hammer), **incus** (IN-cuss; anvil), and **stapes** (STAY-pees; stirrup). As illustrated in **FIGURE 12-17B**, the hammer-shaped malleus abuts the tympanic membrane. When the membrane is struck by sound waves, it vibrates. This causes the malleus to rock back and forth. The malleus, in turn, causes the incus to vibrate, which causes the stapes, the stirrup-shaped bone, to move in and out against the **oval window,** an opening to the inner ear covered with a membrane like the skin on a drum. Thus, vibrations created in the eardrum are amplified as they are transmitted to the inner ear, where the sound receptors are located.

As **FIGURE 12-17** illustrates, the middle ear cavity opens to the pharynx via the **auditory,** or **eustachian, tube** (you-STAY-shun). The eustachian tube serves as a pressure valve. Normally, the eustachian tube is closed. Yawning and swallowing, however, cause it to open, allowing air to flow into or out of the middle ear cavity. This equalizes the internal and external pressure on the eardrum, as you will notice when taking off in an airplane.

Scuba divers and swimmers can sustain considerable damage to an eardrum as they ascend or descend.

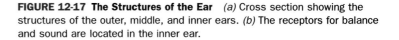
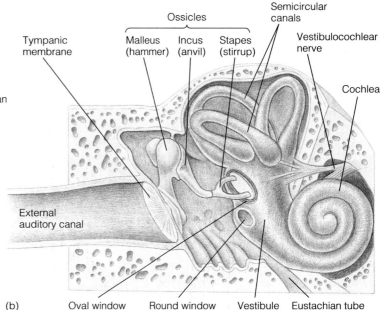

FIGURE 12-17 The Structures of the Ear   *(a)* Cross section showing the structures of the outer, middle, and inner ears. *(b)* The receptors for balance and sound are located in the inner ear.

As a diver descends, for instance, pressure from the water builds, pushing the eardrum inward. To prevent the eardrum from tearing, air must be forced into the middle ear cavity. This can be done by holding one's nose, clamping one's mouth shut, and gently blowing. Air is forced through the eustachian tube into the middle ear cavity. When a diver ascends, just the opposite happens: Air pressure increases inside the middle ear cavity. The diver must release the pressure or else suffer a broken eardrum. This release usually occurs quite naturally, but it can be facilitated by yawning.

The Inner Ear.

The inner ear occupies a large cavity in the temporal bone and contains two sensory organs, the cochlea and the vestibular apparatus. The **cochlea** (COE-klee-ah) is shaped like a snail shell and houses the receptors for hearing. The **vestibular apparatus** consists of two parts: the semicircular canals and the vestibule (**FIGURE 12-17B**). The **semicircular canals** are three ringlike structures set at right angles to one another. They house receptors for body position and movement. The **vestibule** is a bony chamber lying between the cochlea and semicircular canals. It also houses receptors that respond to body position and movement.

▇ Hearing Requires the Participation of Several Structures

As noted previously, sound waves enter the external auditory canal, where they strike the tympanic membrane, or eardrum. The eardrum vibrates back and forth, causing the ossicles of the middle ear cavity to vibrate. The ossicles, in turn, transmit sound waves to the cochlea, which houses the receptor for sound.

Structure and Function of the Cochlea.

The cochlea is a hollow, bony spiral. A cross section through this remarkable structure reveals three fluid-filled canals (**FIGURE 12-18A**). Separating the middle canal from the lowermost one is a flexible membrane called the **basilar membrane** (BAZ-eh-ler). It supports the **organ of Corti,** the receptor organ for sound.

The organ of Corti contains receptor cells, the **hair cells** (**FIGURE 12-18B**). Cellular projections from the hair cells contact the overlying **tectorial membrane** (teck-TORE-ee-al). When sound waves are transmitted from the middle ear to the inner ear, they create pressure waves in the fluid in the uppermost canal of the cochlea, the **vestibular canal.** As shown in **FIGURE 12-19A,** in which the cochlea is unwound, the stirrup transmits vibrations to the drumlike **oval window,** the opening in the cochlea. Fluid pressure waves created in the vestibular canal then travel through the vestibular membrane into the middle canal, the **cochlear duct (canal).** From here, the pressure waves pass through the basilar membrane into the lowermost canal, the **tympanic canal.** Pressure is relieved by the outward bulging of the **round window,** an opening in the bony cochlea, which, like the oval window, is spanned by a flexible membrane.

In their course through the cochlea, the pressure waves cause the basilar membrane to vibrate. This vibration stimulates the hair cells. The hair cells respond by releasing a neurotransmitter, which stimulates the dendrites wrapped around their bases. The nerve impulses travel to the auditory cortex via the **vestibulocochlear**

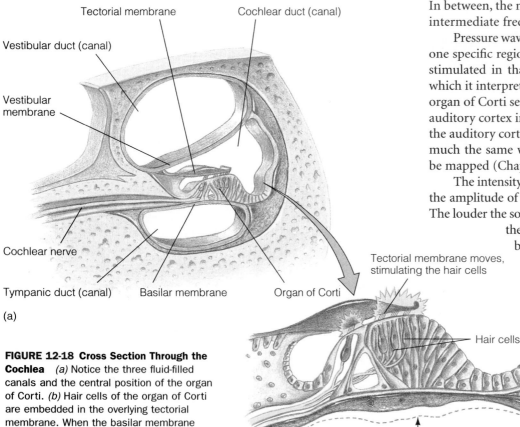

Tectorial membrane

Cochlear duct (canal)

Vestibular duct (canal)

Vestibular membrane

Cochlear nerve

Tympanic duct (canal)   Basilar membrane   Organ of Corti

(a)

Tectorial membrane moves, stimulating the hair cells

Hair cells

Deflection of basilar membrane, because of fluid movements in the cochlea, stimulates the new cells

(b)

FIGURE 12-18 Cross Section Through the Cochlea *(a)* Notice the three fluid-filled canals and the central position of the organ of Corti. *(b)* Hair cells of the organ of Corti are embedded in the overlying tectorial membrane. When the basilar membrane vibrates, the hair cells are stimulated.

In between, the membrane responds to a wide range of intermediate frequencies.

Pressure waves caused by any given sound stimulate one specific region of the organ of Corti. The hair cells stimulated in that region send impulses to the brain, which it interprets as a specific pitch. Each region of the organ of Corti sends impulses to a specific region of the auditory cortex in the temporal lobe of the brain. Thus, the auditory cortex can be mapped according to tone in much the same way the motor and sensory cortex can be mapped (Chapter 11).

The intensity of a sound, or its loudness, depends on the amplitude of the vibration in the basilar membrane. The louder the sound, the more vigorous the vibration of the eardrum. The more vigorous the vibration of the eardrum, the greater the deflection of the basilar membrane in the area of peak responsiveness. The greater the deflection of the basilar membrane, the more hair cells are stimulated. Loud rock music or sirens, however, cause extreme vibrations in the basilar membrane, destroying hair cells over time and causing partial deafness.

Hearing Loss.

As people grow older, many lose their hearing, but hearing loss usually occurs so slowly that most people are unaware of it. In some cases, though, people lose their hearing suddenly. A loud explosion, for example, can damage the hair cells or even break the ossicles.

Hearing losses may be temporary or permanent, partial or complete. Hearing loss falls into one of two categories, depending on the part of the system that is affected. The first is **conduction deafness,** which occurs when the conduction of sound waves to the inner ear is impaired. Conduction deafness may result from excessive earwax in the auditory canal or a rupture of the eardrum. Damage to the ossicles and rupture of the oval window are additional causes.

Conduction deafness most often results from infections in the middle ear. Bacterial infections may result in the buildup of scar tissue that causes the ossicles to fuse and lose their ability to transmit sound. Infections of the middle ear usually enter through the eustachian tube. Thus, a sore throat or a cold can easily spread to the ear, where it requires prompt treatment. Ear infec-

nerve (vess-TIB-you-low-COE-klee-er), one of the 12 cranial nerves.

Distinguishing Pitch and Intensity.

A pitch pipe helps a singer find the note to begin a song. It has a range of notes from high to low. The ear can distinguish between these various pitches, or frequencies, in large part because of the structure of the basilar membrane. The basilar membrane underlying the organ of Corti is stiff and narrow at the oval window, where fluid pressure waves are first established inside the cochlea (**FIGURE 12-19B**). As the basilar membrane proceeds to the apex of the spiral, however, it becomes wider and more flexible. The change in width and stiffness results in marked differences in its ability to vibrate. The narrow, stiff end, for example, vibrates maximally when pressure waves from high-frequency sounds ("high notes") are present (**FIGURE 12-19C**). The far end of the membrane vibrates maximally with low-frequency sounds ("low notes").

FIGURE 12-19 The Transmission of Sound Waves Through the Cochlea The cochlea is unwound here to simplify matters. *(a)* Vibrations are transmitted from the stirrup (stapes) to the oval window. Fluid pressure waves are established in the vestibular canal and pass to the tympanic canal, causing the basilar membrane to vibrate. *(b)* A representation of the basilar membrane, showing the points along its length where the various wavelengths of sound are perceived. Notice that the basilar membrane is narrowest at the base of the cochlea at the oval window end and widest at the apex. *(c)* High-frequency sounds set the basilar membrane near the base of the cochlea into motion. Hair cells send impulses to the brain, which interprets the signals as a high-pitch sound. Low-frequency sounds stimulate the basilar membrane where it is widest and most flexible.

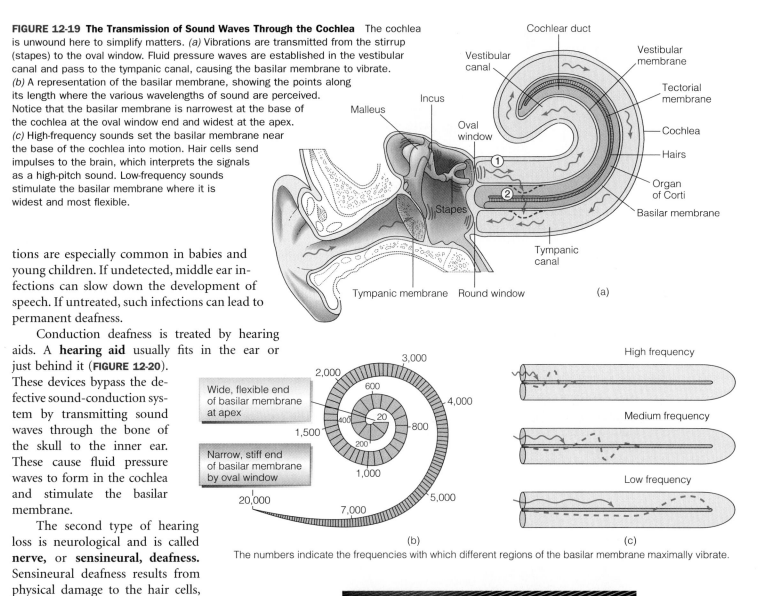

The numbers indicate the frequencies with which different regions of the basilar membrane maximally vibrate.

tions are especially common in babies and young children. If undetected, middle ear infections can slow down the development of speech. If untreated, such infections can lead to permanent deafness.

Conduction deafness is treated by hearing aids. A **hearing aid** usually fits in the ear or just behind it (**FIGURE 12-20**). These devices bypass the defective sound-conduction system by transmitting sound waves through the bone of the skull to the inner ear. These cause fluid pressure waves to form in the cochlea and stimulate the basilar membrane.

The second type of hearing loss is neurological and is called **nerve,** or **sensineural, deafness.** Sensineural deafness results from physical damage to the hair cells, the vestibulocochlear nerve, or the auditory cortex. Explosions, extremely loud noises, and some antibiotics can all damage the hair cells in the organ of Corti, creating partial to complete deafness. The auditory nerve, which conducts impulses from the organ of Corti to the cortex, may degenerate, thus ending the flow of information to the cortex. Tumors in the brain or strokes can destroy the cells of the auditory cortex.

Although the ear is vulnerable to loud noise and other problems, it contains a mechanism to protect itself from damage. This mechanism consists of two tiny skeletal muscles (the smallest in the body) located in the middle ear cavity. One of these muscles inserts on the malleus, and the other attaches to the stapes. As noted earlier, the malleus attaches to the eardrum, and the stapes attaches to the oval window. Loud noises stimulate

FIGURE 12-20 Hearing Aids Worn by people with conduction deafness, hearing aids send sound impulses through the bone of the skull to the cochlea.

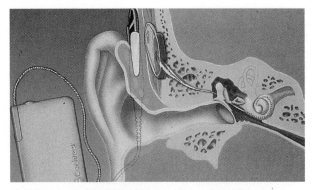

FIGURE 12-21 Cochlear Implant   The cochlear implant can correct for nerve deafness. Electrodes convey electrical impulses from a small microphone mounted in the ear to the auditory nerve.

a reflex contraction of the middle ear muscles, pulling them away from their membrane contacts. Whenever an individual is exposed to loud noise, this reflex reduces the conduction of the noise to the inner ear. Unfortunately, the reflex requires about 40 milliseconds, so it cannot protect the ear from explosions.

Correcting Profound Deafness.

More than 2 million Americans are profoundly deaf. Until recently, this condition was considered virtually untreatable.

Children who are born deaf or are deafened before they begin to speak often fail to mature emotionally.

Even reading comprehension can be impaired. Some profoundly deaf children, in fact, never advance beyond third- or fourth-grade reading levels.

Hearing aids usually cannot help individuals who are born deaf or those who suffer from nerve damage. Researchers, however, have developed a device, called a **cochlear implant**, which simulates the function of the inner ear remarkably (**FIGURE 12-21**). This device picks up sound and transmits it to a receiver implanted inside the skull. The signal then travels to an electrode implanted in the vestibulocochlear nerve. Electrical impulses in the electrode stimulate the nerve, creating impulses that travel to the auditory cortex.

Today, hundreds of adults and children are equipped with cochlear implants, which detect and transmit a wider range of sounds. Recipients of the new models can perceive many distinct words. Some individuals equipped with these devices have apparently developed a remarkable ability to perceive sounds. Cochlear implants also help the deaf monitor and regulate their voices and make lip reading easier.

The Vestibular Apparatus Houses Receptors That Detect Body Position and Movement

The cochlea lies next to the **vestibular apparatus.** As noted earlier, it consists of the two parts, the semicircular canals and the vestibule. The vestibule is a bony chamber between the cochlea and semicircular canals. The vestibular apparatus houses receptors that detect body position and movement (**FIGURE 12-22**).

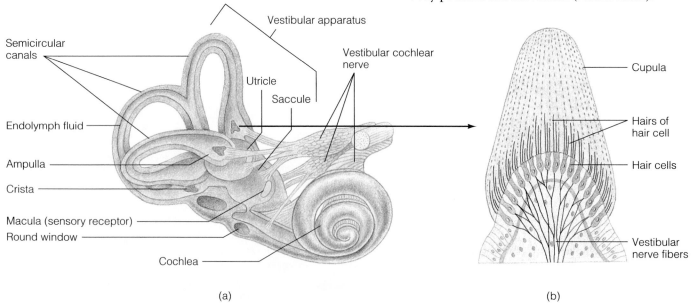

(a)   (b)

FIGURE 12-22 Vestibular Apparatus and Semicircular Canals
(a) This illustration shows the location of the cristae in the ampullae of the semicircular canals. The semicircular canals are filled with endolymph. (b) When the head spins, the endolymph is set into motion, deflecting the gelatinous cupula of the crista, thus stimulating the receptor cells.

The Semicircular Canals.

The three semicircular canals are arranged at right angles to one another. Each canal is filled with a fluid called **endolymph.** As **FIGURE 12-22A** shows, the base of each semicircular canal expands to form the **ampulla** (AM-pew-lah). On the inside wall of each ampulla is a small ridge of tissue, or **crista** (CHRIS-tah). Each crista consists of a patch of receptor cells. Each receptor cell contains numerous microvilli and a single cilium embedded in a cap of gelatinous material, the **cupula** (CUE-pew-lah), shown in **FIGURE 12-22B.** Dendrites of sensory nerves wrap around the base of the receptor cells, and the cupula extends into the cavity of the ampulla, where it is bathed in endolymph.

Rotation of the head causes the endolymph in the semicircular canals to move. The movement of the fluid deflects the cupula, which stimulates the hair cells to release a neurotransmitter. This excites the sensory neurons, which send impulses to the brain, alerting it to the rotational movement of the head and body.

Because the semicircular canals are set in all three planes of space, movement in any direction can be detected. By alerting the brain to rotation and movement, the semicircular canals contribute to our sense of balance. They are therefore important to **dynamic balance—** helping us stay balanced when in movement.

The Utricle and Saccule.

Two additional receptors play a role in balance and the detection of movement, the **utricle** (YOU-treh-kul) and **saccule** (SACK-yule) (**FIGURE 12-22**). These membranous compartments inside the vestibule provide input under two distinctly different conditions: at rest and during acceleration in a straight line. As a result, these receptors contribute dynamic balance and **static balance—**helping us stay balanced when we are not moving.

The utricle and saccule contain small receptor organs akin to the cristae of the semicircular canals. The nerve impulses generated by the receptors in the semicircular canals and the utricle and saccule are sent to a cluster of nerve cell bodies in the brain stem. Here all of the information the receptors generate on position and movement is integrated with input from the eyes and from receptors in the skin, joints, and muscles (earlier in this chapter). From this center, information flows in many directions. One major pathway leads to the cerebral cortex. This path makes us conscious of our position and movement. Another path leads to the muscles of the limbs and torso. Signals to the muscles maintain our balance and, if necessary, correct body position.

In some people, activation of the vestibular apparatus results in motion sickness, characterized by dizziness and nausea. The exact cause of motion sickness is not known.

Health, Homeostasis, and the Environment

Noise Pollution

The nervous system is essential to homeostasis because it regulates many activities aimed at maintaining a stable internal environment. This, in turn, depends on receptors that provide input essential for homeostasis. While it is fairly obvious that the general senses are involved in regulating key physiological functions essential to homeostasis, what about the special senses?

The special senses help us—and other animals— acquire food, defend ourselves against danger, and affect other behavioral responses that are essential to homeostasis. Damage to these senses may therefore have profound effects. Consider noise.

Noise is one of the most widespread environmental pollutants in modern industrial societies. It may be turning industrial nations into countries of the hearing-impaired. Traffic noise, airport noise, loud music, crowd noise, and other loud sounds so common in modern society may be slowly destroying our hearing. By the time many New York City residents reach the age of 20, their hearing has been so impaired that they hear only as well as a 70-year-old African Bushman who has lived a life free of city noises.

For many years, scientists attributed the decline in hearing to middle ear infections, certain antibiotics, and the natural deterioration of the hair cells. New research, however, suggests that noise is probably the principal cause of hearing loss in modern societies.

Like that of any other pollutant, the damage caused by noise is related to two factors: exposure level (how loud a noise is) and the duration of exposure (how long one is exposed to it). In general, the louder the noise, the more damaging it is. In addition, the longer you are exposed to a damaging level of noise, the more hearing loss you will suffer.

The noise to which people are exposed in factories and at construction sites is sufficient to cause gradual hearing loss. A worker may notice a dulled sense of hearing after working in a noisy environment; this is called a **temporary threshold shift.** Over time, the continued assault leads to a **permanent threshold shift,** complete hearing loss in certain frequencies. In most cases, hearing loss occurs so gradually that workers do not notice it until it is too late.

TABLE 12-5

The Decibel Scale

Sound Level, dB	Sound Sources	Effects		
		Perceived Loudness	Damage to Hearing	Community Reaction to Outdoor Noise
180—	Rocket engine			
170—				
160—				
150—	Jet plane at takeoff	Painful	Traumatic injury	
140—			Injurious range irreversible damage	
130—	Maximum recorded rock music			
120—	Thunderclap			
	Auto horn, 1 meter away	Uncomfortably loud		
110—	Riveter		Danger zone; progressive loss of hearing	
	Jet flying over at 300 meters			
100—	Newspaper press			
90—	Motorcyle, 8 meters away			Vigorous action
	Food blender			
	Diesel truck, 80 km/hr, 15 meters	Very loud	Damage begins after long exposure	
80—	Garbage disposal			Threats
70—	Vacuum cleaner			
	Ordinary conversation			
		Moderately loud		Widespread complaints
60—	Air-conditioning unit, 6 meters			
	Light traffic noise, 30 meters			Occasional complaints
50—	Average living room			
40—	Bedroom	Quiet		No action
	Library			
30—	Soft whisper			
20—	Broadcasting studio	Very quiet		
10—	Rustling leaf			
		Barely audible		
0—	Threshold of hearing			

SOURCE: Table from *Environmental Science*, 3rd ed., by Jonathan Turk and Amos Turk. Copyright © 1984 by Saunders College Publishing. Reproduced by permission of the publisher.

The intensity of sound is measured in **decibels** (DESS-seh-bells) or **dB. TABLE 12-5** lists the dB level of some common sounds. Surprising new research shows that continuous exposure to sounds over 55 decibels can result in hearing loss. Light traffic and an air conditioner operating 6 meters (18 feet) from you are 60-decibel sounds.

Besides deafening us to the important sounds of modern life, noise disturbs communications, rest, and sleep. The latter two are particularly important to homeostasis and health. Noise also raises our level of stress, which, in turn, shortens our lives. When hearing is impaired, communication falters, and tensions often rise. People who are losing their hearing complain that they feel inadequate in social situations.

Hearing loss is not inevitable. You can take steps to protect your hearing. Keep your stereo at a reasonable level. Avoid noisy events. Cover your ears when an ambulance or fire engine approaches. Wear ear plugs or ear guards when operating firearms or noisy equipment such as vacuum cleaners, chain saws, or construction equipment. Get treatment if you develop an ear infection. Prevention is the best medicine, because once you lose your hearing, it is gone forever, and so is an important part of your life.

SUMMARY

THE GENERAL AND SPECIAL SENSES

1. The body contains two types of receptors—those that provide general senses and those that provide special senses.

THE GENERAL SENSES

2. The *general senses* are pain, light touch, pressure, temperature, and position sense. Receptors for these senses may be naked or encapsulated nerve endings.

3. Receptors fit into five functional categories: (a) *mechanoreceptors*, (b) *thermoreceptors*, (c) *photoreceptors*, (d) *chemoreceptors*, and (e) *pain receptors (nociceptors)*.

4. Sensory stimuli may be internal or external. Some stimuli cause reflex actions. Still others may stimulate physiological changes.

5. The *naked nerve ending receptors* in the body include those stimulated by pain, light touch, and temperature. The *encapsulated receptors* include those that detect pressure *(Pacinian corpuscles)*, light touch *(Meissner's corpuscles,* Krause's end-bulbs, and Ruffini's corpuscles), and muscle extension *(muscle spindles* and *Golgi tendon organs)*.

6. Many sensory receptors cease responding to prolonged stimuli, a phenomenon called *sensory adaptation*. Pain, temperature, pressure, and olfactory receptors all adapt.

TASTE AND SMELL: THE CHEMICAL SENSES

7. The *special senses* include taste, smell, vision, hearing, and balance, which are made possible by more elaborate receptor organs.

8. Taste receptors called *taste buds* are located principally on the upper surface of the tongue. Taste buds contain receptor cells. Food molecules dissolve in the saliva and bind to the membranes of the microvilli of the receptor cells.

9. Taste buds respond to four flavors: salty, bitter, sweet, and sour.

10. The receptors for smell are located in the *olfactory membrane* in the roof of the nasal cavities. The receptor cells in the olfactory membrane respond to thousands of different molecules, which bind to membrane receptors on the olfactory hairs, stimulating nerve impulses that are transmitted to the brain via the olfactory nerve.

THE VISUAL SENSE: THE EYE

11. The *eye* is the receptor for visual stimuli and is located in the orbit, a bony socket.

12. The wall of the human eye consists of three layers. The outermost layer consists of the *sclera* (the white of the eye) and the *cornea* (the clear anterior structure that lets light shine in).

13. The middle layer consists of the *choroid* (the pigmented and vascularized region), the *ciliary body* (whose muscles control the lens), and the *iris* (which controls the amount of light entering the eye).

14. The innermost layer is the *retina,* the light-sensitive layer, which contains two types of photoreceptors: rods and cones. *Rods* function in dim light and provide black-and-white vision, and *cones* operate in bright light and provide color vision. Cones are also responsible for visual acuity. The highest concentration of cones is found in the *fovea centralis.*

15. Light is focused on the retina by the cornea and *lens*. The cornea has a fixed refractive power, but the lens can be adjusted to bend light according to need. The muscles of the ciliary body play an important role in this process. Focusing on near objects is aided by pupillary constriction.

16. The lens may become cloudy with old age or because of exposure to excess ultraviolet radiation. This condition, called *cataracts,* can be corrected by surgically removing the lens and replacing it with a plastic one.

17. The eye is divided into two cavities. The *posterior cavity* lies behind the lens and is filled with a gelatinous material,

the *vitreous humor*. The *anterior cavity* is filled with a plasmalike fluid called the *aqueous humor,* which nourishes the lens and other eye structures in the vicinity. If the rate of absorption of aqueous humor decreases, however, pressure can build inside the anterior cavity, resulting in *glaucoma*.

18. As people age, the lens becomes less resilient and less able to focus on nearby objects, a condition called *presbyopia*.

19. *Myopia*, or nearsightedness, results from a lens that is too strong (too concave) or from an elongated eyeball. In the uncorrected eye, the image from distant objects comes in focus in front of the retina.

20. *Hyperopia* results from a weak lens (too convex) or from a shortened eyeball. In the uncorrected eye, light rays from an image nearby would come into focus behind the retina.

21. *Astigmatism* is an irregularly curved lens or cornea that creates fuzzy images.

22. Three types of cones are present in the eye: red, green, and blue. Each type responds maximally to one specific color of light. Intermediate colors activate two or more types of cones.

23. *Color blindness* is a genetic disorder, more common in men than women. It results from a deficiency or an absence of one or more types of cones. Red-green color blindness is the most common type.

HEARING AND BALANCE: THE COCHLEA AND MIDDLE EAR

24. The human ear consists of three portions: the outer, middle, and inner ears.

25. The *outer ear* consists of the *auricle* and *external auditory canal,* both of which direct sound to the *eardrum*.

26. The *middle ear* consists of the eardrum and three small bones, the ossicles, which transmit vibrations to the inner ear.

27. The *auditory tube* equilibrates the pressure inside the middle ear cavity.

28. The *inner ear* contains the cochlea where the receptors for sound are located. The inner ear also houses receptors for movement and head position: the semicircular canals, *utricle,* and *saccule.*

29. The *cochlea* is a spiral-shaped, bony structure that contains three fluid-filled canals. Separating the middle canal from the lower one is the flexible *basilar membrane* which supports the *organ of Corti.* Hair cells in the organ of Corti are embedded in the relatively rigid *tectorial membrane.*

30. Sound waves create vibrations in the eardrum and ossicles, which are transmitted to fluid in the cochlea. Pressure waves in the cochlea cause the basilar membrane to vibrate, which stimulates the hair cells.

31. Pressure waves resulting from any given sound cause one part of the membrane to vibrate maximally. The hair cells stimulated in that region send signals to the brain, which it interprets as a specific frequency.

32. Hearing loss may occur as a result of damage or blockage to the conducting system: the external auditory canal, the eardrum, and the ossicles. Damage to the hair cells, the auditory nerve, or the auditory cortex are forms of *nerve deafness.*

33. The *semicircular canals* are three hollow rings filled with a fluid called *endolymph*. The receptors for head movement are located in an enlarged portion at the base of each canal, the *ampulla.*

34. Fluid movement inside the semicircular canals deflects the gelatinous cap *(cupula)* lying over the receptor cells, stimulating them and alerting the brain to head movements.

35. The semicircular canals are set in all three planes of space so that movement in any direction can be detected.

36. Two membranous sacs in the *vestibule,* the *utricle* and *saccule,* contain receptors that respond to linear acceleration and tilting of the head.

HEALTH, HOMEOSTASIS, AND THE ENVIRONMENT: NOISE POLLUTION

37. Receptors that provide for the general and special senses are extremely important to homeostasis.

38. Noise damages the ears. Extremely loud noises can rupture the eardrum or break the ossicles. Less intense noises, however, generally destroy hearing gradually by damaging hair cells. In most people, hearing loss occurs so gradually as to be undetected.

39. Besides deafening us and cutting us off from the important sounds of modern life, noise disturbs communications, rest, and sleep—the last two of which are particularly important to homeostasis and health. Noise also raises our level of stress, which, in turn, shortens our lives.

Critical Thinking

THINKING CRITICALLY— ANALYSIS

This Analysis corresponds to the Thinking Critically scenario that was presented at the beginning of this chapter.

This type of reporting provides anecdotal information—that is, it presents one person's story, which might be the exception to the rule. Anecdotal information does not provide enough information on which to base a decision. It would have been more useful for the news story to have included some statistics—especially the percentages of patients who improve and those who don't. This would give you a better understanding of your chances of benefitting from the operation.

In addition, it might be helpful to look at side effects. Have any patients developed problems as a result of the operation? What was the incidence of infections? Do the incisions heal in all patients? If not, what percentage do not heal?

Answers to these questions might come from physicians, but will they provide accurate, up-to-date informa-

tion? Early in the history of radial keratotomy, reports of the side effects were disturbingly common. In some patients, the incisions never healed. Others developed infections. Has the procedure improved since then? Perhaps the best answers might come from recent long-term studies published in reputable medical journals.

EXERCISING YOUR CRITICAL THINKING SKILLS

In an experiment to determine how chemicals affect vision, two researchers expose rats to varying levels of a toxin, chemical A. They find that at low doses, the chemical has no effect, but that higher doses result in a severe loss of vision and blindness. The researchers immediately ask the Food and Drug Administration to ban the chemical from production for fear it might similarly affect humans.

Imagine that you are the head of the FDA. Using your critical thinking skills, how would you go about considering the request? What factors would help you determine whether the request was valid? What studies might you want to see done?

TEST OF CONCEPTS

1. Define the terms *general senses* and *special senses.*
2. Using your knowledge of the senses and of other organ systems gained from previous chapters, describe the role that sensory receptors play in homeostasis. Give specific examples to illustrate your main points.
3. Make a list of both the encapsulated and the nonencapsulated general sense receptors. Note where each is located and what it does.
4. Define the term *adaptation.* What advantages does it confer? Can you think of any disadvantages?
5. Describe the receptors for taste, explaining where they are located, what they look like, and how they operate.
6. Taste buds detect four basic flavors. What are they? How do you account for the thousands of different flavors that you can detect?

7. Describe the olfactory membrane and the structure of the receptor cells. How do these cells operate? In what ways are taste receptors and olfactory receptors similar? In what ways are they different?
8. Explain the following statement: taste and smell are complementary.
9. Draw a cross section of the human eye and label its parts.
10. Define the following terms: retina, rods, cones, fovea centralis, optic disc, ganglion cells, bipolar neurons, and optic nerve.
11. Compare and contrast rods and cones.
12. When focusing on a nearby object, your eyes go through several changes. Describe those changes and what they accomplish.
13. Define the following terms: myopia, hyperopia, presbyopia, and astigmatism.

14. You walk into a dark movie theater and find that you can barely make out the aisle. After a while, your vision recovers. Explain both phenomena.
15. What is color blindness? What is the most common type? Explain what your world would look like if you were afflicted by red-green color blindness.
16. What is a cataract, and how is it treated?
17. Describe the disease known as glaucoma. What causes it, and how is it treated?
18. Describe the anatomy of the ear and the role of the outer, middle, and inner ears in hearing.
19. How do the semicircular canals operate? How do the utricle and saccule operate?
20. Describe the different types of deafness and how they are treated.

TOOLS FOR LEARNING

Tools for Learning is an on-line student review area located at this book's web site HumanBiology (www.jbpub.com/humanbiology). The review area provides a variety of activities designed to help you study for your class:

Chapter Outlines. We've pulled out the section titles and full sentence sub-headings from each chapter to form natural descriptive outlines you can use to study the chapters' material point by point.

Review Questions. The review questions test your knowledge of the important concepts and applications in each chapter. Written by the author of the text, the review provides feedback for each correct or incorrect answer. This is an excellent test preparation tool.

Flash Cards. Studying human biology requires learning new terms. Virtual flash cards help you master the new vocabulary for each chapter.

Figure Labeling. You can practice identifying and labeling anatomical features on the same art content that appears in the text.

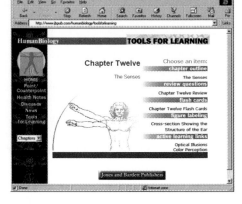

Active Learning Links. Active Learning Links connect to external web sites that provide an opportunity to learn basic concepts through demonstrations, animations, and hands-on activities.

THE SKELETON AND MUSCLES

X ray of hip socket and femur showing steel rods used to repair the broken bone.

Thinking **Critically**

A study on osteoporosis in Australian women involving 120 nonsmokers, ranging in age from 50 to 60, concluded that the best prevention against bone fractures is estrogen-replacement therapy. During the 2-year study, the researchers examined the density of bone in the forearms of the women at 3-month intervals. Initial measurements showed very low levels of calcium, which indicates a high risk of fractures.

The women were then divided into three groups. The first group received a placebo, the second group received 1 gram of calcium per day, and the third group received daily doses of estrogen and progesterone (sex steroid hormones). All the women were encouraged to take two brisk 30-minute walks every day and engage in one low-impact aerobics class each week.

The study showed that bone loss continued in the control group (group 1, exercise only) at a rate of 2.5% per year. Bone loss also continued in group 2, which received a daily dose of calcium, but was slower than in group 1 (0.5%–1.3% per year). In group 3, whose participants exercised and received hormone therapy, bone density increased from 0.8% to 2.7% per year.

Given the potential side effects of hormone therapy, the researchers recommended that only women with lowest bone densities receive hormone therapy and that others should exercise and take calcium to reduce their risk of osteoporosis.

Can you pinpoint any potential flaws in the experiment's design or execution? How could they be corrected? ■

THOMAS POWERS IS A 40-YEAR-OLD COLLEGE professor. Like many of his friends, he likes to downhill ski. But lately, his skiing has been curtailed by chronic knee pain from an old skiing injury. After visiting his doctor, he found that years ago a spill on the slopes tore some cartilage in his knee. Although it never really healed, the tear caused few problems until he recently took up mountain biking. The additional strain combined with some degeneration of the knee joint itself conspired to create an intolerable situation.

This chapter discusses the human skeleton and the muscles that attach to it, the skeletal muscles. It examines the joints, which are essential to movement but also susceptible to injury and wear and tear, like those of Professor Powers.

Together, the bones, joints, and muscles constitute the musculoskeletal system. In adults, bones and muscles make up a remarkable 50%–60% of the body weight.

Section 13-1

Structure and Function of the Human Skeleton

The word **skeleton** is derived from a Greek word that means "dried-up body." Your first impression of bone might not be much different (**FIGURE 13-1**). **Bone,** after all, appears to be a dry, dead structure; appearances aside, however, bone is living, metabolically active tissue. Bone tissue contains numerous cells, known as **osteocytes** (OSS-tee-oh-SITES). These cells are embedded in a calcium-impregnated extracellular material, known as the **matrix** (MAY-tricks). The matrix also consists of an organic component, collagen (COLL-ah-gin), a protein that imparts flexibility. The inorganic component of the matrix, chiefly calcium phosphate crystals, is deposited on the collagen fibers and imparts hardness and strength to the bone.

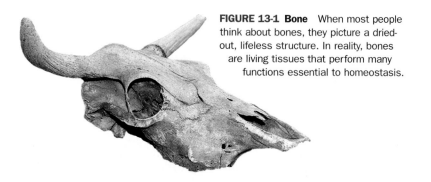

FIGURE 13-1 Bone When most people think about bones, they picture a dried-out, lifeless structure. In reality, bones are living tissues that perform many functions essential to homeostasis.

Bones Serve Many Functions and Play an Important Role in Homeostasis

The human skeleton consists of 206 bones, discrete structures made of bone tissue (**FIGURE 13-2**). Bones provide internal structural support, giving shape to our bodies and enabling us to maintain an upright posture. Some bones protect internal organs. The rib cage, for instance, protects the lungs and heart, and the skull forms a protective shell for the brain. Bones serve as the site of attachment for the tendons of many skeletal muscles whose contraction results in purposeful movements. Bones are home to cells that give rise to red blood cells, white blood cells, and platelets. Bones are a storage depot for fat, needed for cellular energy production at work and at rest. Finally, bones are a reservoir of calcium and other minerals. Bones release and absorb calcium as needed, which helps maintain normal blood levels. Because calcium is essential to muscle contraction, disturbances in blood calcium levels can impair muscle contraction.

The Human Skeleton Consists of Two Parts

The human skeleton consists of the axial and appendicular skeletons (**FIGURE 13-2**). The **axial skeleton** (AXE-ee-al) forms the long axis of the body. It consists of the skull, the vertebral column, and the rib cage. The **appendicular skeleton** (ah-pen-DICK-you-lar) consists of the bones of the arms and legs as well as the bones of the shoulders and pelvis, to which the upper and lower extremities are attached.

All Bones Have a Hard, Dense Outer Layer That Surrounds a Less Compact Central Region

Take a moment to study the skeleton in **FIGURE 13-2**. As you examine it, you may notice that bones come in a variety of shapes and sizes. Some are long, some are short, and some are flat and irregularly shaped. Nevertheless, all bones share some characteristics. Consider one of the long bones of the body, the **humerus** (HUGH-mer-us).

FIGURE 13-3 illustrates the anatomy of the humerus, which is in the upper arm. As shown, this bone, like other long bones of the body, consists of a long, narrow shaft, the **diaphysis** (die-AF-eh-siss), and two expanded ends, the **epiphyses** (ee-PIF-eh-seas). The ends of the epiphyses are covered with a thin layer of hyaline cartilage (Chapter 4), which reduces friction in joints. The protrusions on the bone mark the sites of muscle attachment. Many other long bones have the same features. Like other bones, the humerus consists of an

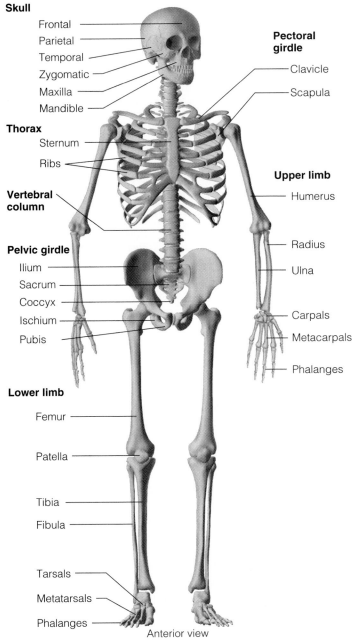

Skull
- Frontal
- Parietal
- Temporal
- Zygomatic
- Maxilla
- Mandible

Thorax
- Sternum
- Ribs

Vertebral column

Pelvic girdle
- Ilium
- Sacrum
- Coccyx
- Ischium
- Pubis

Lower limb
- Femur
- Patella
- Tibia
- Fibula
- Tarsals
- Metatarsals
- Phalanges

Anterior view

Pectoral girdle
- Clavicle
- Scapula

Upper limb
- Humerus
- Radius
- Ulna
- Carpals
- Metacarpals
- Phalanges

FIGURE 13-2 The Human Skeleton Over 200 bones of all shapes and sizes make up the human skeleton. The cartilage is shaded blue.

On the outer surface of the compact bone is a layer of connective tissue, the **periosteum** (PEAR-ee-oss-TEA-um; "around the bone"). The periosteum serves as the site of attachment for skeletal muscles. It also contains osteogenic (OSS-tea-oh-GEN-ick; "bone-forming") cells that participate in the production of new bone during remodeling or repair.

The periosteum is richly supplied with blood vessels, which enter the bone at numerous locations. Blood vessels travel through small canals in the compact bone and course through the inner spongy bone, providing nutrients and oxygen and carrying off cellular wastes such as carbon dioxide. The periosteum is also richly supplied with nerve fibers. The majority of the pain felt after a person bruises

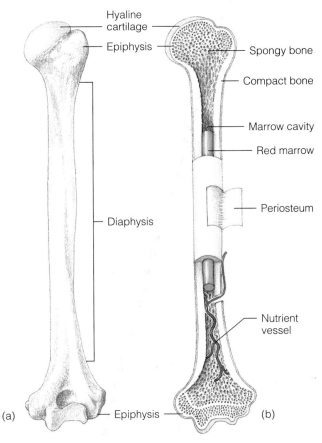

- Hyaline cartilage
- Epiphysis
- Spongy bone
- Compact bone
- Marrow cavity
- Red marrow
- Periosteum
- Diaphysis
- Nutrient vessel
- Epiphysis

(a)   (b)

FIGURE 13-3 Anatomy of Long Bones (a) Drawing of the humerus. Notice the long shaft and dilated ends. (b) Longitudinal section of the humerus showing compact bone, spongy bone, and marrow.

outer shell of dense material called **compact bone.** Inside the humerus is a mass of **spongy bone,** so named because of its spongy appearance. Spongy bone forms an interlacing network and is less dense than compact bone. Spongy bone contains numerous small, adjoining cavities.

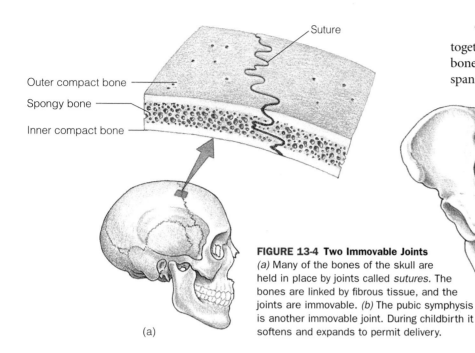

FIGURE 13-4 Two Immovable Joints
(a) Many of the bones of the skull are held in place by joints called *sutures*. The bones are linked by fibrous tissue, and the joints are immovable. (b) The pubic symphysis is another immovable joint. During childbirth it softens and expands to permit delivery.

The bones of the skull shown in **FIGURE 13-4A** are held together by immovable joints. As illustrated, opposing bones in the skull interlock. Fibrous connective tissue spans the space between the interlocking bones, holding them together. Another immovable joint is the pubic symphysis (PEW-bick SIM-fa-siss), which is formed by the two pubic bones (**FIGURE 13-4B**). These bones are held in place by fibrocartilage. Near the end of pregnancy, however, hormones loosen the fibrocartilage of the pubic symphysis, allowing the pelvic outlet to widen enough to permit a baby to pass through—an adaptation of enormous value to our species!

or fractures a bone results from stimulation of the pain fibers in the periosteum.

Inside the shaft of long bones is a large cavity, the **marrow cavity** (MARE-row). The marrow cavities in most of the bones of the fetus and newborn contain **red marrow,** so named because of its color. Red marrow is a blood-cell factory of the human body. It produces RBCs, WBCs, and platelets to replace those routinely lost each day. As an individual ages, most red marrow is slowly "retired" and becomes filled with fat, becoming **yellow marrow.** Yellow marrow begins to form during adolescence and, by adulthood, is present in all but a few bones. Red blood cell formation, however, continues in the bodies of the vertebrae, the hip bones, and a few others. Yellow marrow can be reactivated to produce blood cells under certain circumstances—for example, after an injury.

■ The Joints Permit Varying Degrees of Mobility

A gymnast races across the mat and leaps into space, twirling effortlessly before landing on her feet. As soon as she lands, she takes off again across the mat in a series of elegant and energetic handsprings. These delightful movements are the result of long hours of practice and exercise and are made possible by joints.

Joints are the structures that connect the bones of the skeleton and can be classified by the degree of movement they permit. Those that allow no movement are called *immovable joints,* and those that allow free movement are *freely movable joints.* Consider some examples.

The bodies of the vertebrae are united by slightly movable joints (**FIGURE 13-5**). Each vertebra is separated from its nearest neighbors by an intervertebral disc. The inner portion of the disc acts as a cushion, softening the impact of walking and running. The outer, fibrous portion holds the disc in place and joins one vertebra to its nearest neighbor. The joints between the vertebrae offer some degree of movement, resulting in a fair amount of flexibility. If they did not, we would be unable to bend over to tie our shoes or to curl up on the couch for an afternoon snooze.

The most common type of joint is the freely movable, or **synovial joint** (sin-OH-vee-al). The synovial joints are more complex than other types and permit varying degrees of movement. Although synovial

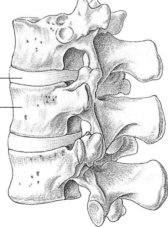

FIGURE 13-5 A Slightly Movable Joint The intervertebral discs allow for some movement, giving the vertebral column flexibility.

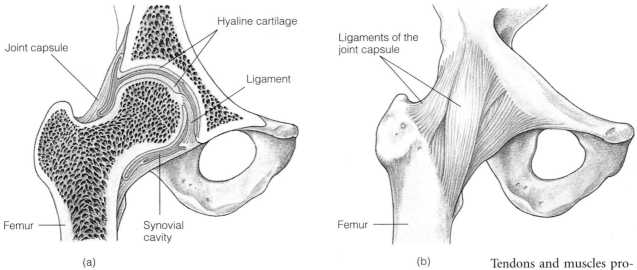

FIGURE 13-6 A Synovial Joint *(a)* A cross section through the hip joint (a ball-and-socket joint) showing the structures of the synovial joint. *(b)* Ligaments in the outer portion of the joint capsule help support the joint.

joints differ considerably in architecture, they share several features. The first commonality is the hyaline cartilage located on the articular (joint) surfaces of the bones (**FIGURE 13-6A**). This thin cap of hyaline cartilage reduces friction and facilitates movement. The second commonality is the **joint capsule,** a structure that joins one bone to another (**FIGURE 13-6A**). The outer layer of the joint capsule consists of dense connective tissue that attaches to the periosteum of adjoining bones. Parallel bundles of dense connective tissue fibers in the outer layer of the capsule form **ligaments,** which run from bone to bone, giving additional support to the joint.

As a rule, ligaments are fairly inflexible. However, some individuals have remarkably flexible ligaments and tendons, which attach muscles to bone. Because of this, some people can extend their thumbs well beyond the 90 degrees possible for most of us. And some can extend their fingers so much that they can touch the back of their hand. These people are said to be "double-jointed."

The inner layer of the joint capsule is the **synovial membrane.** It produces a fairly thick, slippery substance called **synovial fluid.** Synovial fluid provides nutrients to the articular cartilage (the hyaline cartilage on the articular surfaces of the bone) and also acts as a lubricant. Normally, the synovial membrane produces only enough fluid to create a thin film on the articular cartilage. Injuries to a joint, however, can result in a dramatic increase in synovial fluid production, causing swelling and pain in joints.

Tendons and muscles provide additional support in some joints and are commonly associated with synovial joints. In the shoulder, for example, muscles help hold the head of the humerus in the socket (formed by the scapula). Muscles in the hip help hold the head of the femur in place. Because muscles strengthen joints, individuals who are in poor physical shape are much more likely to suffer a dislocation on a ski trip or during exercise than someone who is in good shape. **Dislocation** is an injury in which a bone is displaced from its proper position in a joint due to a fall or some other unusual body movement. In some cases, bones slip out of place, then back in without assistance. In others, the bone can be put back in place only by a trained health care worker.

Synovial joints come in many shapes and sizes and are classified on the basis of structure. Two of the most common are the hinge joint and the ball-and-socket joint. The knee joint is a **hinge joint,** as are the joints in the fingers. Hinge joints open and close like hinges on a door. Such joints provide for movement in only one plane. The hip and shoulder joints are **ball-and-socket joints.** They allow a wider range of motion. (Compare the movements permitted by the shoulder and hip joints to those permitted by the knee joint.)

Arthroscopy Is a Procedure That Permits Surgeons to Repair Injured Joints with a Minimum of Trauma.

Joint injuries are common among athletes and other physically active people (**TABLE 13-1**). A hard blow to the knee of a football player, for instance, can tear the ligaments inside the knee joint or in the joint capsule. A rough fall can dislocate a skier's shoulder. Improperly

TABLE 13-1

Common Injuries of the Joints

Injury	Description	Common Site
Sprain	Partially or completely torn ligament; heals slowly; must be repaired surgically if the ligament is completely torn.	Ankle, knee, lower back, and finger joints
Dislocation	Occurs when bones are forced out of a joint; often accompanied by sprains, inflammation, and joint immobilization. Bones must be returned to normal positions.	Shoulder, knee, and finger joints
Cartilage tears	Cartilage may tear when joints are twisted or when pressure is applied to them. Torn cartilage does not repair well because of poor vascularization. It is generally removed surgically; this operation sometimes makes the joint less stable.	Knee joint

lifting an object can strain the ligaments that join the vertebrae of the back.

Torn ligaments, tendons, and cartilage in joints heal very slowly because they have few blood vessels to bring nutrients needed for repair.[1] For years, joint repair required major surgery that was so traumatic it put patients out of commission for several months. Today, however, new surgical techniques allow physicians to repair joints with a minimum of trauma (**FIGURE 13-7**). Through small incisions in the skin over the joint, surgeons insert a device called an **arthroscope** (ARE-throw-scope). It allows them to view the damage inside the joint and to insert instruments that remove dam-

aged cartilage. The surgeon can thus repair damaged cartilage without opening the joint. The damage caused by large incisions is eliminated, and athletes are usually back on their feet and on the playing field in a matter of weeks. New surgical techniques can rebuild torn ligaments and return joints to nearly their original state.

Osteoarthritis Is a Degenerative Bone Disease Caused by Wear and Tear on the Articular Cartilages.

Virtually every time you move, you use one or more of your joints. Problems in joints, which elicit pain, are therefore often quite noticeable. One of the most common problems is called **degenerative joint disease** or **osteoarthritis.** Degenerative joint disease probably results from wear and tear on a joint. Over time, excess wear may cause the articular cartilage on the ends of bones to flake and crack. As the cartilage degenerates, the bones come in contact and may grind against each other during movement. This causes considerable swelling, pain, and discomfort.

Osteoarthritis occurs most often in the weight-bearing joints: the knee, hip, and spine—regions subject to the most wear over time. Osteoarthritis may also develop in the finger joints.

Osteoarthritis is extremely common. X-ray studies of people over 40 years of age show that most people have some degree of degeneration in one or more joints. Fortunately, many people do not even notice the problem, and the disease rarely becomes a serious medical problem.

Wear and tear on joints is worsened by obesity. The extra pressure wears the cartilage away more quickly.

---

[1]Because most cartilage is avascular, it may not heal at all.

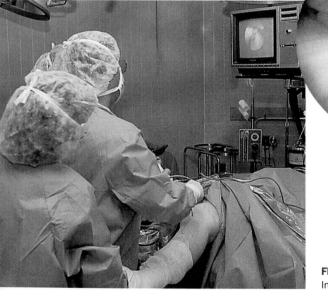

(b)

(a)

FIGURE 13-7 Arthroscopic Surgery   (a) A physician performing arthroscopic surgery. (b) Inside view of the knee joint through an arthroscope.

FIGURE 13-8 Hand Disfigured by Rheumatoid Arthritis Degeneration of the finger joints causes mild to extreme disfiguration.

Rheumatoid arthritis
www.jbpub.com/humanbiology

For individuals with a joint problem, weight control can reduce the rate of degeneration and ease the pain.[2] Painkillers such as aspirin and other anti-inflammatory drugs are used to treat the pain and swelling. Injections of steroids may help reduce inflammation, although repeated injections often damage the joint.

Rheumatoid Arthritis Is an Autoimmune Disease.

Another common disorder of the synovial joint is rheumatoid arthritis. **Rheumatoid arthritis** (RUE-mah-toid) is the most painful and crippling form of arthritis. It is caused by an inflammation of the synovial membrane. The inflammation often spreads to the articular cartilages. If the condition persists, rheumatoid arthritis can wear through the cartilage and cause degeneration of the underlying bone. In such cases, the thickening of the synovial membrane and degeneration of the bone often disfigure the joints, reduce mobility, and cause considerable pain (**FIGURE 13-8**). Afflicted joints can be completely immobilized. In severe cases, the bones become dislocated, causing the joints to collapse.

Rheumatoid arthritis generally occurs in the joints of the wrist, fingers, and feet. It can also affect the hips, knees, ankles, and neck.

Research suggests that rheumatoid arthritis results from an **autoimmune reaction**—that is, an immune response to the cells of one's own synovial membrane (Chapter 8). Rheumatoid arthritis occurs in people of all ages, but most commonly appears in individuals between the ages of 20 and 40. Rheumatoid arthritis is usually a permanent condition, although the degree of severity varies widely. Patients suffering from it can be treated with physical therapy, painkillers, anti-inflammatory drugs, and surgery.

Diseased joints can also be replaced by **prostheses** (pross-THEE-seas), artificial joints that restore mobility and reduce pain. Plastic joints are used to replace the finger joints, greatly improving the appearance of the hands by eliminating the gnarled, swollen joints. Moreover, patients regain the use of previously crippled fingers. Day-to-day chores (buttoning a shirt) that often required assistance become noticeably easier. Once impossible tasks—such as opening screw-top jars—are again feasible. Severely damaged knee and hip joints are replaced with special steel or Teflon substitutes, which, if fitted properly, may last 10–15 years (**FIGURE 13-9**).

Most of the Bones of the Human Skeleton Start Out as Hyaline Cartilage

During embryonic development, hyaline cartilage forms in the arms, legs, head, and torso where bone will eventually be (**FIGURE 13-10**). In short order, the cartilage is converted to bone. This process is known as **endochondral ossification** (en-doh-CON-drall).

Bones Are Constantly Remodeled in Adults to Meet Changing Stresses Placed on Them

Bones are dynamic structures that undergo considerable remodeling in response to changes in our lives. This process begins early in life. In a newborn baby, for example, the bones of the leg (the tibia and fibula) are quite bowed. Cramped inside the mother's uterus, the baby's bones do not grow very straight. During the first 2 years of life, however, as the child begins to walk and run, the leg bones generally straighten (**FIGURE 13-11**). The bones are remodeled to meet the markedly different stresses placed on them by upright posture.

Remodeling occurs throughout adult life as well. During sedentary periods, for example,

FIGURE 13-9 Artificial Joints
An artificial knee joint (a) and hip joint (b).

(a)   (b)

---

[2]Weight control is also a preventive measure that helps people avoid the problem in the first place.

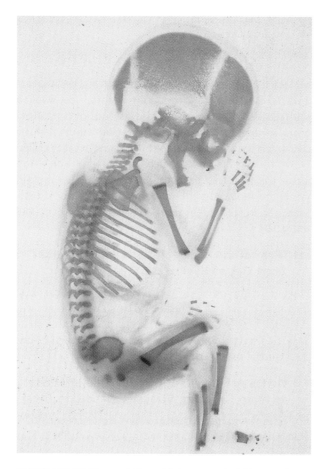

FIGURE 13-10 Human Fetus Eighteen-week-old human fetus showing forming bones. The fetus is six to nine inches in length.

compact bone decreases in thickness. Increasing one's level of activity, however, causes compact bone to thicken, thus helping the bone withstand stresses placed on it by walking, running, or standing. Spongy bone also undergoes considerable remodeling. Thus, even the internal architecture of a bone changes to meet new stresses.

Two cells are responsible for bone remodeling: osteoclasts and osteoblasts. **Osteoclasts** (OSS-tea-oh-KLASTS) are akin to the wrecking crew that comes into a house to tear out walls before a remodeling job. **Osteoblasts** (OSS-tea-oh-BLASTS) lay down new bone during the remodeling phase like the drywallers and carpenters who rebuild walls after they have been torn down. When an osteoblast becomes sur-

Tibia
Fibula

Birth

20 months

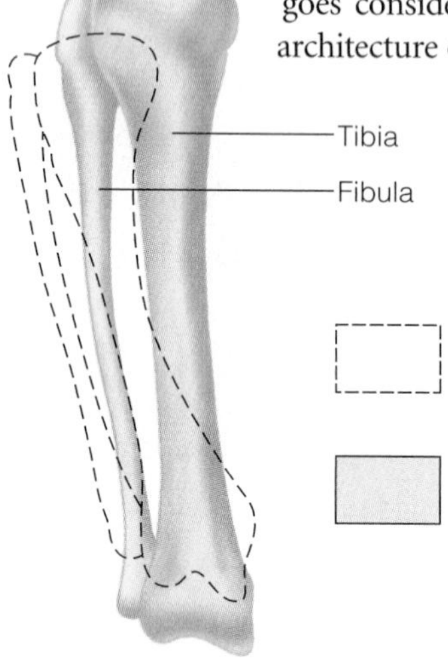

FIGURE 13-11 Remodeling of a Baby's Leg Bones Notice how dramatically the bones change to meet the changing needs of the toddler.

rounded by calcified extracellular material, it is referred to as an osteocyte.

Bone Is a Homeostatic Organ That Helps Maintain Proper Levels of Calcium in the Body

In addition to their role in remodeling bone, osteoblasts and osteoclasts help control blood calcium levels and are therefore part of a homeostatic mechanism. They are controlled by two hormones: parathormone and calcitonin.

When blood calcium levels fall, the parathyroid glands (small glands embedded in the thyroid glands in the neck) release **parathormone** (PAIR-ah-THOR-moan) or **PTH** into the bloodstream. PTH travels throughout the body. When it reaches the bone, it stimulates the osteoclasts, causing them to digest bone in their vicinity. The calcium released by the activity of the osteoclasts replenishes blood calcium levels.

When calcium levels rise—for example, after a meal—the thyroid releases **calcitonin** (CAL-seh-TONE-in). This hormone inhibits osteoclasts, stopping bone destruction. It also stimulates osteoblasts, causing them to deposit new bone. The inhibition of osteoclasts and the formation of new bone help lower blood calcium levels, returning them to normal.

Bone Fractures Are Repaired by Fibroblasts and Osteoblasts

If you're like most people, chances are that sometime in your life you'll break a bone. Bone fractures can vary considerably in their severity. Some only involve hairline cracks, which mend fairly quickly. Others involve considerably more damage—a complete break, for instance—that takes much longer to repair. In some instances, the broken ends must be realigned. A plaster cast is then applied to hold the limb in position and keep the ends together to optimize healing. Severe fractures may require surgeons to insert steel pins to hold the bones together.

Bones, like other tissues, are capable of self-repair. As **FIGURE 13-12** illustrates, after a fracture, blood from broken blood vessels in the periosteum and marrow cavity pours into the fracture and forms a blood clot. Within a few days, the clot is invaded by fibroblasts, connective tissue cells from the periosteum. Fibroblasts secrete collagen fibers, thus forming a mass of cells and fibers, the **callus** (CAL-us), which bridges the broken ends internally and externally.

The callus is next invaded by osteoblasts from the periosteum. Osteoblasts slowly convert the callus to bone, thus "knitting" the broken ends together. As **FIGURE 13-12** shows, the callus is initially much larger than

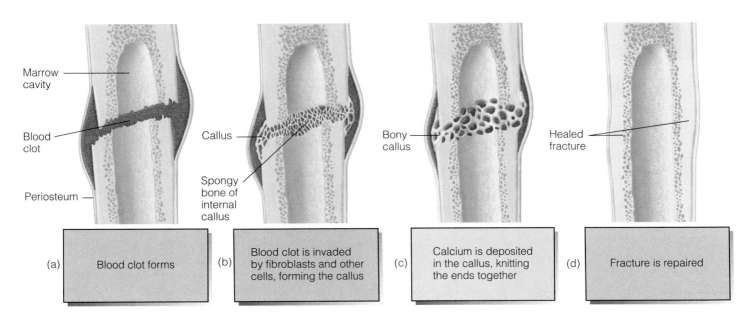

Marrow
cavity

Blood
clot

Periosteum

Callus

Spongy
bone of
internal
callus

Bony
callus

Healed
fracture

(a) Blood clot forms

(b) Blood clot is invaded by fibroblasts and other cells, forming the callus

(c) Calcium is deposited in the callus, knitting the ends together

(d) Fracture is repaired

FIGURE 13-12 Stages in Fracture Repair

the bone itself. But the excess bone is gradually removed by osteoclasts, leaving little, if any, evidence that a break has occurred.

▪ Osteoporosis Involves a Loss of Calcium, Which Results in Brittle, Easy-to-Break Bones

One of the most common problems adults face is **osteoporosis** (OSS-tea-oh-puh-ROE-siss; "porous bone"), a condition characterized by a progressive loss of bone calcium (**FIGURE 13-13**). In some individuals, calcium loss is so severe that bones become extremely brittle—so much

so that even normal activities such as getting out of bed in the morning or doing housework cause fractures.

Osteoporosis occurs most often in postmenopausal women. **Menopause** (MEN-oh-pause) occurs between 45 and 55 years of age and results from a cessation of ovarian estrogen production. Although estrogen is primarily a reproductive hormone, it also helps to maintain bone.

Osteoporosis also occurs in people who are immobilized for long periods. Hospital patients who are restricted to bed for 2 or 3 months, for example, show signs of osteoporosis.

Osteoporosis may also result from environmental factors. In the 1960s, women living along the Jintsu

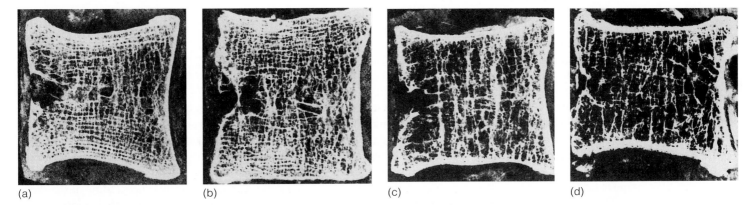

(a)   (b)   (c)   (d)

FIGURE 13-13 Osteoporosis The loss of estrogen or prolonged immobilization weakens bone. In these situations, bone is dissolved and becomes brittle and easily breakable. (a) A section of the body of a lumbar vertebra from a 29-year-old woman. (b) Some thinning is evident in a vertebra of a 40-year-old woman. (c) Bone loss is severe in an 84-year-old woman. (d) Bone loss is most severe in a 92-year-old woman. Osteoporosis is not inevitable and can be prevented by exercise, calcium supplements, estrogen supplements, and other measures.

Health Note 13-1

Preventing Osteoporosis: A Prescription for Healthy Bones

Twenty million Americans, mostly women, suffer from a painful, debilitating disease called **osteoporosis** (**FIGURE 1**). Caused by a gradual deterioration of the bone, osteoporosis results in nagging pain and discomfort. In more severe cases, bones fracture easily.

If current trends continue, one of every two American women will develop postmenopausal osteoporosis. Each year, nearly 60,000 Americans—most of them women—will die from complications resulting from this common malady. The three most common are hemorrhage; fat embolisms (globules), sometimes released from the yellow marrow of broken bones; and shock. Unfortunately, most women do not realize they have the disease until it has progressed quite far.

FIGURE 1 Bone Deterioration An elderly woman suffering from osteoporosis. Notice the hunched back due to the collapse of vertebrae.

Recent research shows that osteoporosis begins much earlier than researchers once thought—by the time a woman reaches her mid-20s. Bone demineralization occurs very rapidly. In fact, by age 30, many women have lost one-third of their bone calcium! Between the ages of 30 and 50, many women's bones continue to deteriorate, becoming extremely brittle.

Calcium loss begins so early in American women because many weight-conscious women in their mid-20s abstain from fatty foods such as whole milk, cheese, and ice cream to control their weight. Although these foods are fatty, they are also a major source of calcium. Milk products are also shunned by many adults who develop an intolerance to lactose (a sugar) in milk and other dairy products. Lactose intolerance results from a sharp reduction in the production and secretion of the intestinal enzyme lactase as one ages, which makes it difficult to digest lactose. Because of

River in Japan developed a painful bone disease known as itai-itai (which literally means "ouch, ouch"). The women lived downstream from zinc and lead mines that released large amounts of the heavy metal cadmium into the river. They used river water for drinking and for irrigating rice paddies. Even though men, young women, and children were exposed to cadmium, 95% of the cases occurred in postmenopausal women.

Researchers hypothesized that the accelerated bone loss in the postmenopausal Japanese women was due to cadmium and tested this hypothesis by feeding mice diets containing various levels of cadmium chloride. The group receiving the highest dose showed significant reductions in bone calcium when their ovaries had been removed.

These findings may also explain why older women who smoke are more likely to develop osteoporosis and why they experience more bone fractures and tooth loss than nonsmokers. Cadmium is one of several

harmful substances present in cigarette smoke. Smoking, however, may also act by decreasing estrogen levels, making it doubly dangerous to a woman's health. For a discussion of ways to prevent osteoporosis, see Health Note 13-1.

Section 13-2

The Skeletal Muscles

Purposeful movement is one of the most distinctive features of animal life. As noted in the introduction to this chapter, in humans, skeletal muscles acting on bones result in movement. Most skeletal muscles are under the control of the nervous system.[3]

FIGURE 13-14 on page 304 shows some of the skeletal muscles of the body. As illustrated, most skeletal

---

[3]Chapter 4 discussed three types of muscle: skeletal, cardiac, and smooth.

these and other factors, adult women often consume only about one-half of the 1000–1500 milligrams of calcium they need every day.

Osteoporosis may be prevented and even reversed by eating calcium-rich foods, such as spinach, milk, cheese, shrimp, oysters, and tofu (soybean curd). Calcium supplements can also help retard or even stop bone deterioration and restore calcium levels. Because vitamin D increases the absorption of calcium in the intestines vitamin supplements can help. (A word of caution, however: excessive vitamin D—five times the RDA—can be toxic.)

Studies also suggest that osteoporosis can be prevented by exercise. Aerobics, jogging, walking, and tennis in conjunction with the dietary changes all help prevent the disease.

Osteoporosis can be *reversed* by exercise even after the disease has reached the dangerous stage. Forty-five minutes of moderate exercise (walking) 3 days a week, for example, greatly decreases the rate of calcium loss in older individuals. In addition, this exercise regime stimulates the rebuilding process, replacing calcium lost in previous years. Continued exercise increases bone calcium levels and decreases the rate of bone fractures and fatal complications.

Another effective treatment for postmenopausal women is estrogen. Low doses of estrogen halt bone demineralization and promote bone formation. Because women who are given estrogen suffer an increased risk of cancer of the uterine lining, physicians often prescribe a mixed dose of estrogen and progesterone, which reduces the likelihood of this type of cancer.

Studies have also shown that high doses of calcium fluoride stimulate bone development. Calcium fluoride treatment increases bone mass approximately 3%–6% per year and decreases bone fractures. The average patient in one study experienced one fracture every 8 months before treatment. After treatment, that figure dropped to one fracture every 4.5 years.

Unfortunately, large doses of fluoride may erode the stomach lining, causing internal bleeding. They may also stimulate abnormal bone development and cause pain and swelling in joints. To offset these problems, researchers have developed a pill that releases the fluoride gradually.

For millions of young women, early detection and sound preventive measures, including exercise, vitamin D, dietary improvements, and fluoride treatments, can prevent this unnecessary disease.

Visit *Human Biology's* Internet site, www.jbpub.com/humanbiology, for links to web sites offering more information on this topic.

www.jbpub.com/humanbiology

---

muscles cross one or more joints. When they contract, they produce movement.

As a rule, muscles generally work in groups to bring about various body movements. Groups of muscles are often arranged in such a way that one set produces one movement and another set on the opposite side of the joint causes an opposing movement. For example, the biceps (BUY-seps) is a muscle in the front of the upper arm. When it and other members of its group contract, they cause the arm to bend, a movement called *flexion* (FLECK-shun). On the back side of the upper arm is another muscle, the triceps (TRY-seps). When it contracts, it causes the arm to straighten, or extend. Such opposing muscles are called **antagonists.** When one muscle contracts to produce a movement, the antagonistic muscles relax.

Not all skeletal muscles make bones move. Some muscles simply steady joints, allowing other muscles to act. These stabilizing muscles are called **synergists** (SIN-er-gists). These muscles help to maintain our posture, permitting us to stand or sit upright despite the constant pull of gravity.

Another group of muscles that don't move bones are the muscles of the face. Anchored to the bones of the skull and to the skin of the face, these muscles allow us to wrinkle our skin, open and close our eyes, and move our lips.

Muscles also produce enormous amounts of heat as a by-product of metabolism. Working muscles produce even more heat—so much, in fact, that you can cross-country ski in freezing weather wearing only a light sweater.

Skeletal Muscle Cells Are Known as Muscle Fibers and Are Both Excitable and Contractile

Skeletal muscles consist of long, unbranched **muscle fibers.** Each fiber contains many nuclei derived from individual cells that fuse during embryonic development to form a fiber. Viewed with the light microscope, skeletal

FIGURE 13-14 The Skeletal Muscles

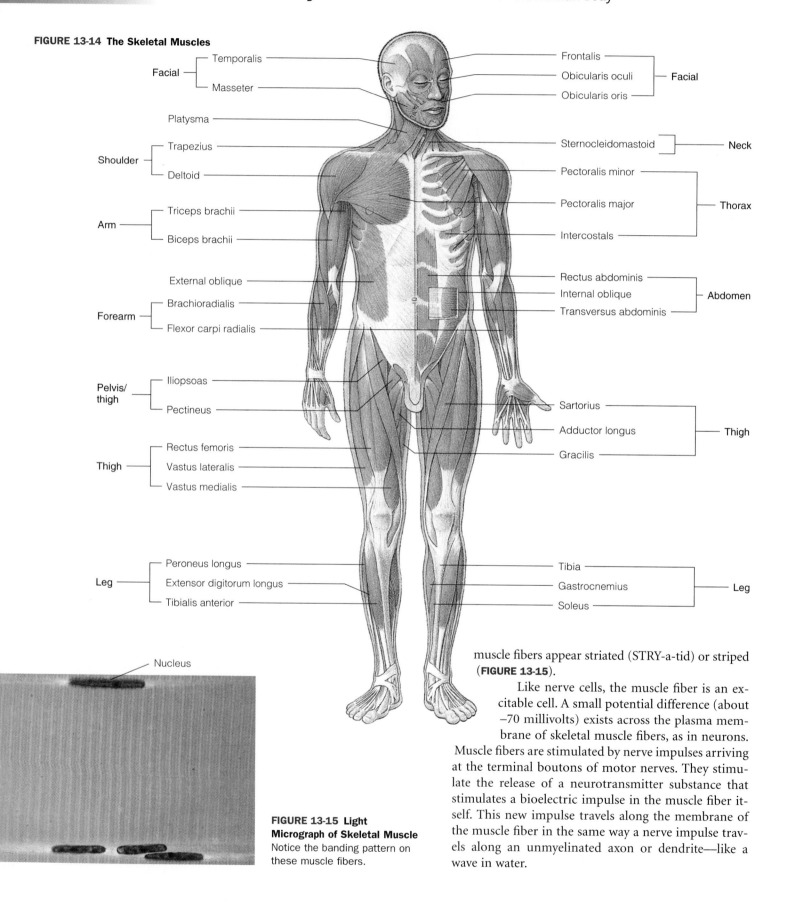

Facial
- Temporalis
- Masseter

Platysma

Shoulder
- Trapezius
- Deltoid

Arm
- Triceps brachii
- Biceps brachii

Forearm
- External oblique
- Brachioradialis
- Flexor carpi radialis

Pelvis/thigh
- Iliopsoas
- Pectineus

Thigh
- Rectus femoris
- Vastus lateralis
- Vastus medialis

Leg
- Peroneus longus
- Extensor digitorum longus
- Tibialis anterior

Frontalis
Obicularis oculi — Facial
Obicularis oris

Sternocleidomastoid — Neck

Pectoralis minor
Pectoralis major — Thorax
Intercostals

Rectus abdominis
Internal oblique — Abdomen
Transversus abdominis

Sartorius
Adductor longus — Thigh
Gracilis

Tibia
Gastrocnemius — Leg
Soleus

Nucleus

FIGURE 13-15 Light Micrograph of Skeletal Muscle Notice the banding pattern on these muscle fibers.

muscle fibers appear striated (STRY-a-tid) or striped (**FIGURE 13-15**).

Like nerve cells, the muscle fiber is an excitable cell. A small potential difference (about −70 millivolts) exists across the plasma membrane of skeletal muscle fibers, as in neurons. Muscle fibers are stimulated by nerve impulses arriving at the terminal boutons of motor nerves. They stimulate the release of a neurotransmitter substance that stimulates a bioelectric impulse in the muscle fiber itself. This new impulse travels along the membrane of the muscle fiber in the same way a nerve impulse travels along an unmyelinated axon or dendrite—like a wave in water.

Muscle fibers are also contractile. When stimulated, the contractile proteins inside the fibers cause the cells to shorten. Muscle fibers are also elastic and therefore capable of returning to normal length after a contraction has ended.

Muscle Fibers Contain Many Small Bundles of Contractile Filaments Known as Myofibrils

Understanding how a muscle contracts first requires an understanding of its structure. To begin, imagine that you could tease a single skeletal muscle fiber (cell) free from a skeletal muscle (**FIGURE 13-16A**). Under the microscope, you would find that each muscle fiber is a long cylinder encased by plasma membrane and containing many nuclei. Each muscle fiber is further characterized by a series of dark and light bands—hence its striated appearance (**FIGURE 13-16B**).

Take the tissue to an electron microscope, and you will find that inside each muscle fiber are lots of threadlike filaments. They exist in tiny bundles called **myofibrils** (MY-oh-FIE-brills) (**FIGURE 13-16B**). Myofibrils contain contractile filaments and, as **FIGURE 13-16C** illustrates, are striated (banded).

The light and dark bands of the myofibril are arranged in a uniform pattern that gives the myofibril and muscle fiber its striated appearance. The details are shown in **FIGURE 13-16C**.

FIGURE 13-16D shows that the myofibrils contain thick and thin filaments. The thick filaments consist of the protein **myosin** (MY-oh-sin). The thin filaments are composed primarily of the protein **actin** (ACK-tin).

During Muscle Contraction, the Actin Filaments Slide Inward, Causing the Sarcomeres to Shorten

When a muscle contracts, it shortens. This is achieved by the sliding of actin filaments over the myosin filaments. In actuality, the actin filaments do not passively slide to the

middle; they are pulled inward by the myosin molecules. How is this accomplished?

The contraction of muscle has to rank as one of the most fascinating biological processes. As **FIGURE 13-17** shows, each myosin filament actually consists of numerous golf-club-shaped myosin molecules arranged with their "club ends," or heads, projecting toward the actin filaments. During muscle contraction, the heads of the myosin molecules attach to actin filaments, forming **cross bridges,** which tug the filaments inward.

FIGURE 13-16 Structure of the Skeletal Muscle Fiber, Myofibril, and Sarcomere (a) A single muscle fiber teased out of the muscle. (b) Each muscle fiber consists of many myofibrils. (c) Note the banded pattern of the myofibril. (d) Sarcomeres consist of thick and thin filaments, as shown here. (e) Molecular structure of the thick (myosin) and thin (actin) filaments.

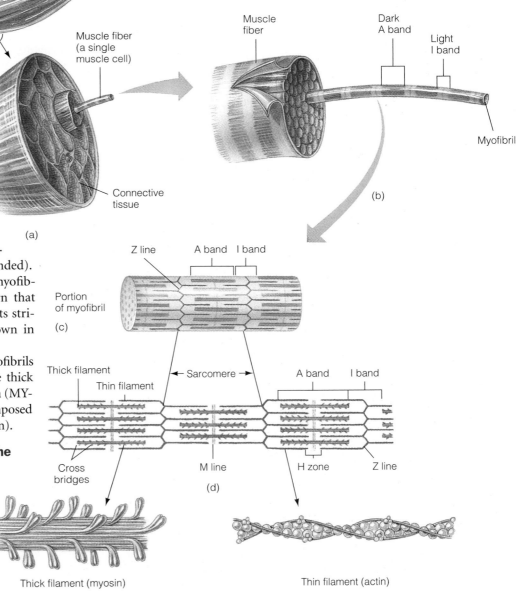

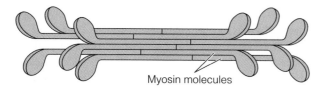

FIGURE 13-17 Structure of Myosin Filaments Myosin molecules join to form a myosin filament. Note the presence and orientation of the heads of the myosin molecules.

The attachment of the myosin molecules to the actin molecules is stimulated by the release of calcium inside muscle cells. Calcium is stored in the smooth endoplasmic reticulum, which forms a rather extensive network inside the muscle cells. The release of calcium, in turn, is stimulated by electrical impulses from the nerves. These impulses reach the muscle cell via motor nerves, which form synapses with the muscle cells. Neurotransmitters released from the terminal boutons of the motor nerves cause impulses to be generated in muscle cells. These impulses travel along the membrane of muscle cells and enter the interior via small infoldings of the plasma membrane. These infoldings come in close contact with the smooth endoplasmic reticulum. When an impulse penetrates into the interior, it stimulates calcium release. This causes the heads of the myosin molecules to attach to the actin filaments. ATP in the muscle provides the energy needed to pull the actin filaments inward.

Muscle Cells Use Enormous Quantities of ATP Daily, Which Come From Several Sources.

ATP is in relatively short supply in cells, and must be recycled over and over again in rapid succession to meet cellular demands. In muscles, some of the ATP is regenerated by a substance called **creatine phosphate** (CREE-ah-teh-NEEN). This molecule contains a high-energy bond, indicated by the squiggly line (below). Stored in muscle in high concentrations, creatine phosphate reacts with ADP as follows:

$$\text{creatine} \sim P + ADP \rightarrow ATP + \text{creatine}$$

Creatine phosphate replenishes ATP used during muscle contraction.

ATP is also generated by glycolysis, the citric acid cycle, and the electron transport system (Chapter 3). The electron transport system is the main source of ATP. But this process requires oxygen. During vigorous muscle contraction, oxygen supplies inside muscle cells can fall precipitously. If the circulatory system cannot replace oxygen used as quickly as it is being used, the citric acid cycle and electron transport system shut down.

To generate ATP, the cell must resort to fermentation—the metabolic breakdown of glucose in the absence of oxygen. Besides being inefficient, this process results in the buildup of lactic acid. The shortage of ATP and the buildup of lactic acid that occurs during vigorous exercise result in **muscle fatigue.** Muscle fatigue also results from a depletion of glycogen stores in skeletal muscle. Glycogen, you may recall, is a polysaccharide found in muscle and liver cells and is composed of thousands of glucose molecules.

Glycogen broken down in muscles during exercise is replaced during rest. Oxygen depleted during exercise must also be replaced. This replacement occurs rather quickly. The muscle oxygen deficiency, called the **oxygen debt,** is often largely replaced right after you exercise, explaining why you keep breathing hard for a while after you stop exercising.

Individual Skeletal Muscle Fibers Contract After Being Stimulated by an Action Potential

The Strength of Muscle Contraction Can Be Increased by Stimulating (Recruiting) Additional Muscle Fibers to Contract.

Individual skeletal muscle fibers contract when activated by an action potential. A single contraction, followed by relaxation, is called a **twitch.**

Generating the force needed to move arms and legs or eyelids requires the action of many muscle cells. The engagement of additional muscle fibers during muscle contraction is called **recruitment.**

To understand how recruitment occurs in muscle, we first take a look at **FIGURE 13-18**. The drawing shows that the axons of motor neurons form many branches upon reaching the muscles they supply. A single motor neuron, in fact, may end on dozens of individual muscle fibers in a skeletal muscle. A motor neuron and the muscle fibers it supplies constitute a **motor unit.** When the neuron supplying a motor unit is activated, all of the muscle fibers in the unit are stimulated.

Motor units account for the degree of control that various muscles exert. The fewer muscle fibers in a motor unit, the finer the control. Such is the case in the muscles of the fingers, which are capable of very fine movements. In contrast, the more muscle fibers in a motor unit, the cruder the control. Thus, the neurons that supply the muscles of the leg, which produce a crude but strong propulsive force, may end on as many as 2000 muscle fibers. Each time an impulse is delivered to the motor unit, it stimulates all 2000 muscle fibers.

How does this process relate to recruitment? To increase the force of contraction in a skeletal muscle, more motor neurons are stimulated and therefore more mus-

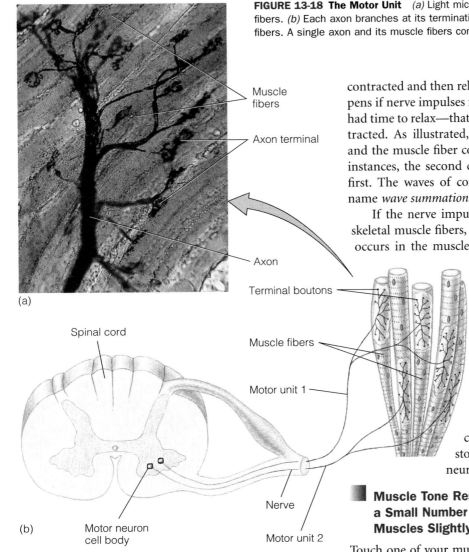

FIGURE 13-18 The Motor Unit   *(a)* Light micrograph of axon branching to terminate on many muscle fibers. *(b)* Each axon branches at its termination, supplying a few dozen to many thousand muscle fibers. A single axon and its muscle fibers constitute a motor unit.

Muscle fibers

Axon terminal

Axon

Terminal boutons

Muscle fibers

Motor unit 1

Spinal cord

(a)

(b)

Motor neuron cell body

Nerve

Motor unit 2

contracted and then relaxed. Panel (b) shows what happens if nerve impulses reach a muscle fiber before it has had time to relax—that is, while it is still somewhat contracted. As illustrated, this creates additional tension and the muscle fiber contracts more forcefully. In such instances, the second contraction "piggybacks" on the first. The waves of contract are added up; hence the name *wave summation.*

If the nerve impulses arrive frequently at enough skeletal muscle fibers, a smooth, sustained contraction occurs in the muscle (**FIGURE 13-19C**). This is called **tetanus** (not to be confused with the serious, often fatal bacterial infection of the same name).[4] Tetanic contractions occur in your arm muscles when you carry a bag of groceries. In such instances, the muscles contract to support the weight and remain contracted throughout the activity. Tetanic contractions eventually cause muscle fatigue and muscles stop contracting, even though the neural stimuli may continue.

Muscle Tone Results from the Contraction of a Small Number of Muscle Fibers That Keep Muscles Slightly Tense

Touch one of your muscles. Even if you are not in peak physical condition, you will notice that the muscle is firm. This firmness is called **muscle tone.** Muscle tone is essential for maintaining posture. Without it, you would fall into a heap on the floor. Muscle tone also generates heat in warm-blooded animals like us.

Muscle tone results from the contraction of muscle fibers during periods of inactivity. But not all fibers contract—just enough of them to keep the muscles slightly tense. Muscle tone is maintained, in part, by the muscle spindle, a receptor that monitors muscle stretching (Chapter 12). The spindle "alerts" the brain and spinal cord to the degree of stretching. When muscles relax, signals travel to the spinal cord, and then back out to motor axons, which stimulate a low level of muscle contraction, maintaining muscle tone.

cle fibers are called into action. Therefore, if you were to lift a piece of Styrofoam, only a small fraction of the motor neurons going to the arm muscle would be used. If you were to lift a 40-pound bag of dog food, your brain would recruit many more motor neurons and muscle fibers.

Additional Tension May Be Created in a Muscle Fiber if a Nerve Impulse Arrives While the Muscle Fiber Is Still Contracted.

The strength of muscle contraction can also be increased by a process called **wave summation.** Let me explain what this means.

Each time a muscle fiber is stimulated, it contracts to produce a twitch. **FIGURE 13-19A** shows the strength of twitch (contractions) resulting from two separate nerve impulses, the second arriving after the muscle fiber has

---

[4]The bacterial infection is called *tetanus* because the toxins from the bacterial infection cause the muscle to lock.

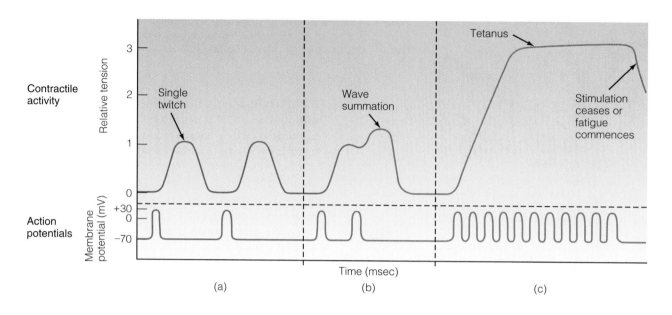

The duration of the action potentials is not drawn to scale, but is exaggerated.

FIGURE 13-19 Graph of Muscle Contraction *(a)* The force generated by two separate stimuli (action potentials from motor neuron). *(b)* The force generated by two closely timed stimuli. *(c)* The force generated by many closely spaced stimuli.

Two Types of Muscle Fibers Are Found in Skeletal Muscle, Slow- and Fast-Twitch

Physiological studies have revealed the presence of two types of skeletal muscle fibers: fast-twitch and slow-twitch fibers.[5] **Slow-twitch muscle fibers** contract relatively slowly but have incredible endurance. It is not surprising, then, that the muscles of endurance athletes (for example, long-distance runners) contain a high proportion of slow-twitch muscle fibers (**FIGURE 13-20**). These permit such athletes to perform for long periods without tiring. The flight muscles of birds, which can travel long distances without stopping, are similarly made.

Fast-twitch muscle fibers contract swiftly. The muscles of sprinters and other athletes whose performance depends on quick bursts of activity contain a high proportion of fast-twitch fibers.

Skeletal muscles generally contain a mixture of slow- and fast-twitch fibers, giving each muscle a wide range of performance abilities. However, a muscle that performs one type of function more often than another tends to have a disproportionately higher number of fibers corresponding to the type of activity it performs.

FIGURE 13-20 Built to Last An abundance of slow-twitch fibers gives the Olympic long-distance runner endurance.

[5]In truth, there are three types—one kind of slow-twitch and two types of fast-twitch. One type of fast-twitch fiber fatigues rather easily, whereas the other type is fatigue resistant.

The muscles of the back, for example, contain a larger number of slow-twitch fibers. These muscles operate throughout the waking hours to maintain posture. They do not need to contract quickly, but they must be resistant to fatigue. In contrast, the muscles of the arm are used for many quick actions—waving, playing tennis, or grasping falling objects. Fast-twitch fibers are more common in the muscles of the arm.

Genetic Differences May Account for Differences in Athletic Performance.

New research suggests that one of the reasons some people excel in certain sports may lie in the relative proportion of fast- and slow-twitch fibers in their muscles. In fact, a study of the skeletal muscles of world-class long-distance runners suggests that their physical endurance is primarily due to a high proportion of slow-twitch fibers, a trait that may be genetically determined. Biochemical studies also show that the muscle cells in endurance athletes have a higher level of ATP, both at rest and during exercise, thus providing more energy for muscle contraction. Endurance athletes start out with a larger storehouse of energy and maintain a larger supply throughout exercise.

Exercise Builds Muscles and Increases Endurance

Milo of Croton was a champion wrestler in ancient Greece who stumbled across a revolutionary way to build muscle. His method was simple. Milo borrowed a newborn calf, then proceeded to carry the animal around every day. Day after day, he faithfully followed this routine so that by the time the calf had become a bull, so had he. Milo's program worked, for he went on to win six Olympic championships in wrestling.

The Greek wrestler discovered a principle of muscle physiology familiar to weight lifters today: When muscles are made to work hard, they respond by becoming larger and stronger. The increase in size and strength results from an increase in the amount of contractile protein inside muscle cells.

Unfortunately, muscle protein is quickly made and quickly destroyed. In fact, about half of the muscle you gain in a weight-lifting program is broken down 2 weeks after you stop exercising. In order to build muscle, then, the rate of formation must exceed the rate of destruction. If you don't keep up with your exercise program, your newly developed biceps will disappear quickly.

High-intensity exercise, such as weight lifting, builds muscle, and it takes surprisingly little exercise to have an effect. Working out every other day for only a few minutes will result in noticeable changes in muscle mass. According to some sources, 18 contractions of a muscle (in three sets of six contractions each) are enough to increase muscle mass, if the contractions force your muscles to exert over 75% of their maximum capacity.

Low-intensity exercise, such as aerobics and swimming, tends to burn calories but not build muscle bulk. Aerobic exercise therefore helps increase endurance—that is, it increases one's ability to sustain muscular effort. Stamina or endurance results from numerous physiological changes. One of the most important changes occurs in the heart.

The heart responds to exercise like any other muscle—it grows stronger and enlarges. A well-exercised heart beats more slowly but pumps more blood with each beat, as explained in Chapter 6. The net result, then, is that the heart works more efficiently, delivering more oxygen to skeletal muscles.

Increased endurance also results from improvements in the function of the respiratory system. For example, exercise increases the strength of the muscles involved in breathing. These muscles become stronger and can operate longer without tiring. Breathing during exercise becomes more efficient.

Increased endurance may also be attributed to an increase in the amount of blood that occurs as a result of exercise. An increase in blood volume, in turn, results in an increase in the number of RBCs, which increases the amount of oxygen available to cells. This improvement, combined with others, allows an individual to work out longer without growing tired.

When you set out on an exercise program, it is important to establish your goals first. If you are interested in increasing your endurance, you should pursue an exercise regime that works the heart and muscles at a lower intensity over longer periods such as riding a stationary bicycle. If you are after bulk, the answer lies in high-intensity exercises such as weight lifting that result in increased muscle mass.

Many health clubs and university gyms offer advice and exercise programs to help you achieve your goals. And many offer a variety of machines to help you build muscles or simply tone them up. Most of the exercise machines work one particular muscle group—for example, the muscles of the upper arm or the muscles of the chest. In most gyms, a half-dozen or more machines are usually placed in a line so you can go down the line, working one set of muscles after another until you have exercised your entire body.

Exercise machines are popular because they are safer than free weights (barbells and dumbbells). Progressive resistance machines eliminate the chances of your dropping a weight on your toes—or someone else's! Moreover, it is almost impossible to strain your back if you make a mistake using one. In contrast, lifting free weights requires care and training as well as brawn.

These machines also reduce the amount of time a person needs to exercise by about half because they've been designed to require work when a joint is both flexed and extended. The biceps machine, for example, requires you to pull the weights up, then return them slowly. Your muscles are being forced to work in both directions.

Such machines also permit individuals to lift heavier weights. The reason for this is simple. With free weights, you can only lift a weight that can safely be moved through the part of the exercise where your muscle is the weakest. Any more, and you can tear a muscle or drop a weight. One popular brand of progressive-resistance machine, Nautilus, alters the resistance automatically as you perform an exercise. In the weakest phase of the exercise, the machine reduces the resistance to prevent damage. Throughout the rest of the exercise, however, the resistance is full.

Health, Homeostasis, and the Environment

Athletes and Steroids

The skeleton and muscles provide many obvious benefits, the most evident one being mobility. These two systems also contribute to homeostasis. The skeleton, for instance, serves as a calcium reservoir, essential for controlling blood levels of calcium, which are crucial for normal muscle function. Muscle helps provide the heat necessary for the normal operation of cells.

This section is a departure from previous Health, Homeostasis, and the Environment sections, which have shown how health is affected by environmental pollution or adverse environmental conditions that upset homeostatic workings of cells and various organ systems. Here we will discuss how an intentional action (taking anabolic steroids) to improve a body system (the muscles) can upset other systems.

Anabolic steroids are synthetic hormones that resemble the male sex hormone, testosterone. When taken in large doses, anabolic steroids stimulate muscle formation. They increase muscle size and strength by stimulating protein synthesis in muscle cells. High doses may also reduce the inflammation that frequently results from heavy exercise, allowing athletes to work out harder and longer.

When it comes to building muscle, steroids and exercise are an unbeatable combination. Some users claim that steroids even increase aggression, which may be helpful to competitive athletes such as football players.

Despite the benefits of steroids, most physicians view them as a dangerous proposition. For example, steroids can result in psychiatric (mental) and behavioral problems. One set of researchers interviewed 41 athletes who used steroids in doses 10–100 times greater than those used in medical studies. The athletes also reported using as many as five or six steroids simultaneously in cycles lasting 4–12 weeks, a practice known as "stacking." In this study, researchers found that one-third of the athletes developed severe psychiatric complications.

Athletes in the study reported episodes of severe depression during and after steroid use. Some reported feelings of invincibility. One man, in fact, deliberately drove a car into a tree at 40 miles per hour while a friend videotaped him. Some subjects reported psychotic symptoms in association with steroid use, including auditory hallucinations (hearing voices). Withdrawal from steroids resulted not only in depression, but also in suicidal tendencies.

Other studies have shown that steroids used in excess may damage the heart and kidneys and reduce testicular size in men. In women, steroids deepen the voice and may cause enlargement of the clitoris. Steroids also cause severe acne and liver cancer. Unfortunately, there are no scientific studies on the long-term health effects of steroids.

Despite the fact that anabolic steroids are banned by the National Football League, the International Olympic Committee, and college athletic programs, athletes continue to use them. Most steroids used in the United States are imported illegally from Mexico and Europe. Because of a federal crackdown on the importation of steroids, some experts believe that the inflow may be slowing. Others are not so optimistic, contending that the $100-million-a-year black market will not be easily deterred.

A recent survey of 46 public and private high schools across the United States involving over 3000 teenagers suggests that steroid use is especially prevalent in high school seniors. About 1 of every 15 senior boys reported taking anabolic steroids. The study also showed that the use of anabolic steroids begins in junior high school. Two-thirds of the students surveyed said that they had used steroids by age 16. Nearly half the users said they took the drugs to boost athletic performance, and 27% said their primary motive was to improve their appearance. Researchers say that adolescents who use steroids may be putting themselves at risk of stunted growth, infertility, and psychological problems.

Steroid use in the United States illustrates our dependence on quick fixes. It is a tragic result of the almost

obsessive focus on performance and achievement in our highly competitive society, a social environment that may be endangering the health of our children and our athletes.

SUMMARY

STRUCTURE AND FUNCTION OF THE HUMAN SKELETON

1. *Bones* serve many functions. They provide internal support, allow for movement, and help protect internal body parts. They also produce blood cells and platelets, store fat, and help regulate blood calcium levels.
2. Most bones have an outer layer of *compact bone* and an inner layer of *spongy bone*. Inside the bone is the *marrow cavity*, filled with either fat cells (*yellow marrow*) or blood cells and blood-producing cells (*red marrow*) or with combinations of the two.
3. The *joints* unite bones. Three types of joints are found: immovable, slightly movable, and freely movable. The movable joints allow for flexion, extension, and other important movements.
4. Joints are vulnerable to injury and disease. Torn ligaments and ripped cartilage are common injuries among athletes. Wear and tear on some joints can cause the articular cartilage to crack and flake off, resulting in degenerative joint disease, or *osteoarthritis*. *Rheumatoid arthritis* results from an autoimmune reaction that produces a painful inflammation and thickening of the *synovial membrane*, disfiguring and stiffening joints.
5. Most of the bones form from hyaline cartilage in a process called *endochondral ossification*.
6. Bone is constantly remodeled after birth to accommodate changing stresses. During bone remodeling, *osteoclasts* destroy bone. *Osteoclasts* are stimulated by *parathormone*, produced by the parathyroid glands.
7. Osteoclasts also participate in the homeostatic control of blood calcium levels. When activated, these cells free calcium from the bone, raising blood calcium levels.
8. *Osteoblasts* are bone-forming cells stimulated by *calcitonin* from the thyroid

gland. Calcitonin secretion decreases blood calcium levels and increases the amount of calcium in bones.
9. *Osteoporosis* is a disease of the bone caused by progressive decalcification. The bones become brittle and easily broken. Osteoporosis is most common in postmenopausal women and results from the loss of the ovarian hormone estrogen. It also occurs in people who are immobilized for long periods.
10. Osteoporosis can be prevented and reversed by exercise, calcium supplements, calcium-rich foods, vitamin D, and estrogen therapy.

THE SKELETAL MUSCLES

11. *Skeletal muscles* are involved in body movements; they help maintain our posture and produce body heat both at rest and while we are working or exercising.
12. Each skeletal muscle consists of many long, unbranched, multinucleated cells known as *muscle fibers*. Muscle fibers are both excitable and contractile.
13. Inside each muscle fiber are numerous *myofibrils*, bundles of the contractile filaments *actin* and *myosin*.
14. Contraction occurs when the heads of the myosin filaments attach to binding sites on the actin filaments, then pull the actin filaments inward.
15. Muscle contraction is stimulated by nerve impulses from motor neurons that cause the release of acetylcholine.
16. The energy for muscle contraction comes from ATP. ATP is replenished by creatine phosphate, glycolysis, and cellular respiration.
17. Individual muscle fibers contract when stimulated by an action potential, producing a *twitch*. Contractions of varying strength (graded contractions) can be generated in whole muscles by recruitment and by wave summation.
18. *Recruitment* results from the engagement of many motor units. A *motor

unit* is a motor axon and all of the muscle cells it innervates.
19. *Wave summation* is a piggybacking of muscle fiber contractions that occurs when stimuli arrive before the fiber relaxes.
20. *Muscle tone* is the rigidity of resting muscle caused by low-level contraction of some muscle fibers.
21. The body contains two types of skeletal muscle fibers: *slow-twitch* and *fast-twitch fibers*. Skeletal muscles generally contain a mixture of the two types, but fast-twitch fibers are found in greatest number in muscles that perform rapid movement. Slow-twitch fibers are found in muscles such as those of the back that perform slower motions or are involved in maintaining posture.
22. Muscle mass can be increased by certain forms of exercise, such as weight lifting. An increase in muscle mass results from an increase in the amount of contractile protein (actin and myosin) in muscle fibers.
23. Endurance can be increased by other forms of exercise such as aerobics.
24. Endurance is a function of at least three factors: the condition of the heart, the condition of the muscles of inspiration, and the blood volume. Improvement in all three factors increases the efficiency of oxygen delivery to muscles.

HEALTH, HOMEOSTASIS, AND THE ENVIRONMENT: ATHLETES AND STEROIDS

25. Many athletes are using synthetic anabolic steroids to improve performance and build muscle.
26. Unfortunately, massive doses of steroids have many harmful effects. They can increase aggression, cause psychiatric imbalance such as severe depression and result in damage to the heart and kidneys.

Critical Thinking

THINKING CRITICALLY—ANALYSIS

This Analysis corresponds to the Thinking Critically scenario that was presented at the beginning of this chapter.

If you noted that no mention was made of dietary calcium intake by the various groups, you're correct. Unfortunately, the diets of the women in this study were not monitored to determine how much calcium was being ingested through food and beverages. Second, this study focused on the bones of the forearm, not the load-bearing bones that are at greatest risk of fractures. Walking would probably influence the bones of the legs, hips, and back. To accurately assess the effects of the exercise regime prescribed in this study, the researchers probably should have looked at bone density in all three sites, not the arms. Given these inadequacies, do you think this study's conclusions are reliable?

EXERCISING YOUR CRITICAL THINKING SKILLS

A friend is selling an all-natural supplement guaranteed to give you more energy and help you sleep better at night. She cites a number of examples of users who feel energized and are sleeping better. She also notes that the developer of the product has a Ph.D. in nutrition. She wants you to try the product, which costs $40 a bottle. Using the critical thinking rules you learned in Chapter 1, describe what you might do in this situation. How would you evaluate the evidence your friend has given supporting the use of this supplement? What additional information would you want?

TEST OF CONCEPTS

1. Describe the functions of bone. In what ways does bone participate in homeostasis?

2. The synovial joints move relatively freely. What structures support the joint, helping keep the bones in place?

3. A young patient comes to your office with swollen joints and complains about pain and stiffness in the joints. Friends have suggested that the boy has arthritis, but his parents argue that he is too young. Only old people get arthritis, they say. How would you answer them?

4. Describe the process of bone formation. Be sure to define the following terms: primary and secondary centers of ossification, osteoclasts, perichondrium, osteoblasts, and osteocytes.

5. Bone is constantly remodeled, from infancy through adulthood. Explain when bone remodeling occurs and what cells and hormones participate in the process.

6. Using what you know about bone, explain why an office worker who exercises very little is more likely to break a bone on a skiing trip than a counterpart who works out every night after work.

7. Describe how a bone heals after being fractured.

8. A 30-year-old friend of yours who smokes, exercises very little, and avoids milk products because of her diet says: "Why should I worry about osteoporosis? That's a disease of old women." Based on what you know about bone, how would you respond to her?

9. Describe the major functions of skeletal muscle.

10. Describe the detailed structure of a skeletal muscle fiber.

11. Describe the molecular events involved in muscle contraction.

12. Define the term *graded contraction,* and describe how they are achieved.

13. Describe the two types of skeletal muscle fibers, and explain how they differ.

14. A friend comes to you complaining that he can't seem to lose weight. He works out on barbells three times a week for an hour or so each time. His diet hasn't changed much, but with the increase in exercise, he thinks he should be losing weight, rather than staying even. Why do you think he isn't losing weight?

15. A friend is thinking about using anabolic steroids to improve his performance in gymnastics. He argues that he is young and will be taking the drug only for a year or so. He points out that other young men are using steroids and they really help build muscle and endurance. What advice would you give him?

TOOLS FOR LEARNING

www.jbpub.com/humanbiology

Tools for Learning is an on-line student review area located at this book's web site HumanBiology (www.jbpub.com/humanbiology). The review area provides a variety of activities designed to help you study for your class:

Chapter Outlines. We've pulled out the section titles and full sentence sub-headings from each chapter to form natural descriptive outlines you can use to study the chapters' material point by point.

Review Questions. The review questions test your knowledge of the important concepts and applications in each chapter. Written by the author of the text, the review provides feedback for each correct or incorrect answer. This is an excellent test preparation tool.

Flash Cards. Studying human biology requires learning new terms. Virtual flash cards help you master the new vocabulary for each chapter.

Figure Labeling. You can practice identifying and labeling anatomical features on the same art content that appears in the text.

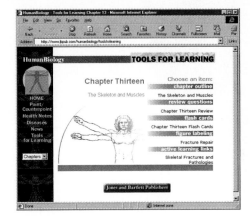

Active Learning Links. Active Learning Links connect to external web sites that provide an opportunity to learn basic concepts through demonstrations, animations, and hands-on activities.

THE ENDOCRINE SYSTEM

Thinking Critically

One of the hormones you'll learn about in this chapter is growth hormone. As its name implies, it stimulates growth. Some individuals' bodies produce inadequate amounts, resulting in mild to severe stunted growth.

With the advent of genetic engineering, growth hormone is now commercially available. As a result, physicians can now treat patients whose bodies produce inadequate amounts, permitting them to grow more or less normally. Some parents and doctors are treating children who might be normally small. Should growth hormone be given to these "normal" children to enhance their growth—for example, to make them better basketball players? ■

Special image of the human thyroid gland.

ONE DAY WHILE ON HIS MORNING JOG, PRES-ident George Bush began to feel weak. His heartbeat became irregular. He felt breathless. Fearing a heart attack, his security entourage rushed the president to the hospital where a blood test revealed elevated levels of a hormone called *thyroxine*. Produced by the thyroid gland (THIGH-roid), thyroxine accelerates all chemical reactions in the body, affecting a number of mental as well as physical processes. This condition, called **hyperthyroidism** (HIGH-per-THIGH-roid-izm), or **Grave's disease** (after its discoverer, not the final outcome), is rare and can occur at any age. It is just one of many diseases caused by a hormonal imbalance. This chapter discusses hormones of the endocrine system and presents some interesting examples of homeostatic imbalance caused by endocrine disorders.

Section 14-1

Principles of Endocrinology

The human **endocrine system** consists of numerous small glands scattered throughout the body (**FIGURE 14-1**). These highly vascularized glands produce and secrete chemical substances known as hormones (HOAR-moans). (The word *hormone* comes from the Greek *hormon*, which means "to stimulate" or "excite.") A **hormone** is a chemical produced and released by cells or groups of cells that constitute the **endocrine (ductless) glands**.

Hormones travel in the blood to distant sites, where they elicit certain response(s). The hormone insulin, for example, is produced by the pancreas and travels in the blood to skeletal muscle and other body cells, where it stimulates glucose uptake and glycogen synthesis. The cells affected by a hormone are called its **target cells**. For a discussion of some of the early research that contributed to our understanding of hormones, see Scientific Discoveries 14-1.

Hormones function in five areas: (1) homeostasis; (2) growth and development; (3) reproduction; (4) energy production, storage, and use; and (5) behavior. At any one moment, the blood carries dozens of hormones. The cells of the body are therefore

exposed to many different chemical stimuli. How does a cell keep from responding to all of these signals?

▮ Target Cells Contain Receptors for Specific Hormones

Despite the fact that they're literally flooded with hormonal signals, target cells respond only to specific ones. In other words, target cells are selective. Selectivity is conferred by protein **receptors**. In some target cells, the receptors are embedded in the plasma membrane; in others, they're located in the cytoplasm. (More

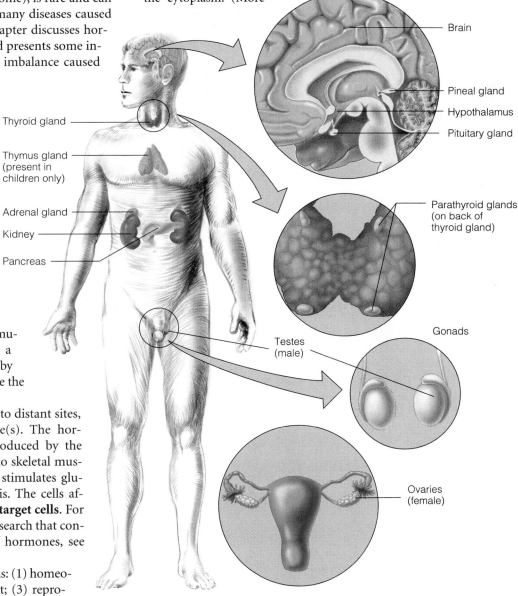

Brain

Thyroid gland

Thymus gland (present in children only)

Pineal gland

Hypothalamus

Pituitary gland

Adrenal gland

Kidney

Pancreas

Parathyroid glands (on back of thyroid gland)

Gonads

Testes (male)

Ovaries (female)

FIGURE 14-1 The Human Endocrine System The endocrine system consists of a scattered group of glands that produce hormones, chemicals that regulate growth and development, homeostasis, reproduction, energy metabolism, and behavior.

Scientific Discoveries that Changed the World 14–1

Pancreatic Function: Is It Controlled by Nerves or Hormones?

Featuring the Work of Bayliss and Starling

The concept of chemical control of body functions, like many other important biological ideas, is rooted in numerous experiments.

Of the many experiments crucial to our understanding of endocrinology, one by two British physiologists, W. M. Bayliss and E. H. Starling, stands out. Published in 1902, this experiment set to rest a debate about the mechanism controlling the release of pancreatic secretions.

Several experiments before the publication of this work suggested that hydrochloric acid in chyme (produced in the stomach) stimulated nerve endings in the small intestine. This, in turn, was supposed to trigger a reflex that caused the release of pancreatic juices containing sodium bicarbonate. In these studies, researchers injected substances into the small intestine of anesthetized dogs to see what stimulated the release of pancreatic secretions. Inter-

estingly, even though severing the nerves that supplied the small intestine did not affect the results, researchers still argued that a nervous system reflex was involved in the control. They hypothesized that clumps of nerve cells (ganglia) and their network of interconnected fibers in the wall of the intestine participated in local reflex arcs.

Bayliss and Starling repeated these experiments but were careful to sever all nerve connections to the small in-

on this later.) Each cell contains receptors for the hormones it's genetically programmed to respond to.

▎ Hormones Stimulate the Synthesis and Release of Other Hormones or Activate Cellular Processes

Hormones fall into two broad categories. The first are the trophic or more commonly today, tropic hormones (TROW-pick; "to nourish"). **Tropic hormones** stimulate the production and secretion of hormones by other endocrine glands. An example is **thyroid-stimulating hormone** (**TSH**). Produced by the pituitary gland, TSH travels in the blood to the thyroid gland located in the neck on either side of the larynx. TSH stimulates the cells of the thyroid, causing them to release a hormone known as **thyroxine** (thigh-ROX-in).

Thyroxine, in turn, circulates in the blood and stimulates metabolism in many types of body cells. Thyroxine is a nontropic hormone. **Nontropic hormones** stimulate cellular growth, metabolism, or other functions.

Hormones can also be classified into three groups according to their chemical composition: (1) steroids, (2) proteins and polypeptides, and (3) amines.

Steroid hormones are derivatives of cholesterol. **FIGURE 14-2** illustrates two common steroid hormones. Only minor differences in the chemical structure of these molecules exist, but they produce profound functional differences. **Protein** and **polypeptide hormones** are the largest group of hormones. Proteins and polypeptides are polymers of amino acids (Chapter 3). Growth hormone and insulin are two examples. **Amine hormones** are derivatives of the amino acid tyrosine (TIE-row-seen). **FIGURE 14-3** shows the structure of two of the four amine hormones produced in the body.

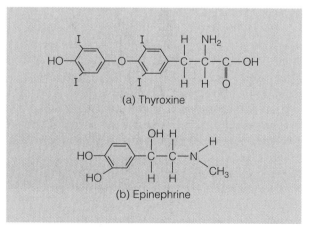

(a) Thyroxine

(b) Epinephrine

FIGURE 14-3 Representative Amine Hormones *(a)* Thyroxine from the thyroid and *(b)* adrenalin (epinephrine) from the adrenal medulla.

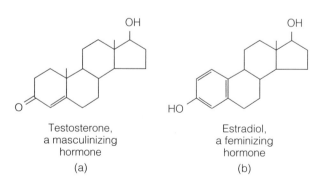

Testosterone, a masculinizing hormone

(a)

Estradiol, a feminizing hormone

(b)

FIGURE 14-2 Two Common Steroids

testine. They then tied off a piece of the small intestine and injected small amounts of hydrochloric acid into it. This treatment resulted in the production of pancreatic juices. Because it was previously known that acid introduced into the bloodstream had no effect on the release of pancreatic secretions, the researchers concluded that "the effect was produced by some chemical substance finding its way into the veins of the loop of jejunum in question. . . ." This substance, they wrote, was "carried in the bloodstream to the pancreatic cells."

To verify their hypothesis, the researchers tried another experiment. In this one, they scraped off the cells lining the section of the small intestine, exposed them to acid, ground them up, and then extracted the fluid, which they injected into the bloodstream. After a brief latent period, the pancreas began secreting, a clear indication that these cells were producing a chemical substance that stimulated the pancreas.

Bayliss and Starling referred to the mystery substance as "secretin," be-cause it stimulated pancreatic secretion. This name remains in use today.

Besides settling the debate over the control of pancreatic function, Bayliss and Starling helped clarify our understanding of endocrinology, furnishing definitions for some of the most important terms in the field. One good example is the term *endocrine gland*, which they defined as an organ that secretes specific substances into the blood that affect some other organ or process located at a distance from the gland.

Hormone Secretion Is Often Controlled by Negative Feedback Mechanisms

Hormones help to control many homeostatic mechanisms. Not surprisingly, their production and release are generally controlled by negative feedback loops (Chapter 4). Consider the hormone glucagon (GLUE-ka-gone).

Glucagon is secreted (seh-CREET-ed) by cells in the pancreas when glucose concentrations fall—for example, between meals. After being released into the bloodstream, glucagon travels to the liver, where it stimulates the breakdown of glycogen. Glucose released from this process then enters the bloodstream, restoring normal levels. When normal glucose levels are achieved, the glucagon-producing cells in the pancreas cease secretion.

Not all feedback loops are as simple as this one. Some involve intermediary compounds. Nevertheless, all operate on the same basic principle.

Positive feedback loops are also encountered in the endocrine system, but only rarely. A **positive feedback loop** results when the hormonal product of a cell or organ stimulates the production of another hormone. This, in turn, causes more secretion and so on and so on. (The noise in a restaurant is a good example of a positive feedback. As the noise level increases, people talk louder. This increases the overall noise, forcing people to talk even louder.)

Positive feedback loops perform specialized functions. Ovulation (OV-you-LAY-shun), the release of the ovum from the ovary, for example, is stimulated by a positive feedback loop discussed in Chapter 19. Fortunately, each positive feedback loop has a built-in mechanism that ends the escalating cycle, preventing the response from getting out of hand.

Most Hormones Undergo Periodic Fluctuations in Their Release

The fact that hormones are controlled by negative feedback loops does not mean that hormone concentrations in the blood are constant 24 hours a day, 365 days a year. In fact, virtually all hormones undergo periodic fluctuations in their release, causing many body functions to also fluctuate. These natural fluctuations in body function are called **biological cycles**, or **biorhythms**, and were described in Chapter 4.

Biological cycles vary in length. Some hormones are released in hourly pulses, or cycles. Others are released in daily cycles. Cortisol (CORE-te-zol) is a steroid hormone produced by the adrenal cortex in a 24-hour, or **circadian rhythm** (sir-KAY-dee-an) (**FIGURE 14-4**). Cortisol has many functions, one of which is to increase

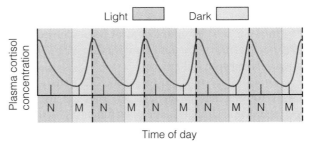

N is noon; M is midnight.

FIGURE 14-4 Cortisol Secretion Cortisol secretion follows a diurnal (daily) rhythm with highest levels occurring in the night just before waking. N = noon and M = midnight. Adapted with permission from George A. Hedge, Howard D. Colby, and Robert L. Goodman, *Clinical Endocrine Physiology* (Philadelphia: W. B. Saunders Company, 1987), Figure 4-4, p. 80.

levels of blood glucose. Its secretion increases during the night, reaching a peak just before one awakens, then falls sharply during the day.

Other hormones are released in monthly cycles. Estrogen and progesterone, for instance, are released in a 28-day menstrual cycle. Still other hormones are released in seasonal cycles. In humans, thyroid hormone is released in greater amounts during the winter than during the summer. Thyroid hormone stimulates metabolism, which raises body temperature.

Most hormonal cycles are controlled by biological clocks, those regions of the brain that regulate biological cycles (Chapter 4). Seasonal cycles, however, are usually controlled by environmental conditions—for example, temperature or photoperiod (day length). Such an interplay permits an organism to function in sync with environmental conditions.

The Chemical Nature of a Hormone Determines How It Is Transported in the Blood and How It Acts on Cells

Protein and polypeptide hormones are water-soluble and, therefore, readily dissolve in the plasma of the blood. In contrast, steroid hormones are lipids and do not dissolve in water. To be transported, they must be bound to much larger plasma proteins like albumin.

When they arrive at their target cells, protein and polypeptide hormones bind to receptors in the plasma membrane. This binding of hormone to receptor stimulates a series of metabolic changes that activate the target cell. Steroid hormones, being lipid-soluble, pass through the plasma membrane of target cells. Inside the cytoplasm, they bind to their receptors.

The Endocrine System and the Nervous System Are Both Control Systems, But They Operate in Markedly Different Ways

In the last two chapters, you studied the nervous system and the ways it controls basic body functions. The endocrine system is the second control system; however, it operates in quite a different fashion. Consider some of the similarities and differences.

Both the nervous and the endocrine systems send signals to cells that initiate changes. They are also both involved in coordinating basic body functions—creating a smooth operating system essential for homeostasis.

Despite these similarities, the nervous system and endocrine system differ in several key respects. Perhaps the most important is that the nervous system often elicits a rapid, short-lived response (for example, a muscle contraction). Although there are a few notable exceptions to this rule, the endocrine system generally brings about much slower, longer lasting responses

(such as a change in body temperature). The differences can be attributed to the type of signal found in each system. For example, messages in the nervous system are conveyed by bioelectric impulses that travel along the nerves of the body. In the endocrine system, messages are chemical in nature and are transmitted through the bloodstream. Despite these differences, the endocrine and nervous systems both promote homeostasis. As you will soon see, these systems sometimes work in concert.

Section 14-2

The Pituitary and Hypothalamus

Attached to the underside of the brain by a thin stalk is the **pituitary gland** (peh-TWO-eh-TARE-ee) (**FIGURE 14-5**). About the size of a pea, the pituitary lies in a depression in the base of the skull.

The pituitary gland is one of the most complex of all the endocrine organs. It is divided into two major parts: the anterior pituitary and the posterior pituitary. Together, they secrete a large number and variety of hormones, affecting a great many of the body's functions. The anterior pituitary, for example, produces seven protein and polypeptide hormones, six of which are discussed in this chapter (**TABLE 14-1**).

The release of hormones from the anterior pituitary is controlled by a region of the brain, the **hypothalamus**, lying just above the pituitary gland. The hypothalamus contains receptors that monitor blood levels of hormones, nutrients, and ions. When activated, the receptors stimulate specialized nerve cells within the hypothalamus. These cells are **neurosecretory neurons**—so named because they synthesize and secrete hormones. The hormones they produce act on the anterior pituitary (**FIGURE 14-5**). Some hypothalamic hormones stimulate the release of anterior pituitary hormones and are appropriately called **releasing hormones,** designated RH. Others inhibit the release of hormones from the anterior pituitary and are called **inhibiting hormones,** designated IH. How do these hormones travel to the pituitary?

The releasing and inhibiting hormones are produced in the cell bodies of the neurosecretory neurons. They reach the pituitary via what may seem like a fairly circuitous route. First, the hormones travel down the axons of the neurosecretory cells, which terminate in the lower part of the hypothalamus, just above the pituitary gland. The hormones are then stored in the axon terminals until needed.

When released, the hormones diffuse into nearby capillaries. As shown in **FIGURE 14-5**, these capillaries

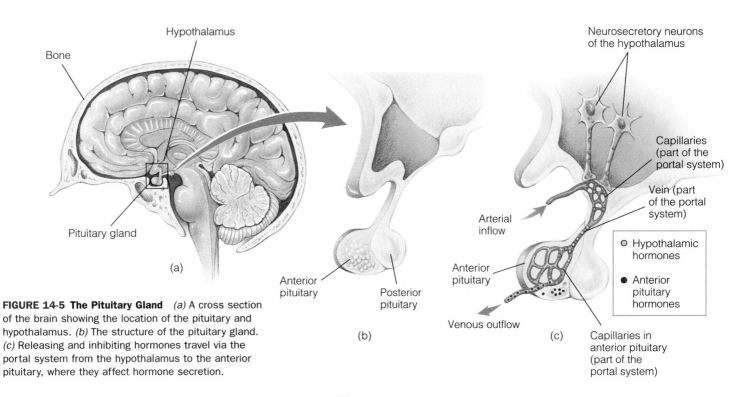

FIGURE 14-5 The Pituitary Gland *(a)* A cross section of the brain showing the location of the pituitary and hypothalamus. *(b)* The structure of the pituitary gland. *(c)* Releasing and inhibiting hormones travel via the portal system from the hypothalamus to the anterior pituitary, where they affect hormone secretion.

drain into a series of veins in the stalk of the pituitary, which drain into a capillary network in the anterior pituitary. From here, the hormones are free to diffuse out and reach their target cells. This unusual arrangement of blood vessels where a capillary bed drains to a vein, which then drains into another capillary bed, is called a **portal system.**[1]

_____

[1]Another portal system is associated with the liver and digestive tract.

The Anterior Pituitary Secretes Seven Hormones with Widely Different Functions

The pituitary produces a number of hormones that profoundly influence our bodies. We will now discuss the major ones produced by the anterior pituitary gland.

Growth Hormone Stimulates Cell Growth, Primarily Targeting Muscle and Bone.

It never fails to amaze me how much people differ. One of the most noticeable differences is in height and body build. What causes such variation?

TABLE 14-1

Hormones Secreted by the Pituitary Gland

Hormone	Function
Anterior pituitary	
Growth hormone (GH)	Stimulates cell growth. Primary targets are muscle and bone, where GH stimulates amino acid uptake and protein synthesis. It also stimulates fat breakdown in the body.
Thyroid-stimulating hormone (TSH)	Stimulates release of thyroxine and triiodothyronine.
Adrenocorticotropic hormone (ACTH)	Stimulates secretion of hormones by the adrenal cortex, especially glucocorticoids.
Gonadotropins (FSH and LH)	Stimulate gamete production and hormone production by the gonads.
Prolactin	Stimulates milk production by the breast.
Melanocyte-stimulating hormone (MSH)	Function in humans is unknown.
Posterior pituitary	
Antidiuretic hormone (ADH)	Stimulates water reabsorption by nephrons of the kidney.
Oxytocin	Stimulates ejection of milk from breasts and uterine contractions during birth.

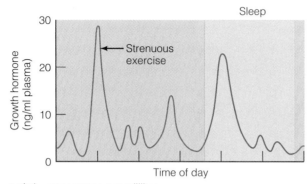

ng/ml = nanograms per mililiter

FIGURE 14-6 Growth Hormone Secretion in an Adult Growth hormone stimulates muscle and bone growth and is released during exercise and at night.

These differences are largely attributable to **growth hormone** (**GH**) a protein hormone produced by the anterior pituitary. As its name implies, growth hormone stimulates growth in the body. It does so by promoting cellular enlargement (hypertrophy; HIGH-per-TROW-fee) and an increase in the number of cells through division (hyperplasia; HIGH-per-PLAY-za).

Although growth hormone affects virtually all body cells, it acts primarily on bone and muscle. In muscle, it stimulates the uptake of amino acids and protein synthesis. In bone, growth hormone stimulates cell division and protein synthesis in bone cells, resulting in an increase in both the length and width of bone. Recent evidence shows that growth hormone actually operates via intermediaries—several polypeptide hormones produced largely by the liver. Known as **somatomedins** (so-MAT-oh-ME-dins), these hormones directly stimulate bone and cartilage cells.

As a rule, the more growth hormone produced during the growth phase of an individual's life cycle, the taller and heftier he or she will be.[2] In men, body growth is also stimulated by testosterone, an anabolic steroid produced by the testes, which stimulates bone and muscle growth, explaining why men are generally taller and more massive than women.

Growth hormone secretion undergoes a diurnal (daily) cycle. As **FIGURE 14-6** shows, the highest blood levels are present during sleep and during strenuous exercise. It is no wonder that sleep is so important to a growing child. Growth hormone secretion declines gradually as we age.

The secretion of growth hormone is controlled by a releasing hormone (GH-RH) and an inhibiting hormone (GH-IH). Both are produced by the hypothalamus. Growth hormone is controlled by a negative feedback loop. Its release can also be stimulated directly through the nervous system. Stress and exercise, for example, stimulate the hypothalamus to release GH-RH.

Deficiencies in growth hormone can result in dramatic changes in body shape and size, depending on when the deficiency occurs. Undersecretion, or **hyposecretion** (HIGH-poe-seh-KREE-shun), occurring during the growth phase of a child is one cause of stunted growth (dwarfism) (**FIGURE 14-7A**). Oversecretion, or **hypersecretion** (HIGH-per-seh-KREE-shun), results in **giantism** (GIE-an-tizm) if the excess occurs during the growth phase (**FIGURE 14-7B**). If the pituitary begins producing excess growth hormone after growth is complete, the result is a relatively rare disease called **acromegaly** (AH-crow-MEG-al-ee). Facial features become coarse, and hands and feet continue to grow throughout adulthood (**FIGURE 14-8**). Growth of the vertebrae results in a hunched back. The tongue, kidneys, and liver often become quite enlarged in patients with acromegaly.

Thyroid-Stimulating Hormone Stimulates the Thyroid Gland to Produce Thyroxine.

Thyroid-Stimulating Hormone (TSH) is a protein hormone produced by the anterior pituitary and is controlled by **TSH-RH** (**TSH-releasing hormone**) from the hypothalamus. TSH-RH secretion is stimulated by cold and stress.

(a)

(b)

FIGURE 14-7 Disorders of Growth Hormone Secretion (a) Pituitary dwarves. (b) Pituitary giant.

[2]Growth hormone production levels are ultimately determined by the genes.

FIGURE 14-8 Acromegaly Hypersecretion of growth hormone in adults results in a gradual thickening of the bone, which is especially noticeable in the face, hands, and feet. There is no sign of the disorder at age 9 *(a)* or age 16 *(b)*. Symptoms are evident at age 33 *(c)* and age 52 *(d)*.

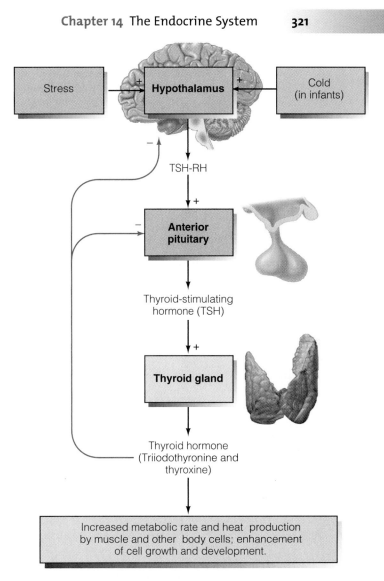

FIGURE 14-9 Negative Feedback Control of TSH Secretion Triiodothyronine (T_3) and thyroxine levels are detected by receptors in the hypothalamus. They also feedback on the anterior pituitary. When levels of T_3 and thyroxine are low, the hypothalamus releases TSH-RH, which stimulates the pituitary to release TSH. TSH, in turn, stimulates the thyroid gland to secrete T_3 and thyroxine. Other factors such as stress and cold influence the release of TSH via the hypothalamus. (A + denotes stimulation; a − denotes inhibition.)

Day-to-day levels, however, are primarily regulated by the level of thyroxine in the blood in a classical negative feedback loop relying on receptors in the hypothalamus that detect the level of circulating thyroxine. When circulating levels are low, these receptors signal the hypothalamus to release TSH-RH. When the level of thyroxine increases, TSH-RH secretion declines (**FIGURE 14-9**).

TSH-RH travels to the anterior pituitary, where it stimulates the production and release of TSH. TSH then travels in the blood to the thyroid gland, located in the neck (**FIGURE 14-1**). TSH stimulates the production and release of two thyroid hormones, thyroxine (thigh-ROX-in) and triiodothyronine (try-eye-OH-doe-THIGH-row-neen).

These hormones circulate in the bloodstream and influence many body cells. One of their chief functions is to stimulate the breakdown of glucose by body cells. Because this process produces energy and heat, increased levels of these hormones also raise body temperature—in the same way that adding wood to a fire increases its heat output.

ACTH Stimulates the Release of Hormones from the Adrenal Cortex.

The **adrenal cortex** is the outer layer of the adrenal gland and is under the control of the anterior pituitary via a polypeptide hormone known as ACTH (adrenocorticotropic hormone; ad-REE-no-core-tick-oh-TRO-pick). ACTH secretion, in turn, is controlled by the hypothalamus via ACTH-RH (**FIGURE 14-10**). Its release will be discussed shortly. For now, let us concentrate on ACTH and the adrenal gland.

As shown in **FIGURE 14-10**, ACTH stimulates the production and release of a group of steroid hormones known as the **glucocorticoids** (GLUE-co-CORE-teh-KOIDS). Glucocorticoids (like the hormone cortisol) increase blood glucose levels, thus helping maintain homeostasis.

ACTH-RH secretion is controlled by at least three factors, two of which are discussed here. The first is the level of circulating glucocorticoids, which participate in a negative feedback loop (**FIGURE 14-10**). ACTH-RH secretion is also controlled by stress, acting through the nervous system. A stress-stimulated increase in ACTH-RH results in a rise in ACTH release. This stim-ulates an increase in the release of glucocorticoids by the adrenal cortex, which increases blood glucose levels. This mechanism is an adaptation that ensures additional energy for cells (especially muscle) when the body is under stress.

The Gonadotropins Are Hormones That Stimulate the Ovaries and Testes.

Reproduction in both males and females is under the control of the anterior pituitary. It produces two hormones that affect the gonads, appropriately known as **gonadotropins** (go-NAD-oh-TROW-pins). Because these hormones are discussed in detail in the next chapter, a few words will suffice for now.

The pituitary's gonadotropins are follicle-stimulating hormone and luteinizing hormone. **Follicle-stimulating hormone** (**FSH**) promotes gamete formation in men and women. **Luteinizing hormone** (LEW-tin-IZE-ing), or **LH**, stimulates gonadal hormone production. In men, LH stimulates the production of testosterone, the male sex steroid, by the testes. In women, LH stimulates progesterone secretion by the ovaries. Both FSH and LH are under the control of **gonadotropin releasing hormone**, produced and secreted by the hypothalamus.

Prolactin Stimulates Milk Production in the Breasts of Women.

Prolactin (pro-LACK-tin) is a protein hormone produced by the anterior pituitary. In women, prolactin is secreted at the end of pregnancy and acts on the mammary glands (breasts), stimulating milk production.

Prolactin secretion is controlled by a **neuroendocrine reflex,** a reflex involving both the nervous and endocrine systems. As **FIGURE 14-11** shows, suckling stimulates sensory fibers in the breast. Nerve impulses travel to the hypothalamus via sensory neurons. In the hypothalamus, these impulses stimulate the release of prolactin releasing hormone (PRH). This hormone, in turn, travels to the anterior pituitary in the bloodstream, where it stimulates the secretion of prolactin. As long as suckling continues, prolactin secretion will continue.

As a side note, the neuroendocrine reflex is the basis of commercial milk production. When a cow gives birth, it produces milk to feed its calf. On dairy farms, though, calves are separated from their mothers fairly early in life. Milk production in their mothers continues, however, because manual or machine milking stimulates the neuroendocrine response.

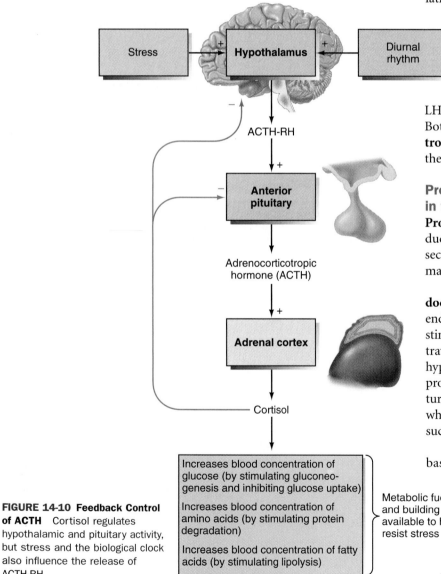

FIGURE 14-10 Feedback Control of ACTH Cortisol regulates hypothalamic and pituitary activity, but stress and the biological clock also influence the release of ACTH-RH.

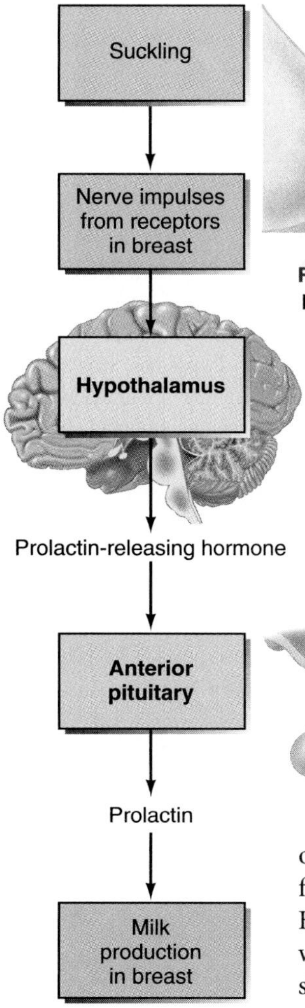

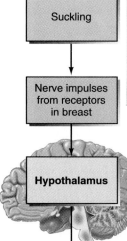

Suckling

↓

Nerve impulses from receptors in breast

↓

Hypothalamus

↓

Prolactin-releasing hormone

↓

Anterior pituitary

↓

Prolactin

↓

Milk production in breast

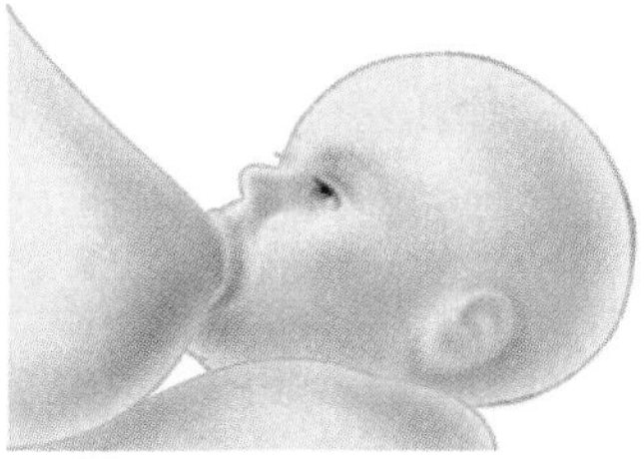

FIGURE 14-11 Neuroendocrine Reflex and Prolactin Secretion Suckling stimulates prolactin release by the anterior pituitary. Prolactin stimulates milk production by the breast. Milk release requires another hormone, oxytocin, from the posterior pituitary.

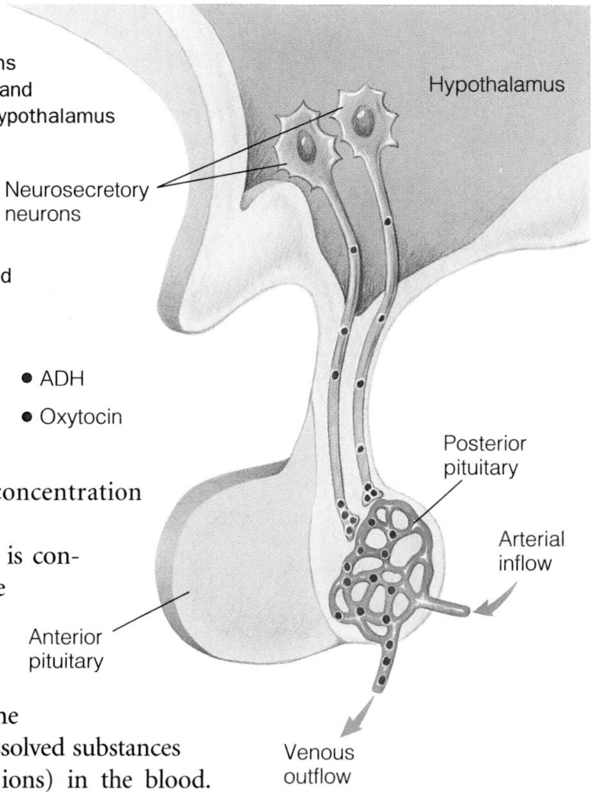

FIGURE 14-12 The Posterior Pituitary Neurosecretory neurons that produce oxytocin and ADH originate in the hypothalamus and terminate in the posterior pituitary. Hormones are produced in the cell bodies of the neurons and are stored and released into the bloodstream in the posterior pituitary.

● ADH
● Oxytocin

Hypothalamus

Neurosecretory neurons

Posterior pituitary

Arterial inflow

Anterior pituitary

Venous outflow

The Posterior Pituitary Secretes Two Hormones

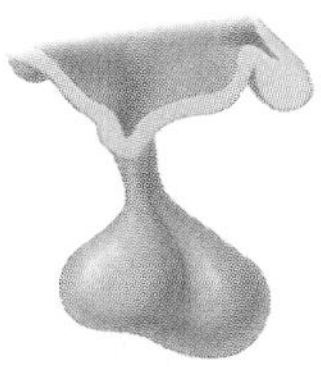

Have you ever wondered why alcoholic beverages make a person urinate so much or what actually causes milk to flow from a woman's breast? Have you ever wondered why a woman's uterus contracts when she delivers a baby?

These phenomena can be attributed to two hormones from the posterior pituitary. Like the hypothalamus, the **posterior pituitary** is a neuroendocrine gland—that is, a gland made of neural tissue that produces hormones. Derived from brain tissue during embryonic development, the posterior pituitary remains connected to the brain throughout life.

The posterior pituitary secretes two hormones: oxytocin (OX-ee-TOE-sin) and antidiuretic hormone (an-tie-DIE-yur-ET-ick) (ADH). Both hormones consist of nine amino acids. As shown in **FIGURE 14-12**, ADH and oxytocin are produced in the cell bodies of the neurosecretory cells located in the hypothalamus. They then travel down the axons into the posterior pituitary. The posterior pituitary itself consists of the axons and terminal ends of neurosecretory cells in which these hormones are stored until being released into the surrounding capillaries.

Antidiuretic Hormone Increases Water Absorption in the Kidney.

ADH regulates water balance in humans by increasing water absorption in the nephrons of the kidneys (**FIGURE 14-13**). As a result, water reenters the bloodstream, increasing blood volume and maintaining the normal osmotic concentration of the blood.

ADH secretion is controlled by a negative feedback loop (**FIGURE 14-13**). **Receptors** in the hypothalamus monitor the concentration of dissolved substances (especially sodium ions) in the blood. When the concentration exceeds the normal level—for example, during dehydration—ADH is released into the bloodstream. ADH stimulates the reabsorption of water by the nephrons, thus diluting the blood. When the osmotic concentration of the blood approaches homeostatic levels, ADH secretion ceases.

As Chapter 10 notes, ADH is inhibited by ethanol in alcoholic beverages. The reduction in ADH decreases water reabsorption in the kidneys and increases urine output. This, in turn, can lead to dehydration—the dry mouth and intense thirst a person may experience as part of a hangover.

ADH secretion may also decline as a result of traumatic head injuries that damage the hypothalamus. Such injuries can cause a sharp decline in ADH output. In the absence of ADH, the kidneys produce several gallons of urine a day. To keep up with the water loss and to prevent dehydration, an individual must drink enormous quantities of water. This condition is known as **diabetes insipidus** (Chapter 10).

Because of its ability to increase arterial blood pressure, ADH is sometimes referred to as **vasopressin** (VEY-zoh-PRESS-in). Under most conditions, however, it is doubtful that enough ADH is secreted to increase blood pressure. ADH secretion in levels sufficient to exert a vasopressive effect occurs when an individual suffers an extreme loss of blood. When released in massive quantities,

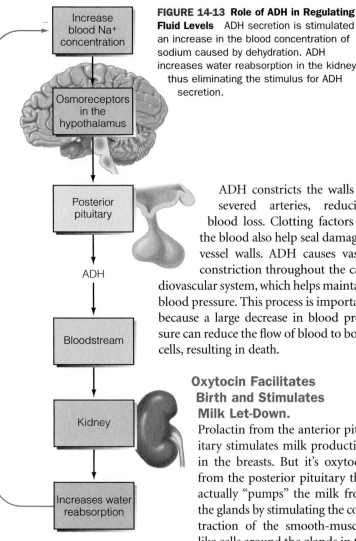

FIGURE 14-13 Role of ADH in Regulating Fluid Levels ADH secretion is stimulated by an increase in the blood concentration of sodium caused by dehydration. ADH increases water reabsorption in the kidney, thus eliminating the stimulus for ADH secretion.

ADH constricts the walls of severed arteries, reducing blood loss. Clotting factors in the blood also help seal damaged vessel walls. ADH causes vasoconstriction throughout the cardiovascular system, which helps maintain blood pressure. This process is important because a large decrease in blood pressure can reduce the flow of blood to body cells, resulting in death.

Oxytocin Facilitates Birth and Stimulates Milk Let-Down.

Prolactin from the anterior pituitary stimulates milk production in the breasts. But it's oxytocin from the posterior pituitary that actually "pumps" the milk from the glands by stimulating the contraction of the smooth-muscle-like cells around the glands in the breast.

Oxytocin release is stimulated by suckling. Sensory fibers in the breast conduct nerve impulses to the hypothalamus, triggering the release of oxytocin. The hormone travels in the blood to the breast, where it stimulates **milk let-down**—the ejection of milk from the glands soon after suckling begins. Milking machines and hand milking stimulate the reflex in dairy cattle.

At the end of pregnancy, in the hours prior to delivery, the walls of the uterus and cervix stretch. Impulses from the stretch receptors in the uterus travel to a region of the neurosecretory cells in the hypothalamus that produce oxytocin. Impulses from the stretch receptors cause the neurosecretory cells to release the hormone. Oxytocin travels in the blood to the uterus, where it stimulates smooth muscle contraction, aiding in the expulsion of the baby.

The Thyroid Gland

On a 50° day in the fall, you bundle up in a sweater to feel warm. However, 50° in the spring feels warm. You may not even need a sweater. Why the difference?

The answer lies in the thyroid gland and two of its hormones. The **thyroid gland** is a U- or H-shaped gland (it varies from one person to the next) located in the neck (**FIGURE 14-14A**). The thyroid gland produces three hormones: (1) thyroxine (tetraiodothyronine, or T_4), (2) a chemically similar compound, triiodothyronine (T_3), and (3) calcitonin. The first two are involved in controlling metabolism and heat production; the last helps regulate blood levels of calcium.

As **FIGURE 14-14B** shows, the thyroid gland consists of large, spherical structures known as **follicles** (FOLL-ick-als). Each follicle consists of a central region containing a gel-like glycoprotein called **thyroglobulin** and a surrounding layer of cuboidal **follicle cells** that produce thyroglobulin. Thyroxine and triiodothyronine are relatively small molecules (amines) derived from much larger thyroglobulin molecules in each follicle.

To make T_3 and T_4, the thyroid requires large quantities of iodide, I^-, which is actively transported into the cell. A shortage of iodide in the diet results in a decline in the levels of thyroxine and triiodothyronine in the blood. The hypothalamus responds by producing more TSH. TSH causes the thyroid to increase thyroglobulin production. Without sufficient iodide, however, the thyroid gland cannot produce thyroxine and triiodothyronine. As a result, the thyroid gland continues to produce thyroglobulin. This causes the gland to enlarge, sometimes forming softball-sized enlargements on the neck. This condition is known as **goiter** (GOY-ter) (**FIGURE 14-15**).

Goiter was once common in areas of Europe and the United States where iodine had been leached out of the soil by rain—for example, near the Great Lakes. Crops grown in iodine-poor soils failed to provide sufficient quantities. To counteract this deficiency, iodine was added to table salt. Today, few people in industrialized nations develop goiter because they eat iodized salt or ingest sufficient quantities of iodine in their food or through mineral supplements.

Thyroxine and Triiodothyronine Accelerate the Breakdown of Glucose and Stimulate Growth and Development

Thyroxine and triiodothyronine are virtually identical and, as a result, have nearly identical functions. Both

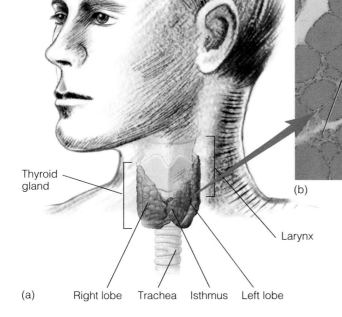

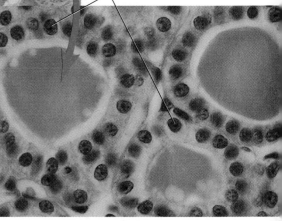

Follicles   Thyroglobulin

Parafollicular cell

(b)

(c)

FIGURE 14-14 The Thyroid Gland *(a)* The thyroid gland is located in the neck on either side of the larynx. *(b)* The thyroid follicles produce thyroxine and triiodothyronine. They are stored in the colloidal material called *thyroglobulin.* *(c)* Enlargement showing calcitonin-producing cells.

Thyroid gland

Larynx

(a)   Right lobe   Trachea   Isthmus   Left lobe

hormones, for example, accelerate the rate of glucose breakdown in most body cells. Thyroid hormones also stimulate cellular growth and development. Bones and muscles are especially dependent on them during the growth phase. Even normal reproduction requires these hormones. A deficiency of T_3 and T_4, for example, delays sexual maturation in both sexes. In children, depressed thyroid output stunts mental as well as physical growth. If the deficiency is not detected and treated, the effects are irreversible.

In adults, reduced thyroid activity, or **hypothyroidism**, is less severe. Hypothyroidism decreases the metabolic rate, making a person feel cold much of the time. People suffering from hypothyroidism also feel tired and worn out. Even simple mental tasks become difficult. Their heart rate may slow to 50 beats per minute. Hypothyroidism is treated by pills containing artificially produced thyroid hormone.

Excess thyroid activity, **hyperthyroidism**, in adults results in elevated metabolism, excessive sweating (due to overheating), and weight loss, despite increased food intake. The increase in thyroid hormone levels results in increased mental activity, resulting in nervousness and anxiety. People suffering from hyperthyroidism often find it difficult to sleep. Their heart rate may accelerate, as in the case of former President Bush, and they may lose their

sensitivity to cold. Some people suffering from hyperthyroidism exhibit a condition called **exophthalmos** (EX-oh-THAL-mus), or bulging eyes. In such individuals, the eyes may protrude so much that the eyelids cannot close completely. Exophthalmos can cause double vision or blurred vision.

Hyperthyroidism is treated in a number of ways. Patients may be given antithyroid medications, drugs that block the effects of thyroid hormones. Surgery may be required to remove part or all of the gland if it has become cancerous. The most common treatment for hyperthyroidism, however, is radioactive iodine. Iodine is

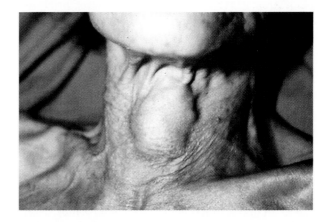

FIGURE 14-15 Goiter An enlargement of the thyroid gland most often results from a lack of iodine in the diet.

concentrated in the thyroid gland. Radioactive iodine therefore accumulates in the cells of the thyroid follicle, damaging them and reducing their output of thyroid hormones. This procedure, while effective, may lead to other problems (notably cancer) later in life.

◼ Calcitonin Decreases Blood Levels of Calcium

Large, round cells in the perimeter of the thyroid follicles produce **calcitonin**, or **thyrocalcitonin**, a polypeptide hormone. Calcitonin lowers the blood calcium level (**FIGURE 14-14C**) in one of three ways: (1) by inhibiting osteoclasts, bone-resorbing cells, thus reducing the release of calcium from bone. (2) by stimulating bone-forming cells, osteoblasts, causing calcium to be deposited in bone. (3) by increasing the excretion of calcium (and phosphate) ions by the kidneys.

Calcitonin is involved in a simple negative feedback loop with calcium ions in the blood. When the calcium-ion concentration increases, calcitonin secretion increases. As calcium concentrations fall, calcitonin secretion falls.

Section 14-4
The Parathyroid Glands

The **parathyroid glands** are four small nodules of tissue embedded in the back side of the thyroid gland. These glands, once mistaken by anatomists for undeveloped thyroid tissue, are actually independent endocrine glands. They produce a polypeptide hormone known as **parathyroid hormone**, or **parathormone** (**PTH**).

Parathyroid hormone increases blood calcium levels. Its secretion is stimulated when calcium levels in the blood drop. PTH restores blood calcium levels by (1) increasing intestinal absorption of calcium, (2) stimulating bone destruction by osteoclasts, and (3) increasing calcium reabsorption in the kidney.

Calcium homeostasis is also influenced by vitamin D. This compound increases calcium absorption in the intestine when blood levels fall. Vitamin D also increases the responsiveness of bone to PTH. In calcium homeostasis, then, PTH plays a major role that's supported by vitamin D.

As in other glands, the parathyroid may malfunction. **Hyperparathyroidism**, excess secretion of parathyroid hormone, is the most common condition and can result from a tumor of the parathyroid glands, which causes the secretion of excess PTH. Excess PTH results in a loss of calcium from the bones and teeth. Excess PTH also upsets several metabolic processes, resulting in indigestion and depression. Because bones contain enormous amounts of calcium, most symptoms of hyperparathyroidism (except high blood calcium) do not appear until 2–3 years after the onset of the disease. Therefore, by the time the disease is discovered, kidney stones may already have formed from calcium, cholesterol, and other substances, and bones may have become more fragile and susceptible to breakage. To prevent further complications, parathyroid tumors must be removed.

Section 14-5
The Pancreas

A young girl's parents complain to the daughter's doctor of frequent urination, bladder infections, fatigue, and weakness. Tests show that she is suffering from **diabetes mellitus**, a disorder of the pancreas. Diabetes mellitus has several causes, but in young people it is generally caused by a lack of insulin production.

Insulin stimulates the uptake of glucose by body cells. It also stimulates the synthesis of glycogen in liver and muscle cells. As noted in Chapter 5, glycogen is the storage form of glucose. Supplies of glycogen in the liver increase immediately after meals, then decline in the period between feedings as the body uses up its storage depots to supply body cells with glucose. Muscles use glycogen supplies principally during exercise.

Insulin is produced by the pancreas (PAN-kree-us), a dual-purpose organ located in the abdominal cavity and described in Chapter 5 (**FIGURE 14-16A**). The head of the pancreas lies in the curve of the duodenum, and its tail stretches to the left kidney. Most of the pancreas consists of tiny clumps of cells called *acini* (A-sin-eye) that produce digestive enzymes and sodium bicarbonate. Scattered throughout these enzyme-producing cells are small islands of endocrine cells, the **islets of Langerhans** (EYE-lits of LANG-er-hahns) as illustrated in **FIGURES 14-16B** and **C**. The islets are composed of four types of cells, two of which will be discussed in this chapter, the alpha cell (AL-fah) and the beta cell (BAY-tah).

◼ Insulin Is a Glucose-Storage Hormone and Is Produced by the Beta Cells

The **beta cells** produce insulin, a protein hormone released within minutes after glucose levels in the blood begin to rise. Insulin affects a number of cellular processes and a number of different cells. Its principal targets, however, are skeletal muscle cells, liver cells, and fat cells. In this chapter we examine a few of its major functions, beginning with skeletal muscles.

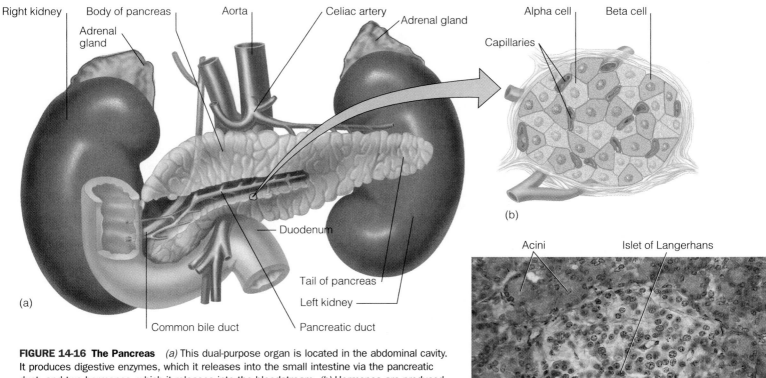

FIGURE 14-16 The Pancreas *(a)* This dual-purpose organ is located in the abdominal cavity. It produces digestive enzymes, which it releases into the small intestine via the pancreatic duct, and two hormones, which it releases into the bloodstream. *(b)* Hormones are produced by small clumps of cells called *islets of Langerhans*. *(c)* The islets of Langerhans, located among the acini of the pancreas, produce two hormones: glucagon, which increases blood glucose, and insulin, which lowers blood glucose.

Skeletal muscle cells are virtually impermeable to glucose in the absence of insulin. When insulin is present, however, the glucose transport into muscle cells increases dramatically. Glucose uptake rises because insulin stimulates facilitated diffusion. Insulin also increases glycogen synthesis in skeletal muscle cells, which not only helps store glucose for later use but also lowers intracellular concentrations, thus accelerating diffusion. In addition, insulin increases the uptake of amino acids by muscle cells and stimulates protein synthesis in them, thus promoting muscle formation.

In contrast, the plasma membrane of liver cells is quite permeable to glucose, so glucose enters with great ease whether or not insulin is present. Nevertheless, insulin still increases the uptake of glucose by liver cells by stimulating the addition of phosphate groups to glucose molecules that have entered the cytoplasm. Because phosphorylated glucose cannot diffuse through the plasma membrane, glucose is trapped in the liver cell. Insulin also increases glycogen formation, creating glucose stores for times of need (**FIGURE 14-17**). Phosphorylation and glycogen synthesis decrease the cytoplasmic concentrations of glucose, thus maintaining the con-centration gradient between the blood and the cytoplasm, which ensures a steady influx of glucose into the liver cell.

In fat cells, insulin increases glucose uptake and also stimulates lipid synthesis, creating food stores for times of need.

Glucagon Increases Blood Levels of Glucose, Thus Opposing the Actions of Insulin

A successful homeostatic mechanism often requires antagonistic—or opposing—controls. (In a car, we have an accelerator and brakes to control speed. In Congress, we have Democrats and Republicans.) In glucose homeostasis, the antagonistic control is provided by **glucagon**.

Produced by the **alpha cells** of the islets of Langerhans, glucagon is a polypeptide hormone that increases blood levels of glucose by stimulating the breakdown of glycogen in liver cells (**FIGURE 14-17**). This process, called **glycogenolysis** (GLY-co-jeh-NOL-i-sis), helps maintain proper glucose levels in the blood between meals. One

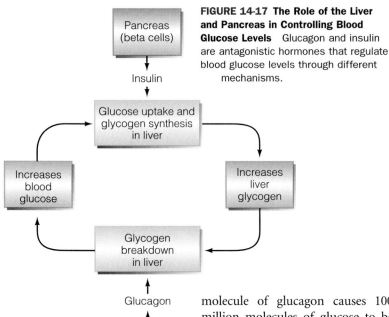

FIGURE 14-17 **The Role of the Liver and Pancreas in Controlling Blood Glucose Levels** Glucagon and insulin are antagonistic hormones that regulate blood glucose levels through different mechanisms.

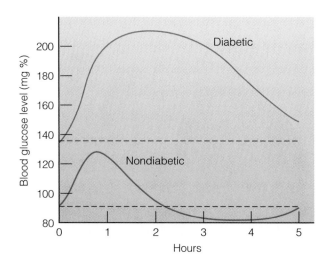

FIGURE 14-18 **Glucose Tolerance Test** In a diabetic (pink line), blood glucose levels increase rapidly and remain high after ingestion of glucose. In a normal person (blue line), blood glucose levels rise but are quickly reduced by insulin.

molecule of glucagon causes 100 million molecules of glucose to be released.

Glucagon also elevates blood glucose levels by stimulating **gluconeogenesis** (GLUE-coe-KNEE-oh-GEN-eh-siss). Although its name may be forbidding, it is merely the synthesis of glucose from amino acids and glycerol molecules derived from the breakdown of triglycerides. Gluconeogenesis occurs in the liver.

Glucagon secretion, like that of insulin, is regulated by a negative feedback mechanism stimulated by glucose concentrations in the blood. When glucose levels fall, the alpha cells release the hormone into the bloodstream. When glucose levels rise, glucagon secretion declines. Glucagon release is also stimulated by an increase in the concentration of amino acids in the blood. Because glucagon stimulates gluconeogenesis, this process ensures that excess amino acids in high-protein meals are used to produce glucose.

Diabetes Mellitus Is a Disease Resulting from an Insulin Deficiency or a Decrease in Tissue Sensitivity to Insulin

FIGURE 14-18 shows a graph of blood glucose levels in nondiabetic and diabetic people obtained during a **glucose tolerance test**. During this test, physicians give their patients an oral dose of glucose, then check blood levels at regular intervals. The blue line on the graph in **FIGURE 14-18** represents the blood glucose levels in a normal person with a healthy pancreas. As illustrated, glucose levels increase slightly after the administration of glucose at time zero. However, within 2 hours, glu-

cose levels have decreased to normal, thanks to insulin secretion.

The pink line illustrates the response in a diabetic person, who produces insufficient amounts of insulin or whose cells have become unresponsive to insulin. Glucose levels rise considerably in these people after they've been given an oral dose of glucose and they remain elevated for 4–5 hours. A similar response is seen in untreated diabetics after they have eaten a meal. In both instances, blood glucose levels remain high so long because there are few places for the glucose to go. In other words, the absence of insulin in diabetics means that virtually no glucose can enter skeletal muscle cells and that liver uptake is greatly reduced. Glycogen synthesis in liver and muscle cells is also greatly reduced.

Glucose levels eventually decline as this nutrient is used by brain cells, red blood cells, and others. The kidneys also excrete excess glucose. Unfortunately, this wastes a valuable nutrient and causes the kidneys to lose large quantities of water. Excess urination and constant thirst are the chief symptoms of untreated diabetes. Diabetics may have to urinate every hour, day and night.

Diabetes Is Really Two Diseases with Similar Symptoms But Different Causes.

Physicians recognize two types of diabetes (**TABLE 14-2**). **Type I diabetes** occurs mainly in young people and results from a deficiency in insulin production. Also called **juvenile diabetes** or **early-onset diabetes,** Type I diabetes is believed to be caused by damage to the insulin-producing

Diabetes

TABLE 14-2

Type I and Type II Diabetes

Characteristic	Type I Diabetes	Type II Diabetes
Level of insulin secretion	None or almost none	May be normal or exceed normal
Typical age of onset	Childhood	Adulthood
Percent of diabetics	10%–20%	80%–90%
Basic defect	Destruction of beta cells	Reduced sensitivity of insulin's target cells
Associated with obesity?	No	Usually
Genetic and environmental factors important in precipitating overt disease?	Yes	Yes
Speed of development of symptoms	Rapid	Slow
Development of ketosis	Common if untreated	Rare
Treatment	Insulin injections; dietary management	Dietary control and weight reduction; occasionally oral hypoglycemic drugs

cells of the pancreas. A leading hypothesis suggests that it is the result of an autoimmune reaction. Another holds that viral infection of the pancreas may be the cause.

Insulin production varies in patients with Type I diabetes. In some, it is only slightly reduced; in others, it is completely suppressed. In the latter, the absence of insulin starves body cells for glucose. Energy must be provided by the breakdown of fats.

Type II diabetes usually occurs in people over 40 and is also called **late-onset diabetes** (TABLE 14-2). In this disease, the beta cells produce normal or above-normal levels of insulin. However, because of a decline in the number of insulin receptors in target cells, cells are unresponsive to insulin.

Type II diabetes is commonly associated with obesity. Heredity may also be an important contributor, as it is in Type I diabetes. In fact, studies show that about one-third of the patients with Type II diabetes have a family history of the disease. Critical thinking rules in Chapter 1 suggest the need to look for other explanations. Perhaps it may be the obesity that is inherited, rather than the tendency to develop diabetes.

Although they have different causes, both forms of diabetes exhibit similar symptoms. Excess urination and thirst are generally the first signs of trouble. Patients often feel tired, weak, and apathetic. Weight loss and blurred vision are also common. Excess glucose in the urine may result in frequent bacterial infections in the bladder.

Although Symptoms are Similar, Treatments for Type I and Type II Diabetes Are Quite Different.

Early-onset diabetes (Type I) is treated with insulin injections and therefore is often called **insulin-dependent diabetes.** Patients receive regular injections of insulin—usually two to three times per day. Patients are also required to eat meals and snacks at regular intervals to maintain constant glucose levels in the blood and to ensure that regular insulin injections always act on approximately the same amount of blood glucose.

Insulin injections are tailored to an individual's life-style and body demands. To mimic the body's natural release, medical researchers have developed a device called an **insulin pump**. This device delivers predetermined amounts of insulin. Because the pump cannot actually measure blood glucose levels, it may deliver inappropriate amounts at certain times. Medical researchers are also experimenting with ways to transplant healthy beta cells into the pancreas of a diabetic.

In a recent study, researchers from the University of Florida reported findings that may help prevent Type I diabetes. The scientists withdrew blood from 5000 young children who showed no symptoms of early-onset diabetes. Next, they analyzed the blood for antibodies produced by an autoimmune reaction and followed the children's health status over time to determine which ones developed diabetes. During the study, 12 of the subjects developed early-onset diabetes. Blood samples from these individuals all contained an antibody that attacks a protein on the surface of the beta cells of the pancreas. The antibody was found in blood samples taken as many as 7 years before the onset of diabetes.

The antibody to the surface protein of the beta cell may provide a biochemical marker, an advance notice of the disease. Eventually, researchers hope to find ways to attach chemicals to the surface proteins of the beta cells that would neutralize or destroy the antibodies that bind to them, thus preventing the disease from developing. If this research proves fruitful, regular blood tests could be used to screen children. When the antibodies are found, special treatment could be used to annihilate them before they have time to attack the islet cells. This could free millions of people from a lifetime of insulin injections and the risk of blindness and other serious medical problems later in life.

Type II diabetes is also called **insulin-independent diabetes** because insulin injections are useless, given the fact that the disease very likely stems from a lack of insulin receptors in target cells. In many patients, Type II diabetes can be eliminated by weight loss. In others, the disease can be controlled by diet and exercise. For example, physicians restrict carbohydrate intake of their patients and instruct them to eat small meals at regular intervals during the day. Candy, sugar, cakes, and pies are strictly off-limits. Glucose must be supplied by complex carbohydrates, such as starches.

The treatments described above have dramatically changed the prognosis for diabetics. At one time the disease was fatal. Today, patients can live fairly normal lives. However, risks are still present. Type I diabetics, for example, may suffer from diabetic comas, or unconsciousness, when they receive insufficient amounts of insulin or if they go without an insulin injection for a prolonged period. Without insulin,

the body cells become starved for glucose (even though blood levels are high) and begin breaking down fat. Excessive fat catabolism releases toxic chemicals (called *ketones*) that cause the patient to lose consciousness.

Diabetics may also suffer from **insulin shock** caused by an overdose of insulin. This reduces blood glucose levels, creating hypoglycemia. In mild cases, symptoms include tremor, fatigue, sleepiness, and the inability to concentrate. These symptoms result from a lack of glucose in the brain. In severe cases, unconsciousness and death may occur.

Early- and late-onset diabetics may also suffer from loss of vision, nerve damage, and kidney failure 20–30 years after the onset of the disease, even if they are being treated. Damage to the circulatory system can cause gangrene (GANG-green), requiring amputation of limbs, especially the lower extremities. These serious complications result from the inevitable elevations in blood glucose levels that occur periodically over the years.

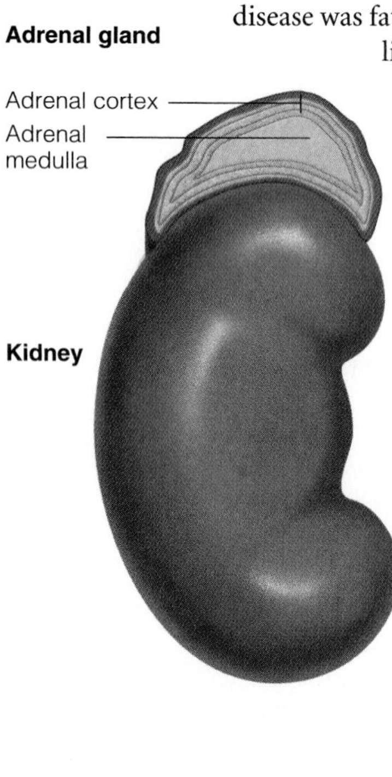

Section 14-6

The Adrenal Glands

You stand on the banks of a raging river. Your raft is tied to a tree and floating in the calm water of an eddy above the tumultuous white water. As you watch kayakers and rafters head into the rapids, your heart starts to race, and your intestines churn in excitement (and fear).

This natural response results from the secretion of two of the many hormones produced by the **adrenal glands** (ah-DREE-nal). As **FIGURE 14-19** shows, the adrenal glands perch atop the kidneys, and each consists of two zones. The central region, or **adrenal medulla**, produces the hormones that increase the heart rate and accelerate breathing when a person is excited or frightened. The outer zone, the **adrenal cortex**, produces a number of steroid hormones discussed in more detail shortly.

The Adrenal Medulla Produces Stress Hormones

The adrenal medulla produces two hormones: **adrenalin** (epinephrine) and **noradrenalin** (norepinephrine). In humans, about 80% of the adrenal medulla's output is adrenalin. Helping us meet the stresses of life, these hormones are instrumental in the fight-or-flight response—the physiological reactions that take place when we are threatened and that facilitate our ability to fight or flee. Adrenalin and noradrenalin are secreted under all kinds

Adrenal gland

Adrenal cortex
Adrenal medulla

Kidney

FIGURE 14-19 **Adrenal Gland** The adrenal glands sit atop the kidney and consist of an outer zone of cells, the adrenal cortex, which produces a variety of steroid hormones, and an inner zone, the adrenal medulla. The adrenal medulla produces adrenalin and noradrenalin, the secretion of which is controlled by the autonomic nervous system.

of stress—for example, when a careless driver cuts in front of you in heavy traffic or as you wait outside a lecture hall, anticipating an exam. Nerve impulses traveling from the brain to the adrenal medulla trigger the release of adrenalin and noradrenalin. The response is part of the autonomic nervous system (Chapter 11).

The physiological changes these hormones cause are many. For example, adrenalin and noradrenalin elevate blood glucose levels, making more energy available to cells, particularly skeletal muscle cells. They also increase breathing rate, which provides additional oxygen to skeletal muscles and brain cells. These hormones cause heart rate to accelerate as well. This increases circulation and ensures adequate glucose and oxygen for cells that might be called into action. In addition, adrenalin and noradrenalin cause the bronchioles in the lungs to dilate, permitting greater movement of air in and out of them. Furthermore, these hormones cause blood vessels in the intestinal tract to constrict, putting digestion on temporary hold, while the blood vessels in skeletal muscles dilate, increasing flow through them. Mental alertness increases as a result of increased blood flow and hormonal stimulation. You are ready to fight or flee.

The Adrenal Cortex Produces Three Types of Hormones with Markedly Different Functions

Surrounding the medulla is the adrenal cortex (**FIGURE 14-19**). The adrenal cortex produces three types of steroid hormones, each of which has a different function. The first group, the **glucocorticoids**, affect glucose metabolism and maintain blood glucose levels. The second group, the **mineralocorticoids** (MIN-er-al-oh-CORE-teh-KOIDS), regulate the ionic concentration of the blood and tissue fluids. The final group, the **sex steroids**, are identical to the hormones produced by the ovaries and testes. In healthy adults, the secretion of adrenal sex steroids is insignificant compared to the amounts produced by the gonads.

Glucocorticoids.

The prefix *gluco* reflects the fact that these steroid hormones affect carbohydrate metabolism. (Glucose, of course, is one of many carbohydrates in living things.) The secretion of glucocorticoid is governed by ACTH from the anterior pituitary, as noted earlier. Several chemically distinct glucocorticoids are secreted, the most important being **cortisol**. Cortisol increases blood glucose by stimulating gluconeogenesis (the synthesis of glucose from amino acids and glycerol). This makes more glucose available for energy production in times of stress. Cortisol also stimulates the breakdown of pro-

teins in muscle and bone, freeing amino acids that can then be chemically converted to glucose in the liver.

In pharmacological doses (levels beyond those seen in the body), cortisol inhibits inflammation, the body's response to tissue damage. Pharmacological doses of glucocorticoids also depress the allergic reaction. Cortisol brings about its effects by inhibiting the movement of white blood cells across capillary walls, thus impeding their migration into damaged tissue. Cortisol also reduces the number of circulating lymphocytes by destroying them at their site of formation.

Because they reduce inflammation, cortisol and other glucocorticoids (particularly cortisone, a synthetic glucocorticoid-like hormone) can be used to treat inflammation resulting from diseases such as rheumatoid arthritis or physical injuries. However, the benefits must be weighed carefully against the damage that can be caused by upsetting homeostasis.

Mineralocorticoids.

The mineralocorticoids are involved in electrolyte or mineral salt balance. The most important mineralocorticoid is **aldosterone**. It is the most potent mineralocorticoid and constitutes 95% of the adrenal cortex's hormonal output.

Although mineralocorticoids regulate the level of several ions, their main function is to control sodium- and potassium-ion concentrations, and their chief target is the kidney. As noted in Chapter 10, aldosterone increases the movement of sodium ions out of the nephron and into the blood, a process called *tubular reabsorption*. Aldosterone also stimulates sodium-ion reabsorption in sweat glands and saliva and potassium excretion by the kidney. When sodium is shunted back into the blood in the kidney and the skin, water follows. Aldosterone therefore helps conserve body water.

Aldosterone secretion is controlled by the sodium-ion concentration in a negative feedback loop. As sodium levels fall, aldosterone secretion increases. As sodium levels are restored, aldosterone secretion declines. As you might expect, aldosterone secretion is also stimulated when potassium levels climb and when blood volume and blood pressure decline.

Diseases of the Adrenal Cortex.

FIGURE 14-20 shows a patient with Cushing's syndrome. **Cushing's syndrome** generally results from pharmacological doses of cortisone, a synthetic glucocorticoid used to treat rheumatoid arthritis or asthma. In a few instances, the disease may be caused by a pituitary tumor that produces excess ACTH or a tumor of the

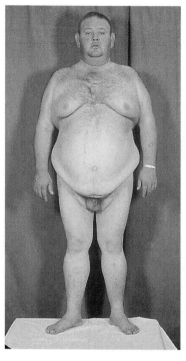

FIGURE 14-20 Cushing's Syndrome
This disease results from an excess of glucocorticoid hormone, either cortisol or cortisone. It is most often caused by cortisone treatment for allergies or inflammation. The most common symptoms are a round face due to edema and excess fat deposition.

Addison's disease
www.jbpub.com/humanbiology

adrenal cortex that secretes excess glucocorticoid.

Patients with Cushing's disease often suffer persistent **hyperglycemia**—high blood sugar levels—because of the presence of high levels of glucocorticoid. Bone and muscle protein may also decline sharply, because glucocorticoids stimulate the breakdown of protein. Individuals complain of weakness and fatigue. Loss of bone protein increases the ease with which bones fracture.

Water and salt retention are also common in Cushing's patients, resulting in tissue edema (swelling). Because of this, Cushing's patients typically have a rounded "moon face." Swelling occurs because glucocorticoids, in high concentrations, have mineralocorticoid effects.

Because most cases of Cushing's syndrome result from steroids taken for health reasons, treatment is a simple matter: by gradually reducing the glucocorticoid dose. Tumors in the pituitary and the adrenal cortex are treated with radiation or surgery.

Another disease of the adrenal cortex is **Addison's disease**. Most cases of Addison's disease are thought to be autoimmune reactions in which the cells of the adrenal cortex are recognized as foreign and are then destroyed by the body's own immune system. Addison's disease is characterized by a variety of symptoms caused by the loss of hormones from the adrenal cortex. These include loss of appetite, weight loss, fatigue, and weakness. Although insulin and glucagon are still present, the absence of cortisol upsets the body's homeostatic mechanism for controlling glucose. The body's reaction to stress is also impaired. The lack of aldosterone results in electrolyte imbalance. Because aldosterone maintains sodium levels and blood pressure, patients with Addison's disease have low blood sodium levels and low blood pressure. Addison's disease can be treated with steroid tablets that replace the missing hormones. Treatment allows patients to lead a fairly normal, healthy life.

Health, Homeostasis, and the Environment

Herbicides and Hormones

Hormones orchestrate an incredible number of body functions, and help to create a dynamic balance that is vital for good health. Hormones influence homeostasis primarily by controlling the rate of various metabolic reactions and by regulating ionic balance. When this balance is altered, human health suffers (**TABLE 14-3**). The endocrine system, like other systems, is sensitive to outside factors. Stress, for example, can lead to an imbalance in adrenal hormones, resulting in high blood pressure and other complications. High blood pressure, in turn, puts strain on the heart.

Hormonal balance can also be upset by toxic pollutants. One example is dioxin (die-OX-in), a contaminant found in some herbicides and in paper products. Infamous for its presumed carcinogenic properties, dioxin was found in the chemical known as Agent Orange, which was used in the Vietnam War to kill vegetation to reduce hiding spots for enemy soldiers. Oil spread on dirt roads in Times Beach, Missouri, also contained dioxin, deliberately added to the oil by a less-than-honest waste disposer. In 1983, after considerable public debate, the U.S. government bought up the town for $35 million, moved the residents out, and fenced it off.

New evidence suggests, however, that dioxin may be carcinogenic primarily at high levels—those in the workplace and those associated with chemical spills. The low levels found in Times Beach, in fact, may have little effect on human cancer.

Before you write your congressional representative to complain about the costly buyout, consider recent studies that suggest that at low levels dioxin may suppress immune functions. Scientists believe that it binds to plasma membranes and cytoplasmic receptors to which hormones normally attach, upsetting body functions. Thus, some researchers are calling dioxin an "environmental hormone." One form of dioxin, TCDD, suppresses the immune system in mice at least 100 times more effectively than corticosterone, one of the body's glucocorticoids.

TCDD also causes a variety of other direct biological effects. In some cells, it causes rapid cell growth. In others, it may alter cellular differentiation. Nonetheless, its effect on the immune system may be far more important than its impact on cancer.

Dioxin is not the only dangerous chemical in the environment. Studies suggest that a number of com-

TABLE 14-3

Summary of Some Endocrine Disorders

Disease	Cause	Symptoms
Giantism	Hypersecretion of GH starting in infancy or early life	Excessive growth of long bones
Dwarfism	Hyposecretion of GH in infancy or early life	Failure to grow
Acromegaly	Hypersecretion of GH after bone growth has stopped	Facial features become coarse; hands and feet enlarge; skin and tongue thicken.
Hyperthyroidism	Overactivity of the thyroid gland	Nervousness; inability to relax; weight loss; excess body heat and sweating; palpitations of the heart
Hypothyroidism	Underactivity of the thyroid gland	Fatigue, reduced heart rate; constipation; weight gain; feel cold; dry skin
Hyperparathyroidism	Excess parathyroid hormone secretion, usually resulting from a benign tumor in the parathyroid gland	Kidney stones; indigestion; depression; loss of calcium from bones
Hypoparathyroidism	Hyposecretion of the parathyroid glands	Spasms in muscles; numbness in hands and feet; dry skin
Diabetes insipidus	Hyposecretion of ADH	Excessive drinking and urination; constipation
Diabetes mellitus	Insufficient insulin production or inability of target cells to respond to insulin	Excessive urination and thirst; poor wound healing; urinary tract infections; excess glucose in urine; fatigue and apathy
Cushing's syndrome	Hypersecretion of hormones from adrenal cortex or, more commonly, from cortisone treatment	Face and body become fatter; loss of muscle mass; weakness; fatigue; osteoporosis
Addison's disease	Gradual decrease in production of hormones from adrenal gland; most common cause is autoimmune reaction.	Loss of appetite and weight; fatigue and weakness; complete adrenal failure

mon herbicides (the thiocarbamates) may upset the thyroid's function and result in the formation of thyroid tumors. These herbicides are chemically similar to thyroid hormone and therefore block the secretion of TSH, resulting in goiter. At higher levels, they may cause thyroid cancer.

The overall impact on human health of toxic chemicals released into the environment is probably small, especially compared with the impact of tobacco smoke. However, not everyone agrees. The Point/Counterpoint in Chapter 17 debates this subject, which is an issue worth thinking about. This is an important debate, and it will be around for years.

SUMMARY

PRINCIPLES OF ENDOCRINOLOGY

1. The *endocrine system* consists of a widely dispersed set of highly vascularized ductless glands that produce *hormones*.
2. Hormones affect five vital aspects of our lives: (a) homeostasis; (b) growth and development; (c) reproduction; (d) energy production, storage, and catabolism; and (e) behavior.
3. Hormones act on specific cells. Specificity results from the presence of hormone receptors on target cells. Steroid hormone receptors are generally located in the cytoplasm, and protein and polypeptide hormone receptors are located in the plasma membrane.
4. Three types of hormones are produced in the body: (a) steroids, (b) proteins and polypeptides, and (c) amines.
5. Hormone secretion is controlled principally by negative feedback loops, many of which involve the hypothalamus.
6. The endocrine and nervous systems are similar in several respects. They both send signals that regulate cell structure and function. Both help coordinate body functions.
7. The endocrine and nervous systems differ in several key respects. The nervous system elicits rapid, generally short-lived responses. The endocrine system elicits slower, longer-lasting responses.

THE PITUITARY AND HYPOTHALAMUS

8. The *pituitary* is a pea-sized gland suspended from the *hypothalamus* by a thin stalk. It consists of two parts: the anterior pituitary and the posterior pituitary.

9. The *anterior pituitary* produces seven protein and polypeptide hormones. Their release is controlled by *releasing* and *inhibiting hormones* produced by the hypothalamus, which are transported to the anterior pituitary via a portal system.

10. The hypothalamic hormones are produced by *neurosecretory neurons*. Their release is controlled by chemical stimuli and nerve impulses.

11. The hormones of the anterior pituitary and their functions are summarized in **TABLE 14-1**.

12. The *posterior pituitary* is derived from brain tissue during embryonic development and consists of axons and terminal ends of neurosecretory cells whose cell bodies are located in the hypothalamus. Hormones are produced in the cell bodies and transported down the axons to the posterior pituitary, where they are stored in the axon terminals until released.

13. The posterior pituitary produces two hormones, *antidiuretic hormone* (*ADH*) and *oxytocin*, whose functions are also summarized in Table 14-1.

THE THYROID GLAND

14. The *thyroid gland* is located in the neck, on either side of the trachea near its junction with the larynx. The thyroid produces three hormones: thyroxine (T_4), triiodothyronine (T_3), and calcitonin.

15. *Thyroxine* and *triiodothyronine* accelerate the rate of glucose breakdown in most cells, increasing body heat. These hormones also stimulate cellular growth and development.

16. Thyroxine and triiodothyronine both require iodine for their synthesis. A deficiency of this element in the diet results in *goiter*, an enlargement of the thyroid gland.

17. *Calcitonin* lowers blood calcium levels by inhibiting osteoclasts, thus reducing bone destruction and the release of calcium from bone. Calcitonin also increases formation of bone by osteoblasts and increases the excretion of calcium in the kidneys.

THE PARATHYROID GLANDS

18. The *parathyroid glands* are located on the back of the thyroid gland and produce a polypeptide hormone called *parathyroid hormone* (*PTH*), or *parathormone*.

19. PTH is released when calcium levels in the blood drop. This hormone increases blood calcium levels by stimulating bone reabsorption by osteoclasts, increasing intestinal absorption, and increasing renal reabsorption of calcium.

THE PANCREAS

20. The *pancreas* produces two hormones, insulin and glucagon, from the *islets of Langerhans*.

21. *Insulin* is the glucose-storage hormone. It stimulates the uptake of glucose by many body cells and stimulates the synthesis of glycogen in muscle and liver cells. Insulin also increases the uptake of amino acids and stimulates protein synthesis in muscle cells.

22. *Glucagon* is an antagonist to insulin. It raises glucose levels in the blood between meals by stimulating glycogen breakdown and the synthesis of glucose from amino acids and fats (gluconeogenesis).

23. *Diabetes mellitus* is a disease involving insulin and blood glucose. It has two principal forms: Type I and Type II. *Type I*, also called *early-onset diabetes*, occurs early in life and may be caused by an autoimmune reaction that destroys the beta cells of the pancreas. It can be treated by insulin injections.

24. *Type II*, or *late-onset*, *diabetes*, results from a reduction in the number of insulin receptors on target cells. It may be caused by obesity and genetic factors and can often be treated successfully by dietary management.

THE ADRENAL GLANDS

25. The *adrenal glands* lie atop the kidneys and consist of two separate portions: the adrenal medulla, at the center, and the adrenal cortex, a surrounding band of tissue.

26. The *adrenal medulla* produces two hormones under stress: adrenalin and noradrenalin. These hormones stimulate heart rate and breathing, elevate blood glucose levels, constrict blood vessels in the intestine, and dilate blood vessels in the muscles.

27. The *adrenal cortex* produces three classes of hormones: glucocorticoids, mineralocorticoids, and sex steroids.

28. The *glucocorticoids* affect carbohydrate metabolism and tend to raise blood glucose levels. The principal glucocorticoid is cortisol.

29. In pharmacological doses, cortisol inhibits the immune system and allergic reactions and is used to treat allergies or inflammation caused by injury and infections. High doses, however, have many adverse impacts on the body.

30. The chief *mineralocorticoid* is *aldosterone*. It acts on the kidneys, sweat glands, and salivary glands, causing sodium and water retention and potassium excretion.

HEALTH, HOMEOSTASIS, AND THE ENVIRONMENT: HERBICIDES AND HORMONES

31. Hormones influence an incredible number of body functions, fostering homeostasis, a dynamic balance necessary for good health. When homeostasis is altered, our health suffers.

32. The endocrine system, like others, is sensitive to upset from stress and environmental factors such as pollutants.

33. Dioxin, a contaminant in some herbicides and paper products, may exert most of its effects through the endocrine system.

34. A number of common herbicides (the thiocarbamates) upset the thyroid's function and may even cause thyroid tumors.

Critical Thinking

THINKING CRITICALLY— ANALYSIS

This Analysis corresponds to the Thinking Critically scenario that was presented at the beginning of this chapter.

This ethical dilemma hinges on several key questions—for example, should people be tinkering with nature? It is also a medical dilemma. For example, will growth hormone injections over a long period have any adverse effects?

As it turns out, no one knows the answers to the medical questions, but some studies suggest that prolonged usage may result in diabetes or high blood pressure. A recent study showed that growth hormone therapy in normal children, while greatly increasing height and muscle mass, dramatically reduced body fat. Children appeared gangly and raw-boned (emaciated). What effects these might have in the long term, no one knows. Some scientists believe that subtle abnormalities may develop in the cells, tissues, and organs of these individuals.

Clearly, the jury is still out on the long-term effects of growth hormone therapy in children. Parents who are subjecting their children to such treatment may be essentially creating a generation of human guinea pigs.

EXERCISING YOUR CRITICAL THINKING SKILLS

Find a copy of the *New England Journal of Medicine* in the library with an article on a discovery in endocrinology. How is the article organized? Analyze the article, using your critical thinking skills. Was the study performed on humans or laboratory animals? What were the major conclusions?

In your view, was the study performed correctly? Did the researchers include a control group? Was the control group similar to the experimental group? How many subjects were used? Should more have been used? Why? Can you tell whether the conclusions supported the data? Can you think of any alternative explanations for the data?

TEST OF CONCEPTS

1. Define the following terms: endocrine system, hormone, and target cell.

2. Hormones function in five principal areas. What are they? Give some examples of each.

3. Describe the concept of specificity. How is it created in the nervous system? How is it created in the endocrine system?

4. Compare and contrast the functions of the nervous and endocrine systems.

5. The endocrine system elicits slower, longer-lasting responses than the nervous system. Do you agree or disagree? Explain your reasons.

6. Define the terms *tropic* and *nontropic hormones*, and give several examples of each.

7. Give two examples of negative feedback loops in the endocrine system, a simple feedback mechanism and a more complex one that operates through the nervous system.

8. Define the term *neuroendocrine reflex*, and give some examples.

9. Give several biological reasons for the following observations: (a) the endocrine response tends to be delayed; (b) the endocrine response tends to be prolonged; (c) some hormones perform several different functions.

10. List the hormone(s) involved in each of the following functions: blood glucose levels, growth, milk production, milk let-down, calcium levels, and metabolic rate.

11. Describe the role of the hypothalamus in controlling anterior pituitary hormone secretion.

12. Describe the ways in which the posterior pituitary differs from the anterior pituitary.

13. Explain the hormonal reasons for each of the following: acromegaly, dwarfism, and giantism. Acromegaly and giantism are both caused by the same problem. Why are these conditions so different?

14. ACTH is controlled by levels of glucocorticoid and by a biological clock. How are the controls different?

15. Describe the neuroendocrine reflex involved in prolactin secretion.

16. Offer some possible explanations for the following experimental observation: Milk production occurs late in pregnancy and is thought to be stimulated by the hormone prolactin. A nonpregnant rat is injected with prolactin but does not produce milk.

17. Where is ADH produced? Where is it released? Describe how ADH secretion is controlled. What effects does this hormone have?

18. Where is oxytocin produced? Where is it released? Describe how oxytocin secretion is controlled. What effects does this hormone have?

19. A patient comes into your office. She is thin and wasted and complains of excessive sweating and nervousness. What tests would you run?

20. A patient comes into your office. He is suffering from indigestion, depression, and bone pain. An X-ray of his bone shows some signs of osteoporosis. You think that the disorder might be the result of an endocrine problem. What test would you order?

21. How are the two basic types of diabetes mellitus different? How are they similar? How are they treated, and why are these treatments chosen?

22. Describe the physiological changes that occur under stress. What hormones are responsible for them?

23. What is gluconeogenesis? What hormones stimulate the process?

24. Cortisone depresses the allergic response. A patient comes to your office and asks that you treat her allergies with cortisone. What would you tell her?

25. Aldosterone is a mineralocorticoid. Describe its chief functions. How does it help retain body fluid? Under what conditions is aldosterone secreted?

TOOLS FOR LEARNING

www.jbpub.com/humanbiology

Tools for Learning is an on-line student review area located at this book's web site HumanBiology (www.jbpub.com/humanbiology). The review area provides a variety of activities designed to help you study for your class:

Chapter Outlines. We've pulled out the section titles and full sentence sub-headings from each chapter to form natural descriptive outlines you can use to study the chapters' material point by point.

Review Questions. The review questions test your knowledge of the important concepts and applications in each chapter. Written by the author of the text, the review provides feedback for each correct or incorrect answer. This is an excellent test preparation tool.

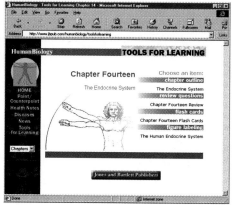

Flash Cards. Studying human biology requires learning new terms. Virtual flash cards help you master the new vocabulary for each chapter.

Figure Labeling. You can practice identifying and labeling anatomical features on the same art content that appears in the text.

Active Learning Links. Active Learning Links connect to external web sites that provide an opportunity to learn basic concepts through demonstrations, animations, and hands-on activities.

CHAPTER 15

CHROMOSOMES, CELL DIVISION, AND CANCER

Human cell arrested in cell division.

Thinking Critically

Since the 1970s, environmentalists have been warning people of the increasing danger of a global calamity caused by the release of a class of chemicals called chlorofluorocarbons or CFCs for short. Used in refrigerators, freezers, and many other products, CFCs drift into the upper atmosphere where sunlight causes them to dissociate, producing a chemical that many scientists believe could seriously deplete the ozone layer, which protects us and many other species from dangerous ultraviolet radiation. Ultraviolet radiation, of course, can cause skin cancer, cataracts, and other problems. As the ozone layer thins, skin cancer rates and mortality from skin cancer could skyrocket.

Against this gloom and doom, a loud dissenting voice was raised on the airways of America, led in large part by radio and television celebrity Rush Limbaugh. He called the ozone threat a "scam." He said scientists were behind it, in part, because they viewed it as a way to enhance their funding. He and others alleged that claims by environmentalists and scientists were wrong.

Environmentalists countered that Limbaugh and others like him were spreading misinformation—or more bluntly, falsehoods. How would you go about analyzing this debate? ■

ELECTRICITY FLOWING THROUGH WIRES PRO-duces a magnetic field. Since the discovery of this phenomenon, scientists have been interested in it mostly as a matter of curiosity. In 1979, though, two researchers from the University of Colorado discovered a potential dark side to this natural phenomenon. They reported the health effects of extremely low frequency (ELF) magnetic waves generated by high-voltage power lines. Their studies suggested that these magnetic fields may increase the incidence of leukemia (a cancer of the white blood cells) in children who live near high-voltage power lines. The cancer death rates in these children were twice what would have been expected in the general public. In November 1986, researchers from the University of North Carolina published reports showing a fivefold increase in childhood cancer (particularly leukemia) in families living 25–50 feet from wires transporting electricity from power substations to neighborhood transformers (**FIGURE 15-1**).

Studies of human health, such as these, are part of the science of **epidemiology** (EP-uh-DEEM-ee-OL-uh-ge), literally, "the study of epidemics." Epidemiological studies depend primarily on statistical analysis to examine potential correlations between cause and effect. Today, they're used to study the impacts of a wide range of substances and physical agents such as magnetic fields on human health. One epidemiological study by itself rarely proves a point. But when several studies report the same correlation, as in the ELF case, the results are more reliable.

Experimental work in the laboratory also lends support to epidemiological studies. In the ELF controversy, additional evidence in favor of the hypothesis came from a researcher who found that ELF fields increased the growth rate of cancer cells in tissue culture. They also rendered the cells 60%–70% more resistant to the immune system's naturally occurring killer cells, which attack cancer cells.

Some studies showed that numerous household items, such as water bed heaters and electric blankets, produce magnetic fields and that these devices increased the likelihood of miscarriage (spontaneous abortion of an embryo or fetus). In October, 1996, however, a panel of scientists convened by the prestigious National Research Council concluded that electromagnetic fields associated with power lines, appliances, and other sources *do not* pose a health risk to humans. Their conclusion was based on an exhaustive analysis of over 500 studies on the subject performed over a 17-year period. Although some studies do show an effect, the bulk of the evidence shows no credible link, they concluded. In those studies that demonstrated a link, scientists suggested that other unmeasured factors may have been responsible.

This chapter provides basic information that will help you understand this debate as well as others on potential cancer-causing agents, or **carcinogens** (car-SIN-oh-GINS). **Cancer** is a disease caused by the uncontrolled replication of cells. It is induced by alterations, or **mutations**, of the genetic information (DNA) contained in the chromosomes.

FIGURE 15-1 Too Close for Comfort?   Some studies suggest that extremely low frequency radiation from power lines may increase the incidence of cancer, especially childhood leukemia, in nearby residents. A recent analysis of the research, however, indicates that this link between cancer and electrical lines is *not* real.

Section 15-1

The Cell Cycle

As noted in Chapter 1, all cells arise from preexisting cells via cell division. This process is essential to all life because it is the basis of reproduction and growth. In humans and other multicellular organisms, cell division also plays an important role in the repair of damaged tissues and organs.

Cell division is a well-orchestrated process, occurring during a distinct part of the life cycle of a cell, commonly referred to as the **cell cycle** (**FIGURE 15-2**). As illustrated, the cell cycle is divided into two parts: cell division and interphase. During **cell division**, the nucleus and the cytoplasm divide, splitting the cell more or less equally, thus forming two new cells. **Interphase** is the period between cell divisions.

Interphase Is Divided into Three Parts

Once thought to be a resting period, interphase is now known to be a time of intense metabolic activity. During interphase, for example, cells replicate their DNA. They also replicate many cytoplasmic components, including organelles, a process that prepares cells for division. Thus, when cells divide, their offspring, or daughter cells, receive an adequate supply of the organelles and molecules needed to function normally.

Interphase is divided into three phases: G_1, S, and G_2 (**FIGURE 15-2** and **TABLE 15-1**). G_1 (**gap 1**) begins immediately after a cell divides. During G_1, the nuclear DNA orchestrates the cell's synthesis of RNA, proteins, and other molecules vital for proper cellular function and growth. These molecules also serve as a stockpile that ensures each daughter cell an adequate supply of nutrients.

The length of G_1 varies considerably from one cell to another. Some cells in an organism spend only a few

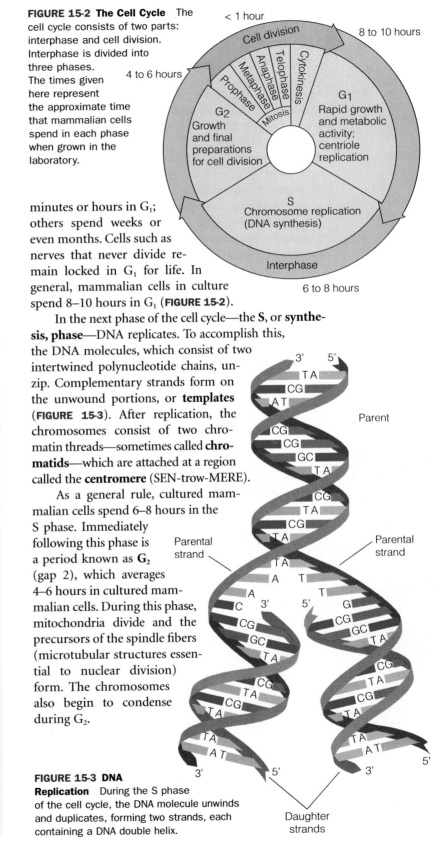

FIGURE 15-2 The Cell Cycle The cell cycle consists of two parts: interphase and cell division. Interphase is divided into three phases. The times given here represent the approximate time that mammalian cells spend in each phase when grown in the laboratory.

minutes or hours in G_1; others spend weeks or even months. Cells such as nerves that never divide remain locked in G_1 for life. In general, mammalian cells in culture spend 8–10 hours in G_1 (**FIGURE 15-2**).

In the next phase of the cell cycle—the **S, or synthesis, phase**—DNA replicates. To accomplish this, the DNA molecules, which consist of two intertwined polynucleotide chains, unzip. Complementary strands form on the unwound portions, or **templates** (**FIGURE 15-3**). After replication, the chromosomes consist of two chromatin threads—sometimes called **chromatids**—which are attached at a region called the **centromere** (SEN-trow-MERE).

As a general rule, cultured mammalian cells spend 6–8 hours in the S phase. Immediately following this phase is a period known as G_2 (gap 2), which averages 4–6 hours in cultured mammalian cells. During this phase, mitochondria divide and the precursors of the spindle fibers (microtubular structures essential to nuclear division) form. The chromosomes also begin to condense during G_2.

FIGURE 15-3 DNA Replication During the S phase of the cell cycle, the DNA molecule unwinds and duplicates, forming two strands, each containing a DNA double helix.

TABLE 15-1	
Phases of the Cell Cycle	
Phase	**Characteristics**
Interphase	
G_1 (gap 1)	Stage begins immediately after mitosis.
	RNA, protein, and other molecules are synthesized.
S (synthesis)	DNA is replicated.
	Chromosomes become double stranded.
G_2 (gap 2)	Mitochondria divide; precursors of spindle fibers are synthesized.
Mitosis	
Prophase	Chromosomes condense.
	Nuclear envelope disappears.
	Centrioles divide and migrate to opposite poles of the dividing cell.
	Spindle fibers form and attach to chromosomes.
Metaphase	Chromosomes line up on equatorial plate of the dividing cell.
Anaphase	Chromosomes begin to separate.
Telophase	Chromosomes migrate or are pulled to opposite poles.
	New nuclear envelope forms.
	Chromosomes uncoil.
Cytokinesis	Cleavage furrow forms and deepens.
	Cytoplasm divides.

Nuclear and Cytoplasmic Division Occur Separately

Interphase is followed by cell division, a rapid event by cell standards, which lasts only about 1 hour in mammalian cells. During cell division, the nucleus and cytoplasm both divide. However, nuclear division, commonly known as **mitosis** (my-TOE-siss), precedes cytoplasmic division.

Mitosis involves a series of rather dramatic structural changes. In order for the replicated chromosomes to divide, they must first condense, becoming compact structures clearly visible in ordinary light microscopic preparations (**FIGURE 15-4**). Condensation facilitates chromosomal segregation. Without it, cell division would be a nightmare. Dividing the 46 unravelled chromosomes in humans would result in a tangled mass of broken chromosomes. Cell division without chromosome compaction would be like trying to separate a plate of spaghetti into two equal piles without breaking a single strand.

In addition, for mitosis to occur smoothly, the chromosomes must line up in the center of the cell. From this position, the two chromatids of each chromosome can easily be drawn apart and then be divvied up, one-half going to each new daughter cell.

Cytoplasmic division, or **cytokinesis** (literally, "cell movement," pronounced SITE-toe-ki-NEE-siss), occurs independently of mitosis and usually toward the end of it.

With this overview of cell division, we now turn our attention to one of the main actors in this exciting cellular drama, the chromosome.

Section 15-2

The Chromosome

As you surely know by now, chromosomes contain the genetic information of cells, DNA, which directs most cellular events. (For a discussion of the scientific work that led to the discovery of DNA, see Scientific Discoveries 15-1; for a review of terminol-

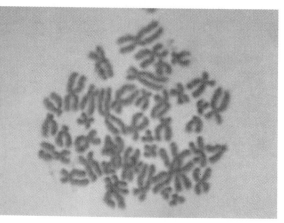

FIGURE 15-4 Metaphase Chromosomes Taken from a human cell during metaphase, these chromosomes are clearly visible under a light microscope.

TABLE 15-2	
Review of Terminology	
Chromatin	General term referring to a strand of DNA and associated histone protein
Chromatin fiber	Strand of chromatin
Chromosome	Structure consisting of one or two chromatin fibers
Chromatid	Generally used to refer to one of the chromatin fibers of a replicated chromosome
Centromere	Region of each chromatid to which a sister chromatid attaches

ogy see **TABLE 15-2**.) All organisms have a set number of chromosomes. All of the cells in your body, for example, contain 46 chromosomes, except for the gametes (or germ cells—the sperm and egg), which contain half that number.

Each type of chromosome contains very specific genetic information and is different from every other type. Chromosomes also differ in several other key respects, such as size, location of the centromere, and banding patterns (**FIGURE 15-5**). By studying these features, **geneticists**, scientists who study chromosomes, genes, and heredity, have found that chromosomes in

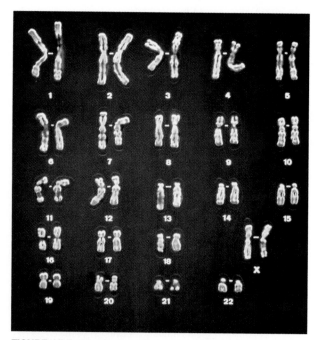

FIGURE 15-5 Karyotype Chromosomes from a normal human female arranged in order of decreasing size. Note that there are 46 chromosomes—or 23 pairs—in somatic cells. Each pair is similar in size, banding pattern, and location of the centromere.

FIGURE 15-6 Chromosome Structure
(a) The DNA helix wraps around histone molecules to form a chromatin fiber. During prophase, the chromatin fiber coils tightly, forming the densely packed chromosome. *(b)* Transmission electron micrograph of a metaphase chromosome.

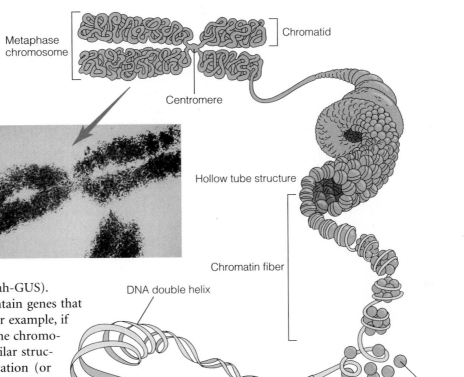

most of the cells of the body exist in pairs. In humans, for example, geneticists can distinguish 23 structurally similar pairs. As shown in **FIGURE 15-5**, each pair has the same length, banding pattern, and centromere position. Such chromosomes are described as **homologous** (ha-MOLL-ah-GUS).

Homologous chromosomes contain genes that control the same inherited traits. For example, if a gene for hair color is located on one chromosome of a pair, another gene of similar structure will be found at the same location (or **locus**) on the other member of the pair.

The presence of homologous chromosome pairs results from sexual reproduction—the uniting of a sperm and an ovum. During sexual reproduction, a new individual is created when the sperm and egg combine. Both the sperm and egg contain 23 chromosomes.

Cells that contain the full complement of chromosomes—that is, 46 chromosomes—are said to be **diploid** (diploid = twofold). In humans, all body cells, like those of skin and muscle, are diploid. Such cells are often called **somatic cells** to distinguish them from **germ cells**, the sperm and ovum, which carry paternal and maternal chromosomes, respectively. Germ cells are **haploid** because they contain half the chromosomes of somatic cells. Germ cells are produced by a special kind of cell division known as **meiosis** (my-OH-siss), which occurs in the gonads (ovaries and testes). Meiosis is discussed in Chapter 16.

Chromosomes Condense After Replication, Forming Tightly Coiled Structures in the Nucleus

Soon after replication, the chromosomes of the cell begin to coil, compacting in much the same way that a stretched phone cord shortens and compacts when the tension is removed. Unlike a phone cord, which functions just as well when it is stretched or compacted, chromosomes in the condensed state are metabolically inactive—that is, unable to produce either RNA or DNA.

The structure of a condensed chromosome is shown in **FIGURE 15-6B**. To understand this very complex structure, begin at the bottom and work up. As illustrated, each chromatid consists of a DNA in a double helix and associated proteins called **histones** (HISS-tones). The histones are globular proteins, which are thought to play a role in regulating the DNA's activities (Chapter 17).

As illustrated in **FIGURE 15-6**, loops of the double helix of DNA encircle the histones, forming small clusters. The small clusters along the DNA molecule compact to form hollow tubules. These tubules, in turn, form coils like those of a phone cord. When the coils are compacted, they form the body of the chromosomes.

Chromosomal condensation is of great importance to cells, as pointed out earlier. It also has many practical applications. For example, it provides physicians and geneticists an opportunity to search for genetic defects. This is especially useful to expectant parents who are concerned about possible genetic defects in their babies.

In order to acquire fetal cells, physicians extract fluid from the liquid-filled cavity surrounding the

Scientific Discoveries that Changed the World 15-1

Unraveling the Structure and Function of DNA

Featuring the Work of Miescher, Griffith, Avery, Watson, Crick, Wilkins, and Franklin

In a word-association test, when the tester says "hot," you'd probably answer "cold." If he or she said "DNA," a biologist would probably answer "Watson and Crick." James Watson and Francis Crick, to be exact.

In 1953, Watson and Crick published a brief paper that proposed a model (hypothesis) for the structure of DNA (**FIGURE 1**). They suggested that DNA was a double helix and explained how the nucleotides fit together.

Like many other scientific advances that changed our understanding of the world and, in some cases, our way of life, the discovery of the structure of DNA resulted from the efforts of many scientists over many years. In fact, work on DNA began in the 1860s, when Frederick Miescher launched an inquiry into the chemical composition of white blood cells. In 1868, Miescher separated the nuclei of white blood cells from their cytoplasm. He then began to study the chemical nature of the material he had extracted. Research by other scientists using his techniques revealed the presence of two types of molecules in the nucleus: RNA and DNA.

Over the next 80 years, evidence that DNA played a role in heredity steadily accumulated. In 1920, for example, Frederick Griffith discovered a chemical substance in bacteria he called the "transforming factor." When transmitted from one bacterium to another, this substance caused specific structural changes in

the recipient. In 1944, Oswald Avery and his colleagues identified the transforming factor as DNA. They showed that DNA extracted from one strain of bacteria produced an altered capsule in a recipient strain. Furthermore, this trait was passed to all offspring of the recipient.

To confirm the hypothesis that DNA was the transforming agent, Avery and co-workers treated the extracts from the first bacteria with enzymes that destroy protein and RNA. They found that removing protein and RNA did not affect transformation. To test their hypothesis that DNA was the hereditary molecule, the researchers treated the preparation with an enzyme that destroys DNA. As they had expected, this blocked transformation altogether.

This work led some scientists to conclude that DNA is the genetic material and that it controls the synthesis of specific cellular products. Prior to this time, most scientists believed that proteins were the most likely candidate for hereditary molecules. In fact, this notion persisted until the mid-1940s because many

FIGURE 1 The Double Helix Model for DNA Watson and Crick and their double helix model.

growing fetus through a long needle inserted through the mother's abdomen (**FIGURE 15-7**). This process is called **amniocentesis** (AM-knee-oh-cen-TEE-siss).

Fetal skin cells present in the fluid are separated from it, and then grown in culture dishes. There, they proliferate, dividing by mitosis. After the number of cells has increased, the tissue culture is treated with a chemical substance that arrests cell division. The cells are then removed from the culture and placed on a glass slide. A small piece of glass or plastic is placed over them and compressed, causing the cells to flatten out. The chromosomes are then stained and photographed through a camera mounted on the microscope. Chromosomes are then cut out of the photos, and arranged

in homologous pairs with the largest first, as shown in **FIGURE 15-5**. The resulting display is called a **karyotype** (CARE-ee-oh-TYPE).

Nowadays, computers can also be used to produce a karyotype. A camera mounted on the microscope projects the image onto a computer screen. Technicians then isolate the chromosomes on the screen and assign each one a number; the computer then creates the karyotype.

Lining the chromosomes up like this helps geneticists locate obvious structural defects—for example, extra chromosomes or missing segments. More subtle changes cannot be detected this way. As the next chapter explains, gross defects in chromosomes can have serious consequences. In many cases, chromosomal

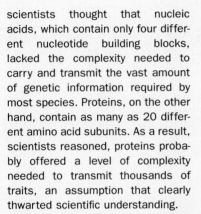

scientists thought that nucleic acids, which contain only four different nucleotide building blocks, lacked the complexity needed to carry and transmit the vast amount of genetic information required by most species. Proteins, on the other hand, contain as many as 20 different amino acid subunits. As a result, scientists reasoned, proteins probably offered a level of complexity needed to transmit thousands of traits, an assumption that clearly thwarted scientific understanding.

Once it was established that DNA was the molecule responsible for heredity, scientists turned their attention to unraveling its structure. This work began in the mid-1940s and culminated in the publication of Watson and Crick's paper in 1953. Watson and Crick used available information from other researchers. One vital piece came from Rosalind Franklin, who worked in the laboratory of British researcher Maurice Wilkins. Franklin produced X-ray photographs of crystals of highly purified DNA (**FIGURE 2**). Her remarkable photographs suggested that the DNA molecule was helical.

Another vital piece of evidence came from the laboratory of Erwin Chargaff and his colleagues, who had spent years studying the chemical nature of DNA. Chargaff's lab showed that the amount of purine in DNA always equals the amount of pyrimidine and that the ratios between adenine and thymine and between cytosine and guanine are always 1:1.

Armed with these and other vital details, Watson and Crick devised an ingenious model that fit the chemical and physical data. They published their results in 1953 in the journal *Nature,* and 2 months later they published a second paper proposing that the complementary strands of their model could explain replication. In addition, Watson and Crick hypothesized that genetic information could be stored in the sequence of bases and that mutations might result from a change in that sequence.

In the ensuing years, numerous scientists have provided hard evidence that supports the model proposed by Watson and Crick. And in 1962, Watson, Crick, and Wilkins were awarded the Nobel Prize for

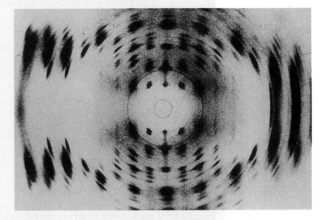

FIGURE 2 Discovering DNA's Helical Shape The famous X-ray diffraction photograph of DNA by Rosalind Franklin.

their discovery. Interestingly, although much of the X-ray data on which the Watson-Crick model was based had come from Franklin, she was not included in the celebrations. She couldn't be. Unfortunately, Rosalind Franklin died of cancer in 1958, and Nobel Prizes are awarded only to living scientists. Had she been alive, it is likely that she too would have shared in this coveted prize and been a household name as well.

analysis assures parents that their child has no obvious genetic defects, alleviating worry during a period of heightened anxiety. It has the added benefit of helping parents learn the sex of their baby, which, of course, helps them determine the color to paint the baby room, illustrating the importance of science to our daily lives.

Although Somatic Cells Contain a Set Number of Chromosomes, the Number of Chromatids Varies, Depending on the Stage of the Cell Cycle

The study of genetics can be quite confusing at first, because cells contain different numbers of chromosomes and because chromosomes may contain one or two chromatids, depending on the stage of the cell cycle. For example, as noted earlier, during G_1 of interphase, each chromosome consists of a single DNA molecule (a double helix) and associated protein. The 46 unreplicated chromosomes in the human cell are packed loosely in the nucleus in an apparently random fashion. During the S phase, however, the DNA strands replicate (**FIGURE 15-3**). Each chromosome now consists of two identical chromatids held together at the centromere. Each chromatid is a molecule of DNA, arranged in a double helix, with its associated proteins. When a cell divides, these chromosomes are split in half, with one chromatid going to each daughter cell.

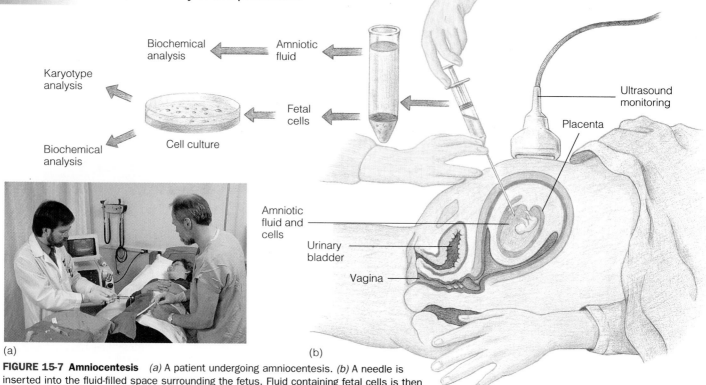

FIGURE 15-7 Amniocentesis *(a)* A patient undergoing amniocentesis. *(b)* A needle is inserted into the fluid-filled space surrounding the fetus. Fluid containing fetal cells is then withdrawn. Cells are subjected to chromosomal and biochemical analyses. Ultrasound monitors are used to direct the needle and prevent injury to the fetus.

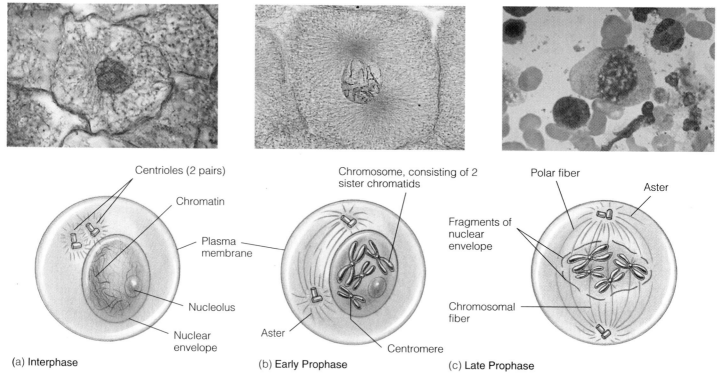

(a) Interphase (b) Early Prophase (c) Late Prophase

FIGURE 15-8 Mitosis

Cell Division: Mitosis and Cytokinesis

With this background information in mind, let's explore some of the changes occurring during cell division, beginning with mitosis, nuclear division.

Mitosis Is Divided Into Four Stages

The four stages of mitosis are prophase, metaphase, anaphase, and telophase (**FIGURE 4-2**).[1]

Prophase.

Prophase begins immediately after interphase, a time during which the chromosomes replicate. During prophase, the replicated chromosomes shorten and thicken considerably, forming compact structures (**FIGURES 15-8A** and **15-8B**). In addition, the nucleoli, regions of active rRNA synthesis, gradually disappear.

Two events of great importance occur in the cytoplasm during prophase. The first is the division of the

cell's centrioles. Shown in **FIGURE 15-9**, each **centriole** consists of two small, cylindrical structures identical to the basal bodies (Chapter 3). Like other organelles, centrioles replicate during interphase. During prophase, the centrioles separate, migrating to opposite ends of the nucleus.

The second cytoplasmic event of importance is the formation of the mitotic spindle. The **mitotic spindle** is an elaborate array of microtubules responsible for subsequent movement of the chromosomes. As biologist Norman Wessels writes, "Just as a dancing puppet is operated by moving strings, the dance of the chromosomes is choreographed and controlled by the action of the spindle fibers."

As shown in **FIGURE 15-10**, two types of fibers are found in the mitotic spindle: **polar fibers**, which extend from the spindle pole inward toward the center of the cell; and **chromosomal fibers**, which also extend from the poles inward, but attach to individual chromosomes via their centromeres. Each fiber is actually a bundle of microtubules composed of tubulin molecules. These molecules are derived from the disassembly of the cell's cytoskeleton during mitosis.

Also associated with the centrioles in animal cells is a star burst of microtubules called the **aster**. Its role in

---

[1]You can use the acronym PMAT to remember these steps.

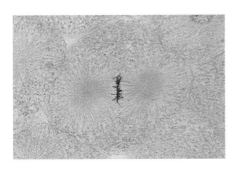
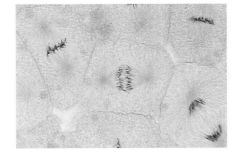
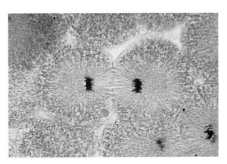

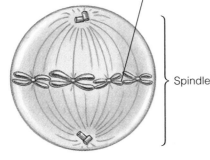

Metaphase plate

Spindle

(d) Metaphase

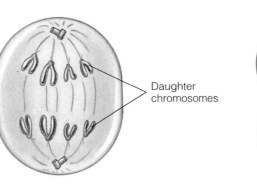

Daughter chromosomes

(e) Anaphase

(f) Telophase and cytokinesis

FIGURE 15-8 Mitosis—continued

FIGURE 15-9 Centriole *(a)* The centriole, like the basal body, consists of nine sets of microtubules with three in each set. Centrioles are found in the cytoplasm and replicate during interphase. *(b)* They migrate to opposite poles of the nucleus during mitosis.

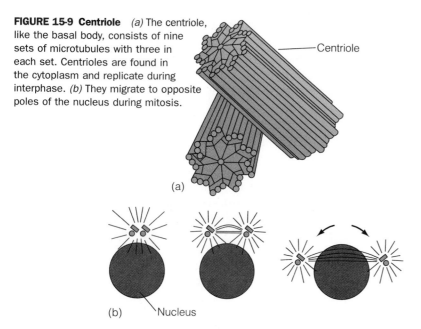

(a)

(b) Nucleus

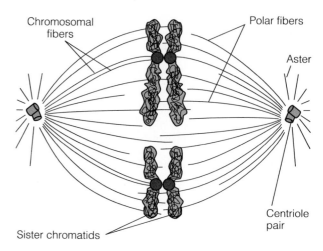

FIGURE 15-10 Mitotic Spindle The mitotic spindle, made of microtubules, forms during prophase in the cytoplasm of the cell. Chromosomal fibers connect to the chromosomes, and polar fibers extend from a pole to the equatorial region of the spindle. In ways not currently understood, the mitotic spindle separates each double-stranded chromosome during mitosis.

cell division remains unclear, although it may anchor the spindle to the plasma membrane.

Late in prophase (**FIGURE 15-8C**), the nuclear envelope begins to disintegrate. This permits the chromosomal fibers of the mitotic spindle to attach to the chromosomes, setting the stage for the next step, metaphase.

Metaphase.

Before chromosomes can be divided equally, they are lined up in the center of the cell an equal distance from the two poles of the mitotic spindle with their centromeres located along the equatorial plane (**FIGURE 15-8D**). This alignment occurs during metaphase and greatly facilitates the separation of chromatids. It's not unlike the process you and a roommate might engage in at the end of the semester to be sure you've equally divided your things as well as any communal property.

Anaphase.

When the chromatids of each chromosome begin to separate, the cell enters anaphase. The shortest part of mitosis, anaphase lasts only a few minutes (**FIGURE 15-8E**). During anaphase, the chromatids of the homologous chromosomes are drawn toward opposite poles of the mitotic spindle and therefore toward opposite ends of the cell. As shown in **FIGURE 15-8E**, the chromosomes are dragged centromere-first, with their arms trailing behind. All of the chromosomes begin to separate simultaneously.

Telophase.

The final stage of nuclear division is telophase (**FIGURE 15-8F**). Telophase begins when the chromosomes

complete their migration to the poles, and it involves a series of changes in the nuclei that is essentially the reverse of prophase.

During telophase, nuclear envelopes form around the chromosomes of each daughter cell. New nuclear envelopes are produced by the endoplasmic reticulum. They soon fuse into one. Also during telophase, the spindle fibers disappear, and the chromosomes uncoil, regaining their threadlike appearance. Nucleoli reappear. Telophase ends when the nuclei of the daughter cells appear to be in interphase.

Cytokinesis Is the Division of the Cytoplasm

As **FIGURE 15-8** shows, cytokinesis usually begins late in anaphase or early in telophase. Cytokinesis is brought about by a dense network of contractile fibers, or microfilaments, that lies beneath the plasma membrane. These microfilaments, which are part of the cytoskeleton of the cell, are composed of the same proteins found in muscle cells.

When the time comes for the cytoplasm to divide, the microfilaments along the midline of the cell contract and pull the membrane inward (**FIGURE 15-11**). The membrane furrows as if it were being constricted by a thread tied around the cell, then tightened. As cytokinesis proceeds, the furrow deepens, eventually pinching the cell in two. This results in two daughter cells with more or less equal amounts of cytoplasm. With this equal allocation of cytoplasm to the daughter cells comes a sharing of the cell's organelles.

Control of the Cell Cycle

What stimulates the replication of DNA and the division of organelles? What causes a cell to divide?

Despite years of intensive research, our understanding of the controls of the cell cycle is far from complete. Some experiments suggest that chemical messages from the cytoplasm control the cell cycle. If cytoplasmic signals control the nucleus, the next logical question is, What controls the cytoplasm?

Unfortunately, no one knows for sure. Research shows that certain external signals may exert some control over cytoplasmic events. Studies of **cell cultures**—that is, cells grown in the laboratory under precise conditions—have shown that normal cells migrate and divide until they spread evenly across the bottom surface of culture dishes. At this point, the cells stop dividing, even though nutrients are plentiful. The contact of cells with one another apparently inhibits further growth; this process is called **contact inhibition**. Contact inhibition also occurs in the tissues of the body.

Interestingly, one of the reasons cancer cells grow uncontrollably is that they lose contact inhibition. Thus, in a culture dish, cancer cells proliferate wildly, growing on top of one another and quickly utilizing nutrients in the nutrient bath they're immersed in. In the body, cancer cells grow in a similar way, forming large growths, called **tumors**, that may eventually destroy vital organs and cause death.

Hormones affect the cell cycle as well. During each menstrual cycle, the hormone estrogen released from a woman's ovaries stimulates the growth of breast tissue and the uterine lining in preparation for pregnancy. Several other hormones also stimulate the division of body cells.

Growth-promoting and growth-inhibiting factors likewise play a role in controlling the cell cycle. Skin cells, for example, produce a growth-inhibiting factor that prevents cell division. However, damage to the skin—say, a cut or burn—reduces the number of skin cells and presumably reduces the amount of inhibiting factor at the site of injury. This, in turn, releases cells from inhibition. Cells proliferate and repair the damaged area. When the tissue is repaired, the rate of cell division returns to normal.

Obviously, there is no single answer to the question of what controls the cell cycle. Many factors play a role, depending on the type of cell. In time, other factors are sure to be discovered. These factors may act directly on the cytoplasm but more likely induce changes in the

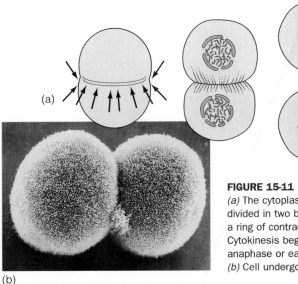

(a)

(b)

FIGURE 15-11 Cytokinesis *(a)* The cytoplasm of a cell is divided in two by the contraction of a ring of contractile microfilaments. Cytokinesis begins late in anaphase or early in telophase. *(b)* Cell undergoing cytokinesis.

genes in the nucleus that regulate the cell cycle. Activation of these genes may, for example, lead to the production of chemicals in the cytoplasm that diffuse back into the nucleus, thus causing the nuclear changes of mitosis.

Cancer: Understanding a Deadly Disease

In the early 1970s, Americans were shocked by reports of hazardous wastes seeping into their drinking water, pesticides contaminating the food they ate, and pollution fouling the air they breathed. Many feared that these conditions would lead to an increase in cancer. In fact, some experts estimated that 90% of all cancers were caused by environmental and workplace pollution and by other agents such as X-rays, ultraviolet light, and viruses. That conjecture caused a storm of controversy and a flurry of research into cancer, a disease that kills over 550,000 Americans each year.

More recent studies suggest that the early estimates were probably in error. Today, in fact, many health experts believe that only 20%–40% of all cancers arise from environmental and workplace pollutants. Most of the rest are caused by smoking, diet, and natural causes.

No matter what the cause, cancer is a disease that most of us will have close experience with. According to national statistics, one of every three Americans will contract cancer, and one of every four will die from it. For a debate on whether America is in a cancer epidemic see the Point/Counterpoint discussion.

Are We in a Cancer Epidemic?

THE MYTHS OF THE CANCER EPIDEMIC

David L. Eaton

David L. Eaton *is a professor of Environmental Health and Environmental Studies, and director of Toxicology at the University of Washington. He has an active research program on the mechanisms by which chemicals cause cancer.*

There is no debate that cancer is a devastating and deadly disease. One in three people living in the United States today will contract some form of cancer in his or her lifetime, and one in four will die from it, if current rates continue. But are cancer rates increasing in epidemic proportions?

The total number of people and the percentage of all deaths caused by cancer have increased dramatically in the past 50 years. However, cancer is largely a disease of old age, and thus it is necessary to adjust such statistics for changes in the age distribution of our population. A 1988 report from the National Cancer Institute states that "the age adjusted mortality rates for all [types of] cancers combined, except lung cancer, have been declining since 1950 for all age groups except 85 and above." Statistics from the American Cancer Society yield the same conclusion. In contrast, the incidence of some childhood leukemias and brain tumors has increased significantly in the past decade. How much of this increase is a result of better diagnosis and reporting, rather than a "true" increase, remains controversial.

We *are* in an "epidemic" of lung cancer. For most of the first half of this century, lung cancer mortality was not even in the "top five" types of cancer-related deaths. Lung cancer is now the leading cause of cancer-related deaths in both men and women. About 85–90% of all lung cancers in men, and perhaps 70% in women, is directly attributable to smoking. *Per capita* consumption of cigarettes increased fivefold in men from 1900 to 1960, and with it a concomitant increase in lung cancer. The risks of several other types of common cancers are also increased by smoking (for example,

cancers of the bladder and esophagus). Approximately one-third of *all* cancer deaths could be eliminated by eliminating smoking from our society.

Of the variety of environmental factors other than smoking, dietary factors are now generally thought to represent the largest source of cancer risk, perhaps 30%–40% of all cancers. Although synthetic chemicals such as industrial pollutants and pesticides present in trace amounts in our food supply may contribute to dietary risk, recent studies suggest that this contribution is trivial relative to other "nonpollutant" factors. For example, the risk of breast cancer in women (second only to lung cancer in incidence and mortality) is significantly increased by high fat diets, and the amount of fiber in the diet substantially influences the risk of colon cancer, a major site of cancer in both men and women.

The largest source of exposure to cancer-causing chemicals may not be industrial pollution, but chemicals that occur naturally in our diet. All plants produce toxic chemicals to protect against insects, fungi, and animal predators. It has been estimated that we ingest in our diet about 10,000 times more of "nature's pesticides" than human-made chemical residues. Many of these chemicals are potent mutagens and carcinogens and are frequently present at levels thousands of times higher than the synthetic pesticide residues and industrial chemicals sometimes found in foods. Taken together, the dietary risk factors from natural sources, often present in relatively high amounts, are far more important than the pesticide residues and industrial chemicals that can often be detected at exceedingly small concentrations in our diets. Unfortunately, because of the relatively high exposure to carcinogens from natural sources, the complete elimination of synthetic industrial chemicals from our diet, if it were possible, would not likely have any significant beneficial effect on cancer incidence and mortality.

Finally, recent advances in the understanding of the biology of can-

cer suggest that "spontaneous" or "background" alterations in DNA may explain much of the cause of cancer. Studies show that DNA is inherently unstable and can be altered by normal errors in DNA replication. DNA is subject to extensive damage from processes associated with normal cellular metabolism. Within our life span, our cells undergo about 10 million billion cell divisions. Spontaneous errors in this process, which lead to mutations and cancer, accumulate with age. It is not surprising then that cancer seems to be a frequent outcome of old age.

The view that we are in a cancer epidemic largely due to industrial chemicals is not supported by the majority of cancer researchers throughout the world. Unfortunately, it will take time for the political arena to come to grips with the fact that further reduction in public exposure to synthetic chemicals will not have much impact on cancer incidence, and that such results will come only at great social and economic expense. The United States is currently spending about $80 billion per year on pollution reduction, about nine times the total budget for all basic scientific research. I believe that much of this is justified to enhance the quality of our environment and to ensure the habitability of our planet for humans and other species. However, there is also little question that huge sums of money are spent each year to reduce what is in all likelihood a trivial cancer risk, with few other environmental benefits.

If our society is truly concerned about reducing the human tragedy from cancer, more efforts should be focused on eliminating smoking and alcohol abuse, on better research and education on dietary risk factors, on more research into the biochemical and molecular events that lead to cancer (which, in turn, will lead to more effective preventive and curative measures), and on continued identification and reduction in occupational exposures to those chemicals that pose a significant cancer risk.

THE POISONING OF A NATION: AMERICA'S EPIDEMIC OF CANCER

Lewis G. Regenstein

America is in the throes of an unprecedented cancer epidemic, caused in large part by the pervasive presence in our environment and food chain of deadly, cancer-causing pesticides and industrial chemicals.

Today, significant levels of hundreds of toxic chemicals known to cause cancer, miscarriages, birth defects, and other health effects, are found regularly in our food, our air, our water, and our own bodies. Accompanying this widespread pollution has been a dramatic and alarming rise in the cancer rate in recent decades.

Each year, over a million Americans (an estimated 1,130,000 for 1992, not including 450,000 skin cancers) are diagnosed as having cancer—over 3000 people a day! The disease now strikes almost one American in three, and kills over a thousand of us *every day!* This means that of the Americans now alive, some 70–80 million people can expect to contract cancer in their lifetimes. More Americans die of cancer *every year* (an estimated 520,000 in 1992) than were killed in combat in World War II, Korea, and Vietnam combined.

And cancer has now become a common disease of the young as well as the old, with incidence of childhood cancers, especially leukemia and brain tumors, mounting sharply in recent years.

Human-made chemicals are also depleting the Earth's protective ozone layer, which makes life on the planet possible by shielding us from most of the sun's ultraviolet rays. The U.S. Environmental Protection Agency (EPA) has projected that the increase in radiation hitting the Earth will cause Americans to suffer 40 million cases of skin cancer, 800,000 deaths in the next 88 years, and 12 million incidences of eye cataracts.

In 1978, the President's Council on Environmental Quality (CEQ) reported unequivocally that "most researchers agree that 70% to 90% of all cancers are caused by environmental influences and are hence theoretically preventable."

Evidence demonstrating that toxic chemicals are heavily contributing to the cancer epidemic continues to mount. In general, the most polluted areas of the country have the highest cancer rates. Heavily industrialized New Jersey has the greatest concentration of chemical and petroleum facilities in the United States, and in July, 1982, the University of Medicine and Dentistry of New Jersey released a study showing a correlation between the presence of toxic waste dumps and elevated cancer death rates (up to 50% above average) in certain areas of the state.

In February, 1984, a report by researchers at the Harvard School of Public Health demonstrated a link between the consumption of chemically contaminated well water near Woburn, Massachusetts, and the extraordinary incidence of childhood leukemia, stillbirths, birth defects, and disorders of kidneys, lungs, and skin among local residents.

Dr. Samuel Epstein of the University of Illinois Medical Center, perhaps the foremost authority on the subject, points out that apart from AIDS, "cancer is the only major killing disease which is on the increase," with incidence rising by at least 2% a year, and death rates rising at 1% annually over the last decade. He concludes that "the facts show very clearly that we are in a cancer epidemic now," in large part because of "the carcinogenizing of our environment, the increasing contamination of our air and our water and our food and our workplace."

Today, every American is exposed to a variety of health-destroying chemicals. Dozens of pesticides used on our food are known or thought to cause cancer in animals. By the time restrictions were placed on some of the deadliest chemicals, such as DDT, dieldrin, BHC, and PCBs, these carcinogenic poisons were being found in the flesh tissues of literally 99% of all Americans tested, as well as in the food chain and even mother's milk. In fact, breast milk is heavily contaminated with high levels of banned, cancer-causing chemicals. And virtually all Americans carry in their bodies traces of dioxin (TCDD), the most deadly synthetic chemical known.

The response of the U.S. government has been largely weak or nonexistent enforcement of the nation's health and environmental protection laws. For example, with few exceptions, the EPA has refused to carry out its legal duty to ban or restrict pesticides known to cause cancer. Nor has the government adequately implemented or enforced the laws regulating hazardous waste, which is being generated at a rate of up to 250 million metric tons a year—over a ton for every man, woman, and child in the nation. Much of this is disposed of in a manner that will ultimately threaten the health of nearby residents.

Thus, we are even now sowing the seeds for the cancer epidemics of the future. Only time will tell what will be the effect on this generation, and future ones, of Americans—the chemical industry's ultimate guinea pigs.

Lewis G. Regenstein, *an Atlanta writer and conservationist, is the author of* How to Survive in America the Poisoned *and director of the Interfaith Council for the Protection of Animals and Nature, an affiliate of the Humane Society of the United States.*

SHARPENING YOUR CRITICAL THINKING SKILLS

1. Regenstein asserts that America is in the midst of a cancer epidemic caused in large part by pesticides and industrial chemicals. Eaton argues the opposite. He says that the death rate for all types of cancer except lung cancer is actually declining. Summarize the supporting data of each author.

2. Using the critical thinking skills described in Chapter 1, analyze each argument.

Visit *Human Biology's* Internet site, www.jbpub.com/humanbiology, to research opposing web sites and respond to questions that will help you clarify your own opinion. (See Point/Counterpoint: Furthering the Debate.)

www.jbpub.com/humanbiology

349

Health Note 15-1

Breast Cancer: Early Detection Is the Best Cure

Joan Lowden is 47 years old. She is a successful realtor and the mother of two children. Last week, she noticed a lump in one of her breasts. An examination by her doctor indicated that the lump was breast cancer, a potentially lethal disease that afflicts over 150,000 women every year in the United States. Each year, breast cancer claims 40,000 lives in this country alone.

Accounting for 29% of all cancers in women, breast cancer is the leading cause of death in women under the age of 54. One of every eight women alive today will be diagnosed with breast cancer sometime during her life—most commonly between the ages of 45 and 65. Men also develop breast cancer, but it is extremely rare (fewer than 1% of all cases).

In the early stages of development, breast cancer can only be detected by a special type of X-ray known as a mammogram. In later stages, a distinct swelling or lump

can be palpated (felt). Patients may complain of a vague discomfort (not pain) in the breast. The nipple may retract and the contour of the breast may become distorted. The skin covering the breast may become dimpled or reddened. On rare occasions, blood may be discharged from the nipples.

Breast cancer afflicts the nipple or tissue of the breast and often spreads to nearby lymph glands in the armpit. Thus, swollen lymph nodes in the armpit are also a symptom of breast cancer.

Breast cancer can also spread to the lungs, bone (especially in the skull), and the liver. Breast tumors grow very slowly. One the size of the tip of your little finger may have taken 8 to 10 years to reach that size, the size at which they are often detected by self-examination.

The earlier breast cancer is detected, the more successful the outcome of treatment. For this reason,

doctors recommend that women over the age of 20 practice monthly breast self-examination (BSE). This is best carried out one week after the beginning of menstruation. A pamphlet that describes BSE is available from the American Cancer Society. From the ages of 20 to 40, women should also have their breasts examined by a doctor every three years. From ages 40 to 49, the American Cancer Society recommends a doctor's exam every year and a mammogram every one to two years. After age 49, a mammogram should be administered once a year. Although X-rays may increase a women's risk of developing breast cancer, doctors consider the risk from this exposure in women over age 35 to be minimal. Because mammograms permit early detection, there is much to be gained by this procedure.

All women are at risk for developing breast cancer, but some are at a higher level than others. Those at

(b)

FIGURE 15-12 Cancer in Plants
Cancers occur in a number of different species, not just humans. *(a)* A tumor on a tomato plant. *(b)* A cancerous tumor on a tree.

(a)

Fortunately, new drugs have helped boost the survival rate of individuals suffering from certain types of cancer, such as childhood leukemia. Surgery also works effectively for some cancers, notably certain types of skin cancer. Radiation treatment and chemotherapy are successful in still others. Despite an outpouring of cancer treatments, one of the most productive advances in the treatment of cancer is not a drug or radiation; it is early detection. The earlier a cancer is detected, the greater one's chances of survival. The importance of early detection of breast cancer is noted in Health Note 15-1. Even so, the war on cancer seems to be a losing battle—with only marginal gains in survival rates after years of intensive research.

Although we often think of cancer as a human disease, in actuality, cancer occurs in a wide range of animals and plants (**FIGURE 15-12**). This section looks at cancer in humans. It will familiarize you with terms and concepts that you will encounter throughout your lifetime.

greatest risk are women who come from a family with a history of breast cancer. Women who have not had children or women whose first pregnancy occurred after age 30 are also at a higher risk of developing breast cancer. High-fat diets, obesity, and alcohol use also appear to increase the risk of developing breast cancer, as does the use of oral contraceptives and the use of estrogen therapy in postmenopausal women. Early menstruation (before the age of 12) and late menopause (after 55) are also related to a higher risk of developing breast cancer.

After a tumor has been detected, physicians typically perform a biopsy (BY-op-see). In this procedure, a needle is used to aspirate a small piece of the tumor, which is then examined under a microscope. Surgeons can also make a small incision to excise a small piece of the tumor. If the lump is a cancerous tumor and not a benign growth or a cyst, it is surgically removed.

Basically, two types of surgery are now performed: a lumpectomy and a modified radical mastectomy (mas-TECK-toe-me). A lumpectomy or breast conservation surgery is a procedure in which the tumor and a small amount of surrounding tissue is removed. The lymph nodes in the armpit in which the tumor has typically spread are also removed.

Modified radical mastectomy is the removal of the entire breast including the nipple. Lymph nodes in the armpit are also excised. This procedure is often accompanied by surgery to place a breast implant in the skin, creating an artificial breast.

Radiation treatment, chemotherapy, and hormonal therapy or combinations of them are also given to women to destroy tumor cells that may have invaded the surrounding tissues. The exact treatment varies with the type of tumor—there are 14 different types of breast cancer—and the stage at which it was detected.

The later the stage, the more vigorous the treatment.

Although breast cancer is common in women, it is also curable—if diagnosed and treated early. As in many instances, prevention is important. Eating a well-balanced, low-fat diet can reduce a woman's risk of breast cancer. Pregnant women might consider breast feeding (Health Note 8-1), as women who have breast fed their infants are at a slightly lower risk of developing breast cancer. A great deal of information is available for women through the American Cancer Society, in libraries, and on the Internet.

Visit *Human Biology's* Internet site, www.jbpub.com/humanbiology, for links to web sites offering more information on this topic.

www.jbpub.com/humanbiology

Two Types of Tumors Are Encountered in Humans

As mentioned, cancer is a disease in which cells divide uncontrollably, often invading other parts of the body. But not all abnormal cellular proliferation is cancerous. Some cells, in fact, form small masses that reach a certain size, then stop growing. These are called **benign tumors**. As a rule, they pose no significant medical problems unless they put pressure on nerves or block blood vessels.

Of great concern, however, are the **malignant tumors,** which continue to grow and often spread to other parts of the body, where they may destroy vital organs. The term *cancer* is reserved for malignant tumors.

Malignant tumors spread when individual cells or clusters of cells break loose from the tumor. They then can travel through the body in the **circulatory system,** or blood vessels, and in the **lymphatic system**, a network of vessels that drains excess fluid from body tissues (Chapter 6). These cells may settle in other sites where they can divide uncontrollably to form secondary tumors. The spread of cells from one region of the body to another is called **metastasis** (ma-TASS-tah-SISS).

Primary tumors, from which secondary tumors arise, can often be removed surgically or destroyed by radiation. The removal of the original tumor, however is fruitless unless secondary tumors are destroyed. Unfortunately, finding secondary tumors is extremely difficult, for there are literally thousands of places where the cells can become established. Cancer cells from the breast, for example, may lodge in the brain or lungs or in **lymph nodes**, organs interspersed along the lymphatic system to filter the fluid coursing through the vessels. Physicians can take tissue samples of lymph nodes to assess the spread of the cancer and can remove cancerous nodes. But when a secondary tumor is difficult to remove,

Health Note 15-2

Cancer: Ultimately, You May Hold the Key to a Cure

Hardly a year goes by without headlines announcing several promising new treatments for cancer.

ANTiCANCER VACCINE

In 1988, for example, researchers reported preliminary results of a clinical study using a "vaccine" against lung cancer, a disease that annually kills 150,000 Americans. The new vaccine, said proponents, could double survival rates in patients whose lung cancers were diagnosed early.

The lung cancer vaccine is not really a vaccine in the traditional sense, because it does not prevent the disease. This treatment is actually an immune system boost, which helps fight off an existing disease. In the study, 34 patients with early-stage lung cancer were treated with a plasma membrane protein extracted from lung cancer cells. When given to patients in early stages of lung cancer, the protein appears to stimulate the patient's immune system. The immune system, in turn, mounts an assault on the foreign protein in the blood. But because the protein is also attached to the plasma membrane of lung cancer cells, the immune system attacks and kills the cancer cells, especially those that have spread to other sites.

In clinical studies, this experimental therapy greatly increased the subjects' 5-year survival rates. Normally, only 1 patient in 10 who is diagnosed with lung cancer is alive 5 years after diagnosis and conventional treatment (surgery followed by chemotherapy). Five years after the immune system boost, 5 of 10 patients were still alive.

MICROSPHERES AND MONOCLONAL ANTIBODIES

In recent years, some scientists have expressed their enthusiasm for lipid microspheres (liposomes) containing cancer-killing chemicals, or oncotoxins. Liposomes can be designed to target cancer cells, in the process delivering small doses of the lethal drugs to cancer cells and sparing the rest of the body. Targeting cancer cells is the job of antibodies, special proteins incorporated in the membrane of the liposome. These antibodies bind to proteins in the plasma membrane of cancer cells, allowing the "toxic bullet" to find its target.

Oncotoxins may also be attached directly to monoclonal antibodies, which are synthesized by the immune system of animals that have been injected with human cancer cells. These antibodies can be chemically bonded to oncotoxins, and then administered to patients. Inside the body of the patient, the antibody-oncotoxin complex binds to the cell surface of cancer cells and destroys them.

TREATING CANCER WITH LIGHT

Scientists have found that certain drugs, when activated by laser light, can kill cancer cells. In one therapy, patients with tumors in the lining of the esophagus and the bronchi of the lung are given the drug Photofrin (porfimer sodium). After the drug circulates in the blood and bathes the cancer cells, the researchers insert a small fiberoptic fiber in the affected area. A laser beam is trained on the cancer and, when turned on, activates the chemical, which selectively destroys the cancer cells. This treatment has been approved in the United States for certain forms of advanced esophageal cancer. Companies are now (November, 1997) seeking approval for other uses. One of the chief advantages is that photodynamic therapy, as it is called, poses little risk to adjacent tissues and can be repeated several times and combined with other therapies.

PREVENTION

Despite the promising new developments in cancer treatment, many experts stress the importance of prevention. By quitting smoking or not

physicians generally use radiation or **chemotherapeutic drugs**, highly toxic drugs that attack rapidly dividing cells of tumors. This shotgun approach has many side effects, including nausea, hair loss, and general sickness, but it may be a patient's only hope. Some new developments in cancer treatment are noted in Health Note 15-2.

■ The Conversion of a Normal Cell to a Cancerous One Involves Two Steps

Few questions interest cell biologists more than what causes a cell to begin dividing uncontrollably. A growing body of research suggests that the production of a malignant tumor (cancer) actually involves two steps that may take place over a long period of time: (1) conversion and (2) development and progression (**FIGURE 15-13**).

The **conversion** of a normal cell to a cancerous one typically begins with a **mutation**, a change in the DNA caused by a chemical, physical, or biological agent (a virus, for example). Those agents that cause the initial transformation are known as **co-carcinogens**. Mutating just any gene won't do, though. In order to produce a mutation that sets the stage for a cancerous growth, the

taking it up in the first place, by limiting exposure to X-rays and UV light, and by choosing a safe occupation and dwelling, you can greatly reduce your likelihood of contracting cancer.

As noted in Chapter 5, proper diet may also help prevent cancer. Dietary fiber in grains and vegetables, for example, may help reduce your chances of contracting colon cancer, a leading killer of men (**FIGURE 1**). Other substances in food may also help reduce the risk of cancer. Recent research suggests that a chemical substance called ellagic acid, which is found in raspberries, blackberries, strawberries, and other fruit and nuts, may destroy carcinogenic molecules in the body, reducing the likelihood of developing cancer.

Scientists studying carcinogenic agents found in tobacco smoke, auto exhaust, and foods noted that in laboratory animals, ellagic acid reduced DNA damage caused by one prevalent carcinogen by 45–70%.

Ellagic acid has to be added before or during exposure to a carcinogen, and researchers have not yet identified the mechanism by which this substance works in people. They suspect that it competes for binding sites on the DNA molecule to which carcinogens attach. Studies are currently underway in humans to determine if ellagic acid is an effective anticancer agent.

Emotions and Cancer

Recent studies indicate that there may be a link between one's emotional state and the likelihood of developing cancer. Researchers at Johns Hopkins University studied medical students subjected to personality tests between 1948 and 1964 and compared the results of the tests with subsequent health records. They found that over a 30-year period students characterized as "loners," who suppressed emotions beneath a bland exterior, were 16 times more likely to develop leukemia and several other cancers than a group that gave vent to emotions and took active measures to relieve anger and frustration. Why this might be so is anyone's guess.

Corroborating these results, a 17-year study of nearly 7000 people in Alameda County, California, showed that two types of social isolation (having few close friends and feeling alone even in the presence of friends) elevated the risk of certain cancers in women. A study in Yugoslavia also showed that repression and denial of emotions both on a regular basis and in response to stress were related to an increase in cancer and heart disease.

Despite these gains from years of research and billions of dollars spent studying cancer, modern science may actually be losing the war against

FIGURE 1 Reducing Cancer by Eating Right Fruits, vegetables, and grains provide fiber that helps reduce the incidence of colon cancer.

many types of cancer. For example, long-term survival rates in cancer patients over the past decade or so have increased only 4.2% compared with an overall increase of 5.1% in survival rates for all other diseases. Because of this statistic, many agree that the cheapest and most effective cure is prevention.

Visit *Human Biology's* Internet site, www.jbpub.com/humanbiology, for links to web sites offering more information on this topic.

www.jbpub.com/humanbiology

FIGURE 15-13 Stages Leading to Cancer.

DNA-reactive carcinogen

Genetic mutation in growth control gene

Hormonal imbalance, immune system alteration, or tissue injury

Normal cell

Epigenetic carcinogen

Conversion

Promoter

Mutated cell

Development and Progression

Cancer

mutation must occur in one of the 50 genes that control the replication of cells. This, in turn, may partially or completely release a cell from the normal growth controls.

Partially released cells often remain dormant for long periods, presumably controlled by tissue factors. This homeostatic control keeps the cancer cell from dividing and thus provides an important measure of protection. Unfortunately, this safeguard can be overridden. In some instances, for example, certain hormones produced by the body can stimulate the growth of these cells. Such cancers are said to be hormone-dependent.

The second stage leading to cancer is called **development and progression**. It occurs when a chemical substance stimulates the growth of a mutated cell. Those chemical substances that influence the second stage of carcinogenesis are called **promoters**. Promoters stimulate cells to divide uncontrollably, as do errors in DNA replication. Once released, cells begin a frenzy of growth that, if unchecked, usually kills the host.

Mutations, like those that result in cancer, occur with great frequency. Fortunately, though, cells have evolved an intricate mechanism to repair genetic damage. Thus, if a mutation that transforms a cell into a cancerous one can be repaired before the cell divides, the cancer is eliminated. These mechanisms must also be considered part of the body's homeostatic weaponry. The immune system can also locate and destroy tumor cells.

Unfortunately, not all cancerous mutations are repaired in time or eliminated by the immune system. Given the large number of mutations occurring every day in our bodies and the longer lives we now lead, it is little wonder that one out of every three of us will develop some form of cancer in our lifetime.

Some Carcinogens Exert Their Effect Outside of the DNA

As mentioned above, many chemical carcinogens induce cancer by altering the DNA of cells. These are generally referred to as **DNA-reactive carcinogens**. A growing body of evidence, however, indicates that many chemical substances act outside the DNA—that is, they do not react with DNA. These carcinogens, known as **epigenetic carcinogens**, operate by eliciting other biological effects. For example, some may cause a hormonal imbalance that leads to cellular proliferation. Others may suppress the immune system, which, in turn, permits cellular proliferation. Still others may cause chronic (persistent) tissue injury. Evidence suggests that some epigenetic carcinogens stimulate changes in the DNA indirectly and may facilitate tumor development from already-converted cells.

Interestingly, seemingly innocuous chemicals can turn into dangerous substances inside the body. A good example is the nitrites. Although normally harmless, nitrites are converted to carcinogenic nitrosamines inside the human body. (See the Health, Homeostasis, and the Environment section in Chapter 2.) The enzymes responsible for this conversion reside in the smooth endoplasmic reticulum (SER) of liver cells and possibly others. Although the liver's SER is usually an ally in detoxifying chemical substances, in this instance it converts a harmless substance into a potentially lethal one.

Cancers Develop Many Years After the Initial Exposure to a Carcinogen

As a rule, mutations do not manifest themselves immediately in cancer. In fact, many tumors form 20 or 30 years after the mutation. A few cancers such as leukemia occur within 5 years. The period between exposure and the emergence of a cancerous tumor is called the **latent period**. Because the latent period is long, it has been extremely difficult for medical researchers to determine the exact causes of many cancers, a fact that has hindered public health research considerably.

Several Physical and Biological Agents Are Also Responsible For Producing Cancer

Chemicals are not the only cause of cancer. X-rays that physicians use to diagnose disease, ultraviolet radiation from the sun or a tanning salon, and radon gas given off by soils in some parts of the country can lead to mutations that result in cancer.

Biological agents also cause cancer. In 1911, Peyton Rous discovered that a certain virus produced cancerous growths in chickens. Since that time, researchers have discovered many viruses that cause cancer in a wide variety of animals, including humans. Unlike physical and chemical carcinogens, which alter DNA structure, carcinogenic viruses contain genes that, when inserted into the genetic material of the cells, may cause uncontrollable cell division.

Health, Homeostasis, and the Environment

Depleting the Ozone Layer

Encircling the Earth, 12–30 miles above your head, is a layer of air rich in ozone (O_3) (**FIGURE 15-14**). It is called the **ozone layer**. Although the ozone layer may seem far removed from the discussion at hand, it's not. It is vital

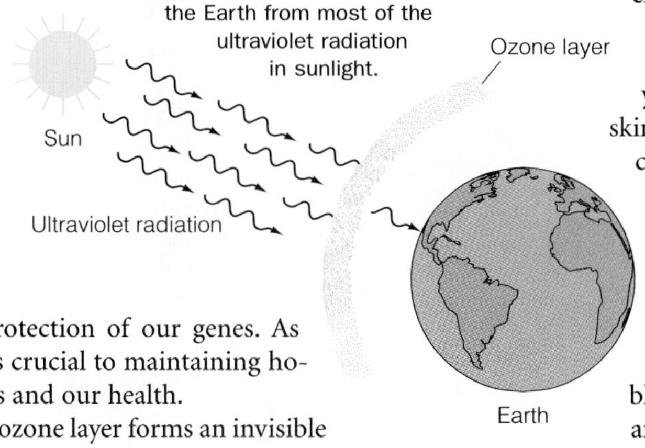

FIGURE 15-14 **The Ozone Layer**   High above our heads lies the ozone layer of the atmosphere. The ozone layer shields the Earth from most of the ultraviolet radiation in sunlight.

to the protection of our genes. As such, it is crucial to maintaining homeostasis and our health.

The ozone layer forms an invisible shield that filters out most (99%) of the ultraviolet (UV) radiation emitted from the sun. Consequently, it protects all living things—from plants to humans—from damaging UV light, which can cause mutations in cells that may lead to cancer and death. The formation of the ozone layer over 1 billion years ago, in fact, was probably necessary for the evolution of life on land (Chapter 21). Without it, terrestrial life could probably not exist or would have evolved thick, protective shields.

Today the ozone layer is disappearing. In 1974, two chemists warned that chemical substances used as spray-can propellants and refrigerants, known as the **chlorofluorocarbons (CFCs),** were eroding the ozone layer. Over the next decade and a half, numerous estimates of ozone depletion were presented, but because of conflicting results and because the ozone concentrations in the ozone layer fluctuate naturally from year to year, atmospheric scientists could not tell for certain whether the projections were accurate.

Then, in 1988, a group of 100 scientists, working under the auspices of the National Aeronautics and Space Administration (NASA), gathered to review the data of nearly 20 years of atmospheric monitoring. They agreed that the ozone layer was indeed declining and that levels over the poles had fallen by as much as 10% since 1969. The decline over the heavily populated regions of North America and Europe over the same period was 3%. More recent studies show that an overall decline of 3% (over the entire panet), and a 13–14% decline over the northern hemisphere since the early 1970s.

Some scientists predict that each 1% decrease in the ozone layer will increase the incidence of skin cancer by at least 2%. If this prediction proves correct, the 3% decline over North America could increase the skin cancer rate by 6%.[2] This would result in an additional 20,000–60,000 cases of skin cancer each year in the United States alone. Given that 4% of all skin cancers are fatal, an additional 800–2400 Americans could die each year from skin cancer. A 13% decline would lead to a far greater incidence of skin cancer. Scientists haven't seen this increase yet in large part because of air pollution, which blocks the incoming ultraviolet light.

Increased UV light may also harm other animals and plants. Plants are particularly vulnerable. High UV radiation interrupts photosynthesis and can result in death.

In September 1987, about a year before the NASA report, representatives from 24 nations met in Montreal to work out a treaty to restrict the use of most CFCs and a few related chemicals also thought to destroy the ozone layer. The nations agreed to cut their use to one-half of the 1986 levels by 1999. Given the accumulation of CFCs in the upper atmosphere since the 1950s, many critics warned that this reduction would not save the ozone layer.

Much to the surprise of the critics, Du Pont, a world leader in CFC production, agreed in 1988 that further cuts were necessary and called for a total worldwide ban on CFC production. The company argued that stronger regulations could eliminate CFC production by the end of the century. Two years after the Montreal Treaty, the signatories met in London and agreed to a complete phase-out of CFCs and other ozone-depleting chemicals by 2000.

In 1992, another treaty was signed in Copenhagen, calling for an acceleration of the phaseout of CFCs and other ozone-depleting chemicals. The three treaties have resulted in a dramatic decline in the production and release of ozone-depleting chemicals. While that is cause for celebration, many scientists point out that the ozone layer will continue to deteriorate because massive amounts of ozone-depleting chemicals can still be found in the stratosphere and because these chemicals last for decades. According to one estimate, at least 100 years will be required before the ozone layer is back to normal.

---

[2]More recent estimates suggest that each 1% decline in the ozone layer will result in a 4% increase in skin cancer.

SUMMARY

1. Researchers have found a disturbing correlation between extremely low frequency (ELF) magnetic radiation and leukemia in children. At least one study shows that ELF radiation accelerates the growth of cancer cells and increases their resistance to the body's natural killer cells.
2. ELF magnetic fields are generated by power lines and by electric appliances, water bed heaters, and electric blankets.
3. An analysis of all published studies, by a prestigious group of scientists, however, concludes that there is no correlation between ELF magnetic radiation and cancer.

THE CELL CYCLE

4. The life cycle of a cell is called the *cell cycle* and consists of two parts: interphase and cell division.
5. *Interphase* is a period of active cellular synthesis and is divided into three phases: G_1, S, and G_2.
6. During G_1, the cell produces RNA, proteins, and other molecules; G_1 is the period during which a cell carries out its normal activities. It is also a period of preparation for division.
7. During the S phase, the DNA replicates. After replication, each chromosome in the nucleus of the cell contains two chromatin fibers, or chromatids.
8. G_2 is a much shorter period and is relatively inactive.
9. During interphase, the cell replicates its organelles and molecules needed by the two daughter cells.
10. *Cell division* follows interphase. It requires two separate but related processes: *mitosis*, or nuclear division, and *cytokinesis*, or cytoplasmic division.

THE CHROMOSOME

11. Each organism has a set number of chromosomes. All body cells, except the germ cells, are called *somatic cells*; they contain a full complement of chromosomes and are described as *diploid*.
12. *Germ cells* or *gametes* contain half the number of chromosomes of somatic cells and are referred to as *haploid cells*.
13. Gametes are produced by a special type of cell division known as *meiosis*, which occurs in the gonads (ovaries and testes).
14. The chromosomes are loosely arranged in the nucleus during interphase but condense during prophase. Condensation facilitates chromosome separation.
15. Chromosomal condensation also allows physicians to study the morphology of chromosomes. Chromosomes arranged according to size form a *karyotype*, which allows geneticists to count the chromosomes and to locate gross abnormalities.

CELL DIVISION: MITOSIS AND CYTOKINESIS

16. During cellular division, the cell divides its chromosomes equally and distributes its cytoplasm and organelles more or less evenly between the two daughter cells.
17. Cell division is essential to reproduction and growth. It is the basis of tissue repair and provides replacements for cells lost through normal wear and tear.
18. *Mitosis*, nuclear division, is divided into four stages: prophase, metaphase, anaphase, and telophase. **TABLE 15-1** summarizes the major changes in each stage.
19. Successful mitosis depends on the presence of the *mitotic spindle*, an array of microtubules that forms during prophase. Some of the spindle fibers attach to the chromosomes and help pull them apart during anaphase.
20. *Cytokinesis*, the division of the cytoplasm, begins in late anaphase or early telophase. In human cells, cytokinesis results from the contraction of microfilaments lying beneath the plasma membrane.

CONTROL OF THE CELL CYCLE

21. Research suggests that the cell cycle is controlled in part by substances produced in the cytoplasm. These chemicals cause changes in the nucleus of the cell.
22. External controls, such as hormones, growth regulators, cell contact, and others, are also imposed on the cell.

CANCER: UNDERSTANDING A DEADLY DISEASE

23. Cancer is a disease characterized by uncontrolled cell division. Current estimates suggest that 20%–40% of all cancers arise from environmental and workplace pollutants. The remaining cases are caused by smoking, diet, and natural causes.
24. No matter what the cause, cancer is a disease that most of us will have close experience with. One out of three Americans will contract the disease, and one out of four will die from it.
25. Cells that divide uncontrollably form *tumors*. Tumors that fail to grow and spread are *benign* and rarely cause medical problems.
26. *Malignant tumors* are those that continue to grow, often spreading to other parts of the body and invading vital organs. The term *cancer* is reserved for malignant tumors.
27. Malignant tumor cells spread (*metastasize*) through the body via the circulatory and lymphatic systems, often becoming established in distant sites, where they form *secondary tumors*.
28. Tumors can be removed surgically or destroyed with radiation, but successful treatment depends on the location and destruction of secondary tumors.
29. Patients can also be treated with chemotherapeutic agents, chemical substances that attack rapidly dividing cells.
30. The transformation of a normal cell into a cancerous one often results from a mutation, a change in the DNA resulting from a chemical, biological, or physical agent. These agents are called *co-carcinogens*.
31. The initial mutation, caused by a co-carcinogen, may or may not result in a cancerous growth. A second factor, a *promoter*, however, may stimulate the cell to proliferate uncontrollably.

HEALTH, HOMEOSTASIS, AND THE ENVIRONMENT: DEPLETING THE OZONE LAYER

32. The ozone layer, found in the upper atmosphere, filters out most of the incoming ultraviolet light, protecting many life-forms from harmful burns and damaging mutations that can lead to cancer.
33. Unfortunately, certain chemicals called *chlorofluorocarbons* (CFCs) are destroying the ozone layer. In industrialized nations, CFCs were used as spray-can propellants, refrigerants, coolants, cleaning agents, and blowing agents for foam.
34. CFCs have caused a 3% decline in ozone levels globally and a 13% decline over North America.
35. CFC production and release have been nearly eliminated thanks to international treaties.

Critical Thinking

THINKING CRITICALLY— ANALYSIS

This Analysis corresponds to the Thinking Critically scenario that was presented at the beginning of this chapter.

First and foremost, one should examine the source of information and look for hidden bias. First, you will find that Rush Limbaugh is not a scientist and has little knowledge of the atmospheric science necessary to understand ozone depletion. Furthermore, Limbaugh is a powerful advocate of free enterprise and takes a dim view of government regulation. In other words, he's not exactly an unbiased observer.

Furthermore, if you read his work, you'll find that he gets most of his facts from the late Dixie Lee Ray, also an avowed anti-environmentalist. According to author Robert Boyle, she, in turn, gets much of her information from Rogelio Maduro, an associate editor of *21st Century*. Maduro coauthored a book entitled *The Holes in the Ozone Scare*. It asserts that efforts to ban CFCs stem in large part from a corporate plot led by Du Pont, the world's leading manufacturer of CFCs. According to Maduro, they are eager to profit from substitutes for CFCs, which they developed.

Unfortunately, this characterization is in error. Du Pont and other chemical companies actually opposed bans on ozone-depleting CFCs for many years. Only when the evidence that CFCs were damaging the ozone layer became irrefutable did Du Pont back down and support a ban.

Articles in *Science* and *Chemical and Engineering News* have systematically refuted the claims made by Limbaugh, Ray, and Maduro. Unfortunately, few people will ever read these journals to explore the scientific treatment of this subject.

This controversy, like others, illustrates the importance of examining the source of information and looking for hidden bias.

EXERCISING YOUR CRITICAL THINKING SKILLS

Professor Arthur Kronberg has spent a lifetime exploring the secrets of cells, and with great success. In 1956, for example, he discovered DNA polymerase, the enzyme cells required to synthesize DNA. In 1959, he was awarded the Nobel Prize for his contributions. In 1967, Kronberg synthesized an entire viral DNA chain, which successfully replicated inside host cells.

Kronberg has focused much of his research on cellular enzymes. In much of his work, he grinds up cells and extracts the liquid contents. He then manipulates the extract to perform the specific process in which he is interested. After that, he begins removing enzymes from the extract until the process stops. When it stops, he knows he has isolated the enzyme responsible for catalyzing the function he is studying.

Kronberg has used this technique to find out what stops a cell from dividing—in other words, what factor or factors are responsible for turning off DNA replication in nondividing cells. Kronberg and his colleagues began this search for this "off switch" by extracting enzymes from the cell extracts. Through this work, they discovered several enzymes and proteins that stop replication. Unfortunately, the researchers found that the proteins work by destroying the DNA, obviously not the way nature works.

Although you're not a trained cell biologist, can you see anything wrong with the way Kronberg approached this problem?

One of Kronberg's postdoctoral students did. In 1988, Deog Su Hwang, who was working in Kronberg's lab, began to study the same problem, but from a different angle. He had observed that two intertwined strands of DNA separate slightly just before replication. This separation takes place at a site along the DNA strands known as the "origin." Many enzymes involved in DNA replication attach to the origin. This separation forms a kind of bubble that promptly causes the strands to separate further. The nucleotides necessary for making complementary strands are then assembled by DNA polymerase. Hwang thought that this separation site was a likely place for the off switch and concentrated on this region. In other words, he located a probable location for the switch, then began the search for the protein that binds to that site, hoping that it would somehow block DNA replication. His approach focused more narrowly on the problem and very probably increased his chance of success.

Hwang used enzymes to slice off portions of the DNA chain and associated protein around the origin. These fragments were then isolated and studied. In late 1989, he succeeded in isolating a pure sample of the protein that inhibited DNA replication when added before initial separation of the DNA helix. How did the experimental procedures differ?

Kronberg's enzyme purification procedure searched for an enzyme that catalyzed a specific cellular process. In a sense, he was searching for a needle in a haystack. Hwang's method involved searching for the protein that fit a specific site and then testing it to determine its function. The two procedures are obviously quite different. What is to be learned from this exercise?

With all due respect to Arthur Kronberg, question the experts and their assumptions. At times, even Nobel-Prize-winning scientists can embark on paths of scientific inquiry that yield little information. In this case, Kronberg's line of inquiry was based on techniques that worked in other situations but weren't applicable to the problem at hand.

TEST OF CONCEPTS

1. Describe the cell cycle, and explain what happens during each part.
2. Discuss the factors that control the cell cycle. Is it likely that many factors play a role in determining when and how often a cell divides?
3. Draw a diagram of the cell cycle, and note how many chromosomes are found in the cell at each stage and whether they are single- or double-stranded.
4. List the stages of mitosis, and describe the major nuclear changes occurring in each stage. When does cytokinesis occur? What are the major cytoplasmic changes?
5. In what ways are cancer cells different from normal cells? What events lead to a cancer? Why don't more cells become cancerous? What is the difference between a benign and a malignant tumor?
6. The Health, Homeostasis, and the Environment section argues that chemical substances produced by modern society—notably, the chlorofluorocarbons—may be making our environment hazardous to human health. Describe how.
7. Cancer is often considered a disease of aging. Given the fact that cancer is caused by DNA mutation, why is this assertion correct? Can you think of some reasons why the incidence of some cancers has risen in the past 20 years?

TOOLS FOR LEARNING

www.jbpub.com/humanbiology

Tools for Learning is an on-line student review area located at this book's web site HumanBiology (www.jbpub.com/humanbiology). The review area provides a variety of activities designed to help you study for your class:

Chapter Outlines. We've pulled out the section titles and full sentence sub-headings from each chapter to form natural descriptive outlines you can use to study the chapters' material point by point.

Review Questions. The review questions test your knowledge of the important concepts and applications in each chapter. Written by the author of the text, the review provides feedback for each correct or incorrect answer. This is an excellent test preparation tool.

Flash Cards. Studying human biology requires learning new terms. Virtual flash cards help you master the new vocabulary for each chapter.

Figure Labeling. You can practice identifying and labeling anatomical features on the same art content that appears in the text.

Active Learning Links. Active Learning Links connect to external web sites that provide an opportunity to learn basic concepts through demonstrations, animations, and hands-on activities.

CHAPTER 16

PRINCIPLES OF HUMAN HEREDITY

Scanning electron micrograph of human chromosomes.

Thinking Critically

More than a hundred years after his death, President Abraham Lincoln has become the subject of a scientific controversy. The debate among some scientists is not over who killed him or why but, rather, over the possibility that he suffered from a genetic disorder called *Marfan's syndrome.*

Marfan's syndrome is an extremely rare disorder that affects the skeletal system, the eyes, and the cardiovascular system. It is difficult to diagnose and is usually identified only after its victims die. Flo Hyman, the star of the 1984 U.S. women's Olympic volleyball team had Marfan's syndrome. In 1986, Hyman collapsed on the volleyball court and died from a ruptured aorta, the huge blood vessel that carries blood away from the heart to the body.

Marfan's victims typically have exceedingly long arms and legs, which make them excellent volleyball and basketball players. In normal individuals, the arm span (the distance from fingertip to fingertip) is equal to their height. In Marfan's syndrome, however, arm span exceeds height.

Nearsightedness and lens defects are also common among Marfan's patients. The most serious problem, however, is enlargement and weakening of the aortic arch, caused by a degeneration of the connective tissue in the wall of this vessel. If untreated, the aorta can burst, causing instant death.

Evidence suggests that Lincoln had Marfan's syndrome. He was tall and lanky, for example, and he wore glasses. And, his great-great-grandfather had the disease, making it possible that Abe had received the Marfan's gene. What additional evidence might you look for to confirm this hypothesis? What critical thinking rule(s) do(es) this exercise illustrate? ∎

FOR MOST OF US, LIFE IS AN EMOTIONAL ROLLER coaster. That is, unless you're numb to the world, your emotions tend to fluctuate with changing circumstances like the cars on a roller coaster. Some events bring elation, while others evoke anger, fear, or other deep-seated emotions. Such peaks and troughs are quite normal. But for some people, the emotional roller coaster ride is intense—their feelings swing erratically for little or no reason. One minute they are in a state of elated overactivity (mania), the next they fall into a state of depressed inactivity. This condition, known as **manic-depression,** is a psychological disorder characterized by cyclic mood swings that are unrelated to external events.

Manic-depression varies widely from one person to the next. In some, it persists for only a short time. In others, it continues for years. During mania, the animated and highly energetic state, manic-depressives often act irrationally or obnoxiously. Such behavior can ruin relationships and can lead to financial disaster, especially if it interferes with decision making. When in a state of deep depression, victims sometimes threaten suicide, although their lack of energy often prevents them from following through. In some cases, however, the self-destructive yearning persists after the period of depression and results in suicide.

Manic-depression occurs in about 3% of the U.S. population and tends to run in families, a fact that suggests a genetic link. Researchers, in fact, recently located a gene near the tip of chromosome 11 that may predispose its bearers to manic-depression and possibly severe depression.

Genes on chromosomes determine the structure and function of cells and organs. Growing evidence suggests that they even influence behavior. This chapter examines how genes are passed from parent to offspring—that is, the inheritance of traits. We begin with an overview of meiosis, a special kind of cell division responsible for gamete formation in humans and other sexually reproducing organisms. We then focus our attention on some basic principles of heredity.

Meiosis and Gamete Formation

For successful reproduction to occur, sexually reproducing organisms, including humans, must have some mechanism to produce gametes with half the number of chromosomes of somatic cells. Thus, when the male and female gametes unite, the offspring will have the normal (diploid) number of chromosomes. This process of reduction is called meiosis (my-OH-siss). Technically, **meiosis** is a type of nuclear division that occurs in germ cells and yields haploid gametes. Germ cells are formed in the reproductive organs, the ovaries (OH-var-ees) of women and testes (TESS-tees) of men.

As mentioned in Chapter 15, human somatic cells contain two sets of chromosomes, one maternal in origin, the other paternal. All told, human somatic cells contain 46 chromosomes, 23 from the mother and 23 from the father. These chromosomes occur in pairs, called homologous pairs, because they are the same length, have the same appearance, and contain the same genes. During meiosis, the homologous pairs are split so each gamete ends up with one chromosome of each pair.

Meiosis Involves Two Cellular Divisions

As noted above, meiosis involves two nuclear divisions. The two nuclear divisions are referred to as meiosis I and II.

Meiosis I Is a Reduction Division

Meiosis I is often called a reduction division. **FIGURE 16-1** illustrates the process. One of the first things you will note is that meiosis contains the same steps as mitosis: prophase, metaphase, anaphase, and telophase. Although meiosis I has the same steps, there are some important differences. As shown during prophase I, the chromosomes condense, but unlike meiosis, the homologous chromosomes "pair up." During metaphase I, the homologous pairs line up along the equatorial plate. In mitosis, chromosomes line up singly on the equatorial plate. During anaphase I, the pairs are separated. Thus, during telophase, each daughter cell ends up with half the number of chromosomes of the parent cell from which it was formed. Because of this, meiosis II is called a reduction division. Note, however, that each chromosome contains two chromatids.

Meiosis II Is Akin to Mitotic Division

The second division, **meiosis II,** also resembles mitosis, except for one difference: The cells start out as haploid. As illustrated in **FIGURE 16-1**, the chromosomes of the haploid cells condense during prophase II, then line up in single file along the equatorial plate during metaphase II. During anaphase II, the chromatids dissociate, one going to each pole. The result is two cells, each containing a haploid number of single-stranded (unreplicated or one-chromatid) chromosomes. These cells give rise to the gametes.

In summary, in meiosis I the cells begin with 46 chromosomes—each with two chromatids. During meiosis I, these cells divide. Each daughter cell ends up with 23 chromosomes. Each chromosome has two chromatids. During meiosis II, the 23 chromosomes split apart, and each daughter cell ends up with 23 unreplicated chromosomes—each with one chromatid.

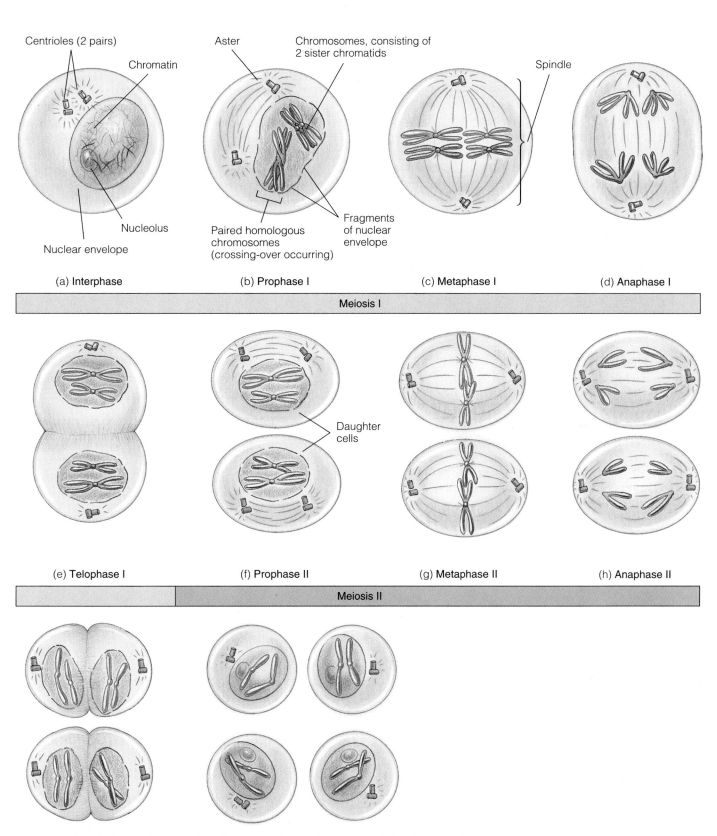

Centrioles (2 pairs)

Chromatin

Nucleolus

Nuclear envelope

(a) Interphase

Aster

Chromosomes, consisting of 2 sister chromatids

Paired homologous chromosomes (crossing-over occurring)

Fragments of nuclear envelope

(b) Prophase I

Spindle

(c) Metaphase I

(d) Anaphase I

Meiosis I

(e) Telophase I

Daughter cells

(f) Prophase II

(g) Metaphase II

(h) Anaphase II

Meiosis II

FIGURE 16-1 Meiosis. Note that meiosis is divided into two phases: meiosis I and meiosis II.

▮ In Males, Meiosis Produces Four Gametes; In Females, It Produces Only One

Meiosis differs between males and females. In males, the process is much as I have described above. Thus, a single diploid cell in the gamete-producing testis (singular; pronounced TESS-tiss) gives rise to four sperm cells. During this process, the cytoplasm of the original cells that give rise to the sperm is divided equally among the four germ cells. During the final stages of sperm development, much of the cytoplasm is discarded to produce streamlined sperm capable of swimming up the female reproductive tract (**FIGURE 16-2**).

In females, the single diploid cell gives rise to only one egg (**FIGURE 16-3**). Thus, all of the cytoplasm is retained in one cell, and the extra chromosomal material is systematically discarded during the two meiotic divisions. This difference reflects the fact that the egg is fairly sedentary. It's the sperm that must do most of the work during fertilization! This early division of labor also ensures that adequate cellular organelles (supplied by the female) are available for the fertilized ovum, or **zygote** (ZYE-goat).

Section 16-2

Principles of Heredity: Mendelian Genetics

By now, you know that the genes on human chromosomes are largely responsible for who we are—what we look like and, at least partly, how we behave. The gene-bearing chromosomes

are divvied up during meiosis into germ cells. The germ cells combine during fertilization to form a new being. With these basics in mind, we turn our attention to heredity—the transmission of genes from parents to offspring.

Heredity is one aspect of the study of genetics. The formal study of genetics began with a look at heredity at the organismic level carried out by a 19th-century monk, Gregor Mendel, working in his garden (**FIGURE 16-4**).

Born in Eastern Europe in 1821, Mendel entered an Augustinian monastery in Brno, in what is now the Czech Republic, at the age of 21.[1] After completing his studies, Mendel enrolled at the University of Vienna, where he studied botany, mathematics, and other sciences. Returning to the monastery, he began a series of experiments on garden peas—studies that would reveal several fundamental principles of genetics.

During Mendel's time, many scientists believed that the traits of a child's parents were blended in the offspring, producing a child with intermediate characteristics. In addition, because the ova of sexually reproducing organisms are much larger than the sperm, some scientists believed that the female had a greater influence on the characteristics of the offspring than the male.

---

[1]Brno is now the city of Brunn.

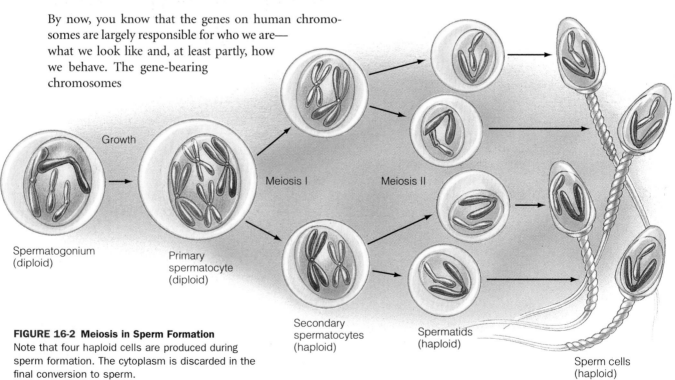

FIGURE 16-2 Meiosis in Sperm Formation
Note that four haploid cells are produced during sperm formation. The cytoplasm is discarded in the final conversion to sperm.

Spermatogonium
(diploid)

Growth

Primary
spermatocyte
(diploid)

Meiosis I

Secondary
spermatocytes
(haploid)

Meiosis II

Spermatids
(haploid)

Sperm cells
(haploid)

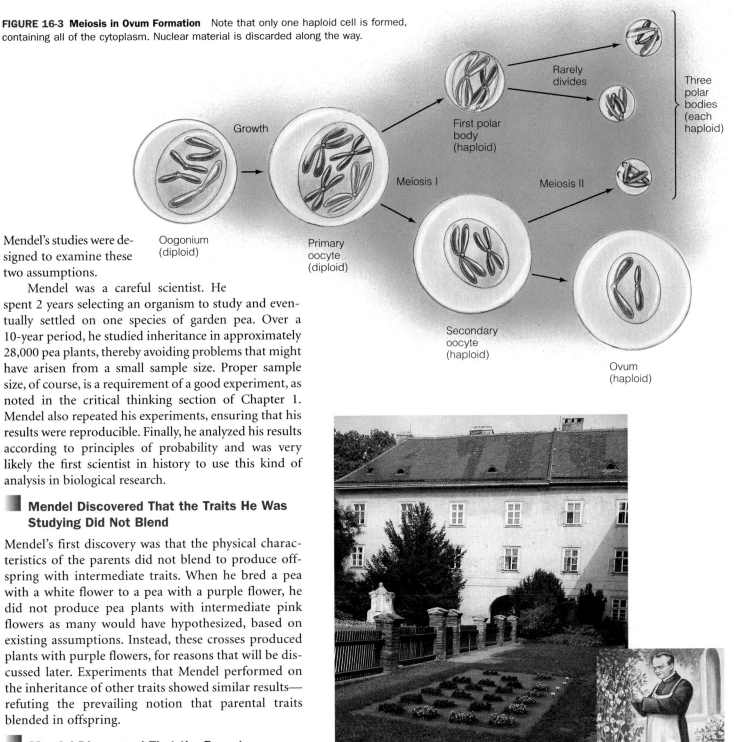

FIGURE 16-3 Meiosis in Ovum Formation Note that only one haploid cell is formed, containing all of the cytoplasm. Nuclear material is discarded along the way.

Growth

Oogonium (diploid)

Primary oocyte (diploid)

Meiosis I

Meiosis II

First polar body (haploid)

Rarely divides

Three polar bodies (each haploid)

Secondary oocyte (haploid)

Ovum (haploid)

Mendel's studies were designed to examine these two assumptions.

Mendel was a careful scientist. He spent 2 years selecting an organism to study and eventually settled on one species of garden pea. Over a 10-year period, he studied inheritance in approximately 28,000 pea plants, thereby avoiding problems that might have arisen from a small sample size. Proper sample size, of course, is a requirement of a good experiment, as noted in the critical thinking section of Chapter 1. Mendel also repeated his experiments, ensuring that his results were reproducible. Finally, he analyzed his results according to principles of probability and was very likely the first scientist in history to use this kind of analysis in biological research.

Mendel Discovered That the Traits He Was Studying Did Not Blend

Mendel's first discovery was that the physical characteristics of the parents did not blend to produce offspring with intermediate traits. When he bred a pea with a white flower to a pea with a purple flower, he did not produce pea plants with intermediate pink flowers as many would have hypothesized, based on existing assumptions. Instead, these crosses produced plants with purple flowers, for reasons that will be discussed later. Experiments that Mendel performed on the inheritance of other traits showed similar results—refuting the prevailing notion that parental traits blended in offspring.

Mendel Discovered That the Parents Contributed Equally to the Characteristics of Their Offspring

Mendel's research led him to conclude that adult plants contain pairs of hereditary factors for each trait (for example, seed color) and that each pair governed the

(b)

FIGURE 16-4 Mendel and His Garden *(a)* Gregor Mendel worked out some of the basic rules of inheritance in the mid-1800s in his research on garden peas. *(b)* Mendel's garden at the monastery in Czechoslovakia as it appears today.

(a)

inheritance of a single trait. Today, we refer to Mendel's hereditary factors as the genes, which we know are located in the DNA of the chromosomes.

Mendel reasoned that, because a male gamete and a female gamete combine to form a new organism and an adult contains only two hereditary factors for any given trait, each gamete must contain only one hereditary factor (gene) for each trait. Mendel's studies therefore suggested that the paired hereditary factors of the mother and father must separate during gamete formation so that each gamete contributes one and only one hereditary factor to the zygote. The separation of hereditary factors during gamete formation is known as the **principle of segregation.**

Because the gametes of the parents combine to produce an offspring and because each gamete contains one hereditary factor for each trait, Mendel concluded that the contributions of the parents must be equal. That is, each parent contributes one hereditary factor, or gene, for each trait. These conclusions may seem unremarkable to us today, especially with our knowledge of meiosis and gamete formation. However, remember that Mendel performed his experiments in the 1850s and 1860s, long before chromosomes had been discovered! In fact, he hypothesized the presence and behavior of genes long before scientists knew even the most basic facts of cell biology and inheritance.

Mendel Also Discovered the Principle of Dominance

Mendel postulated that hereditary factors (genes) are either **dominant** or **recessive** (**TABLE 16-1**). When a dominant factor and a recessive factor are present in a pea, Mendel concluded, the dominant factor is always expressed. A recessive factor is expressed only when the dominant factor is missing. An alternative form of the

TABLE 16-1

Genetic Traits Studied by Mendel

Structure Studied	Dominant Trait	Recessive Trait
Seeds	Smooth	Wrinkled
	Yellow	Green
Pods	Full	Constricted
	Green	Yellow
Flowers	Axial (along stems)	Terminal (top of stems)
	Purple	White
Stems	Long	Short

same gene is called an **allele** (ah-LEEL). Thus, dominant and recessive factors are known today as dominant and recessive alleles. Dominant alleles are designated by capital letters; recessive alleles are signified by lowercase letters. In Mendel's peas, for instance, the gene for flower color is the *P* gene. The dominant form is *P* (purple flowers), and the recessive form is *p* (white flowers). The *P* and *p* genes are both alleles of the flower-color gene.

For genes with two alleles, three combinations are possible during fertilization. The first consists of two dominant genes—for example, *PP*. This individual is said to be **homozygous dominant** (hoe-moe-ZYE-gus) for that particular trait. The second consists of two recessive alleles—in this example, *pp*—and the individual is said to be **homozygous recessive** for that trait. The third occurs when dominant and recessive alleles are present (*Pp*), and the individual is **heterozygous.**

The Genotype of an Organism Is Its Genetic Makeup; the Phenotype Is Its Appearance

As noted earlier, Mendel determined that blending did not occur in the inheritance of traits, such as flower color. Thus, a pea plant with purple flowers (*PP*) bred with a pea plant with white flowers (*pp*) produced offspring with purple flowers and not pink flowers, which would occur if blending took place. Mendel proposed that the purple-flowered offspring of this cross were heterozygous—*Pp*. In such cases, he argued, the recessive hereditary factors (genes) are masked by the dominant hereditary factors (genes). Because of dominance, Mendel noted, the outward appearance of a plant—its **phenotype** (FEEN-oh-type)—does not always reflect its genetic makeup, or **genotype** (JEAN-oh-type). In fact, in this example, two different genotypes lead to the same phenotype (*PP* = purple and *Pp* = purple). As a general rule, a homozygous dominant individual and a heterozygous individual are indistinguishable on the basis of phenotype.

Genotypes and Phenotypes Can Be Determined by the Punnett Square

To track the genotypes and phenotypes of breeding experiments, such as those that Mendel performed, geneticists use a relatively simple tool known as the **Punnett square** (**FIGURE 16-5**). To illustrate the process, consider one of the traits that Mendel studied—seed-coat texture. The gene for seed-coat texture has two alleles: one that codes for smooth seeds (*S*) and one that codes for wrinkled seeds (*s*). As indicated by the capital *S*, the smooth-seed allele is dominant.

Suppose that you bred a homozygous-dominant plant (*SS*) with a homozygous-recessive plant (*ss*).

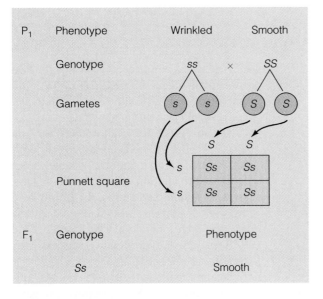

FIGURE 16-5 Mendel's Early Experiments When studying the inheritance of seed conformation, Mendel first crossed homozygous-recessive (*ss*) and homozygous-dominant *(SS)* plants. The offspring were all heterozygotes *(Ss)*.

FIGURE 16-5 lists the genotypes of each parent and the possible genotypes of the gametes. Determining the genotype is a rather simple affair. If a diploid parent is *SS*, its haploid gametes are all *S*. If a parent is *ss*, the gametes are all *s*.

To determine the outcome of crossing these two plants, all of the possible gametes produced from one plant are listed along the top of the Punnett square, and all of the possible gametes from the other are listed along the side. The gametes are then combined as shown, producing all of the possible genotypes present in the offspring.

As this example illustrates, a cross between a homozygous-dominant (*SS*) individual and a homozygous-recessive (*ss*) individual produces only heterozygous (*Ss*) offspring. Thus, the offspring are phenotypically uniform, and all of them resemble the homozygous-dominant parent—they have a smooth seed coat.

In the language of genetics, when one organism is bred to another to study the transmission of a single trait, the procedure is called a **monohybrid cross** (MON-oh-HYE-brid). In a monohybrid cross, the organisms that are bred initially constitute the P₁ generation (P for parents). The first set of offspring constitutes the F₁ generation (first filial generation; *filialis* is Latin for "of a son or daughter"). In this example, the F₁ offspring could be bred to produce additional offspring, the F₂ generation (second filial generation).

To determine the genotypes and phenotypes of such a cross, the Punnett square can be used once again. As **FIGURE 16-6** shows, the gametes of the F₁ generation are *S* and *s* because all of the parents are heterozygous

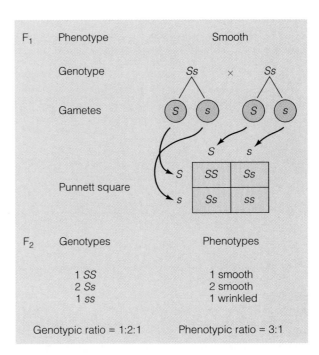

FIGURE 16-6 Monohybrid Cross In Mendel's early work, he crossed offspring from his F₁ generation *(Ss)* and found a variety of genotypes and phenotypes.

(*Ss*). All of the possible gametes from the father are placed along the top of the Punnett square, and all of the possible gametes of the mother are placed along the side. The gametes are combined as before in the appropriate boxes.

As shown in **FIGURE 16-6**, this cross produces three different genotypes: *SS*, *Ss*, and *ss*. The ratio of genotypes, or the genotypic ratio of the F₂ generation, is one *SS* to two *Ss* to one *ss*, or 1 : 2 : 1. Thus, if 100 offspring are produced, you would expect 25 of them to be *SS* (homozygous dominant), 50 to be *Ss* (heterozygous), and 25 to be *ss* (homozygous recessive). Because of dominance, though, this cross yields only two phenotypes: plants with smooth seeds (*SS* and *Ss*) and plants with wrinkled seeds (*ss*). The ratio of phenotypes, or the phenotypic ratio, is three smooth to one wrinkled, or 3 : 1.

Genes on Different Chromosomes Segregate Independently of One Another During Gamete Formation

In his later experiments, Mendel tracked two traits at the same time, a procedure geneticists refer to as a **dihybrid cross** (DIE-HIGH-brid) (**FIGURE 16-7**). This work led to the **principle of independent assortment**— the idea that genes located on different chromosomes segregate independently during meiosis. To understand

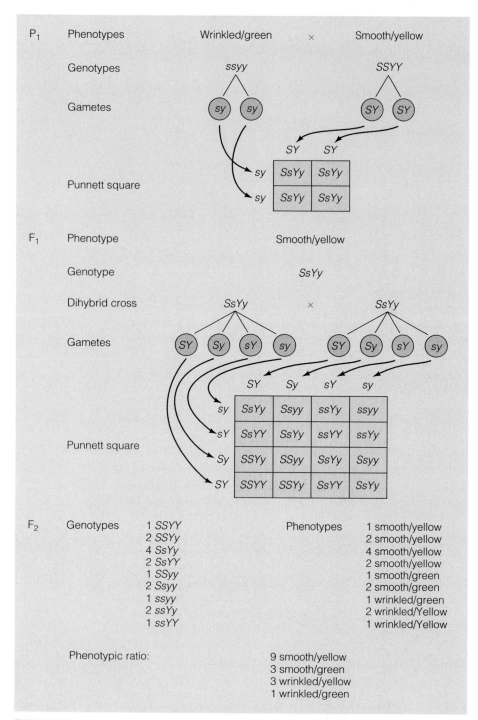

FIGURE 16-7 Dihybrid Cross The dihybrid cross examines the inheritance of two traits. Independent assortment of genes produces a large number of genotypes.

and the recessive form is *s* (wrinkled). Peas also contain a gene for seed color; the dominant form (*Y*) yields yellow seeds, and the recessive form (*y*) yields green seeds.

To study the inheritance of these two genes, Mendel crossed homozygous-dominant plants with smooth, yellow seeds (*SSYY*) with homozygous-recessive plants (*ssyy*). As **FIGURE 16-7** shows, the genotype of the F_1 generation was *SsYy*. Consequently, the offspring were phenotypically identical, and all displayed the dominant characteristics (smooth, yellow seeds).

When Mendel crossed two of the F_1 off-spring (genotype = *SsYy*), however, he got a mixture of phenotypes (**FIGURE 16-7**). To produce this combination of offspring, Mendel concluded, the hereditary factors *S* and *s* must have segregated independently of *Y* and *y* during gamete formation. Translated into modern terms, this means that the *Y* and *S* genes (and their alleles *y* and *s*) were on different chromosomes and separated independently of each other during gamete formation (meiosis). Because of independent assortment, the *gametes* of the F_1 generation contain all combinations of the alleles in equal proportions: *SY, Sy, sY,* and *sy.* Fertilization also occurs at random, giving rise to 16 possible genetic combinations (see F_2 genotypes in **FIGURE 16-7**.

This research led Mendel to propose the principle of independent assortment, which states that the segregation of the alleles of one gene on one chromosome is independent of the segregation of the alleles of another gene on a second chromosome during gamete formation.

Mendel may have been one of the luckiest scientists in the world—or maybe he was divinely blessed. We'll never know. In either case, he fortuitously chose to study seven traits, listed in **TABLE 16-1**, each of which scientists now believe is on a different chromosome. (This assumption has not yet been proved.) If two genes under study were located on the same chromosome, Mendel might have never discovered his principle of independent assortment because such genes tend to be inherited together. All this is to say, as you'll soon see, that independent assortment is not a universal truth. It holds only for genes located on different chromosomes.

what this principle means and how Mendel arrived at it, consider another example that examines seed-coat texture and seed color.

As noted above, peas contain a gene for seed-coat texture, the *S* gene. The dominant form is *S* (smooth),

Mendelian Genetics in Humans

What do studies on garden peas have to do with humans? It's simple. Much of what Mendel discovered in garden peas pertains to human genes and human inheritance. **TABLE 16-2** lists a number of human traits and diseases whose inheritance follows basic Mendelian principles. This section describes the inheritance of those traits, using several common diseases as examples.

Autosomal-Recessive Traits Are Expressed Only When Both Alleles Are Recessive

Human cells contain 23 pairs of chromosomes. They can be divided functionally into one pair of sex chromosomes and 22 pairs of autosomes (AU-toe-zomes).

The **sex chromosomes** are involved in sex determination—that is, determining whether we turn out male or female—as well as a few other functions. Two types of sex chromosomes exist, X and Y. Females have two homologous X chromosomes; their genotype is therefore XX. Males have a nonhomologous pair, consisting of one X chromosome and one Y chromosome. Their genotype is XY.

The remaining 22 pairs of chromosomes are called the **autosomes.** The autosomes contain numerous genes that control a variety of traits. This section examines several **autosomal-recessive traits.** These are traits expressed only when both recessive alleles are present.

Over 600 traits in humans have been identified as autosomal recessive, and another 800 are strongly suspected of being autosomal recessive. Some autosomal-recessive traits are the cause of abnormalities, among them albinism and cystic fibrosis.

Albinism.

Albinism is one of the most common genetic defects known to science. It occurs in 1 of every 38,000 Caucasian births and in 1 of every 22,000 African-American births. Among the Hopi and Navajo Indians of the American Southwest, the incidence is 1 in every 200.

Albinism in humans and other animals as well (rabbits and rats, for example) occurs when two recessive genes are inherited from one's parents. The recessive genes result in a deficiency in the metabolic pathways leading to the production of melanin. **Melanin** is the brown pigment responsible for coloration of the eyes, skin, and hair. Individuals who are homozygous-recessive for albinism have no melanin at all or may have reduced levels (**FIGURE 16-8**). Consequently, the skin of an albino is pale, and the hair is white. The eyes of albino children are pink, because there is no pigment in the iris or retina.

In most of us, melanin in the skin protects against the effects of ultraviolet light. Its absence in albinos makes them highly susceptible to sunburn and skin cancer. Moreover, the lack of pigment in the eyes may result in damage in the light sensitive region, the retina, which is essential to vision. This, in turn, may result in blindness.

TABLE 16-2

Traits and Diseases Carried on Human Chromosomes

Autosomal recessive

Albinism	Lack of pigment in eyes, skin, and hair
Cystic fibrosis	Pancreatic failure, mucus buildup in lungs
Sickle-cell anemia	Abnormal hemoglobin leading to sickle-shaped red blood cells that obstruct vital capillaries
Tay-Sachs disease	Improper metabolism of a class of chemicals called gangliosides in nerve cells, resulting in early death
Phenylketonuria	Accumulation of phenylalanine in blood; results in mental retardation
Attached earlobe	Earlobe attached to skin
Hyperextendable thumb	Thumb bends past 45° angle

Autosomal dominant

Achondroplasia	Dwarfism resulting from a defect in epiphyseal plates of forming long bones
Marfan's syndrome	Defect manifest in connective tissue, resulting in excessive growth, aortic rupture
Widow's peak	Hairline coming to a point on forehead
Huntington's disease	Progressive deterioration of the nervous system beginning in late twenties or early thirties; results in mental deterioration and early death
Brachydactyly	Disfiguration of hands, shortened fingers
Freckles	Permanent aggregations of melanin in the skin

FIGURE 16-8 Albinism Albinism is an autosomal-recessive trait. An albino man with his wife in India.

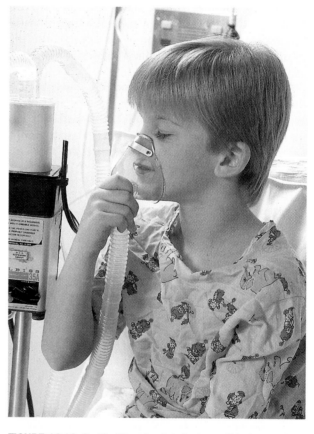

(a)   (b)

FIGURE 16-9 Comparison of a Pancreas from Patient with Cystic Fibrosis with a Normal Pancreas Cystic fibrosis is a disease caused by an autosomal-recessive gene. *(a)* It results in a blockage of the ducts draining the pancreas, leading to cysts in the pancreatic tissue. The tissue degenerates and is replaced by fibrous connective tissue. *(b)* Normal pancreas.

Cystic Fibrosis.

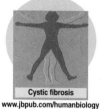

Cystic fibrosis
www.jbpub.com/humanbiology

Cystic fibrosis (SISS-tick fie-BRO-siss) is an autosomal-recessive disease that leads to early death. The presence of two recessive alleles for the disease alters the function of sweat glands of the skin, mucous glands in the respiratory system, and the pancreas. Defective sweat glands, for example, release excess amounts of salt, a marker that helps physicians diagnose the disease. The most significant symptoms, however, occur in the pancreas and lungs.

In individuals with cystic fibrosis, the ducts that drain digestive enzymes from the pancreas into the small intestine become clogged. This not only impairs digestion, but also causes a buildup of enzymes in the organ that results in the formation of cysts. Over time, the pancreas begins to degenerate, and fibrous tissue replaces glandular tissue—hence, the name *cystic fibrosis* (**FIGURE 16-9**).

Despite adequate nutritional intake, victims of cystic fibrosis often show signs of malnutrition. To enhance digestion, patients must eat powdered or granular extracts of animal pancreases containing digestive enzymes. Massive doses of vitamins and nutrients must also be taken.

The respiratory systems of most victims of cystic fibrosis produce copious amounts of mucus. Mucus blocks the passages, making breathing difficult. Thus, patients must be treated several times a day to remove the mucus (**FIGURE 16-10**). Despite antibiotic therapy and other treatments, most cystic fibrosis patients live only into their late teens or early twenties.

Cystic fibrosis is one of the most common genetic diseases known to medical science. Surprisingly, 1 of every 22 Caucasians carries a gene for this disease, and

FIGURE 16-10 Cystic Fibrosis Inhalants, antibiotics, and physical therapy are used to treat patients with cystic fibrosis. To remove mucus from the lungs, parents or physical therapists must treat the patient two or three times a day. Pounding on the rib cage with a cupped hand (clopping) loosens the mucus.

(a)

(b)

FIGURE 16-11 Widow's Peak (a) The widow's peak is a dominant trait carried on one of the autosomes. (b) A straight hairline is a recessive trait. Simple Mendelian genetics can be used to determine the genotype of the offspring.

approximately 1 of every 2000 Caucasians born in the United States suffers from it. In the African-American population, only about 1 in 100,000–150,000 individuals is a carrier.

Autosomal-Dominant Traits Are Expressed in Heterozygous and Homozygous-Dominant Individuals

Many human traits are autosomal dominant; that is, they are carried on the autosomes and are expressed in heterozygous (Aa) and homozygous-dominant (AA) individuals. To date, nearly 1200 human traits have been identified as autosomal dominant, and 1000 others are suspected. The absence of dermal ridges (which give rise to fingerprints), short fingers and toes, freckles, cleft chin, and drooping eyelids are all autosomal-dominant traits. Marfan's syndrome, discussed in the Thinking Critically section in this chapter, is an autosomal-dominant disease. This section discusses two other examples: widow's peak and achondroplasia.

Widow's Peak.

Take a moment to examine the hairlines of your friends and classmates. In some individuals, the hairline runs straight across the forehead. In others, it juts forward in the center, forming a "widow's peak" (**FIGURE 16-11**). Widow's peak is determined by an autosomal-dominant gene, indicated by W. Because the W allele is dominant, this phenotype is expressed in homozygous-dominant individuals (WW) and also heterozygotes (Ww). Individuals with the genotype ww have a straight hairline.

Achondroplasia.

The boy shown in **FIGURE 16-12** has a genetic disease called **achondroplasia** (a-CON-drow-PLAY-zee-ah),

one form of dwarfism. Individuals with this disease have short, stubby legs and arms but a relatively normal-sized trunk. As a rule, they do not grow taller than 4 feet. Surgeons have developed a technique to lengthen the arms and legs of people with this disorder. In this procedure, the leg and arm bones of the person are fractured (under anesthesia), and the patient is fitted with a traction device. Traction is applied to the broken bones, causing them to separate. New bone is formed in the gap, and the bones slowly elongate.

Achondroplasia afflicts about 1 of every 10,000 children born in the United States and results from an autosomal-dominant gene. Most cases are believed to arise from spontaneous mutations, because the majority of children with the condition are born to phenotypically normal parents.

Section 16-4

Variations in Mendelian Genetics

Mendel presented his work at meetings of the Natural Science Society in Czechoslovakia in 1865 and published his results the following year. Like many ideas ahead of their time, Mendel's conclusions went largely unnoticed. It was not until 1900, 16 years after he died, that Mendel's studies received the attention they deserved. At that time, the publication of three other studies confirmed Mendel's findings. More studies followed, and excitement began to mount. The scientific community realized that Mendel's principles pertained to a great many organisms. Since that time, researchers have uncovered additional modes of inheritance and gene expression.

Incomplete Dominance Results in Intermediate Traits—That Is, a Kind of Blending of Traits

One example is a phenomenon called incomplete, or partial, dominance. **Incomplete dominance** occurs when heterozygous offspring exhibit intermediate

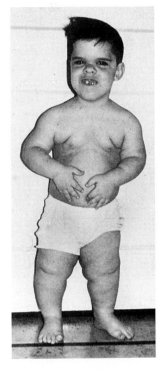

FIGURE 16-12 Achondroplasia Achondroplasia, one form of dwarfism, is an autosomal-dominant disorder.

phenotypes. In other words, incomplete dominance produces F₁ offspring with phenotypes intermediate to the parental phenotypes. This is the kind of blending that Mendel said didn't exist!

Incomplete dominance occurs in a plant called *Mirabilis*. **FIGURE 16-13** shows a cross between a plant with red flowers (*RR*) and one with white flowers (*rr*). Even though the red color is dominant, this cross produces offspring with pink flowers, an intermediate phenotype (**FIGURE 16-13**). That's because the *R* gene does not exert complete dominance. Interestingly, incomplete dominance also occurs in a number of human disorders, including sickle-cell anemia.

Sickle-Cell Anemia.

Sickle-cell anemia affects the blood, and chiefly afflicts African Americans and Caucasians of Mediterranean descent. Sickle-cell anemia is caused by a genetic defect that leads to abnormal hemoglobin (HEME-oh-GLOW-bin) formation. Hemoglobin is a protein found in red blood cells, which carry oxygen to body cells.

Sickle-cell anemia occurs in individuals that are homozygous recessive. The abnormal hemoglobin causes red blood cells to convert to sickle-shaped cells when they encounter low oxygen levels in the blood—for example, when blood cells flow through capillaries (small blood vessels) in metabolically active tissues.

Sickle-shaped red blood cells clog capillaries and reduce oxygen flow to brain cells, heart cells, and other organs. In homozygous-recessive individuals, sickle-cell anemia is usually lethal. In fact, most individuals with the disease die by their late twenties.

Individuals who are heterozygous for the trait are referred to as **carriers,** because they can pass the gene on to their children. Carriers generally lead relatively normal lives, but they are subject to occasional problems. Although their red blood cells appear normal, they actually contain 50% normal and 50% abnormal hemoglobin. Moderate sickling occurs when these cells are exposed to low oxygen.

Approximately 1 in every 500 African Americans born in the United States is homozygous recessive, and about 1 of every 12 is heterozygous (a carrier) for the sickle-cell trait. Why is this allele so prevalent?

In tropical climates, from which African Americans come, the sickle-cell allele protects carriers and homozygous-recessive individuals from malaria, a deadly disease prevalent in humid, tropical regions. Although homozygous-recessive individuals die earlier, the selective advantage that carriers enjoy has caused the allele to increase in the population.

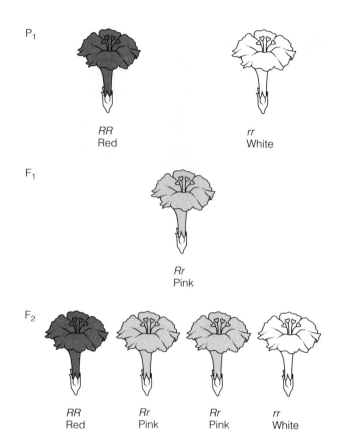

FIGURE 16-13 Incomplete Dominance Incomplete dominance involves two alleles, neither of which is dominant over the other. The result is an intermediate phenotype as shown here in the flower of the plant *Mirabilis*.

Malaria is caused by a microscopic parasite called *Plasmodium* (plaz-MOW-dee-um), which is transmitted from one person to the next by certain mosquitoes. Inside the body, the parasites invade and colonize the liver, where they multiply rapidly. New parasites leave the liver and enter the bloodstream, where they attack and destroy red blood cells. Many of the parasites remain in the liver, however, continuing to reproduce and periodically release new offspring. Unless treated, a victim of malaria suffers from repeated attacks of chills followed by fever. Each attack corresponds with the release of a new batch of *Plasmodia* from the liver.

How does the sickle-cell allele protect people? As noted earlier, the hemoglobin in the red blood cells of carriers and individuals who are homozygous recessive is altered by the sickle-cell allele. For reasons not entirely clear, the presence of altered hemoglobin changes the plasma membrane of red blood cells and prevents the parasite from entering. As a result, both homozygous-recessive individuals and carriers (heterozygotes) are relatively immune to the parasite.

Because the sickle-cell allele confers a selective advantage on those who have it and because heterozygotes suffer relatively few problems, in East Africa, where the malaria parasite is quite common, 45% of all blacks are carriers of the trait (heterozygotes). In the United States, where malaria is virtually nonexistent, the frequency of the sickle-cell allele has decreased considerably.

Some Genes Have Multiple Alleles

Mendel studied seven characteristics of peas. Each characteristic is determined by a gene with two alleles—two alternative forms. In human and other species, however, researchers have shown that some genes can have more than two alleles; such a gene is said to have **multiple alleles.**

Consider blood types. One gene that controls blood type is called the *I* **gene** (for isoagglutinin; pronounced EYE-sa-AH-glue-TEH-nin). The *I* gene is located at a particular site (or locus) on one pair of chromosomes. But unlike the genes Mendel studied, which had only two alleles, the *I* gene exists in one of three distinct forms. The three alleles are *I*^{A}, *I*^{B}, and *I*^{O}. Note, however, that even though there are three possible alleles in human beings, an individual can have only two of the alleles in his or her **genome** (genome refers to all of one's genes), one on each homologous chromosome. It's like having three mittens. You can only wear two at a time.

Four possible blood types exist: A, B, AB, and O. **TABLE 16-3** lists the four blood types and the six different genotypes that give rise to them. Take a moment to study them.

As shown, the A blood type occurs when an individual has two *A* alleles of the *I* gene or an *A* and an *O*. Type B occurs when an individual has two *B* alleles of the *I* gene or a *B* and an *O*. Type AB has both *A* and *B* alleles. Type O occurs when two *O* alleles of the *I* gene are present.

The *I* gene is responsible for the production of glycoproteins (proteins with carbohydrate attached) that project from the surface of red blood cells. Thus, the *A* and *B* alleles of the *I* gene produce A and B glycoproteins, respectively. The *O* allele, on the other hand, produces neither type of cell-surface glycoprotein. If you know this, you can determine the type of glycoproteins in the different phenotypes. For example, AA individu-

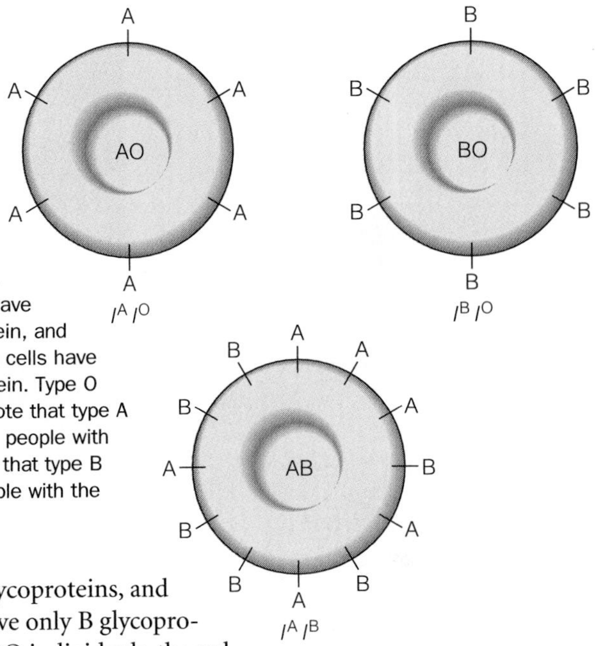

FIGURE 16-14
Codominance
Codominant genes are expressed fully when present in the same cell. In blood type AB, both *A* and *B* genes produce their characteristic glycoproteins. Type A (genotype AO) cells have only type A glycoprotein, and type B (genotype BO) cells have only type B glycoprotein. Type O cells have neither. Note that type A blood may also be in people with the AA genotype and that type B blood may be in people with the BB genotype.

als have only A glycoproteins, and BB individuals have only B glycoproteins. In AO and BO individuals, the red blood cells contain only type A glycoproteins and B glycoproteins, respectively (top, **FIGURE 16-14**). Type AB individuals have both A and B glycoproteins.

Both A and B are dominant genes, and the *O* allele is recessive. Therefore, the *I*^{A} and *I*^{B} genes are said to be codominant. **Codominant genes** are expressed fully and equally. So not only are blood types an example of multiple genes, they are also an example of codominance, another type of inheritance.

Incomplete dominance (blending), multiple alleles, and codominance are three exceptions to Mendel's principles. But don't be too quick to dismiss Mendel. These examples do not negate his discoveries. They are simply additional modes of gene expression. Moreover, Mendelian principles can be used to predict the genotypic ratios and patterns of inheritance in these alternative forms of inheritance.

Some Traits Are Determined by More Than One Gene Pair

For many years, geneticists believed that each gene controls a single trait. In humans and other animals, however, many traits are controlled by a number of genes—from a few to perhaps hundreds. Skin color, for example, is controlled by as many as eight genes. This type of inheritance is called **polygenic inheritance** (polly-JEAN-ick). Polygenic inheritance results in incredible phenotypic variation.

To see how polygenic inheritance leads to such wide phenotypic variation, consider an example using two genes for skin color, designated *A* and *B*. In this

TABLE 16-3	
Phenotype and Genotype in ABO system	
Phenotype (blood type)	**Genotypes**
Type A	*I*^{A}*I*^{A}, *I*^{A}*I*^{O}
Type B	*I*^{B}*I*^{B}, *I*^{B}*I*^{O}
Type AB	*I*^{A}*I*^{B}
Type O	*I*^{O}*I*^{O}

FIGURE 16-15 Polygenic Inheritance Skin color is probably determined by at least two, perhaps as many as eight, genes, resulting in a wide range of phenotypes: *(a)* black, *(b)* dark, *(c)* mulatto, *(d)* light, and *(e)* white.

TABLE 16-4

Possible Skin-Color Genotypes and Phenotypes with Two Skin-Color Genes*

Genotype	Phenotype	Number of Recessive Genes
AABB	Black	0
AABb	Dark	1
AaBB	Dark	1
AaBb	Mulatto	2
AAbb	Mulatto	2
aaBB	Mulatto	2
Aabb	Light	3
aaBb	Light	3
aabb	White	4

*Skin color probably involves many more genes.

example, we will say that the genotype of an African American is *AABB* and the genotype of a Caucasian is *aabb*. **TABLE 16-4** lists all of the possible genotypes in this example with possible phenotypes; **FIGURE 16-15** shows what they look like. Take a moment to study the table and the figure.

Polygenic inheritance is also responsible for height, weight, intelligence, and a number of behavioral traits. But in each instance, the genes are not the only factors controlling these traits. As a rule, the genotype establishes the range in which a phenotype will fall, but environmental factors (for example, diet) determine exactly how much of the potential will be realized for many genes. For example, although genes determine height, this trait is also influenced by one's diet. The better one eats from infancy to adolescence, the taller he or she will be.

Genes Located on the Same Chromosome Are Said To Be Linked

As pointed out earlier, Mendel found that during gamete formation, the seven genes under study in his pea plants were segregated independently. Also noted earlier, independent assortment generally occurs *only* when the genes under study are on different chromosomes. As a rule, if two genes are on the same chromosome, they do not segregate independently.

Humans have an estimated 100,000 genes on their 46 chromosomes. Those genes located on the same chromosome tend to be inherited together and are said to be **linked.**

To illustrate the concept, imagine that you are studying two traits. Designate the dominant form of the first trait *A* and the recessive allele *a*. The dominant form of the second trait is designated *B* and the recessive form is *b*. Suppose we cross a homozygous-dominant individual (*AABB*) with a homozygous-recessive mate (*aabb*). This scenario is shown in **FIGURE 16-16A.** Take a moment to study the top part of the figure.

As shown, if the *A* and *B* genes are *not* linked, this cross produces an F_1 generation that's entirely heterozygous (*AaBb*). Furthermore, F_1 offspring will produce the four genetically different gametes: *AB, Ab, aB,* and *ab.* If an F_1 offspring mates with another F_1 offspring (of the opposite sex of course), the result is an F_2 generation with nine distinct genotypes and four different phenotypes. You can take my word for this, or examine **FIGURE 16-16A** for yourself.

If the *A* and *B* genes are on the same chromosome (they're linked) and the *a* and *b* genes are on the other (also linked), the outcome changes dramatically. First, as **FIGURE 16-16B** shows, only two gametes are produced: *AB* and *ab.* Second, as **FIGURE 16-16B** also shows, all members of the F_1 generation are heterozygous, *AaBb.* This is the same as if the genes were unlinked. However, because the *A* and *B* genes are on the same chromosome, the F_1 generation produces only two types of gametes, *AB* and *ab.* Consequently, only two phenotypes are produced in the F_2 generation.

So can we conclude that independent assortment does not occur when genes are linked? Yes—at least most of the time. When two genes are on the same chromo-

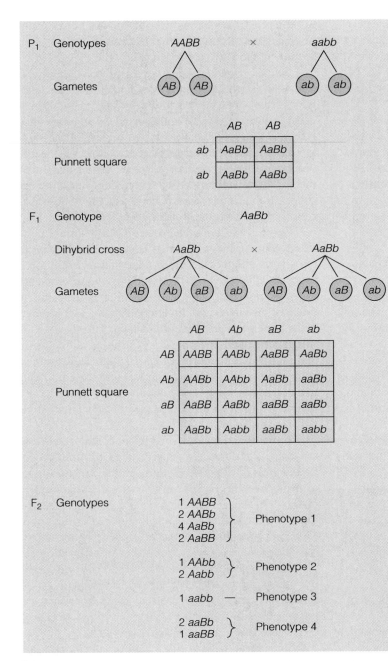

(a)

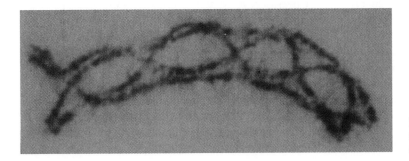

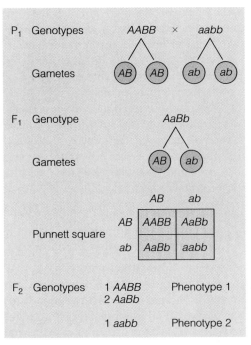

(b)

FIGURE 16-16 Linkage *(a)* Hypothetical dihybrid cross with no linkage. *(b)* Hypothetical dihybrid cross with linkage. Linkage reduces the number of genotypes in the offspring.

some, they tend to be inherited together. However, there is one exception—a phenomenon called crossing-over.

Crossing-over occurs during meiosis: when the homologous chromosomes come together, or "pair up," in prophase I. During this process, many chromosomes exchange segments of their chromatin with chromatids on homologous chromosomes (**FIGURE 16-17**). When this occurs, genes that were linked are no longer linked and are not necessarily inherited together.

Crossing-over "unlinks" genes and increases genetic variation in gametes; this, in turn, leads to genetic variation in offspring. (Other factors also contribute to genetic variation.) The more variation, the more genotypes and possible phenotypes in a population. As Chapter 21 points out,

FIGURE 16-17 Crossing-Over The crossing-over shown here increases genetic variation in gametes and offspring and occurs during meiosis.

variation is essential to evolution. Genetic variants may have characteristics that give one organism an advantage over another. In a very general sense, then, crossing-over provides variation that ensures the survival of populations in changing environmental conditions. It's a protective mechanism of sorts.

Crossing-over can occur anywhere along the length of a chromosome. However, the greater the distance between two genes on the same chromosome, the more likely it is that a crossover will occur between them. This fact is helping scientists map the human **genome,** a monumental project now under way.

The Human Genome Project Seeks To Determine the Sequence of Bases in Human DNA and the Location of All of the Genes

Scientists the world over have embarked on what is conceivably the largest unified biological research project of its kind in human history. It is designed to determine the location of all of the genes on all 46 chromosomes and the precise sequence of bases of the human genes and intervening DNA in the chromosomes. With an estimated 100,000 genes and three billion bases in the human DNA, this monumental task could cost $3 billion dollars and was estimated to take up to 15 years or more to complete.

The **Human Genome Project,** as it is called, began in earnest in 1991. Scientists found that the first part of the project—mapping the location of the 100,000 or so genes—occurred at a much faster pace than anticipated. At this writing (November, 1998), geneticists have pinpointed the location of 95% of the human genes. The remaining 5% may take proportionately longer. Sequencing the DNA—that is, determining the sequence of the bases—is also occurring at a faster-than-anticipated rate.

Why is it important to know the location of human chromosomes? First, it adds to our overall knowledge of human genetics. Second, it provides scientists a way to isolate and synthesize human genes. For example, if a gene controls the production of useful products (hormones, blood-clotting factors, etc.), it could be synthesized in the lab and used to mass produce the product to treat various illnesses. Chapter 18 discusses this process in more detail.

A third practical, but controversial, application of this research is that it could help medical geneticists locate disease-producing genes. This would allow geneticists to screen unborn fetuses to determine if they contain any life-threatening or debilitating disorders. If results show this to be so, parents may choose to abort their offspring. For those for whom abortion is not an option, genetic screening with this kind of power could help prepare parents for the challenges that await them with their specially challenged child.

Fourth, knowing the location of disease genes also leads some scientists to speculate on the possibility of genetic repair—for example, finding ways to splice normal genes in place of defective ones, avoiding problems altogether. A parent, faced with the possibility of having a severely physically impaired child because of a genetic disorder, for example, might opt to have gene replacement therapy on the unborn fetus in hopes of having a healthy child. Although this prospect may be years away from successful completion, some early research seems promising.

The Human Genome Project is not without opposition. Some critics argue that the project is too costly and that the high dollar investment, in an era of limited research dollars, reduces the amount of money available for other research. On a related issue, it turns out that much of the DNA is not functional; it's a kind of genetic "filler." Spending millions of dollars to find the code of this genetic filler, say some scientists, is a waste of time and money. Another criticism is that knowledge gained from knowing the human genome intimately could lead to a potentially tricky situation in which a physician can diagnose an illness through genetic means long before cures are available.

Some individuals are concerned that insurance companies could access a person's records and, by examining his or her genetic code, would be in a position to deny coverage. Finally, some critics worry about the possibilities the Human Genome Project will present for manipulation of the human genes. Is it opening doors to genetic manipulation that should be better left to evolution? Are we playing God with our own genome?

These and other controversies, especially the debate over the need to sequence all the useless genes, have slowed progress in this project. But proponents are pushing ahead while policymakers and others engage in the ethical and economic debates.

One upshot of this project has been the mapping of chromosomes in nonhuman species, a step that could help bring important genetic improvements, in the case of livestock, or in the case of disease organisms could help scientists find better tools to combat them. Scientists have already started to explore the genetic code of the organism thought to be responsible for infections in patients with cystic fibrosis, the most common fatal genetic disease in the United States, which currently afflicts approximately 30,000 children and young adults. Knowing the complete genetic code of this organism, which successfully evades the human immune system could help scientists find new ways to penetrate its defenses.

Section 16-5

Sex-Linked Genes

Earlier, I noted that the sex chromosomes determine an individual's sex. The truth of the matter is that the sex of an individual is determined by the Y chromosome. If the Y chromosome is present, an individual becomes a male. The absence of the Y chromosome results in the development of a female phenotype.

The Y chromosome exerts its effect during embryonic development. Interestingly enough, males and females start out looking alike. That is, early in the embryo's development, the gonads, which develop in the abdominal cavity, are identical. When the Y chromosome is present, however, the embryonic gonad becomes a testis—thanks, at least in part, to the presence of the t gene (testis-determining gene). The testis, in turn, produces testosterone and other androgens, which are responsible for the male secondary sex characteristics—hair growth, body shape and size, and so on (Chapter 19). The absence of the Y chromosome results in the development of an ovary with its hormones and the resultant female phenotype.

The X and Y chromosomes also carry genes that determine many other traits. Genes situated on the X and Y chromosome are therefore known as **sex-linked genes.** So far, the majority of the sex-linked genes discovered by geneticists are located on the X chromosome. These are known as **X-linked genes.** The following sections describe the inheritance of some common sex-linked genes.

Recessive X-Linked Genes Are the Best Understood of the Sex-Linked Genes

At least 124 genes have been assigned to the X chromosome. At least 160 more are also thought to be located on it. Color blindness, discussed in detail in Chapter 12, and certain forms of hemophilia are examples of recessive traits carried on the X chromosome.

In order for a female (XX) to display a recessive sex-linked trait, each of her X chromosomes must carry the recessive gene. For males (XY), however, only one recessive allele is required. Why? Because the Y chromosome is not genetically equivalent to the X chromosome.[2] Thus, in males, only one recessive gene is needed to exhibit a recessive X-linked trait.

FIGURE 16-18 shows four possible genetic combinations leading to color blindness. This illustration also introduces you to a genetic-tracking system used to follow traits in families, known as a **pedigree.** In a pedigree, squares represent men and circles represent women. (We assume the square for males is not social commentary.) The horizontal line linking a square to a circle indicates a mating. Offspring are shown below their parents. When a square is lightly shaded, the

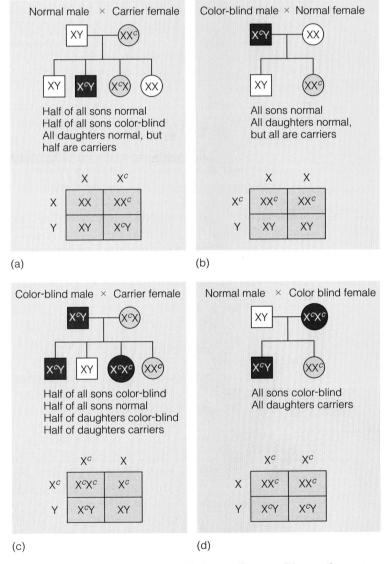

(a)

(b)

(c)

(d)

FIGURE 16-18 Inheritance of Color Blindness Four possible genetic ways a sex-linked recessive gene such as color blindness can be passed to offspring. Males are indicated by boxes and females by circles. In this scheme, green boxes and circles represent men and women with color blindness. Light blue boxes represent men and women who are carriers. White boxes and circles represent men and women without the color blindness gene. The notation X^c indicates an X chromosome carrying the gene for color blindness.

[2]The X and Y chromosomes are believed to share few genes, and only part of the Y chromosome is homologous with the X chromosome.

individual is a carrier of the gene. He or she does not have the disease, but does carry the recessive gene and can pass it on to his or her offspring. A darkly shaded square or circle indicates the person has the trait.

In **FIGURE 16-18A**, for example, a man and a woman have four children, two boys and two girls. The woman (light blue circle) is a carrier of color blindness. Her cells contain one X chromosome with a recessive allele for color blindness (designated X^c) and another X chromosome with the normal allele. Consequently, half of her ova will contain the recessive allele, and the other half will contain the normal allele. As shown, the woman's husband (white square) is not color-blind. He produces sperm containing either an X or a Y chromosome. When one of his X-bearing sperm unites with one of her ova bearing an X chromosome with the allele for color blindness, the result is a daughter who is a carrier, indicated by a light blue circle. When one of his X-bearing sperm unites with an ovum carrying a normal X chromosome, the result is a daughter who is neither a carrier nor color blind (white circle).

Now what about male children of these parents? Males are produced when a Y-bearing sperm unites with an ovum carrying an X chromosome. If the X chromosome carries the recessive allele for color blindness, the boy is color-blind (green square). If the X chromosome is normal, the boy's color vision is unimpaired (white square). Take a moment to study the other possibilities in **FIGURE 16-18**.

Dominant X-Linked Genes Are Relatively Rare

Recessive X-linked genes are the most common sex-linked trait in humans. However, there are a few noteworthy examples of dominant X-linked genes. One of the best understood is a disorder with a tongue-twisting name of **hypophosphatemia** (high-poe-FOS-fuh-TEEM-ee-uh). This disease is characterized by low phosphate levels in the blood and tissues of the body. This, in turn, results in a form of **rickets** (RICK-its), or bowleggedness (**FIGURE 16-19**). Rickets usually results from a dietary deficiency of vitamin D

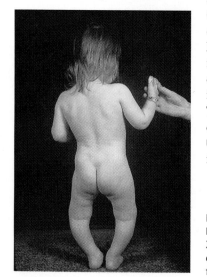

FIGURE 16-19 Hypophosphatemia
People with hypophosphatemia, a dominant X-linked genetic disorder, resemble this child with rickets, which usually results from inadequate vitamin D intake.

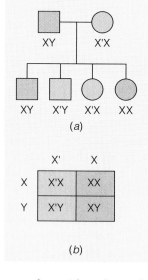

FIGURE 16-20 Inheritance of a Sex-Linked Dominant Gene (*a*) Pedigree. (*b*) Corresponding Punnett square.

or insufficient exposure to sunlight (Chapter 5). Alleviating the dietary deficiency usually solves the problem. In this genetic disease, however, vitamin D cannot reverse the symptoms.

Hypophosphatemia occurs when either the male (XY) or the female (XX) has one X chromosome containing the dominant gene, indicated as X' (X prime). **FIGURE 16-20** illustrates the pattern of inheritance when a woman who is heterozygous (X'X) for the trait mates with a man who does not carry the trait (XY). Study the figure and see if you can determine any rules that apply to the inheritance of X-linked dominant genes.

What you should have discovered is that, whenever the dominant gene is present, it is expressed—regardless of the sex of the individual.

Both Dominant and Recessive Y-Linked Genes Are Always Expressed

The genes that control gonadal differentiation, sperm production, and male secondary sex characteristics are thought to be on the Y chromosome. The Y-linked genes have a simple and distinct pattern of inheritance. Because only males have Y chromosomes, Y-linked traits only appear in males. Obviously, Y-linked genes can only be transmitted from fathers to sons. Because the X and Y chromosomes in men are not homologous, each gene on the Y chromosome has only one allele. Thus, both dominant and recessive Y-linked genes are always expressed.

The Action of Some Genes Is Influenced by the Sex of an Individual

In one of the most interesting twists on basic genetics, it turns out that certain autosomal genes behave differently in the two sexes. In one sex, for example, an allele will be dominant; in the other sex, it will be recessive. These genes are known as **sex-influenced genes.**

The best-known example is the gene for pattern baldness. **Pattern baldness** is the loss of hair that often begins in a man's twenties (**FIGURE 16-21**). The name of this condition results from the fact that affected individuals do not go completely bald; they retain a rim of hair on the temples and back of the head.

The gene for pattern baldness is present in men and women, but in men the gene acts as if it is autosomal-dominant. That is, it is expressed in both heterozygous and homozygous-dominant individuals, which means that lots of men end up with pattern baldness. In contrast, in women the allele acts as if it is autosomal-recessive. Only women who are homozygous recessive for the trait exhibit baldness, which means very few end up with this affliction. For more on the causes and cures of baldness, see Health Note 16-1.

What accounts for the different behavior of these genes in men and women? Geneticists believe that the genes are influenced by testosterone, a sex steroid hormone found in far greater concentration in the blood of men than women. In this case, the genotype and hormonal environment interact to determine the expression of the pattern-baldness gene.

Section 16-6

Chromosomal Abnormalities and Genetic Counseling

A young couple wait in their doctor's office for the results of a genetic test on cells of their unborn baby,

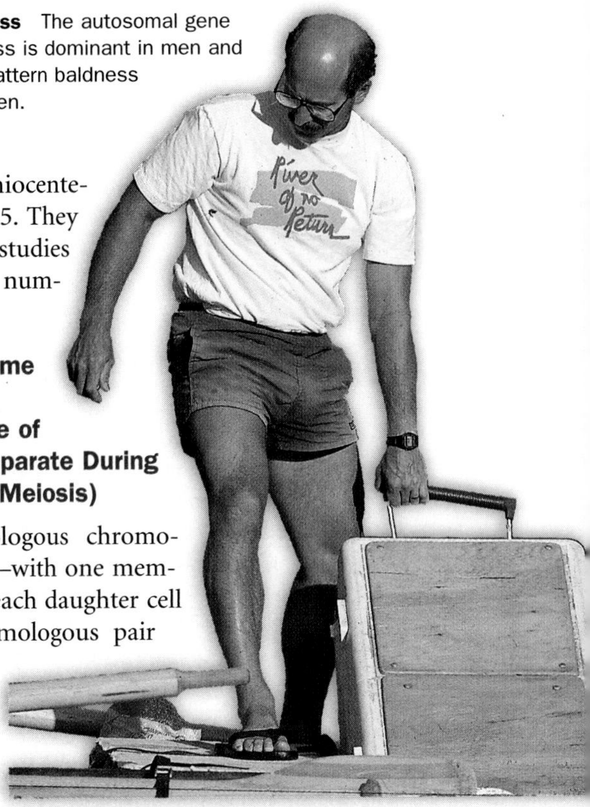

FIGURE 16-21 Pattern Baldness The autosomal gene responsible for pattern baldness is dominant in men and recessive in women, making pattern baldness much more common among men.

which were drawn via amniocentesis, described in Chapter 15. They will soon learn that the studies have revealed an abnormal number of chromosomes.

Abnormal Chromosome Numbers Generally Result from a Failure of Chromosomes to Separate During Gamete Formation (Meiosis)

During meiosis I, homologous chromosomes pair, then separate—with one member of each pair going to each daughter cell (**FIGURE 16-22A**). If a homologous pair fails to separate during meiosis, however, one of the new cells will end up with an extra chromosome (**FIGURE 16-22B**).

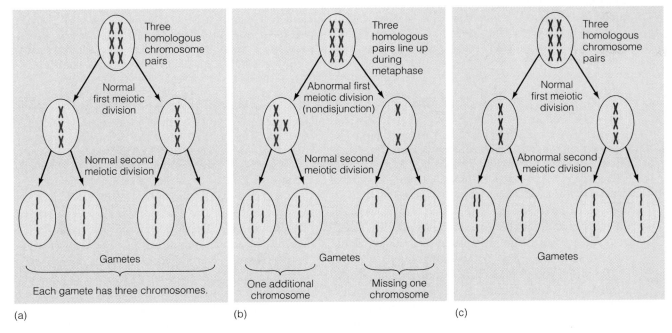

(a)   (b)   (c)

FIGURE 16-22 Meiosis and Abnormal Chromosome Numbers
(a) A simplified version of meiosis. During the first meiotic division, the homologous pairs line up, then separate, producing daughter cells with one-half the number of chromosomes. In the second meiotic division, the chromosomes line up single file and separate, with one chromatid going to each daughter cell.

(b) Nondisjunction in the first meiotic division. A chromosome pair may fail to separate during meiosis I, resulting in abnormal gametes. Half are missing a chromosome, and the other half have an extra chromosome. *(c)* Nondisjunction in second meiotic division, resulting in two normal gametes, one gamete with two chromosomes, and one gamete with four.

Health Note 16-1

Going, Going . . . Gone: The Causes and Cures of Baldness in Men and Women

Americans and others in Western society spend billions of dollars a year on their hair—styling it, coloring it, cutting it, and, yes, preventing its loss. Hair loss, however, is as inevitable as death itself. All of us lose hair as we age, but some people lose a lot more than others—and, for some, hair loss begins early in life, even in childhood.

Doctors recognize three types of hair loss. Partial hair loss involves the loss of small, isolated patches of hair. The second type, pattern baldness, involves the loss of most or all of the hair on the head. Total body hair loss is the complete loss of hair from every part of the body: head, eyelids, eyebrows, arms, legs, and so on.

Hair loss strikes both sexes, all ethnic groups, and all age groups. In fact, two of every three American men will develop some form of balding. An even larger percentage of men and women will lose some hair.

Hair loss is not a life-threatening condition, but can have severe emotional impacts on individuals. Young children are especially devastated by hair loss. Men whose hair begins to thin in their twenties often feel self-conscious and anxious in public. They may feel isolated from their cohorts, for hair loss often makes them look

five to ten years older than their peers. Friends may poke fun at them, further adding to their self-conscious feelings. Even in older men, whose friends are balding, hair loss can result in depression, feelings of low self-esteem, and a general sense of inadequacy. In women, hair loss can have even more devastating effects. What causes hair loss?

A surprising number of factors can cause hair to fall out. One of the most widely recognized is chemotherapy—chemical treatments used to fight cancer. These drugs attack rapidly dividing cells in the body such as cancer cells and the cells in hair follicles. High fever, severe infections, and severe cases of the flu also cause hair loss that mysteriously begins four weeks to twelve months after the illness. Many prescription drugs also cause hair loss. Thyroid disease—either an overactive or an underactive thyroid—can cause hair to fall out. Birth control pills, iron deficiencies, and protein-deficient diets can have a similar effect in some patients. Many women report excessive hair loss after pregnancy. But the most common cause of all is genetic.

Hereditary balding (pattern baldness) or thinning, described in the

chapter, results from genes that arise from either the mother's or the father's side of the family. In men, this condition results in near total or complete loss of hair. Women, on the other hand, tend to experience excessive thinning, but rarely go completely bald. What can be done to combat balding?

Two lines of attack are possible: medicinal and surgical. Consider the medicinal approach first. Several medicines, some of which are over the counter, are available. One of the most widely publicized is Rogaine. In a study of 2300 patients with male pattern baldness, Rogaine treatment resulted in moderate to marked hair growth in 39% of the patients, compared to 11% of those in the control group. For Rogaine to be effective, however, a man must be prepared for a lifetime of work because Rogaine must be applied twice a day every day of one's life. Stopping treatment causes bald spots to reappear, and reverting to once a day results in continued hair loss. A one-year supply of Rogaine costs $600 to $1000 per year, depending on the size of one's bald spot. Rogaine also creates some discomfort in the form of itching, headaches (in 40% of the patients), dizzy spells, and an irregular

---

The other cell will be short one chromosome. The failure of homologous chromosomes to separate is called **nondisjunction** (non-diss-JUNK-shun).

Nondisjunction can also occur in the second meiotic division (**FIGURE 16-22C**). In this division, you may recall, the 23 replicated chromosomes split apart, with one chromatid going to each daughter cell. If a chromosome fails to separate into its two chromatids, the result is the same as nondisjunction in meiosis I—a daughter cell with an extra chromosome and another daughter cell missing one chromosome.

When a gamete with one extra chromosome unites with a normal gamete, the zygote produced will contain

47 chromosomes. The zygote may be able to divide successfully by mitosis, producing an embryo all of whose cells will have an additional chromosome. Thus, instead of the normal 23 chromosome pairs, each cell in the embryo contains 22 pairs and one triplet. This condition is called **trisomy** (TRY-sew-mee; literally, "three bodies").

Gametes with a missing chromosome can also unite with normal gametes. This results in individuals with 45 chromosomes—22 chromosome pairs and a chromosome singlet. This condition is called **monosomy** (MON-oh-SO-me).

Surprisingly, one of every two conceptions contains an abnormal chromosome number. Most of

heartbeat in some patients. Rogaine is recommended for younger men from ages 20 to 30 who have begun to lose hair within the last five years. Men who are completely bald are poor candidates for treatment.

Antiandrogens can be given to women. These drugs inhibit the binding of androgens (male hormones found in women's blood) to the hair follicles. For reasons not well understood, this treatment can result in a complete reversal of hair loss, but only if it is begun within two years of the onset of hair loss. Men can also be treated with antiandrogens, but not without significant problem, for these drugs cause a loss of sex drive and an undesirable elevation of the voice that would qualify men under treatment to sing in a girl's choir.

For those who do not respond to drugs, surgery is an option. But surgery can be quite expensive, costing as much as $15,000. Despite its potentially high price tag, an estimated 250,000 American males elect to have one of several different types of surgical procedures each year.

One of the most common is hair transplantation (**FIGURE 1**). In this operation, small plugs of hair are taken from the sides and back of the scalp where hair grows thickly. These plugs—usually 50 to 60 at a time—are placed in the bald spot. Depend-

ing on the size of the bald spot, up to four sessions may be required. To make the graft look more natural, surgeons often place single-hair follicle or several-hair follicle grafts along the hair line.

Another common procedure is scalp reduction. In this procedure, surgeons remove well-defined bald spots (up to 2 by 7 inches) in the top of the scalp. The edges of the incision are then drawn together, thus reducing the area of baldness. Hair transplants may also be performed in conjunction with this procedure to fill in the remaining area.

Another option is a wig or toupee. Although wigs or toupees have improved dramatically and well-crafted hair pieces made from real or synthetic hair are very difficult to distinguish from the real thing, there are enough bad ones around that this option is often looked upon with disfavor by many men. Hair pieces can even be sutured in place.

Certain cosmetic remedies are also available. For example, cutting one's hair short tends to make baldness less obvious. Dark hair can be lightened so that it is more closely matched in color with the scalp, thus making the contrast between hair and scalp less obvious.

People who are experiencing baldness should seek medical attention

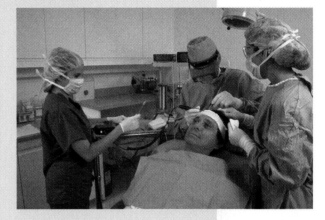

FIGURE 1 Patient undergoing hair transplant surgery.

quickly to determine if it is transient, as in the case of post-pregnancy; medicinally induced loss; or heredity. With their doctor's help, they can plot a strategy, if they so desire, to ward off balding. Or, they can simply accept their fate and learn to live with—or even appreciate—it.

Visit *Human Biology's* Internet site, www.jbpub.com/humanbiology, for links to web sites offering more information on this topic.

these embryos and fetuses die in utero. This condition is believed to be responsible for 70% of all early embryonic deaths and 30% of all fetal deaths. It is also associated with an increased miscarriage rate in older mothers.

Down Syndrome Is Trisomy 21.
One of the most common trisomies is **Down syndrome,** or **trisomy 21.**[3] Approximately 1 of every 700

babies born in the United States has Down syndrome. Down syndrome children are typically short with round, moonlike faces (**FIGURE 16-23**). Their tongues protrude forward, forcing their mouths open, and their eyes slant upward at the corners. These children have IQs that are rarely over 70. A significant number of Down syndrome babies die from heart defects and respiratory infections in the first year of infancy. Modern medical care, especially antibiotics, has reduced early death, and many individuals with Down syndrome live to age 20 or beyond.

The incidence of Down syndrome (and many other monosomies and trisomies) increases with maternal age.

---

[3]Contemporary geneticists generally refer to this as Down syndrome, rather than Down's syndrome.

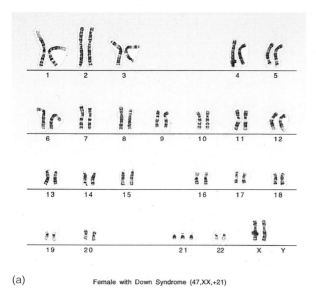

(a)   Female with Down Syndrome (47,XX,+21)

FIGURE 16-23 Down Syndrome   *(a)* Karyotype of Down syndrome girl. Note trisomy of chromosome 21. *(b)* Notice the distinguishing characteristics described in the text.

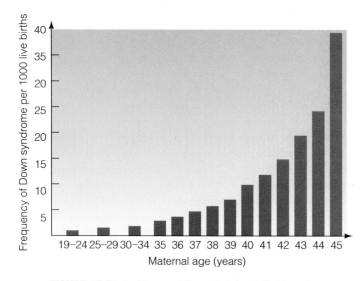

(b)

Why does the incidence of Down syndrome increase with age? A new study suggests that the reason for the rise in Down syndrome children later in life may be related to the mother's ability to carry such a fetus to term. This study suggests that the number of trisomy 21 conceptions is the same in all women regardless of age. The embryos are simply less likely to be naturally aborted in older women. Why? No one knows.

Nondisjunction of the Sex Chromosomes.

Nondisjunction can occur in any chromosome—that is, in both autosomes and sex chromosomes. Nondisjunction of the sex chromosomes can lead to a variety of nonlethal genetic disorders. One of the most common is Klinefelter syndrome.

Klinefelter syndrome occurs when an ovum with an extra X chromosome is fertilized by a Y-bearing sperm, resulting in an XXY genotype. This condition occurs in about 1 of every 700 to 1000 newborn males.[4] Although Klinefelter syndrome patients are males, masculinization is incomplete. The males' external genitalia and testes are unusually small, and about 50% of them develop breasts (**FIGURE 16-25**). Spermatogenesis is abnormal, and Klinefelter patients are generally sterile.

Another common disorder resulting from nondisjunction of the sex chromosomes is **Turner syndrome,** a monosomy. Turner syndrome results when an ovum lacking the X chromosome is fertilized by a X-bearing sperm. It may also result when a genetically normal ovum is fertilized by a sperm lacking an X or Y chromosome. In either case, the result is an offspring with 22 pairs of autosomes and a single, unmatched X chromosome (XO).

Turner syndrome patients are phenotypically female and are characteristically short with wide chests and a prominent fold of skin on their necks (**FIGURE 16-26**). Because their ovaries fail to develop at puberty, Turner syndrome patients are sterile, have low levels of estrogen, and have small breasts.[5] For the most part, they lead fairly nor-

As **FIGURE 16-24** shows, a woman's chances of having a Down syndrome baby increase dramatically after age 35. For this reason, many couples choose to have their children earlier in the woman's child-bearing years. If the woman does get pregnant after age 35, many couples choose to have amniocentesis. If the test shows that the fetus has Down syndrome, a couple may decide to opt for an abortion. Amniocentesis also enables parents for whom abortion is not an option to prepare psychologically and to seek education that will help them care for their child.

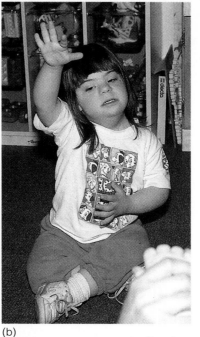

FIGURE 16-24 Incidence of Down Syndrome Babies at Various Maternal Ages   The incidence of Down syndrome rises quickly after the maternal age of 35.

---

[4]Nondisjunction can also occur in sperm development, resulting in an XY sperm. The XY sperm can fertilize a normal ovum, resulting in an XXY individual.

[5]Note that some estrogen is released from the adrenal cortex, but not enough to permit breast development.

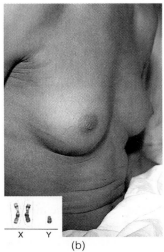

FIGURE 16-25 Klinefelter Syndrome *(a)* Notice that the sex chromosomes of a person with Klinefelter's syndrome include two X and one Y. *(b)* Breast development in a male with Klinefelter syndrome.

(a) X   Y   (b)

TABLE 16-5	
Chromosome Deletions	
Syndrome	**Phenotype**
Wolf-Hirschhorn syndrome	Growth retardation, heart malformation, cleft palate; 30% die within 24 months.
Cri-du-chat syndrome	Infants have catlike cry, some facial anomalies, severe mental retardation.
Wilm's tumor	Kidney tumors, genital and urinary tract abnormalities
Retinoblastoma	Cancer of eye, increased risk of other cancers
Praeder-Willi syndrome	Infants are weak and grow slowly; children and adults are obese and are compulsive eaters.

mal lives. Mental retardation is not associated with the disorder, although some studies suggest that Turner patients are not as capable at numerical skills and spatial perception as genetically normal children. Turner syndrome occurs in 1 of every 10,000 female births. The rarity of this condition, compared with Klinefelter's syndrome, is due to the fact that the XO embryo is more likely to be spontaneously aborted.

Genetic Disorders May Also Result from Variations in Chromosome Structure

Another defect in chromosomes involves alterations in chromosome structure, the most common of which are (1) **deletions,** the loss of a piece of chromosome, and (2) **translocations,** breakage followed by reattachment elsewhere.

Deletions.

Most deletions are deleterious, and embryos with them are usually eliminated early in pregnancy. Nevertheless, some embryos whose cells contain deletions do survive. **TABLE 16-5** lists a few disorders caused by deletions. One of the more striking is called Praeder-Willi syndrome (PRAY-der Will-ee).

Praeder-Willi syndrome occurs in 1 in 10,000–25,000 births. It is caused by the loss of one of

the arms of chromosome 15 during gamete formation and is characterized by slow infant growth, compulsive eating, and obesity. Babies born with the syndrome are weak. They have a poor suckling reflex and do not feed well. By age 5 or 6, however, these children become compulsive eaters. Parents must lock their cupboards and refrigerators. Neighbors must be warned to discourage begging and must keep their garbage cans under lock and key. The urge to eat results in obesity, which often leads to diabetes. If food intake is not restricted, victims can literally eat themselves to death. Researchers believe that the eating disorder may result from an endocrine imbalance caused by the deletion—illustrating how a genetic defect can lead to an upset in homeostasis.

Translocations.

Translocations occur when a segment of a chromosome breaks off and reattaches to another site on the same chromosome or to another chromosome. Although this might at first seem innocuous, movement of a segment of a chromosome to another site can upset the delicate balance of gene expression and alter homeostasis. Translocations, for example, may be the cause of certain forms of leukemia.

Genetic Screening Allows Parents to Determine Whether They Will Have a Genetically Normal Baby

Chapter 15 noted that parents can now find out the sex of their child and the presence of certain genetic defects well before birth via amniocentesis. This procedure permits geneticists to search for abnormal chromosome numbers, as well as deletions and translocations. Biochemical tests can also be run to pinpoint metabolic disease, although only a dozen or so are routinely screened.

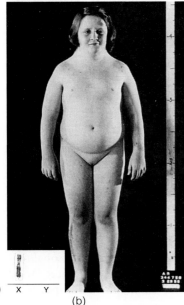

FIGURE 16-26 Turner Syndrome *(a)* In a person with Turner syndrome only one X chromosome is present. *(b)* Characteristic physical features of a Turner syndrome girl.

(a) X   Y   (b)

Amniocentesis increases the risk of spontaneous abortion by about 1%. It also slightly increases the risk of maternal uterine infection. Therefore, this procedure is usually recommended only if one or more of the following conditions are met: (1) a woman is over 35, (2) she has already delivered a baby with a genetic defect, (3) she is a carrier of an X-linked genetic disorder, or (4) she or the father has a known chromosomal or genetic abnormality.

As a rule, amniocentesis is usually not performed until the sixteenth week of pregnancy.[6] Before this time, there is not enough fluid surrounding the fetus, and the needle could damage the fetus. Analysis of the fetal cells withdrawn from the amnion requires 10–15 additional days. If a serious defect is observed, a couple may elect to have an abortion. However, the risk of an abortion to the mother is slightly greater at this time than it is during the 12- to 16-week period, and some state laws do not permit abortion after 16 weeks.

To permit earlier detection of genetic defects, a new procedure, known as **chorionic villus biopsy** (CORE-ee-on-ick VILL-us BYE-op-see), has been developed. Chorionic villi are composed of embryonic tissue that form the fetal portion of the placenta (plah-SEN-tah), a structure that nourishes the growing fetus. To perform this procedure, physicians first insert a device into the uterus through the vagina. It is used to remove a small sample of a villus. The cells of the villus are then examined for chromosomal abnormalities, much the same way as those removed during amniocentesis.

Because a biopsy can be performed much earlier than amniocentesis, abortions can be performed between 8 and 12 weeks of gestation; these abortions are safer than those performed between 16 and 20 weeks. Although chorionic villus biopsy allows for earlier detection, it poses a slightly higher risk to the mother and her fetus.

▉ DNA Abnormalities May Also Occur in Mitochondria, Resulting in Diseases

Mitochondria contain a tiny fraction (about 0.3%) of a cell's DNA. As noted in Chapter 3, the DNA inside a mitochondrion may be an evolutionary remnant from the time when the mitochondrion was a free-living organism. Until recently, though, the significance of mitochondrial DNA was not well understood.

A few years ago, however, researchers discovered a rare form of blindness in humans linked to a defect in the mitochondrial DNA. This finding confirmed suspicions that defective mitochondrial genes result in genetic defects, and it also suggests an additional mechanism for inheriting genetic diseases.

The defective gene in the mitochondria that leads to blindness codes for a protein required in the first step of ATP production. The absence of this protein in the neurons of the optic nerve results in their death, which, in turn, leads to blindness usually by age 20.

Because mitochondria are passed on by the mother, all of the children of a woman with the defective gene will inherit it. Only a small fraction of the children who inherit the defective gene actually go blind, so it is thought that the mutation only predisposes people to blindness. Other factors may also contribute to blindness.

Several rare genetic diseases may also result from mitochondrial DNA defects. Researchers suggest that even some of the more common diseases may be caused by genetic defects in mitochondrial DNA. Some cases of heart, kidney, and central nervous system failure, whose causes are now unknown, may one day be linked to defective mitochondrial DNA.

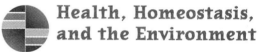

Health, Homeostasis, and the Environment

Nature Versus Nurture

Throughout this book, the central theme—the idea that homeostasis and human health are profoundly influenced by our environment—has emerged repeatedly. Put in other words, the environment, which includes our social, psychological, and physical environments, has a profound impact on our internal environment (homeostasis) and our health.

In this section, we examine a controversy that has raged in science for decades over the connection between our genes and our environment. We will look at two basic questions: How do our genes contribute to our personalities? What is the role of our environment in determining our personality and behavior and, perhaps, even our health?

At one time, psychologists viewed the baby as a blank slate. A child's personality, they said, develops through interaction with its environment—its parents, friends, teachers, and so on. However, extensive research now suggests that our personalities are also influenced by our genes. Thus, our genes and our environment probably operate together in determining personality.

Michael Lewis, a researcher at the Robert Wood Johnson Medical School, studies infant response to

---

[6]Improvements in this technique now allow physicians to perform the procedure as early as the fourteenth week, although risks of spontaneous abortion increase somewhat.

stress. His research shows that newborn babies differ markedly in how they respond to the stress of a blood test performed well before their environment could have affected their personality. Lewis found that some children wail when poked with a needle during routine blood tests in the first few days of life; others hardly seem to notice. Of the newborns who cry, some quickly dampen their response. Others seem to go on forever, much to their parents' chagrin.

Lewis notes that a child's reaction to stress at this time is likely to be repeated 3 months later when the child receives an inoculation. He believes that the difference in response to stress is genetically based and that the inherent differences will persist.

Lewis has also found that babies differ in how they react to frustration. He performed a series of experiments to test infant response to a frustrating situation. Most children responded with anger. Some, however, showed no response at all, and others displayed sadness. These innate differences in behavior, occurring too early to stem from differences in upbringing, probably result from genetic differences. They may help account for the profound differences in responses to stress seen in adults.

Psychologist Nathan Fox has performed some intriguing studies on shy and extroverted children. His studies show that shy children cling to their mothers when a clown suddenly appears; extroverted children eagerly engage the clown in play. Fox has found that shy children show greater electrical activity in the right part of the brain; extroverted children show a higher level of activity in the left side. Because the children's personalities differ at birth and because the personality differences are reflected in differences in brain activity, genetics may be playing a powerful role in personality development.

Allison Rosenberg, a researcher at the National Institutes of Health, has also studied shy and outgoing children. Her work shows marked differences between these two groups in heart rate and the release of the hormone cortisol, further supporting the notion that there are inherent physiological differences in children from the outset. These are related to early differences in personality and, very possibly, to differences in genetic makeup.

Similar findings have been reported in other primates. Infant monkeys, for instance, display a wide range of personality traits early in life. Research shows that timid animals differ physiologically from their braver counterparts. Stephen Soumi, a researcher at the National Institutes of Health, believes that the individual differences seen in personality and in response to stress

are so pronounced that they must be genetically based. Furthermore, he and his colleagues have found that shyness and extroversion tend to run in families, further suggesting a link between early behavior and genetics.

Soumi notes, however, that environmental effects, especially events very early in life, can modify genetically programmed behavior. If a shy baby monkey is paired with an exceptionally nurturant mother, he finds, the shy animal develops rapidly and actively explores its environment.

Another example of behavioral modification by environment comes from research on thrill seeking. Psychologists believe that some individuals are naturally born thrill seekers, or type T people (**FIGURE 16-27**). Some researchers believe that thrill seeking may be genetically based. One theory is that risk takers, the type T or "big T" individuals, may be hard to excite and, therefore, may attempt extraordinary feats for excitement. At the opposite end of the spectrum are "small t" people, risk avoiders, who are easily aroused. They seek to avoid stimulation. Presumably, there's a whole spectrum of folks in the middle.

Some researchers believe that type T individuals have an imbalance of a neurotransmitter known as monoamine oxidase (MAO) in the brain. Thrill seeking supposedly increases the MAO levels in the brain, creating a feeling of exhilaration.

Type T behavior can be modified by an individual's upbringing and turned in a negative or positive direction, say some psychologists. A positive direction might lead an individual to play for the Green Bay Packers football team. A negative direction might lead that same individual, under different environmental conditions, into gang fighting and crime in the streets.

Peers, teachers, relatives, ministers, parents, and others make up the environment. Their influence may turn the type T child to healthy, constructive opportunities or unhealthy, destructive ends. Environmental influences, then, may affect health in a roundabout way. Children who respond abnormally to stress, for instance, may be steered toward mechanisms that reduce stress, resulting in a more healthy life-style. Homeostasis is, therefore, broadly affected by the psychological environment.

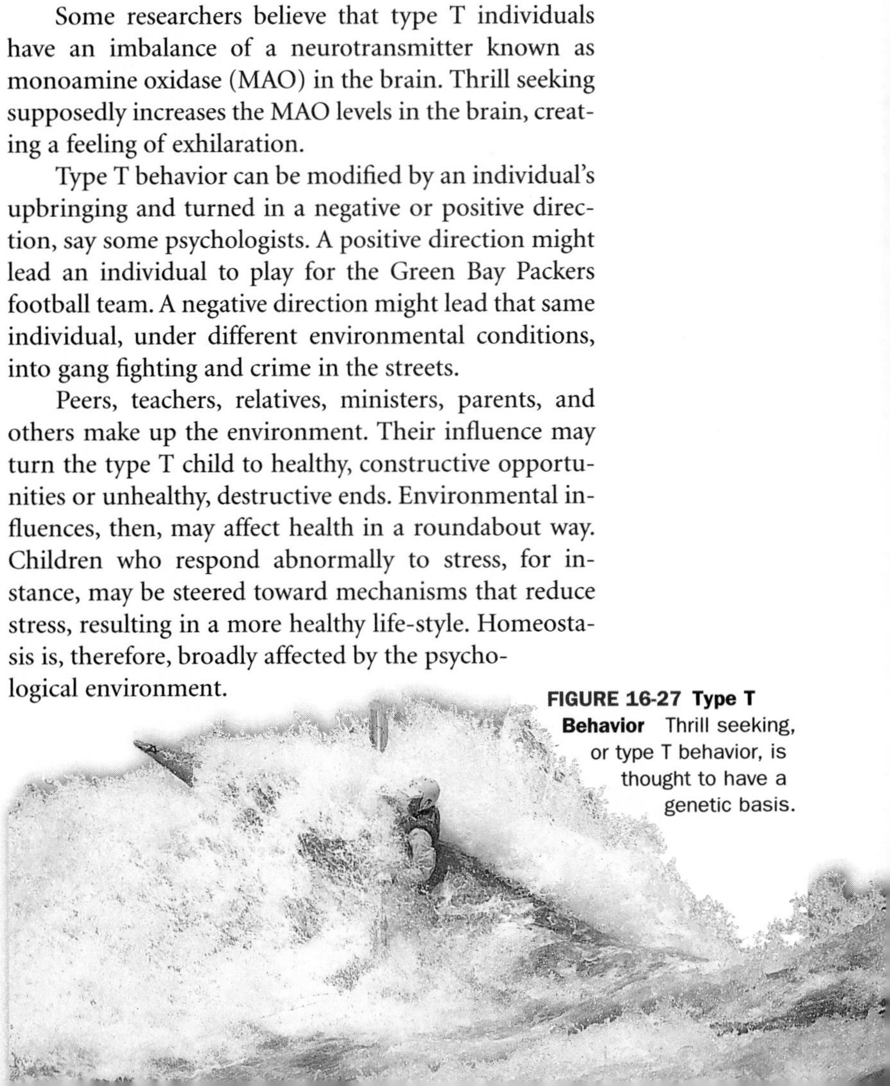

FIGURE 16-27 Type T Behavior Thrill seeking, or type T behavior, is thought to have a genetic basis.

SUMMARY

MEIOSIS AND GAMETE FORMATION

1. Sexually reproducing organisms halve the number of chromosomes in germ cells through *meiosis*, a type of nuclear division found only in germ-cell production in the gonads of sexually reproducing animals.

2. Meiosis involves two nuclear divisions. During the first division, *meiosis I*, the chromosome number is halved. Thus, a diploid cell produces two haploid cells.

3. The second meiotic division, *meiosis II*, is virtually identical to mitosis, except for the fact that the cells are haploid. When haploid cells divide in meiosis II, they produce two new cells, each containing a haploid number of single-stranded (unreplicated, or one-chromatid) chromosomes.

4. In males, meiosis produces four gametes, but in females it produces only one.

PRINCIPLES OF HEREDITY: MENDELIAN GENETICS

5. Gregor Mendel, a nineteenth-century monk, derived several important principles of inheritance from his work on garden peas. Mendel determined that, in peas, traits do not blend as was commonly thought at the time. He also postulated that each adult has two hereditary factors for a given trait and that these factors (genes) separate during gamete formation. He called this notion the *principle of segregation*.

6. Mendel also postulated that a hereditary factor might be either dominant or recessive. A *dominant factor* masks a recessive factor. A *recessive factor* is expressed only when the dominant factor is missing. The dominant and recessive genes are alternative forms of the gene, or *alleles*.

7. Three genetic combinations are possible for a given trait: *heterozygous*, *homozygous dominant*, and *homozygous recessive*.

8. The genetic makeup of an organism is called its *genotype*. The physical appearance, which is determined by the genotype and the environment, is the *phenotype*.

9. From his studies, Mendel concluded that the hereditary factors were separated independently of one another during gamete formation. This is the principle of *independent assortment* and holds true only for nonlinked genes.

MENDELIAN GENETICS IN HUMANS

10. Human cells contain 23 pairs of chromosomes: 22 pairs of *autosomes* and 1 pair of *sex chromosomes*. Chromosomes carry dominant and recessive traits, and inheritance of these traits is consistent with Mendel's principles of inheritance, although additional mechanisms are at work in humans and other organisms.

11. Sickle-cell anemia, cystic fibrosis, and albinism are *autosomal-recessive traits* and are expressed only in homozygous-recessive individuals.

12. Widow's peak, achondroplasia, and Marfan's syndrome are *autosomal-dominant traits* and are expressed in heterozygous and homozygous-dominant genotypes.

VARIATIONS IN MENDELIAN GENETICS

13. Genetic research since Mendel's time has turned up several additional modes of inheritance. One mode is *incomplete dominance*. It occurs when an allele exerts only partial dominance, producing intermediate phenotypes.

14. Some genes have more than two possible alleles. *Multiple alleles* result in more genotypes and phenotypes in a population. Because chromosomes exist in pairs, individuals can have only two of the possible alleles.

15. *Codominance* occurs in multiple-allele genes. Codominant genes are expressed fully and equally.

16. Some traits are controlled by many genes. This phenomenon is referred to as *polygenic inheritance*.

17. Genes that are found on the same chromosome are said to be *linked*. If crossing-over does not occur, these genes are inherited together.

SEX-LINKED GENES

18. The X and Y chromosomes are commonly referred to as the sex chromosomes. However, studies suggest that the real determinant of sex is the Y chromosome. If it is absent, the individual (XX) is a female. If it is present, the individual (XY) is a male.

19. The sex chromosomes also carry genes that determine physical traits. A trait determined by a gene on a sex chromosome is a *sex-linked trait*. Most sex-linked traits occur on the X chromosome. Both dominant and recessive sex-linked traits can be present.

CHROMOSOMAL ABNORMALITIES AND GENETIC COUNSELING

20. Abnormalities in the human genome arise from mutations (changes in DNA structure), abnormalities in chromosome number (aneuploidy and polyploidy), and alterations in chromosome structure (deletions and translocations).

21. Alterations in the number of chromosomes result chiefly from errors in gamete formation when chromosomes fail to separate during meiosis. This process is called *nondisjunction*.

22. Variations in chromosome structure result from two occurrences: *deletions*, or the loss of a piece of chromosome, and *translocations*, or breakage followed by reattachment elsewhere.

23. Embryos with abnormal chromosome numbers or abnormal chromosome structure are likely to die and be aborted spontaneously.

HEALTH, HOMEOSTASIS, AND THE ENVIRONMENT: NATURE VERSUS NURTURE

24. At one time, psychologists thought that a child's personality developed principally through interaction with the environment—parents, friends, and teachers. New research suggests, however, that our personalities are also influenced by our genes.

Critical Thinking

THINKING CRITICALLY— ANALYSIS

This Analysis corresponds to the Thinking Critically scenario that was presented at the beginning of this chapter.

To begin, reread the previous material in this exercise and make a list of the symptoms characteristic of Marfan's syndrome.

They are aortic weakening, abnormally long legs and arms, excellence on the basketball and volleyball court, and nearsightedness.

Although you won't be able to find any information about aortic weakening, you could possibly find information on Lincoln's health—perhaps old medical files. In fact, Lincoln's medical records are available, and they show that the former president displayed no signs of cardiovascular disease. That doesn't really prove anything, by itself.

As for the president's skill on the basketball court or volleyball court . . . we'll never know. The sports hadn't even been invented yet. But, more seriously, it just so happens that Lincoln's lanky limbs were within the normal dimensions of tall people. As for the president's eyesight, it turns out that Lincoln was farsighted, not nearsighted.

This evidence strongly suggests that Abe Lincoln did not have this genetic disorder. This exercise illustrates the importance of a close examination of the facts.

EXERCISING YOUR CRITICAL THINKING SKILLS

The sex of a child is determined by the presence or absence of the Y chromosome. As you learned in this chapter, the sex chromosomes in males are XY; in females, they are XX. Thus, the presence of a Y chromosome yields a male; its absence produces a female.

Since 1959, geneticists have been searching for the specific gene on the Y chromosome responsible for male characteristics. In 1983, a group of scientists was studying a rare condition—males with XX chromosomes. This anomaly occurs in approximately 1 of every 20,000 males. When they examined the X chromosomes of these males, the researchers discovered that the X chromosomes (presumably from the father) carried a small fragment from a Y chromosome. Maleness, they concluded, comes from the presence of a gene on that segment exchanged with the X chromosome.

The magazine article in which I read this report noted that the researchers had studied several dozen XX males with these fragments. After examining the fragments, they isolated a specific gene and identified it as the possible sex determiner. This astounding finding was reported with much fanfare by the news media in 1987.

When I went to the original research publication, however, I found that the researchers had not reported how many XX males in their study had the fragment from the Y chromosome. Was it all XX males or just a few? (The editors should have insisted on full disclosure.) Given the conclusions of the study and my faith in scientific publications, I surmised (uneasily) that all of the XX males studied had the Y fragment—that is, that this phenomenon occurred in 100% of the XX males.

A few years later, though, a second report hit the press. In this study, researchers found that three XX males in their study group did not have the much-heralded gene, suggesting that in the earlier research the same phenomenon might have occurred.

The second study clearly illustrates the importance of good follow-up research, which allows scientists to check the results of earlier work and find weaknesses. As pointed out in the Thinking Critically section in Chapter 1, follow-up research is essential to good science! Critical thinkers must demand it.

Continuing the saga, a more recent study uncovered another gene that researchers think may be the sex determiner. Their studies suggest that this gene may be something of a master gene—that is, it may control other genes involved in sexual development. The sex-determining gene controls sexual development by inducing testicular development. Subsequent male sexual differentiation is a consequence of testosterone, a hormonal product of the testis. Interestingly, the sex-determining gene appears to work just prior to the development of male sex organs in the embryo.

If this gene is indeed responsible for sex determination, it would be expected to be found in all mammals. Research shows that an almost identical gene is present in many mammals, including mice, rabbits, chimpanzees, horses, and tigers.

The research team that discovered the "master gene" is hesitant to announce that it has found the sex determiner gene even with all this evidence. Because any announcement would be met with at least some skepticism, the evidence must be concrete before definite conclusions can be made. In other words, more studies are needed.

Using your knowledge of genetics, can you think of any way to test whether this gene is the sex determiner? Pause for a few minutes and jot down your ideas.

Here's what the research team is doing: They are using a procedure that allows them to insert the gene into mouse embryos with two X chromosomes. These embryos would normally develop into females. If the gene is the sex determiner, the researchers assert, the embryos should develop into males.

TEST OF CONCEPTS

1. Explain the process of meiosis in general terms. Where does it occur? What does it accomplish?

2. Draw a diagram showing the various stages of meiosis I and meiosis II. Make a note of the number of chromosomes at each stage and their condition—that is, whether they have one chromatid or two. Which division is the reduction division?

3. How is mitosis different from meiosis? How is it similar?

4. Mendel's research was designed to answer two basic questions. What were the questions and what were his findings?

5. Define the following terms: principle of segregation, principle of independent assortment, allele, phenotype, genotype, heterozygous, homozygous, monohybrid cross, and dihybrid cross.

6. Freckles are an autosomal-dominant trait. A woman with freckles (*Ff*) marries and has a baby by a man without freckles (*ff*). What are the chances that their children will have freckles?

7. Two freckled adults marry and have children. The first baby has no freckles. What are the genotypes of the parents?

8. Attached earlobes (*A*) are an autosomal-dominant trait. The *A* allele is dominant over the *a* allele, which produces unattached earlobes in homozygous-recessive individuals. A woman with freckles and attached earlobes (*FfAa*) marries a man who has freckles and attached earlobes (*FfAa*). Draw a Punnett square showing the various gametes as well as the genotypes of the offspring. List all possible phenotypes and the genotypes that correspond to them.

9. An albino guinea pig with the genotype *ccBB* is mated to a brown guinea pig with the genotype *CCbb*. Using a Punnett square, determine the genotype and phenotype of the F₁ generation.

10. What is sickle-cell anemia? What causes it? Why can a person be a carrier of the disease but not display outward symptoms?

11. How do incomplete dominance and codominance differ? Give examples of each.

12. Assuming that two genes (A and B) control height, list all of the possible genotypes, and indicate the phenotype associated with each.

13. Describe how crossing-over works. What impact does it have on the genotype of a person's gametes?

14. Color blindness is a recessive, X-linked gene. A color-blind man and his wife have four children, two boys and two girls. One boy and one girl are color-blind, and the other two are normal. What is the genotype of the woman?

15. What is a sex-influenced gene? Give some examples. What criteria would you use to assess whether a trait was sex-influenced?

16. Explain how each of the following genetic defects could arise: trisomy 18, monosomy 10, triploidy, and tetraploidy.

TOOLS FOR LEARNING

www.jbpub.com/humanbiology

Tools for Learning is an on-line student review area located at this book's web site HumanBiology (www.jbpub.com/humanbiology). The review area provides a variety of activities designed to help you study for your class:

Chapter Outlines. We've pulled out the section titles and full sentence sub-headings from each chapter to form natural descriptive outlines you can use to study the chapters' material point by point.

Review Questions. The review questions test your knowledge of the important concepts and applications in each chapter. Written by the author of the text, the review provides feedback for each correct or incorrect answer. This is an excellent test preparation tool.

Flash Cards. Studying human biology requires learning new terms. Virtual flash cards help you master the new vocabulary for each chapter.

Figure Labeling. You can practice identifying and labeling anatomical features on the same art content that appears in the text.

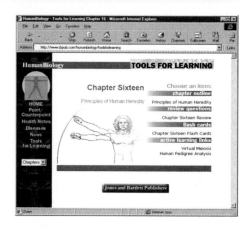

Active Learning Links. Active Learning Links connect to external web sites that provide an opportunity to learn basic concepts through demonstrations, animations, and hands-on activities.

MOLECULAR GENETICS

HOW GENES WORK AND HOW GENES ARE CONTROLLED

Thinking Critically

One of America's leading drug companies is spending about $300 million to construct a facility that will produce synthetic growth hormone to be sold to dairy farmers. The hormone will be produced by bacteria that contain growth-hormone genes isolated and transplanted from dairy cattle. In other words, they're genetically engineered bacteria. Given to cows, the hormone from these bacteria dramatically increases milk production. Business economists believe that an increase in milk production will reduce the cost of milk to the consumer. Based on this information, does introducing synthetic growth hormone seem like a good idea? ■

Tunneling electron micrograph of DNA.

N 1988, A U.S. MILITARY COURT SENTENCED A SER-viceman in Korea to 45 years in prison for rape and attempted murder. Ten days later, a Florida court convicted a man on two counts of first-degree murder. What makes these two convictions important is that prosecutors relied heavily on the results of a new technique known as "DNA fingerprinting."

In this procedure, which today is being used with more and more frequency, criminologists analyze the composition of the genetic material DNA in samples of hair, semen, or blood found at the scene of the crime. Then they compare the DNA in these samples with the DNA of the accused. If it matches, prosecutors can make a strong case for guilt. In fact, without this procedure, prosecutors believe, the two convictions cited above would not have been won.

Many people believe that DNA fingerprinting could revolutionize criminal investigations. King County, Washington, in fact, has already begun taking DNA samples from all convicted sex offenders. Like fingerprints, the results of DNA analysis will be kept on file, readily available for future cases. The FBI is also using this technique in its work.

DNA fingerprinting holds great promise, say supporters, because there is only 1 chance of a mistaken identity in 4 or 5 trillion. In contrast, the best conventional methods, such as blood typing and blood-enzyme analysis, run the risk of error in about 1 in every 1000 cases.

Although DNA fingerprinting sounds promising, not all geneticists believe that it is as reliable as many think. In December 1991, in fact, two prominent geneticists published an article in the

prestigious journal *Science* in which they assert that the probabilities (noted above) are based on improper assumptions (violating one of the critical thinking rules) and comparatively little scientific data on the genetic composition of populations. The Exercising Your Critical Thinking Skills section at the end of this chapter offers a case in point.

DNA fingerprinting is one of many offshoots of research in genetics. This chapter covers some of the basic research that led to the discovery of DNA fingerprinting and offers additional information—notably, how genes work and how they are controlled—that helped scientists develop other practical spin-offs such as genetic engineering.

DNA and RNA: Macromolecules with a Mission

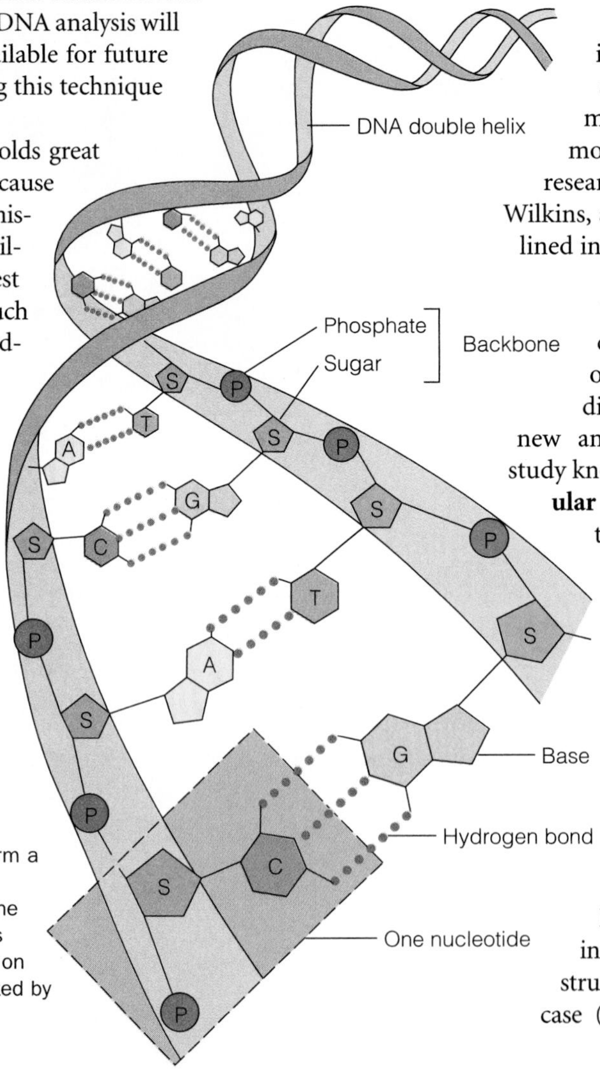

FIGURE 17-1 DNA DNA consists of two intertwined polynucleotide chains that form a double helix. The sugars and phosphates form the backbone of each chain, with the bases projecting inward. The bases on opposite strands are connected by hydrogen bonds.

In 1953, James Watson, an American biologist, and Francis Crick, a British biologist, proposed a model for the structure of the DNA molecule. This model was based on research by Rosalind Franklin, Maurice Wilkins, and many other scientists, as outlined in Scientific Discoveries 15-1.

In 1962, Watson, Crick, and Wilkins, a British biophysicist, received the Nobel Prize in Physiology and Medicine for their model, a discovery that opened the doors to a new and infinitely fascinating field of study known as molecular genetics. **Molecular genetics** is the study of the structure and function of RNA and DNA at the molecular level. It brings us as close as we can get to the roots of inheritance. We turn first to the structure and the synthesis of DNA.

DNA Is the Molecular Basis of the Gene and Consists of a Double Helix Held Together by Hydrogen Bonds

DNA consists of two strands that intertwine to form a double helix, a structure that resembles a spiral staircase (**FIGURE 17-1**). Each strand of the double helix contains millions of smaller molecules, or nucleotides. Like several other important biological molecules, then, DNA is a polymer.

Each nucleotide consists of three smaller molecules: a nitrogenous base, a phosphate group, and a monosaccharide, deoxyribose (**FIGURE 17-2**). The nucleotides join by covalent bonds to form polynucleotide chains. As shown in **FIGURE 17-1**, nucleotides unite by covalent bonds between the sugar molecules and phosphates.

The two polynucleotide chains of DNA are held together by hydrogen bonds that form between the bases. As shown in **FIGURE 17-1**, the bases of the nucleotides of each chain project inward and therefore lie inside the helix. Hydrogen bonds between the bases are indicated by the dotted lines in **FIGURE 17-1**. These bonds are much weaker than covalent bonds and can be easily broken so the molecule can unzip for replication.

DNA contains two types of bases: purines (PURE-eens) and pyrimidines (pa-RIM-a-DEENS) (**FIGURE 17-2**). The nitrogenous base of **purines** consists of two fused rings; the nitrogenous base of **pyrimidines** contains only one ring. In DNA, two purines are found: **adenine** (A) (AD-ah-neen) and **guanine** (G) (GUAN-neen). The pyrimidines are **cytosine** (C) (sigh-toe-seen) and **thymine** (T) (thigh-mean). My genetics prof came from the agricultural college of my university, and evidently proud of his roots, gave us the mnemonic "pure AG" to remember these.

Purines on one strand bind (via hydrogen bonds) to pyrimidines on the opposite strand. But the relationship between the two is even more specific. If you look at **FIGURE 17-3B**, you will see that the adenine binds only to the pyrimidine thymine. Guanine binds only to the pyrimidine cytosine. Adenine and thymine are therefore said to be complementary bases, as are guanine and cytosine. As you will soon see, this unalterable coupling, called **complementary base pairing,** ensures the accurate replication of DNA and the reliable transmission of the genetic information from a parent cell to its daughter cells during cell division.

▮ DNA Replication Takes Place on the Individual Polynucleotide Strands After the Double Helix "Unzips"

As noted in Chapter 3, before a cell can divide, it must first make an exact copy of all of its DNA. DNA replication ensures that a cell about to divide has two identical sets of genetic information, one for each daughter cell.

During the S phase of interphase, cells replicate their DNA by unzipping each DNA double helix along the hydrogen bonds that unite the complementary bases. Each polynucleotide strand then serves as a template (described

FIGURE 17-2 Nucleotides Containing Purine and Pyrimidine Bases All nucleotides consist of three subunits: a phosphate group; an organic, nitrogen-containing base; and a five-carbon sugar. DNA nucleotides contain the sugar deoxyribose. Two types of bases are found: purines and pyrimidines. *(a)* The purines are adenine and guanine. *(b)* The pyrimidines are cytosine and thymine.

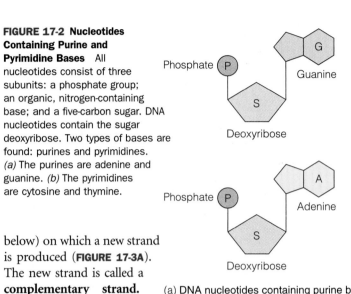

(a) DNA nucleotides containing purine bases

below) on which a new strand is produced (**FIGURE 17-3A**). The new strand is called a **complementary strand.** The template is often called the **original strand.**

DNA replication is referred to as a **semiconservative process** because each polynucleotide chain of the DNA double helix serves as a template for the production of a new strand of DNA.

DNA replication during the S phase begins when special enzymes start to pull apart, or unzip, the DNA double helix. As the two strands are separated, the bases of the polynucleotide chains are exposed. They are then free to form hydrogen bonds with complementary nucleotides in the nu-

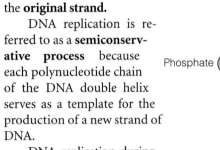

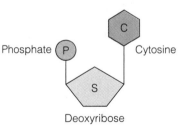

(b) DNA nucleotides containing pyrimidine bases

cleoplasm (**FIGURE 17-3**). Because adenine binds only to thymine and guanine binds only to cytosine, the original strands, the templates, are said to direct the synthesis of new strands. Synthesis on the DNA templates occurs one base at a time, and the accuracy of replication is ensured by complementary base pairing.

During replication, incoming nucleotides must first be aligned properly so that the hydrogen bonds can form between complementary bases and so that the phosphate group of the incoming nucleotide can bond to the sugar of the nucleotide already in place, as illustrated in **FIGURE 17-4**. Alignment and attachment are aided by an enzyme known as **DNA polymerase** (PUL-yi-merr-ace) (**FIGURE 17-4**). DNA polymerase slides

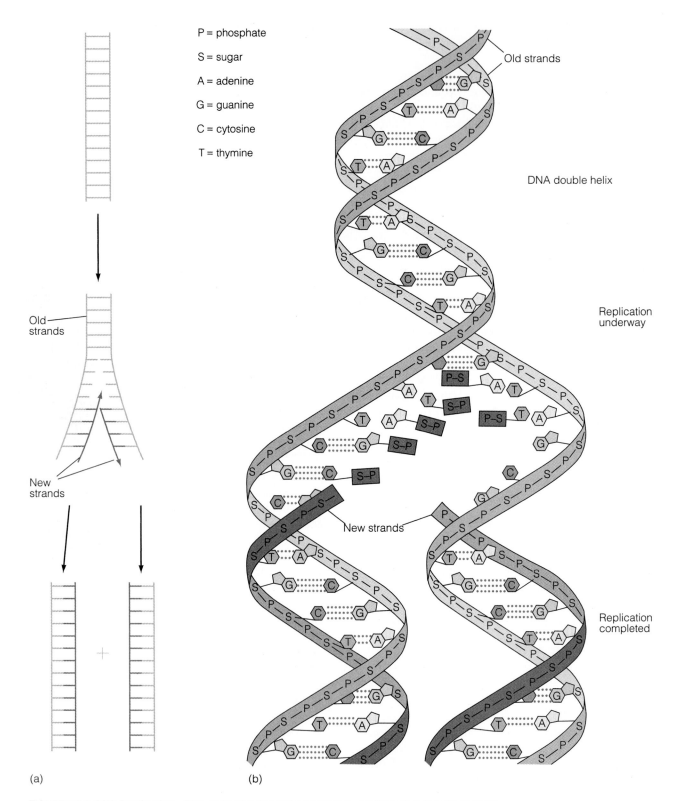

P = phosphate
S = sugar
A = adenine
G = guanine
C = cytosine
T = thymine

Old strands

DNA double helix

Old strands

Replication underway

New strands

New strands

Replication completed

(a)

(b)

FIGURE 17-3 DNA Replication   DNA replication is semiconservative. *(a)* Each double helix unwinds, and each half of the helix serves as a template for the production of a new strand of DNA. When replication is complete, each new helix contains one old and one new strand. *(b)* Nucleotides attach to the template one at a time and are joined together with the aid of enzymes.

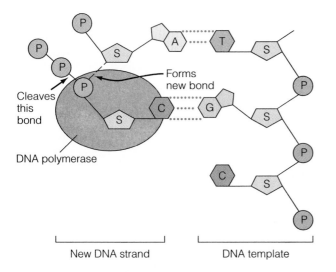

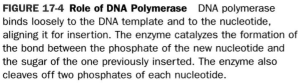

New DNA strand · DNA template

FIGURE 17-4 Role of DNA Polymerase DNA polymerase binds loosely to the DNA template and to the nucleotide, aligning it for insertion. The enzyme catalyzes the formation of the bond between the phosphate of the new nucleotide and the sugar of the one previously inserted. The enzyme also cleaves off two phosphates of each nucleotide.

along the template, aligning nucleotides one at a time, then linking each new nucleotide to the one already in place. As noted above, the enzyme joins the phosphate group of one nucleotide to the deoxyribose molecule (sugar) of its neighbor.

When DNA synthesis is finished, two new DNA molecules exist, each of which consists of a double helix. Each double helix, in turn, contains one strand from the original molecule and one new polynucleotide chain. The chromosome, once a single molecule of DNA and protein, now consists of two molecules with associated proteins. As noted in Chapter 15, each strand of DNA and associated protein is called a chromatid. The two chromatids of each chromosome are joined at their centromeres after synthesis.

Three Types of RNA Exist, Each of Which Is Involved in Protein Synthesis

DNA contains the genetic information, which determines the structure and controls most of the functions of cells. Like a commander in the army, DNA does not exert its influence directly. Its work is carried out by other molecules, notably RNA. Three types of RNA are involved in this process: ribosomal RNA (rRNA), messenger RNA (mRNA), and transfer RNA (tRNA). Each has a unique function in protein synthesis (**TABLE 17-1**). Despite major functional differences in these molecules, all three RNA molecules are biochemically similar. In humans, for example, all RNAs are single-stranded molecules, and all are polynucleotides (**FIGURE 17-5**). RNA nucleotides consist of three molecules: a sugar, a nitrogenous base, and a phosphate group. Unlike the DNA nucleotides, however, the RNA nucleotides contain the sugar ribose instead of deoxyribose and the pyrimidine

---

TABLE 17-1

Role of RNA Molecules

Molecule	Role
Messenger RNA (mRNA)	Carries the genetic information that is needed to make proteins in the cytoplasm from the nucleus
Transfer RNA (tRNA)	Binds to specific amino acids, transports them to the mRNA, and inserts them in the correct location on the mRNA
Ribosomal RNA (rRNA)	Component of the ribosome

FIGURE 17-5 RNA RNA is a single-stranded molecule consisting of many RNA nucleotides. A small section of an RNA molecule is shown here.

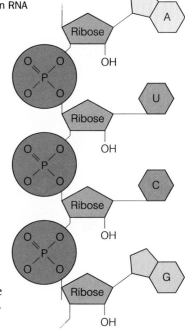

uracil instead of thymine. **TABLE 17-2** summarizes the key differences between RNA and DNA.

RNA Synthesis Is Called Transcription and Takes Place on a DNA Template in the Nucleus of the Cell

Unlike DNA, which is self-replicating, all three types of RNA must be synthesized on DNA templates. The synthesis of RNA on a DNA template permits the genetic information coded in the DNA to be transferred from the DNA molecule to RNA, notably, to the messenger RNA. Because messenger RNA is free to leave the nucleus and DNA is not, it serves as a kind of shuttle service or messenger that transfers the genetic information

---

TABLE 17-2

Differences Between RNA and DNA in Prokaryotic and Eukaryotic Cells

DNA	RNA
Double-stranded	Single-stranded
Contains the sugar deoxyribose	Contains the sugar ribose
Contains adenine, guanine, cytosine, and thymine	Contains adenine, guanine, cytosine, and uracil
Functions primarily in the nucleus	Functions primarily in the cytoplasm

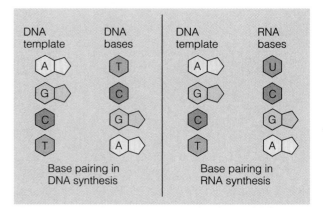

FIGURE 17-7 DNA and RNA Base Pairing The sequence of bases on the DNA template determines the sequence of bases on the complementary strand of DNA and RNA.

FIGURE 17-6 RNA Synthesis on DNA Template. RNA is synthesized on a DNA template. As shown, the DNA double helix unwinds, but only one strand serves as a template for the production of RNA. Complementary base pairing determines the exact sequence of bases in the RNA molecule.

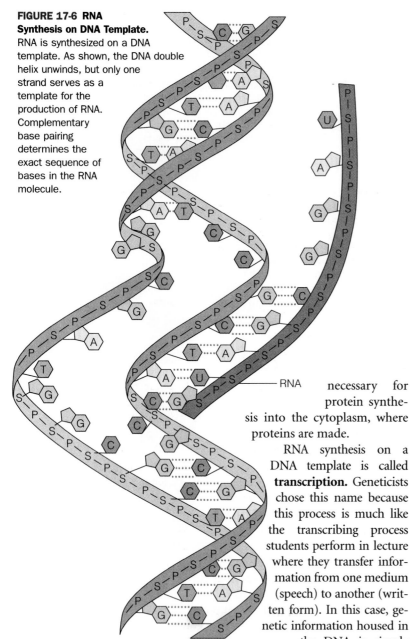

A strand of RNA produced from DNA template

RNA necessary for protein synthesis into the cytoplasm, where proteins are made.

RNA synthesis on a DNA template is called **transcription.** Geneticists chose this name because this process is much like the transcribing process students perform in lecture where they transfer information from one medium (speech) to another (written form). In this case, genetic information housed in the DNA is simply written in another form, RNA.

Transcription occurs during all three phases of interphase of the cell cycle at more or less the same rate. During transcription, small sections of the DNA helix unzip temporarily with the aid of special enzymes from RNA nucleotides. This creates a DNA template on which RNA can be made (**FIGURE 17-6**). But RNA is produced on only one of the DNA strands, and only a small portion (less than 1%) of a cell's DNA is used to make RNA. (This helps explain why most mutations in the DNA have no effect on cell function.)

During RNA synthesis, an enzyme called **RNA polymerase** helps align the RNA nucleotides on the DNA template in much the same way that DNA polymerase functions during DNA synthesis. RNA polymerase also catalyzes the formation of covalent bonds between ribose and phosphate groups, forming a polynucleotide chain. When the synthesis is complete, the RNA molecule is released from the DNA template, and the two strands of the DNA molecule reunite, reforming the double helix. RNA is then free to leave the nucleus.

During the synthesis of RNA, base pairing ensures the proper sequence of nucleotides on the RNA strand. **FIGURE 17-7** shows which bases pair up during RNA synthesis. The only difference between RNA and DNA synthesis is that adenine on the DNA template pairs with uracil.

Section 17-2

How Genes Work: Protein Synthesis

The information required to synthesize protein is transported out of the nucleus on one type of RNA, known as messenger RNA (mRNA). In the cytoplasm, mRNA serves as a template for protein synthesis, which may occur either on the endoplasmic reticulum or free within the cytoplasm, as noted in Chapter 3. The synthesis of protein on an mRNA template is called **translation** (**FIGURE 17-8**). In effect, the RNA message (the genetic code) is translated into a new "molecular language"—that of the protein.

Protein synthesis requires two additional "players," transfer RNA (tRNA) and ribosomes. **Transfer RNA** molecules are relatively small strands of RNA that bind to specific amino acids in the cytoplasm and transport them to the mRNA, where they are incorporated into

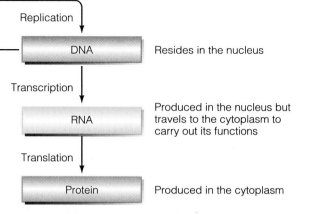

FIGURE 17-8 The Central Tenet of Molecular Genetics The DNA controls the cell through protein synthesis—that is, the production of enzymes and structural proteins. RNA serves as an intermediary, carrying genetic information to the cytoplasm, where protein is synthesized.

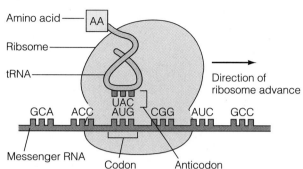

FIGURE 17-9 Messenger RNA The mRNA, either free within the cytoplasm or bound to the endoplasmic reticulum, serves as a template for protein synthesis. The codons, each consisting of three bases on the mRNA, determine the sequence of amino acids by binding to complementary anticodons on the tRNA molecules. Each tRNA, in turn, delivers a specific amino acid to the mRNA template. The ribosome provides binding sites for tRNA–AA molecules, catalyzes the formation of the peptide bonds, and slides down the mRNA template to permit chain elongation.

the protein (**FIGURE 17-9**). A tRNA molecule bound to an amino acid is generally written like this: tRNA–AA.

Ribosomes, mentioned briefly in Chapter 3, are composed of **ribosomal RNA** (**rRNA**) and protein and consist of two subunits, a small one and a large one. The subunits are often detached in the cytoplasm, joining for protein synthesis on the mRNA template.

Protein synthesis consists of three stages: (1) chain initiation, (2) chain elongation, (2) and chain termination. During chain initiation, the small subunit of the ribosome attaches to the mRNA at a specific site called the **initiator codon** (COE-dawn). The initiator codon is a sequence of three bases on the mRNA molecule that marks where protein synthesis should begin. After the small subunit attaches, the large subunit of the ribosome links up. As shown in **FIGURE 17-10A**, the ribosome now contains two binding sites for tRNA–AA.

Soon after the ribosome attaches to the mRNA, a tRNA bearing a specific amino acid enters the first binding site. But how does the cell know which amino acid to insert? At the base of all tRNA molecules is a sequence of three bases. Called an **anticodon,** it contains complementary bases that pair with the bases of the initiator codon of the mRNA template.

After the first tRNA–amino acid is in place, a second tRNA–AA enters the scene, inserting itself into the second binding site on the ribosome. The second binding site is located above the next codon, the next three bases on the mRNA template (**TABLE 17-3**).

In the example shown in **FIGURE 17-10**, the codon CGG binds to the tRNA with the anticodon GCC. This tRNA, in turn, binds to only one amino acid, arginine. The cell therefore ensures the proper sequence of amino acids through two mechanisms: (1) complementary base pairing between codons (mRNA) and anticodons

(tRNA), and (2) the specificity (matching) of tRNA molecules for (with) amino acids.[1]

Once the two amino acids are in place, the next step is to link them via a peptide bond. This occurs with the assistance of an enzyme in the ribosome (**FIGURE 17-10C**). After the peptide bond is formed, the first tRNA (minus its amino acid) leaves the binding site. It is then free to retrieve another amino acid for use later on. As illustrated in **FIGURE 17-10C**, the dipeptide formed during the first reaction is now attached to the second tRNA attached at the second binding site.

In order for the chain to grow, the ribosome must move down the mRNA strand. This is accomplished with the aid of a contractile protein in the ribosome. This protein permits the ribosome to slide along mRNA one codon at a time.

As illustrated in **FIGURE 17-10D**, after the first tRNA is released, the ribosome moves down the mRNA, and the dipeptide is shifted to the first binding site. This frees up the second site, permitting it to accept another tRNA–AA. A new tRNA–AA then enters the vacant site and is joined to the dipeptide, thus forming a tripeptide.

Chain elongation takes place by the addition of one amino acid at a time, but this does not mean that

---

[1]This process is a little more complicated than presented here, but it has been simplified for ease of comprehension. In reality, there are 64 codons that can be read from the mRNA, but only 40 distinct types of tRNA molecules. The reason is that some tRNA anticodons can pair with two or three different mRNA codons that specify the same amino acid.

FIGURE 17-10 Protein Synthesis

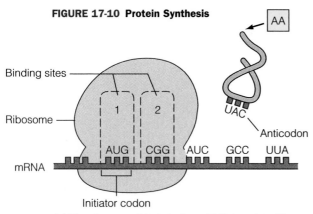

(a) The ribosome binds to the mRNA template. The tRNA binds to a specific amino acid in the cytoplasm.

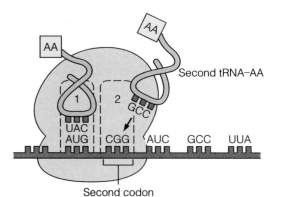

(b) The tRNA–amino acid (tRNA–AA) complex binds to the first codon and is held in place by the first binding site. A second tRNA–AA complex binds to the second codon.

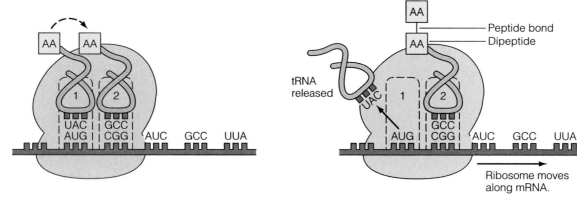

(c) The ribosome contains an enzyme that catalyzes the formation of a peptide bond between the two amino acids. The dipeptide is then attached to the second tRNA. This frees up the first tRNA, which vacates the first binding site.

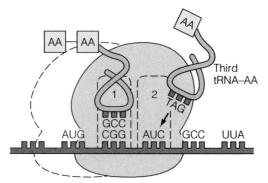

(d) The ribosome next slides down the mRNA, transferring the tRNA–dipeptide to the first binding site and opening up the second binding site to a third tRNA–amino acid.

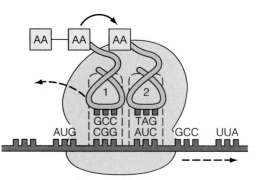

(e) The dipeptide is then linked by a peptide bond to the third amino acid, forming a tripeptide. This frees the tRNA in the first binding site. The ribosome next slides down one more codon, exposing the second binding site and freeing it up for the addition of another tRNA–AA. This process repeats itself until the terminator codon is reached.

TABLE 17-3

Codons on mRNA and Their Corresponding Amino Acids

Codon	Amino Acid	Codon	Amino Acid	Codon	Amino Acid	Codon	Amino Acid
AAU AAC	Asparagine	CAU CAC	Histidine	GAU GAC	Aspartic acid	UAU UAC	Tyrosine
AAA AAG	Lysine	CAA CAG	Glutamine	GAA GAG	Glutamic acid	UAA UAG	Terminator codon*
ACU ACC ACA ACG	Threonine	CCU CCC CCA CCG	Proline	GCU GCC GCA GCG	Alanine	UCU UCC UCA UCG	Serine
AGU AGC AGA AGG	Serine Argenine	CGU CGC CGA CGG	Arginine	GGU GGC GGA GGG	Glycine	UGU UGC UGA UGG	Cysteine Terminator codon* Tryptophan
AUU AUC AUA AUG	Isoleucine Methionine	CUU CUC CUA CUG	Leucine	GUU GUC GUA GUG	Valine	UUU UUC UUA UUG	Phenylalanine Leucine

*Terminator codons signal the end of the formation of a polypeptide chain.

protein synthesis is a slow process. Quite the contrary, protein synthesis occurs with remarkable speed. In bacteria, many proteins containing numerous amino acids are synthesized in 15–30 seconds. In humans, the large alpha and beta chains of the very large hemoglobin molecule are synthesized in 3 minutes.

As the peptide chain is formed, hydrogen bonds between amino acids on different parts of the chain cause it to bend and twist, forming the secondary structure of the protein or peptide, described in Chapter 3 (FIGURE 17-11). When the ribosome reaches the terminator codon, the sequence of bases that signals the end of the protein chain, the protein is released. If the protein is produced on mRNA on the surface of the rough endoplasmic reticulum, it is transferred into its interior. The protein may then be chemically modified and packaged into transport vesicles, which pinch off and transport their cargo to the Golgi complex. If protein synthesis occurs on mRNA molecules in the cytoplasm, the protein is released into the cytoplasm.

Protein synthesis is quite involved and perhaps a little confusing at first glance, but once you've mastered it, you'll probably agree: It's one of the most fascinating processes you've ever encountered.

FIGURE 17-11 Protein Synthesis As the protein is synthesized on the mRNA, it begins to coil and bend, forming its secondary structure. Several ribosomes may "work" a single strand of mRNA simultaneously.

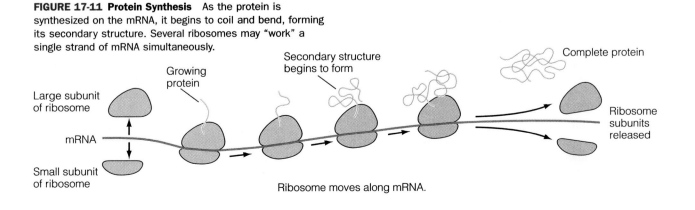

Scientific Discoveries that Changed the World 17-1

Unraveling the Mechanism of Gene Control

Featuring the Work of Jacob and Monod

One of the questions that intrigued geneticists for many years—and still does—is how the genetic information contained in a cell's DNA is controlled. Many experiments have been performed over the years in an at-

FIGURE 1 Jacob and Monod These two French scientists proposed the operon hypothesis.

tempt to answer this question, but none so important as those of two French molecular geneticists, Francois Jacob and Jacques Monod, on the common intestinal bacterium *E. coli* (**FIGURE 1**). Today, much of what is known about gene regulation in bacteria and most of the terms that describe the process come from their studies.

Working at the Pasteur Institute in the late 1950s, Jacob and Monod focused much of their attention on the bacterial uptake and breakdown of lactose, a disaccharide also known as milk sugar. *E. coli* absorb lactose and catabolize it with the aid of an enzyme known as galactosidase (ga-lack-TOSE-uh-DACE). The catabolic breakdown of lactose results in the release of energy and carbon atoms useful in bacterial biosynthesis.

E. coli living in a lactose-free medium, however, contain very little galactosidase. It is only when lactose is added that the enzyme is synthesized.

The introduction of lactose also increases the production of a carrier molecule (galactoside permease) that resides in the plasma membrane and transports lactose into the bacterium. In addition, the presence of lactose increases the intracellular concentration of another enzyme, which may play a role in lactose catabolism, but which is not essential to the process. The carrier protein and two enzymes are part of an inducible enzyme system—that is, their production is induced by the presence of lactose.

Mapping studies have shown that three adjacent genes on the bacterial DNA are responsible for the produc-

---

Section 17-3

Controlling Gene Expression

Hard as it is to imagine, there's yet another fascinating bit of molecular wizardry involved in controlling genes—that is, in controlling the production of messenger RNA and protein synthesis. This section examines this process.

After decades of research, molecular geneticists have a remarkably good understanding of the molecular mechanism responsible for gene regulation, although it is still not completely understood. But why, you ask, is this so important to know?

Besides the fact that it's likely to be covered on your next exam, understanding the gene control is important because it could help scientists discover new ways to treat, or even cure, diseases such as cancer that originate in the genes of an organism. To begin this study, we first look to the bacteria. Insight gained here will provide a basis for understanding human gene regulation.

Bacterial Genes Consist of Three Interdependent Segments of DNA: the Regulator Gene, Control Regions, and Structural Genes

As explained in Scientific Discoveries 17-1, two French researchers, Francois Jacob and Jacques Monod, worked out the details of bacterial gene control based on their studies of a common intestinal bacterium with a tongue-twisting name, *Escherichia coli* (ESH-ur-EE-shee-ah KOLL-eye), or *E. coli* for short (**FIGURE 17-12**). *E. coli* inhabits the large intestine of humans and other mammals, where it lives off undigested and unabsorbed food matter—not a glamorous existence for an organism that has contributed so much to science and our well-being.

Jacob and Monod's research allowed them to develop a model of bacterial gene control. It states that bacterial genes consist of three parts: structural genes, regulator genes, and control regions. (I use the letters SRC to remember them.)

tion of these proteins. These studies led Jacob and Monod to hypothesize that the three genes belong to a single unit, which they called an *operon*. They defined an operon as a cluster of genes with related functions that is regulated in such a way that all the genes in the group are activated and inactivated simultaneously.

In a paper published in 1961, the researchers proposed a mechanism by which these genes might be controlled, creating the model you studied in this chapter. This intriguing model is "one of the truly important conceptual advances in biology," according to University of Wisconsin cell biologist Wayne Becker. The bulk of the evidence in support of the model came from genetic analysis of mutant bacteria.

Extensive studies of the mutants and how they responded to lactose helped Jacob and Monod piece together a picture of gene regulation in *E. coli*. In these studies of the inducible *lac* operon, the researchers found that structural gene mutations resulted in an alteration of one specific protein. If the bacterium had a mutation in the *y* gene, one of the structural genes, the permease molecule would not be produced when the inducer was added to the system. Galactosidase was formed, but very little lactose could enter the cell. If it had a mutation in the *z* gene (another structural gene), galactosidase would not be produced, but permease would be. The bacterial cell could absorb lactose; it just could not do much with it.

In contrast, mutations in the regulatory regions affected all of the structural genes and proteins they produced. A defective operator site, which binds the repressor, for example, resulted in the production of galactosidase and permease whether or not the inducer, lactose, was present. (This response was due to the fact that RNA polymerase was free to move down the *lac* operon.)

Bacteria with a mutation in the regulator gene, the segment that codes for the production of mRNA that gives rise to the repressor protein, also produced galactosidase and permease. They did so because this mutation blocked all repressor protein production, letting RNA polymerase wander freely down the *lac* operon, transcribing the genes.

Data from these and other mutants helped the scientists piece together a model. As a testimony to the thoroughness of their work, the original model has undergone very little change in over 30 years.

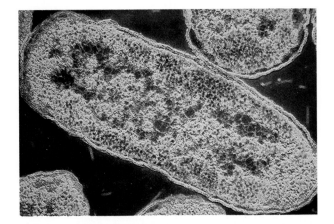

FIGURE 17-12 *E. coli* A micrograph of the bacterium *E. coli*, a common inhabitant of the large intestine of mammals.

Structural genes, shown in light blue in **FIGURE 17-13**, code for the production of mRNA that is used to synthesize enzymes and other proteins. **Regulator genes** are shown in a slightly darker shade of blue.

They and the **control regions** are shown in yellow and green. In the model proposed by Jacob and Monod, the structural genes do the work—that is, produce the mRNA—and the regulator genes and control regions, as their names imply, govern the operation of the structural genes. It's a little like a government work crew, with one set of workers and two sets of supervisors.

Jacob and Monod's research led them to hypothesize that the control regions (yellow and green) and structural genes (light blue) responsible for producing the enzymes of a single metabolic pathway were located side by side on the chromosome (**FIGURE 17-13**). Together, the structural genes and the control regions form a functional unit known as an **operon** (OP-ur-on). The regulator genes are located some distance from the operon. Using our analogy of government workers, the regulator gene is the big boss, who stays in the office downtown. The control region is the job foreman.

Based on the results of many experiments, Jacob and Monod proposed two types of operons: inducible and repressible. Let's look at each one in more detail.

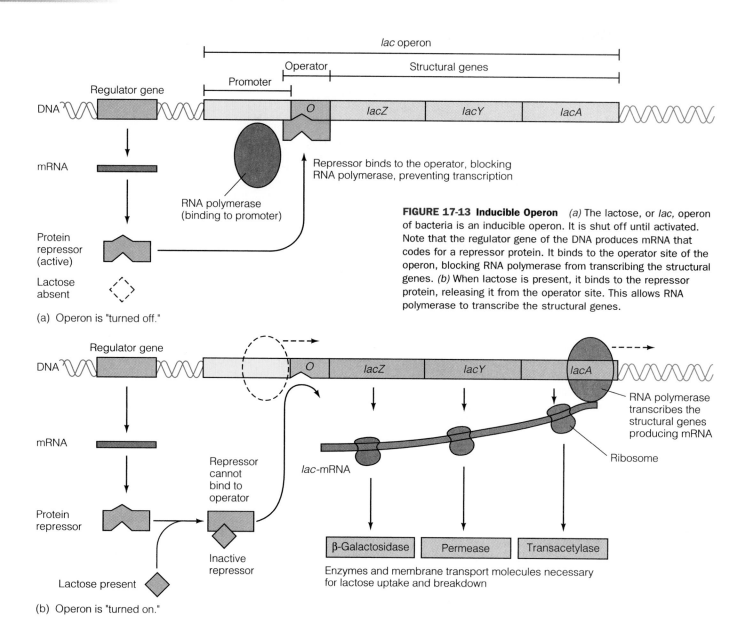

FIGURE 17-13 Inducible Operon *(a)* The lactose, or *lac,* operon of bacteria is an inducible operon. It is shut off until activated. Note that the regulator gene of the DNA produces mRNA that codes for a repressor protein. It binds to the operator site of the operon, blocking RNA polymerase from transcribing the structural genes. *(b)* When lactose is present, it binds to the repressor protein, releasing it from the operator site. This allows RNA polymerase to transcribe the structural genes.

Inducible Operons Are Switched On When Enzymes Are Needed.

Most people don't turn on lights in a room unless they're needed. Many bacterial genes operate by the same efficiency principle. For example, although *E. coli* is capable of producing many enzymes needed to digest leftover foodstuffs in our intestines, the organism only produces these enzymes when they're needed. That is, it produces them on demand. It does so by activating inducible operons. An **inducible operon** is a set of structural genes and accompanying control regions that remains off until activated or induced.

The operon we will examine to understand induction is called the **lac operon.** It is responsible for the production of three different proteins associated with the uptake and metabolism of lactose (milk sugar) in *E. coli.*

When lactose is absent, this operon is inactive. (Like a light bulb in a room, it is kept off until needed.) When lactose is present in the digestive system, small quantities diffuse into the bacteria. These molecules activate the *lac* operon. Lactose is therefore said to be an inducer. **Inducers** activate the operon, which then produces mRNA that acts as a template for the production of proteins involved in the transport and metabolism of lactose.

Repressible Operons Provide a Steady Supply of Enzymes and Other Proteins and Remain On Until Switched Off.

Bacteria require a steady supply of some enzymes to produce molecules, such as amino acids, necessary for cellular metabolism. During evolution, another strategy for controlling genes evolved to meet this demand. It is the repressible operon. Unlike the inducible operon, the **repressible operon** remains active unless turned off. The repressible operon is like the inducible operon. In repressible operons, however, end products of chemical reactions build up inside cells, turning off the genes that control the production of mRNA needed to make proteins that catalyze chemical reactions in question.

■ In Humans, Genetic Expression Is Controlled at Four Levels

Humans are a bit more complicated than your average bacterium and, as you might expect, the control of genes in human cells is also a bit more involved. Although inducible and repressible systems are present in human cells, most human genes are inducible. But several other mechanisms are also present. For simplicity, these can be divided into four major categories: (1) control at the chromosome level, (2) control at transcription, (3) control after transcription before translation, and (4) control at translation (**FIGURE 17-14**).[2]

Control at the Chromosome: Access to the Genes Is Controlled by Coiling and Uncoiling of the Chromosomes During Interphase.

Chapter 15, you may recall, noted that chromatin fibers in the nucleus condense and become inactive during prophase of mitosis. In the condensed state, they cannot produce RNA or new DNA. Condensation essentially shuts down the DNA as the cell prepares for nuclear division.

Cells also selectively inactivate some of their chromatin during interphase. Why? Most cells have far more DNA than they need. Muscle cells, for instance, don't need all of the genes that a liver cell needs, and vice versa. But since it can't discard this unnecessary DNA, it inactivates it.

Inactivated chromatin in the interphase nucleus is tightly coiled, or compacted, and appears as dark clumps called **heterochromatin.** Some heterochromatin re-

---

²Bacteria also control gene expression at these levels.

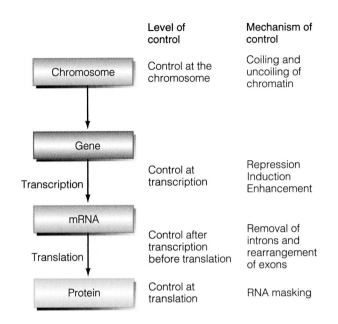

Level of control	Mechanism of control
Control at the chromosome	Coiling and uncoiling of chromatin
Control at transcription	Repression Induction Enhancement
Control after transcription before translation	Removal of introns and rearrangement of exons
Control at translation	RNA masking

FIGURE 17-14 Gene Expression In humans, gene expression is regulated at four levels.

mains in this state throughout the cell cycle, except during replication (S phase). The remaining chromatin in the interphase nucleus is diffusely organized—consisting of fine threads unraveled in the nucleus. It is called **euchromatin** ("true chromatin"). Euchromatin is metabolically active.

Chromosomal condensation or coiling provides a crude way of regulating genetic expression. More precise controls occur at other levels.

Three Control Mechanisms Operate at the Level of Transcription: Induction, Repression, and Enhancement.

In humans, transcription (RNA production) can be controlled by the induction and repression of operons, similar to that which occurs in bacteria. Human cells also control gene activity via a third mechanism, known as enhancement.

Enhancement is a process in which already active genes increase their production of mRNA. This occurs as a result of the action of nearby segments of the DNA called **enhancers.** Research suggests that certain protein molecules bind to enhancer regions of the chromosome. These proteins, in turn, increase RNA production by the structural gene. Thus, enhancers do not turn nearby structural genes on and off like inducers or corepressors, they simply amplify their activity. An enhancer is therefore a little like the accelerator in your car—it doesn't turn the engine on, it simply speeds it up.

Genetic Expression Can Be Controlled by Altering the Structure of mRNA.

Genetic studies show that eukaryotic DNA contains two types of genetic material: sections that are used to produce mRNA and sections that are not. As **FIGURE 17-15** shows, the noncoding segments of DNA are called **introns** and are interspersed within the functional segments of DNA (functional genes), the **exons.** The name *intron* signifies that these are *inter*vening segments of noncoding DNA. Exons are *ex*pressed segments—sections of the DNA that can be used to produce the mRNA that is used to make protein.

Both the introns and exons are transcribed during the cell cycle, producing an RNA transcript that is a complete "readout" of the DNA. Eukaryotic cells process the transcript, cutting out the introns, then join the exons together to produce a functional mRNA molecule. Although this seems like a lot of work, it is necessary because in humans and other eukaryotes the structural genes that code for the enzymes of a metabolic pathway are not lined up next to one another as they are in bacteria. In order for a functional mRNA molecule to be synthesized, it must be "cut and pasted" together.

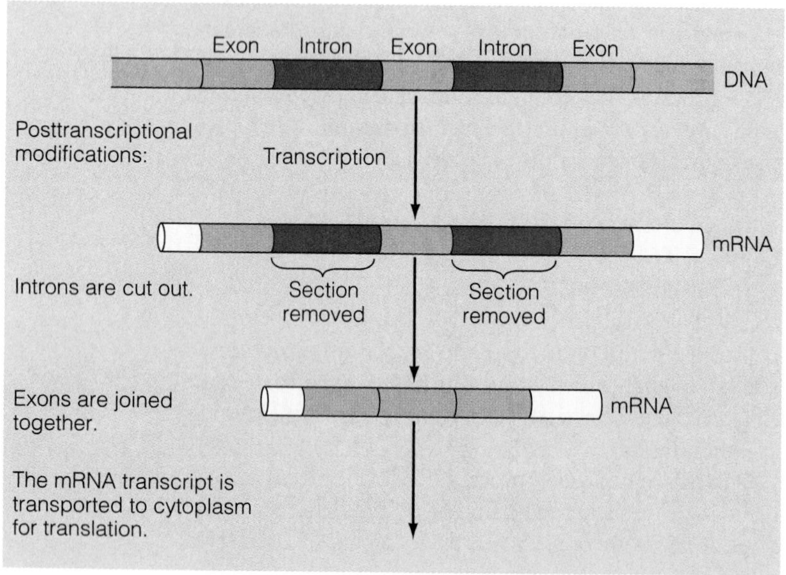

FIGURE 17-15 Posttranscriptional Control of Human Genes One key mechanism of control occurs after transcription. DNA contains noncoded segments (introns) and useful segments (exons). An mRNA transcript is produced from the DNA, and the segments produced by introns are removed. The final product is a strand of mRNA containing only RNA copies of the exons. The exon copies can be linked differently, producing slightly different products.

Control at Translation Occurs by Masking mRNA.

Eukaryotic cells can also control gene expression at the level of translation (protein synthesis). Messenger RNA, for example, which is produced in the nucleus, may be transferred to the cytoplasm in an inactive state. The mRNA is said to be **masked.** Masking allows the cell to build up large supplies of mRNA in preparation for a sudden burst of protein synthesis. When the time is right, the mRNA is activated and begins producing large quantities of protein. This phenomenon occurs in the human ovum, the female sex cell produced in the ovaries. The ovum produces an enormous amount of masked mRNA as it awaits fertilization. When the sperm arrives and fertilizes the ovum, the mRNA that has accumulated in the cytoplasm is unmasked, and there is an explosion of protein synthesis necessary for subsequent cell divisions.

In closing, gene expression in humans and other eukaryotes is a complex phenomenon. Like the cell cycle, it is no doubt under the influence of many controlling factors.

Section 17-4

Applications of Biology: Oncogenes—The Seeds of Cancer Within Us

Chapter 15 introduced you to the disease called *cancer* and noted that it was caused by numerous biological, chemical, and physical agents, known as **carcinogens.** It also noted that most cancers arise from mutations, alterations of parts of the genetic material. Broadly defined, mutations include (1) alterations of the DNA, such as the deletion or addition of a purine or pyrimidine base; (2) alterations of the chromosome, such as missing segments; and (3) alterations in the chromosome number—that is, too many or too few chromosomes.

Mutations may occur in germ cells, the sperm and ovum, or in body cells. Germ cell mutations may result in genetically defective offspring or cancers in early life. This section concerns itself with mutations in body cells, or somatic cells, that involve changes in the structure of the DNA.

Mutations in the DNA (1) can improve cellular function; (2) may be neutral, having neither a positive nor a negative effect; or (3) may affect vital sections of the genome, killing the cell outright or releasing the cell on the rampage of growth called cancer. It is the mutations that lead to cancer that will concern us here. In most instances, mutations in the DNA of somatic cells

FIGURE 17-16 Hazardous Wastes The careless dumping of hazardous materials in rivers and lakes, in abandoned fields, along highways, and in abandoned warehouses creates a health hazard to humans and many other species and results in costly cleanup efforts. Thankfully, new laws have put tighter controls on hazardous waste disposal, although some industries and waste disposal companies still violate the law.

are rapidly repaired by enzymes in the nucleus. This helps ensure cellular homeostasis. Despite the presence of remarkably efficient repair mechanisms, some somatic mutations simply escape repair.

Researchers have made a startling discovery about such mutations that may explain the cause of most cancers. They have found that humans and other organisms contain a group of genes called **proto-oncogenes** (pro-toe-ON-co-JEANS). When mutated and not repaired, these genes unleash the cell from normal growth controls.

Proto-oncogenes control functions related to cellular replication, among them cell adhesion and the production of the plasma membrane receptors that bind to growth factors or hormones. When mutated, these genes produce uncontrolled cellular proliferation, cancer. (For a discussion of methods used to assess carcinogenicity, see the Point/Counterpoint.)

Certain viruses also cause cancer. Cancer-causing viruses fall into two groups. Some viruses stimulate cancer by activating proto-oncogenes in body cells. Others contain cancer-causing genes themselves, which they insert into the DNA of the nuclei of infected cells. These genes are known as **oncogenes.** After being inserted into a cell's DNA, oncogenes turn on the cell's genes that control division.

The discovery of proto-oncogenes and viral oncogenes has resulted in a quantum leap in our understanding of cancer. Some scientists believe that the key to treating cancer may ultimately lie in finding ways to turn off activated proto-oncogenes, thus ending the cancerous proliferation of cells. Who knows, maybe someday you or someone you know will benefit directly from this idea. (For more on cancer treatment, see Health Note 15-2.)

Health, Homeostasis, and the Environment

Hazardous Wastes and Mutations

Thousands of mutations occur in each cell of your body every day, many of which could lead to serious problems, including cancer. As noted earlier, the cells of your body fortunately repair much of the damage. This repair is brought about by special enzymes in the nucleus that "snip off" damaged sections of the DNA, then rebuild the molecule, thus helping to ensure normal cell function.

Remaining healthy, though, requires that we not stretch that resiliency to the breaking point. And yet some individuals fear that modern industrial society may be doing just that. Today, 60,000 chemical substances are in commercial use. The National Academy of Sciences notes that few of these substances have been adequately tested for their ability to cause mutations, cancer, and birth defects.

The extensive use of chemicals and the widespread publicity over the ill effects of some have created a chemical paranoia—sometimes referred to as chemophobia—in our society. New research suggests that while the threat is real, many people are overreacting to it. In fact, numerous studies suggest that naturally occurring chemicals in our food, like one type of mold that grows on peanut butter, probably cause more cancer than pesticide residues. (For a debate over chemical contamination and the relative importance of pesticides and other industrial chemicals on human health, see the Point/Counterpoint in Chapter 18.)

Some critics argue, however, that the debate over human health is misleading, for it ignores the potentially widespread effects of toxic chemicals on plants, animals, and microorganisms that share this planet with us. Nowhere has the impact on humans and other organisms been more noticeable than around toxic waste dumps, where for decades companies have disposed of highly dangerous substances, often in cardboard containers or steel barrels that rust within a few years of burial (**FIGURE 17-16**).

Point/Counterpoint

Are Current Procedures for Determining Carcinogens Valid?

ANIMAL TESTING FOR CANCER IS FLAWED

Philip H. Abelson

Philip Abelson is the Deputy Editor of Science. *This essay is excerpted from "Testing for Carcinogens with Rodents,"* Science 249: September 21, 1990. Copyright © 1990 by the American Association for the Advancement of Science.

The principal method of determining potential carcinogenicity of substances is based on studies where huge doses of chemicals are administered daily to inbred rodents for their lifetime. Then by questionable models, which include large safety factors, the results are used to extrapolate the effects of minuscule doses in humans. Resultant stringent regulations and attendant frightening publicity have led to public anxiety and chemophobia. If current ill-based regulatory levels continue to be imposed, the cost of cleaning up phantom hazards will be in the hundreds of billions of dollars with minimal benefit to human health. In the meantime, real hazards are not receiving adequate attention.

The current procedures for gauging carcinogenicity are coming under increasing scrutiny and criticism. A leader in the examination is Bruce Ames, who with others has amassed an impressive body of evidence and arguments. Ames and Gold summarized some of their recent data and conclusions in *Science* (31 August 1990, p. 970). Three articles in the *Proceedings of the National Academy of Sciences* provide an elaboration of the information with extensive bibliographies. The articles also provide data about other pathologic effects of natural chemicals.

A limited number of chemicals tested, both natural and synthetic, react with DNA to cause mutations. Most chemicals are not mutagens, but when the maximum tolerated dose (MTD) is administered daily to rodents over their lifetime, about half of the chemicals give rise to excess cancer, usually late in the normal life span of the animals. Experiments in which synthetic industrial chemicals were administered in the MTD to both rats and mice resulted in 212 of 350 chemicals being labeled as carcinogens. Similar experiments with chemicals naturally present in food resulted in 27 of 52 tested being designated as carcinogens. These 27 rodent carcinogens have been found in 57 different foods, including apples, bananas, carrots, celery, coffee, lettuce, orange juice, peas, potatoes, and tomatoes. They are commonly present in quantities thousands of times as great as the synthetic pesticides.

The plant chemicals that have been tested represent only a tiny fraction of the natural pesticides. As a defense against predators and parasites, plants have evolved a large number of chemicals that have pathologic effects on their attackers and consumers. Ames and Gold estimate that plant foods contain 5,000–10,000 natural pesticides and breakdown products. In cabbage alone, some 49 natural pesticides have been found. The typical plant contains a total of a percent or more of such substances. Compared with the amount of synthetic pesticides we consume, we eat about 10,000 times more of the plant pesticides.

It has long been known that virtually all chemicals are toxic if ingested in sufficiently high doses. Common table salt can cause stomach cancer. Ames and others have pointed out that high levels of chemicals cause large-scale cell death and replacement by division. Dividing cells are much more subject to mutations than quiescent cells. Much of the activity of cells involves oxidation, including formation of highly reactive free radicals that can react with and damage DNA. Repair mechanisms exist, but they are not perfect. Ames has stated that oxidative DNA damage is a major contributor to aging and to cancer. He points out that any agent causing chronic cell division can be indirectly mutagenic because it increases the probability of DNA damage being converted to mutations. If chemicals are administered at doses substantially lower than MTD, they are not likely to cause elevated rates of cell death and cell division and hence would not increase mutations. Thus, a chemical that produces cell death and cancer at the MTD could be harmless at lower dose levels.

Diets rich in fruits and vegetables tend to reduce human cancer. The rodent MTD test that labels plant chemicals as cancer-causing in humans is misleading. The test is likewise of limited value for synthetic chemicals. The standard carcinogen tests that use rodents are an obsolescent relic of the ignorance of past decades. At that time, extreme caution made sense. But now tremendous improvements of analytical and other procedures make possible a new toxicology and far more realistic evaluation of the dose levels at which pathological effects occur.

CURRENT METHODS OF TESTING CANCER ARE VALID

Devra Davis

The vast majority of the scientific community endorses the conduct of experimental studies in order to try to identify those materials that could cause disease and prevent harmful exposures. In the typical toxicologic study of 50 rodents, each animal is a stand-in for 50,000 people. Because rodents live about 2 years on average, they are usually exposed to amounts of the suspect agent that approximate what a human would encounter in an average lifetime of 70 years.

Some have argued that the use of the maximum tolerated dose (MTD) in these studies produces tissue damage and cell proliferation, which, in turn, lead to cancer. This is biological nonsense that ignores a fundamental characteristic of cancer biology that has been known for more than 50 years: cancer is a multi-stage disease, with multiple causes, which arises by a stepwise evolution that involves progressive genetic changes, cell proliferation, and clonal expansion. Thus, swamping of tissues with high doses alone may well kill an animal or damage its tissues, but high doses alone are not sufficient to cause cancer. A 2-year study at the National Institute of Environmental Health Sciences (NIEHS) by David Hoel and colleagues provides good evidence on this point. They looked for signs of damage in tissues taken from rats and mice used in cancer studies. Cancers occurred in organs that did not show apparent damage, and some damaged organs were completely free of tumors.

In addition, most compounds tested do not cause cancer only in the highest dose group tested but typically produce a dose-response relationship, where the amount of cancer developed is proportional to the dose administered. Some chemicals cause toxicity only, others cause only cancer. Not all of those that cause cancer do so through organ toxicity. In fact, almost 90% of the substances shown to cause cancer in the National Toxicology Program do so without producing any increased cellular toxicity, and they also cause cancer at both lower and higher doses.

Animal studies are evolving and being further refined, as is our understanding of differences between species that need to be taken into account in conducting these studies. Every compound known to cause cancer in humans also produces cancer in animals, when adequately tested. For 8 of the 54 known human cancer-causing agents, evidence of carcinogenicity was first obtained in laboratory animals; in many cases, the same target organs and doses have been involved in producing cancer in both animals and humans.

The NIEHS has carried out nearly 400 long-term rodent studies, which have been published following peer review by specialists in the field. Chemicals nominated for testing usually represent a sample of potentially "problematic" materials. About 40% of the "suspect" pesticides evaluated to date have been found to cause cancer. As to the role of so-called natural pesticides, rodent diets are also loaded with many of these materials. Nevertheless, a number of test compounds added to these diets markedly increase the amount of tumors produced. Thus, animals are more sensitive to certain synthetic compounds than to the background level of natural materials. Humans are also likely to have acquired some resistance to these natural materials through evolution.

To date, only about 20% of all synthetic organic chemicals have been adequately tested for their potential human toxicity. Those who must set public policy on the use of chemicals need a rational basis on which to stake their actions. Continued advances in animal testing provide an important contribution to environmental health sciences and to public health efforts to predict, and then prevent, the development of environmentally caused disease.

In summary, the current system should be used until it can be replaced by a demonstrably better one. There is no scientific basis for rejecting the MTD as capable of inducing cancer.

Devra Davis *is a senior adviser in the office of the Assistant Secretary for Health, Department of Health and Human Services.*

SHARPENING YOUR CRITICAL THINKING SKILLS

1. Summarize each author's main points and supporting data. Do you see any inconsistencies or examples of faulty logic in either essay? If so, where?

2. Why is it that two scientists can disagree on an issue such as this?

3. Given the disagreement, what course of action would you recommend?

Visit *Human Biology's* Internet site, www.jbpub.com/humanbiology, to research opposing web sites and respond to questions that will help you clarify your own opinion. (See Point/Counterpoint: Furthering the Debate.)

Love Canal in the city of Niagara Falls, New York, was the scene of one of the nation's worst toxic nightmares. In an abandoned canal, Hooker Chemical Company dumped over 20,000 metric tons of highly toxic and carcinogenic wastes from 1947 to 1952.[3] In 1952, the city of Niagara Falls began condemnation proceedings on the site. But their intention was not to shut down the operation. It was a legal maneuver to seize the land to permit the construction of a school and housing. Under pressure from the city, Hooker sold the land to the city for $1 in exchange for a release from future liability.

A few years later, bulldozer operators preparing the site for construction of the school removed the protective clay lid that Hooker had placed over the site, but they didn't remove the hazardous wastes. In the late 1950s, after the construction was complete, rusty barrels began to work their way to the surface. Toxic chemicals oozed out of the site, killing trees and gardens and causing chemical burns in children. Some children even died (**FIGURE 17-17**).

The problem came to a head in the 1970s. After a period of heavy rainfall, toxic wastes began to leak into basements of local residents, and the chemical stench became unbearable. Over 80 different chemical substances turned up in studies of water, air, and soil. Many of the chemicals were known or suspected carcinogens. A New York State Health Department study showed that one of every three pregnant women who lived in the area had miscarried. The miscarriages could have been caused by chemically induced mutations in the early embryos or in the germ cells of the parents. Birth defects were present in one of every five children, far in excess of the expected rate. Birth defects may have resulted from chemically induced mutations.

The chemical substances emitted from the old dump site irritated lungs, gave residents headaches, and resulted in convulsions in some people. Genetic studies showed mutations in the chromosomes of residents.

Nearly 1000 families were evacuated from the vicinity over the years, and now many of the homes sit idle, boarded up, in solemn tribute to human carelessness. New laws and concerted efforts by industries have cut back on the reckless disposal of hazardous wastes.

Cleaning up past mistakes like Love Canal has proved costly. Over $40 million has been spent so far to clean up the Love Canal area. What is even more startling, though, is that some experts estimate that there are nearly 10,000 hazardous waste sites, some worse than Love Canal, in the United States in need of cleanup. Government estimates put the cost of cleaning up the mess at $100 billion. Critics believe that the actual cost will be far greater. Adding to the problem however, are the more than 230 million metric tons of hazardous materials that American industry generates annually—nearly a ton of hazardous waste for every man, woman, and child each year!

Improper hazardous waste disposal now pollutes groundwater in many areas. That concerns public health officials because half the people in the United States get their drinking water from wells. By one estimate, more than 10 million Americans drink tap water contaminated with pollutants in excess of EPA standards. Because of years of careless waste disposal, groundwater contamination is expected to grow worse. Even if we stop polluting now, the problem will linger for many years.

A healthy population requires clean water. But how do we get it? Many changes are needed. Especially important are ways to reduce hazardous waste production. By redesigning chemical processes, for example, manufacturers can make significant reductions in their waste output. They can also reuse and recycle hazardous wastes. A waste product from one process may actually become the raw material for another process. These steps are preventive and can help avoid the problems of disposal altogether.

Individuals can also help reduce hazardous waste by reducing unnecessary consumption. Environmentally safe cleaning products, for example, help reduce pollution. Individual actions can have a profound effect on the production of toxic waste and the health of the environment and us. Using energy and other resources more frugally and recycling household products, paper, aluminum, glass, and plastics can also make substantial inroads into hazardous waste production. The health of the planet and the health of people depends on it.

---

[3]A metric ton is 2240 pounds.

FIGURE 17-17 A Victim of Love Canal

SUMMARY

DNA AND RNA: MACROMOLECULES WITH A MISSION

1. *DNA* houses the genetic information. DNA is a molecule that consists of two polynucleotide strands joined by hydrogen bonds between purine and pyrimidine bases on the opposite strands. The strands produce a *double helix.*

2. Each nucleotide in the DNA molecule consists of a purine or pyrimidine base, the sugar deoxyribose, and a phosphate group. The nucleotides are joined by covalent bonds between phosphate groups and deoxyribose molecules.

3. *Complementary base pairing* is an unalterable coupling in which adenine on one strand of the DNA molecule always binds to thymine on the other and guanine always binds to cytosine.

4. Complementary base pairing ensures accurate replication of the DNA and accurate transmission of genetic information from one cell to another and from one generation to another.

5. To replicate, DNA must first unwind with the aid of a special enzyme. After unwinding, the strands provide templates for the production of complementary DNA strands, a process aided by the enzyme *DNA polymerase.*

6. The cell contains three types of *RNA:* transfer RNA, ribosomal RNA, and messenger RNA. All play a role in the synthesis of protein in the cytoplasm.

7. All three RNA molecules are single-stranded polynucleotide chains. RNA nucleotides contain the sugar ribose instead of deoxyribose, which is found in DNA. RNA nucleotides contain four bases: adenine, guanine, cytosine, and uracil (instead of thymine).

8. RNA is synthesized in the nucleus on a template of DNA. The synthesis of RNA is called *transcription.*

HOW GENES WORK: PROTEIN SYNTHESIS

9. The genetic information contained in the DNA molecule is transferred to *messenger RNA.* Messenger RNA molecules carry this information to the cytoplasm, where proteins are synthesized.

10. Messenger RNA serves as a template for protein synthesis. Ribosomes are required to produce proteins on the mRNA template.

11. *Transfer RNA* molecules deliver amino acid molecules to the mRNA and insert them in the growing chain. Each tRNA binds to a specific amino acid and delivers it to a specific *codon,* a sequence of three bases on the mRNA. Thus, the sequence of codons determines the sequence of amino acids in the protein.

12. Proteins are synthesized by adding one amino acid at a time.

13. During protein synthesis, the ribosome first attaches to the mRNA at the *initiator codon.* Soon after, the large subunit attaches. A specific tRNA bound to an amino acid binds to the initiator codon and the first binding site of the ribosome. A second tRNA–amino acid then enters the second site.

14. An enzyme in the ribosome catalyzes the formation of a peptide bond between the two amino acids. After the bond is formed, the first tRNA (minus its amino acid) leaves the first binding site.

15. The ribosome moves down the mRNA, shifting the tRNA bound to its two amino acids to the first binding site and opening the second site for another tRNA–amino acid. This process repeats itself many times in rapid succession.

16. As the peptide chain is formed, hydrogen bonds begin to form between the amino acids, and the chain begins to bend and twist, forming the secondary structure of the protein or peptide. When the ribosome reaches the terminator codon, the peptide chain is released.

CONTROLLING GENE EXPRESSION

17. Bacterial DNA contains functional units called *operons,* which consist of three parts: structural genes, regulator genes, and control regions. *Structural genes* code for the production of enzymes and other proteins and are controlled by the *regulator genes* and the *control regions.*

18. Two types of operons are present: inducible and repressible.

19. *Inducible operons* are switched off until needed. In an inducible operon, the regulator gene produces a *repressor protein.* The repressor protein binds to the *operator site* next to the structural genes and prevents the transcription of the structural genes by blocking RNA polymerase. When an inducer substance is present, it binds to the repressor, causing it to release from the operator site and permitting the transcription of the DNA.

20. *Repressible operons* are generally continuously activated but can be switched off by the presence of a chemical substance, usually an end product of a metabolic pathway.

21. Gene control in humans occurs at four levels: at the chromosome, at transcription, after transcription but before translation, and at translation.

22. At the chromosome: Condensation, or coiling, of the chromatin fibers inactivates genes. How the cell condenses a segment of the chromatin is not clearly understood.

23. At transcription: Research shows that inducible and repressible systems are present in human cells. Research also shows that certain segments of the DNA, called *enhancers,* can greatly increase the activity of nearby genes. Geneticists think that protein molecules bind to enhancer regions of the chromosome and increase gene activity.

24. After the production of RNA before translation: Human DNA contains far more genetic material than it needs. Those segments of the DNA used to produce the RNA that will be used to make protein are called *exons.* The intervening segments of noncoding DNA are called *introns.*

25. Both the introns and exons are transcribed during the cell cycle. The cell, however, removes the RNA from the introns and joins the RNA fragments produced by exons to create a functional mRNA molecule.

26. The exon-produced RNA can be spliced together in different ways; the resulting mRNAs produce different proteins.

27. At translation: Messenger RNA may be transferred to the cytoplasm in an inactive, or masked, state. Masking permits the cell to build up large supplies of mRNA in preparation for a sudden burst of protein synthesis.

28. Most cancers arise from mutations. Mutations include three basic changes:

(a) alterations of the DNA itself, (b) alterations of the chromosome, and (c) alterations in the chromosome number.

APPLICATIONS OF BIOLOGY: ONCOGENES—THE SEEDS OF CANCER WITHIN US

29. Humans and other organisms contain specific genes, called *proto-oncogenes*, that when mutated lead to cancerous growth. Proto-oncogenes are normal genes that code for cellular structures and functions, such as cell adhesion, mitotic proteins, and the production of plasma-membrane receptors for growth factors or hormones.

30. Chemical, physical, and biological agents can mutate these genes, resulting in uncontrolled cellular proliferation (cancer).

31. Viruses can also cause cancer. Some viruses, for example, possess *oncogenes*, which enter human body cells and are incorporated in the cell's genome, stimulating uncontrolled cellular division. Other viruses carry genes that stimulate cancer by activating human proto-oncogenes.

HEALTH, HOMEOSTASIS, AND THE ENVIRONMENT: HAZARDOUS WASTES AND MUTATIONS

32. Numerous chemicals are released into the environment from factories, farms, automobiles, and other sources. Many chemicals now in common use have not been tested to determine their ability to cause cancer and birth defects.

33. New research suggests that people may be overreacting to the threat of low-level chemical pollutants. Naturally occurring chemicals in our food, in fact, may cause more cancer than pesticide residues.

34. In some locations, such as toxic waste dumps, chemicals are present in extremely high concentrations. Decades of improper waste disposal have left a legacy of pollution, ill health, and contaminated groundwater.

35. Reducing the contamination of our environment by hazardous wastes will require concerted efforts on the part of governments, businesses, and individuals. Especially important are ways to reduce the production of toxic waste by redesigning processes, recycling and reusing wastes, and finding substitutes.

Critical Thinking

THINKING CRITICALLY— ANALYSIS

This Analysis corresponds to the Thinking Critically scenario that was presented at the beginning of this chapter.

After you have thought about it for a while, consider some additional facts. The U.S. dairy industry already produces an excess of milk. The federal government buys the surplus, dehydrates some of it, and makes cheese out of the rest. The government stockpiles dehydrated milk and cheese at considerable cost to taxpayers. Does this change your view?

Before you draw a firm conclusion, however, remember that cheese and dehydrated milk are given to the needy. The food is not going to waste. Will increasing the surplus produce more food for the poor?

Again, before you draw any conclusions, consider another factor—the effects on small dairy operations of increasing the surplus of milk. Many owners of small dairy herds believe that the large producers will be the primary beneficiaries of the hormone. It will allow them to increase their milk production. As noted earlier, this increase will probably drive down the cost of milk and milk products, and it could put many small dairy farmers out of business. This loss could adversely affect the economies of many rural regions.

Some consumer groups are concerned about the potential health effects of using synthetic growth hormone. A trace of growth hormone is present in milk produced normally. Will the use of synthetic growth hormone increase this concentration in milk? What effect will that have on your health or the health of children?

Make a list of the pros and cons of using synthetic growth hormone. Then make a list of questions—those that you can answer and those that you cannot answer. Do you need more information to decide?

If you are able to make up your mind, what factors swayed your opinion? Did the concerns of one group outweigh the concerns of another? What critical thinking rules did you use in this exercise?

EXERCISING YOUR CRITICAL THINKING SKILLS

A man was accused of raping and murdering his grandmother in a small town in Texas. Unfortunately, prosecutors had very little physical evidence to link the man to the crime—except for the DNA in his semen. As it turned out, the DNA fingerprint proved to be pivotal to the case. Prosecutors claimed that there was a 96 million to 1 probability that the suspect was guilty and told the jury that there was absolutely no way anyone other than the suspect could have committed the crime. Based on this evidence and this statement, the man was convicted of murder and is now awaiting execution on Death Row.

Justice was served, right? Consider some more facts. First, the man who allegedly committed the crime was an unskilled laborer and lacked the financial resources required to hire a top-notch attorney or scientific experts to analyze the DNA evidence. Second, some time after his conviction, a team of geneticists reanalyzed the data and concluded that the DNA match was improperly performed. They argued that the statements made to the jury by the prosecuting attorney were grossly exaggerated. According to them, the probability of the man being the killer were really closer to 100 to 1. Would this conflicting view sway your opinion if you were a juror? In other words, would there be a reasonable doubt in your mind about the responsible party?

Consider some more facts. The town where the crime was committed and where the suspect had lived since birth contained only 300 people. Many of the residents have been genetically related for several generations. In other words, for about 200 years people in the town had been intermarrying, greatly increasing the chances that many individuals shared the same DNA fingerprint. According to the geneticists, any of the suspect's male relatives, of which there were many, could have shared the same DNA. Does this change your view? What critical thinking rule(s) does this exercise rely on?

TEST OF CONCEPTS

1. What is the genetic code? How is the genetic code, housed in the DNA, translated into instructions the cell can understand?

2. Describe how DNA is synthesized. How does the cell ensure the accurate replication of DNA?

3. List the three types of RNA, and briefly describe their function in protein synthesis.

4. In what ways are RNA and DNA similar? In what ways are they different?

5. Describe, in detail, the production of protein on mRNA. Be sure to note the enzymes involved and the role of the ribosome.

6. Discuss how bacterial cells control their genes. In what way does an end product of a metabolic pathway control genes? In what way does a starting material control the genes?

7. Discuss the various levels at which human genes are controlled. Give an example of each.

8. What is a proto-oncogene? How is it affected by a mutagen, a chemical or physical agent that causes mutation?

TOOLS FOR LEARNING

Tools for Learning is an on-line student review area located at this book's web site HumanBiology (www.jbpub.com/humanbiology). The review area provides a variety of activities designed to help you study for your class:

www.jbpub.com/humanbiology

Chapter Outlines. We've pulled out the section titles and full sentence sub-headings from each chapter to form natural descriptive outlines you can use to study the chapters' material point by point.

Review Questions. The review questions test your knowledge of the important concepts and applications in each chapter. Written by the author of the text, the review provides feedback for each correct or incorrect answer. This is an excellent test preparation tool.

Flash Cards. Studying human biology requires learning new terms. Virtual flash cards help you master the new vocabulary for each chapter.

Figure Labeling. You can practice identifying and labeling anatomical features on the same art content that appears in the text.

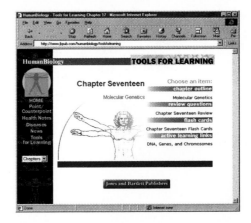

Active Learning Links. Active Learning Links connect to external web sites that provide an opportunity to learn basic concepts through demonstrations, animations, and hands-on activities.

GENETIC ENGINEERING AND BIOTECHNOLOGY
SCIENCE, ETHICS, AND SOCIETY

Thinking Critically

A chef at a famous restaurant in New York City organized fellow chefs in the city to protest genetically engineered produce, such as tomatoes. His main concern is that the government has refused to require tomato producers who are growing genetically engineered tomatoes to label their products. Consumers, among them chefs, therefore won't know if they're getting a genetically engineered vegetable or one produced naturally. How would you analyze this issue? What critical thinking rules might apply? ■

Recombinant DNA insulin.

N THE MOVIE *JURASSIC PARK*, SCIENTISTS RE-
moved DNA from dinosaur blood found in the gut
of mosquitoes preserved in amber (hardened sap of
pine trees) during the age of the dinosaurs. They took
this DNA and inserted it into the nuclei of frog cells and
produced living, breathing, lawyer-devouring dinosaurs.

Although this scenario is pure science fiction, it illustrates the kinds of genetic manipulations that scientists are engaged in. This procedure, in which a segment
of the DNA of one species is cut out and transferred to
another species, is popularly referred to as **genetic engineering.** For reasons explained shortly, scientists typically refer to it as **recombinant DNA technology.**

Although we like to think of genetic engineering as
a modern invention, it is not new at all. In fact, plants,
animals, and microorganisms have been "running" their
own genetic experiments for at least 3.5 billion years.
They've been undergoing mutations, exchanging segments of genes (via crossing-over), and "conducting"
other genetic experiments. These manipulations were
obviously not the product of conscious thought, but
were the product of natural processes. Nonetheless, the
outcome has been the same: new genetic combinations.
Over the course of evolution, nature's genetic tinkering
has resulted in a diverse array of life-forms.

Humans took genetic experimentation a step further by **selective breeding**—intentionally crossing certain plants or animals to produce offspring with desirable traits. Domestic animals, such as dogs, came about
by selective breeding, as have cattle, cats, fancy pigeons,
and other animals (**FIGURE 18-1**). Crops, such as corn
and tomatoes, are here today thanks to the efforts of
early geneticists who crossed plants with desirable traits
to produce offspring with good-tasting fruits and seeds.

FIGURE 18-1 Fancy Pigeon This beautiful bird barely
resembles its relative the rock dove, or common pigeon.
Selective breeding produced this bird, as it has many other
new genetic combinations.

Some consider genetic engineering, the newest
method of genetic manipulation, to be nothing more
than an advanced form of the genetic tinkering that occurs in nature and in human societies through selective
breeding. Others see it as an intrusion into the natural
process of evolution. Still others view it as a dangerous
gamble, a perilous flirtation with ecological disaster.
Opponents warn that genetically engineered organisms
could spread throughout the world, causing disease and
ecological disruption.

Yet another growing controversy has to do with a
process called **cloning.** In this technique, cells from an
organism are used to develop another, genetically identical organism—a clone. In 1997, much furor arose
when a Scottish researcher reported the birth of a
cloned sheep. The clone, named Dolly, was derived from
a cell taken from the mammary gland of another sheep.
Although there is some suspicion about this being a true
clone, the announcement stirred considerable controversy, as people began to wonder whether this technique
would be used for humans. In 1998, scientists announced yet another clone—this one in mice.

This chapter examines recombinant DNA technol-
ogy and its real and potential applications. It also outlines
the basic controversy over the use of this technology.

Recombinant DNA Technology: Slicing, Splicing, and Cloning Genes

Genetic engineering got its start over 20 years ago as
the result of research by Stanley Cohen at Stanford
University and Herbert Boyer at the University of California at San Francisco. In 1973, these scientists cut segments of DNA from one bacterium using a special
enzyme and spliced them into the genetic material of
another. The recipient bacterium then multiplied and,
in the process, produced many copies of the altered genetic material. The process Cohen and Boyer pioneered
still provides the basis for much of the genetic engineering research and development being done today.
To understand how it works, let's look at the individual
steps, beginning with the slicing of a piece of DNA
from a chromosome.

Enzymes Are Used to Excise Segments of Genes and Insert Them into Foreign DNA Molecules

Genetic engineering was made possible by the discovery
of a group of enzymes that snip off segments of the

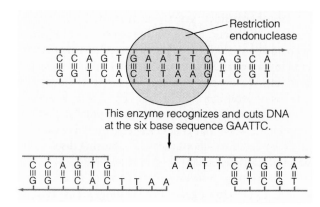

This enzyme recognizes and cuts DNA at the six base sequence GAATTC.

FIGURE 18-2 DNA Scissors Restriction endonucleases extracted from bacteria selectively slice through different segments of the DNA double helix, producing a staggered cut as shown here.

DNA molecule. Known as **restriction endonucleases,** these enzymes cut through both strands of the DNA, as shown in **FIGURE 18-2.**

Restriction endonucleases are found in many species of bacteria, where they serve a protective function, destroying the DNA of viruses that enter bacteria and preventing these cellular pirates from taking over. In fact, the term *restriction endonuclease* simply refers to the fact that the enzyme is found in the nucleus (endonuclease) and that it restricts the reproduction of viruses. To date, scientists have discovered hundreds of different restriction endonucleases, each one specific for a particular sequence of bases on the DNA.

As illustrated in **FIGURE 18-2,** these enzymes do not make a nice neat cut across the double helix; they create staggered cuts. Fortunately for geneticists, this "sloppy job" of cutting creates a gap in the DNA into which another piece of DNA can be inserted. All that's necessary for a segment of another DNA molecule to fit in is for its bases to match with those exposed by the enzyme.

Scientists can cut DNA from a bacterium and any other organism they'd like—for example, a human cell. If they use the same enzyme in both instances, they'll end up with segments of human DNA whose cut ends contain complementary base pairs to those in the bacterial DNA. Thus, the human segments can be inserted into bacterial DNA.

After a slice of DNA is added to the bacterial DNA, a new enzyme is used to seal it in place. This enzyme is called **DNA ligase.** At this point, the DNA splicing is complete. Geneticists have formed a **recombinant DNA molecule**—a molecule that contains DNA from two different organisms (a combination of DNAs).

Plasmids Can Be Used to Clone DNA Fragments

As noted earlier, recombinant DNA technology began with experiments in bacteria. As noted in Chapter 3, bacteria are prokaryotes with a single, circular "strand" of DNA (**FIGURE 18-3**). But bacteria also contain smaller circular strands, called **plasmids** (PLAZ-mids). Plasmids carry a few genes—such as those that confer antibiotic resistance to bacteria. They also replicate independent of the chromosome. Plasmids, being smaller than bacterial DNA and also relatively easy to extract, are often used in genetic engineering as a vector—a carrier of foreign genes.

As **FIGURE 18-4** illustrates, plasmids can be sliced with restriction endonucleases. This opens up the circular DNA. Foreign genes can be inserted into the gap in the plasmid and sealed in place with DNA ligase. This creates recombinant plasmids—plasmids containing foreign genes.

Plasmid carrying foreign genes can then be added to culture dishes containing bacteria or yeast cells. In culture, the plasmids are taken up by the new hosts. Then, as the host cells divide, the plasmids replicate. This produces multiple copies of the gene, which is then said to be **cloned.** Cloning produces numerous identical copies of the foreign gene. These can be used to produce useful products, such as hormones.

Cloning Is One Type of Gene Amplification

Geneticists can produce multiple copies of DNA in other ways, too. One common way of doing this is by extracting messenger RNA from cells and making copies of DNA from them. But wait a minute, you say, that's impossible! RNA can't produce DNA, it works the other way around!

FIGURE 18-3 Plasmids Rupturing a bacterium releases circular chromosome and numerous plasmids, indicated by arrows.

In one of those interesting quirks of nature, there is an enzyme that can make DNA from messenger RNA. This process is called **reverse transcription** because it is the reverse of transcription. It relies on an enzyme called **reverse transcriptase,** which comes from certain viruses.

To make DNA, then, geneticists remove mRNA from a cell and add reverse transcriptase (**FIGURE 18-5**). This enzyme then builds a single strand of DNA on the mRNA template. The hybrid DNA/RNA molecule is then treated with a chemical that destroys the RNA, leaving a single strand of DNA. Additional enzymes convert the single-stranded DNA to a double-stranded molecule. DNA molecules copied from mRNA are referred to as **cDNA.**

Perhaps the most widely used method of gene amplification is called the **polymerase chain reaction.** Basically, it boils down to this: Geneticists extract a segment of the DNA double helix from a cell's nucleus. They then split the DNA molecules into two strands by heating them. Next, they cool the solution and add DNA polymerase enzymes. These enzymes catalyze the formation of complementary DNA molecules.

Scientists then split the DNA molecules with heat and add more DNA polymerase. The new polymerase triggers another round of DNA replication. This procedure, which is carried out by machine, permits scientists to produce a million or more copies of the original DNA.

Mass production permits scientists to produce large quantities of DNA from organisms needed to chemically analyze DNA to determine the exact sequence of bases that comprise genes. This procedure, then, is helping scientists map the entire 100,000 or more genes of the human **genome** (JEAN-ome), as discussed in Chapter 16. An interesting finding of this study is that the human genome contains many genes found in other organisms, such as bacteria, mice, and snakes. This finding supports evolutionary theory.

Knowing the sequence of bases in a gene permits scientists to synthesize genes. Manufacturing genes permits geneticists to create normal copies of human genes, which can be inserted into human cells to correct genetic defects.

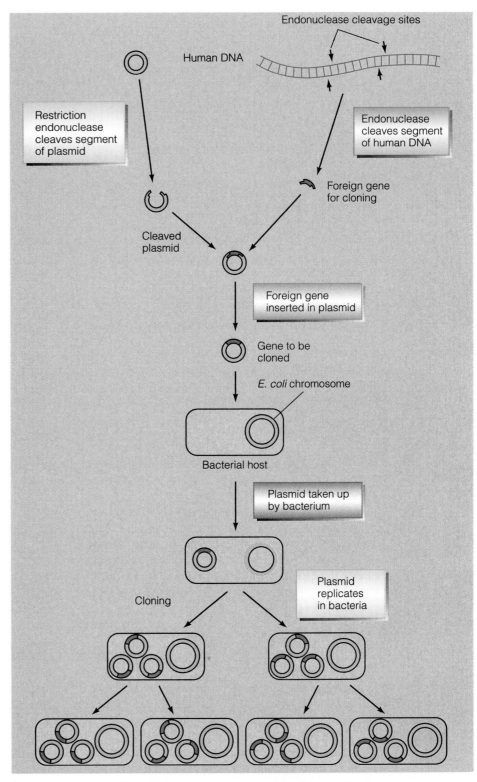

FIGURE 18-4 Genetic Engineering Simplified Genes can be snipped from chromosomes and inserted in plasmids. The recombinant plasmids can then be reinserted into bacteria, where they are replicated in large quantity. This process is known as cloning.

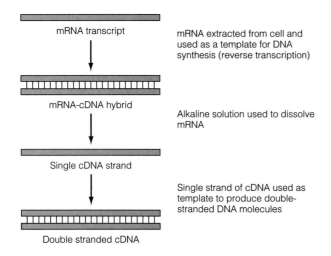

mRNA transcript

mRNA extracted from cell and used as a template for DNA synthesis (reverse transcription)

mRNA-cDNA hybrid

Alkaline solution used to dissolve mRNA

Single cDNA strand

Single strand of cDNA used as template to produce double-stranded DNA molecules

Double stranded cDNA

FIGURE 18-5 Making Genes from mRNA In this process, mRNA is extracted and then used as a template for the production of a complementary strand of DNA (cDNA), a process called reverse transcription. The mRNA molecule is then destroyed, leaving behind a single-stranded molecule of cDNA, which is then used to produce complementary cDNA.

Mass-producing genes also gives scientists the needed material to mass-produce gene products—notably, proteins. This topic is discussed in the next section.

Section 18-2

Applications of Recombinant DNA Technology

Recombinant DNA technology and other related techniques have resulted in a number of practical, though sometimes controversial, applications. This section looks at several potentially useful applications, beginning with one with great practical importance to medicine: the large-scale production of hormones and other proteins to treat diseases.

Hormones and Other Proteins Mass-Produced in Microorganisms Can Be Used to Treat a Variety of Disorders

By removing the human genes that code for hormones and transferring those genes into bacterial cells, scientists can create biological factories that generate huge quantities of hormones and other human proteins (such as blood clotting factors) missing in some individuals. This technique has resulted in a dramatic increase in the supply of proteins and hormones such as insulin and growth hormone for medical use. **TABLE 18-1** lists some of the products and their uses. So what's so important about this?

Consider an example: Human growth hormone used to treat people with glandular defects was once extracted

TABLE 18-1	
Some Useful Products of Genetic Engineering	
Products	**Used to**
Growth hormone	Treat patients with dwarfism
Insulin	Treat patients with diabetes
Tumor necrosis factor	Treat patients with cancer
tPA (tissue plasminogen activator)	Prevent clotting in heart attacks and strokes
Erythropoietin	Stimulates blood cell formation for patients with some forms of anemia
Clotting factors	Treat people with hemophilia
Surfactant	Treat babies with blue baby syndrome, caused by collapse of air sacs in lungs, which is caused by a lack of surfactant
Interferons	Treat some forms of cancer and some viral infections
Vaccines	To prevent hepatitis A, B, and C, as well as AIDS*, malaria*, and herpes*
DNA probes	Map human chromosomes, perform DNA fingerprinting, detect infectious diseases, and detect genetic disorders

*Under development, but not currently available.

from the pituitary glands of human cadavers. In order to extract enough growth hormone for one dose, technicians had to remove glands from 50 human cadavers. This process was not only costly, but time-consuming.

Recombinant DNA technology is also safer than previous methods because it avoids potential immune reactions caused by injecting proteins from pigs or other animals into humans. Although these proteins stimulate the desirable biological effect, they are chemically dissimilar enough to evoke an immune reaction in some patients. Genetic engineering produces biochemically identical proteins so you get both the desired biological effects and no immune response!

Protein hormones, like growth hormone, can also be produced for other uses, notably for livestock production. Cows, for instance, can be injected with bovine growth hormone produced by genetically engineered bacteria. This treatment greatly increases milk production. (For a discussion of the pros and cons of this, see the Thinking Critically section in Chapter 17.)

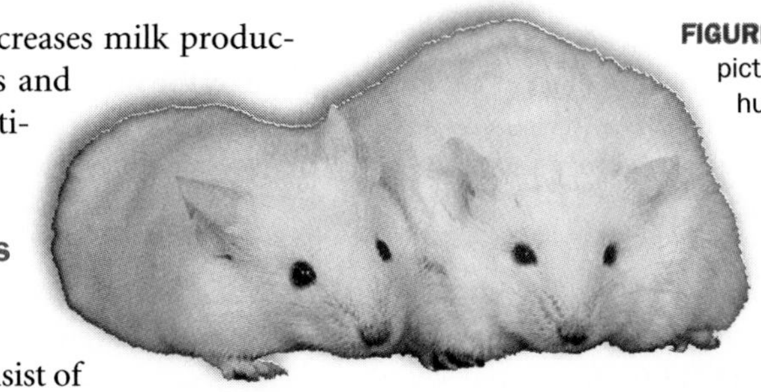

Recombinant DNA Techniques Can Be Used to Produce Vaccines.

As noted in Chapter 8, vaccines consist of inactivated bacteria or viruses. When injected in an individual, a vaccine stimulates an immune response. This, in turn, protects a person for long periods from the infectious form of the bacterium or virus. Using genetic engineering, scientists can produce vaccines in large quantity and at a lower cost than is possible by conventional means.

At present, only one vaccine, for an infectious disease that strikes the liver (hepatitis B), has been produced via this method. But geneticists hope to develop others, perhaps even one for the AIDS virus, which, if successful, could stop the spread of this deadly disease.

Gene Splicing Can Be Used to Transfer Genes from One Organism to Another.

In what some people view as a major intrusion into the evolution of life, scientists are now able to transfer single genes from one organism to another, creating a **transgenic organism.** These organisms don't differ much from their original form; they may, in fact, display only one new trait, such as a larger body size. One of the most successful attempts to date involves the transplantation of the human gene that codes for growth hormone into pigs and mice (**FIGURE 18-6**). When successful, this results in offspring that are much larger than their siblings who didn't receive the gene.

Researchers are looking for ways to re-engineer cows so that the milk they produce will be chemically more similar to human milk. Scientists, for example, have transplanted a human gene that codes for the production of a chemical found in mother's milk, called *lactoferin,* into cows. This substance helps babies fight off infections.

One of the most exciting developments is the production of mice containing genes identical to those that cause human diseases. This could advance medical research on diseases and speed the development of treatments and cures. In 1997, scientists from the Medical College of Georgia announced the successful transplantation of genes that code for a green fluorescing protein into zebra fish. It turned the red blood cells green under blue light, allowing scientists to study blood cell formation in the zebra fish embryos. The scientists had performed the same process with nerve cells, and hope that

FIGURE 18-6 Supermice The mouse on the right, pictured with its littermate, has received a gene for human growth hormone.

this will allow them to study how nerve cells form connections in the developing brain.

Although such techniques sound simple, they're actually quite difficult. One can't just inject a foreign gene into a baby pig or mouse or cow. In order to successfully transplant a gene from one species into another, the gene generally has to be introduced at the embryonic stage when the organism is in the one-cell stage. This is accomplished by injecting genes into the nucleus of the ovum via tiny needles. Some nuclei incorporate the foreign gene, which is then transmitted from one cell to the next during embryonic development. Even so, there's no guarantee that the gene will end up in the proper location.

Transgenic Plants Are Easier to Produce Than Transgenic Animals, and Many Successes Have Already Occurred In This Area.

Plants are easier to perform gene transplants on than animals. One of the first success stories involved the transplantation of a gene that codes for an enzyme called luciferase (lew-SIFF-er-ace) into a tobacco plant. This enzyme is normally found in fireflies and is responsible for catalyzing a light-generating reaction. In the tobacco plant, the enzyme didn't do much unless the plant was sprayed with a chemical called luciferin. At this point, the plant glowed (**FIGURE 18-7**).

This experiment was not performed to create tobacco plants that glow in the dark so they could be harvested at night, but only to show that foreign genes

FIGURE 18-7 Glow Little Tobacco Plant, Glimmer Glimmer This tobacco plant contains firefly genes for the enzyme luciferase. When the plant is sprayed with luciferin, the enzyme breaks it down, releasing light.

could indeed be transplanted from one species to another. It opened up the door to more practical experiments—for instance, transplanting genes that make plants produce more nutritious seeds and fruit or that permit the roots of some plants to synthesize their own nitrogen fertilizer.

One of the more promising developments involves the introduction of genes that allow oats to grow in salty soil. Transplanting these genes into commercial crop species could allow farmers to use vast acreages in the United States and other countries idled by the buildup of salts. This is caused by the irrigation of poorly drained soils with water high in salts.

Researchers are also working on ways to make crop plants resistant to herbicides, chemicals applied to crops to control weeds. Although herbicides generally act only on weed species, they sometimes impair the growth of crop species. Proponents hope that herbicide-resistant crops will help farmers increase yields. Critics argue that if herbicide-resistant crops are used, farmers may apply larger amounts of herbicide to their fields. These chemicals, in turn, could be washed into nearby waterways or could poison soils, killing important bacteria and other microorganisms essential for the long-term health of the soil. (For a discussion, see the Point/Counterpoint in this chapter.)

Geneticists are also working on ways to alter plants themselves to increase their resistance to pests. By transferring the genes that protect wild species from insects into crop species, scientists may be able to produce many varieties that require little, if any, chemical pesticides.

Transgenic Microorganisms Are Perhaps Some of the Most Controversial of All Applications.

Genetic engineering began with microorganisms, and work in this area continues today. For example, scientists have successfully transplanted a bacterial gene that produces a toxin that kills insects into a species of bacteria that lives on the roots of crop plants. Root-eating insects ingest the lethal bacteria when they feed on roots of treated plants. This kills the insects and protects crops. This technique benefits the farmer, who has to use fewer chemicals to control root-eating insects, and also benefits the environment, because it reduces the use of potentially dangerous chemicals that could leach into groundwater or be washed into nearby lakes and streams.

Genetic researchers have also developed a bacterium that retards the formation of frost on plants. The genetic variant is found naturally in the environment, but not in sufficient quantities to protect crops. Thus, researchers have cloned the bacterium and have begun testing its efficiency in the field. Should it prove successful and safe, the bacterium could be sprayed on crops late in the growing season when frosts can be devastating, potentially saving farmers millions of dollars a year. Some environmentalists, however, worry that this bacterium and other genetically engineered creations could escape into the environment and displace normally occurring bacteria. Who knows what detrimental effects this might have on the function of natural systems?

Recombinant DNA Technology Can Be Used to Cure Genetic Disease, a Treatment Called Gene Therapy

Scientists are also tinkering with ways in which genetic engineering can be used to insert normal human genes into genetically defective body cells, thereby curing serious genetic diseases. This technique is called **gene therapy.**

Approximately 1 of every 100 children born in the United States suffers from a serious genetic defect, such as sickle-cell anemia or hemophilia. Thanks to advances in genetic engineering, scientists may soon be able to replace their defective genes with normal genes. Instead of simply finding a treatment, this technique will, if successful, permit doctors to cure diseases previously considered incurable. Many of these diseases cost our society millions of dollars and account for enormous human suffering. The largest obstacle, however, is getting the genes into body cells where they are missing. For instance, to cure a genetic defect in brain cells, the genes must somehow be delivered to the cells of the brain.

One promising means of delivering genes to sites where they're required is the bone marrow transplant. Consider an example. Krabbe's disease is a rare genetic disorder resulting from a deficiency in one enzyme in human brain cells. The absence of this enzyme allows fat to accumulate in the nervous system, causing nerve cells to degenerate. Seizures and visual problems occur early in life, and most victims die within the first 2 years of life.

To test a means of correcting this disease, scientists used a strain of mice that suffers from a similar condition. They injected them with bone marrow cells containing the gene that codes for the missing enzyme. Scientists found that the transplanted cells became established in the lungs and liver and restored enzymatic activity. Some even found their way into the brain!

Another potentially promising technique involves the use of microspheres—tiny lipid spheres. Called liposomes, these tiny lipid spheres can be packed with genes and coated with antibodies for specific target cells. If successful, this will allow the liposomes to deliver their contents to the correct cells.

Researchers have also developed a novel transplantation technique that could enable surgeons to introduce genetically engineered cells into specific organs in the human body. Scientists injected genetically altered liver cells into a foamlike material. The foam was then transplanted into rats after being impregnated with a hormone that stimulates the growth of blood vessels from nearby larger vessels. Within a week after implantation, the artificial tissue was riddled with a network of blood vessels. This technique could be used to introduce genetically altered cells into numerous tissues and organs. It could provide a way to cure diabetes or Parkinson's disease.

DNA Probes Permit Analysis of DNA from Crime Scenes, Detection of Infectious Organisms and Genetic Disorders.

Imagine that you're a police officer investigating a gruesome murder. The victim is a 29-year-old man who was found dead in his apartment. Unfortunately, the physical evidence is scant. There were no eye witnesses, no fingerprints, and no fragments of the perpetrator's clothing in the apartment. However, there is blood on the carpet and on the couch, which turns out to belong to two different individuals—the victim and possibly the perpetrator. Some days later, an anonymous tip points you in the direction of a family member, a cousin. You take him into custody for questions, but he says he wasn't in town at the time and has several people who will attest to his whereabouts. After you sample the blood of the cousin, though, you find that it matches that found in the apartment. Then you have the lab perform a genetic analysis of the alleged perpetrator's blood and find

that it matches samples at the crime scene as well. In a court of law, this evidence alone may be enough to convict the cousin.

This technique of DNA matching is called **DNA fingerprinting,** and was introduced in Chapter 17. DNA fingerprinting relies on techniques derived from genetic engineering. In particular, it depends on **DNA probes,** small fragments of single-stranded DNA produced by machines called DNA synthesizers. Probes bind to complementary base pairs of DNA from samples taken at the scene of a crime and from suspected perpetrators. Several probes are used to compare samples of blood as well as hair or semen.

DNA Probes Can Even Be Used to Help Geneticists Map Human Chromosomes

Geneticists have long been interested in mapping human chromosomes—that is, in determining which genes are on which chromosomes. For years, the mapping process has been rather crude, relying on studies of patterns of linked traits in families.

Another, more refined technique has also been used. In this procedure, scientists fuse a human cell with a mouse cell (**FIGURE 18-8**). The new hybrid cell containing 96 chromosome divides, but when the cell divides, it begins to lose human chromosomes. The daughter cells divide again and lose more. Eventually, cells are produced with all mouse chromosomes and

FIGURE 18-8 Mapping by Fusion This technique permits scientists to determine the chromosomal address of certain genes. Human and mouse cells are fused, creating a new cell with excess chromosomes. This cell divides and, in so doing, loses most of the excess chromosomes. Cells with one or two human chromosomes can be grown in culture. By determining the unique human proteins being synthesized in these cells and the chromosomes present, geneticists can determine the chromosomal "address" of certain genes. More precise techniques are needed to determine the exact location on the chromosome.

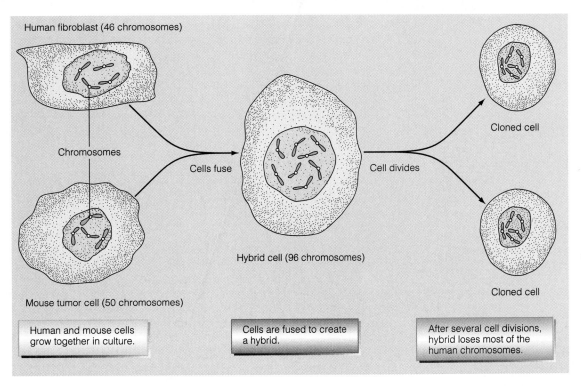

Human fibroblast (46 chromosomes)

Chromosomes

Cells fuse

Hybrid cell (96 chromosomes)

Cell divides

Cloned cell

Cloned cell

Mouse tumor cell (50 chromosomes)

Human and mouse cells grow together in culture.

Cells are fused to create a hybrid.

After several cell divisions, hybrid loses most of the human chromosomes.

Point/Counterpoint

Controversy over Herbicide Resistance Through Genetic Engineering

THE BENEFITS OF GENETICALLY ENGINEERED HERBICIDE RESISTANCE
Charles J. Arntzen

Dr. Charles J. Arntzen is a professor in the Alkek Institute of Biosciences and Technology in the Texas Medical Center in Houston, Texas. Dr. Arntzen was elected to the U.S. National Academy of Sciences in 1983 as a result of his pioneering research in photosynthesis and plant molecular biology. He was the chairman of the National Biotechnology Policy Board of the National Institutes of Health from 1991–1993, and is a member of the Editorial Board of Science. This article is adapted by the author from a 1991 publication entitled The Genetic Revolution, Scientific Prospects and Public Perceptions *with the permission of the publisher, The Johns Hopkins University Press.*

One of the principal outcomes of plant genetic engineering has been a rapid expansion in our understanding of plant cell biology, genetics, and molecular controls over the structure and function of plants. This understanding is now helping to solve practical problems such as developing insect- and disease-resistant crops, thereby decreasing needs for chemical pesticides. Other approaches include creating herbicide-resistant plants, which will allow farmers more choices in weed control, and to switch to newer, rapidly biodegradable herbicides.

Herbicides are chemicals sprayed on crops to control weeds. Their effect is selective—that is, they kill weeds without significantly harming crops. All herbicides currently available to farmers began with an evaluation of the sensitivity of weeds and crops to experimental chemicals. This evaluation is accomplished by applying potentially useful chemicals to test samples of weed and crop seedlings. Compounds that kill weeds but not crops are then subjected to animal toxicology studies. Those deemed acceptable are developed for farmers' use.

The success of herbicides is based on the fact that crop species contain enzymes that convert chemical compounds, including herbicides, to inactive forms. These enzymes tend to be crop-specific. For instance, the enzymes that render soybeans immune to 30 or more commercial herbicides are different from the enzymes that make corn resistant to other herbicides. This can be a benefit when corn or soybeans are grown in rotation. In such instances, both the unwanted "volunteer" corn and weeds that germinate in a soybean field can be controlled by the soybean herbicide.

Interestingly, certain weeds have the same enzymes as the crop species they invade. This is often true when the weeds and crops are somewhat related, such as with wild oats in a field of wheat or barley. Thus, the oats are resistant to the same herbicides as the wheat or barley. Consequently, the farmer currently has few or no choices in chemical weed control since herbicides that would kill the weeds would also kill the crop.

Through genetic engineering, researchers have devised ways to make crops resistant to new classes of herbicides. That is, they have found ways to give crops such as soybeans additional enzymes to augment their natural herbicide resistance. By transferring genes that code for enzymes conferring herbicide resistance, scientists can give farmers a broader set of options when selecting herbicides. If successful, these options will solve problems where there is currently no weed control, or where the herbicides available are prohibitively expensive or environmentally damaging.

This line of research may also give the farmer a simpler means to control weeds through the use of a single, more effective herbicide selected on the basis of several traits, including herbicide resistance, reduced cost, and, ideally, greater safety in the environment.

This explanation of the benefits of genetic research aimed at making crops more resistant to herbicides may not satisfy the reader who is asking, "Why don't they simply stop using all herbicides?" The main answer is economics. Weed control is essential for all crops. Left unchecked, weed populations can reduce crop yields to near zero. Prior to the 1950s, weeds were controlled strictly by mechanical means—that is, by cultivation or hand pulling. This was a labor-intensive and costly process. By the 1980s, more than 95% of all major row crops in the United States were produced using herbicides. Farmers based this decision (to apply herbicides rather than using a cultivator or a hoe) on costs. An application of herbicides at the time of planting, costing as little as $10 per acre, is many times cheaper than mechanical cultivation. Weed control through herbicides is a factor in the low food prices we enjoy in the United States. It also saves on energy use and, when properly used, does not harm the environment.

With the continuing development of herbicide-resistant crops as part of an integrated weed-control strategy, crop production costs can be held at low levels while new methods for weed control are being developed.

THE PERILS OF GENETICALLY ENGINEERED HERBICIDE RESISTANCE
Margaret Mellon

Environmentalists have been active in the debate about genetic engineering since the beginning. The current focus of environmental interest in genetic engineering is based on the prospect of commercial production of a broad range of genetically engineered organisms—for example, bacteria and viruses. These organisms could have direct and adverse impacts on our environment and our health.

Public concern over genetic engineering is also focused on indirect environmental consequences, the best example of which is herbicide-resistant plants. The development of major crops that are tolerant of chemical herbicides seems likely to lead to increased or prolonged use of these dangerous agricultural chemicals.

Herbicide-resistant crops, like most genetically engineered agricultural products, are being developed by large, transnational chemical companies. Worldwide, at least 28 enterprises have launched more than 65 research programs to develop herbicide-resistant crops. All of the major crop plants are involved, including cotton, corn, soybeans, wheat, and potatoes. Engineered crops being made resistant to a company's own herbicides will surely increase that company's market share in chemicals. Thus, industry analysts expect herbicide-resistant plants to be big business, mainly for pesticide and herbicide producers.

Herbicide-resistant products run directly counter to the promise of a reduced dependence on agricultural chemicals put forward by biotechnology advocates. Rather than weaning agriculture from chemicals, these crops will be shackled to herbicide use for the foreseeable future. Herbicides represent an estimated 65% of the chemical pesticides used in agriculture. By continuing to promote the use of herbicides, the biotechnology industry greatly restricts its potential for reducing overall chemical use.

While herbicide-resistant crops offer no hope of environmental benefits in agriculture, there are alternatives to these products that do promise substantial reduction in overall pesticide use. These alternatives fall under the rubric of sustainable agriculture.

Sustainable agriculture employs a variety of agricultural practices such as crop rotation, intercropping, and ridge tillage (that produce long ridges of soil on which plants are grown). These practices control pests without the application of synthetic pesticides and fertilizers. Used by knowledgeable farm managers, these techniques work to make farms more profitable and environmentally sound. Growing different crops in successive growing seasons, for example, dramatically reduces pests by sequentially removing the hosts on which the pests depend. With fewer weeds or insects to contend with, the farmers reduce the need for costly chemical pesticides. Once thought impractical, sustainable agriculture is rapidly gaining support in national policy forums, including the National Academy of Sciences.

The environmental advantage of such an approach is clear. Crop rotations that reduce pests reduce the need for pesticides now and into the future. Thus, developing sustainable practices should be the highest priority for agriculture as it moves toward the twenty-first century.

Although a minor part of the sustainable agriculture picture up to now, biotechnology could help by developing new crop varieties for use in sustainable systems—for example, faster germinating, cold-tolerant crop varieties could enhance low-input sustainable systems by enabling more effective weed control. Engineered products such as these would fulfill the promise of biotechnology, and would benefit farmers, the environment, and the rural economy alike.

Margaret Mellon is the director of the Agriculture and Biotechnology Program at the Union of Concerned Scientists (UCS). UCS works at the interface of technology and society to promote sustainable agriculture and evaluate applications of biotechnology. Dr. Mellon lectures widely on biotechnology issues and has appeared frequently on television and radio talk shows.

SHARPENING YOUR CRITICAL THINKING SKILLS

1. State the main thesis of each author in your own words.
2. List the data or arguments used to support each hypothesis and the data used by the authors to refute the other's point of view, if any.
3. Can you determine whether there are any flaws in the reasoning in either essay?
4. Which viewpoint do you agree with? Why?

Visit *Human Biology's* Internet site, www.jbpub.com/humanbiology, to research opposing web sites and respond to questions that will help you clarify your own opinion. (See Point/Counterpoint: Furthering the Debate.)

one or two human chromosomes. By studying the resulting protein products, geneticists can associate certain genes with certain chromosomes.

Recombinant DNA technology has permitted yet another dramatic refinement in chromosome mapping. The DNA probe is helping geneticists determine not only what genes are on what chromosomes, but also their exact location on the chromosome.

Section 18-3

Controversies over Genetic Engineering— Ethics and Safety Concerns

Although genetic engineering has spawned a great deal of enthusiasm, it also has its critics. And, despite the promises, genetic engineering is fraught with controversy over safety and ethics. In October, 1997, for instance, several scientists issued a worldwide alert asking all governments to ban import of a genetically engineered soybean from the United States. The soybeans were genetically altered to be resistant to a herbicide known as Roundup. The scientists are concerned that the use of this herbicide will increase the level of plant estrogens in soybeans, thus increasing the levels of these potentially harmful chemicals in humans.

At this time, many safety questions remain unanswered. As noted earlier, perhaps the greatest safety concern is the possibility of unleashing genetically altered bacteria or viruses. Some critics fear that a genetically altered strain could spread through the environment, wreaking havoc on ecosystems and, possibly, human populations. Once unleashed, it would be impossible to retrieve. The genetically altered bacterium that retards frost formation, for example, could enter the atmosphere on dust particles, reducing cloud formation and altering global climate. No one knows for sure how serious this threat really is. The Point/Counterpoint in Chapter 2 discusses the safety of genetically engineered food.

Other critics object to genetic tinkering, especially the transfer of genes from one species to another, on ethical grounds. Do humans have the right, they wonder, to interfere with the course of evolution?

Proponents of genetic tinkering argue that livestock and plant breeders have been selectively breeding hardy animals to produce genetically superior livestock and crops for hundreds, if not thousands, of years. In so doing, people have been performing a kind of genetic engineering—albeit a slow one—for millennia. Genetic engineering, say proponents, merely offers a quicker way of achieving the same goals.

To understand this debate, it is important to exercise one's critical thinking faculties by first defining the terms very carefully. When you do this, you find that genetic engineering and selective breeding do indeed achieve the same end points in some instances. Selective breeding simply alters gene frequencies—that is, the percentage of organisms in a population carrying a specific allele. Genetic engineering does the same, as in the case of the frost-retarding bacteria. In other words, geneticists are simply producing large quantities of a mutant already found in nature. But that in and of itself may be reason for concern. In nature, for reasons not well understood, the mutant is found in small quantity. Could shifting the allele frequency cause some ecological catastrophe?

In other instances, the two techniques produce quite different results. That is, in some cases, genetic engineering introduces new genes into species, producing genetic combinations never before encountered in life. Introducing the human growth hormone gene into cattle embryos is an example of this form of tinkering. This manipulation violates a fundamental law of nature: Different species do not interbreed. Is it ethical to blur nature's naturally maintained boundaries? Who can say? And can we put a lid on our ambition? Will we know when to stop?

Unfortunately, experience with genetic engineering is too limited to answer the safety concerns of critics. Preliminary work suggests that the dangers have been exaggerated and that genetically engineered bacteria are not a threat to ecosystem stability. Still, further research is needed to be certain that in our zeal to make life better, we don't do ourselves in. Ethical questions, like political questions that have been plaguing some countries for centuries, may never be answered to the satisfaction of everyone.

Another point, rarely mentioned in the debate, is the potential for wrong-doing—for creating superinfectious organisms that could wipe out huge numbers of people. Imagine for instance, the catastrophe that might arise if some mad scientist introduced the deadly AIDS virus genes into the genome of the flu virus, which is rapidly transmitted around the world.

Whatever the outcome of the debate, you can bet that genetic engineering is here to stay. Future disasters, should they occur, may compel us to prohibit certain forms of genetic tinkering, but in cases where genetic engineering reduces suffering, advances agriculture, and facilitates environmental cleanup, public support may be hard to thwart.

SUMMARY

1. Genetic engineering, or *recombinant DNA technology*, is a procedure by which geneticists remove segments of DNA from one organism and insert them into the DNA of another.

2. Genetic manipulation is not a new phenomenon. Nature has been tinkering with genes for at least 3.5 billion years, yielding multitudes of new genetic combinations during evolution.

RECOMBINANT DNA TECHNOLOGY: SLICING, SPLICING, AND CLONING GENES

3. Genetic engineering got its start over 20 years ago when scientists cut segments of DNA from one bacterium using a special enzyme and spliced them into the genetic material of another that was then permitted to replicate.

4. Geneticists use an enzyme called *restriction endonuclease* to excise segments of genes and insert them into foreign DNA molecules.

5. After a segment of DNA is transplanted, a second enzyme, *DNA ligase*, is used to seal it in place. At this point, geneticists have formed a *recombinant DNA molecule*—a molecule that contains DNA from two different organisms.

6. Bacteria contain small, circular strands of DNA, called *plasmids*, which are separate from the chromosome. Foreign genes can be spliced into plasmids, and the plasmid carrying a foreign gene can be reinserted into bacteria or yeast cells by adding them to culture dishes containing these microorganisms.

7. In culture, the plasmids are rapidly taken up by the new hosts. As these cells divide, the plasmids replicate. This produces multiple copies of the gene, which is then said to be *cloned*.

8. Cloning produces numerous identical copies of the foreign gene, which can be used for further studies in genetics or to produce useful products, such as hormones.

9. Geneticists can produce multiple copies in other ways, too. One of the most common ways of doing this is by extracting messenger RNA from cells and making copies of DNA from them using the enzyme *reverse transcriptase.*

10. The most widely used method of gene amplification is called the *polymerase chain reaction* in which DNA is alternately heated to split the double helix, then cooled. DNA polymerase enzymes are added to catalyze the formation of complementary DNA molecules.

APPLICATIONS OF RECOMBINANT DNA TECHNOLOGY

11. Recombinant DNA technology and other related techniques have resulted in a number of practical and sometimes controversial applications.

12. Hormones and other proteins mass-produced in genetically engineered microorganisms can be used to treat a variety of disorders. Recombinant DNA techniques can be used to produce vaccines.

13. Gene splicing can be used to transfer genes from one organism to another, creating *transgenic organisms.* These organisms rarely differ much from their original form; they may, in fact, only display one new trait, such as a larger body size.

14. Transgenic plants are easier to produce than transgenic animals, and many successes can be enumerated. Attempts are being made to make plants that produce more nutritious products or are more resistant to herbicides and pests.

15. Recombinant DNA technology can be used to cure genetic disease, a treatment called *gene therapy.* In this procedure, scientists attempt to insert normal human genes into genetically defective body cells.

16. Although promising, gene therapy has two major barriers: delivering the gene to the proper cell and ensuring its in-

corporation into the DNA of that cell. Several techniques are under development to overcome these barriers.

17. Another practical outcome of research in recombinant DNA technology is the DNA probe. *DNA probes* are tiny segments of genes that bind to complementary base pairs of sample DNA—for example, hair, blood, or semen taken from crime scenes. DNA probes can also be used to detect infectious organisms and genetic disorders.

18. DNA probes are used in *DNA fingerprinting.*

19. DNA probes are being used to map human chromosomes—that is, to determine the location of genes on human chromosomes.

CONTROVERSIES OVER GENETIC ENGINEERING—ETHICS AND SAFETY CONCERNS

20. Although genetic engineering has spawned a great deal of enthusiasm, it does have its critics who are concerned with safety and ethical issues.

21. Perhaps the greatest safety concern is the possibility of intentionally or unintentionally releasing potentially dangerous genetically altered bacteria or viruses into the environment.

22. Some critics object to genetic tinkering, especially the transfer of genes from one species to another, on ethical grounds, wondering if we have the right to interfere with the course of evolution.

23. Unfortunately, experience with genetic engineering is too limited to answer the safety concerns of critics. Preliminary work suggests that the dangers have been exaggerated and that genetically engineered bacteria are not a threat to ecosystem stability. Further research is needed to be certain.

Critical Thinking

THINKING CRITICALLY— ANALYSIS

This Analysis corresponds to the Thinking Critically scenario that was presented at the beginning of this chapter.

Before you accept or reject an argument, such as the one made by the chef, it's important to gather all the facts. Reading this chapter will give you some perspective on genetic engineering. I'd also suggest reading some specific articles on genetically engineered tomatoes. An excellent one was published in *Science News,* November 28, 1992. This article explains the ways that geneticists are trying to alter tomatoes, using genetic engineering. Upon careful analysis, it appears that several of the improvements they're seeking could have been acquired by selective breeding. In other words, they could have achieved the same results by cross-breeding tomatoes with desired traits with those with other traits, producing a "super" tomato with good taste, long shelf life, and other desirable features. In this instance, there's nothing magical about genetic engineering. It's just a faster way of getting desirable genes into a tomato.

But you'll also learn that geneticists are transplanting bacterial genes into tomato plants. One gene in particular retards the production of ethylene gas, which normally stimulates ripening. When the bacterial gene is present, it slows ripening, allowing farmers to leave the tomato on the vine several extra days. This increases the tomato's flavor with obvious benefits to anyone who has eaten a store-bought tomato and had the opportunity to compare it with one grown in a garden.

The questions here might be: Will this harm consumers? Can you think of any other concerns and ways to address them? Are the chefs merely reacting to the phrase "genetic engineering"? Could this be one of those thought stoppers described in the Critical Thinking section in Chapter 1?

EXERCISING YOUR CRITICAL THINKING SKILLS

In an attempt to increase the supply of blood for surgery and medical emergencies and to avoid potential contamination from the AIDS virus, scientists are looking for blood substitutes. One group of scientists has successfully transplanted the human hemoglobin gene into pigs. About 10% of their hemoglobin is chemically identical to human hemoglobin. Using your critical thinking skills, analyze the potential of this discovery. Would you invest in this company? What obstacles lie in the way?

TEST OF CONCEPTS

1. Describe the methods of recombinant DNA technology. Your answer should explain how genes are removed, spliced, and cloned.
2. Describe three methods of gene amplification.
3. What barriers lie in the way of creating transgenic organisms?
4. List and describe three practical applications of recombinant DNA technology.
5. List the major concerns for genetic engineering and describe each one.
6. Debate the statement: Genetic engineering is morally wrong and should be discontinued. It violates the laws of nature and gives humans a power beyond their control.

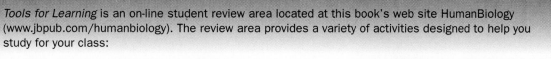

TOOLS FOR LEARNING

Tools for Learning is an on-line student review area located at this book's web site HumanBiology (www.jbpub.com/humanbiology). The review area provides a variety of activities designed to help you study for your class:

www.jbpub.com/humanbiology

Chapter Outlines. We've pulled out the section titles and full sentence sub-headings from each chapter to form natural descriptive outlines you can use to study the chapters' material point by point.

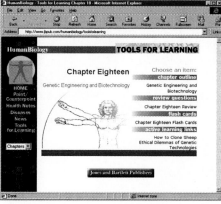

Review Questions. The review questions test your knowledge of the important concepts and applications in each chapter. Written by the author of the text, the review provides feedback for each correct or incorrect answer. This is an excellent test preparation tool.

Flash Cards. Studying human biology requires learning new terms. Virtual flash cards help you master the new vocabulary for each chapter.

Figure Labeling. You can practice identifying and labeling anatomical features on the same art content that appears in the text.

Active Learning Links. Active Learning Links connect to external web sites that provide an opportunity to learn basic concepts through demonstrations, animations, and hands-on activities.

HUMAN REPRODUCTION

Thinking Critically

Imagine that you are a journalist for a major urban newspaper. One day, you receive a press release from a local medical school announcing that one of its researchers has discovered that a chemical found in a common household cleaning agent reduces fertility in rats and mice. Large doses were given to both males and females before conception. The results showed a statistically significant decline in the litter size as well as several physical abnormalities—misshapen legs and malformed skulls. The press release quotes the researcher who says that the chemical should be banned from use in homes. Using your critical thinking skills, what questions would you ask before writing your article? What other information would you seek out? ■

Human sperm, one of the most highly specialized cells in the body.

SANDRA COLLINS WOKE ONE DAY WITH A TERRI-
ble pain in her abdomen that persisted throughout the morning. Instead of calling her doctor,
though, she shrugged off the pain and went Christmas
shopping with her husband. A few hours later, while she
was browsing through a bookstore, the pain grew worse
and she blacked out. Her husband rushed her to the
emergency room, where it was found that Sandra was
suffering from severe internal bleeding caused by an
ectopic (eck-TOP-ick) pregnancy—that is, a fertilized
ovum that had developed in the upper part of her reproductive tract, outside the uterus.

Fortunately, physicians were able to counteract her
blood loss with a transfusion. They then whisked her
into the operating room, where they surgically removed
the fetus, placenta, and surrounding tissue and repaired
torn blood vessels.

Reproduction is one of the most basic body functions. However, as this account shows, it doesn't always
operate smoothly. Like many other body functions, reproduction is controlled by the endocrine and nervous
systems. This chapter describes human reproduction.
The information gained here will help you understand
the material presented in the next chapter, which deals
with fertilization and development.

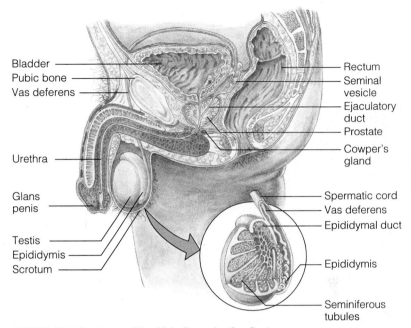

FIGURE 19-1 Anatomy of the Male Reproductive System

The Male Reproductive System

In humans, the male reproductive system consists of
seven parts: (1) the two **testes** (TESS-teas), which produce sex steroid hormones and sperm; (2) the two
epididymides (EP-eh-DID-eh-mid-ees), which store
sperm produced in the testes; (3) a pair of ducts, each
known as a **vas deferens** (vass DEAF-er-ens), that conducts sperm from the epididymis of each testis to the
urethra; (4) **sex accessory glands**, which produce secretions that make up the bulk of the ejaculate; (5) the **urethra** (yur-REETH-rah), which conducts sperm to the
outside; (6) the **penis** (PEA-nuss), the organ of copulation; and (7) the **scrotum** (SCROW-tum), a sac that
houses the testes (**FIGURE 19-1; TABLE 19-1**).

The Scrotum Helps Keep the Testes Cool

The testes are suspended in a pouch known as the **scrotum**. As **FIGURE 19-1** shows, the scrotum is attached to
the body below the attachment of the penis.

The scrotum provides just the right temperature
to permit normal sperm production. Inside the body
cavity, however, the temperature is too high for sperm
development.

The influence of body heat on sperm development
is illustrated by the plight of male long-distance runners
who run hundreds of miles a month. This level of exercise elevates their body temperature, and even though
the testes are suspended in the scrotum, scrotal temperature may be so high that sperm formation declines.
Thus, long-distance runners may become temporarily
sterile. Tight-fitting pants or shorts can have the same
effect.

Sperm Are Produced in the Seminiferous Tubules and Stored in the Epididymis

As **FIGURE 19-2** shows, each testis is surrounded by a dense
layer of connective tissue. This layer is invested with numerous pain fibers, a fact to which most men will attest.

TABLE 19-1

The Male Reproductive System

Component	Function
Testes	Produce sperm and male sex steroids
Epididymes	Store sperm
Vasa deferentia	Conduct sperm to urethra
Sex accessory glands	Produce seminal fluid that nourishes sperm
Urethra	Conducts sperm to outside
Penis	Organ of copulation
Scrotum	Provides proper temperature for testes

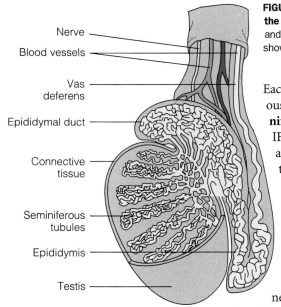

FIGURE 19-2 Interior View of the Testis The epididymal duct and seminiferous tubules are shown.

Nerve

Blood vessels

Vas deferens

Epididymal duct

Connective tissue

Seminiferous tubules

Epididymis

Testis

Each testis contains numerous highly convoluted **seminiferous tubules** (SEM-in-IF-er-uss), in which sperm are formed. Stretched end to end, the seminiferous tubules of each testis would extend for about half a mile.

Sperm produced in the seminiferous tubules empty into a network of connecting tubules in the "back" of the testes. These tubules, in turn, empty into the **epididymal duct** (ep-eh-DID-eh-mal), where sperm are stored until released during **ejaculation**, the ejection of sperm. As **FIGURE 19-2** shows, the epididymal duct forms the epididymis. The epididymal duct empties into the **vas deferens** (plural, **vasa deferentia**, pronounced VAH-sah DEAF-er-en-she-ah). These ducts pass from the scrotum into the body cavity through the **inguinal canals,** small openings in the wall of the abdomen (**FIGURE 19-3A**). As shown in **FIGURE 19-1**, the vasa deferentia empty into the urethra. Sperm therefore travel from the seminiferous tubules to the epididymal duct to the vas deferens to the urethra. The inguinal canals are potential weak spots in the abdominal wall. Loops of intestine can bulge through weakened canals of some men (**FIGURE 19-3B**).

Semen Consists Mostly of Fluids Produced by the Sex Accessory Glands.

During ejaculation, sperm are joined by fluids produced by the **sex accessory glands**. The sex accessory glands are located near the neck of the urinary bladder and include two **seminal vesicles,** the **prostate gland,** and two **Cowper's glands** (**FIGURE 19-1**). The secretions produced by the sex accessory glands make up 99% of the volume of the **ejaculate,** or **semen** (SEA-men). Semen is a

fluid containing sperm and sex accessory gland secretions. The semen contains fructose, a monosaccharide that is used by sperm mitochondria to generate the energy needed to help propel the sperm through the female reproductive tract. Semen also contains a buffer that neutralizes the acidic secretions of the female reproductive tract. Yet another component of the semen is prostaglandin, a chemical that causes the muscle of the uterus to contract, which is believed to be primarily responsible for the movement of sperm up the female

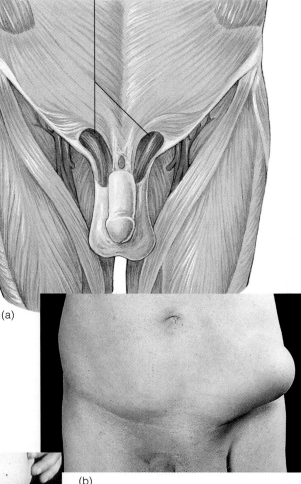

Inguinal canals

(a)

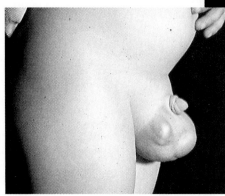

(b)

(c)

FIGURE 19-3 The Inguinal Canal and Hernia *(a)* During development, the testis descends through the inguinal canal, an opening through the musculature in the lower abdominal wall. In adults, the inguinal canal provides a route for the vas deferens, blood vessels, and nerves that supply each testis. *(b)* Loops of intestine may push through the weakened musculature surrounding the inguinal canal. *(c)* In some instances, large sections of the intestine may push into the scrotum.

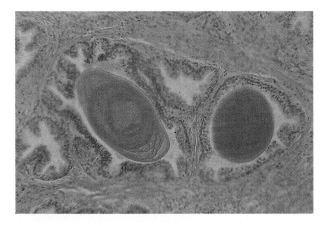

FIGURE 19-4 Prostatic Nodules

tract. Interestingly, lymphocytes "patrol" the testes and seminal vesicles. Some lymphocytes are expelled with the ejaculate, explaining why AIDS is transmitted in the semen of men.

The paired **seminal vesicles** (SEM-in-al) empty into the vasa deferentia and produce the largest portion of the ejaculate. The **prostate gland** (PROS-tate) surrounds the neck of the bladder and empties its contents directly into the urethra. Routine medical examinations of men over

the age of 45 show that nearly all of them have enlarged prostates. This condition results from the formation of small nodules inside the gland. These nodules form by the condensation of prostatic secretions inside the gland (**FIGURE 19-4**). Although they usually cause no trouble, in some cases the nodules grow quite large and can block the flow of urine, making urination painful. In such cases, the prostate can be reamed out by a device inserted through the penis or be removed through surgery. The prostate is also a common site for cancer in men and should therefore be checked every year by a physician in men over 40. The **Cowper's glands** (COW-perz), a pair of small glands located below the prostate on either side of the urethra, are the smallest of the sex accessory glands.

Sperm Are Formed from Stem Cells Known as Spermatogonia.

Sperm are produced in the seminiferous tubules (**FIGURE 19-5A**). The lining of the wall of the tubule is known as the **germinal epithelium** because its cells give rise to sperm, the male germ cells. The formation of sperm is

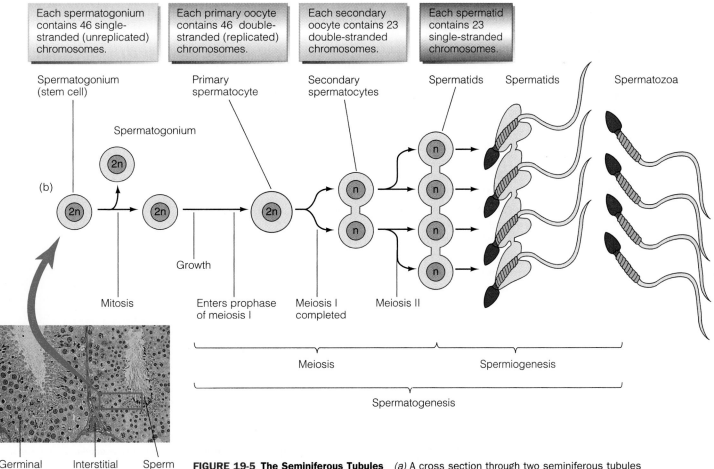

Each spermatogonium contains 46 single-stranded (unreplicated) chromosomes.

Each primary oocyte contains 46 double-stranded (replicated) chromosomes.

Each secondary oocyte contains 23 double-stranded chromosomes.

Each spermatid contains 23 single-stranded chromosomes.

Spermatogonium (stem cell)

Primary spermatocyte

Secondary spermatocytes

Spermatids   Spermatids   Spermatozoa

Spermatogonium

(b)

Growth

Mitosis

Enters prophase of meiosis I

Meiosis I completed

Meiosis II

Meiosis

Spermiogenesis

Spermatogenesis

(a)

Germinal epithelium   Interstitial cells   Sperm

FIGURE 19-5 The Seminiferous Tubules *(a)* A cross section through two seminiferous tubules showing the germinal epithelium and the interstitial cells. *(b)* Details of spermatogenesis.

known as **spermatogenesis** (sper-MAT-oh-GEN-eh-siss) and involves two subprocesses: (1) a special type of cell division known as **meiosis** and (2) **spermiogenesis** (SPERM-ee-oh-GEN-eh-siss), a process of cellular differentiation (**FIGURE 19-5B**).

Spermatogenesis begins with spermatogonia (sper-MAT-oh-GO-nee-ah). Located in the periphery of the seminiferous tubule in the germinal epithelium, the **spermatogonia** divide mitotically and therefore ensure a constant supply of sperm-producing cells. Some of the spermatogonia formed during cellular division, however, enlarge and become **primary spermatocytes** (sper-MAT-oh-sites). Two meiotic divisions follow in the formation of sperm.

As you may recall from Chapter 16, the first division in meiosis is meiosis I. During meiosis I, primary spermatocytes divide to form two **secondary spermatocytes** (**FIGURE 19-5B**). Each secondary spermatocyte contains 23 double-stranded chromosomes. During **meiosis II**, secondary spermatocytes divide, forming four **spermatids**. Each spermatid contains 23 single-stranded (unreplicated) chromosomes.

Spermatids soon develop into sperm (**FIGURE 19-5**). This process is called **spermiogenesis**. During spermiogenesis, the nuclear material of the spermatid condenses and most of the cytoplasm is shed to streamline the cell. The sperm tail also forms during this process from the centriole. Also during this amazing transformation, the mitochondria of the spermatid congregate around the first part of the tail, where they can provide energy for propulsion. The Golgi apparatus enlarges and forms an enzyme-filled cap that fits over the condensed nucleus like a stocking cap. This structure, the **acrosome** (ACK-row-sohm), will help the sperm digest its way through the coatings surrounding the ovum during fertilization. The product of spermiogenesis is the **spermatozoan** (sper-MAT-oh-ZO-an), or mature **sperm**. It is a marvel of biological architecture—a true testimony of the marriage of structure and function. Rid of excess cytoplasmic baggage, it is streamlined for relatively swift movement.

On average, men produce 200–300 million sperm every day. The average 3-milliliter ejaculate contains 240 million or more—nearly as many people as there are in the U.S. population. Such large numbers no doubt evolved because many sperm are required to ensure fertilization. In fact, nearly all sperm are eliminated as they travel through the female reproductive tract.

In humans, each sperm formed during meiosis contains 23 single-stranded (unreplicated) chromosomes—half the number in a normal somatic cell. Thus, when the sperm unites with an ovum (also containing 23 unreplicated chromosomes), they produce a zygote containing 46 single-stranded chromosomes. One-half of its chromosomes come from each parent.

Interstitial Cells Produce the Male Sex Steroid Testosterone

The spaces between the seminiferous tubules contain clumps of large cells known as **interstitial cells** (in-ter-STISH-al) (**FIGURE 19-6**). These cells produce **androgens**, steroids that exert a masculinizing effect. The most important androgen is **testosterone**. Testosterone diffuses out of the interstitial cells and into the seminiferous tubules, where it stimulates spermatogenesis (prophase I) and spermiogenesis, the maturation of sperm. In the absence of testosterone, sperm cell production declines, then stops, and the walls of the seminiferous tubules shrink.

Testosterone is also transported in the bloodstream throughout the body, where it affects a variety of target cells. For example, it stimulates cellular growth in bone and muscle and accounts in part for the fact that men are generally taller and more massive than women. In addition, testosterone promotes facial hair growth and thickening of the vocal cords, typically giving men deeper voices than women. Testosterone stimulates growth of the laryngeal cartilage, producing the prominent bulge called the *Adam's apple*. It also stimulates cell growth in the skin, making most men's skin slightly thicker than women's.

Testosterone affects the hair follicles on the heads of genetically predisposed men, causing pattern baldness. It is not the absence of testosterone, as some believe, but the presence of testosterone and certain genes that lead to this condition.

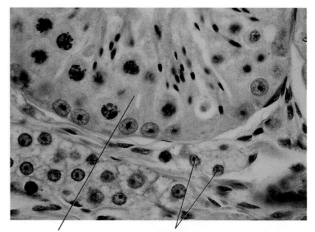

Seminiferous tubule   Interstitial cells

FIGURE 19-6 Interstitial Cells   Cross section of the wall of a seminiferous tubule showing the interstitial cells.

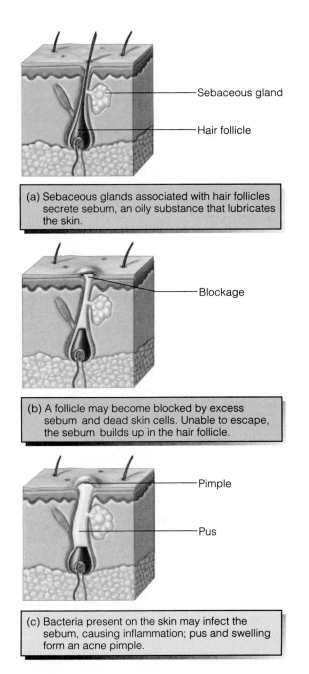

(a) Sebaceous glands associated with hair follicles secrete sebum, an oily substance that lubricates the skin.

— Sebaceous gland

— Hair follicle

— Blockage

(b) A follicle may become blocked by excess sebum and dead skin cells. Unable to escape, the sebum builds up in the hair follicle.

— Pimple

— Pus

(c) Bacteria present on the skin may infect the sebum, causing inflammation; pus and swelling form an acne pimple.

FIGURE 19-7 Formation of an Acne Pimple   *(a)* Testosterone stimulates oil production in the sebaceous glands. *(b)* If the outlet is blocked, sebum builds up in the gland and *(c)* the gland may become infected.

Testosterone also stimulates the sebaceous glands of the skin in both sexes. **Sebaceous glands** (seh-BAY-schuss) secrete oil onto the skin, moisturizing it (**FIGURE 19-7A**). During puberty (sexual maturation) in boys, testosterone levels rise dramatically, causing a marked increase in sebaceous gland activity. Dead skin cells may block the pores that normally carry the oil to the skin's surface (**FIGURE 19-7B**). As a result, sebum collects inside the glands. Bacteria on the skin often invade and proliferate in the small pools of oil, resulting in in-

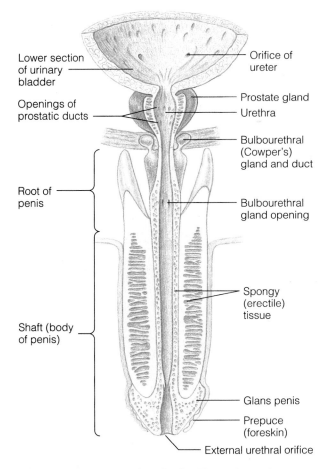

FIGURE 19-8 Anatomy of the Penis   The penis consists principally of spongy tissue that fills with blood during sexual arousal. The urethra passes through the penis, carrying urine or semen.

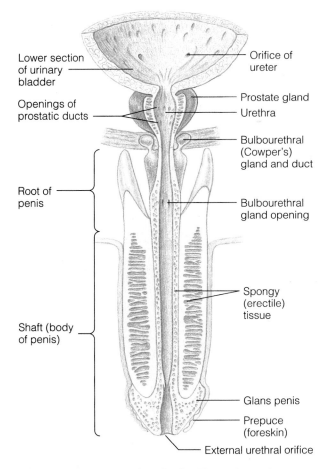

(above image labels)
Lower section of urinary bladder
Openings of prostatic ducts
Root of penis
Shaft (body of penis)
Orifice of ureter
Prostate gland
Urethra
Bulbourethral (Cowper's) gland and duct
Bulbourethral gland opening
Spongy (erectile) tissue
Glans penis
Prepuce (foreskin)
External urethral orifice

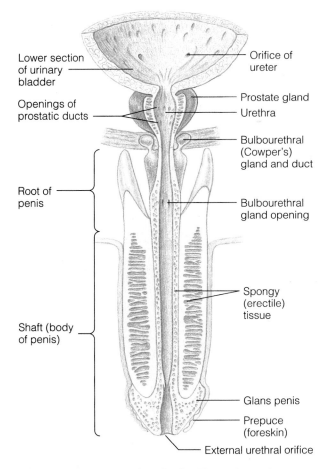

flammation, pus formation, and swelling. The skin protrudes, forming an **acne pimple**.

Mild acne can be treated by washing the skin twice a day with unscented soap. Women should avoid makeup that has an oily base or use a nonoily type of foundation and wash their faces thoroughly each night. Sunlight also helps clear up acne, because it dries the oil on the skin and kills skin bacteria. Severe acne can be treated by special ointments and antibiotics.

The Penis Contains Erectile Tissue That Fills with Blood During Sexual Arousal

Sperm are deposited in the female reproductive tract with the aid of the **penis**. The penis consists of a shaft of varying length and an enlarged tip, the **glans penis** (**FIGURE 19-8**). The glans is covered by a sheath of skin at birth, the **foreskin**. The foreskin gradually becomes separated from the glans in the first 2 years of life. At puberty, the inner lining of the foreskin begins to produce an oily secretion. Bacteria can grow in the protected, nutrient-rich environment created by the foreskin, so special precautions must be taken to keep the area clean.

Because of potential health problems or religious reasons, many parents opt to have the foreskin removed in the first few days of their son's life. The operation, called **circumcision** (sir-come-SIZH-un; literally, "to cut around"), may help reduce penile cancer in men and may also reduce cervical cancer in the wives or sexual partners of circumcised men, as explained in the Point/Counterpoint in this chapter.

For successful copulation, the penis must become rigid, or erect. During sexual arousal, nerve impulses in the parasympathetic division of the autonomic nervous system cause arterioles in the penis to dilate. Blood flows into a spongy **erectile tissue** (eh-REK-tile) in the shaft of the penis, making it harden. The growing turgidity (swelling) compresses a large vein on the dorsal surface of the penis, blocking the outflow of blood and further stiffening the organ.

Coursing through the penis is the urethra, a duct that carries urine from the bladder to the outside of the body during urination. The urethra also transports semen—sperm, and secretions of the sex accessory glands during ejaculation.

Some men lose their ability to become erect or to sustain an erection. This condition, known as **impotence** (IM-poe-tense), may be caused by psychological, physical, or physiological problems. Marital conflict, stress, fatigue, and anxiety all contribute to impotence. If the problem is psychological, therapy is often advised. Patients with nerve damage, however, are likely to suffer permanent impotence. Nerve damage may result from diabetes mellitus or from a traumatic accident.

Smoking is also a major contibutor to impotence. In fact, 52% of all men from 40 to 60 years of age suffer from impotence; smokers are twice as likely to be impotent as nonsmokers because smoking causes arterioles in the penis to constrict, reducing blood flow. Smoking also causes the buildup of atherosclerotic plaque, which permanently blocks blood flow to this organ.

For patients with irreversible impotence, urologists can surgically insert an inflatable plastic implant in the penis. The implant is attached to a small, fluid-filled reservoir in the scrotum. The fluid is manually pumped into the implant, making the penis erect upon demand, thus permitting sexual intercourse. Other types of implants are also available. A number of drug treatments are also available including the much-talked-about drug Viagra, which promotes erection when men are sexually aroused.

Ejaculation Is a Reflex Mechanism

When sexual stimulation becomes intense, sensory nerve impulses traveling to the spinal cord activate motor neurons there. These neurons send impulses to the smooth muscle in the walls of the epididymis and vasa deferentia, causing them to contract. This is spinal cord reflex. Muscle contraction, in turn, propels sperm into the urethra. Nerve impulses from the spinal cord also stimulate the smooth muscle in the walls of the sex accessory glands to contract, causing these glands to empty their secretions into the vasa deferentia and the urethra. The sperm and secretions from the sex accessory glands form the semen.

Semen is propelled onward by smooth muscle contractions in the walls of the urethra, which cause the sperm to be released in spurts.

The Male Reproductive System Is Controlled by Three Hormones: Testosterone, Luteinizing Hormone, and Follicle-Stimulating Hormone

As noted earlier, the testes produce sex steroid hormones—notably, testosterone. Testosterone secretion by the interstitial cells is controlled by **luteinizing hormone** (**LH**). LH in males is also known as **interstitial cell stimulating hormone** (**ICSH**). ICSH secretion is controlled by a releasing hormone produced by the hypothalamus, known as **gonadotropin releasing hormone (GnRH)**.

As **FIGURE 19-9** shows, the secretion of GnRH and ICSH is controlled by testosterone levels in the blood in a classic negative feedback loop. Accordingly, a decline in testosterone levels in the blood signals an increase in GnRH secretion, resulting in an increase in ICSH secretion. But when testosterone levels return to normal, GnRH and ICSH release subsides.

The pituitary also produces the gonadotropin **follicle-stimulating hormone** or **FSH**. Like testosterone, FSH stimulates spermatogenesis (**FIGURE 19-9**). However, FSH does not act directly on the spermatogenic cells. Instead, it exerts its influence through another cell in the germinal epithelium of the seminiferous tubule, the Sertoli cell (ser-TOLL-ee). **Sertoli cells**, shown in **FIGURE 19-10**, are large "nurse cells." The spermatogenic cells (spermatogonia, spermatocytes, and spermatids) divide and differentiate within folds in the plasma membrane of the Sertoli cell, moving slowly to the surface of the germinal epithelium. The spermatids produced during spermatogenesis remain attached to the Sertoli cells, where they differentiate into sperm.

FSH stimulates the Sertoli cells to produce a cytoplasmic receptor protein that binds to testosterone. Called **androgen-binding protein**, this receptor concentrates testosterone within the Sertoli cell. Testosterone, in turn, stimulates spermatogenesis.

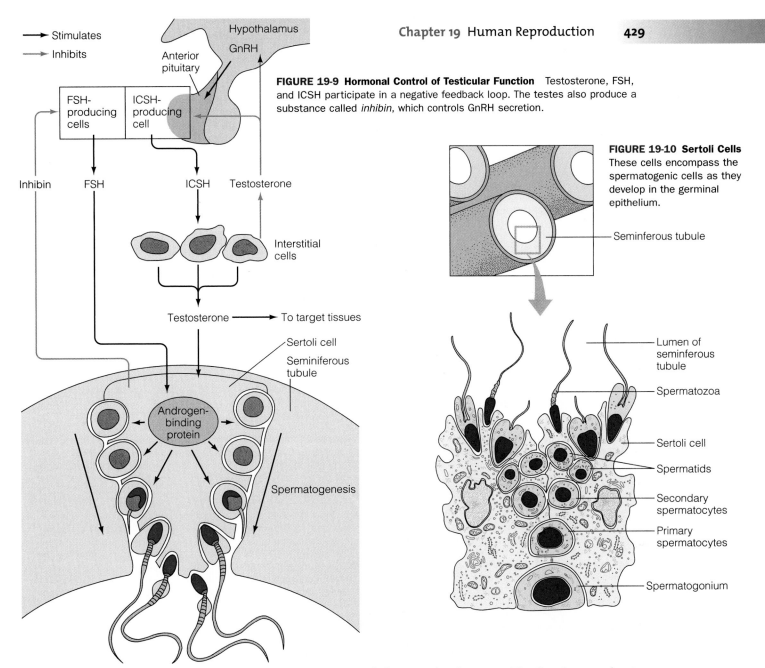

Stimulates
Inhibits

FIGURE 19-9 Hormonal Control of Testicular Function Testosterone, FSH, and ICSH participate in a negative feedback loop. The testes also produce a substance called *inhibin*, which controls GnRH secretion.

Hypothalamus
GnRH
Anterior pituitary

FSH-producing cells
ICSH-producing cell

Inhibin   FSH   ICSH   Testosterone

Interstitial cells

Testosterone → To target tissues

Sertoli cell
Seminiferous tubule

Androgen-binding protein

Spermatogenesis

FIGURE 19-10 Sertoli Cells These cells encompass the spermatogenic cells as they develop in the germinal epithelium.

Seminferous tubule

Lumen of seminferous tubule

Spermatozoa

Sertoli cell

Spermatids

Secondary spermatocytes

Primary spermatocytes

Spermatogonium

FSH secretion is controlled by GnRH and a peptide hormone called **inhibin** (**FIGURE 19-10**), which is produced by Sertoli cells. Inhibin gets its name from the fact that it inhibits the activity of FSH-secreting cells in the anterior pituitary and thus blocks the action of GnRH. When inhibin levels are high, FSH secretion is low.

The Female Reproductive System

FIGURE 19-11 illustrates the **female reproductive system,** which consists of two parts: the external genitalia

and the reproductive tract. The female reproductive tract consists of four structures: (1) the ovaries, (2) the uterine tubes, (3) the uterus, and (4) the vagina. **TABLE 19-2** summarizes the role of each structure.

TABLE 19-2

The Female Reproductive System

Component	Function
Ovaries	Produce ova and female sex steroids
Uterine tubes	Transport sperm to ova; transport fertilized ova to uterus
Uterus	Nourishes and protects embryo and fetus
Vagina	Site of sperm deposition, birth canal

The Medical Debate over Circumcision

AN UNNECESSARY AND COSTLY PRACTICE

Dr. Thomas Metcalf

Dr. Thomas J. Metcalf *is a clinical associate professor of pediatrics at the University of Utah and practices at Willow Creek Pediatrics in Salt Lake City.*

Routine neonatal circumcision continues to be performed on the majority of American boys. I remain convinced that routine circumcision of the newborn is not needed if adequate hygiene of the uncircumcised penis is maintained throughout life. Circumcision costs the American public a significant amount of money and remains a culturally motivated operation without a valid medical *raison d'etre.*

Circumcision for nonreligious reasons began in the United States in the 1870s when genital surgery of both sexes became established as a preventive and/or cure for masturbation. Though genital surgery of females declined, circumcision of males persisted in the United States, even as it fell into disfavor in Europe. In 1949, Gairdner and other authors began to speak out against routine newborn circumcision. The medical community gradually came to the position that "newborn circumcision has potential medical benefits and advantages as well as disadvantages and risks" (1989, American Academy of Pediatrics).

Eighty percent of the world remains uncircumcised. In the United States, however, 60%–90% of 1.9 million male newborns are circumcised annually. The cost for a newborn circumcision in Salt Lake City, Utah is $110; in other areas of the U.S. the cost is 2–3 times higher. Thus, the total annual cost to the American public for newborn circumcisions is between $209 million and $627 million. Are there compelling reasons to circumcise newborn boys at such an expense, in an era of burgeoning health care costs? Is circumcision a public health measure, an "immunization"?

Wiswell and others, in retrospective studies, have documented a greater incidence of urinary tract infections (UTIs) in uncircumcised

male infants, roughly 1%–2% versus 0.1% in circumcised infants. A prospective, controlled study has not been published. The higher rate in uncircumcised males could be due to more frequent and heavier colonization of the periurethral area by pathogens. However, circumcision cannot be thought of as a protective "immunization" against UTI. If a male infant presents with a febrile illness for which no obvious source of infection is found, the physician must consider the possibility of a UTI in *either* a circumcised or uncircumcised male. Circumcision does not obviate the need for this evaluation, which currently costs $99 in Salt Lake City. Given a UTI incidence of 1%–2% in uncircumcised infants, the cost to prevent one UTI by performing 100 circumcisions is $11,000.

Circumcision largely prevents cancer of the penis, which has a mortality rate of up to 25%. This would seem a strong argument for neonatal circumcision as a preventive. However, other factors play a role. According to a report by the Task Force on Circumcision, in developed countries where circumcision is not routinely performed (and parents and boys are used to caring for the uncircumcised penis), the incidence of penile carcinoma ranges from 0.3 to 1.1 per 100,000 men per year, about half the incidence in uncircumcised U.S. men, but greater than that in circumcised men. In developing countries with lower standards of hygiene, the incidence is from 3 to 6 per 100,000 men per year. Thus, good hygiene may make up for the effect of circumcision, at a fraction of the cost. A recent case-controlled study in Washington and British Columbia showed only a 3.2 times greater risk for uncircumcised relative to circumcised men.

Sexually transmitted diseases (STDs) have been shown in some studies to be more prevalent in uncircumcised men. Parker et al. showed a significantly higher risk of four types of STDs among uncircumcised men in Australia. However, these authors did not recommend circumcision to prevent STDs. They stated that "if these findings are confirmed in other studies, it would seem that attention should be di-

rected to the improvement of personal hygiene among uncircumcised men." In the final analysis, physicians cannot promote circumcision as an effective way to prevent STDs.

Lack of circumcision appears to be associated with an increased risk of transmission of the AIDS virus. This correlation has been shown in both clinic-based studies and in a statistical study of male circumcision status in 37 African capital cities. The authors of the study suggest that lack of circumcision is a cofactor in HIV infection, not causative. They state that uncircumcised African males "are apparently at increased risk of developing chancroid and other genital-ulcer disease," which, in turn, facilitate infection with HIV; they also write that perhaps the intact foreskin enhances viral survival and, finally, that more frequent infection of the glans of the penis increases susceptibility to HIV. Again, while all this may be true, circumcision would not be an effective way of solving the AIDS epidemic and should not be promoted as such.

Complications of routine newborn circumcision are indeed infrequent, and most are easily treated. The use of xylocaine for local anesthesia, while inflicting its own pain, renders the remainder of the circumcision procedure painless in 80% of cases.

Studies in the United States and New Zealand have shown more problems in caring for the uncircumcised penis in the first few years of life, but data from Europe suggest that the uncircumcised penis presents few problems for parents and boys used to dealing with it. In any event, problems that arise during care of the circumcised *or* uncircumcised penis are generally minor, requiring only one medical visit for correction.

In summary, there is an increasing body of evidence for the medical benefits of newborn circumcision. However, this data is insufficient to support advocating circumcision, and its cost, as a national public health measure. Even in an individual case, a physician could never say, "Circumcise your newborn and he'll not have to worry about STDs, UTIs, penile cancer, or AIDS." Thus, newborn circumcision remains a costly, painful, and unnecessary procedure.

A SAFE AND BENEFICIAL PROCEDURE

Dr. Thomas Wiswell

Sixty percent to 90% of newborn boys (1.2 to 1.8 million) are circumcised annually in the United States. Several issues have convinced me of the benefits of this procedure: (1) the prevention of urinary tract infections (UTIs) and complications from them; (2) the prevention of penile cancer; (3) the lower incidence of sexually transmissible diseases in uncircumcised males; (4) the low risk for complications from the operation; (5) the greater incidence of penile problems among "intact" boys; and (6) recent evidence showing that circumcision protects against AIDS.

Circumcised boys are 10–39 times less likely to have UTIs during infancy. From population studies involving more than 600,000 children, there are nine investigations that have confirmed the increased risk for UTIs among uncircumcised male infants. In addition, three other studies have found older boys and adult males who are uncircumcised to be more likely to develop these infections. Urinary tract infections are not benign. More than 36% of 88 boys below 1 month of age with a UTI had concurrent infection in their bloodstream. Furthermore, three of these infants had concomitant meningitis, two had renal failure, and two died. Littlewood has reported that 11% of children with a UTI during the first month of life may die. There are longer-term effects of UTIs in children. In infected infants, 10%–15% will subsequently demonstrate kidney scarring. Approximately 10% of these infants will develop high blood pressure, and 2%–3% will ultimately require dialysis or kidney transplantation.

Penile cancer is the only malignancy that can be prevented categorically by a prophylactic procedure, neonatal circumcision. Of the more than 60,000 cases of penile cancer that have occurred in the United States since 1930, fewer than 10 have been in circumcised men. More than 1000 men develop penile cancer each year, and 225–317 die from it annually. The basic therapy for this malignancy is amputation of the penis.

Virtually all sexually transmissible diseases (STDs) have been found to occur more frequently among uncircumcised men. There are more than 70 references that have found STDs to occur more often among "intact" individuals. I am struck by the paucity of contrary reports. There is only one report of a venereal disease (nongonococcal urethritis) being more common in circumcised men. However, in this population, more than 60% of the cases of another STD (gonorrhea) occurred in uncircumcised men.

Serious complications from routine foreskin removal are infrequent and relatively minor. We have recently examined a population of more than 100,000 circumcised boys and found complications in fewer than 2 per 1000. Two other large investigations reported complications from circumcision in 0.06% and 0.20% of circumcised boys, respectively. The majority of the complications are easily treated bleeding and minor infections. Atypical complications of the procedure (glans loss, staphylococcal scalded skin syndrome, etc.) occur infrequently and receive note due to their uniqueness. Death rarely occurs as a complication of circumcision. To date, there have been a total of three reported deaths in the United States since 1954 that can be ascribed to neonatal circumcision. This contrasts sharply with the potentially preventable 7500–11,500 deaths from penile cancer that occurred during the same period.

Herzog and Alvarez found uncircumcised boys aged 4 months to 12 years to be more likely to have "penile problems" than were circumcised boys. Fergusson et al. similarly described uncircumcised boys as having more problems than their circumcised counterparts during the first 8 years of life. In both investigations, the "problems" largely consisted of balanitis (infection in and around the head of the penis) and phimosis (an abnormal constriction of the foreskin that prevents urine from being excreted). Recent investigations have confirmed that adult uncircumcised men will have more "penile problems" (the majority of which are balanitis and phimosis) than their circumcised counterparts. Finally, we found the risks from circumcision during the first month of life to be fewer than the risks from the uncircumcised state.

Over the past six years, at least eight investigations from the United States and Africa have found uncircumcised men to be 4 to 12 times more likely to become infected with the human immunodeficiency virus (HIV—the AIDS virus).

As a pediatrician, I am a child advocate. I have pondered this issue for many years. I understand that we would have to circumcise "the many" to protect "the few." However, we have no way of identifying "the few." Neonatal circumcision is a rapid and generally safe procedure that must be performed by experienced caretakers. With the low complication rate and the many benefits of the procedure, I personally believe we should routinely circumcise newborn boys.

Dr. Thomas E. Wiswell *is a professor of pediatrics and director of Neonatal Research at Thomas Jefferson University in Philadelphia.*

▎SHARPENING YOUR CRITICAL THINKING SKILLS

1. Summarize the key points of each author, then list the data they use to support their main points.

Visit *Human Biology's* Internet site, www.jbpub.com/humanbiology, to research opposing web sites and respond to questions that will help you clarify your own opinion. (See Point/Counterpoint: Furthering the Debate.)

www.jbpub.com/humanbiology

FIGURE 19-11 Anatomy of the Female Reproductive Tract *(a)* Frontal view. *(b)* Midsagittal view.

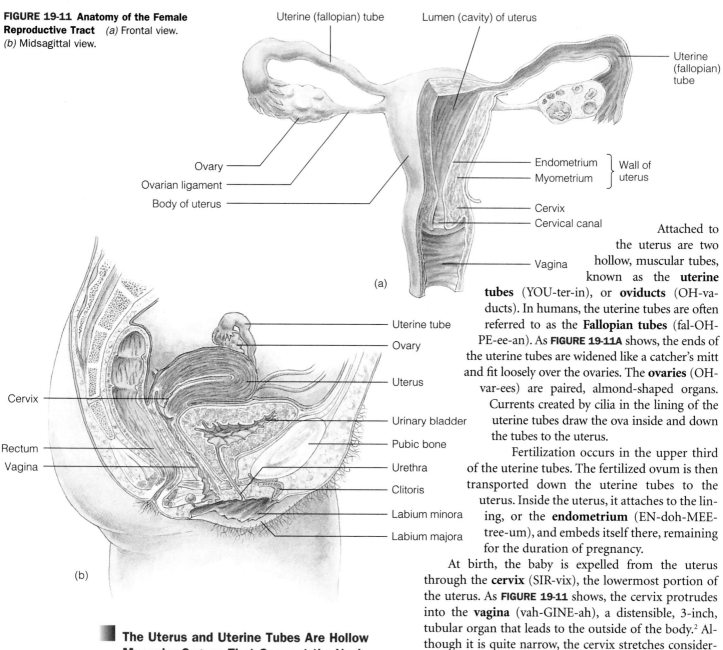

Uterine (fallopian) tube

Lumen (cavity) of uterus

Uterine (fallopian) tube

Ovary

Ovarian ligament

Body of uterus

Endometrium
Myometrium
} Wall of uterus

Cervix
Cervical canal

Vagina

(a)

Uterine tube
Ovary
Uterus
Urinary bladder
Pubic bone
Urethra
Clitoris
Labium minora
Labium majora

Cervix
Rectum
Vagina

(b)

■ **The Uterus and Uterine Tubes Are Hollow Muscular Organs That Connect the Vagina, the Site of Sperm Deposition, with the Ovaries Where the Ova Are Produced**

The **uterus** is a pear-shaped organ about 7 centimeters (3 inches) long and about 2 centimeters (less than 1 inch) wide at its broadest point in nonpregnant women.[1] The wall of the uterus contains a thick layer of smooth muscle cells, the **myometrium** (MY-oh-ME-tree-um). The uterus houses and nourishes the developing fetus.

Attached to the uterus are two hollow, muscular tubes, known as the **uterine tubes** (YOU-ter-in), or **oviducts** (OH-va-ducts). In humans, the uterine tubes are often referred to as the **Fallopian tubes** (fal-OH-PE-ee-an). As **FIGURE 19-11A** shows, the ends of the uterine tubes are widened like a catcher's mitt and fit loosely over the ovaries. The **ovaries** (OH-var-ees) are paired, almond-shaped organs. Currents created by cilia in the lining of the uterine tubes draw the ova inside and down the tubes to the uterus.

Fertilization occurs in the upper third of the uterine tubes. The fertilized ovum is then transported down the uterine tubes to the uterus. Inside the uterus, it attaches to the lining, or the **endometrium** (EN-doh-MEE-tree-um), and embeds itself there, remaining for the duration of pregnancy.

At birth, the baby is expelled from the uterus through the **cervix** (SIR-vix), the lowermost portion of the uterus. As **FIGURE 19-11** shows, the cervix protrudes into the **vagina** (vah-GINE-ah), a distensible, 3-inch, tubular organ that leads to the outside of the body.[2] Although it is quite narrow, the cervix stretches considerably at birth to allow the passage of the baby into the vagina. The vagina also serves as the receptacle for sperm during sexual intercourse. To reach the ovum, sperm must travel through a tiny opening and narrow canal of the cervix that leads into the uterus. From here, sperm move up both uterine tubes.

The **external genitalia** consist of two flaps of skin on either side of the vaginal opening (**FIGURE 19-12**). The outer folds are the **labia majora** (LAY-bee-ah ma-JOR-ah). These large folds of skin are covered with hair on the outer surface and contain numerous sebaceous glands

---

[1]The uterus is slightly larger in women who have had children, and it enlarges considerably during pregnancy to accommodate the growing fetus.

[2]The vagina is often called the *birth canal*.

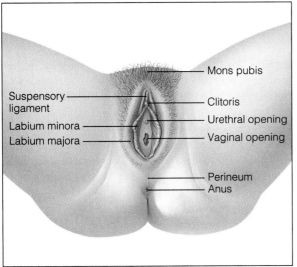

Mons pubis

Suspensory ligament

Clitoris

Urethral opening

Labium minora

Vaginal opening

Labium majora

Perineum

Anus

FIGURE 19-12 Female External Genitalia

One is the female germ cells—immature ova or oocytes. **Oocytes** develop from **oogonia** (oh-oh-GO-knee-uh), cells akin to the spermatogonia of the seminiferous tubules. Oogonia, however, are only found in fetal ovaries. Well before birth, they enter meiosis I but do not complete it. At birth, all oogonia have therefore been converted into primary oocytes. Each primary oocyte contains 46 chromosomes and undergoes two meiotic divisions. This process is called **oogenesis** (oh-oh-GEN-eh-siss).

Meiosis is similar in oogenesis and spermatogenesis. However, oogenesis produces only one ovum,

on the inside. The inner flaps are the **labia minora** (meh-NOR-ah). Anteriorly, they meet to form a hood over a small knot of tissue called the **clitoris** (CLIT-er-iss). The clitoris consists of erectile tissue and is a highly sensitive organ involved in female sexual arousal. It is formed from the same embryonic tissue as the penis. In fact, some women are born with greatly elongated clitorides (plural).

The Ovaries Produce Ova and Release Them During Ovulation

During each menstrual cycle, one ovary releases an ovum, the female gamete. It is drawn into the uterine tube. The release of an ovum is called **ovulation** (OV-you-LAY-shun).[3] Ovulation occurs approximately once a month in women during their reproductive years—from puberty (age 11–15) to menopause (age 45–55). Ovulation is temporarily halted when a woman is pregnant and may be suppressed by emotional and physical stress.

The structure of an ovary is shown in **FIGURE 19-13**. As illustrated, several ovarian landmarks are visible.

---

[3]The release of the ovum is not an explosive event, although many women feel a sharp pain when it occurs.

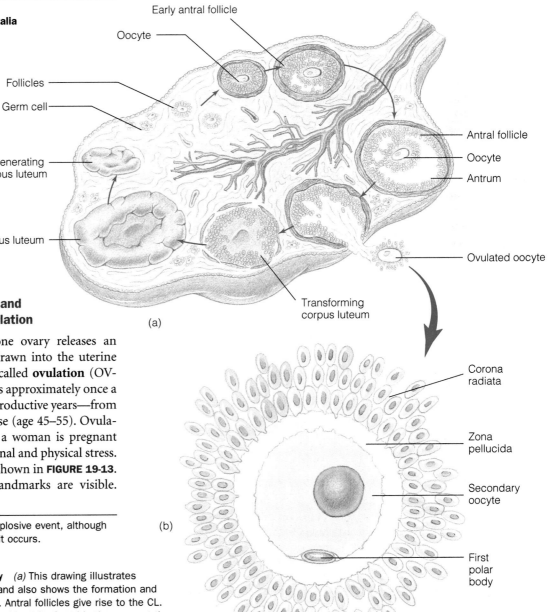

Early antral follicle

Oocyte

Follicles

Germ cell

Degenerating corpus luteum

Corpus luteum

Antral follicle

Oocyte

Antrum

Ovulated oocyte

Transforming corpus luteum

(a)

Corona radiata

Zona pellucida

Secondary oocyte

First polar body

(b)

FIGURE 19-13 Structure of the Ovary *(a)* This drawing illustrates the phases of follicular development and also shows the formation and destruction of the corpus luteum (CL). Antral follicles give rise to the CL. A fully formed CL and antral follicle would not be found in the ovary at the same time. *(b)* A recently ovulated follicle showing the corona radiata and zona pellucida.

whereas spermatogenesis produces four sperm. How does this happen?

During each meiotic division in oogenesis, the nucleus divides in half, but the cytoplasm doesn't (**FIGURE 19-14**). Thus, the first meiotic division produces only one cell, the **secondary oocyte**, and a small package of discarded nuclear material containing 23 double-stranded chromosomes. This structure is called the **first polar body** and contains a tiny amount of cytoplasm.

During the second meiotic division, the nucleus divides again, but the cytoplasm doesn't. This "unequal division" results in the formation of an ovum, containing 23 single-stranded chromosomes and yet another "nuclear discard," the **second polar body.**[4]

Germ cells are housed in the ovary in special structures called *follicles* (**FIGURE 19-13A**). A **follicle** consists of a primary oocyte surrounded by one or more layers of **follicle cells**, which are derived from the loose connective tissue of the ovary.

FIGURE 19-14 (right side) illustrates the growth and development of follicles. Each month, a dozen or so follicles begin to develop. During early development, the oocyte enlarges. The follicle cells divide and grow, first forming a complete layer around the oocyte, then forming many layers.

In the largest follicles, a clear liquid begins to accumulate between the follicle cells. The fluid creates small spaces among the follicle cells, which enlarge as additional fluid is generated. Eventually, the cavities coalesce, forming one central cavity. At this point, the follicle is called an **antral follicle** (AN-tril). Although a dozen or so follicles begin developing during each cycle, as a rule only one makes it to ovulation. The rest stop growing and degenerate.

The follicle (or follicles) that survives continues to enlarge by accumulating more fluid. As the fluid builds up, the antral follicle begins to bulge from the surface of the ovary, not unlike a pimple. The pressure exerted on the outside of the ovary causes the ovary's surface to stretch. Blood vessels supplying the tissue may be compressed, resulting in a region of cellular necrosis (neh-CROW-siss; "death"). This weakens the wall. Enzymes released from ovarian cells in the region then begin to digest the tissue at the weak point. Eventually, the wall of the follicle breaks down, and the oocyte is released.

Around the time of ovulation, the primary oocyte in the antral follicle completes the first meiotic division (**FIGURE 19-14**). It is then called a **secondary oocyte**. As

FIGURE 19-13B shows, the secondary oocyte is surrounded by a fairly thick layer of gel-like material called the **zona pellucida** (pell-LEW-seh-dah) (**FIGURE 19-13B**). It is surrounded by a layer of follicle cells.

During ovulation, the oocyte and surrounding cells are expelled from the ovary, and the ovulated follicle collapses. The collapsed follicle forms the **corpus luteum** (CORE-puss LEU-tee-um; "yellow body") or **CL** for short—so named because of the yellow pigment it contains in cows and pigs (**FIGURE 19-15**).

The CL is a transient endocrine gland that produces two sex hormones, estrogen and progesterone. The fate of the CL ultimately depends on the fate of the oocyte. If the oocyte is fertilized, the CL remains active for several months, producing the estrogen and progesterone needed for a successful pregnancy. If fertilization does not occur, the CL soon disappears. (More on this in Chapter 20.)

Cyclic Changes in Pituitary Hormones in Women Are Responsible for the Menstrual Cycle

Women of reproductive age undergo a series of interdependent hormonal, ovarian, and uterine changes each month known as the **menstrual cycle** (MEN-strell). The length of the menstrual cycle varies from one woman to the next. In some it lasts 25 days, and in others it may last 35 days. The length of the menstrual cycle may also vary from month to month in the same woman. On average, however, the cycle repeats itself every 28 days. Ovulation usually occurs approximately at the midpoint of the 28-day cycle, or about 14 days before the onset of menstruation.

As noted above, the menstrual cycle involves three interdependent cycles (**FIGURE 19-15**, page 436). The first is a hormonal cycle. The hormonal cycle, in turn, produces cyclic changes in the ovary (the ovarian cycle) and the uterus (the uterine cycle). Understanding the menstrual cycle is easiest if we begin with the hormonal and ovarian cycles.

The Hormonal and Ovarian Cycles.

The first half of the menstrual cycle is known as the follicular phase, because it is during this time that the follicles grow toward ovulation. The second half is called the **luteal phase** (LU-tee-al), so named because the corpus luteum forms during this time. To understand what drives these changes, let's turn our attention first to the pituitary.

The pituitary releases two gonadotropic hormones, FSH and LH. These hormones peak in the middle of the menstrual cycle, just before ovulation. As its name implies, FSH stimulates follicular development

---

[4]Note that the second meiotic division occurs only after a sperm penetrates the secondary oocyte. Additionally, in humans and virtually all other animals, the first polar body usually does not divide.

FIGURE 19-14 Oogenesis and Follicle Development

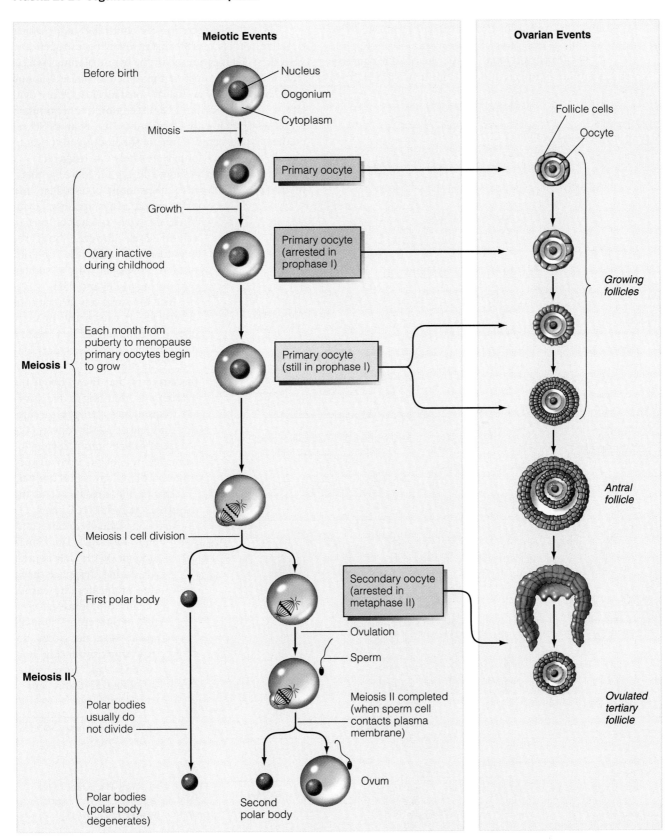

by stimulating the mitotic division of the follicle cells. FSH's counterpart, LH, stimulates estrogen production by follicles. Estrogen, in turn, stimulates mitotic division of follicle cells, promoting growth.

As illustrated in **FIGURE 19-16**, thecal cells convert cholesterol to androgen under the influence of LH. An-drogen, in turn, diffuses into the follicle cells where it is converted to estrogen.[5]

LH and FSH secretion are controlled in a classical negative feedback mechanism involving estrogen and GnRH, which is produced by the hypothalamus. As a result, throughout most of the follicular phase, LH and FSH levels are low and fairly constant. Just before ovulation, however, LH and FSH secretion by the pituitary increases dramatically. These surges in LH and FSH secretion are the result of one of the body's rarest events, a positive feedback loop. Here's how it is triggered.

As **FIGURE 19-15A** shows, during the follicular phase of the menstrual cycle, the amount of estrogen in a woman's blood creeps up fairly slowly. When estrogen reaches a certain critical level, however, both the hypothalamus and the anterior pituitary respond with a sudden outpouring of LH and FSH.

The LH surge has at least four effects: (1) It causes the primary oocyte to complete its first meiotic division, forming a secondary oocyte. (2) It stimulates the release of the enzymes that break down the ovarian wall, resulting in ovulation. (3) It stimulates estrogen production and release. (4) It converts the collapsed follicle into a corpus luteum. The role of the preovulatory surge of FSH, if any, is not known.

During the second half of the menstrual cycle, the luteal phase, LH secretion gradually declines (**FIGURE 19-15A**). The small amount of LH that is present during this stage, however, stimulates the corpus luteum to produce estrogen and progesterone. If pregnancy does not occur, the CL stops producing hormones and degenerates (**FIGURE 19-15B**). Only a

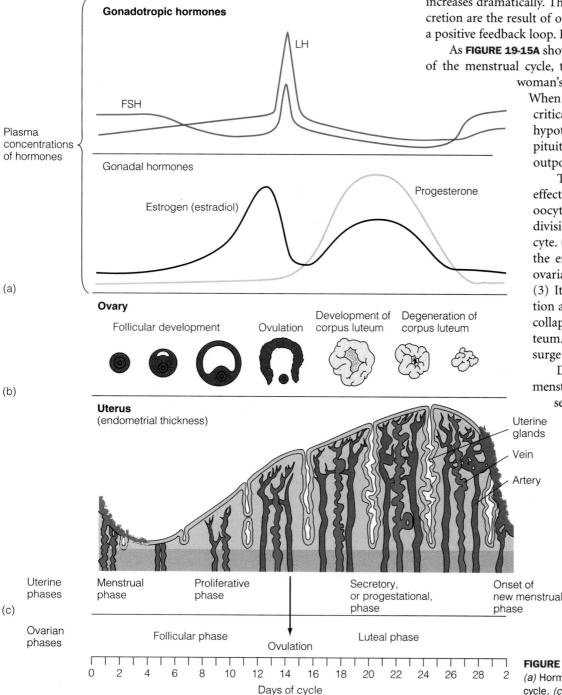

(a)
Plasma concentrations of hormones

Gonadotropic hormones

LH

FSH

Gonadal hormones

Estrogen (estradiol)

Progesterone

(b)
Ovary

Follicular development   Ovulation   Development of corpus luteum   Degeneration of corpus luteum

(c)
Uterus
(endometrial thickness)

Uterine glands

Vein

Artery

Uterine phases

Menstrual phase   Proliferative phase   Secretory, or progestational, phase   Onset of new menstrual phase

Ovarian phases

Follicular phase   Ovulation   Luteal phase

0   2   4   6   8   10   12   14   16   18   20   22   24   26   28   2

Days of cycle

[5]FSH is also necessary for this conversion.

FIGURE 19-15 The Menstrual Cycle
(a) Hormonal cycles. (b) The ovarian cycle. (c) The uterine cycle.

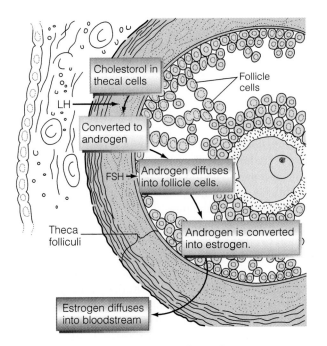

Cholestorol in thecal cells

Follicle cells

LH

Converted to androgen

FSH → Androgen diffuses into follicle cells.

Theca folliculi

Androgen is converted into estrogen.

Estrogen diffuses into bloodstream

FIGURE 19-16 Estrogen Production in the Ovary

hormonal signal (discussed later) from the newly formed embryo can save it.

The Uterine Cycle.

The uterine lining, or endometrium, also undergoes cyclic changes during the menstrual cycle (**FIGURE 19-15C**). These changes result from cyclic changes in ovarian hormones, which, as just discussed, are controlled by changes in pituitary hormone secretion. Before we examine the details of this process, take a moment to study **FIGURE 19-15C**. As illustrated, the endometrium thickens throughout much of the cycle in preparation for pregnancy. In the absence of fertilization, most of the thickened endometrium is shed, a process called **menstruation** (MEN-strew-A-shun).

To understand how the endometrium responds to hormonal changes, we begin on day 1 of the average 28-day menstrual cycle. Day 1 of the cycle marks the first day of menstruation. During the next 4 or 5 days, the uterine lining is shed—that is, the tissue that formed in the previous menstrual cycle sloughs (sluffs) off from the lining. It then passes out of the uterus into the vagina along with a considerable amount of blood—on average, about 50–150 milliliters. (Incidentally, the loss of blood during menstruation is the reason that women are more prone to develop anemia than men.)

As soon as the endometrium has been shed, the lining of the uterus begins to rebuild—to prepare for the possibility of a pregnancy in the new cycle. Initial regrowth in the follicular phase is stimulated by ovarian estrogen. Estrogen stimulates the growth of glands (**uterine glands**) in the endometrium and also promotes cell division in the basal layer (deepest layer) of

the endometrium—all that is left after menstruation. During the first half of the cycle, also called the **proliferative phase**, the uterine glands begin to fill with a nutritive secretion that will nourish an embryo should fertilization occur.

After ovulation the endometrium continues to thicken under the influence of estrogen and progesterone. The uterine glands become distended with a glycogen-rich secretion. The last half of the uterine cycle is therefore called the **secretory phase**.

If fertilization does not occur, the uterine lining starts to shrink approximately 4 days before the end of the cycle, then begins to slough off, starting menstruation. The shedding of the uterine lining (menstruation) is triggered by a decline in estrogen and progesterone concentrations in the blood. Progesterone acts as a uterine tranquilizer, inhibiting smooth muscle contraction in the myometrium. Thus, when progesterone levels fall, the uterus begins to undergo periodic contractions. These contractions propel the detached endometrial tissue out of the uterus and are responsible for the cramps that many women experience during menstruation.

If fertilization occurs, the newly formed embryo produces a hormone called **human chorionic gonadotropin** (KO-ree-ON-ick), or **HCG**. HCG is an LH-like hormone that stimulates the corpus luteum, maintaining its structure and function. When HCG is present, estrogen and progesterone continue to be secreted by the CL. As a result, the uterine lining remains intact. When the newly formed embryo arrives in the uterus, it attaches to the lining. It then embeds in the thickened endometrium from which it derives its nutrients.

HCG maintains the CL for approximately 6 months and shows up in detectable levels in a woman's blood and urine about 10 days after fertilization. Pregnancy tests available through a doctor's office or drugstore detect HCG in a woman's urine. The tests use a commercially prepared antibody to HCG, which binds to the hormone (**FIGURE 19-17**). The home pregnancy tests are relatively inexpensive, fairly reliable, and fast.

Estrogen and Progesterone Exert Numerous Effects.

Like testosterone in boys, estrogen secretion in girls increases dramatically at puberty. As the levels of estrogen in the blood increase, the hormone begins to stimulate follicle development in

FIGURE 19-17 Home Pregnancy Test

the ovaries. Estrogen stimulates the growth of the external genitalia, the breasts, the uterus, uterine tubes, and vagina.

Estrogen's influence extends far beyond the reproductive system. For example, estrogen promotes rapid bone growth in the early teens. Because estrogen secretion in girls usually occurs earlier than testosterone secretion in boys, girls typically experience a growth spurt earlier than boys. However, estrogen also stimulates the closure of the epiphyseal plates, which puts an end to the female growth spurt fairly early. Thus, most girls reach their full adult height by the age of 15–17. Boys experience their most rapid growth later in adolescence and continue growing until the age of 19–21. Finally, estrogen stimulates the deposition of fat in women's hips, buttocks, and breasts, giving the female body its characteristic shape.

Premenstrual Syndrome Is a Condition Afflicting Many Women

For reasons not yet fully understood, many women suffer from irritability, depression, fatigue, and headaches just before menstruation. Many also complain of bloating, tension, joint pain, and swelling and tenderness of the breasts.[6] These complaints are symptomatic of a condition known as **PMS**, or **premenstrual syndrome.**

Premenstrual syndrome is a clinically recognizable condition characterized by one or more of the symptoms noted above. Four of every 10 women of reproductive age experience PMS in varying degrees. Despite the prevalence of PMS, it may be years before scientists can pinpoint the cause or causes of PMS. Nevertheless, dozens of "cures," ranging from massive doses of progesterone to vitamin B-6, have been prescribed by clinics specializing in PMS. Buyers beware, however, for very little good scientific evidence is available to indicate which, if any, of the "cures" really work. Most of the evidence consists of testimonials—individual accounts. The critical thinking skills you learned in Chapter 1 suggest that anecdotal information such as this is no substitute for controlled studies.

Fortunately, work is now under way to test various treatments to see if any consistently bring relief. In the meantime, physicians recommend that women suffering from PMS see their family doctor to be certain that the symptoms are not caused by some other medical problem.

---

[6]All told, women report more than 150 different physical and psychological symptoms that emerge before menstruation begins.

Menopause Is the Cessation of Menstruation

The menstrual cycle continues throughout the reproductive years. However, after a woman reaches 20, her ovaries very gradually begin to become less responsive to gonadotropins. As responsiveness declines, estrogen levels gradually decline (**FIGURE 19-18**). In time, ovulation and menstruation become increasingly erratic and eventually stop. This complete cessation of these functions is called **menopause.**

Menopause is attributed to a reduction in the number of ovarian follicles. At about age 45, most of the follicles that were in the ovary at puberty have been stimulated to grow and have either degenerated or ovulated. Consequently, FSH and LH from the pituitary have no target cells to stimulate. Not surprisingly, the production of ovarian estrogen plummets.

Menopause generally occurs between the ages of 45 and 55, but can occur earlier. The dramatic alteration in the hormonal climate results in several important physiological changes. For example, the decline in estrogen secretion causes the breasts and reproductive organs such as the uterus to begin to atrophy (shrink). Vaginal secretions often decline, and, in some women, sexual intercourse becomes painful.

The decline in estrogen levels may also result in behavioral disturbances. Many women, for instance, become more irritable and suffer bouts of depression. Very noticeable physical changes also occur. Three quarters of all women suffer "hot flashes" and "night sweats" induced by massive vasodilation of vessels in the skin. Fortunately for all people concerned, these symptoms usually pass.

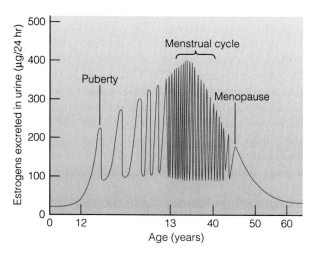

FIGURE 19-18 Ovarian Hormone Secretion Notice that at about age 20, ovarian estrogen secretion begins a gradual decline. Adapted with permission from A. C. Guyton, *Textbook of Medical Physiology,* 7th ed. (Philadelphia: W. B. Saunders Company, 1986), Figure 81–8, p. 979.

As noted in Chapter 13, declining estrogen levels also accelerate osteoporosis. To counter osteoporosis and other impacts of the decline in ovarian function, physicians sometimes prescribe pills containing small amounts of estrogen and progesterone, as well as a program of exercise and a diet rich in calcium and vitamin D (see Health Note 13-1).

Section 19-3
Birth Control

Few topics in modern society generate as much controversy as birth control. **Birth control** is any method or device that prevents births. Birth control measures fit into two broad categories: (1) **contraception**, ways of preventing pregnancy, and (2) **induced abortion**, the deliberate expulsion of a fetus.

Contraceptive Measures Help Prevent Pregnancy

FIGURE 19-19 summarizes the effectiveness of the most common means of contraception. Effectiveness is expressed as a percentage. A 95% effectiveness rating means that 95 women out of 100 using a certain method in a year will not become pregnant.

Abstinence.

Not listed in the figure is a form of birth control that many of us forget to talk about, **abstinence,** refraining from sexual intercourse. This form of birth control is appropriate for many people and should not be overlooked as a strategy of reducing unwanted pregnancy and preventing the transmission of AIDS and other diseases (discussed later).

Sterilization.

Except for complete abstinence, sterilization and the pill (discussed shortly) are the most effective birth control measures (**FIGURE 19-19**). In 1982, sterilization became the leading method of contraception practiced by married couples in the United States.

In women, sterilization is performed by cutting the uterine tubes. This technique is called **tubal ligation** (TWO-bal lie-GAY-shun) (**FIGURE 19-20A**). Surgeons usually make

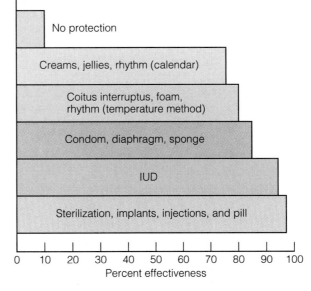

No protection

Creams, jellies, rhythm (calendar)

Coitus interruptus, foam, rhythm (temperature method)

Condom, diaphragm, sponge

IUD

Sterilization, implants, injections, and pill

Percent effectiveness

FIGURE 19-19 Effectiveness of Contraceptive Measures Percent effectiveness is a measure of the number of women in a group of 100 who will not become pregnant in a year.

two small incisions in the abdomen just beneath the navel. An instrument called a **laparoscope** (LAP-er-ah-scope) is inserted through each incision. With the aid of a specially lighted viewing lens in the laparoscope, the surgeon locates and cuts the uterine tubes. The cut ends are either tied off or cauterized—that is, burned using an electrical current. In some cases, surgeons clamp the uterine tubes shut with plastic or metal rings.

Male sterilization requires a far less traumatic surgical procedure called a **vasectomy** (vah-SECK-toe-me), which can be carried out in a physician's office under local anesthesia (**FIGURE 19-20B**). To perform a vasectomy, a physician makes a small incision in the scrotum. Each vas deferens is exposed, then cut, and the free ends are tied off or cauterized.

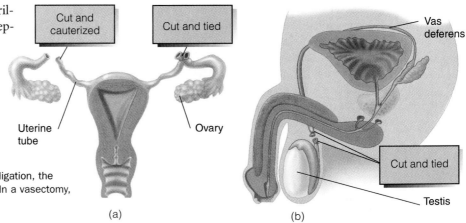

Cut and cauterized

Cut and tied

Uterine tube

Ovary

Vas deferens

Cut and tied

Testis

FIGURE 19-20 Sterilization Methods *(a)* In a tubal ligation, the uterine tubes are cut, then tied off or cauterized. *(b)* In a vasectomy, the vasa deferentia are cut, then tied off.

(a)

(b)

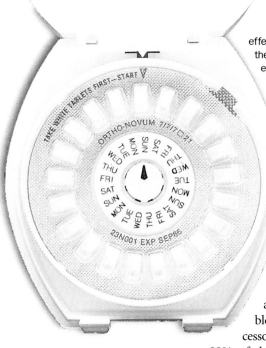

FIGURE 19-21 The Birth Control Pill One of the most effective means of birth control, the pill consists of a mixture of estrogen and progesterone, which is taken throughout the menstrual cycle to block ovulation. Birth control pills are packaged in numbered containers to help women keep track of them.

Vasectomies only prevent the sperm from passing into the urethra during ejaculation. They do not impair sex drive, and because they do not block the flow of the sex accessory glands, which produce 99% of the volume of the ejaculate, they have virtually no effect on ejaculation.

Vasectomy and tubal ligation are essentially irreversible. However, special surgical methods, called **microsurgery**, can be used to reconnect the uterine tubes and the vasa deferentia. This procedure is costly and not always successful.

The Pill.

The **birth control pill** is the most effective temporary means of birth control available (**FIGURE 19-21**). Birth control pills come in several varieties, but the most common is the combined pill. It contains a mixture of synthetic estrogen and progesterone, which collectively inhibit the release of LH and FSH by acting on the pituitary and hypothalamus. This, in turn, inhibits follicle development and ovulation. A minipill containing progesterone alone is also available. Even though it is less effective than the combined pill, the minipill is more suitable for some women because it results in fewer side effects.

Birth control pills must be taken throughout the menstrual cycle. Skipping a few days may release the pituitary and hypothalamus from the inhibitory influences of estrogen and progesterone, resulting in ovulation and possible pregnancy.

Although effective, birth control pills have some adverse health effects worth noting. Even though the incidence of these adverse side effects is small, a woman considering different birth control options should study them carefully before making a decision.

One rare side effect is death. **TABLE 19-3** compares the risk of death from taking birth control pills (and using other contraceptives) with a number of common risk factors. As shown, the risk of a nonsmoker dying from taking birth control pills is 1 in 63,000 in any given year, whereas the risk of dying in an auto accident is 1 in 6000.

Deaths from the use of birth control pills may result from heart attacks, strokes, or blood clots. The incidence of these life-threatening side effects is lowest in nonsmoking women under the age of 30. To reduce this risk even more, pharmaceutical companies have dramatically lowered the estrogen content of the combined pill, because estrogen is responsible for most of the adverse side effects.

Early studies showed a positive correlation between the use of birth control pills and cancers of the breast and cervix. More recent studies suggest that the new generation of low-estrogen pills is less likely to cause cancer of the breast or cervix. Even with reduced estrogen levels, however, women who take birth control pills are more likely to develop cervical cancer than women

TABLE 19-3

Risks Involved in Some Voluntary Activities in the United States

Activity	Annual Risk of Death
Smoking	1 in 200
Motorcycling	1 in 1000
Automobile driving	1 in 6000
Power boating	1 in 6000
Rock climbing	1 in 7500
Using tampons (toxic shock syndrome)	1 in 350,000
Contracting reproductive tract infections through sexual intercourse	1 in 50,000
Preventing pregnancy:	
Oral contraception—nonsmoker	1 in 63,000
Oral contraception—smoker	1 in 16,000
Using intrauterine devices (IUDs)	1 in 100,000
Using barrier methods	None
Using natural methods	None
Undergoing sterilization:	
Laparoscopic tubal ligation	1 in 20,000
Hysterectomy	1 in 1600
Vasectomy	None
Pregnancy:	1 in 10,000
Nonlegal abortion	1 in 3000
Legal abortion:	
Before 9 weeks	1 in 400,000
Between 9 and 12 weeks	1 in 100,000
Between 13 and 16 weeks	1 in 25,000
After 16 weeks	1 in 10,000

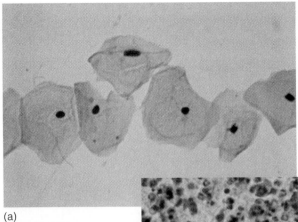

(a)

FIGURE 19-22 The Pap Smear *(a)* A photomicrograph of a normal Pap smear showing large, flattened cells. *(b)* A photomicrograph of a cancerous smear, showing many small cancer cells.

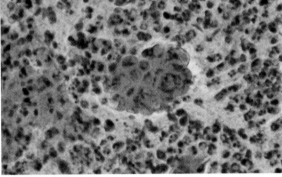

(b)

who do not. Physicians therefore recommend annual Pap smears for women on the pill. During a **Pap smear**, the cervical lining is swabbed. The swab picks up cells sloughed off by the epithelium, which are later examined under a microscope for signs of cancer (**FIGURE 19-22**). This procedure helps physicians in diagnosing cervical cancer early, which increases a woman's chances of survival.

Smoking increases the likelihood of side effects from birth control pills. If a woman is a smoker and takes the pill, for example, she is four times more likely to die from a heart attack or stroke than a nonsmoker. The risk of side effects also increases with age. To reduce the chances of developing serious side effects, women over the age of 35 who smoke should either use an alternative method of birth control or give up smoking. Birth control pills are also not advised for women with a medical history of blood clots, high blood pressure, diabetes, uterine cancer, and cancer of the breast.

Birth control pills do have beneficial effects, not the least of which is that they prevent pregnancy. National statistics show that one of every 10,000 women who becomes pregnant and delivers will die from complications, usually during delivery. Thus, even with the risks associated with the pill, using this mode of contraception is six times safer than pregnancy.

Birth control pills also reduce the incidence of ovarian cysts, breast lumps, anemia, rheumatoid arthritis, osteoporosis, and pelvic infection. Although birth control pills may increase the risk of cervical and breast cancer, they apparently protect a woman from cancer of the ovary and uterus.

Intrauterine Device.

The next most effective means of birth control is the **intrauterine device (IUD)** (**FIGURE 19-23**). The IUD is a small plastic or metal object with a short string attached to it. IUDs are inserted into the uterus by a physician, usually during menstruation, because the cervical canal is widest then and because menstrual bleeding indicates that the woman is not pregnant.

No one knows exactly how the IUD works. Some researchers think that the IUD increases uterine contractions, making it difficult for the early embryo to attach and implant in the wall of the uterus. Others think that the IUD creates a local inflammatory reaction in the uterine lining, resulting in an inhospitable environment for a newly formed embryo. As a result, implantation is blocked. It is possible that both mechanisms are operating.

Like other forms of birth control, IUDs have adverse effects. In some cases, the uterus expels the device, leaving a woman unprotected. Expulsion usually occurs within a month or two of insertion. The IUD may also cause slight pain and increase menstrual bleeding. These effects, however, are minor compared with two much rarer complications: uterine infections and perforation (a penetration of the uterine wall by an IUD). Women with IUDs are more likely to develop uterine infections than women practicing other forms of birth control. If not treated quickly, infections can spread to the uterine tubes, where scar tissue develops and blocks the transport of sperm and ova, causing sterility. Perforation of the uterus is a life-threatening condition requiring surgery to correct.

The Diaphragm, Condom, and Sponge.

The next most effective means of birth control are the barrier methods—the diaphragm, condom, and vaginal sponge—all of which prevent the sperm from entering the uterus. The **diaphragm** (DIE-ah-FRAM) is a rubber cup that fits over the end of the cervix (**FIGURE 19-24**). To increase its effectiveness, a spermicidal (sperm-killing) jelly, foam, or cream should be applied to the rim and inside surface of the cup.

Smaller versions of the diaphragm, called **cervical caps**, are also available. Fitting over the very end of the cervix, the cervical cap most often used is held in place by suction. When used with spermicidal jelly or cream, the caps are as effective as full-sized diaphragms.

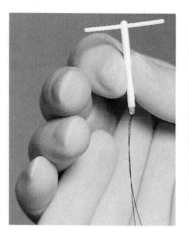

FIGURE 19-23 The IUD IUDs come in a variety of shapes and sizes and are inserted into the uterus, where they prevent implantation. Only one type is currently legal in the United States.

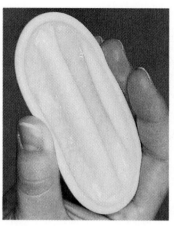

FIGURE 19-24 The Diaphragm Worn over the cervix, the diaphragm is coated with spermicidal jelly or cream and is an effective barrier to sperm.

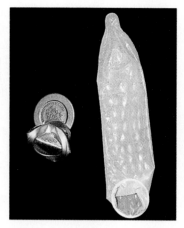

FIGURE 19-25 The Condom Worn over the penis during sexual intercourse, it prevents sperm from entering the vagina.

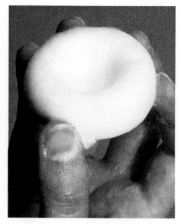

FIGURE 19-26 The Vaginal Sponge Impregnated with a spermicidal chemical, the vaginal sponge is inserted into the vagina and is effective for up to 24 hours.

Condoms are thin, latex rubber sheaths that fit onto the erect penis (**FIGURE 19-25**). Sperm released during ejaculation are trapped inside (in a small reservoir at the tip of the condom) and are therefore prevented from entering the vagina. Besides preventing fertilization, condoms also protect against sexually transmitted diseases, a benefit not offered by any other birth control measure except abstinence.

The newest condom on the market is for women. It fits into the vagina and is recommended by physicians to protect women against AIDS and other sexually transmitted diseases.

Yet another barrier method is the **vaginal sponge** (**FIGURE 19-26**). This small absorbent piece of foam is impregnated with spermicidal jelly. Inserted into the vagina, the sponge is positioned over the end of the cervix. The sponge is effective immediately after placement and remains effective for 24 hours. Cervical sponges can be purchased without a doctor's prescription. Like the condom, one size fits all.

Withdrawal.

One of the oldest, but least successful, means of birth control is **withdrawal,** or **coitus interruptus** (COE-ee-tus), disengaging before ejaculation. This method requires tremendous willpower and frequently fails, for three reasons: because caution is often tossed to the wind in the heat of passion, because the penis is withdrawn too late, or because of preejaculatory leakage—the release of a few drops of sperm-filled semen before ejaculation. Accordingly, it is not an advisable method of birth control.

Spermicidal Chemicals.

As mentioned earlier, spermicidal jellies, creams, foams, and films contain chemical agents that kill sperm but are apparently harmless to the woman. Spermicidal preparations are most often used in conjunction with diaphragms, condoms, and cervical caps. They can also be used alone but are only about as effective as withdrawal.

The Rhythm Method.

Abstaining from sexual intercourse around the time of ovulation—the **rhythm,** or **natural, method**—can help couples reduce the likelihood of pregnancy. If a couple knows the exact time of ovulation, they can time sexual intercourse to prevent pregnancy more precisely.

To practice the natural method successfully, couples must first determine when ovulation occurs. One popular approach is the **temperature method.**[7] A woman's body temperature varies throughout the menstrual cycle, as shown in **FIGURE 19-27**. In most women, body temperature rises slightly after ovulation. By taking her temperature every morning before she gets out of bed, a woman can pinpoint the day she ovulates. By keeping a temperature record over several menstrual cycles, she can determine the length of her cycle and the time of ovulation. Once the length of the cycle and the time of ovulation have been determined, days of abstinence can be determined. Erring on the safe side, some doctors

---

[7]The temperature method can also be used by women who want to get pregnant because it allows them to determine the time of ovulation.

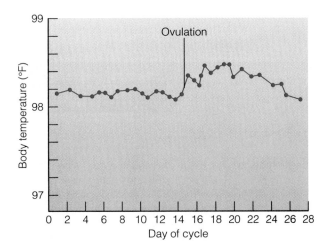

FIGURE 19-27 Body Temperature Measurements During the Menstrual Cycle

recommend that couples refrain from sexual intercourse from the first day of menstruation until 4 days after ovulation. Translated, that means no sex for about 17 days of the 28-day cycle.

Another method used to time ovulation involves taking samples of the cervical mucus, which varies in consistency during the menstrual cycle. By testing its thickness on a daily basis, a woman can tell fairly accurately when she has ovulated.

Because ova remain viable 12–24 hours after ovulation and sperm may remain alive in the female reproductive tract for up to 3 days, abstinence 4 days before and 4 days after the probable ovulation date should provide a margin of safety (**FIGURE 19-28**). As most people practice it, then the natural method of birth control requires about 8 days of abstinence during each menstrual cycle. This minimizes the chances of a viable sperm reaching a viable ovum. Unfortunately, some women experience the greatest sexual interest around the time of ovulation. Sexual intercourse after a period of abstinence may also advance the time of ovulation. (For a discussion of new methods of birth control, see Health Note 19-1.)

■ Abortion Is the Surgical Termination of Pregnancy

Some couples may elect to terminate pregnancy through **abortion**. In the United States, approximately 1 million abortions are performed every year by physicians.

Abortion is not suitable or morally acceptable to all people. Pro-life advocates

argue that abortion should be outlawed or severely restricted—that is, allowed only in cases of rape, incest, and threat to the life of the mother. These individuals advise unmarried women to abstain from sexual intercourse or, if they become pregnant, to give birth and either keep the baby or put it up for adoption.

Pro-choice advocates, on the other hand, support abortion. They argue that women should have the freedom to choose whether to terminate a pregnancy or have a child. Abortion, they say, reduces unwanted pregnancies and untold suffering among unwanted infants, especially in poor families, and gives women more options than motherhood. Although pro-choice advocates view abortion as a legitimate means of family planning, many of them are quick to point out that it should not be practiced as a primary means of birth control. Contraception is less costly, less traumatic, and more morally acceptable.

In the first 12 weeks of pregnancy, abortions can be performed surgically in a doctor's office via **vacuum aspiration**. In this procedure, the cervix is first dilated by a special instrument. Next, the contents of the uterus are drawn out through an aspirator tube.

Vacuum aspiration is a fairly simple and relatively painless procedure. Usually no anesthesia is given. Although women bleed for a week or so after the procedure, they generally experience few complications.

Most abortions are performed by the end of the twelfth week of pregnancy. Vacuum aspiration may also be performed later—between 13 and 16 weeks. In such instances, physicians use a larger aspirator tube. In

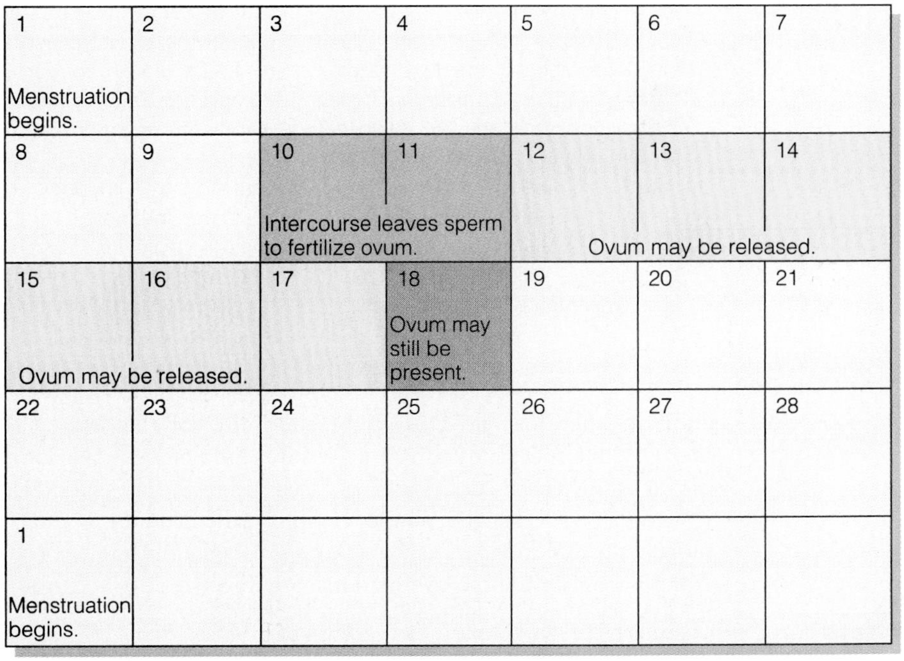

FIGURE 19-28 The Natural Method The yellow-shaded areas indicate an unsafe period for sexual intercourse, assuming ovulation occurs at the midpoint of the cycle.

1	2	3	4	5	6	7
Menstruation begins.						
8	9	10	11	12	13	14
		Intercourse leaves sperm to fertilize ovum.		Ovum may be released.		
15	16	17	18	19	20	21
Ovum may be released.			Ovum may still be present.			
22	23	24	25	26	27	28
1						
Menstruation begins.						

Health Note 19–1

Advances in Birth Control—Responding to a Global Imperative

Every day of the year, 220,000 people join the world population. This translates into an annual rate of growth of about 80 million per year! Faced with resource depletion, environmental destruction, and a host of other problems, many countries recognize the need to stabilize population growth.

Concerns for providing adequate birth control are present in virtually all nations today. In wealthy, industrialized countries, many parents have already taken steps to limit their family size. For many parents in these nations, the decision to limit their family size is based on economics. Raising a single child to college age can cost as much as $85,000 in 1994 dollars. A recent study estimated that another $85,000 will be needed just to send a child born today to a public university. Considerably more is needed to send a student to a private university. Parents may also limit their family size to be able to provide more personal care for their offspring.

Some parents choose to have fewer children for environmental reasons. Consider some startling statistics: a child born in the United States today will require 16 pounds of coal, 3.6 gallons of oil, and 240 cubic feet of natural gas each day of his or her life. This heavy dependence on resources translates into an enormous amount of pollution and environmental damage. Clearly, child rearing is an environmentally taxing activity.

Although most people tend to think of the need for family planning in the poorer, nonindustrial nations, it is just as important, perhaps even more so, in the United States and other developed countries because their impact on the environment is so great. Each American, for example, uses 20–40 times as much of the Earth's resources as a resident of India. In terms of environmental impact, then, the United States' 270 million people are equal to 5–11 billion Indians.

This is not to say that birth control is unimportant in Third World nations. Approximately, 90% of the 80 million people added to the global population each year are born into the Third World, where hunger and starvation abound. And these countries face environmental problems of epic proportions.

Family planning is making headway in Third World nations, but the task has only just begun. To facilitate this process, researchers are trying to develop safer, more convenient, and even more effective methods of birth control. For example, tests are under way on the effectiveness and safety of nasal spray contraceptives for women. Research is also proceeding quickly on transdermal contraceptive patches. These small, Band-Aid-like patches are impregnated with a blend of hormones and hormone analogs, including estrogen and progesterone, and are worn by a woman for a week, then replaced.

In many countries outside the United States, slow-acting, injectable contraceptives are now being used. Women are given a shot of crystalline progesterone under the skin. The crystals dissolve over a period of 3 months, blocking ovulation.

Another novel approach involves matchstick-sized capsules containing an even more potent progesterone implanted under the skin of a woman's arm; these prevent pregnancy for up to 5 years (**FIGURE 1**). Contraceptive implants have been approved for use in 13 countries, including the United States. Population experts hope that they will be widely used in Third World nations because they require virtually no effort on the part of the couple.

Researchers are also experimenting with biodegradable implants. Clinical trials are under way on a biodegradable material impregnated with progesterone that may prevent pregnancy for 18 months or more. The biodegradable material is broken down and gradually disappears.

Experimentation is continuing on "morning-after" pills, which can be taken after sexual intercourse to prevent pregnancy. At least two morning-after pills exist. One contains a synthetic estrogen called *DES* (diethylstilbestrol). DES stimulates muscle contraction in the uterus and uterine tubes, expelling the fertilized ovum from the reproductive tract. DES is sometimes used in cases of incest

addition, they may need to scrape the lining of the uterus with a special instrument to ensure complete removal of the fetal tissue.

After 16 weeks, abortions are more difficult and more risky. Solutions of salt, urea, or prostaglandins, which stimulate uterine contractions, are injected into the sac of fluid surrounding the fetus to induce premature labor. The hormone oxytocin may be administered to the woman with the same effect.

In France, women can use a pill that induces abortion soon after the embryo implants. Called *RU486*, this pill is illegal in the United States, but many believe that it will soon be available. (For more on RU486, see Health Note 19-1.) Although RU486 is not available, there is a chemical treatment of a similar nature available to U.S. women. This pill restricts blood flow to the uterus. Two days after the initial pill is given, a second pill that induces abortion is taken.

or rape, but it is generally avoided because it causes vomiting and nausea.

A synthetic steroid called *RU486* is also in use in France and certain other countries, but not yet in the United States. RU486 binds to progesterone receptors in the uterus, blocking progesterone from binding. Because progesterone inhibits muscular contractions of the uterus, RU486 probably has the same effect as DES, causing an expulsion of the fertilized ovum. Vocal pro-life forces in the United States have taken a strong stand against the use of this method, calling it an abortion pill, but many people believe that RU486 will be approved for use soon.

You may have noticed that when it comes to birth control, there are few options for men. Why not develop a pill for men and shift some of the contraceptive burden to them?

To be effective, a pill would have to shut down spermatogenesis. Testosterone injections would do the job because this hormone blocks the release of pituitary FSH and LH. FSH is required for spermatogenesis, and its absence would depress sperm production. Unfortunately, testosterone injections may also cause aggressive behavior. Complicating matters, excess androgen in males is converted to estrogen, causing feminizing side effects. Finally, androgen treatments may depress spermatogenesis, but not enough to lower sperm count to a level where a couple would feel confident.

Another route is to selectively inhibit FSH secretion. As noted in this chapter, the seminiferous tubules produce a substance called *inhibin,* so named because it inhibits the production of FSH by the pituitary gland. If inhibin could be produced and administered to men, it might give them a better chance to participate in birth control. It could be administered in contraceptive nasal sprays.

One thing is certain, the world's population problem will not be solved through new contraceptive technologies alone. What is required is a change in the attitudes of millions of men and women throughout the world. Controlling family size must be a conscious decision followed by conscientious action.

In Africa, where population is doubling in some countries every 17 years, contraceptive use is a paltry 10%–20%. Worldwide, only about half the women of reproductive age are using contraceptives. It is crucial to involve more people as the world races to stem this tidal wave, which, if unchecked, could add 5 billion people to the world population in the next 40 years. Funds are needed to pay for education, contraception, and other family planning. When one condom costs more than the average person spends on medical care in a year, we can hardly expect widespread use. Many Third World countries, however, divert enormous amounts of money to pay for weapons and almost nothing to family

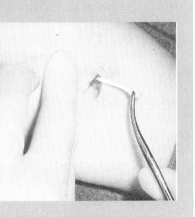

FIGURE 1 Subcutaneous Progesterone Implant Inserted under the skin, this tiny device releases a steady stream of progesterone, blocking ovulation for months.

planning. If the world population is to stabilize, if our children are to inherit a world worth living in, many experts agree that contraceptive use must increase.

Controlling population growth also requires improvements in education and job opportunities for men and women. Small-scale, sustainable economic development will give men and women options other than child bearing.

Visit *Human Biology's* Internet site, www.jbpub.com/humanbiology, for links to web sites offering more information on this topic.

In the state of Washington, it is now legal for pharmacists to dispense emergency "morning-after" pills without a doctor's prescription.

In the controversy over abortion, it is important to remember that abortion is rarely an easy choice for anyone. Contrary to what many think, however, studies show that the majority of women who choose abortion do not suffer lasting emotional harm, especially if they have had counseling.

Section 19-4

Sexually Transmitted Diseases

Certain bacteria and viruses can be transmitted by sexual contact. These organisms can penetrate the lining of the reproductive tracts of men and women and thrive in the moist, warm environment of the body. These organisms cause **sexually transmitted diseases (STDs).** Most of the infectious agents that cause STDs are spread by vaginal intercourse, but other forms of sexual contact such as anal and oral sex are responsible for their transmission. AIDS, for example, can be transmitted by anal sex as well as vaginal and oral sex (Chapter 8). Syphilis is caused by a bacterium that is spread by oral, anal, and vaginal sex.

Although STDs pass from one person to another during sexual contact, the symptoms are not confined to the reproductive tract. In fact, several STDs, including syphilis and AIDS, are primarily systemic diseases—that is, they affect entire body systems.

One complicating factor in controlling STDs is that occasionally some diseases such as gonorrhea produce no obvious symptoms in many men and women. As a result, the disease can be transmitted without a person knowing he or she is infected. In others such as AIDS, symptoms may not appear for weeks or even years after the initial infection. Thus, sexually active individuals who are not monogamous can transmit the AIDS virus to many people before they are aware that they are infected. In this section we will examine the most common STDs, except AIDS, which was discussed in Chapter 8.

Gonorrhea Is a Bacterial Infection That Can Spread to Many Organs

Gonorrhea (GON-or-REE-ah; referred to colloquially as the "clap") is caused by a bacterium that commonly infects the urethra of men and the cervical canal of women. Painful urination and a puslike discharge from the urethra are common complaints in men. Women may experience a cloudy vaginal discharge and lower-abdominal pain. If a woman's urethra is infected, urination may be painful. Symptoms of gonorrhea usually appear about 2–8 days after sexual contact.

If left untreated, gonorrhea in men can spread to the prostate gland and the epididymis. Infections in the urethra lead to the formation of scar tissue. This may narrow the urethra and make urination even more difficult. In some women, bacterial infection spreads to the uterus and uterine tubes, causing the buildup of scar tissue. In the uterine tubes, scar tissue may block the passage of sperm and ova, resulting in infertility.[8] Gonorrheal infections can also spread into the abdominal cavity through the opening of the uterine tubes. If the infection enters the bloodstream in men or women, it can travel throughout the body. Fortunately, gonorrhea can be treated by antibiotics, but early diagnosis is essential to limit the damage.

---

[8]Sexually transmitted diseases are a leading cause of infertility in women.

Syphilis Is Caused by a Bacterium and Can Be Extremely Debilitating if Untreated

Syphilis is a deadly STD caused by a bacterium that either penetrates the linings of the oral cavity, vagina, and penile urethra or enters through breaks in the skin. If untreated, syphilis proceeds through three stages. In stage 1, between 1 and 8 weeks after exposure, a small, painless red sore develops, usually in the genital area. Easily visible when on the penis, these sores often go unnoticed when they occur in the vagina or cervix. The sore heals in 1–5 weeks, leaving a tiny scar.

Approximately 6 weeks after the sore heals, individuals complain of fever, headache, and loss of appetite. Lymph nodes in the neck, groin, and armpit swell as the bacteria spreads throughout the body. This is stage 2.

As a rule, the symptoms of stage 2 syphilis disappear for several years. Then, without warning, the disease flares up again. This is stage 3. During stage 3, an autoimmune reaction occurs. Patients experience a loss of their sense of balance and a loss of sensation in their legs. As the disease progresses, patients experience paralysis, senility, and even insanity. In some cases, the bacterium weakens the walls of the aorta, causing aneurysms (Chapter 6).

Syphilis can be successfully treated with antibiotics, but only if the treatment begins early. Suspicious sores in the mouth and genitals should be brought to the attention of a physician. In stage 3, antibiotics are useless. Tissue or organ damage is permanent.

Chlamydial Infections Are Extremely Common Among College Students

One of the most common sexually transmitted diseases, affecting 3 to 10 million people each year, and many of them college students, is known as **chlamydia** (clam-ID-ee-ah). Caused by a bacterium, this disease is characterized, in men, by a burning sensation during urination and a discharge from the penis. Women also experience a burning sensation during urination and a vaginal discharge. If the bacterium spreads, it can cause more severe infection and infertility. Like other STDs, many people experience no symptoms at all and therefore risk spreading the disease to others. Children born to mothers with chlamydia can develop eye infections and pneumonia.

Chlamydia bacteria often migrate to the lymph nodes, where they cause considerable enlargement and tenderness in the affected area. Blockage of the lymph nodes may result in tissue swelling in the surrounding tissue. Doxycycline and other antibiotics are effective in treating this disease.

Genital Herpes Is Caused by a Virus; It Is Extremely Common, and Essentially Incurable

Genital herpes (HER-peas) is another common sexually transmitted disease. Contracted by 200,000–300,000 people each year, genital herpes is caused by a virus that enters the body and remains there for life. The first sign of viral infection is pain, tenderness, or an itchy sensation on the penis or female external genitalia, occurring 6 days or so after contact with someone infected by the virus. Soon afterward, painful blisters appear on the external genitalia, thighs, buttocks, and cervix, or in the vagina (**FIGURE 19-29**).

The blisters break open and become painful ulcers that last for 1–3 weeks, then disappear. Unfortunately, the herpes virus is a lifelong resident of the body, and new outbreaks can occur from time to time, especially when an individual is under stress. Recurrent outbreaks are generally not as severe as the initial one, and, in time, the outbreaks generally cease.

Unlike other STDs, herpes can be transmitted to other individuals during sexual contact only when the blisters are present or (as recent research suggests) just beginning to emerge. When the virus is inactive, sexual intercourse can occur without infecting a partner.

Although herpes cannot be cured, physicians can suppress outbreaks with drugs such as *acyclovir* (A-sigh-CLOE-ver). These drugs not only reduce the incidence of outbreaks, but also accelerate healing of the blisters.

Herpes is not a particularly dangerous STD, except in pregnant women. These women run the risk of transferring the virus to their infants at birth. Because the virus can be fatal to newborns, these women are often advised to deliver by cesarean section (an incision made just above the pubic bone) if the virus is active at the time of birth.

Nongonococcal Urethritis Is an Extremely Common Disease Caused by Several Types of Bacteria

Nongonococcal urethritis (YUR-ee-THRIGHT-iss), or **NGU** for short, is the most common sexually transmitted disease. Moreover, NGU is one of several STDs whose incidence is steadily rising in the United States. Caused by any of several different bacteria, this infection is generally less threatening than gonorrhea or syphilis, although some infections can result in sterility.

Many men and women often exhibit no symptoms whatsoever and can therefore spread the disease without knowing it. In men, when symptoms occur, they resemble those of gonorrhea—painful urination and a cloudy mucous discharge from the penis. In women,

FIGURE 19-29 Genital Herpes Blisters on the External Genitalia and Inner Thigh

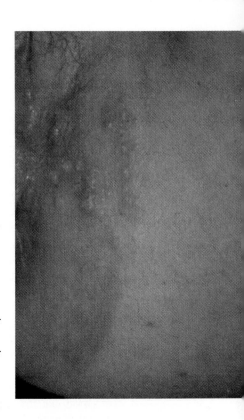

urination becomes painful and more frequent. NGU can be treated by antibiotics, but individuals should seek treatment quickly to avoid the spread of the disease and more serious complications.

Genital Warts Are Caused by Human Papillomavirus (HPV).

The vast majority of Americans carry a virus known as **human papillomavirus** (pap-ILL-oh-mah) or HPV. Transmitted by sexual contact, this virus can cause **gential warts,** benign growths that appear on the external genitalia and around the anuses of men and women (**FIGURE 19-30**). Warts also grow inside the vagina of women. Warts generally occur in individuals whose immune systems are suppressed, for example, after long periods of stress.

These warts can remain small or can grow to cover large areas, creating cosmetically unsightly growths. They may cause mild irritation, and certain strains of HPV are associated with cervical cancer in women. Genital warts can be treated with chemicals or removed surgically—although rates of recurrence are quite high. In 20% to 30% of the cases, genital warts disappear spontaneously. Getting rid of the virus, however, is impossible, for it resides in the body forever.

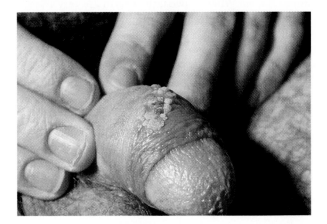

FIGURE 19-30 Genital Warts Genital warts on the penis are caused by infection of the skin by papilloma viruses. Genital warts can be removed by a variety of treatments but sometimes recur.

Section 19-5

Infertility

A surprisingly large percentage (about one in six) of American couples cannot conceive. The inability to conceive (to become pregnant) is called **infertility**. According to some statistics, in about 50% of the couples, infertility results from problems occurring in the woman. Approximately 30% of the cases are due to problems in the man alone, and about 20% are the result of problems in both partners. Can anything be done in such instances?

Fortunately, the answer is yes. If after a year of actively trying to conceive a couple remains unsuccessful, they can consult a fertility specialist who will first check obvious problems such as infrequent or poorly timed sex. Obviously, only intercourse around the time of ovulation is successful. If timing is not the problem, and it rarely is, the physician tests the man's sperm count. A low sperm count is one of the most common causes of male infertility.

A low sperm count may result from overwork, emotional stress, and fatigue. Excess tobacco and alcohol consumption can also contribute to the problem. Tight-fitting clothes and excess exercise, both of which raise the scrotal temperature, also tend to reduce the sperm count. The testes are also sensitive to a wide range of chemicals and drugs that reduce sperm production.

If infertility appears to be caused by a low sperm count, a couple may choose to undergo artificial insemination, using sperm from a sperm bank. These sperm are generally acquired from anonymous donors and are stored frozen. When thawed, the sperm are reactivated, then deposited in the woman's vagina or cervix around the time of ovulation.

If sperm production and ejaculation appear normal, a physician checks the woman's reproductive tract. First on line is a test of ovulation. If ovulation is not occurring, **fertility drugs** may be administered. Several kinds of drugs are available. One of the more common is HCG, which, as noted earlier, is an LH-like hormone that induces ovulation. Unfortunately, fertility drugs often result in superovulation (the ovulation of many fertilizable ova), leaving couples with a "litter" of 4–6 babies, instead of the one child they had hoped for. (Most of the multiple births you hear about on the news are the result of fertility drugs.)

If tests show that ovulation is occurring normally, the physician examines the uterine tubes to determine whether they are obstructed. In some instances, a previous gonorrheal or chlamydial infection that spread into the tubes may have caused scarring that obstructed the passageway. In such instances, couples may be advised to adopt a child or to try *in vitro* (in VEE-trow) fertilization. During *in vitro* fertilization, ova are surgically removed from the woman and fertilized by the partner's sperm outside her body. The fertilized ovum is then implanted in the uterus of the woman where it can grow and develop successfully. Besides being expensive and time-consuming, this procedure has a low success rate and is not widely available. It also places heavy emotional demands on the couple.

Health, Homeostasis, and the Environment

The Sperm Crisis?

The human reproductive system is not essential to homeostasis or the survival of an individual. In the words of physiologist Lauralee Sherwood, "it is essential for sustaining the thread of life from generation to generation." Like other systems, the reproductive system is susceptible to environmental conditions—to pollution and social and psychological conditions that create stress. In this section we will explore one crucial link between the environment and reproduction.

In September 1979, Professor Ralph Dougherty, a chemist at Florida State University, announced findings from a study of 130 healthy male college students. In his test group, the researcher found extremely low sperm counts of only 20 million per milliliter (ml) of semen, compared with an expected value of 60–100 million per ml. Biochemical analyses of the testes also revealed high levels of four toxic chemicals: DDT, polychlorinated biphenyls (PCBs), pentachlorophenol, and hexachlorobenzene.

An article in the Sierra Club's magazine describing the results proclaimed that America was facing a "sperm crisis" caused by toxic chemicals. But not everyone agrees. Health officials, in fact, have challenged these findings. Some officials contend that the low sperm counts that Dougherty recorded may have resulted from improved counting techniques. That is to say, over the years, advances in technology have allowed scientists to count sperm more accurately. As a result, estimates of the normal sperm concentration have been markedly lowered. Dougherty maintains that such improvements, while real, are not entirely responsible for the decline.

Other health officials argue that Dougherty's findings are not representative of the American public. Floridians, they say, are exposed to high levels of pesticides used on farms that contaminate the water supplies of urban and rural residents.

Additional studies in Florida and other states show that sperm counts in American men have indeed been falling since the 1950s. Prior to 1950, the average sperm count was about 110 million per ml. By 1980 and 1981, sperm counts had dropped to about 60 million per ml. Statistical studies suggest that the decline may be related to growing pesticide use, air pollution, and other factors.

Studies in Hawaii support the hypothesis that certain environmental chemicals are causing a decline in sperm count. These studies show that Hawaiian men have a considerably higher sperm count than men residing in the continental United States. This observation has been attributed to a generally cleaner environment. There are, say researchers, fewer factories on the Hawaiian islands than on the mainland. People are exposed to fewer agricultural chemicals, and frequent winds probably result in cleaner air.

Reductions in sperm count are of concern to many people because a sperm count below 20 million per milliliter is generally insufficient for fertilization. Today, low sperm counts account for a significant percentage of all infertility in U.S. couples.

Human reproduction, like other bodily processes, depends on a healthy environment. Research shows that a wide range of factors—from drugs to radiation to industrial chemicals—are toxic, or potentially toxic, to human reproduction (**TABLE 19-4**). "There has been an explosion of spermatotoxins in the environment," says Dr. Bruce Rappaport, a fertility specialist. "The problem is environmental pollution." Today, at least 20 common

TABLE 19-4	
Some Agents Potentially Toxic to Male and Female Reproduction	
Males	**Females**
Natural and synthetic androgens	Natural and synthetic estrogens
Heat	Natural and synthetic progestins
Radiation	Amphetamine
Dioxin	DDT
PCBs	Parathion (insecticide)
Vinyl chloride	Carbaryl (insecticide)
Ethanol	DES
Benzene	PCBs
DES (diethylstilbestrol)	
EDB (ethylene dibromide)	
Paraquat (herbicide)	
Carbaryl (insecticide)	
Cadmium	
Mercury	

industrial chemicals are known to be reproductive toxins. Ten commonly prescribed antibiotics can reduce sperm count. Tagamet, a drug that is used to relieve stress and treat stomach ulcers reduces sperm count by over 40%. By one estimate, at least 40 commonly used drugs depress sperm production. Thousands of other drugs and environmental pollutants have not been tested.

These facts do not necessarily mean that the United States is in a sperm crisis, but they do suggest the need for further research to determine the potential impacts, if any, of the many thousands of chemicals now commonly used or released into the environment. Research may prove that we need to clean up our act, or it may show that these fears are unwarranted.

SUMMARY

THE MALE REPRODUCTIVE SYSTEM

1. The male reproductive tract consists of seven basic components: (a) testes, (b) epididymis, (c) vasa deferentia, (d) sex accessory glands, (e) urethra, (f) penis, and (g) scrotum.

2. The *testes* reside in the *scrotum*, which provides a suitable temperature for sperm development. Each testis contains hundreds of sperm-producing *seminiferous tubules*.

3. *Sperm* produced in the seminiferous tubules are stored in the *epididymis*. During *ejaculation*, sperm pass from the epididymis to the *vas deferens*, then to the *urethra*. Secretions from the *sex accessory glands* are added to the sperm at this time, forming semen.

4. Between the seminiferous tubules are the *interstitial cells*, which produce the hormone *testosterone*. Testosterone secretion is controlled by *luteinizing hormone* (*LH*) or *interstitial cell stimulating hormone* (*ICSH*) from the anterior pituitary. LH release is regulated in a negative feedback loop with testosterone.

5. Testosterone stimulates *spermatogenesis*, facial hair growth, thickening of the vocal cords, laryngeal cartilage growth, sebaceous gland secretion, and bone and muscle development.

6. The *penis* is the organ of copulation. It contains *erectile tissue*, which fills with blood during sexual arousal, making the penis turgid.

7. Ejaculation is under reflex control. When sexual stimulation becomes intense, sensory nerve impulses travel to the spinal cord. There they stimulate motor neurons in the cord, which send impulses to the smooth muscle in the walls of the epididymis, the vasa deferentia, the sex accessory glands, and the urethra, causing ejaculation.

THE FEMALE REPRODUCTIVE SYSTEM

8. The female reproductive system consists of two basic components: the reproductive tract and the external genitalia.

9. The reproductive tract consists of (a) the uterus, (b) the two uterine tubes, (c) the two ovaries, and (d) the vagina.

10. The *external genitalia* consist of two flaps of skin on both sides of the vaginal opening, the *labia majora* and the *labia minora*.

11. Female germ cells are housed in follicles in the ovary. A *follicle* consists of a germ cell and an investing layer of follicle cells.

12. A dozen or so follicles enlarge during each menstrual cycle, but most follicles degenerate. In humans, usually only one follicle makes it to ovulation.

13. The oocyte and an investing layer of follicle cells are released during ovulation and drawn into the uterine tubes.

14. The *menstrual cycle* consists of a series of changes occurring in the ovaries, uterus, and endocrine system of women.

15. The first half of the menstrual cycle is called the *follicular phase*. It is during this period that FSH from the pituitary stimulates follicle growth and development. LH stimulates estrogen production.

16. Estrogen levels rise slowly during the follicular phase, then trigger a positive feedback mechanism that results in a preovulatory surge of FSH and LH, which triggers ovulation.

17. The oocyte is expelled from the *antral follicle* at ovulation. The follicle then collapses and is converted into a *corpus luteum* (*CL*), which releases estrogen and progesterone.

18. In the absence of fertilization, the CL degenerates. If fertilization occurs, however, HCG from the embryo maintains the CL for approximately 6 months.

19. During the menstrual cycle, ovarian hormones stimulate growth of the uterine lining, which is necessary for successful implantation. If fertilization does not occur, the uterine lining is sloughed off during *menstruation*, which is triggered by a decline in ovarian estrogen and progesterone.

20. Like testosterone levels in boys, estrogen levels in girls increase at puberty. Estrogen promotes growth of the external genitalia, the reproductive tract, and bone. It also stimulates the deposition of fat in women's hips, buttocks, and breasts.

21. Progesterone works with estrogen to stimulate breast development. It also promotes endometrial growth and inhibits uterine contractions.

22. Many women suffer from *premenstrual syndrome* or *PMS*, which is characterized by irritability, depression, tension, fatigue, headaches, bloating, swelling and tenderness of the breasts, and joint pain.

23. The menstrual cycle continues throughout the reproductive years, but after a woman reaches 20, the ovaries become progressively less responsive to gonadotropins. As a result, estrogen levels slowly decline as a woman ages. Ovulation and menstruation become erratic as a woman approaches 45.

24. Between the ages of 45 and 55, ovulation and menstruation cease. The end of reproductive function in women is known as *menopause*.

25. The decline in estrogen levels results in the atrophy of the reproductive organs and in behavioral disturbances. Many women become irritable, suffer bouts of depression, and experience hot flashes and night sweats induced by intense vasodilation of vessels in the skin.

BIRTH CONTROL

26. Birth control refers broadly to any method or device that prevents births and includes two general strategies: *contraception* (measures that prevent pregnancy) and *induced abortion* (the deliberate expulsion of a fetus).

27. **FIGURE 19-19** summarizes the effectiveness of the various birth control measures.

28. The most effective form of birth control is abstinence. Another highly effective measure is sterilization—*vasectomy* in men and *tubal ligation* in women.

29. The pill is a highly effective means of birth control. The most common pill in use today contains a mixture of estrogen and progesterone that inhibits ovulation. In some women, however, estrogen causes adverse health effects.

30. The *intrauterine device* or *IUD* is a plastic or metal coil that is placed inside the uterus, where it prevents the fertilized ovum from implanting.

31. The diaphragm, condom, and vaginal sponge are less effective than the measures described above. The *diaphragm* is a rubber cap fitted over the cervix. To

be fully effective, it must be coated with a spermicidal jelly, foam, or cream.

32. The *condom* is a thin, latex rubber sheath worn over the penis during sexual intercourse that prevents sperm from entering the vagina.

33. The *vaginal sponge* is a tiny, round sponge worn by the woman. It is impregnated with a spermicidal chemical.

34. One of the oldest, but least effective, methods of birth control is *withdrawal*, removing the penis before ejaculation. *Spermicidal chemicals* used alone are about as effective as withdrawal.

35. Abstaining from sexual intercourse around the time of ovulation, known as the *rhythm method*, is another way to prevent pregnancy. Statistics on effectiveness show that the natural method is one of the least successful of all birth control measures.

36. Some couples may elect to terminate pregnancy through an abortion.

SEXUALLY TRANSMITTED DISEASES

37. Certain viruses and bacteria can be transmitted from one individual to another during sexual contact. Infections spread in this way are called *sexually transmitted diseases (STDs)*.

38. *Gonorrhea* is caused by a bacterium that commonly infects the urethra in men and the cervical canal in women. Overt symptoms of the infection are frequently not present, so people can spread the disease without knowing it. If left untreated, gonorrhea can spread to other organs, causing considerable damage.

39. *Syphilis* is a serious STD caused by a bacterium that penetrates the linings of the oral cavity, vagina, and penile urethra. If untreated, syphilis proceeds through three stages. It can be treated with antibiotics during the first two stages, but in stage 3, when damage to the brain and blood vessels is evident, treatment is ineffective.

40. Chlamydial infections, an extremely common bacterial STD, resemble gonorrhea.

41. *Genital herpes* is also a very common STD. It is caused by a virus. Once the virus enters the body, it remains for life. Blisters form on the genitals and sometimes on the thighs and buttocks. The blisters break open and become painful ulcers. At this stage, an individual is highly infectious. New outbreaks of the virus may occur from time to time, especially when an individual is under stress.

42. *Nonspecific urethritis (NSU)*, the most common sexually transmitted disease, is caused by several different bacteria, but most commonly by chlamydia. It is less threatening than gonorrhea or syphilis. Many men and women show no symptoms of NSU and can therefore spread the disease without knowing it. Symptoms, when they occur, resemble those of gonorrhea.

43. Genital warts occur in men and women whose immune systems are suppressed and are caused by HPV.

INFERTILITY

44. The inability to conceive is called *infertility*. Infertility can result from a variety of problems in men and women: poorly timed sex, low sperm count, failure to ovulate, or obstruction in the uterine tubes.

HEALTH, HOMEOSTASIS, AND THE ENVIRONMENT: THE SPERM CRISIS?

45. A variety of drugs and chemical pollutants affect sperm development and may be causing a decline in the sperm count of American men.

Critical Thinking

THINKING CRITICALLY— ANALYSIS

This Analysis corresponds to the Thinking Critically scenario that was presented at the beginning of this chapter.

Before writing your article, you might want to check the experimental design. Were adequate sample sizes used in the experiments? Did the researcher have proper control groups? Were the animals particularly prone to these abnormalities? What dosages were given? Are these at all similar to dosages people might be exposed to?

You might also want to see if these results have been reported elsewhere. Moreover, you would probably want

to see whether there is any research data linking this chemical to human abnormalities. Having done this, you would have a strong base from which to piece together a good article.

EXERCISING YOUR CRITICAL THINKING SKILLS

Devise an experiment or set of experiments to test the hypothesis that the United States is in the midst of a sperm crisis, a decline in male sperm production caused by chemicals in the environment. What type of evidence would support or refute your hypothesis? How can you devise your experiment to avoid bias?

TEST OF CONCEPTS

1. You are a fertility specialist. A young woman arrives in your office complaining that she has been trying to get pregnant for 2 years but to no avail. Describe how you would go about determining whether the problem was with her, her husband, or both of them.

2. Describe the anatomy of the male reproductive system. Where are sperm produced? Where are they stored? What structures produce the semen?

3. Describe the process of spermatogenesis, noting the cell types and the number of chromosomes in each type.

4. Why is the first meiotic division called a *reduction division*?

5. List the hormones that control testicular function. Where are they produced, and what effects do they have on the testes?

6. You are a family doctor. A man comes to your office complaining of impotence. What are the possible causes? How would you go about testing for the causes?

7. Trace the pathway for a sperm from the seminiferous tubule to the site of fertilization.

8. Describe the process of ovulation and its hormonal control.

9. What is the corpus luteum? How does it form? What does it produce? Why does it degenerate at the end of the menstrual cycle if fertilization does not occur?

10. What is menstruation? What triggers its onset?

11. Describe the effects of estrogen and progesterone on the reproductive tract and the body.

12. A woman comes to your office. She is 47 years old and complains of irritability and depression. She asks for the name of a reliable psychiatrist who could help her. She says that she wakes up in the middle of the night in a sweat. Would you give her the name of a psychiatrist? Why or why not? If not, what would you do?

13. Describe each of the following birth control measures, explaining what they are and how they work: the pill, IUD, diaphragm, cervical cap, condom, spermicidal jelly, and natural method.

14. Describe ways to prevent the spread of sexually transmitted diseases.

www.jbpub.com/humanbiology

TOOLS FOR LEARNING

Tools for Learning is an on-line student review area located at this book's web site HumanBiology (www.jbpub.com/humanbiology). The review area provides a variety of activities designed to help you study for your class:

Chapter Outlines. We've pulled out the section titles and full sentence sub-headings from each chapter to form natural descriptive outlines you can use to study the chapters' material point by point.

Review Questions. The review questions test your knowledge of the important concepts and applications in each chapter. Written by the author of the text, the review provides feedback for each correct or incorrect answer. This is an excellent test preparation tool.

Flash Cards. Studying human biology requires learning new terms. Virtual flash cards help you master the new vocabulary for each chapter.

Figure Labeling. You can practice identifying and labeling anatomical features on the same art content that appears in the text.

Active Learning Links. Active Learning Links connect to external web sites that provide an opportunity to learn basic concepts through demonstrations, animations, and hands-on activities.

CHAPTER 20

HUMAN DEVELOPMENT AND AGING

Thinking Critically

"Thanks to improvements in medicine, people are living longer." You have heard this statement dozens of times in one form or another. The assertion is that advances in medicine are actually increasing how long you and I will live. This statement is repeated so often that most of us believe it implicitly. But is it true? ■

Human embryo showing the webbed fingers.

I N A HOSPITAL ROOM IN A MAJOR CITY, TWO PAR-ents bring a newborn child into the world with the assistance of a nurse midwife. Three floors up in the geriatric ward, an eldery man quietly passes away. Life and death are two of biology's most dramatic acts, and the subject of this chapter. More specifically, we will focus on fertilization, which creates life, and early intrauterine development in humans. We will also briefly discuss aging and death.

Section 20-1

Fertilization

In sexually reproducing animals like us, life begins when the sperm and egg unite, forming a **zygote.** In humans, the sperm and the ovum unite in the upper third of the uterine tube.

Sperm face many obstacles which eliminate most of them. In humans, sperm are deposited in the vagina and quickly make their way up the reproductive tract (**FIGURE 20-1**). Within a few minutes of ejaculation, they enter the cervical canal; 30 minutes later, they arrive at the junction of the uterus and uterine tubes. Sperm then travel to the upper portion of the uterine tubes, where they may encounter an oocyte released from the ovary. Studies based on laboratory animals suggest that only 1 in 3 sperm makes it through the cervical canal and 1 in 1000 makes it to the uterine tubes (**FIGURE 20-1**). The rest perish along the way.

Although sperm are motile, their movement through the female reproductive tract is primarily due to muscular contractions in the walls of the uterus and uterine tubes. Some evidence suggests that these contractions are stimulated by prostaglandins, hormonelike substances in the semen.

Sperm reach the site of fertilization, the upper third of the uterine tubes, an hour or so after ejaculation, but they cannot fertilize an oocyte until they have been in the female reproductive tract for 6–7 hours. The time spent inside the female reproductive tract before fertilization is required to remove a layer of cholesterol deposited on the plasma membranes of sperm from the secretions of the sex accessory glands. The cholesterol coat stabilizes the sperm plasma membranes and protects sperm on their journey through the female reproductive tract. Removal of this coating renders the sperm's membranes fragile and disruptible, a prerequisite for fertilization.

After this process, the plasma membrane over the head of the sperm fuses with the outer membrane of the acrosome, a caplike structure filled with digestive enzymes. Tiny openings develop at the points of fusion, allowing acrosomal enzymes to leak out (**FIGURE 20-2A**).

Before a sperm can fertilize an oocyte, it must dissolve its way through the layer of follicle cells surrounding the oocyte. But sperm do not work in isolation. Rather, many sperm swarming around the oocyte (like so many bees around a hive) release enzymes from their acrosomes, dissolving the material that binds the follicle cells together (**FIGURE 20-2B**). After passing through this layer, sperm must digest their way through the zona pellucida, the gel-like layer immediately surrounding the oocyte. Acrosomal enzymes play a part in this as well.

Sperm cells that traverse the zona pellucida enter the space between the plasma membrane of the oocyte and the zona. As a rule, the first sperm cell to come in contact with the plasma membrane of the oocyte will fertilize it; all other sperm are excluded.

Sperm are excluded by two mechanisms. The first is called the **fast block to polyspermy** (POLL-ee-SPERM-ee; "many sperm"). The fast block to polyspermy occurs when the sperm cell contacts the plasma membrane of the oocyte, triggering membrane depolarization, a change in the potential difference across the plasma membrane. This is caused by an influx of sodium ions. For reasons not

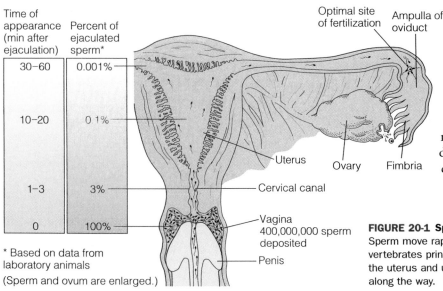

FIGURE 20-1 Sperm Transport in the Female Reproductive System
Sperm move rapidly up the female reproductive tract of humans and other vertebrates principally as a result of contractions in the muscular walls of the uterus and uterine tubes. Notice the rapid decline in sperm number along the way.

well understood, depolarization blocks other sperm from fusing with the oocyte.

Sperm are also excluded by a **slow block to polyspermy**. When a sperm contacts the plasma membrane of a secondary oocyte, it triggers the release of enzymes from membrane-bound vesicles lying beneath the plasma membrane of the oocyte (**FIGURE 20-2A**). These enzymes cause the zona pellucida to harden, blocking other sperm from reaching the oocyte. These secretions may also cause the "extra" sperm that have attached to the plasma membrane of the oocyte to detach.

These two mechanisms are important evolutionary developments, ensuring that only one sperm penetrates an egg. Without them, fertilized eggs might quickly overload with extra nuclear material, a condition that would likely impair subsequent cell divisions and result in embryonic death.

Sperm contact with the plasma membrane of the oocyte also triggers the second meiotic division, thus converting the secondary oocyte into an ovum. But the life of an ovum is short-lived, for once the sperm nucleus enters, the ovum is called a zygote.

Once inside the ovum, the sperm cell

nucleus swells. At this stage, the nuclei of the ovum and sperm are called the male and female **pronuclei** (**FIGURE 20-3**). The chromosomes in the pronuclei replicate as the pronuclei move toward the center of the ovum. A mitotic spindle also assembles. After chromosome replication is complete, the chromosomes of the male and female pronuclei condense, and the nuclear envelopes of the pronuclei disintegrate. Some of the spindle fibers attach to the chromosomes, and the chromosomes line up on the equatorial plate. The zygote is now ready for the first mitotic division.

FIGURE 20-2 Fertilization *(a)* The plasma membrane of the sperm and the outer membrane of the acrosome fuse and the membranes break down, releasing enzymes that allow the sperm to penetrate the corona radiata. Sperm digest their way through the zona pellucida via enzymes associated with the inner acrosomal membrane. Sperm are engulfed by the oocyte plasma membrane. Cortical granules are released when the sperm cell contacts the membrane. These granules cause other sperm in contact with the membrane to detach. *(b)* Although many sperm gather around the egg (this egg is from a clam), only one will enter. Fertilized sperm are phagocytized by the egg.

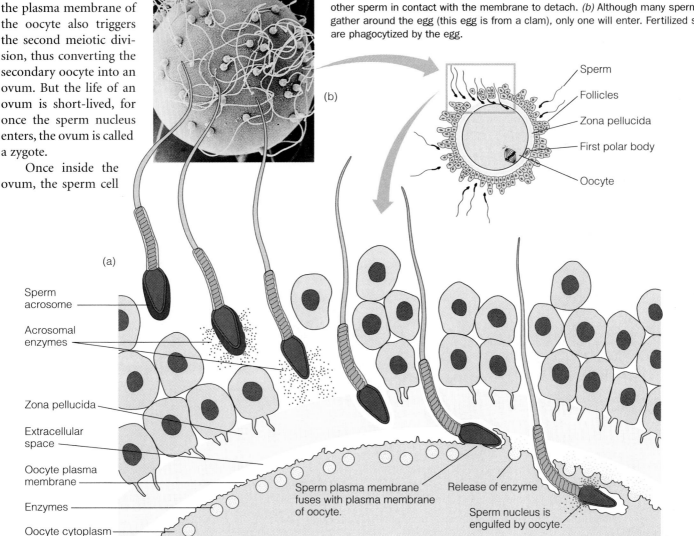

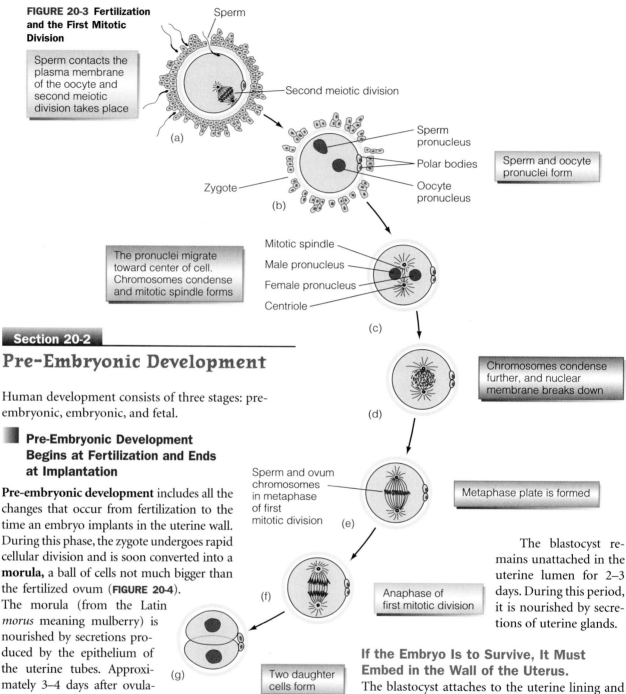

FIGURE 20-3 Fertilization and the First Mitotic Division

Sperm

Sperm contacts the plasma membrane of the oocyte and second meiotic division takes place

(a)

Second meiotic division

Sperm pronucleus

Polar bodies

Oocyte pronucleus

Sperm and oocyte pronuclei form

Zygote

(b)

The pronuclei migrate toward center of cell. Chromosomes condense and mitotic spindle forms

Mitotic spindle

Male pronucleus

Female pronucleus

Centriole

(c)

Chromosomes condense further, and nuclear membrane breaks down

(d)

Sperm and ovum chromosomes in metaphase of first mitotic division

Metaphase plate is formed

(e)

Anaphase of first mitotic division

(f)

Two daughter cells form

(g)

Section 20-2

Pre-Embryonic Development

Human development consists of three stages: pre-embryonic, embryonic, and fetal.

Pre-Embryonic Development Begins at Fertilization and Ends at Implantation

Pre-embryonic development includes all the changes that occur from fertilization to the time an embryo implants in the uterine wall. During this phase, the zygote undergoes rapid cellular division and is soon converted into a **morula,** a ball of cells not much bigger than the fertilized ovum (**FIGURE 20-4**). The morula (from the Latin *morus* meaning mulberry) is nourished by secretions produced by the epithelium of the uterine tubes. Approximately 3–4 days after ovulation, the morula enters the uterus.

Fluid soon begins to accumulate in the morula, converting it into a **blastocyst,** a hollow sphere of cells (**FIGURE 20-4**). The blastocyst consists of a clump of cells, the **inner cell mass,** which will become the embryo, and a ring of flattened cells, the **trophoblast** ("to nourish the blastocyst"). The trophoblast gives rise to the embryonic portion of the **placenta,** an organ that supplies nutrients to, and removes wastes from, the embryo.

The blastocyst remains unattached in the uterine lumen for 2–3 days. During this period, it is nourished by secretions of uterine glands.

If the Embryo Is to Survive, It Must Embed in the Wall of the Uterus.

The blastocyst attaches to the uterine lining and digests its way into the endometrium. This process, called **implantation,** begins 6–7 days after fertilization (**FIGURE 20-5**).

Most embryos implant high on the back wall of the uterus. The cells of the trophoblast first contact the endometrium, then adhere to it, but only if the uterine lining is healthy and properly primed by estrogen and progesterone. If the endometrium is not ready or is "unhealthy"—for example, because of the presence of an

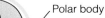

FIGURE 20-4 Formation of the Morula and Blastocyst During Pre-embryonic Development

Zona pellucida

Polar body

(a) 2-cell stage

(b) 4-cell stage

(c) 8-cell stage

(d) Morula

Zona pellucida begins to degenerate.

(e) Blastocyst (early)

Inner cell mass

Trophoblast

(f) Blastocyst (late)

IUD or an endometrial infection—the blastocyst cannot implant. Blastocysts may also fail to implant if their cells contain certain genetic mutations. In either case, the blastocyst perishes and will be reabsorbed (phagocytized by the cells of the endometrium) or expelled during menstruation.

In cases where implantation occurs, the cells of the endometrium at and around the point of contact enlarge (hypertrophy). Enzymes released by the cells of the trophoblast digest a small hole in the thickened endometrium, and the blastocyst "bores" its way into the deeper tissues of the uterine lining (**FIGURE 20-5B**). During this process, the blastocyst literally "eats" its way into the layer of enlarged endometrial cells, feeding on nutrients liberated from the endometrial cells it destroys. Nutrients gained during this process sustain the blastocyst before the placenta forms.

By day 14, the uterine endometrium grows over the blastocyst, walling it off from the uterine lumen (cavity). Endometrial cells respond to the invasion of the blastocyst by producing prostaglandins,

which stimulate an increase in the development of uterine blood vessels. This ensures an ample supply of blood and nutrients for the blastocyst.

In yet another of life's most amazing feats, endometrial and embryonic tissue form the **placenta**, an organ that nourishes the embryo and fetus and gets rid of wastes throughout its long sojourn in the uterus (**FIGURE 20-6**).

Placental Hormones Are Essential to Reproduction.

Besides providing nutrients and getting rid of embryonic wastes, the placenta acts as a temporary endocrine gland, producing several hormones needed to maintain pregnancy. The placenta, therefore, is a respiratory, nutritive, excretory, and endocrine organ. This section discusses three placental hormones: human chorionic gonadotropin (HCG), estrogen, and progesterone (**TABLE 20-1**).

FIGURE 20-5 Implantation *(a)* The blastocyst fuses with the endometrial lining. Endometrial cells proliferate, forming the decidua. *(b)* The blastocyst digests its way into the endometrium. Cords of trophoblastic cells invade, digesting maternal tissue and providing nutrients for the developing blastocyst. *(c)* The blastocyst soon becomes completely embedded in the endometrium.

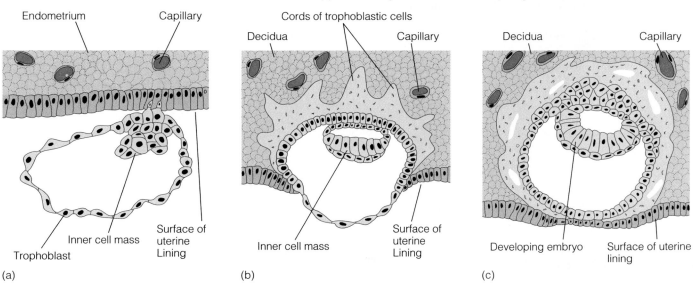

(a)
Endometrium   Capillary
Inner cell mass
Trophoblast
Surface of uterine Lining

(b)
Cords of trophoblastic cells
Decidua   Capillary
Inner cell mass
Surface of uterine Lining

(c)
Decidua   Capillary
Developing embryo   Surface of uterine lining

FIGURE 20-6 The Placenta
This organ, made from maternal and fetal tissue, helps nourish the developing fetus and remove wastes. It also produces important hormones.

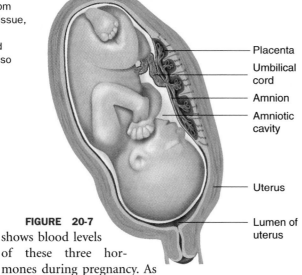

Placenta

Umbilical cord

Amnion

Amniotic cavity

Uterus

Lumen of uterus

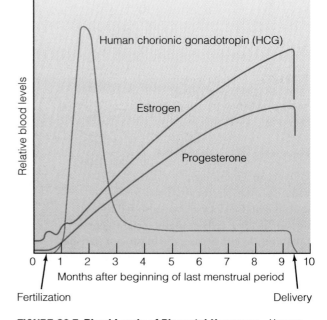

FIGURE 20-7 shows blood levels of these three hormones during pregnancy. As illustrated, HCG levels peak at the beginning of the second month of pregnancy, then drop off by the end of the third month. In contrast, levels of estrogen and progesterone rise dramatically throughout gestation (the period between fertilization and childbirth).

As noted in Chapter 19, HCG is an LH-like hormone produced by the embryo early in pregnancy. One of its main functions is to prevent the corpus luteum (CL) in the ovary from degenerating when an embryo is present. HCG also stimulates estrogen and progesterone production by the CL. These hormones are essential to maintaining pregnancy. However, as illustrated, HCG secretion from the embryonic tissue lasts only about 10 weeks. Because of the natural decline in HCG, the CL degenerates. At this time, estrogen and progesterone production shifts to the placenta.

Many women experience nausea (morning sickness) during the first 2–3 months of pregnancy. Al-

FIGURE 20-7 Blood Levels of Placental Hormones Human chorionic gonadotropin levels peak in the second month of pregnancy, then drop off by the end of the third month. Levels of estrogen and progesterone, produced chiefly by the placenta, continue to rise.

though it often occurs in the morning hours, "morning" sickness in some women can last all day. The exact cause is not known, but some researchers think that HCG may be the culprit. They believe that HCG stimulates the brain directly, creating nausea. Other researchers blame the high levels of estrogen and progesterone. It may be high levels of all three hormones.

Estrogen and progesterone serve a number of functions essential for pregnancy. Together, estrogen and progesterone stimulate the growth of the uterine lining (endometrium), which is essential to successful pregnancy. Estrogen by itself also stimulates growth of the smooth muscle cells of the myometrium; this allows the uterus to expand to many times its original size during pregnancy. Smooth muscle growth not only accommo-

TABLE 20-1

Hormones Produced by the Placenta

Hormone	Function
Human chorionic gonadotropin (HCG)	Maintains corpus luteum of pregnancy
	Stimulates secretion of testosterone by developing testes in XY embryos
Estrogen (also secreted by corpus luteum of pregnancy)	Stimulates growth of myometrium, increasing uterine strength for parturition (childbirth)
	Helps prepare mammary glands for lactation
Progesterone (also secreted by corpus luteum of pregnancy)	Suppresses uterine contractions to provide quiet environment for fetus
	Promotes formation of cervical mucous plug to prevent uterine contamination
	Helps prepare mammary glands for lactation
Human chorionic somatomammotropin	Helps prepare mammary glands for lactation
	Believed to reduce maternal utilization of glucose so that greater quantities of glucose can be shunted to the fetus
Relaxin (also secreted by corpus luteum of pregnancy)	Softens cervix in preparation of cervical dilation at parturition
	Loosens connective tissue between pelvic bones in preparation for parturition

dates the growing baby, it also provides additional propulsive force needed to expel the child at birth.

Progesterone has additional functions as well. For example, it calms the uterine musculature during pregnancy, preventing the premature expulsion of the embryo and fetus. Progesterone also stimulates mucous production by the cervix and the formation of a mucous plug that prevents bacteria from entering the uterus and infecting the growing embryo.

The Amnion Forms from the Inner Cell Mass.

As the placenta begins to form, the inner cell mass (ICM) of the blastocyst undergoes some remarkable changes. Early in development, a layer of cells separates from the ICM to form the **amnion** (AM-knee-on). A small cavity, the **amniotic cavity**, forms between the ICM and the amnion. The amniotic cavity fills with a watery fluid called **amniotic fluid**. It forms a cushion around the baby during development, helping to protect it from injury[1] (**FIGURE 20-6**).

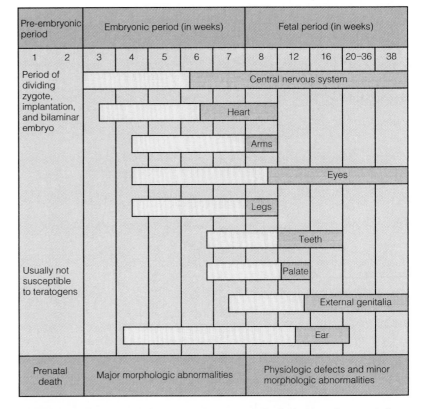

FIGURE 20-8 Organogenesis Human development is divided into three periods, or stages: pre-embryonic, embryonic, and fetal. Organogenesis occurs during the embryonic stage. Each bar indicates when an organ system develops. The yellow-shaded area indicates the periods most sensitive to teratogenic agents (agents that can cause birth defects).

Section 20-3

Embryonic Development

After the amnion forms, the cells of the inner cell mass differentiate, forming three distinct germ cell layers, the ectoderm, mesoderm, and endoderm. These are known as the **primary germ layers**. The formation of the primary germ layers marks the beginning of **embryonic development**.

▍ Embryonic Development Begins with Implantation and Ends When the Organs Are More or Less Formed

The primary germ layers of the embryo give rise to the organs of the body (**TABLE 20-2**) in a process called

---

[1]A portion of the amniotic fluid is produced by the fetus. The rest apparently comes from the amniotic membranes.

organogenesis, which begins about the third week of pregnancy (**FIGURE 20-8**). Organogenesis is the main event of embryonic development.

One of the first events of organogenesis is the formation of the central nervous system (the spinal cord and brain) from ectoderm. Early in embryonic development, the ectoderm invaginates (folds inward) and forms a long trench, the **neural groove**, which runs the length of the

---

TABLE 20-2

End Products of Embryonic Germ Layers

Ectoderm	Mesoderm	Endoderm
Epidermis	Dermis	Lining of the digestive system
Hair, nails, sweat glands	All muscles of the body	Lining of the respiratory system
Brain and spinal cord	Cartilage	Urethra and urinary bladder
Cranial and spinal nerves	Bone	Gallbladder
Retina, lens, and cornea of eye	Blood	Liver and pancreas
Inner ear	All other connective tissue	Thyroid gland
Epithelium of nose, mouth, and anus	Blood vessels	Parathyroid gland
Enamel of teeth	Reproductive organs	Thymus
	Kidneys	

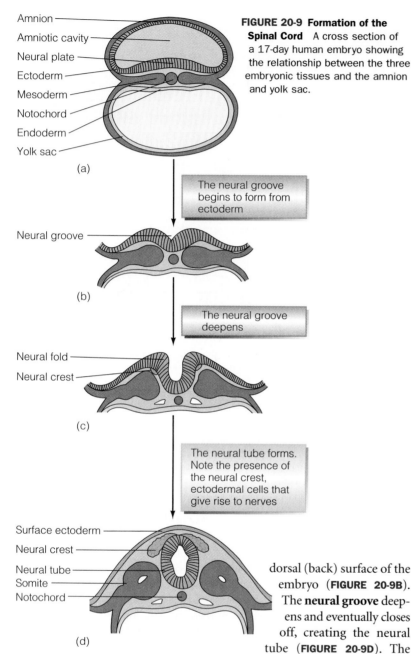

Amnion
Amniotic cavity
Neural plate
Ectoderm
Mesoderm
Notochord
Endoderm
Yolk sac

(a)

FIGURE 20-9 Formation of the Spinal Cord A cross section of a 17-day human embryo showing the relationship between the three embryonic tissues and the amnion and yolk sac.

The neural groove begins to form from ectoderm

Neural groove

(b)

The neural groove deepens

Neural fold
Neural crest

(c)

The neural tube forms. Note the presence of the neural crest, ectodermal cells that give rise to nerves

Surface ectoderm
Neural crest
Neural tube
Somite
Notochord

(d)

The middle germ layer, the mesoderm, gives rise to deeper structures—the muscle, cartilage, bone, and others. Much of the mesoderm first aggregates in blocks, called the **somites** (SO-mights), situated alongside the neural tube (**FIGURE 20-9D**). The somites form the vertebrae (the backbone) and the muscles of the neck and trunk. Mesoderm lateral to the somites becomes the dermis of the skin, connective tissue, and the bones and muscles of the limbs.

The endoderm, the "lowermost" germ layer of the ICM, forms a large pouch under the embryo, the **yolk sac** (yoke sack) (**FIGURES 20-9A** and **20-10**). In birds, reptiles, and amphibians, the yolk sac nourishes the growing embryo, but in humans, nourishment comes from the placenta. The human yolk sac gives rise to blood cells and also primitive germ cells, known as **primordial germ cells**. During organogenesis, these cells migrate from the wall of the yolk sac via amoeboid motion to the developing testes and ovaries near the kidneys. Here the primordial germ cells become spermatogonia or oogonia, depending on the sex of the embryo. Finally, as shown in **FIGURE 20-10C**, the uppermost part of the yolk sac becomes the lining of the intestinal tract.

Section 20-4

Fetal Development

Fetal Development begins in the eighth week of gestation and ends at birth. It involves two basic processes: (1) continued organ development and growth and (2) changes in body proportions—for example, elongation of the limbs.

Fetal Development Begins After the Organs Have Formed

The fetus grows rapidly during the fetal period, increasing from about 2.5 centimeters (1 inch) to 35–50 centimeters (14–21 inches) and increasing in weight from 1 gram to 3000–4000 grams. The fetus also undergoes considerable change in physical appearance. Unfortunately, space limitations prevent a discussion of the development of each organ system, a fascinating subject that you may want to study on your own.

Section 20-5

Ectopic Pregnancy and Birth Defects

Pregnancy and embryonic development are rather complicated, involving hundreds of different processes, any

dorsal (back) surface of the embryo (**FIGURE 20-9B**). The **neural groove** deepens and eventually closes off, creating the neural tube (**FIGURE 20-9D**). The walls of the neural tube thicken, forming the spinal cord. Anteriorly, the neural tube expands to form the brain.

The nerves that attach to the spinal cord (spinal nerves) and brain (cranial nerves) develop from small aggregations of ectodermal cells (the **neural crest**) lying on either side of the neural tube throughout most of its length (**FIGURES 20-9C** and **20-9D**). These cells sprout processes that extend into the body and attach to organs, muscle, bone, and skin, among others. The ectoderm also gives rise to the outer layer of skin (the epidermis).

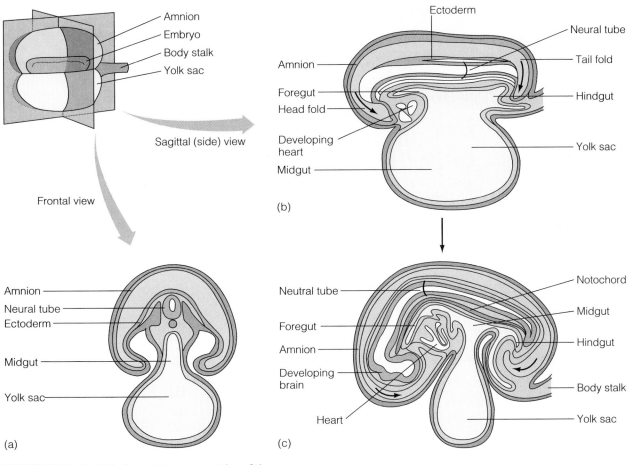

FIGURE 20-10 The Yolk Sac *(a)* A cross section of the embryo. The yolk sac forms from embryonic endoderm. *(b)* A longitudinal section showing how the upper end of the yolk sac forms the embryonic gut, *(c)* which will become the lining of the intestinal tract.

one of which can go wrong. Two fairly common and potentially serious problems are ectopic pregnancy (ECK-TOP-ick) and birth defects.

Most Ectopic Pregnancies Occur in the Uterine Tubes

An **ectopic pregnancy** occurs when a fertilized ovum fails to be transported down the uterine tubes and instead develops outside the uterus, usually somewhere in the uterine tube itself.[2] Because of this, most ectopic pregnancies are also referred to as **tubal pregnancies**.

In a tubal pregnancy, the zygote actually implants in the lining of the uterine tube. However, placental development typically damages the uterine tubes, causing internal bleeding and severe abdominal pain. Because

the uterine tube cannot sustain the embryo, a tubal pregnancy cannot generally proceed to term. Surgery is required to remove the embryo.

About 1 in every 200 pregnancies is ectopic. Usually diagnosed within the first 2 months, tubal pregnancies generally occur as a result of defects in the uterine tubes or scar tissue from a previous infection, which impede the transport of the blastocyst to the uterus.

Embryonic and Fetal Development Can Be Altered by Outside Influences, Resulting in Birth Defects and Miscarriages

By several estimates, 31% of all conceptions (successful fertilizations) end in a miscarriage—that is, a spontaneous abortion. Fortunately, two of every three of these miscarriages occur before a woman is even aware that she is pregnant. Why such a high rate? Biologists hypothesize that early miscarriage is nature's way of "discarding" defective embryos. In other words, it is an evolutionary mechanism that helps ensure a healthier population.

Because nature generally eliminates defects, you might expect that most children born into the world

---

[2]Ectopic pregnancies can also occur in the abdominal cavity.

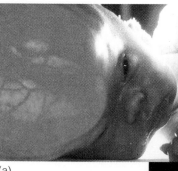

(a)

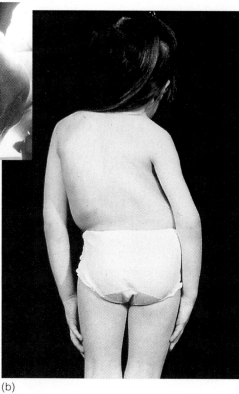

(b)

FIGURE 20-11 Common Birth Defects *(a)* Hydrocephalus. *(b)* Scoliosis, a lateral curvature of the spine.

TABLE 20-3	
Known and Suspected Human Teratogens	

Known Agents	Possible or Suspected Agents
Progesterone	Aspirin
Thalidomide	Certain antibiotics
Rubella (German measles)	Insulin
Alcohol	Antitubercular drugs
Irradiation	Antihistamines
	Barbiturates
	Iron
	Tobacco
	Antacids
	Excess vitamins A and D
	Certain antitumor drugs
	Certain insecticides
	Certain fungicides
	Certain herbicides
	Dioxin
	Cortisone
	Lead

would be free of defects. Unfortunately, this is not the case. Humans experience a surprisingly high rate of **birth defects**, physical or physiological abnormalities. By various estimates, 10%–12% of all newborns have some kind of birth defect, ranging from minor biochemical or physiological problems, which are not even noticed at birth, to gross physical defects (**FIGURE 20-11**).[3]

The study of birth defects is called **teratology** (TARE-ah-TOL-eh-gee). The word *teratology* comes from the Greek *teratos*, meaning "monster" and reflects some of the more gruesome or disfiguring birth defects. The term is an unfair characterization of most birth defects. Many defects are minor.

Although there is much to be learned about the causes of birth defects, scientists believe that most arise from chemical, biological, and physical agents, collectively known as **teratogens** (tah-RAT-uh-gens). **TABLE 20-3** lists known and suspected teratogens in humans.

The effect of a teratogenic agent on the developing embryo is related to three factors: (1) the time of exposure, (2) the nature of the agent, and (3) the dose. Con-

sider time first. Because the organ systems develop at different times, the timing of exposure determines which systems are affected by a given teratogen. Organ systems are usually most sensitive to potentially harmful agents early in their development, as indicated by the yellow bars in **FIGURE 20-8** (page 459). The central nervous system, for example, begins to develop during the third week of gestation, or pregnancy. Because most women do not know they are pregnant for 3–4 weeks after fertilization, the central nervous system is at risk. In contrast, the teeth, palate, and genitalia do not begin to form until about the sixth or seventh week of pregnancy. Exposure to teratogens during the seventh week might therefore affect them but have little effect on the CNS, which has entered a less sensitive phase.[4]

The nature of the teratogen also influences the outcome of exposure. Some teratogens are broad-spectrum agents—that is, they affect a variety of developing systems. Others are narrow-spectrum agents, affecting only one system. Ethanol in alcoholic beverages, for example, is a broad-spectrum teratogen. Consequently, children who are born to alcoholic mothers—or even women who have consumed one or two drinks early in pregnancy—may have a variety of birth defects, including facial, heart, and skeletal defects. Behavioral problems and learning disabilities are also common in children of

---

[3]About 2% of all births show gross malformations.

[4]Note that the central nervous system is by no means fully formed at this time. Some teratogens, such as the rubella virus, can affect development through the first 3 months of gestation.

alcoholic mothers. These symptoms are part of a condition called **fetal alcohol syndrome**.

In contrast, other teratogenic agents are more selective, targeting only one system. Methyl mercury found in fish and shellfish, for instance, damages the CNS, creating abnormalities in the brain and spinal cord, but has little effect on other systems.

Finally, like most toxic substances, teratogens generally follow a dose-response relationship: The greater the dose, the greater the effect.

One of the best-known teratogens is not a chemical substance, but, rather, a biological agent. It is the virus that causes German measles or rubella (rue-BELL-ah). Rubella is less common and less contagious than ordinary measles. If a pregnant woman contracts the disease during the first 3 months of pregnancy, however, she has a one in three chance of giving birth to a baby with a serious birth defect. Deafness, cataracts, heart defects, and mental retardation are common. Moreover, 15% of all babies with congenital rubella syndrome die within one year of birth.

Many birth defects can be avoided. German measles vaccine given to young girls usually protects them throughout their childbearing years. Vaccinating boys also reduces the risk to society. As with many other diseases, prevention is the best policy.

Other precautions also help. For instance, during the first 8 weeks of pregnancy, women should avoid alcohol and other potential teratogens to protect the developing organ systems. As noted earlier, though, most women do not know that they are pregnant until 2–4 weeks after conception. Thus, a woman who is trying to get pregnant should carefully control what she eats and drinks.

Once the period of high sensitivity has passed, good nutrition and a healthy environment of course remain essential. This is because fetal development is also influenced by a number of physical and chemical agents in our homes and places of work. Toxins can stunt fetal growth and, in higher quantities, can even kill a fetus, resulting in a stillbirth. Proper nutrition is essential because nutrient, vitamin, and mineral deficiencies can retard growth.

Section 20-6

Childbirth and Lactation

For most expectant parents, the birth of a child is one of the most exciting events of their lives. Childbirth, also appropriately called *labor*, begins with mild uterine contractions. During labor, contractions increase in strength and frequency until the baby is born. What stimulates this process?

▊ Uterine Muscle Contractions Are Stimulated by a Change in Hormonal Levels

Studies suggest that several factors play a role in stimulating uterine contractions. For instance, during the last few weeks of pregnancy, estrogen levels in the blood of a woman increase dramatically. This rise in estrogen levels is thought to stimulate the production of oxytocin receptors in the smooth muscle cells of the uterus. The increase in the number of receptors in the smooth muscle cells renders them increasingly responsive to the small amounts of oxytocin. But where does the oxytocin come from?

Interestingly, the fetal pituitary gland releases small quantities of oxytocin right before birth. Fetal oxytocin travels across the placenta and circulates in the mother's bloodstream. When it arrives at the sensitized uterine musculature, fetal oxytocin stimulates muscle contraction. Fetal oxytocin also stimulates the release of placental prostaglandins that act on smooth muscle. Together, these hormones stimulate more frequent and powerful uterine contractions.

Maternal oxytocin also plays an important role in childbirth. Scientists believe that uterine contractions and discomfort create stress that triggers the release of maternal oxytocin. It, in turn, augments muscle contractions caused by fetal oxytocin and prostaglandins (**FIGURE 20-12**). As uterine muscle contraction increases, maternal oxytocin release increases. This stimulates even stronger contractions, resulting in additional oxytocin release, a positive feedback loop that continues until the baby is delivered.

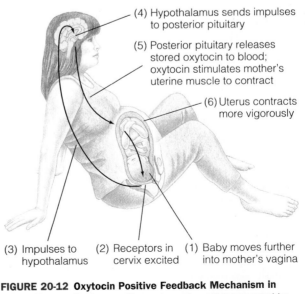

(4) Hypothalamus sends impulses to posterior pituitary

(5) Posterior pituitary releases stored oxytocin to blood; oxytocin stimulates mother's uterine muscle to contract

(6) Uterus contracts more vigorously

(3) Impulses to hypothalamus

(2) Receptors in cervix excited

(1) Baby moves further into mother's vagina

FIGURE 20-12 Oxytocin Positive Feedback Mechanism in Birth The mechanism continues to cycle until interrupted by the birth of the baby.

Research also suggests that the high levels of estrogen at the end of pregnancy block the "calming" influence of placental progesterone. Consequently, the uterus begins to contract at irregular intervals.

Physicians (and expectant mothers) recognize two types of labor contractions. The first are **false labor contractions**. These irregular uterine contractions usually begin a month or two before childbirth and are also known as **Braxton-Hicks contractions**. As the due date approaches, false labor contractions occur with greater frequency, causing many anxious couples to race to the hospital only to be told to go home and wait for a few weeks for the real thing. In contrast, **true labor contractions** are more intense than those of false labor and occur at regular intervals.

Childbirth requires the hormone **relaxin**, which is produced by the ovary and the placenta. Released near the end of gestation, relaxin softens the fibrocartilage uniting the pubic bones. This allows the pelvic cavity to widen and thus greatly facilitates childbirth. Relaxin also softens the cervix, allowing it to expand to allow the baby to pass. Without this remarkable adaptation, delivery, which is painful enough as it is, would be a problematic event indeed!

Childbirth Occurs in Three Stages

FIGURE 20-13 illustrates the three stages of childbirth, dilation, expulsion, and placental.

Stage 1, the **dilation stage**, gets its name from the dilation of the cervix. This phase begins when uterine contractions start and generally lasts 6–12 hours; but it can last much longer.

At the beginning of stage 1, uterine contractions typically last only 30 seconds and may come every half hour or so. As time passes, however, contractions become more frequent and powerful. Uterine contractions generally rupture the amnion early in the dilation phase, causing the release of the amniotic fluid, an event commonly referred to as "breaking the water."

Uterine contractions push the infant's head against the relaxin-softened cervix, causing it to stretch and become thin (**FIGURE 20-13B**). By the end of stage 1, the cervix has dilated to about 10 centimeters (4 inches), approximately the diameter of a baby's head. During

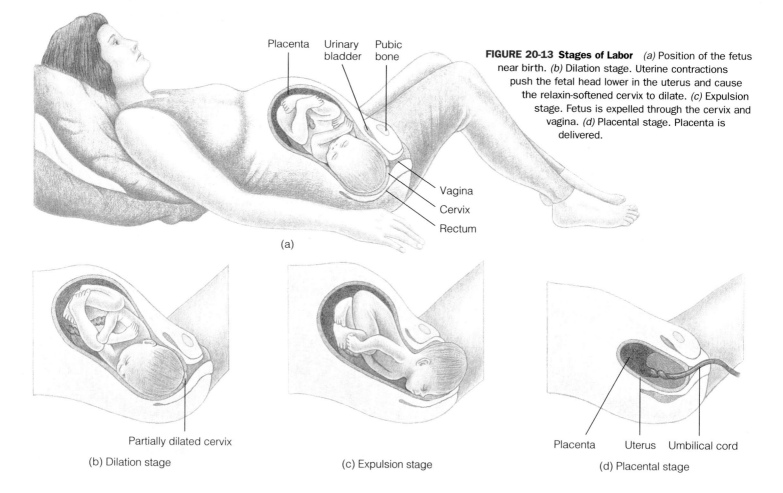

Placenta Urinary bladder Pubic bone

Vagina
Cervix
Rectum

(a)

FIGURE 20-13 Stages of Labor *(a)* Position of the fetus near birth. *(b)* Dilation stage. Uterine contractions push the fetal head lower in the uterus and cause the relaxin-softened cervix to dilate. *(c)* Expulsion stage. Fetus is expelled through the cervix and vagina. *(d)* Placental stage. Placenta is delivered.

Partially dilated cervix

(b) Dilation stage

(c) Expulsion stage

Placenta Uterus Umbilical cord

(d) Placental stage

this phase, the baby is pushed downward by the uterine contractions and descends into the pelvic cavity. When its head is "locked" in the pelvis, the baby is said to be engaged.

Stage 2, the **expulsion stage,** begins after the cervix is dilated to 10 centimeters and the baby is engaged (**FIGURE 20-13C**). At this time, uterine contractions usually occur every 2 or 3 minutes and last 1–1.5 minutes each. For most first-time mothers, stage 2 lasts 50–60 minutes. If a woman is delivering her second child, it lasts about 20-30 minutes on average. The expulsion stage ends when the child is pushed through the vagina into the waiting hands of a doctor, midwife, proud father, or perhaps a bewildered cab driver.

To facilitate the delivery, physicians or midwives often make an incision in the skin to widen the vaginal opening. This procedure, called an **episiotomy** (eh-PEE-zee-AHT-oh-me), is performed when the baby's head enters the vagina in stage 2. The incision prevents unnecessary tearing and allows the infant to pass quickly. The incision is sutured immediately after the baby is born.

Once the baby's head emerges from the vagina, the rest of the body slips out quickly. However, the process isn't over yet. The newborn baby is still attached to the placenta via the umbilical cord. The cord pulsates for a minute or more, continuing to deliver blood to the newborn. Consequently, many health care workers wait about a minute before tying off and cutting the cord to allow the blood remaining in the placenta to be pumped into the newborn.

Most babies (95%) are delivered head first with their noses pointed toward the mother's tailbone (**FIGURE 20-13C**). Occasionally, however, babies may be oriented in other positions—making delivery more difficult, time-consuming, and hazardous for mother *and* baby. The most common alternative delivery is the **breech birth**, in which the baby is expelled rear-end first. Because breech births require more time, they can cause extreme fatigue in the mother and brain damage in the baby. In breech babies, the umbilical cord sometimes wraps around the infant's neck, cutting off its supply of blood, causing brain damage to the baby. To avoid such complications, breech babies are often "turned" by physicians before birth by applying pressure to the woman's abdomen. If a fetus cannot be turned, the baby is usually delivered by a **cesarean section**, a horizontal incision through the abdomen just at the pubic hair line.

Cesarean sections are performed for other reasons as well—for example, if labor is prolonged or if the mother has an active infection caused by herpes or some other sexually transmitted disease that might be transferred to her child as it passes through the vagina.

The final stage of delivery is the **placental stage** (**FIGURE 20-13D**). The placenta remains attached to the uterine wall for a short while, then is expelled by uterine contractions, usually within 15 minutes of childbirth. After the placenta is expelled, the uterine blood vessels clamp shut, preventing hemorrhage, although the mother continues to lose some blood for 3–6 weeks after delivery.

After delivery, the uterus gradually returns to its normal size. Uterine involution (shrinkage) results from the rapid decline in estrogen and progesterone, which had caused the uterus to grow in the first place, and is accelerated by oxytocin released by the posterior pituitary after birth when a woman breast-feeds her baby. Complete involution in women who breast-feed usually occurs within 4 weeks of pregnancy. In women who do not, the process usually takes 6 weeks.

The Pain Associated with Childbirth Can Be Relieved by Drugs and by Special Birthing Methods

The level of pain a woman feels during childbirth varies greatly and is partly governed by her level of fear and tension. Generally, the more tense a woman is, the more pain she feels. For this reason, many hospitals provide relaxation training to expectant mothers and comfortable birthing rooms. Drugs can also be given to reduce tension and pain, but they must be administered well before the birth of the baby to be effective. If given just before delivery, the drug offers little relief to the mother and may impair the baby's breathing.

Painkilling drugs can be injected into the wall of the vagina. This procedure, known as a **pudendal block** (pew-DEN-del), is performed if an episiotomy is likely or if forceps are going to be used to facilitate childbirth.

Many physicians routinely offer **epidural anesthesia** (ep-eh-DU-rel AN-es-THEE-zha). In this procedure, physicians inject an anesthetic agent into the bony canal that houses the spinal cord (just outside the dura mater). The drug temporarily deadens the sensory nerves leading from the vagina and lower body. The drug, however, not only blocks pain impulses, it blocks motor nerve impulses to the muscles of the abdomen, rendering a woman unable to push the baby out. In such cases, forceps may be required to facilitate the delivery, but forceps can damage a newborn and are generally avoided whenever possible.

Many couples and health care workers believe that drugs may be harmful to mothers and their babies. This belief and a desire to do things "naturally" have spawned the natural childbirth movement. **Natural childbirth** means different things to different people. In general, it refers to drug-free childbirth aided by relaxation and/or special breathing techniques.

The most popular natural childbirth method today is the **Lamaze technique** (lah-MAHZZ). Parents learn special breathing techniques to assist in the delivery. Shallow breathing, for example, commences when uterine contractions begin to keep the diaphragm from pressing down on abdominal organs, reducing tension and pain. It also ensures an adequate supply of oxygen to the fetus.

The second most popular method is the Bradley method. The **Bradley method** teaches women how to relax during uterine contractions, reducing pain and the need for drugs. No special breathing is required.

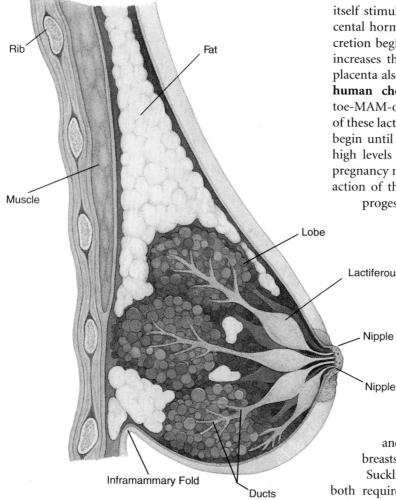

FIGURE 20-14 The Lactating Breast The glandular units enlarge considerably under the influence of progesterone and prolactin. Milk is expelled by contraction of musclelike cells surrounding the glandular units. Ducts drain the milk to the nipple.

Milk Production, or Lactation, Is Controlled by the Hormone Prolactin

A baby emerges into a novel environment at birth. No longer connected to the placental lifeline, the newborn must now begin to breathe to obtain oxygen and dispose of carbon dioxide. It must also find a new way of acquiring food. For most of human evolution, newborns have been fed milk from their mothers' breasts.

The breasts of a nonpregnant woman consist primarily of fat and connective tissue interspersed with milk-producing glandular tissue. The glands are drained by ducts that lead to the nipple, a surface structure where the milk ducts converge (**FIGURE 20-14**). As noted earlier, during pregnancy the ducts and glands proliferate under the influence of estrogen and progesterone from the placenta and ovary. Milk production is itself stimulated by the hormone prolactin and a placental hormone, HCS (described below). Prolactin secretion begins during the fifth week of pregnancy and increases throughout gestation, peaking at birth. The placenta also secretes a mild milk-producing hormone, **human chorionic somatomammotropin** (so-MAT-toe-MAM-oh-trow-pin) or **HCS**. Despite the presence of these lactogenic hormones, milk production does not begin until 2 or 3 days after birth. That's because the high levels of estrogen and progesterone throughout pregnancy necessary for breast development inhibit the action of the prolactin and HCS. When estrogen and progesterone levels decline after birth, HCS and prolactin exert their influence.

Although the breasts do not produce milk for 2 or 3 days, they immediately begin producing small quantities of colostrum. **Colostrum** (co-LOSS-trum) is a fluid rich in protein and lactose (a disaccharide), but lacking fat. Colostrum also contains antibodies that protect the infant from bacteria. A newborn subsists on colostrum for the first few days. (Health Note 8-1 describes the benefits of colostrum and breast-feeding.) Within 2–3 days, the breasts start to produce large quantities of milk.

Suckling stimulates the release of two hormones, both required for successful lactation. When a baby suckles her mother's breast, two things happen: nerve impulses travel to the hypothalamus, causing the release of **prolactin releasing hormone (PRH)**. PRH, in turn, stimulates prolactin secretion, which stimulates milk production. Prolactin levels remain elevated for approximately 1 hour after each feeding. As a result, each surge stimulates the milk production needed for the next feeding. Milk production continues as long as the baby suckles. If nursing is interrupted for 3 or 4 days, however, the breasts stop producing milk. In most women, milk production begins to decline by the seventh to ninth month of lactation, a time at which babies begin to feed on semisolid food and breast-feed less frequently.

In order for milk to reach the nipple, it must be actively propelled from the glandular units and through the ducts. This is the function of oxytocin, the release of which is also controlled by a neuroendocrine reflex stimulated by suckling. Oxytocin causes muscles in the milk glands to contract and eject their milk. Neuroendocrine reflexes are important adaptations that provide hormones on demand. Such systems conserve energy and nutrients and contribute to the evolutionary success of mammals.

Section 20-7

Aging and Death

Aging is a part of life, but it remains one of the great mysteries of biology. Technically, **aging** is defined as a progressive deterioration of the body's structure and function. One function whose deterioration is of extreme importance to us is homeostasis. As homeostatic mechanisms falter, the body becomes less resilient. Healing takes longer. Individuals become more susceptible to disease. Multiple breakdowns accelerate one's physical and mental deterioration.

The most notable changes occurring with age are physical shifts—wrinkling, loss of hair, and stooping. Less obvious is the gradual deterioration of the function of body organs. These often occur so slowly as to seem imperceptible (**FIGURE 20-15**). For example, as people age, their vision and hearing both decline, usually by small increments. Muscular strength also ebbs as a result of a decrease in the number of myofibrils (bundles of contractile filaments) in the skeletal muscles. Bones tend to thin, and joints often show signs of wear. Aging is also accompanied by a gradual reduction in cardiac output and pulmonary function. The decrease in pulmonary function results from a reduction in oxygen absorption and a decrease in the amount of air that can be inhaled and exhaled. The number of functional nephrons in the kidney decreases, as does renal function. The immune system also becomes less able to respond to antigens. The nervous system does not escape the aging process. Memory deteriorates and reaction time decreases. Although these changes are part of the

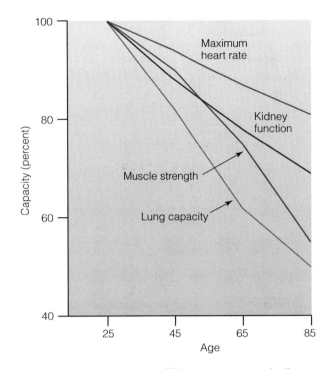

FIGURE 20-15 The Inevitable Slide As we age, our bodies deteriorate. This graph shows the dramatic decline in four essential body functions.

aging process, not all of them begin at the same time. **TABLE 20-4** lists some important age-related changes in various organ systems along with their causes.

Aging May Be Brought About by a Decline in Cell Numbers and a Decline in the Function of Cells

The decline in homeostatic systems and general body function results from at least two factors: a decrease in the number of cells in the organs and a decline in the function of cells.

Cell number is determined by the balance between cell division and cell death. Laboratory experiments show that body cells grown in culture divide a certain number of times, then stop. The number of divisions a cell undergoes is directly related to the age of the organism from which the cell is taken: The older the organism, the fewer divisions are possible from that point in time on.

Laboratory studies such as these have also uncovered a correlation between the number of divisions in culture and the life span of the species from which the cells were taken. In other words, cells from species with long life spans undergo more divisions than cells from species with short life spans. These and other data suggest that the end of cell division and, therefore, aging, may be genetically programmed.

TABLE 20-4

Age-Related Changes in the Body

Organ System	Change	Cause
Skin	Thinning	Reduced cell division in epidermis
	Loss of elasticity	Elastic fibers in dermis replaced by collagen fibers
	Wrinkling	Same as above plus loss of fat in hypodermis
	Drying	Loss of sweat glands and sebaceous glands
	Loss of body hair	Loss of hair follicles
Bones	Become weaker, more porous, and more brittle	Loss of calcium is the primary reason for this, but also caused by loss of collagen fibers
Joints	Often become difficult to move and sometimes arthritic	Deterioration of cartilage on bone surfaces
Muscles	Loss of mass and muscle strength	Loss of motor neurons which stimulate muscles and a loss of muscle protein; fat and collagen fibers tend to invade muscles
Respiratory	Decreased oxygen absorption	Breakdown of alveoli
Cardiovascular	Decreased ability to pump blood so less blood is supplied to other tissues and organs	Reduction in the amount of heart muscle
	Hardening of the arteries and high blood pressure	Loss of elastic fibers in arteries and deposition of calcium, fat, and cholesterol in arterial walls
Nervous	Brain shrinkage	Loss of nerve cells throughout life as a natural process and as a result of alcohol use and trauma
	Deterioration of short-term memory and longer time to process various forms of information	Same as above
Reproductive	Loss of menstrual cycle in women	Dramatic decline in estrogen secretion by ovaries
	Reduced fertility in men	Decline in testosterone production
Urinary	Urinary incontinence	Weakening of the muscles of the urinary sphincter in men and women and weakening of the muscles of the pelvic floor in women who have given birth in earlier years

If cells are programmed to divide a given number of times, why don't all humans live to be the same age? One answer is that people differ genetically, which may contribute to the differences in life span. In addition, people live markedly different life-styles and eat very different diets. Research shows that the better one lives, the longer one lives. Diet, exercise, and stress management are keys to a healthy life.

Scientists are beginning to discover why proper diet and exercise can increase life span. Laboratory studies, for instance, show that by manipulating the chemical environment of cells, one can produce more cell divisions than normal. For example, vitamin E in large quantities increases the number of cell divisions in tissue cultures and increases a cell's life span. Whether vitamin E extends a person's life is not known.

Aging also results from a decrease or deterioration of cellular function. To understand aging, scientists are looking at the cell itself to see what causes its function to deteriorate. Studies show that the decline in cell function results from problems arising in the DNA, RNA, and proteins. Cells are exposed to many potentially harmful factors in the course of a lifetime. Natural and human-produced radiation and other potentially harmful agents (such as chemicals in the home and at work) may damage cells and impair their function. The cells of the body also produce oxygen free radicals, which are oxygen atoms that carry unpaired electrons. These chemical species are natural byproducts of cellular metabolism and can react with (oxidize) proteins, DNA, lipids, and other important molecules, causing damage. Studies of cells of various organs such as the skin, heart, brain, and others show that as they age, they accumulate oxidized lipids and oxidized proteins, including important enzymes, which are essential for normal cell function. Researchers have also found that mitochondrial DNA is particularly susceptible to free radicals. Damage to mitochondria would help explain the loss of cell function with aging.

Additional evidence shows that cells may lose the ability to repair DNA as they age, making matters worse. The impacts of this are many. Cells would, for instance, lose the ability to fix mutations in essential genes—ones that produce important structural and functional proteins such as enzymes. Cancer, the incidence of which increases with age, could result. As the damage accumulates, body function gradually deteriorates.

Some researchers think that aging is largely the result of a gradual decline in immune system function caused by a reduction in cell number and a loss of cell function. Although the immune system does falter, this decline surely cannot explain the other signs of aging.

Aging Is Often Associated with Certain Diseases, But the Likelihood of Contracting a Disease of Old Age Depends in Large Part on Life-Style

Many diseases are associated with old age. Osteoporosis, arthritis, atherosclerosis, Alzheimer's disease, and cancer are examples. Although these diseases are more prevalent in older people, they are emphatically not the inevitable consequence of aging. That is, they are not unavoidable signs of aging like gray hair. As previous chapters have shown, the likelihood of developing most of these diseases can be greatly reduced by exercise, diet, and other life-style adjustments. In other words, people can age in a healthy manner by living healthy lives. Although our bodies will eventually fail, we do not have to suffer chronic illnesses that plague so many people for so long.

One exception is **Alzheimer's disease.** This devastating disease strikes older individuals in about one of every 10 families. Alzheimer's is an incurable nervous system disorder characterized by a progressive loss of memory and other intellectual functions such as reasoning and understanding. Individuals with Alzheimer's become increasingly confused and unaware of their surroundings. Conversations become more and more difficult as mental incapacitation progresses. Afflicted individuals become emotionally unstable and, at the end stages, a person's intellectual abilities and personality may distintegrate entirely.

The cause of Alzheimer's is not known, but researchers believe that it may be a consequence of hardening of the arteries, which restricts blood flow to the brain. This causes brain cells to die, and the brain actually begins to shrink. Nerve tracts become distorted as microfilaments in neurons become tangled and as a plaque-like material called *amyloid* builds up in the brain. Health Note 20-1 offers some additional information on efforts now underway to slow the process of aging.

Death Is the Final Chapter in Our Lives

Aging results in a deterioration of function that eventually leads to death. But death can also result from traumatic injury to the body—for example, severe damage to the brain or a sudden loss of blood—and many other causes. What is death?

When I began to research the subject of death for this book, I thought the task would be easy. After all, death occurs when a person ceases to be alive. What I found was that death is difficult to define precisely. Although we can all agree that death is the cessation of life, trouble begins when people try to define it in instances requiring life-support systems.

In 25 states, a person is considered dead when his or her heart stops beating and breathing ceases. In the remaining 25 states, however, death is more liberally defined as an irreversible loss of brain function. A person in a coma, for example, often shows little or no brain activity other than that needed to keep the heart beating and the lungs working. Is this person alive? In some states, she is; in others, she is not.

Maintaining a person on life support is extremely costly to a family and to society if, for example, the patient is receiving federal benefits through Medicare or Medicaid. The money spent on maintaining a life could arguably be used to save dozens of lives through preventive measures. These issues inevitably lead to another biomedical dilemma, the controversy over euthanasia, a word derived from the Greek meaning "easy death."

Euthanasia (you-thin-A-zah) refers to any act or method of causing death painlessly and may be either passive or active. **Passive euthanasia** consists of decisions and actions to withhold treatment that might prolong life. **Active euthanasia** consists of decisions and actions that actively shorten a person's life—for example, injecting a lethal substance to terminate a patient's life. The Michigan physician, Dr. Jack Kevorkian, has been engaged in active euthanasia.

Opinions on euthanasia vary widely, and the controversy will, no doubt, be debated for many years. (For a debate on physician-administered euthanasia, see the Point/Counterpoint in this chapter.) As a final note,

Health Note 20-1

Can We Reverse the Process of Aging?

Suppose a drug were invented that would help you live to be 100. Would you take it?

If you are like most people, you probably would, as long as your additional years would be healthy and fairly trouble free. Because most of us would like to beat this aging thing that nature imposes on us, we secretly hope for a medical breakthrough that would prolong our lives.

Unfortunately, the search for a "cure" for old age has been fraught with frustration. Over the years, numerous treatments have been tried and have failed. As pointed out in the Critical Thinking section, although the average U.S. life span has increased, we really are not living much longer than our grandparents.

There is some encouraging news, however. Scientists recently discovered a protein called **stomatin** (stow-MA-tin), which they isolated from fibroblasts that have stopped dividing

in tissue culture. Scientists hope that stomatin will stop other cells from dividing as well. If this turns out to be true, researchers may have discovered an important clue in the aging puzzle, the signal that ends cell division. Moreover, they may have found a way to prolong human life.

Their reasoning goes as follows. If this protein signals the end of cellular division, there must be a gene that controls its production. If the stomatin gene can be located, researchers may be able to find a way to inactivate the gene. A drug or genetically engineered chemical, for example, could be injected in people to block the gene. The tissues of these people, researchers think, might continue to regenerate beyond the genetically programmed life span.

Efforts to reverse the aging process have met with some success in another arena as well—slowing down the aging of skin. Because age-

related changes in the skin mirror those occurring in other aging tissues, scientists have long studied the skin to expand their understanding of the aging process. Their studies have shown that skin aging results from two processes: intrinsic or chronological aging—which may be genetically programmed, and extrinsic aging, which results from accumulated environmental damage from sunlight and other factors.

Many skin scientists doubt whether anything can be done about intrinsic aging. Their studies suggest that after a skin cell has lived out its lifetime, its plasma membrane receptors become insensitive to growth factors that stimulate DNA replication and cell division.

New research suggests, however, that extrinsic aging, especially that induced by sunlight, is preventable and even reversible. Unknown to many, sunlight damages epithelial

---

individuals who want to save their relatives emotional and legal turmoil can sign a living will. This is a legal document stipulating conditions under which doctors should allow a person to die.

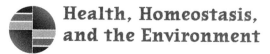 **Health, Homeostasis, and the Environment**

Video Display Terminals

Reproduction has nothing to do with maintaining homeostasis, but homeostasis is absolutely essential to normal reproduction, as this case study on the reproductive effects of computer monitors suggests.

Computer monitors, or video display terminals (VDTs), produce numerous types of radiation with frequencies ranging from X-rays to radio waves. For-

tunately, manufacturers have installed protective shields that prevent most of this radiation from escaping and bombarding the user. It is the lower-frequency radiation (radio frequencies) that has people concerned. Why?

Laboratory experiments show that electromagnetic fields generated by extremely low-frequency radiation can alter fetal development in chickens, rabbits, and swine. In addition, researchers in Oakland, California, recently published results of a medical study on the incidence of miscarriage in nearly 1600 women. The study showed that women who sit in front of a computer monitor for more than 20 hours a week during the first 3 months of pregnancy are nearly twice as likely to miscarry as women in similar jobs not using computers. Researchers suspect that miscarriages result from radiation emitted from the VDTs, but believe that other factors may also be involved—for example, the stress of

cells of the epidermis and the fibroblasts in the underlying dermis. When exposed to sunlight, molecules inside skin cells become electrically excited, and the energy they absorb from the sun drives reactions with other molecules in the cell. These reactions, in turn, lead to the dissociation of chemical bonds, which can damage molecules essential for cellular function. Plasma membranes, in fact, can be damaged within a few minutes of exposure to sunlight (**FIGURE 1**). Sunlight energy can also be absorbed by oxygen in the tissue, resulting in the formation of oxygen free radicals, highly destructive chemicals that are primarily responsible for extrinsic aging.

In 1988, medical researchers found that Retin-A, a derivative of vitamin A used to treat severe cases of acne, reduces sunlight-induced wrinkling. A study of Retin-A that followed patients for 22 months or more after treatment showed that the drug reduces wrinkles, age spots, and roughness for at least 22 months as long as it is applied regularly.

Researchers do not know how the drug works, but studies suggest that it does indeed have several effects that could account for its success. For example, Retin-A stimulates the growth of new blood vessels in the dermis. This, in turn, may nurture the regeneration of damaged skin cells. Retin-A also detoxifies oxygen free radicals in tissues and can inhibit the destruction of collagen fibers in the dermis. Some researchers think that it may regulate genes that play a role in cell growth and differentiation.

Despite the discovery of stomatin and Retin-A, there is as yet no evidence that medical scientists can prevent or even retard the overall rate of aging. The best route to a long and healthy life is to live it well: Learn ways to reduce stress, eat well, exercise regularly, take alcohol in moder-

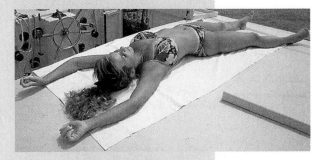

FIGURE 1 Beware: Skin Damage
A gorgeous tan may lead to premature skin aging.

ation or not at all, and avoid harmful practices such as smoking.

Visit *Human Biology's* Internet site, www.jbpub.com/humanbiology, for links to web sites offering more information on this topic.

www.jbpub.com/humanbiology

using a computer. Preferring to err on the conservative side, some scientists advise pregnant women to minimize their exposure to radiation from computer monitors. Should the link between radiation from VDTs and miscarriage be substantiated by further research, it would provide yet another link in the growing case for the importance of a healthy environment to human health.

Physician-Administered Euthanasia

THE RIGHT TO A PHYSICIAN WHO WILL NOT KILL
Rita Marker

Rita L. Marker *is director of the International Anti-Euthanasia Task Force and the author of* Deadly Compassion.

Should physician-administered euthanasia—the direct and intentional killing of a patient by a physician—be legalized? Common sense has always said NO. And laws have wisely banned euthanasia.

Now, however, an effort is underway to change these laws. Those who seek to legalize euthanasia have framed their arguments in terms of personal rights and freedom. They've clearly recognized the importance of words in molding public opinion: carefully crafted verbal engineering is being employed to transform the appalling crime of mercy killing into an appealing matter of patient choice.

Proposals appearing in state after state have such benign titles as "Death With Dignity Act," and the deceptively soothing term "aid-in-dying" is often substituted for "euthanasia."

As efforts of euthanasia activists increase, careful examination of what is at stake has become increasingly important. This examination should include answers to key questions.

Who will be the recipients of new "rights" if euthanasia is legalized? The law would benefit doctors, not patients. Competent adults already have the legal right to refuse medical treatment as well as the right to make their wishes known about such interventions for the future. Additionally, neither attempting nor carrying out the very personal and tragic decision to commit suicide is a criminal offense. Bluntly put, people can refuse medical care, or can kill themselves without a doctor's help.

Doctors, however, are prevented by law from killing their patients. Legalization of euthanasia would give doctors the right to directly and intentionally kill patients. An action that is now considered homicide would become a "medical service."

What about safeguards? Efforts to legalize euthanasia have failed in Washington and California. Opponents pointed out the real dangers inherent in the proposals. Advocates, however, insisted that their measures contained sufficient safeguards. Yet, after each version failed, proponents acknowledged the lack of necessary safeguards. It's as though each and every attempt to legalize euthanasia is treated as a dress rehearsal. Yet we're not dealing with a play or musical event. A "learn as you go" attitude in law and public policy could lead to the deaths of millions of vulnerable people.

What changes in law are now being promoted? Recognizing that the dangers of their initial proposals were too apparent, euthanasia proponents have repackaged their agenda, calling for changes that would allow physician "assisted" death. Often referred to as "assisted suicide" measures, these new proposals would permit a physician to prescribe medication with the express purpose that it be used to end a patient's life. Such a change would lead to classification of such an action as a medical intervention. The reality would be simple: The physician would have prescribed medication to kill.

Logic and common sense make it readily apparent that "prescribing" encompasses far more than writing a prescription that is taken to the corner drug store. Medication prescribed for a patient in a hospital or care facility is often provided by health professionals. The current "change" allowing "only" for the death prescription is yet another deception intended to mask reality in meaningless safeguards.

What has happened where euthanasia has been practiced? One need only look to the experience in Holland for a glimpse of where the euthanasia road leads. Although it is widely practiced in Holland, euthanasia remains technically illegal. Yet, even with "safeguards," which are far tighter than those which have been proposed in the U.S., a 1991 Dutch government study released horrifying information. The study found that in tiny Holland—a country with only one-half the population of the State of California—Dutch physicians deliberately and intentionally end the lives of more than 11,000 people each year by lethal overdoses or injections. And more than half of those killed had *not* requested euthanasia.

What will happen if euthanasia is legalized? With increasing emphasis being placed on cost containment, individuals who are disabled are particularly vulnerable. Leaders in the disability rights community are becoming increasingly alarmed by suggestions such as that of Jack Kevorkian that their "choice" of death would "enhance public health and welfare."

Euthanasia proponents have been clear in describing their goals. For example, Derek Humphry has called for expansion of euthanasia to include those who are physically and mentally disabled. He has also written that there would be a means for handling the dilemma of "terminal old age" once euthanasia is legalized. Jack has outlined plans for designated death zones with special clinics where "planned death" could be carried out.

Those who think that euthanasia, once unleashed, can be controlled are making a deadly mistake. Euthanasia is not a means of dying with dignity, nor is it a matter of liberty. It does not legislate compassion. It only legalizes killing.

THE RIGHT TO CHOOSE TO DIE
Derek Humphry

If we are truly free people and if our bodies belong to ourselves and not to others, then we have the right to choose when and how to die. Death comes to us all in the end, although modern medicine can often help us improve and extend our life spans. Thus, given today's high technology and ruinous cost of medicine, it is wise if we all give advance thought to the manner of our dying. Such decisions should be transferred to paper (a living will and durable power of attorney for health care), because 46 states now legally recognize one's wishes in respect to the withdrawal of life-support equipment.

Advance-declaration documents, of course, deal only with the legal and ethical problems of medical equipment and treatments. Less than half of dying people are connected to such equipment. For many more patients, technology does nothing for their terminal illness, so—in effect—there is no "plug to pull." Therefore, the cutting edge of the right-to-die debate in the 1990s centers on assisted death and voluntary euthanasia.

Hemlock Society supporters feel not only that hopelessly sick people should not only have an unfettered right to accelerate their end to save them pain, distress, and indignity, but also that willing doctors should be able to help them die. [To clarify some terms: *self-deliverance (suicide)* is ending your own life to be free of suffering; *assisted suicide* is helping another die; *euthanasia* is the direct ending of another's life by request. Currently, suicide in any form, for any reason, is legal, but assisted suicide and euthanasia are technically crimes.]

Most people at the close of life do not wish to suffer, desiring a quick and painless demise. The sophisticated pain-management drugs now available, when properly administered, control some 90% of terminal pain. But they do nothing to alleviate the indignities, the psychic pain, and the loss of quality of life associated with some debilitating terminal illnesses.

A complaint that any hastening of the end interferes with God's authority can be answered for many through a belief that *their* God is tolerant and charitable and would not wish to see them suffer. Other people have no faith in God. Pious people differ with this view, of course, and I respect that.

Numerous opinion polls in the United States and other Western countries indicate that two-thirds of people want the right to have a doctor lawfully assist them to die. People in the states of Washington, Oregon, and California have engaged in political action to achieve this law reform. The proposed law in these West Coast states was called the Death With Dignity Act, and the broad outline of its purpose is as follows:

1. The patient wanting physician-assisted dying would have to be a mature adult suffering from a terminal illness likely to cause death within about 6 months.
2. People with emotional or mental illness (especially depression) could not get help to die under this law.
3. The request would have to be in writing, and the signature would have to be witnessed by two independent persons.
4. The physician could decline to help the patient die on grounds of conscience but would then cease to be the treating physician. The patient could then seek a physician who was willing.
5. The family would have to be informed and its views, if any, taken into account. But the family could neither promote nor veto the patient's request to die.
6. The physician would have to be satisfied that the patient was fully aware of his or her condition, had been informed of all possible alternatives, and was competent to make this request.
7. If the physician was unsure of the patient's mental state, a mental-health professional

could be called to make an evaluation.
8. The time and manner of the assisted dying would have to be negotiated between patient and physician, with the patient's wishes paramount.
9. At any time the patient could orally or in writing revoke the request for assisted dying.
10. Any person who pressured another person to get assistance in dying, who forged such a request, or who ignored a revocation would be subject to prosecution.
11. After helping the patient die, the physician would have to report the action in confidence to a state health agency.

The right-to-die movement feels that these conditions, plus others in the Death With Dignity Act too numerous to describe here, are an intelligent and humane approach to euthanasia with the necessary safeguards against abuse.

Dying on one's own terms is not only an idea whose time has come, but also the ultimate civil and personal liberty.

Derek Humphry *was the principal founder of the Hemlock Society and its executive director from 1980 to 1992. He is the author of four books on euthanasia, including the best-seller* Final Exit (1991).

■ SHARPENING YOUR CRITICAL THINKING SKILLS

1. Summarize the key points of each author and their supporting arguments.
2. Which viewpoint corresponds to yours? Why?

Visit *Human Biology's* Internet site, www.jbpub.com/humanbiology, to research opposing web sites and respond to questions that will help you clarify your own opinion. (See Point/Counterpoint: Furthering the Debate.)

SUMMARY

FERTILIZATION

1. The sperm and egg of sexually reproducing animals such as humans unite during *fertilization*.

2. In humans, fertilization usually occurs in the upper third of the uterine tube.

3. Sperm deposited in the vagina reach the site of fertilization with the aid of muscular contractions in the walls of the uterus and uterine tube.

4. Sperm dissolve away the cells surrounding the oocyte and then bore through the zona pellucida and contact the plasma membrane. The first one to contact the membrane fertilizes the oocyte. Further sperm penetration is blocked.

5. Sperm are engulfed by the oocyte. The chromosomes of the sperm and oocyte duplicate and merge in the center of the cell, where mitosis begins.

6. Human development during gestation is divided into three stages: pre-embryonic, embryonic, and fetal.

7. Sperm contact the plasma membrane of eggs and are phagocytized by them. The egg and sperm cell nuclei then fuse to form a diploid *zygote*.

PRE-EMBRYONIC DEVELOPMENT

8. *Pre-embryonic development* begins at fertilization and ends at implantation. The zygote undergoes rapid cellular division and forms a morula.

9. The morula is then converted into a *blastula*, a structure with a hollow cavity, called a *blastocyst* in humans.

10. The morula arrives in the uterus 3–4 days after fertilization and is converted into a blastocyst, which implants in the uterine wall 2–3 days later.

11. The blastocyst consists of a clump of cells, the *inner cell mass* (*ICM*), which becomes the embryo, and the *trophoblast*, which gives rise to the embryonic portion of the placenta.

12. While the placenta is forming, a layer of cells from the ICM of the blastocyst separates from it and forms the *amnion*. The amnion fills with fluid and enlarges during embryonic and fetal development, eventually surrounding the entire embryo and fetus, protecting them during development.

EMBRYONIC DEVELOPMENT

13. After the amnion forms, the cells of the ICM differentiate into the three germ cell layers: ectoderm, mesoderm, and endoderm. The formation of the three primary germ layers marks the beginning of *embryonic development*.

14. The organs develop from the three basic tissues during organogenesis. **TABLE 20-2** lists the organs and tissues formed from each of the layers.

FETAL DEVELOPMENT

15. *Fetal development* begins 8 weeks after fertilization. Because most of the organ systems have developed or are under development, fetal development is primarily a period of growth.

16. The *placenta* produces several hormones that play an important part in reproduction. *Human chorionic gonadotropin* maintains the corpus luteum during pregnancy. Progesterone and estrogen stimulate uterine growth and the development of the glands and ducts of the breast.

ECTOPIC PREGNANCY AND BIRTH DEFECTS

17. About 10%–12% of all newborns enter the world with some form of birth defect. Birth defects arise from chemical, biological, and physical agents known as *teratogens*.

18. The effect of teratogenic agents is related to the time of exposure, the nature of the agent, and the dose. A defect is most likely to arise if a woman is exposed to a teratogen during the embryonic period when the organs are forming.

19. Many physical and chemical agents are toxic to the human fetus and, when present in sufficient quantities, can kill a fetus or retard its growth.

CHILDBIRTH AND LACTATION

20. *Labor* consists of intense and frequent uterine contractions that are believed to be caused by the release of small amounts of fetal oxytocin prior to birth. Fetal oxytocin stimulates the release of prostaglandins by the placenta. Oxytocin and prostaglandins stimulate contractions in the sensitized uterine musculature.

21. Emotional and physical stress in the mother may trigger maternal oxytocin release, augmenting muscle contractions.

22. As uterine contractions increase, they cause more maternal oxytocin to be released, which stimulates stronger contractions and more oxytocin release, a positive feedback that continues until the baby is born.

23. Labor consists of three stages, the *dilation*, the *expulsion*, and the *placental*.

24. The breasts of a nonpregnant woman consist primarily of fat and connective tissue interspersed with milk-producing glandular tissue and ducts.

25. During pregnancy, the glands and ducts proliferate under the influence of placental and ovarian estrogen and progesterone.

26. Milk production is induced by maternal prolactin and *human chorionic somatomammotropin* but does not begin until 2–3 days after birth.

27. Before milk production begins, the breasts produce small quantities of a protein-rich fluid, *colostrum*. A newborn can subsist on colostrum for the first few days; the baby derives antibodies from colostrum that help protect it from bacteria.

28. Suckling causes a surge in prolactin secretion. Each surge stimulates milk production needed for the next feeding.

AGING AND DEATH

29. *Aging* is the progressive deterioration of the body's homeostatic abilities and the gradual deterioration of the function of body organs.

30. These changes result from at least two factors: a decrease in the number of cells in the organs and a decline in the function of existing cells.

31. Death results from aging, traumatic injury, or infectious disease.

Critical Thinking

THINKING CRITICALLY— ANALYSIS

This Analysis corresponds to the Thinking Critically scenario that was presented at the beginning of this chapter.

Medical scientists measure longevity by a statistic called *life expectancy at birth*. Technically, it is the number of years, on average, a person lives after he or she is born. Life expectancy at birth has increased dramatically in the last century. In 1900, for example, on average, white American females lived only 50 years. Today, life expectancy is 81 years. For males, a similar trend is observed. In 1900, for example, the life expectancy of a white American male was 47 years; today, it is 74 years.

Most people take these statistics to mean that men and women are actually living to a much older age. In reality, something very different is happening: The increase in the life expectancy at birth is largely the result of declining infant mortality. That is, thanks to improvements in medicine, more children are living through the first year of life and this has dramatically increased life expectancy. In the early 1900s, 100 of every 1000 babies died during the first year of life. Today, that number has fallen to 12. The impact on average life expectancy is incredible.

This is not to say that all of the gain in life expectancy results from a decline in infant mortality. Medical advances have certainly increased life expectancy after infancy, but these changes are small in comparison to those brought about by lowering infant mortality. In fact, about 85% of the increase in life span in the last century is the result of decreased infant mortality.

EXERCISING YOUR CRITICAL THINKING SKILLS

The frequency of Down syndrome babies increases with maternal age. Researchers attribute 95% of Down syndrome babies to maternal chromosomal nondisjunction (failure of chromosomes to separate during meiosis). Because of this, Down syndrome has long been thought to be the result of the age of the oocytes in a woman's ovaries.

New research, however, suggests that the reason older women have a higher percentage of Down syndrome babies may be that they are more likely to carry one to term than younger women. In an international study conducted by 19 scientists, researchers found that the extra chromosome 21 that is responsible for this genetic disorder is most often incorporated during the first meiotic division. They also found that the frequency of nondisjunction in older and younger women is nearly equal.

If the frequency of nondisjunction is the same in mothers of all ages, one possible reason why older mothers have a higher incidence of Down syndrome babies is that their bodies fail to recognize an abnormal embryo as well as those of younger mothers. As a result, they are more likely to carry a Down syndrome baby to term.

Further research is needed to determine whether this hypothesis is valid. Can you think of any way to test it?

After thinking about it for a while, consider this suggestion: One direct way of testing the hypothesis, of course, would be to compare the ages of women who spontaneously abort Down syndrome fetuses.

TEST OF CONCEPTS

1. Give an overview of the process of development in humans. Describe what happens during each stage.
2. The formation of the gastrula results in the formation of endoderm, mesoderm, and ectoderm and the rearrangement of cells. Why is this process so important to the future development of the embryo?
3. Describe the process of fertilization in humans. Use drawings to elaborate your points, and label all drawings.
4. Define the terms *inner cell mass* and *trophoblast*.
5. Where does the human embryo acquire nutrients before the placenta forms?
6. Describe the formation of the placenta in humans.
7. What are the major functions of the human placenta?
8. Describe the flow of blood from the placenta to the fetus and back. What is the role of the various shunts?
9. List and describe the function of the five placental hormones.
10. Discuss how the nature of a teratogenic agent and the time of exposure affect teratogenesis, the production of birth defects.
11. What factors trigger labor?
12. What factors are responsible for cervical dilation during labor?
13. Define the term *aging*, and describe some of the hypotheses that attempt to explain it.
14. Thanks to modern medicine, men and women are living longer, says a friend. Using your understanding of average life expectancy data and the reason for an increase in life expectancy in the last century, explain why this statement is not quite true.

TOOLS FOR LEARNING

www.jbpub.com/humanbiology

Tools for Learning is an on-line student review area located at this book's web site HumanBiology (www.jbpub.com/humanbiology). The review area provides a variety of activities designed to help you study for your class:

Chapter Outlines. We've pulled out the section titles and full sentence sub-headings from each chapter to form natural descriptive outlines you can use to study the chapters' material point by point.

Review Questions. The review questions test your knowledge of the important concepts and applications in each chapter. Written by the author of the text, the review provides feedback for each correct or incorrect answer. This is an excellent test preparation tool.

Flash Cards. Studying human biology requires learning new terms. Virtual flash cards help you master the new vocabulary for each chapter.

Figure Labeling. You can practice identifying and labeling anatomical features on the same art content that appears in the text.

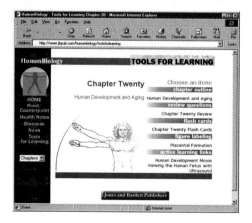

Active Learning Links. Active Learning Links connect to external web sites that provide an opportunity to learn basic concepts through demonstrations, animations, and hands-on activities.

EVOLUTION
FIVE BILLION YEARS OF CHANGE

This mosquito trapped in amber lived when the dinosaurs once ruled the Earth.

Thinking Critically

Nothing stirs the pot of controversy like a discussion of the theories of evolution and creationism. The theory of evolution says, in essence, that all life emerged from a common ancestry. It holds that the Earth is very old (about 4.5 billion years old, in fact) and that over time species changed—that is, evolved—through natural selection. Some species went extinct, and others gave rise to new species or entirely new taxonomic groupings.

Creationism, on the other hand, which proponents refer to as *Scientific Creationism,* says the Earth is very young, only about 6000 years old! This age is based on a study of the *Bible.* Furthermore, creationists argue that all species arose independently and therefore no species is related to any other.

Supporters of creationism argue that it should be taught in science classrooms in public schools alongside discussions of the theory of evolution. Imagine that you're a high school principal, and you get a letter from a parent who insists that creationism be taught along with Darwin's version of evolution. How would you answer this parent? ■

SCIENTISTS BELIEVE THAT THE ATOMS THAT make up both the living world and the nonliving world (rocks, for example) once existed in outer space as a giant cloud of cosmic dust and gas. About 4.5 billion years ago, this huge cloud began to condense. No one knows what triggered the condensation, but some think that a nearby exploding star (supernova) may have been the catalyst. As the huge cloud condensed, scientists speculate, dust particles in the center compacted more rapidly than outlying particles. This region would eventually give rise to the sun. Scientists hypothesize that when the mass of the forming sun reached a critical density, heat and pressure caused hydrogen and helium in it to fuse, forming larger atoms. This process, known as **fusion**, released large amounts of energy in the form of heat, light, X-rays, gamma rays, radio waves, and other radiation.

Cosmic dust lying outside the forming sun also condensed, creating the planets (Greek for "wanderers"). The third planet from the sun was the Earth, the only piece of real estate in our solar system that we know supports life.

When it originated, the Earth was a solid mass of rock and ice. Scientists believe that radioactive decay, intense solar heat, and other sources of heat caused the Earth to melt, turning it into a mass of molten material. As the Earth grew hotter, water and the lighter elements such as hydrogen escaped into the atmosphere.

In the millennia that followed, the Earth gradually cooled, and a thick, rocky crust formed around a molten core. As the Earth cooled, water in the atmosphere began to rain down from the skies, creating lakes and oceans. Today, the oceans cover nearly 70% of the planet. It is in these oceans, or on their shores, that life began.

In this chapter we will trace the emergence of life from the Earth's beginnings and examine evolution, the process responsible for the rich and varied life-forms on Earth.

Section 21-1

The Evolution of Life: An Overview

In the most general terms, **evolution** is a process in which existing life-forms change. As you will soon see, evolution also produces entirely new organisms. The evolution of life on Earth can be divided into three phases. During the first phase organic molecules formed. During the second phase cells evolved. And during the third phase many-celled organisms arose. This section outlines the key events in each phase.

◼ The Formation of Organic Molecules from Inorganic Molecules in the Earth's Primitive Atmosphere Is Called Chemical Evolution

When they first formed, the seas were lifeless bodies of water. The landmasses, so richly carpeted today with grasses, trees, and other plants, were barren rock. How could life have formed from such an unpromising start?

In 1924, a Russian scientist, A. I. Oparin, suggested one hypothesis (**FIGURE 21-1**). He conjectured that inorganic molecules in the Earth's early atmosphere reacted to form the very first organic molecules. They combined to form polymers, which then joined to form the very first primitive cells.

The formation of organic molecules from the Earth's primitive atmosphere that Oparin hypothesized is now referred to as **chemical evolution**. Chemical evolution began about 4 billion years ago or 500–600 million years after the Earth formed. At that time, the Earth's atmosphere contained a mixture of largely inorganic chemicals: water vapor, methane, ammonia, and hydrogen. It may also have included lesser quantities of nitrogen, carbon monoxide, and hydrogen sulfide, but it was devoid of oxygen. These molecules dissolved in rainwater and fell to the Earth. In the shallow waters of the newly formed seas, sunlight, heat from volcanoes, or

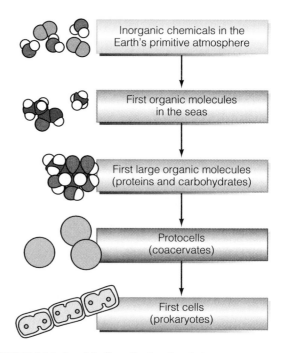

FIGURE 21-1 Oparin's Hypothesis Oparin hypothesized that the organic molecules necessary for life formed from the Earth's primitive atmosphere. Today, this idea is called the *theory of chemical evolution* and is supported by a considerable body of evidence.

lightning energized these molecules, causing them to react with one another in much the same way that molecules in a heated test tube react. These reactions produced a variety of simple organic molecules, including monosaccharides (simple sugars) and amino acids.

According to the theory of chemical evolution, these organic molecules, in turn, began to react, forming small polymers. As a result, primitive proteins and even rudimentary RNA or DNA molecules formed in the shallow waters of the seas. Little by little, and over tens of thousands of years, all of the organic molecules (monomers and polymers) necessary for life began to emerge. The polymers, in turn, combined to form structures known as aggregates, or **protocells** (PRO-toe-cells; literally, "first cells"), the precursors of cells.

The Theory of Chemical Evolution Is Supported by Some Rather Convincing Experiments.

Although chemical evolution may sound like science fiction, a considerable body of research supports this theory. The first direct evidence came from an ingenious American graduate student, Stanley Miller. While studying for his Ph.D. in chemistry at the University of

Chicago in the early 1950s, Miller devised an apparatus to test Oparin's hypothesis (**FIGURE 21-2**). To his closed, sterilized glass apparatus, Miller added three gases thought to have existed in the Earth's primitive atmosphere: methane (CH_4), ammonia (NH_3), and hydrogen (H_2). A sparking device that simulated lightning provided energy. Boiling water created steam that circulated through the device, carrying with it the reactant gases and the products of the reactions occurring inside.

After several days, the water in the apparatus turned brown. A chemical analysis of the brown, soupy liquid revealed the presence of several biologically important organic compounds, including amino acids, urea, and lactic acid (**TABLE 21-1**). In only a few days, Miller had created exciting evidence that supported the hypothesis that organic molecules could form abiotically (in the absence of life).

Although Miller's experiment did not prove Oparin's hypothesis, it created a ground swell of excitement and a flurry of research. Researchers soon found that a variety of organic molecules, including the building blocks of DNA and RNA (purines and pyrimidines), could be created abiotically. This work strongly supports the first part of Oparin's hypothesis, that the small molecules present in the Earth's early atmosphere combined to form the organic building blocks of life.

But what about the next step? Could these molecules join to form polymers? Researchers eagerly sought answers to this intriguing question and, to their delight, soon found that the organic building blocks produced abiotically in the laboratory could indeed assemble into polymers in the absence of living organisms. One notable experiment was performed by Sidney Fox of the University of Miami. He found that when he heated amino acids in air, they joined to form small polymers, which he called **proteinoids** (PRO-teen-oids; meaning "like proteins").

This research led biologists to hypothesize that the dilute solutions of amino acids that formed during chemical evolution may have been splashed onto hot rocks or lava, where the amino acids polymerized. More

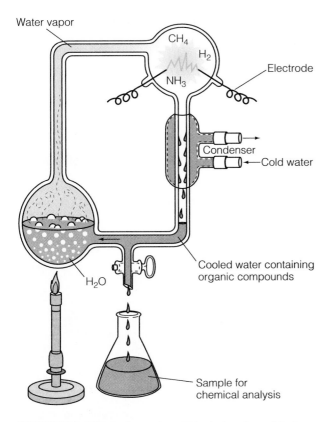

FIGURE 21-2 Miller's Apparatus This device showed that organic molecules could be produced from the chemical components of the Earth's early atmosphere.

TABLE 21-1	
Chemical Products Produced Abiotically in Miller's Experiment	
Glycine	Formic acid
Alanine	Lactic acid
Aspartic acid	Urea
Butyric acid	Succinic acid
Glutamic acid	Aldehyde
Acetic acid	Hydrogen cyanide

recently, scientists have also hypothesized that polymerization may have taken place on clay particles, which tend to concentrate amino acids dissolved in water. Binding to the charged surface of the clay particles, amino acids united via peptide bonds.

Nucleic acids and polymers of carbohydrates have also been produced in similar experiments, thus supporting the second step in Oparin's hypothesis. Scientists next turned their attention to the third and final step, perhaps the most difficult of all to imagine—and to replicate—the formation of living cells from polymers.

The Theory of Cellular Evolution Suggests that the Precursors of Cells Were Aggregations of Polymers

The origin of the very first cells remains an enigma, but experimental evidence suggests a way in which these early cells could have formed from the mixture of organic molecules in the shallow coastal waters. Noteworthy again are the experiments of Professor Fox. He immersed the proteinoids he had produced abiotically (described above) in a boiling salt solution, then cooled the solution. This procedure resulted in the formation of tiny globules of protein, which he called **microspheres** (**FIGURE 21-3**). Much to everyone's surprise, microspheres looked and acted a lot like bacteria. They experienced growth, for example. When a microsphere reached a critical size, tiny protrusions formed and broke off, creating new ones. This process resembles budding, a form of asexual reproduction (Chapter 1).

Other experiments yielded similar results. Scientists found that simple life-like structures emerged. Microspheres and similar structures may have been the precursors of the Earth's first true cells. Some scientists hypothesized that some of the protein or RNA molecules in these precursor cells, or protocells, may have served as primitive enzymes. These primitive enzymes may have allowed the protocells to synthesize some of their own molecules and break down others to generate energy. Researchers also hypothesize that small molecules of DNA that

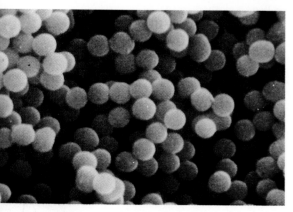

FIGURE 21-3 Microspheres Sidney Fox showed that his abiotically produced proteinoids formed small spherical structures that resemble cells. Could they have acquired genetic material and enzymes to become the first living cells?

formed during chemical evolution may have been incorporated into protocells, providing a primitive mechanism of heredity.

The Very First Cells Probably Acquired Nutrients from the Environment.

Many biologists believe that the protocells gave rise to the first true cells. They also speculate that the first true cells probably received nourishment from organic molecules such as glucose that they absorbed from their environment. The emergence of these cells marks the beginning of the kingdom Monera, which today consists of bacteria (**FIGURE 21-4**).

Many biologists hypothesize that these cells (monerans) gave rise to a group of organisms called *autotrophs*. These organisms are capable of synthesizing their own food. The very first autotrophs to evolve, however, were probably single-celled **chemosynthetic organisms**—organisms that produce their own organic molecules using energy from inorganic molecules in the environment. Some chemosynthetic bacteria can still be found today.[1]

Later on, another group of autotrophs arose—the photosynthetic organisms that synthesize their own food from carbon dioxide, water, and solar energy. Plants and algae are modern-day representatives of this group.

Although early forms of photosynthesis did not release oxygen, later forms did. The emergence of oxygen was a major event in evolution—the very first global pollution disaster. Why? As populations of these oxygen-

---

[1]Chemosynthetic bacteria on the seafloor live on hydrogen sulfide or natural gas.

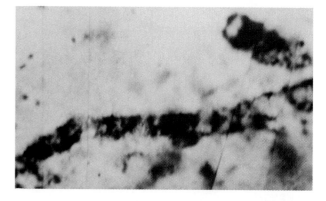

FIGURE 21-4 The First Prokaryote A composite photograph of the first prokaryote fossil, discovered in western Australia. It is believed to be about 3.5 billion years old.

releasing organisms expanded, oxygen levels in the water and atmosphere increased. Many of the anaerobic organisms living at the time were probably unable to cope with the oxygen and perished. Others retreated to oxygen-free environments such as the mud or sediment beneath lakes and oceans, where many of their descendants (the anaerobic bacteria) remain today. Still others evolved mechanisms that allowed them to live in the oxygen-rich atmosphere. These organisms included aerobic monerans (aerobic bacteria) and the eukaryotes (organisms with true nuclei).

Eukaryotes Probably Arose by Endosymbiotic Evolution.

After the emergence of the prokaryotes came the eukaryotes. The very first eukaryotes were members of the kingdom Protista (pro-TEES-tah). (For a timetable of these events, see **FIGURE 21-5**.)

The theory of **endosymbiotic evolution** explains the formation of eukaryotic cells. This theory states that eukaryotes arose from preexisting cells approximately 1.2 billion years ago. In this section, we will consider the origin of mitochondria and nuclei, two distinctive features of eukaryotes.

According to the theory, mitochondria of the eukaryotes came into existence when an anaerobic host cell phagocytized a smaller, oxygen-respiring bacterium. The bacterium set up residence inside the host cell, where it consumed the oxygen that entered its host, much to the benefit of its host. As atmospheric oxygen levels increased, such unions may have conferred a se-

lective advantage on this new biological composite. The theory of endosymbiotic evolution holds that as time went on, the relationship between host cells and their internal partners (the predecessors of mitochondria and chloroplasts) became hereditary. Thus, when host cells divided, they passed the bacteria they contained to their daughter cells.

Unlike the mitochondrion, the nucleus of the eukaryote is believed to have evolved from infoldings of the plasma membrane that came to surround the DNA. The infoldings did not come from endosymbiosis. Evidence of them is found in some species of bacteria. In these species, the plasma membrane folds inward and envelops the DNA, forming packages called **mesosomes.** Although mesosomes remain attached to the plasma membrane, it is thought that similar structures may have formed during cellular evolution and lost their connection, giving rise to the true nucleus.

Like other aspects of the early evolution of life, evidence for endosymbiotic evolution is indirect. The first line of evidence can be classified as "proof by resemblance." Comparisons of mitochondria and free-living bacteria, for example, show several striking similarities. First, mitochondria contain circular DNA, similar to that found in bacteria. Second, mitochondria contain ribosomes structurally similar to those found in bacteria. Third, they contain enzymes needed to produce some of their own proteins. Fourth, mitochondria are capable of dividing. These similarities suggest that mitochondria may have been free-living, bacterialike organisms at one time.

Continuing along this same line, mitochondria also contain two membranes. Some biologists believe that the outer membrane may have been formed when the bacteria that gave rise to this organelle were first phagocytized by the host cells during endosymbiotic evolution many millions of years ago. The outer membrane may have been derived from the membrane of the host cell; the inner membrane may be the phagocytized bacterium's own plasma membrane.

Endosymbiotic evolution is also supported by the fact that endosymbiosis (one organism living inside another) is a rather common occurrence in biology. Termites, for example, contain a microorganism (a flagellate, pronounced FLAJ-eh-LATE) in their gut that digests the wood they eat. The flagellate living in the gut of the termite contains its own internal symbiont—a bacterium that lives inside. Dozens of other examples exist, and many of the internal partners are passed from one generation to another during reproduction. Thus, endosymbiosis is not an anomaly among living things; it's a common strategy for survival.

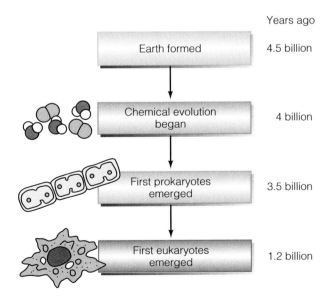

Years ago

Earth formed — 4.5 billion

Chemical evolution began — 4 billion

First prokaryotes emerged — 3.5 billion

First eukaryotes emerged — 1.2 billion

FIGURE 21-5 Summary of the Evolution of the Earth and Life

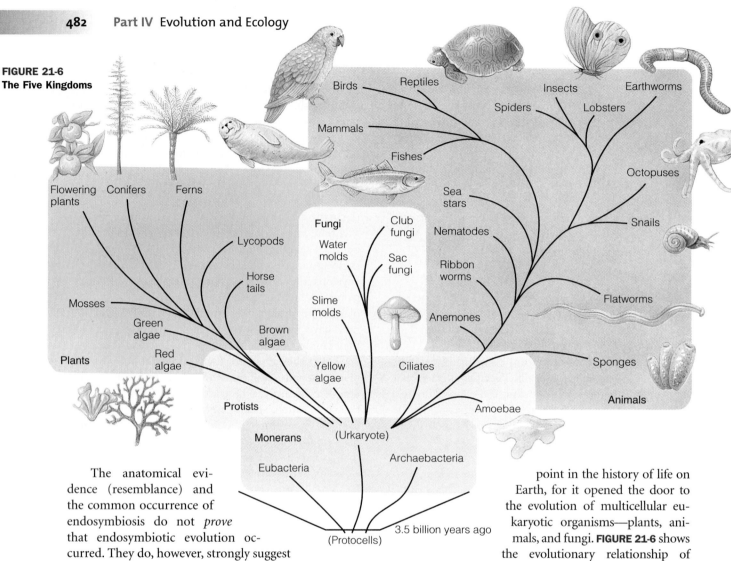

FIGURE 21-6
The Five Kingdoms

The anatomical evidence (resemblance) and the common occurrence of endosymbiosis do not *prove* that endosymbiotic evolution occurred. They do, however, strongly suggest that it is a plausible explanation for the emergence of eukaryotic cells.

However they evolved, the first eukaryotic cells, with their distinct organelles and membrane-bound nuclei, were the free-living, single-celled organisms that compose the kingdom Protista. Common members of the kingdom include such familiar examples as the amoeba and paramecium.

The previous discussion is not meant to suggest that scientists know all there is to know about chemical and cellular evolution. The truth is, we know very little about the origin of life and the emergence of prokaryotic and eukaryotic cells. We have many hypotheses, and this subject is one of the most controversial and open-ended in biology.

The Evolution of Multicellular Organisms Occurred Rather Rapidly Compared with Previous Stages

The evolution of eukaryotes from prokaryotes took an enormous amount of time and marked a major turning point in the history of life on Earth, for it opened the door to the evolution of multicellular eukaryotic organisms—plants, animals, and fungi. **FIGURE 21-6** shows the evolutionary relationship of these three kingdoms to monerans and protistans. The next section offers a brief overview of the evolution of multicellular organisms.

The Evolution of Multicellular Life Began in the Sea.

Because the first eukaryotic cells arose in the oceans and remained there for hundreds of millions of years, it should come as no surprise that the first multicellular plants also evolved in the seas.

The very first aquatic plants were the algae. Three types of algae exist today: brown, red, and green. Scientists believe that each form evolved from a different ancestor. Brown and red algae remained in the water. Green algae also remained in the water but are believed to have given rise to all land plants.

About the same time, between 400–600 million years ago, a variety of multicellular animals also evolved in the oceans. The fossil record suggests that many of them arose independently from early protists (single-celled eukaryotes). One of the first multicellular animals

FIGURE 21-7 The Sponge Probably one of the first multicellular animals, the sponge consists of a central cavity into which it draws nutrients. The sponge exhibits a modest degree of specialization.

may have been the sponge (**FIGURE 21-7**). Sponges are sedentary (immobile) invertebrates (animals without backbones) that attach to rocks and other hard surfaces, usually in shallow water. Another early multicellular invertebrate was a soft-bodied organism that resembled modern-day jellyfish. Molluscs (shellfish) and arthropods (crabs and lobsters) also joined the extraordinary profusion of life (**FIGURE 21-8**).

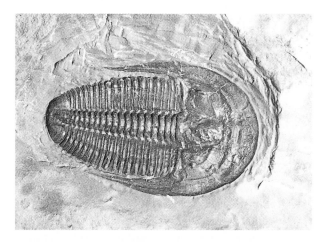

FIGURE 21-8 Early Sea Animals Trilobites were one of the most common life-forms in the ocean 600 to 500 million years ago.

Plants and Animals That Inhabited the Land Arose from the Seas.

While life flourished in the seas, the Earth's continents were barren. Over time, though, plants and animals colonized the land and eventually formed an impressive array of terrestrial life-forms.

Scientists believe that the invasion of the Earth's barren landmasses was prevented for a long time in part by ultraviolet radiation from the Sun, which is lethal to plants and other unprotected organisms. The evolution of life on land may have been made possible in part by the emergence of the **ozone layer**, a thin layer of air containing a slightly higher concentration of **ozone molecules** (O_3) 20–30 miles above the Earth's surface. Ozone molecules absorb ultraviolet light, reducing the amount of harmful radiation that strikes the Earth by about 99%.

The ozone layer probably formed from oxygen released by photosynthesis. As noted earlier, oxygen first came from photosynthetic monerans, then from autotrophic protists, which evolved later. Oxygen molecules produced by organisms drifted into the atmosphere, where they reacted to form ozone molecules. Over time, the concentration of ozone in the upper atmosphere increased.

The colonization of the continents by plants occurred approximately 400–500 million years ago. The first land plants probably evolved from green algae. Soon after the terrestrial plants began to evolve, the first sea animals "invaded" the land. Evolutionary biologists believe that the previous emergence of plants provided the first land animals with an abundant supply of food.

The first animals to come onto land were probably the **arthropods.** Two well-known examples are insects and crustaceans. Arthropods have hard outer skeletons that reduce evaporation and provide support. This feature, which evolved in the ocean most likely as a protective measure, made the transition to terrestrial life much easier. Organisms with characteristics that make them suitable for other environments are said to be **preadapted** to the new environment. The very first terrestrial arthropods were probably the ancestors of the scorpion (**FIGURE 21-9**).

The arthropods had the land and its plants to themselves for millions of years. Meanwhile, in the seas and

FIGURE 21-9 The First Land Animals Arthropods, perhaps the early ancestors of the scorpion, were the first to invade the land.

(a)

FIGURE 21-10 The Dinosaurs
(a) Probably one of the most successful species ever to inhabit the land, the dinosaurs lived on Earth for about 100 million years. Why they disappeared remains a mystery. *(b)* Alligators (shown here) and crocodiles are modern descendants of the dinosaurs.

(b)

lakes of the newly forming world, fishes began to emerge. Evolutionary biologists believe that one group of fishes (the lobe-fins) gave rise to the amphibians. Modern amphibians include organisms such as frogs, toads, and salamanders, all of which are dependent on the land and water for their subsistence. Most amphibians must return to water to lay their eggs.

Reptiles (represented today by snakes, lizards, turtles, and alligators) evolved from amphibians many millions of years later. Ancient reptiles gave rise to the dinosaurs and many modern reptiles (**FIGURE 21-10**). Thanks to their thick, scaly skin and eggs covered by a thick, rubbery shell, reptiles survived nicely on land. One important reptile was the dinosaur. Although the dinosaurs perished after an impressive 125-million-year stay, many reptiles remained, among them alligators, crocodiles, snakes, lizards, and turtles. Birds are believed to have evolved from reptiles about 225 million years ago at a time when the dinosaurs were beginning to diversify. Some scientists think the birds arose directly from early dinosaurs. Others believe they came from a separate reptilian ancestor.

Studies of the fossil record suggest that mammals probably also evolved from reptiles. Many mammal-like reptiles were present about 230 million years ago, around the time of the emergence of dinosaurs. The first truly recognizable mammals appear in rock about 180 million years old. Living among the dinosaurs for millions of years, the earliest mammals were rather small creatures that inhabited trees and were active principally at night.

Mammals are characterized by mammary glands and hair. Over time, body size increased greatly, and the limbs elongated and diversified, a feature that provides many styles of locomotion from flight (bats) to swimming (beavers, whales) to running (cheetahs, deer). Brain size also increased during evolution, and today mammals have the largest brains (relative to body size) of all organisms on Earth.

The primates, which include the apes and humans, are a group of mammals that evolved from a tree-dwelling mammal that lived among the dinosaurs. As you will see in Chapter 22, the early primates persisted long after the dinosaurs perished and gave rise to human beings.

Section 21-2

How Evolution Works

Many life-forms have emerged over the millions of years of the Earth's history. And many have changed, or evolved, to produce a rich and diverse array of species. The theory of evolution presents a plausible explanation as to how life emerged, how it changed, and why it became so diverse. In other words, it explains the mechanisms behind these processes. This next section explains how evolution works.

Genetic Variation Is the Raw Material of Evolution

During evolution, the anatomical, physiological, and even behavioral changes that occur in species take place because of changes in the frequency of genes in populations of organisms.[2] Changes in gene frequency begin with random genetic changes in a population; these arise from one of five processes: (1) mutations, (2) crossing-over, or genetic recombination, (3) the independent assortment of genes during meiosis, (4) new genetic combinations produced during sexual reproduction (that

[2]For our purposes, a species is a group of organisms whose members are anatomically and physiologically similar to one another. The members of a species interbreed successfully; that is, they produce viable, reproductively functional offspring.

is, the combination of genetically different gametes), and (5) gene flow. Of these sources of genetic variation, the only one that creates new forms of genes is mutations. The rest only shuffle existing genes, creating new combinations that may lead to favorable adaptations.

Mutations in the genes of organisms occur naturally and randomly. Many of these errors are corrected by the cells, but some remain. Mutations occur in the body cells (somatic cells) and germ cells. In the evolution of multicellular organisms, somatic mutations are meaningless, for they cannot be passed to future generations. It is germ cell mutations that are of importance.

Like somatic cell mutations, germ cell mutations can be harmful, even lethal. In fact, many human embryos that abort spontaneously are the result of harmful germ cell mutations in the sperm, the ovum, or both (Chapter 20).[3] Other germ cell mutations, however, are beneficial to the offspring. That is, they produce characteristics that give the offspring an advantage over other members of the population.

Genetically based characteristics that increase an organism's chances of passing on its genes are called **adaptations**. For example, a random mutation in the germ cell of a fish may make its offspring more efficient in food digestion. This efficiency, in turn, increases the likelihood that their offspring will survive and reproduce. Members of the same population that do not share this trait are less likely to survive and reproduce and will very likely produce fewer offspring. Over time, the better adapted fish will leave a larger number of offspring. Furthermore, these offspring are more likely than others to survive and reproduce. Over time, their success results in a shift in the gene pool so that future populations contain proportionately more offspring of the fish with the beneficial mutation. On a genetic level, then, the gene that gave some fish an advantage over others increased in frequency in the population.

Beneficial mutations, which produce genetic variation in populations, are often called the raw material of evolution, a fact that is true for sexually as well as asexually reproducing organisms (bacteria). In other words, mutations are a source of new alleles and thus a major source of phenotypic **variation** in a population—that is, differences in anatomical, physiological, and even behavioral characteristics.[4]

In sexually reproducing organisms, such as humans, genetic variation also results from recombination. **Re-**combination is the formation of new combinations of genes during crossing-over in meiosis (Chapter 16). Crossing-over occurs when homologous chromosomes pair up and exchange segments. This reshuffling of genes results in new genetic combinations and thus produces variation in a population. If the new combination of genes produces advantageous adaptations, the genes will usually increase in frequency in the population.

The independent assortment of chromosomes during meiosis may be even more important than crossing-over in producing new genetic combinations and phenotypic variation. Independent assortment, described in Chapter 16, simply means that maternal and paternal chromosomes in the forming germ cells segregate independently of one another during meiosis, creating new genetic combinations.

Still another source of genetic variation in population is the process of sexual reproduction. When male and female gametes combine, they produce new genetic combinations, some of which may provide advantageous adaptations.

The final source of genetic variation is called gene flow. **Gene flow** is the introduction of new alleles to a population by individuals from another population of the same species.[5] This process occurs when members of previously isolated populations mate. A male bird capable of flying long distances while using less energy than other members of the population, for example, might cross a mountain range and mate with a female from a previously isolated population. Her offspring would then carry the gene that permits more efficient flight. Variation has been introduced into the population.

Natural Selection Is a Process by Which Organisms Become Better Adapted to Their Environment

Sociologist Andrew Schmookler once wrote that "evolution employs no author, only an extremely patient editor." What he meant by that is that evolution is not directed. There is no author. Instead, mutations arise spontaneously. Mutations, recombination, gene flow, and new combinations from the independent assortment of genes and from sexual reproduction produce genotypic and phenotypic variants. If beneficial, the genetically based characteristics (adaptations) that arise from these sources tend to persist. The species changes over time, becoming better suited to its environment. In some cases, a whole new species may evolve. The gene pool shifts because of evolution's patient "editor," a

---

[3]A mutation in a somatic cell in a morula or blastocyst can also lead to embryonic or fetal death and miscarriage.

[4]For bacteria and other organisms that reproduce asexually (by dividing or budding), mutation is the only source of variation.

---

[5]The new genes are most likely to have arisen from a mutation.

process called **natural selection**. Charles Darwin, a nineteenth-century British naturalist, and Alfred Wallace independently proposed the idea of natural selection. Darwin described it as a process in which slight variations, if useful, are preserved (**FIGURE 21-11**). Thus, natural selection is a process by which organisms often become better adapted to their environment.

Two principal factors contribute to natural selection: **biotic factors**, or other living organisms, and **abiotic factors**, the physical and chemical environment (temperature, rainfall, and so on). Abiotic and biotic factors influence survival and reproduction by "selecting" the fittest—those best able to reproduce and pass on their genes. If these conditions change, those organisms best adapted to the new conditions tend to remain and pass their genes on to subsequent generations. This process causes a shift in the frequency of certain genes in a population.

In Darwin's time, evolution was widely discussed among naturalists and other scientists, but the mechanism by which it occurred remained a mystery. Darwin dedicated many years to the search for an answer, even traveling around the world by ship and cataloging species (**FIGURE 21-11B**). It was on this lengthy voyage that Darwin reportedly came up with the idea. A careful scientist, though, Darwin spent the next 20 years looking for flaws in his own reasoning. In 1858, much to his surprise, Darwin received a paper from Wallace, a respected naturalist who had proposed the same concept. Darwin sent Wallace's paper to some of his colleagues and suggested that they publish it. Fortunately, Darwin's colleagues, who were aware of his own theory, prevailed on him to write a paper of his own. In 1858, Wallace's and Darwin's papers were both presented to the Linnaean Society. In 1859, Darwin published his now-famous book, *On the Origin of Species by Means of Natural Selection.*

As in the case of Mendel, Darwin's and Wallace's ideas took many years to be understood and accepted. Not until the 1940s, about 60 years after Darwin's death, did natural selection become widely accepted. Today, it is one of the central tenets of biology.

Natural Selection Ensures That the Fittest Organisms in a Population Survive and Reproduce.

Darwin used the phrase "survival of the fittest" to describe how natural selection worked.[6] Survival of the

---

[6]Herbert Spencer is thought to have coined the phrase.

FIGURE 21-11 "Evolutionary Voyage"
(a) Darwin as a young man proposed the theory of evolution by natural selection, helping solve one of the key puzzles of biology: how species evolve. (b) From 1831 to 1836, Darwin made a famous voyage on the *Beagle*, collecting and cataloging thousands of diverse species.

(a)

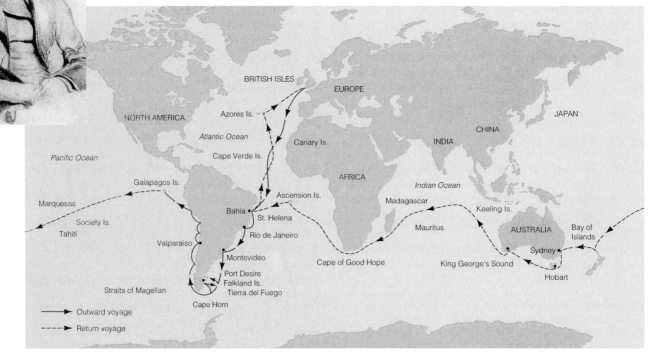

(b)

fittest is commonly interpreted as "survival of the strongest." To a biologist, however, **fitness** is a measure of reproductive success, which can be attributed to many different features, not just strength. Fitness does not necessarily result from speed, either. The ability to hide, digest food more efficiently, or use water more efficiently can be just as important, if not more important, than standard measures of strength.

Ultimately, fitness is a measure of the genetic influence an individual has on future generations. By definition, then, the fittest individuals leave the largest number of descendants. Their influence on the gene pool will be greater than that of their less-fit contemporaries.

Allele Frequencies Can Shift Because of Chance Events

Not all shifts in the gene pool result from natural selection. Some occur quite by chance. Imagine, if you will, a population of mountain goats in which 3 members—out of, say, 30—possess a slightly superior ability to climb extremely steep, rocky slopes. One day those three animals climb a ridge to reach some food. The rest of their herd stays on lower ground. That afternoon a lightning bolt strikes the ridge, killing all three goats.

This chance event causes a shift in the frequency of alleles in the population, wiping out the gene that improved the three goats' climbing ability. A shift in allele frequencies that is due to a chance occurrence is referred to as **genetic drift**. This process is most rapid when a population is small, as in the example above.

If in this example an airplane crashed on lower ground and killed the 27 less-fit animals, leaving only the 3 exceptional climbers, the allele frequency would also shift—this time in favor of improved climbing ability. Over time, a new population would emerge. This extreme case of genetic drift illustrates an important phenomenon, the **founder effect**, because a small number of individuals establish (found) a new and genetically distinct population. The founder effect is extremely important in the colonization of islands and other isolated habitats. It is also important when a few individuals of a species are introduced into new habitat.

Genetic drift may take place in populations that are severely reduced through natural or human causes. The extreme reduction in the number of individuals in a population is called the **bottleneck effect**. In the late 1800s, for instance, the northern elephant seal was nearly hunted to extinction. The 20 or so remaining seals formed the basis of the new population, which today numbers around 30,000. This much larger population has almost no genetic variation compared with other seal populations that have not gone through similar bottlenecks.

Genetic uniformity makes populations highly susceptible because they do not contain the wide genetic base vital to evolutionary change in the event that environmental conditions change. This is one reason why it is prudent not to let wild populations of any species be diminished. For an interesting view of a previous hypothesis on the mechanism of evolution, see Scientific Discoveries 21-1.

Section 21-3

The Evolution of New Species

Evolution is a process by which species change over time, becoming better adapted to their environment. As noted earlier, evolution can also produce new species.

New Species Evolve When Geographic Isolation Results in Reproductive Isolation

The evolution of a new species is called **speciation** (spee-sea-A-shun). New species arise in large part as a result of a phenomenon called **geographic isolation**. Geographic isolation occurs when members of a population of organisms are physically separated by some barrier—a mountain range, a lake, an ocean, a river, or simply a great distance. In their separate habitats, the members of the subpopulations are exposed to different environmental influences. Over time, two distinct species may form, each adapted to its environment. If a population is separated into more than two subgroups, many new species may arise.

One of the best examples of this phenomenon occurred on the Galapagos Islands in the Pacific Ocean off the coast of Ecuador. On these islands, over a dozen species of finches (sometimes called Darwin's finches) evolved from a single population of seed-eating birds that arrived from South America sometime in the past 500,000 years (**FIGURE 21-12**). As the ancestral population settled on different islands, the birds began to evolve in different ways, primarily because of differences in food sources. Major changes took place in body size and the shape of the bill. The differences in beak size and shape correspond to different ways of obtaining food.

Darwin's finches illustrate how new species develop from a common ancestor. This phenomenon is called **divergent evolution**. The term refers to the fact that species diverge genotypically and phenotypically in response to differing environmental conditions.

FIGURE 21-13 (page 489) shows divergent evolution among ancestral reptiles and ancestral placental animals that gave rise to a wide assortment of species, some adapted to life in the water, and others adapted to life on

FIGURE 21-12 Darwin's Finches All but one species (the Cocos Island finch) are believed to have evolved from a single ancestral flock that arrived on the Galapagos Islands within the past 500,000 years.

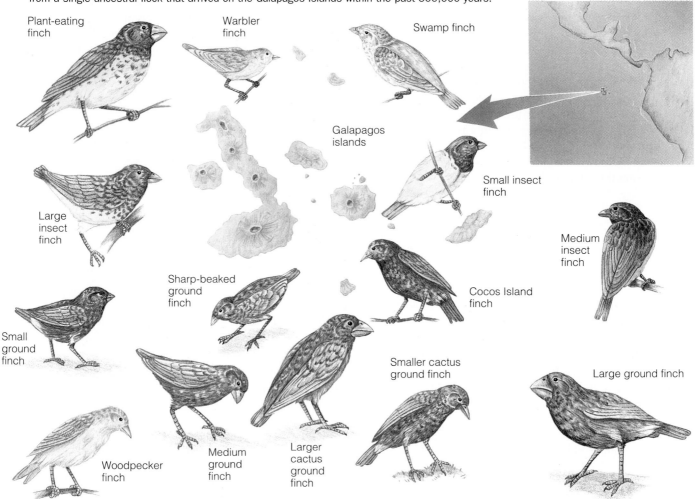

Plant-eating finch

Warbler finch

Swamp finch

Galapagos islands

Large insect finch

Small insect finch

Medium insect finch

Sharp-beaked ground finch

Cocos Island finch

Small ground finch

Smaller cactus ground finch

Large ground finch

Woodpecker finch

Medium ground finch

Larger cactus ground finch

land or in the air. The term **adaptive radiation** is also used to describe the evolution of new species from a common ancestry, each of which is specially adapted to its environment.

Certain genetic lineages may produce a number of new species over tens of millions of years. Others produce numerous species over a relatively short period. But how does geographic isolation "produce" new species?

One of the consequences of geographic isolation is that organisms that arose from the same population often lose the ability to interbreed, for reasons explained shortly. Biologists say that the members of the original population have become **reproductively isolated** from one another. New species emerge when geographical isolation results in reproductive isolation. When new species emerge in geographically separate regions, the process is called **allopatric speciation** (allopatric is derived from the Greek for "other" and "fatherland").

The first step in reproductive isolation in animal populations is the emergence of behavioral differences between the two groups. Courtship patterns, for example, may change so that females from one population no longer respond to males of the other. **TABLE 21-2** lists several additional reproductive isolating mechanisms that fall into one of two broad categories: (1) those factors that prevent mating such as different courtship behaviors and mating times; and (2) those that prevent the production of viable offspring and/or reproduction—for example, the inability of sperm to reach ova.

Interestingly, new species may also form without geographical isolation. This process is common in plants and is known as **sympatric speciation** ("same fatherland"). Sympatric speciation often occurs by polyploidy, the accumulation of one or more additional sets of chromosomes. This, in turn, results from complete nondisjunction of chromosomes during meiosis.

Convergent Evolution Occurs Because Organisms Tend to Evolve Similarly When Faced with the Same Challenge

FIGURE 21-13 illustrates a very interesting phenomenon—notably that unrelated species may adapt to similar environments in similar ways. In this example, ancestral mammals and reptiles both evolved flying forms (in fact, about 20% of all mammal species are bats). In addition, mammals and reptiles both gave rise to carnivores, as represented in the figure by the tyrannosaurus and the lion. The tendency for organisms to develop the same types of adaptations in response to similar environmental conditions is called **convergent evolution**.

Species Often Evolve in Concert with Each Other

So far, the discussion of evolution has focused principally on abiotic factors as agents of natural selection. Abiotic factors such as climate are only part of the picture. Organisms are themselves agents of natural selection. Predatory pressure, for example, may act as an agent of natural selection in the evolution of prey populations. For example, owls hunt mice at night. Changes in the owls' ability to locate mice could have a profound effect on their prey population. If the eyesight of a population of owls improves, for instance, only the quickest mice will survive and reproduce, changing the mouse population over time.

Because biological organisms interact, two populations such as owls and mice evolve in concert. Owls can influence the evolution of mouse populations, and mouse populations, in turn, can influence the evolution of the owl. Changes in the ability of the mice to hide, for example, might eliminate all but the most skilled owl

FIGURE 21-13 Divergent Evolution
From common ancestors, new species emerge, each adapted to its environment.

hunters. Over time, the owl population would "improve." As the owl population became better able to hunt and catch swift mice, the mouse population would change.

When members of two interacting species affect each other's gene pool, the species evolve in concert, a process called **coevolution**. In the example above, coevolution is like an arms race between predator and prey. Each improvement in a predator's ability to catch its prey is followed by an improvement in the prey's ability to avoid or resist attack. Coevolution occurs in relationships other than predator-prey interaction. Defense mechanisms in plants, for instance, may result in coevolutionary changes in insects that eat them. A toxic or noxious-tasting chemical, for example, may evolve in

---

TABLE 21-2

Reasons for Reproductive Isolation

Factors that prevent mating
    Mating occurs at different times.
    Different cues (songs and coloration) develop.
    Different courtship behaviors develop.
    Genital incompatibility develops.
Factors that prevent production of viable offspring and/or reproduction
    Sperm cannot reach ova.
    Hybrid offspring die in utero or shortly after birth.
    Hybrid offspring survive, but are sterile.
    Hybrid offspring survive, but have lower fitness.

Scientific Discoveries that Changed the World 21-1

Debunking the Notion of the Inheritance of Acquired Characteristics

Featuring the Work of Lamarck, Weismann, Castle, and Phillips

The eighteenth-century French naturalist Jean-Baptiste Lamarck argued that life had been created in a relatively simple state. Over time, he said, life-forms gradually improved as a result of an innate drive for perfection. He further hypothesized that this inherent drive was centered in nerve cells and that these cells released a "fluida" (chemical substance) that traveled to different body parts needing improvement.

Consider Lamarck's explanation of the evolution of the giraffe. According to him, the giraffe of years past was a short-necked animal. Over time, the elongation of the giraffe's neck resulted from the simple act of stretching of its neck—a feat required to feed on leaves that were out of the reach of other animals. The act of stretching, said Lamarck, directed fluida to the neck. This stimulus made the neck grow longer. What is more, the slightly stretched neck that an adult acquired was then transmitted to its offspring. The offspring, in turn, stretched their necks

to reach food, causing further elongation. According to Lamarck then, generation after generation of giraffes, each stretching to find food, led to the modern giraffe.

Lamarck proposed an evolutionary theory based on the "use and disuse" of organs. It stated that an individual acquired traits during its lifetime and that such traits were in some way incorporated into the hereditary material and passed to the next generation, thus explaining how a species could change over time. Conversely, disuse of organs led to their disappearance. The inheritance of acquired characteristics was doubted by many of Lamarck's contemporaries, in large part because he failed to support his hypothesis with observations or experiments.

Lamarck's concept is based on what is viewed today as a fundamental fallacy—that is, the inheritance of acquired characteristics supposes that all of the organs of one's parents produce the hereditary factors that form corresponding parts in the

offspring. Thus, any change that occurred in an organ (an acquired characteristic) before one transmitted genetic material to offspring would result in the production of a hereditary factor that would reflect the altered organ. Put another way, if you were lifting weights and had developed extensive musculature before the conception of your child, your child would grow up to be brawny.

After the discovery of fertilization, it was assumed that the organs transmitted parcels of hereditary information to the gametes. In this way, acquired traits were passed on to the offspring.

August Weismann, a German scientist in the late 1800s, attempted to bring some sense to the debate in a series of extraordinary essays, based largely on the work of others, but also on his own ideas. Although Weismann did not perform experiments to disprove Lamarck's notions, his essays remain an important contribution to modern genetics. Among other things, he argued that

some plant species, persisting because it helps ward off hungry insects. In time, some insects may evolve enzymes capable of detoxifying the harmful chemical.

A Modern Version of Evolutionary Theory

Darwin's theory of evolution by natural selection can be summarized in three principles. First, natural variations exist in all species. Second, which members of the species survive to reproduce and reproduce at a greater rate is determined by inherited (genetic) variations (adaptations). Third, natural selection "determines" which organisms survive and reproduce.

For many years, biologists thought that evolutionary changes occurred gradually over many millions of

years, a process called **gradualism**. If gradualism does indeed occur, evolutionary biologists reason, the fossil record should contain many intermediate forms of plants and animals. In most cases, it doesn't.

As a rule, new species in the fossil record appear rather abruptly, persist for several million years, then vanish as abruptly as they arrived on the scene. Based on these and other data, Stephen Jay Gould of Harvard University and Niles Eldredge of the American Museum of Natural History have proposed an alternative hypothesis stating that evolution occurs in spurts. Their hypothesis is called **punctuated equilibrium**.

According to this theory, evolution consists of long periods of relatively little change (equilibrium) interspersed with (punctuated by) brief periods of relatively rapid change, although these periods are still many thousands of years long. During these periods of rapid change, some species become extinct, and new ones

the transmission of traits from one generation to the next depended on the sex cells. The sex cells, he said, are capable of reproducing all of the "peculiarities of the parent body in the new individual."

Furthermore, Weismann asserted that the cells of an organism did not dispatch small hereditary particles to germ cells from which the latter derived their power of heredity. Rather, he contended that heredity was brought about by the transference from one generation to another of a substance with a definite molecular constitution. He called it "germ-plasm" and agreed with others that it was probably found in the nucleus of germ cells. Today, we know it as DNA.

Although Weismann propagated a few erroneous ideas himself, his notion that a chemical substance contained the hereditary information responsible for the faithful transmission of traits profoundly influenced the thinking of many biologists and marked a turning point in our understanding of genetics. It also helped debunk Larmarck's notion of the inheritance of acquired characteristics

and his erroneous view that evolution was driven by such a mechanism.

One of the most rigorous tests of Weismann's hypothesis came about 20 years later in experiments by W. E. Castle and John C. Phillips. Castle and Phillips transplanted ovaries from black guinea pigs (homozygous dominant) into albino guinea pigs (homozygous recessive) whose ovaries had been removed. Later, the albino females were bred to albino males.

Previous studies had shown that matings between albino males and females resulted in 100% albino offspring. Therefore, if acquired characteristics were inherited, Castle and Phillips argued, the offspring of these matings should be white, having acquired their coloration from the host. When bred to an albino male, however, the females produced only black offspring. The young, they said, "are such as might have been produced by the black guinea pig herself, had she been allowed to grow to maturity and been mated with the albino male used in the experiment."

Lamarck did have many good ideas—for instance, that species changed over time and that the environment was a factor in this change. He also contributed to the science of biology in other ways. For example, he was the first scientist to distinguish between animals with backbones (vertebrates) and those without backbones (invertebrates). He went on to classify many invertebrates into the categories of arachnids, crustaceans, and echinoderms, and he also wrote a text on invertebrate systems and a seven-volume treatise on the natural history of invertebrates. But history generally knows Lamarck for his erroneous concepts of inheritance and evolution. In fact, despite the general criticism of Lamarck's views, the twentieth-century Russian agronomist Trofim D. Lysenko adopted a Lamarckian viewpoint and, with Stalin's backing, established it as Soviet doctrine in genetics research and teaching. The result was a significant setback in genetic research in the former Soviet Union during the Stalinist era.

arise. Proponents of punctuated equilibrium believe that species undergo most of their morphological change when they first diverge from their parent species. Thus, a species will appear rather suddenly, then undergo little or no change for long periods. Recent genetic research suggests that only minor genetic changes are necessary to produce species dramatically different from their common ancestor.

The Evidence Supporting Evolution

Evolution by natural selection is one of the central theories of modern biology. Although biologists may argue over some of the details, such as gradualism versus punctuated equilibrium, they agree about the basic tenets.

From the outside, this bickering may appear to be evidence that the theory of evolution is on shaky ground. Nothing could be further from the truth. What is disputed is the time scale and other details, not whether evolution has occurred. The next section examines some of the evidence in support of the theory of evolution.

The Fossil Record Yields Some of the Best Supporting Evidence for the Existence of Evolution

Human knowledge of evolution comes from many sources. One of the most important sources is fossils. **Fossils** consist of the remains (wood from trees, bones and eggs from animals) or impressions of organisms that lived on Earth in times past.[7] Over the past 100 years,

---

[7]The word *fossil* comes from a Latin word meaning something "dug up."

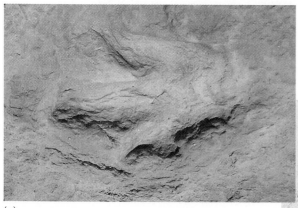

(a)

FIGURE 21-14 Fossils   (a) Dinosaur tracks made in a Texas streambed about 120 million years ago. (b) Imprint of a leaf. (c) Insect embedded in amber about 40 million years ago.

(c)

(b)

bubbles in amber have even been analyzed to determine the composition of the atmosphere in earlier times.

The fossil remains of species no longer in existence have intrigued humans for 200 years and provide evidence in support of the theory of evolution. Because fossils and the rocks they come from can be dated using radioactive techniques, scientists can determine when different species lived.

Studying when a species lived and tracking the types of life-forms present at different times during the Earth's history have permitted scientists to piece together an evolutionary history of the planet. These studies show that the oldest rock contains relatively simple single-celled organisms and that successively younger rock houses fossils representing increasingly more complex life-forms.

To date, scientists have discovered fossils belonging to about 250,000 species, most of them from rock formed within the last 600 million years. In this still-growing record, some lineages are nearly complete. The best records come from prehistoric environments such as shallow seas where sedimentation rates (and thus preservation) were high. Many other lineages are incomplete and may never be fully understood. This is particularly true of animals that lived on dry land. Land animals died and were probably eaten by scavengers. Their bones may have been devoured as well or scattered far and wide by hungry animals. Without sedimentation to bury them, the bones eventually crumbled into oblivion. Even in environments with rapid sedimentation, not all species left their mark. Soft-bodied creatures, without bones, outer skeletons, or shells, perished without a trace.

FIGURE 21-15 shows a much simplified diagram of the evolution of the modern horse. This schematic is called an **evolutionary tree** and is based principally on the study of fossils and the anatomy of modern species. It tracks the progression of life from simple to complex over a period of millions of years. Evolutionary trees also show how various species are related to one another. As shown in **FIGURE 21-15**, horses evolved from dog-sized, woodland browsers (Eohippus) through a series of increasingly larger, plains-dwelling grazers. Structural similarities in the fossil remains of the horse's ancestors suggest that this progression is indeed real.

You will note in **FIGURE 21-15** that the evolution of horses was not a linear progression. A number of new species arose from a common ancestor along the way, as indicated by the circles in the evolutionary scheme. All

thousands of bones of dinosaurs and other animals have been excavated from sites all over the globe. How were these bones preserved?

The most common explanation is that animals that died in regions of rapid sediment deposition (shallow lakes, mud flats, and swamps) were quickly buried and decayed over time, leaving behind sediment-entombed bones. In these sites, the minerals of the bones were often replaced by minerals in the water that seeped through them. (A similar process is responsible for the formation of petrified wood.)

Some creatures such as dinosaurs also left footprints in the mud that hardened into stone to become part of the vast body of fossil evidence of earlier life (**FIGURE 21-14A**). Imprints of leaves make up part of the fossil record (**FIGURE 21-14B**). Intriguingly, some ancient organisms (frogs, insects, and flowers) were preserved in amber, resins released by ancient trees (**FIGURE 21-14C**). Blocks of amber, like plastic blocks containing biological specimens, provide a window to our past. In fact, gas

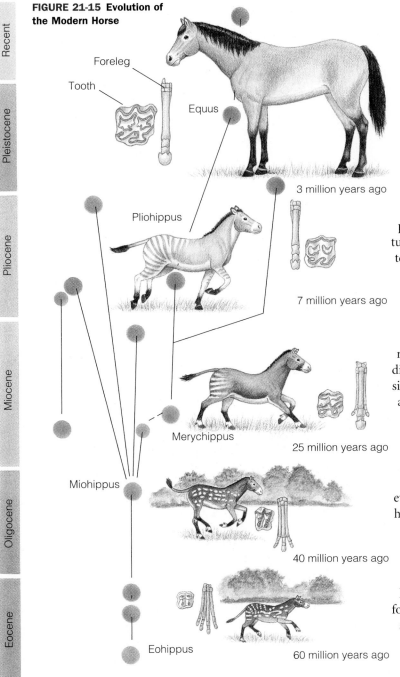

FIGURE 21-15 Evolution of the Modern Horse

Recent

Pleistocene

Foreleg

Tooth

Equus

3 million years ago

Pliocene

Pliohippus

7 million years ago

Miocene

Merychippus

25 million years ago

Oligocene

Miohippus

40 million years ago

Eocene

Eohippus

60 million years ago

are remarkably similar to the fossil remains of the earliest amphibians (**FIGURE 21-16**). Similarly, the early reptiles share many features in common with their amphibian ancestors.

The second line of evidence is chronological. That is, the amphibians, reptiles, mammals, and birds appear in the fossil record in this order. No one has ever found a mammal or a bird that predated the amphibians. Thus, the structural and chronological evidence can be interpreted to support evolutionary theory.

Common Anatomical Features in Different Species Support the Theory of Evolution

Today's organisms are adapted to a wide range of conditions and exhibit a dazzling diversity of appearance. Despite the diversity of life, many organisms exhibit similar anatomical features. Structures thought to have arisen from common ancestors are known as **homologous structures**. The presence of homologous structures among life's diverse forms represents yet another piece of evidence in support of evolution. **FIGURE 21-17** illustrates some homologous structures: wings, flippers, arms, and legs. As shown, the bones in the wing of a bird and bat, the flipper of a whale, the arm of a human being, and the leg of a horse and cat are quite similar, even though these appendages perform quite different functions. Similarities such as these correspond to what you would expect if birds and mammals evolved from a common ancestor.

Creationists, who believe that life on Earth was created by God, argue that the presence of common anatomical features among diverse organisms

of these species became extinct, except, of course, the modern horse, which is doing very well.

The fossil record also suggests the origin of species or how new groups of organisms arose. As noted earlier, during evolution, a primitive fish probably gave rise to the amphibians. Amphibians gave rise to the reptiles. Birds and mammals evolved from reptiles. This progression is supported by two lines of evidence from the fossil record.

The first is structural evidence; that is, species thought to have given rise to new groups bear a remarkable resemblance to their supposed descendants. For example, the fossil remains of primitive fishes (lobe-fins)

FIGURE 21-16 Fossil of Early Amphibian This early amphibian resembles the lobe-fin fishes in many respects, suggesting an evolutionary relationship.

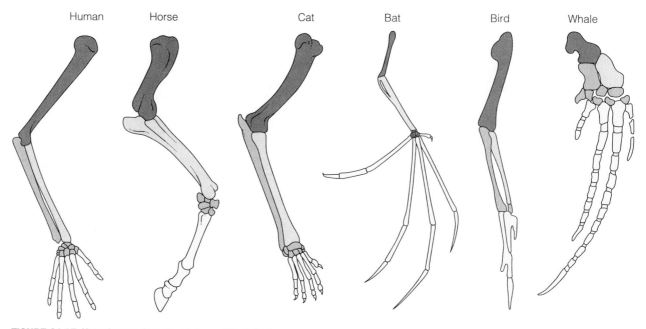

FIGURE 21-17 Homologous Structures Among Vertebrates
The presence of homologous structures in vertebrates and other groups supports the theory of evolution.

and the presence of analogous structures lend support to their beliefs. The common criticism from scientists is that creationism is not a science. It cannot be subjected to experimental method. It is, instead, based on faith.

Vestigial Structures Support Evolutionary Theory.

Vestigial structures (ves-TIJ-ee-al) serve no apparent purpose in an organism. An example is the pelvic bones of whales. The pelvis generally attaches an organism's legs to the axial skeleton, but because whales have no legs, the pelvis is useless. Then why are pelvic bones found in whales? Evolutionary biologists believe that whales evolved from terrestrial mammals that had four legs and an obvious need for pelvic bones. During evolution, whales retained the pelvic bones.

Vestigial structures are "evolutionary baggage." They support the notion of evolution. Had the whale been created through divine intervention, it is doubtful that a pelvis would have been included as standard equipment.

Humans contain many vestigial structures. Hair on the body, muscles that move the ears, the appendix, and the pointed canine teeth are examples. The presence of these structures should come as no surprise because we share so many genes with our ancestors.

Comparative Anatomy Permits Scientists to Determine Evolutionary Relationships

Studies of the anatomy of different species have long been used to determine "relatedness" among organisms, based on the belief that the more similar the internal structures of two species were, the more closely related they must be. Structural and functional similarities among species thus also allow scientists to classify organisms into groups.

The Common Biochemical Makeup of Organisms Also Supports the Theory of Evolution

Organisms as distantly related as roses and rhinos are made of the same basic biochemicals—ATP, DNA, RNA, and protein. All living organisms store genetic information in DNA, and all use ATP to collect, store, and transport energy. In addition, many biochemical pathways in organisms (such as those involved in producing energy) bear a remarkable similarity. Are these coincidental, or are they the result of evolution?

To many biologists, the common biochemistry of the Earth's diverse organisms lends further support to the notion of a common ancestry and thus to the theory of evolution. (To creationists, it supports the notion of a designer using similar blueprints.) Interestingly, although many species have similar enzymes and other proteins, differences do exist. Hemoglobin, for example, is found in vertebrate red blood cells, but differences in the composition of hemoglobin are common among vertebrates. These differences have proven quite useful to researchers. By comparing the amino acid sequence of certain proteins (such as hemoglobin) from different species, evolutionary biologists can determine how closely related they are. Chimpanzees and humans, for instance, share about 98% of the same genes, suggesting that they are indeed very closely related. The more differences, the more distant the evolutionary relationship.

Analysis of the amino acid sequence of proteins and the composition of the DNA have enabled evolutionary biologists to create their own evolutionary trees. Not surprisingly, the evolutionary trees they have developed correspond well to those created by comparative anatomists.

Biochemical analyses of DNA and protein have recently been expanded to extinct organisms (museum specimens) and even fossils. DNA extracted from the cells of extinct species can be cloned via genetic engineering. This process provides sufficient amounts of DNA to determine its sequence and make comparisons with other organisms, both living and dead, thus allowing evolutionary biologists to study the relationships among many more species.

Another Line of Evidence That Supports the Theory of Evolution Comes from Comparative Embryology

Studies show that the embryos of many different groups of organisms develop similarly. Thus, the embryos of chickens, turtles, mice, fishes, and humans—all vertebrates—bear a remarkable resemblance during their early stages of development. For example, each of them has a tail and gill slits, even though only adult fishes and amphibian larvae have gills.

One plausible explanation for the similarity of vertebrate embryos is that all of these species contain the genes that control the development of tails and gills and that these genes were passed on from a common ancestor. In humans, however, these genes are active only during embryological development. The structures they produce either become inconspicuous, as in the case of the tail (all that is visible is the tailbone), or become other structures, as in the case of the gill slits.

Experimental Evidence Supports the Theory of Evolution by Natural Selection

A long list of experiments supports the theory of evolution by natural selection. An exhaustive series of field and laboratory studies, for example, shows how genetic differences in wild populations of fruit flies can be attributed to natural selection.

Health, Homeostasis, and the Environment

The Pesticide Treadmill

Evolution has resulted in the presence of many homeostatic mechanisms—in the bodies of organisms like us and in the environment upon which we depend. As pointed out previously, humans have found many ways to upset those mechanisms. One of these ways—the use of pesticides—has potentially far-reaching impacts.

During World War II, chemical pesticides became the cornerstone of modern agriculture. Chemical pesticides kill insects and other organisms that damage crops. To combat weeds and animal pests worldwide, farmers, homeowners, and others spray 2.5 million metric tons of pesticides on the land each year. These chemicals cause a number of problems. First, many broad-spectrum pesticides kill beneficial insects that control pests naturally. Spider mites were once only a minor crop pest in California, but the heavy use of chemical pesticides to control them has inadvertently killed the mites' natural predators, species that were more sensitive to spraying than the mites. As a result, spider mites today cause twice as much damage in California as all other insect pests combined. Pesticides have thus destroyed an important homeostatic mechanism in California's environment.

A second problem of great concern is genetic resistance. **Genetic resistance** results from genetic variation—the presence of a mutation that makes a small percentage (about 5%) of any insect population resistant to chemical pesticides. Therefore, when farmers spray their fields with chemical pesticides, they kill only the nonresistant insects. Surviving the onslaught, the small subpopulation of genetically resistant insects breeds and produces a new population that is resistant to spraying. To kill it, farmers must apply additional pesticides or switch to another type. Once again, though, they find that a small segment of the population is genetically resistant to the higher dose or the new chemical preparation. That group survives the spraying, is selected for, and produces an even more resistant population, thus continuing an ever-escalating cycle often referred to as the "pesticide treadmill."

The pesticide treadmill is an excellent example of artificial selection at work. Some scientists warn farmers

Health Note 21-1

Well Done, Please: The Controversy over Antibiotics in Meat

FIGURE 1 Feed Lot Cattle   Antibiotics reduce infections in cattle in tight quarters and also accelerate growth. Microbiologists, however, worry that they may be stimulating the evolution of super strains of bacteria.

What does evolution have to do with how you have your meat cooked? Next time you order a steak or cook a hamburger over a grill, think about this: Approximately 40 year ago, livestock growers began adding antibiotics to cattle and pig feed to protect animals confined to pens from disease as they were being fattened for market (**FIGURE 1**). Antibiotics were first given to control disease, which can run rampant under crowded conditions. However, farmers soon found that the drugs had an unanticipated beneficial effect: For reasons still not fully understood, antibiotics acceler-ated the rate of body growth. That meant farmers could turn a higher profit.

Not surprisingly, today 70% of all cattle and 90% of all veal calves and pigs are reared on feed laced with penicillin or tetracycline. Nearly half of the antibiotics sold in the United States, in fact, are used for livestock feed.

The addition of antibiotics to feed has been sharply criticized by microbiologists and health officials who fear that widespread use could promote the evolution of superstrains of bacteria that are immune to antibiotics. In an editorial in the *New England Journal of Medicine*, a Tufts University microbiologist, Stuart Levy, noted that "every animal . . . taking an antibiotic . . . becomes a factory producing resistant strains" of bacteria. Resistant bacteria, in turn, could transfer their resistance to other bacteria, creating highly lethal strains that could infect humans. This is called the *crossover effect*.

Scientists are also concerned that some resistant bacteria in cattle (such as Salmonella) could be trans-

that they can never win in the battle against pests. Thanks to genetic variation in a population, no matter what pesticide they use, or how much they use, there will always be a resistant strain. Ever more powerful and ever more frequent applications at higher doses will be necessary just to stay even. As an example, a few decades ago, farmers in Central America applied pesticides about 8 times a year; today, 30–40 applications are the norm on any given field. Increased spraying pollutes the environment and kills fishes, birds, and other beneficial animals.

Unfortunately, farmers and chemical pesticide manufacturers have failed to take into account the existence of genetic variation and the selection process at work in the populations of insects. As a result, approxi-mately 500 species of insects are resistant to one or more chemical pesticides. Twenty of the worst pests are now resistant to all types of insecticide. Even weeds develop resistance to herbicides.

Knowledge of this phenomenon suggests alternative strategies. One alternative is **crop rotation**. This practice successfully reduces damage by maintaining pest populations at manageable levels. Practiced years ago on many farms but now largely abandoned, crop rotation requires farmers to alternate the crops they plant in a given field. One year corn might be grown in the field, the next year, beans, and the third year, alfalfa. This practice not only maintains soil nutrient levels, but also effectively cuts down on pests. How does it work?

mitted directly to people in meat or milk. The effects could be grave.

Despite these concerns, efforts to reduce the use of antibiotics in feed have been soundly defeated. In 1978, for instance, the Food and Drug Administration, which controls food additives such as antibiotics, proposed cutting back on penicillin and tetracycline use in feeds. Livestock producers, feed producers, and the multimillion-dollar drug industry, however, fought vigorously. They argued that microbiologists' concerns had not been proven.

Scientific evidence is beginning to confirm the suspicions of the medical profession. Dr. Thomas O'Brien of the Harvard University School of Medicine published a study in the *New England Journal of Medicine* in 1982, showing that bacteria that commonly infect humans and other animals share genetic information quite freely through the exchange of plasmids, tiny segments of DNA separate from the bacteria's chromosomes. O'Brien argued that drug resistance could be transferred easily from bacterium to bacterium.

Another study, by researchers at the Centers for Disease Control, pub-

lished in *Science* in 1984, confirmed the suspicion that antibiotic-resistant bacteria artificially selected by adding antibiotics to cattle feed could be transferred directly from meat to humans. This research showed that some of the outbreaks of illness caused by antibiotic-resistant Salmonella in the previous decade can be traced to meat from animals that had been fed antibiotic-treated grains. The research also showed that 20%–30% of the Salmonella outbreaks involved antibiotic-resistant strains. About 4.2% of the people contracting the antibiotic-resistant bacteria died, compared with only 0.2% of the victims of the normal bacteria.

Proponents of antibiotics believe that the link between antibiotics and human disease is still weak and that further research is needed. Even if these findings are substantiated by further research, proponents believe, the benefits of using antibiotics outweigh the potential health effects. Banning antibiotics or cutting back on their use, they say, could have enormous economic impacts that must be weighed against sickness and loss of life. But health experts

hope that the United States, like Europe, which strictly limited the use of antibiotics in animal feed in the early 1970s, will find the political will to end this activity.

Individuals can take direct action, too. You can buy "organically grown" beef—that is, beef fed a diet that contains no hormones or antibiotics. You can also change your eating habits. Instead of asking for rare meat or a half-cooked burger, you may want to consider asking for a well-done piece of meat. If ground beef has been heated in the center to over 160 °F, you are pretty safe; chicken and poultry need to be heated to more than 170 °F or until the fluids run clear.

Visit *Human Biology's* Internet site, www.jbpub.com/humanbiology, for links to web sites offering more information on this topic.

www.jbpub.com/humanbiology

Many insect pests are specialists, preferring one crop over another. A field of corn, for example, provides an abundant supply of food for corn borers. At the end of the season, corn borers lay their eggs in the soil or in organic material on the surface. If the same crop is planted the next year, the newly hatched insects will have an abundant food supply. The insect population will increase quickly, causing considerable damage. If a different crop is planted the second year, however, the corn borer population will decline. Continued rotation helps hold down pests year after year without costly and potentially harmful pesticide applications.

Chemical resistance is a rather common occurrence in the modern world. As noted above, weeds also become resistant to herbicides; even microorganisms develop re-

sistance to antibiotics, which is one reason many physicians prescribe antibiotics sparingly. They fear, and rightly so, that the more antibiotics we use in society, the more likely a resistant strain will emerge. Antibiotics are also used in livestock feed to protect closely housed animals and to stimulate growth; Health Note 21-1 discusses the problems that can arise from this common practice.

In sum, pesticide use upsets natural biological control mechanisms and puts us on a dangerous and costly pesticide treadmill. A little knowledge of genetics and natural selection can help farmers find alternative ways of controlling pests, ways that mimic nature's natural balance. These methods can markedly reduce pesticide use and protect people and the many species that share this planet with us.

SUMMARY

THE EVOLUTION OF LIFE: AN OVERVIEW

1. Scientists believe that all atoms and molecules in our solar system came from the same source, an enormous cloud of cosmic dust and gas that gave rise to the Earth and the sun.

2. The Earth formed about 4.5 billion years ago.

3. The evolution of life probably began in the sea and can be divided into three phases: chemical evolution, cellular evolution, and the evolution of multicellular organisms.

4. *Chemical evolution*, scientists hypothesize, began about 4 billion years ago. At that time, the Earth's primitive atmosphere contained a mixture of water vapor and several gases. As rains drenched the Earth, the oceans formed. Dissolved in the waters of early seas were many inorganic molecules (from the atmosphere) that combined to form small organic molecules. Energy needed to drive the reaction may have come from sunlight or other sources. Over time, the small organic molecules combined to form polymers—small proteins and nucleic acids.

5. The theory of chemical evolution holds that organic polymers combined to form aggregates called *microspheres*. These aggregates exhibit structural and functional characteristics of living organisms and may have been the precursors of cells.

6. *Cellular evolution* is the evolutionary development of prokaryotic cells from cell precursors.

7. The first true cells probably contained primitive enzymes, rudimentary genes, and selectively permeable membranes. These cells may have derived nourishment from organic molecules they absorbed from their environment. They are thought to have been heterotrophic fermenters.

8. Autotrophs, organisms capable of synthesizing their own food, arose next. The earliest autotrophs were probably chemosynthetic bacteria that acquired energy from chemicals in the environment.

9. With the evolution of chlorophyll, photosynthetic autotrophs arose. The first form of photosynthesis, however, did not produce oxygen.

10. Over time, scientists believe, photosynthesis evolved further, and the new photosynthetic organisms began to produce oxygen. Oxygen was toxic to the organisms living in the oxygen-free environment of that time. As a result, many of the anaerobic organisms died. Others retreated to oxygen-free environments. Still others evolved ways to survive in an oxygen-rich atmosphere.

11. The theory of endosymbiotic evolution states that eukaryotes arose from prokaryotes when a host cell acquired an internal symbiotic partner (possibly a smaller, oxygen-respiring or photosynthetic bacterium). As atmospheric oxygen levels increased, these unions persisted, and, in time, the relationship became hereditary.

12. Prokaryotes emerged about 3.5 billion years ago, and eukaryotes evolved about 1.2 billion years ago. The evolution of eukaryotes opened the door for the evolution of multicellular organisms.

13. A variety of multicellular plants and animals evolved from single-celled eukaryotes in the oceans. As the ozone layer developed, life on land became possible.

14. The first species to invade the land were plants that thrived in moist environments. Soon after the land plants invaded, animals came ashore. The very first of these animals were probably scorpionlike creatures, which were preadapted to life on land.

HOW EVOLUTION WORKS

15. Evolution has produced a great diversity of organisms.

16. Evolution takes place because of genetic variation and natural selection.

17. *Genetic variation* in a species arises from mutations, gene flow, recombination, independent assortment, and new combinations resulting from sexual reproduction. Genetic variation results in phenotypic variation. Some phenotypes confer a selective advantage on certain offspring, giving them a better chance of surviving and reproducing and passing the genes on to future generations.

18. Beneficial traits are preserved in a population by *natural selection*. Abiotic and biotic factors are the agents of natural selection.

19. It is important to remember that natural selection affects the survival and reproduction of individuals in a population, but that individuals do not evolve, populations do.

THE EVOLUTION OF NEW SPECIES

20. Natural selection results in a shift in the gene pool of a population and produces organisms that are better adapted to their environment. Dramatic changes in the gene pool may lead to the evolution of entirely new species.

21. *Geographical isolation* is one of the most common mechanisms by which new species evolve. It occurs when a population becomes physically separated. Subject to different environmental conditions, the populations may evolve independently. Over time, they may become *reproductively isolated*—that is, unable to interbreed. When this occurs, two different species are said to have formed.

22. Geographical isolation results in *divergent evolution*—the emergence of two or more species from a common ancestor. Another term that describes this phenomenon is *adaptive radiation*—the evolution of numerous related species, each of which is specially adapted to its environment.

23. Species often evolve in concert with each other, a phenomenon called *coevolution*. Predatory pressure, for example, may act as an agent of natural selection in the evolution of prey populations. Changes in the prey population may also affect the evolution of the predator population.

A MODERN VERSION OF EVOLUTIONARY THEORY

24. Many evolutionary biologists believe that evolution occurs in spurts. According to the *theory of punctuated equilibrium*, long periods of relatively

little change (equilibrium) are punctuated by briefer periods of relatively rapid change, many thousands of years long.

THE EVIDENCE SUPPORTING EVOLUTION

25. The scientific knowledge in support of evolution is rich and varied. The fossil record, anatomical similarities in groups of organisms, the occurrence of vestigial structures, the common biochemical makeup of organisms, similar embryological development among many groups of organisms, and experimental evidence all support the theory.

HEALTH, HOMEOSTASIS, AND THE ENVIRONMENT: THE PESTICIDE TREADMILL

26. Homeostasis at the organismic and environmental levels can be upset by pesticides. Pesticides upset ecological balances by destroying beneficial insects.
27. Pesticides also create a form of artificial selection that can be witnessed today in modern agriculture. Chemical pesticides used to control insects select resistant species. Farmers who spray their fields to kill insects leave behind a genetically resistant subpopulation that breeds and repopulates

farm fields. A second application at a higher dose or an application of another pesticide kills off more susceptible insects, but once again leaves behind another subset that often becomes a further pest, forcing farmers to use higher doses or switch to another pesticide. This escalation in the war against pests is called the *pesticide treadmill*.
28. Chemical resistance is a rather common occurrence in the modern world. Weeds can become resistant to herbicides, and even microorganisms develop resistance to antibiotics.

Critical Thinking

THINKING CRITICALLY— ANALYSIS

This Analysis corresponds to the Thinking Critically scenario that was presented at the beginning of this chapter.

Begin by making a list of basic tenets of creationism and evolution. From this list, you may see that creationism is based on interpretations of the Bible. The Bible, although a powerful basis for a personal belief system, is not a systematic and methodical study of nature. In other words, creationism is based on faith, not scientific fact. Evolution is based on interpretations of hard scientific data—a wide range of methodical and systematic observations and measurements. Although there are gaps in our knowledge of evolution, the data supporting this theory is quite impressive. In fact, there's not one iota of data that contradicts the theory of evolution.

One thing you may note about the creationist hypothesis is that it cannot be tested. You cannot perform experiments or test hypotheses to determine whether creationism existed—another indication that it is based on faith. So, should it be taught in science classrooms? Most scientists believe that religion should not be taught in science classrooms. However, it may deserve some mention—similar to the one in this chapter. And, a comparison of evolutionary theory and creationism can be instructive. Finally, it might be useful to point out that many scientists have no conflict between science and re-

ligion. They do not see evolutionary theory as a refutation of the presence of a divine being.

EXERCISING YOUR CRITICAL THINKING SKILLS

For more than a century, biology teachers have been relating a fascinating story of mimicry among butterflies that involves two colorful species, the monarch and viceroy (**FIGURE 21-18**).

The monarch larvae feed on milkweed and ingest a heart toxin (cardiac glycoside) produced by the plant, which remains in the bodies of the butterflies through adulthood.[8]

Some birds that prey on the monarch die, but most simply become ill and from that point on avoid

---

[8]Cardiac glycosides may have evolved to protect milkweed from herbivores.

FIGURE 21-18 Monarch and Viceroy Butterflies New studies suggest that long-standing beliefs about the evolution of the viceroy butterfly may be wrong.

the monarch. According to common belief, birds also avoid similarly colored viceroy butterflies. Viceroys evolved from the tasty but drab-looking admiral butterfly. Over time, scientists thought, viceroys evolved a coloration similar to the monarch butterfly, and this coloration (mimicry) persisted because it protected the viceroy from predation.

Two researchers at the University of Florida in Gainesville, however, have recently challenged this classic explanation. They conducted an avian taste test, feeding the abdomens of seven different butterfly species, including viceroys and monarchs, to red-winged blackbirds. In their experiments, they found that viceroys were just as unappetizing as monarch butterflies.

This finding throws into question the belief that viceroys are Batesian mimics—otherwise palatable insects that, through evolution, acquired a protective coloration. (Henry Bates was the nineteenth-century British naturalist who studied the phenomenon.) Because the viceroy butterfly feeds on nontoxic willows, the Florida researchers hypothesized that this species must produce its own chemical defense. This hypothesis, if it turns out to be true, could dispel a long-held belief—that all butterflies depend on plant poisons for the protective chemicals they harbor.

This new research suggests that instead of being Batesian mimics—that is, exploiting another's protective mechanism—the viceroy is actually a Mullerian mimic (after zoologist Fritz Muller who first described the phenomenon). This type of mimicry evolves in two equally distasteful butterfly species, which gain greater protection from predators by evolving nearly identical coloration. Mullerian mimicry is therefore mutually beneficial to both species. A bird that eats either a viceroy or a monarch would very probably leave both species alone.

This is another example of a failure in critical thinking—notably, the failure of scientists to question underlying assumptions in the formulation of their theories. One key assumption in this instance was that the viceroy was a palatable species. Scientists knew that the viceroy evolved from the rather tasty admiral butterfly; therefore, if it came from a palatable species, it must be edible. When this assumption was finally tested, a classic biological hypothesis was called into question.

The second assumption that led scientists astray has to do with the evolution of color and wing pattern. Because the viceroy's ancestors are dark-colored and drab, scientists believed that the viceroy probably evolved to mimic the monarch. If the two species turn out to be Mullerian mimics, it may be that they evolved simultaneously, benefiting from similar wing and color patterns.

TEST OF CONCEPTS

1. In 1924, the Russian scientist A. I. Oparin suggested that life arose from nonliving matter. Explain his hypothesis, and discuss the research supporting it.

2. Describe how the first cells may have arisen during evolution. What critical requirements must have been met for life to begin?

3. If life formed abiotically in the shallow waters of the sea, would you expect the process to be occurring now? Why or why not?

4. Describe the theory of endosymbiotic evolution and the evidence supporting it.

5. The development of photosynthesis and the emergence of eukaryotes were pivotal events in evolution. Why?

6. Numerous plants and animals had evolved in the sea before life emerged on land. Why?

7. A critic of evolution says, "Life-forms are too diverse to have come from a common ancestor." How would you respond?

8. What is meant by the phrase "the conservative nature of evolution"?

9. Over the course of the Earth's history, mountain ranges have risen and gradually worn down. Explain what might happen to a species population that was geographically isolated by the emergence of a mountain range.

10. How do random mutations in germ cells contribute to the evolutionary process?

11. Define the following terms: adaptation, variation, natural selection, biotic factors, abiotic factors, selective advantage, and fitness.

12. Discuss the following statement: natural selection is nature's editor.

13. Discuss the evidence supporting the theory of evolution.

14. How do crop rotation and heteroculture keep pest populations in check?

TOOLS FOR LEARNING

www.jbpub.com/humanbiology

Tools for Learning is an on-line student review area located at this book's web site HumanBiology (www.jbpub.com/humanbiology). The review area provides a variety of activities designed to help you study for your class:

Chapter Outlines. We've pulled out the section titles and full sentence sub-headings from each chapter to form natural descriptive outlines you can use to study the chapters' material point by point.

Review Questions. The review questions test your knowledge of the important concepts and applications in each chapter. Written by the author of the text, the review provides feedback for each correct or incorrect answer. This is an excellent test preparation tool.

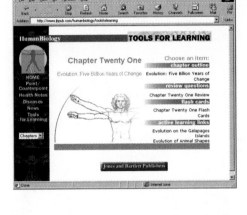

Flash Cards. Studying human biology requires learning new terms. Virtual flash cards help you master the new vocabulary for each chapter.

Figure Labeling. You can practice identifying and labeling anatomical features on the same art content that appears in the text.

Active Learning Links. Active Learning Links connect to external web sites that provide an opportunity to learn basic concepts through demonstrations, animations, and hands-on activities.

TRACING OUR ROOTS

THE STORY OF HUMAN EVOLUTION

Humans have come a long way in their 3.7 million year evolutionary history.

Thinking **Critically**

The emergence of modern humans (people like us, *Homo sapiens sapiens*) is, like many other scientific issues, a matter of considerable debate among scientists. Two basic hypotheses exist: the Multiregional Evolution model (or Regional Continuity model) and the African Origins model.

The Multiregional Evolution model holds that modern humans derive from archaic human populations living in various regions of Europe, Africa, China, and Indonesia. That is to say, these archaic populations gave rise to existing populations. Thus, modern Chinese derive from archaic Chinese (*Homo erectus*). In contrast, the African Origins model holds that, although regional archaic populations existed, they were displaced by a single population of humans that arose in Africa approximately 200,000 years ago. This new species supposedly spread into new territories replacing regional human populations living there.

The African Origins model is based on genetic analyses of mitochondrial DNA. In these analyses, scientists have found that only the African population exists in an unbroken maternal line. Other lineages of archaic peoples seem to have faded into oblivion.

Using your critical thinking skills, make a list of questions that might be useful in analyzing the African Origins model. ■

HUMORIST WILL CUPPY ONCE QUIPPED THAT "all modern men are descended from a wormlike creature, but it shows more in some people." In reality, humans belong to a group or, more correctly, an order known as the primates. The primates are believed to have evolved not from worms, but from an insect-eating mammal that probably resembled the modern-day tree shrew of Southeast Asia (**FIGURE 22-1**).

This chapter examines the evolutionary development of the first primates, beginning with the mammalian insectivores. It then describes how human beings evolved. This chapter also highlights important evolutionary advances in primate evolution and concludes with a brief discussion of human cultural evolution, laying a foundation for an understanding of contemporary environmental problems.

FIGURE 22-1 Look Familiar? An organism resembling this tree shrew is believed to have been the early ancestor of the primates.

Section 22-1

Early Primate Evolution

Before you begin your study of primate evolution, it is important to spend a few minutes on terminology and some basic relationships. First, the order we call **primates** consists of two subgroups or suborders: (1) the prosimians (pro-SIM-ee-ans; "premonkeys") and (2) the anthropoids (AN-throw-POIDS), as illustrated in **FIGURE 22-2**. The **prosimians** were the very first primates to inhabit the Earth (**FIGURE 22-2**). Today, prosimians consist of tree shrews, lemurs, and tarsiers (**FIGURES 22-1** and **22-3**). During the course of evolution, the prosimians gave rise to **anthropoids**. This group includes monkeys, apes, and humans.

Today, most primates—prosimians and anthropoids alike—inhabit tropical or subtropical regions where they live in forests, grasslands, and mixed woodlands (consisting of both grassland and forest). Most primate species are tree-dwellers. Humans are an important exception. Our kind ranges widely over the Earth, inhabiting virtually every available biome.

Primates are characterized by several features, among them grasping hands, which permit them to pick up objects, and forward-directed eyes, which enhance three-dimensional vision. Primates also have the largest brains of all mammals in proportion to their body size.

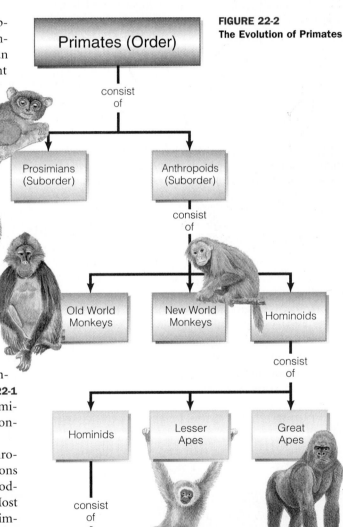

FIGURE 22-2 The Evolution of Primates

The Fossil Evidence of Primate Evolution Is Limited, Making it Difficult to Determine the Exact Progression of Early Human Ancestors

Unlike the dinosaurs, early primates did not live in habitats conducive to fossil preservation. Whereas the mud in swamps or along rivers preserved the bones of many a giant dinosaur, the forests and grasslands in which primates lived were not environments where bones of a dead animal would be readily covered with sediment and preserved. Thus, the primate fossil record is less than complete. Another reason for the scarcity of early primate fossils is that dead or dying animals were probably eaten by scavengers and their bones no doubt scattered across the landscape.

Because the primate fossil record is incomplete, the origin of primates remains speculative. But this has not deterred many scientists from making tentative conclusions about primate evolution. The scheme presented here represents a synthesis of current thinking on the evolution of primates. Bear in mind, however, that new evidence could radically change the present interpretation.

The Prosimians Probably Evolved from Tree-Dwelling Mammals That Lived During the Age of the Dinosaurs

Approximately 80 million years ago, tree-dwelling insectivores (similar to tree shrews) lived in tropical regions. Over time, they gave rise to many nonprimate mammals, among them the ground-dwelling shrews, water shrews, moles, and bats. Although no one knows for sure, the earliest prosimians probably emerged about 65 million years ago. As a group, the early prosimians became quite abundant and geographically widespread and lived for tens of millions of years. Approximately 55 million years ago,

FIGURE 22-3 Prosimians The prosimians (premonkeys) were probably the first primates. The earliest ones probably resembled modern-day (a) tarsiers and (b) lemurs.

(a)

they gave rise to the modern-day prosimians: tree shrews, lemurs, and tarsiers (**FIGURES 22-2** and **22-3**).

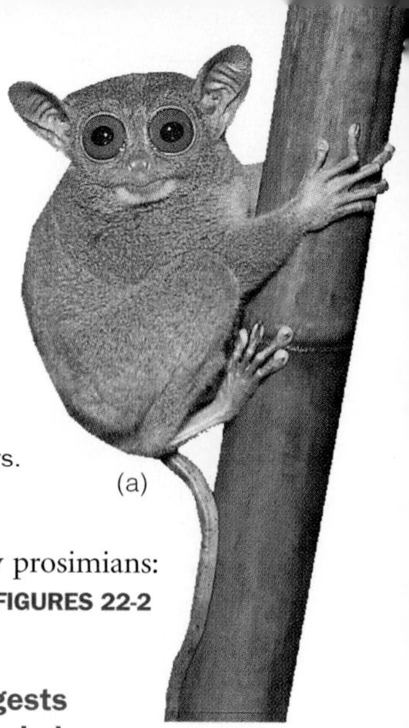

The Fossil Evidence Suggests That One of the Early Prosimian Lines Gave Rise to the Anthropoids

Within the following 15 million years, the **anthropoids** (monkeylike primates) evolved from prosimian stock in Africa and possibly earlier in Asia. Technically, biologists divide the anthropoids into (1) the Old World monkeys; (2) the New World monkeys; and (3) another larger group, the **hominoids** (HAHM-eh-NOIDS). The hominoids consists of great apes (like gorillas), the lesser apes (like the gibbons), and humans (**FIGURE 22-2**).

(b)

The Old World monkeys occupy the tropical regions on the continents of Africa and Asia. They include ground-dwelling as well as arboreal species. The Old World monkeys have grasping hands, but lack prehensile tails. Most monkeys in both groups live in bands and are active during the day. Some familiar examples are the baboon and rhesus monkey (**FIGURE 22-4**).

(a)

(b)

FIGURE 22-4 Old World Monkeys (a) Olive baboon and (b) Rhesus monkey.

FIGURE 22-5 **New World Monkeys** *(a)* Squirrel monkey and *(b)* spider monkey.

The New World monkeys, which inhabit Central and South America, include the familiar squirrel and spider monkeys (**FIGURE 22-5**). All New World monkeys live in the trees and have prehensile tails (pre-HEN-sill; "grasping").

The relationship between Old World and New World monkeys is not clearly delineated. However, most evolutionary biologists believe that Old World monkeys gave rise to the New World monkeys. First emerging in Africa and Asia, the Old World monkeys may have migrated across North America into Central and South America or may have arrived on rafts of logs or floating debris.

The hominoid line, which today consists of great apes, lesser apes, and hominids (all humanlike forms), diverged from the Old World monkeys over 20 million years ago. The very first hominoids were apelike creatures. Two distinct apes were present at this time and are considered to be potential ancestors of humans and modern apes.

The first family, the **dryopithecines** (dry-o-PITH-a-seens), originated about 20 million years ago in Africa and spread to Eurasia about 14 million years ago (**FIGURE 22-6**). Unfortunately, no complete skulls or skeletons of the dryopithecines have been found, and our knowledge of these animals is based on fragments of skulls, jaws, and limb bones. This evidence suggests that dryopithecines had relatively small brains and apelike teeth, jaws, and faces. The structure of their limbs suggests that dryopithecines walked on four legs, much like chimpanzees and gorillas, but spent most of their time in the trees.

The second possible ancestor of apes and humans lived slightly later—from about 17–7 million years ago. Called **ramapithecines** (ram-ah-PITH-o-seens), these creatures were similar to the dryopithecines in many respects.

Dryopithecines and ramapithecines may have been related or may have evolved separately. Unfortunately, fossil evidence is not complete enough to establish the relationship between them or to determine their exact relationship to apes and humans. For now, all we can do is content ourselves with knowing that these creatures existed and hypothesize that they represent a link between early primates and more modern primates, including gorillas, chimpanzees, and humans.

Section 22-2

Evolution of the Australopithecines

Archaeological evidence suggests that the first **hominids,** the first humanlike primates, belonged to the genus *Australopithecus* (awe-STRAY-loh-PITH-eh-cuss; meaning "southern apeman") (**FIGURE 22-6**). The oldest known australopithecine skeleton was unearthed by Donald Johanson, Yves Coppens, and their co-workers in Africa and is about 3.5 million years old. The specimen named Lucy is one of the most complete fossils of early hominids yet found (**FIGURE 22-7**). Known as ***Australopithecus afarensis*** (afarensis means from the Afar region of Ethiopia), these hominids stood only about 3 feet high and had brains only slightly larger than those of apes. The skulls of *A. afarensis* have many apelike features, including massive brow ridges, low foreheads, and forward-jutting jaws. The shape of its pelvis suggests that *A. afarensis* walked erect.

About 3 million years ago, *A. afarensis* vanished and was replaced by another species, *A. africanus*, which lived for about 1.5 million years (**FIGURE 22-6**). *A. africanus* was slightly taller than its predecessor and had a slightly larger brain.

Studies of the fossil record suggest that about 2.3 million years ago another species of *Australopithecus* emerged, *A. robustus*—so named because it was heavier and taller and had a larger brain than *A. africanus* (**FIGURE 22-6**). About 2.2 million years ago, the fourth species, *A. boisei*, appeared. It was even more robust than *A. robustus*.

Members of the genus *Australopithecus* had many common features. For example, they were all **bipedal** (buy-PED-al)—that is, they walked on two legs—and they all

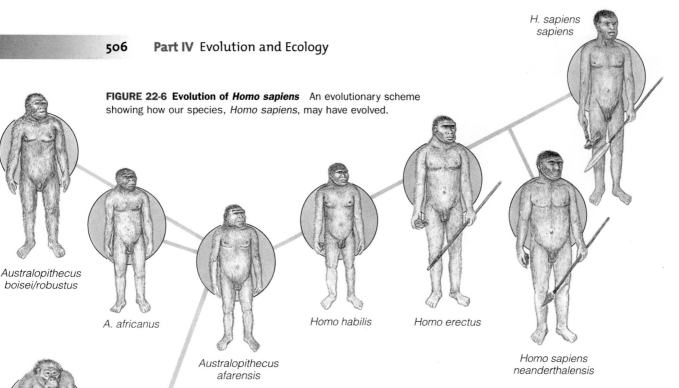

FIGURE 22-6 Evolution of _Homo sapiens_ An evolutionary scheme showing how our species, _Homo sapiens_, may have evolved.

H. sapiens sapiens

Australopithecus boisei/robustus

A. africanus

Australopithecus afarensis

Homo habilis

Homo erectus

Homo sapiens neanderthalensis

Modern apes

Dryopithecus

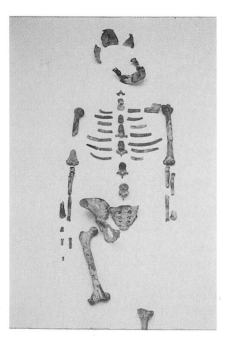

FIGURE 22-7 Skeleton of _Australopithecus afarensis_ This skeleton, estimated to be about 3.5 million years old, is popularly known as Lucy.

ranged in height from 1 meter to 1.7 meters. Their brains were larger than those of chimpanzees, but considerably smaller than modern humans. The differences among the four species are relatively minor, mostly a matter of degree. As the genus evolved, their brains got larger, and their height increased. The enlarged brain is believed to be the result of natural selection.

FIGURE 22-8 indicates when the various species of _Australopithecus_ lived. As illustrated, anthropologists believe that all three "younger" species coexisted, at least for a while, during the time span ranging from 1 million to 3 million years ago. Over 1 million years ago, how-

ever, australopithecines disappeared. Why they vanished no one knows.

Section 22-3

Evolution of the Genus _Homo_

The First Truly Humanlike Creatures Were Called _Homo habilis_

The earliest evidence of the genus _Homo_ is a 1.8 million year old skull and partial skeleton found in Tanzania in 1960 by Mary Leakey. She and her husband, Louis, called this species **_Homo habilis_** ("skillful man"). Archaeological evidence indicates that _Homo habilis_ apparently made primitive tools from fractured rocks and butchered large animals.

What was _Homo habilis_ like physically? The skull of _H. habilis_ is like that of _Australopithecus_, suggesting

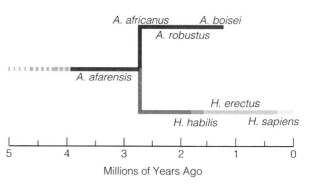

FIGURE 22-8 Australopithecine Evolution Conventional thinking holds that the genus _Homo_ evolved from _Australopithecus afarensis_.

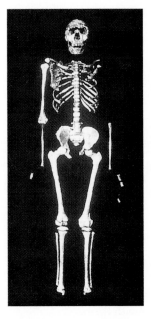

FIGURE 22-9 Skeleton of Homo erectus This, the most complete fossilized skeleton of *H. erectus* ever found, belonged to a boy who lived about 1.6 million years ago.

an evolutionary relationship. However, the brain of *Homo habilis* is 50% larger than its presumed australopithecine relatives. The skeleton of *H. habilis* displays many apelike features, especially the relatively long arms and small body. Because of this, some paleontologists contend that skeletons classified as belonging to *Homo habilis* are really members of the genus *Australopithecus*.

Homo habilis Gave Rise to Homo erectus

Two hundred thousand years after the emergence of *Homo habilis*, the first unmistakable member of the genus *Homo*, known as **Homo erectus** ("upright man"), arose (see **FIGURE 22-6**). Skeletons of *Homo erectus* appear in geological formations 300,000 to 1.6 million years old in the Old World. The most complete fossilized human skeleton to have been found is that of a 12-year-old boy (**FIGURE 22-9**).

Unlike *Australopithecus* and *H. habilis*, which remained in Africa, *H. erectus* first appeared in Africa, then spread to Europe and Asia (India, China, and Indonesia). *Homo erectus* appears to be the first hominid to leave the warmth and abundance of the tropics and subtropics and take up residence in the **temperate zone**, a region characterized by warm summers but cold winters.

Homo erectus stood about 5 feet tall, used fire, and made more sophisticated tools and weapons than its predecessor, *Homo habilis*. With a brain slightly smaller than ours, *Homo erectus* is believed to be the direct ancestor of *Homo sapiens*, the self-proclaimed "thinking man."

Modern Humans Belong to Homo sapiens

Homo sapiens emerged about 300,000 years ago. *Homo sapiens* consists of two subspecies, **Homo sapiens neanderthalensis**, the Neanderthals, and **Homo sapiens sapiens**, modern humans (**FIGURE 22-6**).

Widely distributed in Europe and Asia, the Neanderthals lived in caves and camps. Their name comes from the Neander Valley in Germany, where the first specimens were discovered. Archaeological evidence

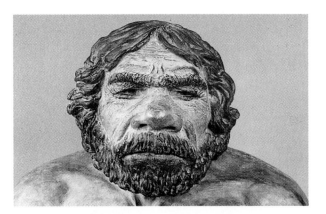

FIGURE 22-10 Neanderthal Notice the projecting brow ridges of this reconstruction of a Neanderthal man.

shows that the Neanderthals gathered fruits, berries, grains, and roots and hunted animals with weapons. They cooked some of their food on fires. Neanderthals stood erect and walked upright. Evidence also indicates that they lived in small clans and buried their dead in elaborate rituals.

Although the skeletons of the Neanderthals resemble those of humans, they differ in several ways. For example, the skulls of Neanderthals were, on average, larger than those of modern humans. Their skeletons were also more massive and heavily muscled than those of modern humans, and they had rather short lower limbs, much like Inuits (Eskimos) and other cold-adapted people.

Many people think of the Neanderthals as dim-witted and brutish, shuffling along bent over like an ape. This view is based on an interpretation of a Neanderthal skeleton found in 1908 in France. At the time of the discovery, however, the researchers failed to recognize that the skeleton under study was bent over because the individual suffered from arthritis of the hip and had diseased vertebrae. It is interesting, though, how this hasty conclusion from one observation spawned such a long-standing myth.

Some anthropologists wonder whether Neanderthals, despite many humanlike behaviors, should really be considered members of *Homo sapiens*. The differences in their physical appearance are quite striking. With broad faces, large projecting brow ridges, and heavily built bodies, they should perhaps be placed in a species all their own (**FIGURE 22-10**). Nonetheless, if one compares skeletons of australopithecines, early species of the genus *Homo*, and modern *Homo sapiens*, the Neanderthal does seem to fit well in the progression from *Homo erectus* to *Homo sapiens sapiens* (**FIGURE 22-11**).

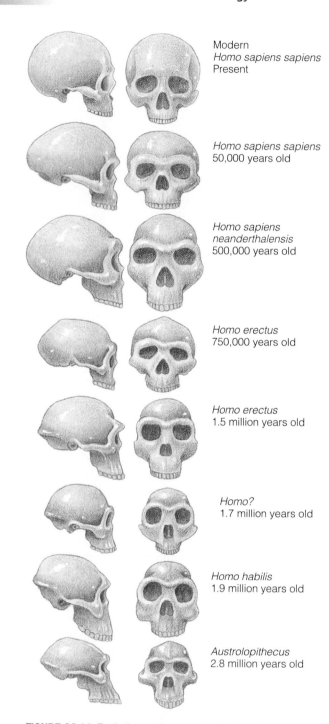

FIGURE 22-11 Evolutionary Progression in Skull Development Notice the changes that occurred over time in the shape of the human skeleton, starting with the apelike *Australopithecus* at the bottom. The dates indicate age of skulls in the drawing, not the evolutionary emergence of each species.

Modern
Homo sapiens sapiens
Present

Homo sapiens sapiens
50,000 years old

Homo sapiens neanderthalensis
500,000 years old

Homo erectus
750,000 years old

Homo erectus
1.5 million years old

Homo?
1.7 million years old

Homo habilis
1.9 million years old

Australopithecus
2.8 million years old

Neanderthals disappeared approximately 40,000 years ago for reasons still not understood. Some archaeologists believe that they were replaced by the **Cro-Magnons** (CROW-MAN-yens), the earliest known

FIGURE 22-12 Cro-Magnon Art

members of *Homo sapiens sapiens*. Cro-Magnons first appeared in Africa and then spread out across Europe and northern Asia, perhaps wiping out the Neanderthals or possibly interbreeding with them.

Archaeological evidence shows that Cro-Magnons used sophisticated tools and weapons, including the bow and arrow, and were highly skilled nomadic hunters, following great herds of animals during their seasonal migrations. They may have had a well-developed language. Cro-Magnons lived in caves and rock shelters, in groups of 50–75 people, and are best known for the elaborate artwork that adorned the walls of their caves (**FIGURE 22-12**).

Approximately 10,000 years ago, truly modern humans emerged, but the changes were not great. In fact, over the past 40,000 years, human evolution has produced little noticeable change in the physical appearance of humans. A Cro-Magnon on the streets of Los Angeles, in fact, would probably go unnoticed. Those 40,000 years have not been without change, however, for during this period, *Homo sapiens* has developed a rich and varied culture, complex language, and extraordinarily sophisticated tools.

Human Races Result from Variations Caused by Geographic Separation

As noted in Chapter 21, species that become subdivided into isolated populations undergo changes in response to their environment. If the changes are profound, the subpopulations often lose the ability to interbreed and produce new species.

Over the course of time, the human population has wandered far and wide on the planet and has fragmented into distinct subpopulations, living in very different environments. This fragmentation has resulted in the formation of a number of phenotypically distinct subpopulations, or **races** (**FIGURE 22-13**). It is important to remember, however, that races are not distinct species, just regional variants of the one species *Homo sapiens*.

Most of us are familiar with regional characteristics. We know, for instance, that Mexicans tend to have

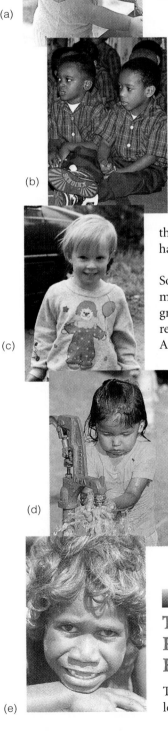

FIGURE 22-13 Human Races
Humans, no matter what they look like, belong to one species, *Homo sapiens*. The various races are phenotypic variants of that species. Many scientists recognize five distinct races: *(a)* Mongoloid, *(b)* Negroid, *(c)* Caucasian, *(d)* Native American, and *(e)* Australian Aborigine.

(a)

(b)

(c)

(d)

(e)

brown eyes, dark skin, and black hair and that Scandinavians tend to have blue eyes and blond hair. Many differences found in the races are undoubtedly adaptations to differing environmental conditions. The darker skin that evolved in tropical and subtropical populations, for instance, may be an adaptation that protects people from harmful ultraviolet light.

How many races are there? Some scientists recognize three main ones: Mongoloids, Negroids, and Caucasians. Others recognize two others: Native Americans and Australian Aborigines (**FIGURES 22-13D** and **22-13E**). Still others recognize 30 different races. What all this tells us is that the races are arbitrary categories. As time goes on, the distinction among various races will probably decrease. Why? As people move about the planet and interbreed, the races are mixing, blurring the lines among the world's peoples.

Section 22-4

Trends in Primate Evolution

The evolutionary path that leads first to primates and then to humans is marked by a number of significant changes. One important change was the evolution of bipedal (two-legged) locomotion, which freed the front legs for other useful tasks. In other words, bipedal locomotion may have, among other things, facilitated the evolution of the highly mobile front limbs we call arms.

In order to understand the importance of highly mobile front limbs, consider the four-legged ancestors of primates, the earliest prosimians. In these quadrupeds, the limbs function primarily as a means of propulsion. The front limbs move back and forth, but cannot be rotated or spread very far from the body. In sharp contrast, the front limbs of monkeys, apes, and humans are highly mobile. In tree-dwelling primates, this evolutionary advance permitted free movement among the branches, and it has enabled humans to use tools and weapons.

The evolution of primates was no doubt accelerated by the increased mobility of the fingers. This mobility permitted monkeys to grasp branches and fruit and apes to manipulate objects and even primitive tools. Chimpanzees, for instance, use sticks to dig termites out of the ground. In humans, manual dexterity permits us to engage in intricate crafts, to play musical instruments, and to perform many other fine motor skills.

Yet another significant advance, related most likely to the arboreal life of the early primates, was the development of eyes on the front of the head. This placement permits three-dimensional vision, essential to the quick pace of life in the trees. Swinging from branch to branch, often jumping long distances, monkeys would be nearly helpless without three-dimensional vision.

Another important development during the evolution of primates to humans was the steady increase in brain size. Larger brains were probably selected for because they gave organisms an advantage over others. Precise hand-eye coordination afforded by a larger brain no doubt facilitated life within the trees. A larger brain has also been a boon to humans. Our large, highly capable brain combined with our extraordinary manipulative abilities have permitted humans to devise elaborately powerful tools to reshape the environment—sometimes with untoward consequences.

 # Health, Homeostasis, and the Environment

Human Cultural Evolution

This book has been organized around the concept of homeostasis, showing how important homeostasis is to human health and how human actions can upset

FIGURE 22-14 Hunters and Gatherers Some anthropologists believe that the early hunters and gatherers acquired most of their food by gathering nuts, fruits, berries, roots, and seeds. Hunting may have provided a supplementary food source.

homeostasis—in our bodies and our environment. As we have seen, upsets in both systems lead to human disease. To better understand why humankind finds itself in a growing environmental crisis, let us take an admittedly simplified look at human cultural evolution.

Human culture has evolved through three general phases: hunting and gathering, agricultural, and industrial. For 99% of human evolutionary history, humans survived by gathering grains, fruits, nuts, berries, and roots and by hunting animals (**FIGURE 22-14**). Recent archaeological evidence suggests that our earliest ancestors were also scavengers, eating animals that died from natural causes. Our modern way of life and our modern agricultural practices, however, are really newcomers in human cultural evolution.

Based on studies of modern-day hunters and gatherers, whom some anthropologists call our "living ancestors," early hunter-gatherers were probably extraordinarily knowledgeable about the environment. For example, they very likely knew hundreds of edible and medicinal plants and were adept at locating insects, grubs, and other foods in nature's vast outdoor kitchen. Some studies also suggest that the lives of many hunters and gatherers, especially in warmer climates, were not as difficult as we often imagine. In many locales, they did not live in constant danger of starvation and did not spend a great deal of time finding food. Their lives were often leisurely and healthy.

Today, remaining hunters and gatherers live in relative harmony with nature, taking what they need and generally causing little ecological disruption, especially compared with the rest of humankind. Because of this, it is assumed that ancient hunting and gathering societies also lived in relative harmony. Their lack of environmental damage can be attributed to at least three facts of life: (1) many were nomadic, wandering in search of food and a favorable climate, (2) their numbers were small, and (3) their tools were primitive.

This is not to say that all hunter-gatherers were environmentally benign. Some North American Indians, for example, set fire to the prairie to drive buffalo off cliffs. Their massive fires killed large numbers of animals. In addition, the extinction of several North American species parallels the migration of ancient hunters and gatherers across the continent.

New archaeological research suggests that many groups of hunters and gatherers grew their own food and raised animals to feed their people earlier than we previously thought. Some may even have engaged in trade with other groups. Over time, of course, more and more of our early ancestors began to cultivate food crops, slowly giving rise to a new form of life, the agricultural society.

Agricultural societies emerged between 10,000 and 6,000 years ago. In the moist rain forests of Southeast Asia, farmers cleared small jungle plots to raise their crops. They grew a variety of vegetables and raised pigs and other domesticated animals for additional food.

Seed crops (grains like wheat) originated in a wider region, extending from China west to India and eastern Africa. In these regions, farmers cleared forests to plant crops and, with the advent of the plow, began to till the rich grassland soils.

The plow allowed for larger fields and higher grain production. Farming, which had generally supplied the immediate needs of a single family, began to change. Now a farmer could produce food for many families and with this development, towns and cities came into existence. People who were no longer needed on the farm congregated in cities and towns, where they began to engage in trades—making clothing, pots, tools, and weapons.

Several important changes in the human-environment interaction were also evident as agriculture grew. First, humans began to drastically modify the natural environment—sometimes on an enormous scale. Poor soil management, overgrazing, and heavy timber cutting destroyed large regions. The rich Tigris-Euphrates river valley, often referred to as the cradle of civilization and now part of Iran and Iraq, was once lush and productive. For several millenia now, much of this landscape has lain parched and unproductive because of human abuse. Further abuse continues today in many parts of the world, threatening long-term food production.

The second major change during the shift to agriculture came with an upsurge in commerce. Commerce

FIGURE 22-15 The Control of Nature Many forms of agriculture attempt to control or dominate the forces of nature, reshaping landscapes to human liking.

demands natural resources—metals, energy, and stone. These materials came from the outlying countryside. The towns and cities, therefore, drew heavily on the surrounding land, often causing considerable environmental damage.

The third change, possibly the most important of all, was a weakening of the link between humans and nature. As noted above, many hunter-gatherers took little from the land and lived within the bounds of nature. Agriculturalists, however, attempted to harness the forces of nature. Humankind began to see itself as separate from the environment, as nature's master (**FIGURE 22-15**). This profound change in attitude followed humans into the next phase of cultural evolution, the industrial society. With the advent of trade and commerce in agricultural societies, humans began to regard the natural world primarily as a source of wealth.

The industrial society is a recent occurrence in human history. In fact, if the Earth's history were condensed into a 1-year-long movie, the Industrial Revolution would occur in the last half-second of the film. The **Industrial Revolution**, the advent of mechanized production, began in England in the 1700s and in the United States in the 1800s. Although machines took over much of the manual production, they required enormous amounts of energy and produced enormous amounts of air and water pollution (**FIGURE 22-16**).

Mechanization swept the farms, too, in the industrial era. As a result, still fewer people were needed to raise food. Cities grew. Pollution increased. Streams once teeming with fish turned putrid with the stench of human and factory wastes. The countryside was often pillaged to provide energy and materials for factories. Pollution and species extinction were two of the principal environmental problems that arose during the Industrial Revolution.

Changes occurring during the agricultural and industrial revolutions planted the seeds of a dramatic increase in human population. Between 1850 and 1999, human population increased from 1 billion to 6 billion. This dramatic rise occurred for many reasons. The most important were (1) modern medicines, which lowered infant mortality; (2) improved sanitation, which stopped the spread of disease in the increasingly crowded urban environment; and (3) greater food production.

Humans' relationship with nature grew even more strained. Philosophers, economists, and some theologians argued that people needed to seek power over nature. Control became the byword. Survival, as many saw it, required complete domination of nature. Today, efforts at domination continue and have spawned serious global environmental problems. Some think they are a warning sign that we have pushed too far—that our culture is on a collision course with its future.

Current problems such as global warming and ozone depletion threaten to alter global climate and to create widespread species extinction. It is no exaggeration to say that the survival of humankind and the other species that inhabit this planet is in danger. Our health and survival on the planet depend on reestablishing a balance with nature. Today many people believe that to meet this challenge we must find sustainable ways of living and conducting business. Our future depends on finding ways to meet our needs without bankrupting the Earth and foreclosing on future generations and the millions of species that share this planet with us. In a sense, we need to construct a way of life that makes sense from both an economic and an environmental standpoint. We need a creative balancing of life-style and economics with the dictates of the Earth's ecological life-support systems.

FIGURE 22-16 A Product of Cultural Evolution

SUMMARY

EARLY PRIMATE EVOLUTION

1. Humans belong to the order called *primates*, which includes two suborders: the *prosimians* (premonkeys) and the *anthropoids* (monkeys, apes, and humans).

2. Today, most primates live in tropic and subtropic forests and are well adapted to an arboreal way of life. The main exception is humans, who inhabit a wide range of habitats.

3. Primates are characterized by grasping hands, forward-looking eyes, and large brains (in proportion to body size).

4. Based on fossil evidence, it appears that the primates evolved from a mammalian insectivore that resembled the modern-day tree shrew and lived about 80 million years ago.

5. The first primates to evolve were the tree-dwelling prosimians. Modern-day prosimians include the lemurs and tarsiers.

6. The prosimians gave rise to the *New World and Old World monkeys*. The *hominoids*, which today consist of the great apes, the lesser apes, and humans, diverged from the Old World monkeys over 20 million years ago.

7. The very first hominoids were apelike creatures. Two distinct apes present at that time are considered potential ancestors of humans and modern apes.

8. The first family, the *dryopithecines*, originated about 20 million years ago in Africa and spread to Eurasia about 14 million years ago. The scant fossil remains of these creatures suggest that they walked about on four legs, much like chimpanzees and gorillas, but spent most of their time in the trees.

9. The second possible ancestor of apes and humans lived slightly later, from about 17 million years ago to 7 million years ago, and are called *ramapithecines*.

EVOLUTION OF THE AUSTRALOPITHECINES

10. The first truly human primate belonged to the genus *Australopithecus* and may have evolved from either the dryopithecines or ramapithecines.

11. The oldest known australopithecine skeleton was unearthed in Africa and is believed to be about 3.5 million years old. It belongs to a group called *Australopithecus afarensis*.

12. *A. afarensis* stood only about 3 feet high and had a brain only slightly larger than an ape's, but it probably walked erect.

13. About 3 million years ago, *A. afarensis* was replaced by another species, *A. africanus*, which was slightly taller than its predecessor and had a slightly larger brain.

14. About 2.3 million years ago *A. robustus* emerged. It was taller and heavier and had a larger brain than its predecessors.

15. About 2.2 million years ago, the fourth species, *A. boisei*, appeared. It was even more robust than *A. robustus*.

EVOLUTION OF THE GENUS *HOMO*

16. Many paleontologists believe that *A. afarensis* also gave rise to the genus *Homo*, the ancestors of modern humans, *Homo sapiens*. Others believe that the genus *Homo* may have evolved from an as-yet-unidentified ancestor.

17. The earliest discovered evidence of the genus *Homo* is a 1.8 million year old skull and partial skeleton found in Tanzania. It belongs to *Homo habilis*.

18. Two hundred thousand years after the emergence of *Homo habilis*, *Homo erectus* arose. Skeletons of *Homo erectus* appear in geological formations 300,000 to 1.6 million years old in the Old World.

19. Unlike *Australopithecus* and *H. habilis*, which remained in Africa, *H. erectus* moved from Africa to Europe and Asia.

20. *Homo erectus* stood about 5 feet tall, used fire, and made more sophisticated tools and weapons than its predecessors. With a brain slightly smaller than ours, *Homo erectus* is believed to be the direct ancestor of *Homo sapiens*.

21. *Homo sapiens* emerged about 300,000 years ago and consists of two subspecies: *Homo sapiens neanderthalensis* (the Neanderthals) and *Homo sapiens sapiens*.

22. The Neanderthals lived in caves and camps in Europe and Asia until approximately 40,000 years ago, when they disappeared. Some archaeologists believe that they were replaced by modern humans, the *Cro-Magnons*, the earliest known members of *H. sapiens sapiens*.

23. The Cro-Magnons first appeared in Africa and then spread across Europe and northern Asia, perhaps wiping out the Neanderthals or possibly interbreeding with them.

24. Over the course of time, the human population has wandered far and wide on the planet and come to inhabit a wide range of climatic zones. This has resulted in the formation of a number of phenotypically distinct subpopulations, or races. Many differences found in the races are thought to be adaptations to differing environmental conditions.

TRENDS IN PRIMATE EVOLUTION

25. The evolutionary history that leads first to primates and then to humans is marked by significant developments, including (1) the evolution of bipedal locomotion, (2) an increase in brain size, (3) modifications of the hands that improve dexterity, and (4) the emergence of stereoscopic vision.

HEALTH, HOMEOSTASIS, AND THE ENVIRONMENT: HUMAN CULTURAL EVOLUTION

26. Over the past 40,000 years, human evolution has provided little noticeable change in physical appearance but tremendous cultural change.

27. Human culture has evolved through three phases: *hunting and gathering*, *agriculture*, and *industry*. During that transition, humans have sought ways to control the environment, sometimes with disastrous consequences.

28. Today, our attempts to dominate nature continue. But current global environmental problems are warning signs that we may have pushed too far, that our society is out of balance with the natural world, threatening our own existence.

Critical Thinking

THINKING CRITICALLY—ANALYSIS

This Analysis corresponds to the Thinking Critically scenario that was presented at the beginning of this chapter.

One of the first questions you might ask is this: Does the archaeological data support the genetic data? In other words, does the fossil record support the African Origins model? Or does it support the Multiregion Evolution model?

Some archaeological evidence suggests that the Multiregion Evolution model might be the most valid hypothesis. For example, studies of facial features of the Chinese skulls illustrate commonalities in modern and archaic Chinese. More importantly, these features are not present in the skulls of early modern Africans. One would expect the skulls of early modern Africans to resemble the Chinese skulls if the modern Chinese populations did indeed come from an ancestral African population. A similar study of skulls of modern and archaic European populations shows no similarities with either the earliest modern Africans or archaic African populations.

Another question that might be useful is this: Is there any archaeological evidence of the spread of African populations into other areas at the proper time? Again, the answer is no. Nor is there evidence of the spread of African culture (as witnessed by stone tools) as one might expect.

Given these facts, can you think of any reason to question the Multiregion Evolution model?

One common criticism of the Multiregion Evolution model is that archaeologists make many subjective judgments and often base their conclusions on small numbers of skulls. Geneticists wonder if such small samples are truly representative of an entire population. In contrast, the genetic data offers quantitative indicators of evolutionary change that are relatively free of biases.

Given the controversy, it is clear that neither hypothesis can be ruled out. More research is needed, and many years will no doubt pass before scientists can agree on which is correct.

EXERCISING YOUR CRITICAL THINKING SKILLS

Some critics make a blanket statement that genetic engineering in humans allows scientists to tinker with our evolutionary process and should therefore be avoided. Using your knowledge of human evolution, genetics, and genetic engineering and your critical thinking skills, how would you analyze this statement?

Hint: You might want to start by examining what genetic engineering in humans attempts to do. Then determine whether it could affect our evolution as a species. Is there any possibility that it will help or hinder our evolution?

TEST OF CONCEPTS

1. Why is the primate fossil record incomplete? What problems does this gap create in tracing the evolutionary history of primates?

2. Draw a flow diagram indicating the progression of primate evolution, beginning with the mammalian insectivores and ending with humans. Briefly describe each organism, and note how it differed from the early prosimians.

3. When you have finished Question 2, describe the major anatomical changes that have taken place over the course of primate evolution.

4. Describe the major changes that took place in human cultural evolution. How have they affected the health of the planet and the future of humankind?

www.jbpub.com/humanbiology

TOOLS FOR LEARNING

Tools for Learning is an on-line student review area located at this book's web site HumanBiology (www.jbpub.com/humanbiology). The review area provides a variety of activities designed to help you study for your class:

Chapter Outlines. We've pulled out the section titles and full sentence sub-headings from each chapter to form natural descriptive outlines you can use to study the chapters' material point by point.

Review Questions. The review questions test your knowledge of the important concepts and applications in each chapter. Written by the author of the text, the review provides feedback for each correct or incorrect answer. This is an excellent test preparation tool.

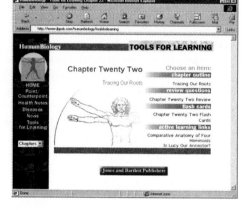

Flash Cards. Studying human biology requires learning new terms. Virtual flash cards help you master the new vocabulary for each chapter.

Figure Labeling. You can practice identifying and labeling anatomical features on the same art content that appears in the text.

Active Learning Links. Active Learning Links connect to external web sites that provide an opportunity to learn basic concepts through demonstrations, animations, and hands-on activities.

CHAPTER 23

PRINCIPLES OF ECOLOGY
UNDERSTANDING THE ECONOMY OF NATURE

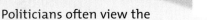

Thinking **Critically**

Politicians often view the environment as just another issue that must compete for limited funds with other issues such as crime. Furthermore, many people view expenditures on the environment as a detriment to economic health. They assert that investments in environmental protection automatically hurt the economy. Using your critical thinking skills, analyze this belief. Is it valid? Should environmental issues compete with other funds, or are they core issues that deserve maximum attention? Are there ways to achieve environmental goals without harming the economy? Or, better yet, are there ways to create a clean, healthy environment while fostering a strong economy? ■

Mount Cook in New Zealand, a cut above the rest.

MOST OF US LIVE OUR LIVES SEEMINGLY apart from nature. We make our homes in cities and towns, surround ourselves with concrete and steel, and drown out the sound of birds with our noise. The closest many of us get to nature is a romp with the family dog on the grass in our backyard.

Raymond Dasmann, a world-renowned ecologist, wrote that despite what many of us may think, a human apart from nature is an abstraction. No such being exists. Human life depends heavily on the environment. The clothes we wear, our morning coffee, and even the breakfast cereal we eat are all products of nature. So is the oxygen we breathe. Without plants, which produce oxygen, humans and other animal species could not survive. Nature provides other free services as well. For example, plants protect the watersheds near our homes, preventing flooding and erosion. Swamps help purify the water we drink. Birds control insect populations, and predators control rodent populations. Clearly, nature "serves" us well. Thus, although we may have isolated ourselves from nature, we have not emancipated ourselves. The ties that bind us cannot be broken without serious repercussions. This chapter will help you understand why.

This chapter discusses ecology. Technically, **ecology** is a branch of science. As ecologist Garrett Hardin once wrote, "ecology takes as its domain the entire living world." But its main focus is on interactions—how organisms interact with one another and how they are affected by the abiotic, or nonliving, components of the environment (like rainfall and temperature). It also looks at ways organisms affect their environment. An understanding of ecology provides many practical lessons. Throughout this chapter, I will point out our personal connections to the living world and ways we can lessen our impact on living systems.

Principles of Ecology: Ecosystem Structure

Ecology, like all disciplines in science, is a body of knowledge and a process of inquiry that seeks to understand the mysteries of nature. However, ecology probably ranks as one of the most misused words in the English language. Banners proclaim, "Save Our Ecology." Speakers argue that "our ecology is in danger," and others talk about the "ecological movement." These common uses of the word *ecology* are incorrect. Why?

Ecology is a scientific field of inquiry. It is not synonymous with the word *environment*. It does not mean the web of interactions in the environment. We can save our ecology department and ecology textbooks, but we cannot save our ecology. Our ecology is not in danger; our environment is. You cannot join the ecology movement, but you would be a welcome addition to the environmental movement.

The Biosphere Is the Zone of Life

The science of ecology, unlike many other branches of scientific endeavor, often focuses on systems. The largest biological system on Earth is the **biosphere** (BUY-oh-sfear). The biosphere can be thought of as the thin skin of life on the planet. As shown in **FIGURE 23-1**, the biosphere forms at the intersection of air, water, and land. All organisms consist of components derived from all three contributing spheres. The carbon atoms in protein come from the carbon dioxide in the atmosphere that is captured by plants. The minerals in your bones come from the soil in which plants grow. Water comes from streams and lakes.

The biosphere extends from the bottom of the ocean to the tops of the highest mountains. Although that may seem like a long way, it's not—at least, in comparison with the size of the Earth. In fact, if the Earth were the size of an apple, the biosphere would be about the thickness of its skin. Although life exists throughout the biosphere, it is rare at the extremes, where conditions for survival are marginal.

The biosphere is a **closed system**, much like a sealed terrarium. By definition, a closed system receives no materials from the outside.[1] The only outside contribution is sunlight, which is vital to the health and well-being of virtually all life.

As noted previously in this book, sunlight powers almost all life on the planet. Even the energy released by the combustion of coal, oil, and natural gas, which we use to light our homes and run our factories, owes its origin to the sunlight that fell on the Earth several hundred million years ago.

Because the Earth is a closed system, all materials necessary for life must be recycled over and over. The carbon dioxide you exhale, for instance, may be used by a rice plant during photosynthesis next week in Indonesia. Those carbon dioxide molecules will be incorporated in carbohydrate produced by the plant and stored in the seed. Consumed by an Indonesian boy, the carbohydrate will be broken back down during cellular respiration, and the carbon dioxide molecules will be

---

[1]Cosmic dust settles on the Earth, but virtually all of the materials necessary for life come from the Earth itself.

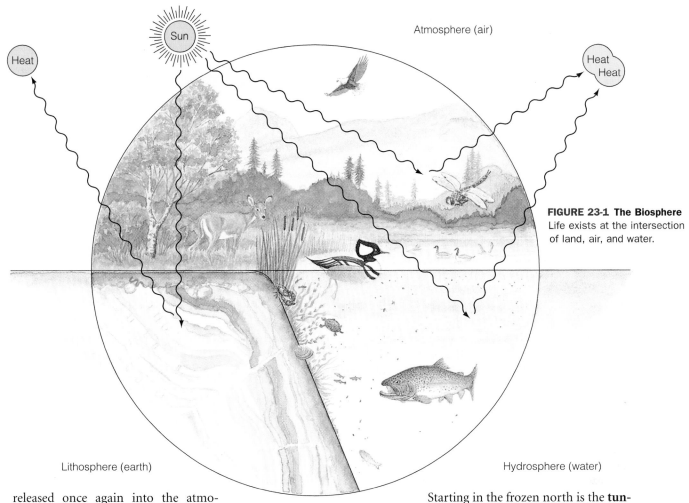

Atmosphere (air)

Heat

Heat
Heat

Sun

FIGURE 23-1 The Biosphere
Life exists at the intersection of land, air, and water.

Lithosphere (earth)

Hydrosphere (water)

released once again into the atmosphere to continue their never-ending cycle. Without this and dozens of other recycling processes, all life on the planet would grind to a halt. Part of the reason for protecting the environment is to protect the components of the global recycling systems on which we all depend.

The Biosphere Can Be Divided Into Distinct Regions Called Biomes and Aquatic Life Zones

Viewed from outer space, the Earth resembles a giant jigsaw puzzle, consisting of large landmasses and vast expanses of ocean. The landmasses, or continents, can be divided into large biological subregions or biomes (**FIGURE 23-2**). A **biome** is a region characterized by a distinct climate and a particular assemblage of plants and animals adapted to it. This section gives an overview of some of the major biomes of the North American continent. As illustrated in **FIGURE 23-2**, the North American continent contains seven biomes, five of which are discussed here.

Starting in the frozen north is the **tundra** (TON-drah), a region of long, cold winters and rather short growing seasons (**FIGURE 23-3A**). The rolling terrain of the tundra supports grasses, mosses, lichens, wolves, musk oxen, and other animals adapted to the bitter cold. Trees cannot grow on the tundra for at least two reasons. First, the growing season is quite short. Second, the subsoil (permafrost) remains frozen year-round, preventing the deep root growth necessary for trees.

Immediately south of the tundra lies the **taiga** (TIE-ga), the northern coniferous, or boreal, forest (**FIGURE 23-3B**). The taiga's milder climate and longer growing season result in a greater diversity and abundance of plant and animal life than exists on the tundra. Evergreen trees, bears, wolverines, and moose are characteristic species.

East of the Mississippi River lies the **temperate deciduous forest biome,** characterized by an even warmer climate and more abundant rainfall (**FIGURE 23-3C**). Broad-leafed trees make their home in this biome. Opossums, black bears, squirrels, and foxes are characteristic animal species.

FIGURE 23-2 The Biomes

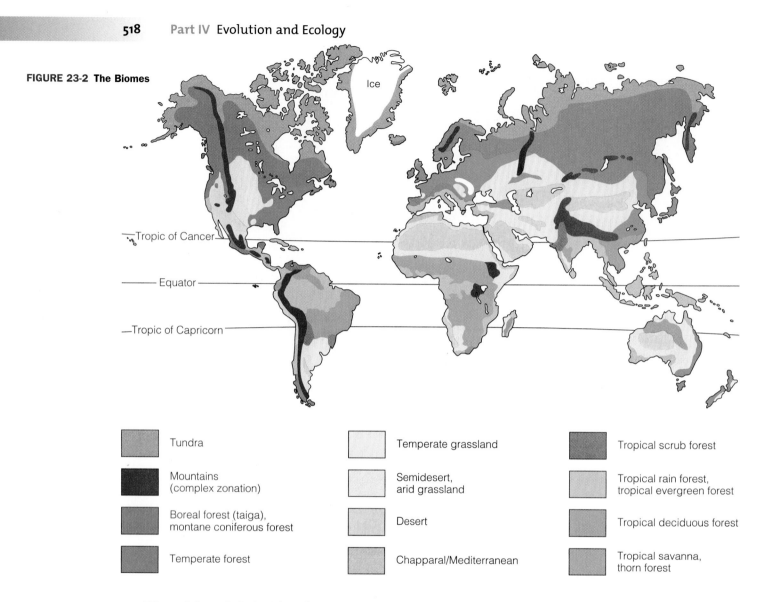

Tundra

Mountains (complex zonation)

Boreal forest (taiga), montane coniferous forest

Temperate forest

Temperate grassland

Semidesert, arid grassland

Desert

Chapparal/Mediterranean

Tropical scrub forest

Tropical rain forest, tropical evergreen forest

Tropical deciduous forest

Tropical savanna, thorn forest

West of the Mississippi lies the **grassland biome** (**FIGURE 23-3D**). Inadequate rainfall and periodic drought prevent trees from growing on the grasslands, except near rivers, streams, and human habitation. Over the years, deep-rooted grasses evolved on the plains. These grasses could tolerate drought and withstand grazing. However, virtually all of the natural grasslands have been plowed under to make room for human crops. Coyotes, hawks, and voles are characteristic animal species.

In the Southwest, where even less rain falls, is the **desert biome** (**FIGURE 23-3E**). Contrary to what many people think, the desert often contains a rich assortment of plants and animals uniquely adapted to the aridity and heat. Cacti, mesquite trees, rattlesnakes, and a variety of lizards all make their home in this environment.

The oceans can also be divided into subregions, known as **aquatic life zones**. Aquatic life zones are the aquatic equivalent of biomes. Like their land-based counterparts, each of these regions has a distinct environment and characteristic plant and animal life adapted to conditions of the zone. Four major aquatic life zones exist: coral reefs, estuaries (the mouths of rivers where fresh and salt water mix), the deep ocean, and the continental shelf.

Humans inhabit all biomes on Earth, a testimonial to our remarkable adaptability. Within these regions, humans benefit in numerous ways from the microbes, plants, animals, soil, water, and air that compose the biome. We tend to think of biomes, though, in terms of their natural resources (coal, timber, oil). Although vital to our economic health, these resources are but a fraction of the services we receive from the biome. Trees, for instance, not only provide oxygen and protect watersheds, they may remove certain pollutants from the air. Soil bacteria recycle nutrients on farms. Lest we forget, many other species also depend on these same services.

(a)

(b)

(c)

(d)

(e)

FIGURE 23-3 North American Biomes
(a) Tundra *(b)* taiga *(c)* temperate deciduous forest *(d)* grassland *(e)* desert.

Ecosystems Consist of Organisms and Their Environment

The biosphere is a global **ecological system**, or **ecosystem**. The term *ecosystem* is used to describe a community of organisms and their physical and chemical environment. Innumerable interactions are possible within an ecosystem. For the sake of convenience, ecologists often limit their study to small ecological systems—for example, a pond, a rotting log, or a grassy meadow. Even in these small ecosystems, the number of organisms present and the number of possible interactions can be astounding. Reduced to a minimum, ecosystems consist of two basic components: abiotic and biotic.

Abiotic Components.

The **abiotic components** of an ecosystem are the physical and chemical factors necessary for life. They include such things as sunlight, precipitation, temperature, and nutrients. Because abiotic conditions within most ecosystems vary, organisms must be able to survive a range of conditions. The range of conditions to which an organism is adapted is called its **range of tolerance.** As **FIGURE 23-4** (page 522) shows, organisms do best in the optimum range. Outside of that are the **zones of physiological stress**, where survival and reproduction are possible but not optimal. Outside of these zones are the **zones of intolerance,** where life for that organism is not possible.

Humans often alter the physical and chemical components of the environment. Changes can lead to serious biological impacts. If conditions change drastically, for instance, some species perish. Dams, for example, create lakes in streambeds, changing the water temperature, water flow, and other abiotic factors. As a result, many fishes that live in the stream perish. Native razorback suckers and other species that once thrived in the Colorado River, for example, are now endangered species (in danger of extinction) because of the large dams constructed along the river (**FIGURE 23-5**, page 522). Cold water released from the bottom of huge reservoirs formed by these dams has changed the temperature of the river and wiped out most of the native fish. The Point/Counterpoint in this chapter gives two opposing views on human-caused extinction.

Although species are sensitive to all of the abiotic factors in their environment, one factor often turns out to be more important than others in regulating growth of an entire ecosystem. This factor, which ultimately regulates the growth of a population, is called a **limiting factor**.

In freshwater lakes and rivers, for example, dissolved phosphate is the limiting factor. Phosphate is required by plants and algae for growth, but phosphate concentrations are naturally low. As a result, plant and algal growth is held in check. When phosphate is added to a body of water, plants and algae proliferate. Algae often form dense surface mats, blocking sunlight (**FIGURE 23-6**, page 522). As a result, plants rooted on the

Point/Counterpoint

Why Worry About Extinction?

HUMANS ARE ACCELERATING EXTINCTION

David M. Armstrong

David M. Armstrong *teaches science for nonscientists at the University of Colorado at Boulder and has written several books on the mammals and ecology of the Rocky Mountain region.*

Evolution is the process of change in gene pools. When one gene pool becomes reproductively independent of another, a new species has formed. Such speciation generates species; extinction takes them away. Simply put, extinction is a failure to adapt to change, the termination of a gene pool, and the end of an evolutionary line.

Extinction is a natural process. Most of the species that have lived on this planet are now extinct. The 3–30 million species on Earth today are no more than 1%–10% of the species that have evolved since life began about 3.5 billion years ago. Given these facts, why are thoughtful people concerned about endangered species? After all, history makes it clear that—given enough time—all species will become extinct.

The basis for concern is that today the natural process of extinction is proceeding at an unnatural rate. Let us estimate by how much human activity has accelerated rates of extinction. The lifespan of species seems to average from 1 million to 10 million years. Assume (to be conservative) that the average longevity of a species of higher vertebrates is 1 million years. In round numbers (to make calculations easy), there are 10,000 species of birds and mammals. So, on average, one species ought to go extinct each century. However, between 1600 and 1980, at least 36 species of mammals and 94 species of birds became extinct. That is about 0.29 species per year, 29 times the natural rate.

What does it mean to increase a rate by 29 times? The speed limit is 55 miles per hour. Exceed the speed limit by 29-fold, and you are moving 1,595 miles per hour, over twice the speed of sound. The difference between natural rates of extinction and present, human-influenced rates is analogous to the difference between a casual drive and Mach 2! Is that a problem? You decide: concern is a moral construct, not a scientific one.

Several human activities have contributed—mostly inadvertently—to accelerating rates of extinction. The dodo and the passenger pigeon were extinguished by overhunting. Wolves and grizzly bears were exterminated over much of their ranges as threats to livestock. The black-footed ferret was driven to the verge of extinction because prairie dogs, its staple food, were poisoned as agricultural pests. The smallpox virus was exterminated in the "wild" (but survives in a half-dozen laboratories). This is the closest that humans have come to deliberate elimination of an organism, and note that thoughtful scientists with the power to destroy smallpox chose not to do so, electing instead to manage it with care.

Habitat change is the most important cause of endangerment and extinction. Clearing forests for agriculture has decimated the lemurs of Madagascar. Chemical pesticides led to the decline of the peregrine falcon. Introducing exotic species (like goats on the Galapagos and mongooses in Hawaii) displaces native animals and plants. Developing the Amazon Basin is a habitat alteration, and a cause of extinction, on an unprecedented scale.

Many urge saving species for their aesthetic value. Whooping cranes are beautiful, and part of the beauty is that they are products of a marvelous evolutionary process. Most concern about accelerated extinction, however, stresses economic value. A tiny fraction of Earth's seed plants are used commercially. Perhaps an obscure plant like jojoba will become a source of oil more reliable than that beneath the sands of Saudi Arabia. Wild grasses have furnished genes that improved disease-resistance in wheat. Numerous wild animals (like musk ox, kudu, and whales) could contribute protein to the human diet. Wild species may have medical value; penicillin, after all, was once merely an obscure mold on citrus fruit. Some sensitive species are useful monitors of environmental quality, and the presence of healthy populations of many species may promote greater stability or resilience of ecosystems. Naturalist Aldo Leopold noted that we humans have a way of "tinkering" with the ecosphere to see how it works. Given that, we ought to remember the first rule of tinkering: never throw away any of the parts.

"Extinction is forever," and extinction impoverishes both Earth's ecosystems and the potential richness of human life. Borrowing again from Aldo Leopold, I believe we should be concerned about unnaturally rapid extinction because such concern is part of a "right relationship" between people and the landscapes that nurture and inspire them.

Biologist Sir Julian Huxley noted that "we humans find ourselves, for better or worse, business agents for the cosmic process of evolution." We hold power over the future of the biosphere, the power to destroy or to preserve. German philosopher George Hegel noted that freedom (including the power to destroy species) implies responsibility (to preserve them). I agree. The question of human-accelerated extinction boils down to a simple ethical question, "Does posterity matter?" Some of us have ethics that are human-centered. We ask simply, "Do my children deserve a life as rich, with as much opportunity, as mine?" If they do, then we have the responsibility to choose restraint. (Perhaps my life has been rich enough without the dodo, but I am reluctant to make that judgment for future generations.)

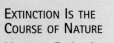

EXTINCTION IS THE COURSE OF NATURE

Norman D. Levine

Evolution is the formation of new species from preexisting ones by a process of adaptation to the environment. Evolution began long ago and is still going on. During evolution, those species better adapted to the environment replaced the less well adapted. It is this process, repeated year after year for millennia, that has produced the present mixture of wild species. Perhaps 95% of the species that once existed no longer exist.

Human activities have eliminated many wild species. The dodo is gone, and so is the passenger pigeon. The whooping crane, the California condor, and many other species are on the way out. The bison is still with us because it is protected, and small herds are raised in semicaptivity. The Pacific salmon remains because we provide fish ladders around our dams so it can reach its breeding places. The mountain goat survives because it lives in inaccessible places. But some thousands of other animal species, to say nothing of plants, are extinct, or soon will be.

Some nature lovers weep at this passing and collect money to save species. They make lists of animals and plants that are in danger of extinction and sponsor legislation to save them.

I don't. What the species preservers are trying to do is to stop the clock. It cannot and should not be done.

Extinction is an inevitable fact of evolution, and it is needed for progress. New species continually arise, and they are better adapted to their environment than those that have died out.

Extinction comes from failure to adapt to a changing environment. The passenger pigeon did not disappear because of hunting alone, but because its food trees were destroyed by land clearing and farming. The prairie chicken cannot find enough of the proper food and nesting places in the cultivated fields that once were prairie.

And you cannot necessarily introduce a new species, even by breeding it in tremendous numbers and putting it out into the wild. Thousands of pheasants were bred and set out year after year in southern Illinois, but in the spring of each year there were none left. Another bird, the capercaillie, is a fine, large game bird in Scandinavia, but every attempt to introduce it into the United States has failed. An introduced species cannot survive unless it is preadapted to its new environment.

A few introduced species are preadapted and some make spectacular gains. The United States has received the English sparrow, the starling, and the house mouse from Europe, and also the gypsy moth, the European corn borer, the Mediterranean fruit fly, and the Japanese beetle. The United States gave Europe the gray squirrel and the muskrat, among others. The rabbit took over in Australia, at least for a time.

The rabbit and the squirrel were successful on new continents because their requirements are not as narrow as those of species that failed. Today, adjustment to human-made environments may be just as difficult as adjustment to new continents. The rabbit and the squirrel have succeeded in adjusting to the backyard habitat, but most wild animals have disappeared.

Human-made environments are artificial. People replace mixed grasses, shrubs, and trees with rows of clean-cultivated corn, soybeans, wheat, oats, or alfalfa. Variety has turned into uniform monotony, and the number of species of small vertebrates and invertebrates that can find the proper food to survive has become markedly reduced. But some species have multiplied in these environments and have assumed economic importance; the European corn borer in this country is an example.

Would it improve Earth if even half of the species that have died out were to return? A few starving, ship-wrecked sailors might be better off if the dodo were to return, but I would not be. The smallpox virus has been eliminated, except for a few strains in medical laboratories. Should it be brought back? Should we bring panthers back into the eastern states? Think of all the horses that the automobile and tractors have replaced, and of all the streets and roads that have been paved and the wild animals and plants killed as a consequence. Before people arrived in America about 10,000 years ago, the animal-plant situation was quite different. What should we do? Should we all commit suicide?

Evolution exists, and it goes on continually. People are here because of it, but people may be replaced someday. It is neither possible nor desirable to stop it, and that is what we are trying to do when we try to preserve species on their way out. It can be done, I think, but should we do it to them all? Or to just a few, as we are doing now?

Norman D. Levine *is a professor emeritus at the College of Veterinary Medicine and Agricultural Experiment Station, University of Illinois at Urbana. His research interests cover parasitology, protozoology, and human ecology. This essay is from "Evolution and Extinction," BioScience 39:38, and is copyright © 1989 by the American Institute of Biological Sciences.*

SHARPENING YOUR CRITICAL THINKING SKILLS

1. Summarize the key points of both authors.
2. Do you see any flaws in the reasoning of either author?
3. Which viewpoint do you adhere to? Why?

Visit *Human Biology's* Internet site, www.jbpub.com/humanbiology, to research opposing web sites and respond to questions that will help you clarify your own opinion. (See Point/Counterpoint: Furthering the Debate.)

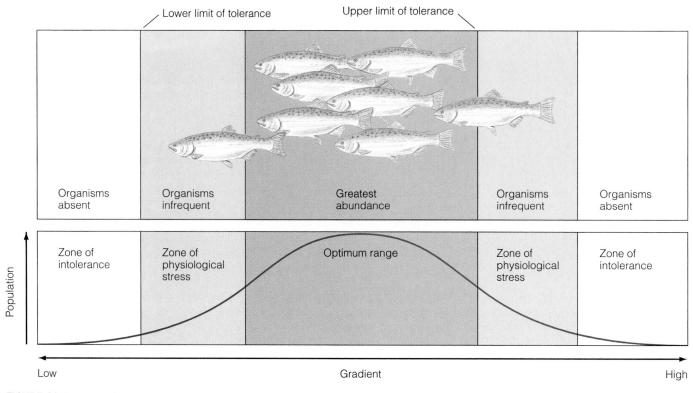

FIGURE 23-4 **Range of Tolerance** Organisms tolerate a range of conditions but thrive within an optimum range.

bottom may die, and oxygen levels in the deeper waters may decline, killing fishes and other aquatic organisms.

Phosphate pollution in lakes and rivers come from several sources. One of the major sources is the sewage-treatment plant. The phosphates released by these plants come from detergents and human waste. People can lessen the problem by using low- or no-phosphate detergents.

On land, precipitation tends to be the limiting factor. At any given temperature, the more moisture that falls, the richer the plant and animal life.

Biotic Components.

The **biotic components** of an ecosystem are the organisms that live there. Within biomes and aquatic life zones, organisms of a given species often occupy very specific regions to which they are well adapted. A group of organisms of the same species occupying a specific

FIGURE 23-5 **Altering Native Fish Life Along the Mighty Colorado** Dams on the Colorado River have changed the water temperature in the river below the dam. Extremely cold water flows out of the bottom of the reservoirs into the river and has nearly eliminated several native fish species, including the razorback sucker.

FIGURE 23-6 **Algal Bloom** This pond is choked with algae due to the abundance of plant nutrients from human sources.

region constitutes a **population**. The members of a population may remain together throughout all or much of the year. Bighorn sheep, for example, live in herds throughout the year. Other organisms such as the grizzly bear are solitary for the most part, keeping to themselves except for mating. In any given ecosystem, several populations exist together and form a **biological community**, an interdependent network of plants, animals, and microorganisms.

Competition May Occur Between Species Occupying the Same Habitat if Their Niches Overlap

If asked to give a brief description of yourself, you would probably begin by describing the place where you live. You would then likely discuss the work you do, the friends you have, and other important relationships that describe your place in society. A biologist would do much the same when describing an organism. He or she would start with a description of the place an organism lives—that is, its **habitat**. Next, he or she would describe how the organism "fits" into the ecosystem, its **ecological niche**, or simply **niche** (nitch). An organism's niche includes its habitat and all of its relationships with its environment. For example, the niche includes what an organism eats, what eats it, its range of tolerance for various environmental factors, and other important facts.

Organisms in a community occupy the same habitat, but most of them have quite different niches. This phenomenon minimizes competition for resources and is a condition favored by natural selection. The fact that organisms occupy separate niches provides for a wider use of an ecosystem's resources, especially food.

Niches do overlap somewhat. For example, two species may feed on some of the same foods. The more two species' niches overlap, however, the more the organisms compete with each other. Just as in the human economy, competition in the natural world can be a good thing, for it is an agent of natural selection that may lead to the evolution of new adaptations. When niches overlap considerably, competition becomes intense, and one species usually suffers. If two species occupy identical niches, competition will eliminate one of them. As a result, two species cannot occupy the same niche for long. This law of nature is called the **competitive exclusion principle**.

The concept of the niche is very important to us. For example, successful control of an insect pest is best achieved through an understanding of the pest's niche. Such an analysis might show that a particular species of bird or insect feeds on the pest. By simply encouraging these beneficial species—say, by providing trees for the birds or a preferred food for the pest-eating insect—farmers can hold pest populations down and save themselves enormous sums of money on chemical pesticides. Individuals interested in saving species have also learned that to protect a plant or animal, you have to protect the ecosystem it lives in, efforts that help protect the species' niche.

Humans Have Become a Major Competitive Force in Nature

Humans compete with many other species for food and living space. However, humans possess a marked advantage provided, in large part, by our technology.

Already, commercial overfishing has depleted dozens of the world's fisheries (**FIGURE 23-7**). Such actions not only reduce our food supply, they also greatly reduce the food supply of other species—notably, seals and other fish-eating animals. On every continent and in every nation, humans are outcompeting the species that share this planet with us. According to estimates by various experts, 40–100 species become extinct every day, largely because of tropical deforestation, but also

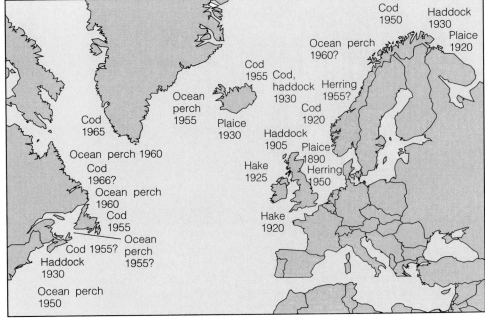

FIGURE 23-7 Depletion of North Atlantic Fisheries The dates on the map indicate the approximate time when various commercial fisheries were depleted by overfishing.

because of a loss of wetlands and coral reefs. Unless we do something, hundreds of thousands of species will become extinct in the next decade.

Saving other species is more than a humanitarian act. It is an act of self-protection. Cutting down the rain forests, for example, could alter global climate, making the Earth hotter. If conditions became bad enough, massive food shortages could occur.

Section 23-2

Ecosystem Function

Life on land and in the Earth's waters is possible principally because of the existence of one group of organisms, the **producers.** These organisms primarily include the algae and plants. These organisms absorb sunlight and use its energy to synthesize organic molecules from atmospheric carbon dioxide and water. This process is called **photosynthesis.**[2] Organic molecules generated by

---

[2]Some producers are chemosynthetic organisms—that is, organisms capable of using methane and hydrogen sulfide as a source of energy to produce organic molecules.

photosynthesis not only nourish the producers, but also feed all the other organisms in the web of life. As a result, producers form the foundation of the living world.

Another large group of organisms is the **consumers.** They reap the benefit of the producers' photosynthetic work. Ecologists place consumers into four general categories, depending on the type of food they eat. Some such as deer, elk, and cattle feed directly on plants and are called **herbivores.** Others such as wolves feed on herbivores and other animals and are known as **carnivores.** Humans and a great many other animal species subsist on a mixed diet of plants and animals and are known as **omnivores.** The final group feeds on animal waste or the remains of plants and animals. These organisms are called **detritivores** (DEE-treh-TEH-vores), or **decomposers.**[3] This important group includes many bacteria and fungi, and insects.

Food and Energy Flow Through Food Chains That Are Generally Part of Much Larger Food Webs in Ecosystems

As scientists began to study the relationships between producers and consumers, they found that specific organisms fed on others. They called these feeding relationships food chains. Technically, a **food chain** is a series of organisms, each one feeding on the organism preceding it (**FIGURE 23-8**). Biological communities consist of numerous food chains. All organisms in the community are members of one or more food chains.

Biologists recognize two general types of food chains: grazer and decomposer. **Grazer food chains** begin with plants and algae, the producers. These organisms are consumed by herbivores, or **grazers**; hence the name *grazer food chain*. Herbivores, in turn, may be eaten by carnivores. **Decomposer food chains** begin with

---

[3]Remember that detritus is waste material.

FIGURE 23-8 Simplified Food Chains
Two grazer food chains are shown, a terrestrial one and an aquatic one.

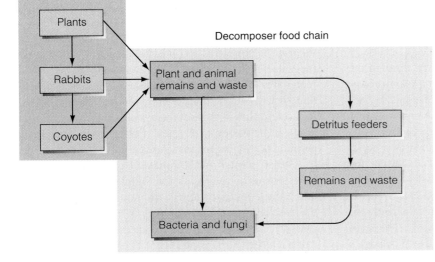

Grazer food chain

Decomposer food chain

FIGURE 23-9 Food Chains A grazer food chain and a decomposer food chain, showing the connection between the two.

During cellular respiration, much of the energy stored in organic food molecules is lost as heat. Heat escaping from plants and animals is radiated into the atmosphere and then into outer space. It cannot be recaptured and reused by the Earth's organisms. Because all solar energy is eventually converted to heat, energy is said to flow unidirectionally through food chains and food webs. Put another way, energy cannot be recycled.

In sharp contrast, nutrients flow cyclically; that is, they are recycled over and over. Nutrients in the soil, air, and water are first incorporated into plants and algae, then passed from plants to animals in various food chains. Nutrients in the food chain eventually reenter the environment by one of three paths: (1) the excretion of wastes, (2) the decomposition of dead organisms, and (3) the decomposition of waste.

Every time you exhale, for example, you release carbon dioxide. It reenters the atmosphere for reuse. Thus, through the act of breathing,

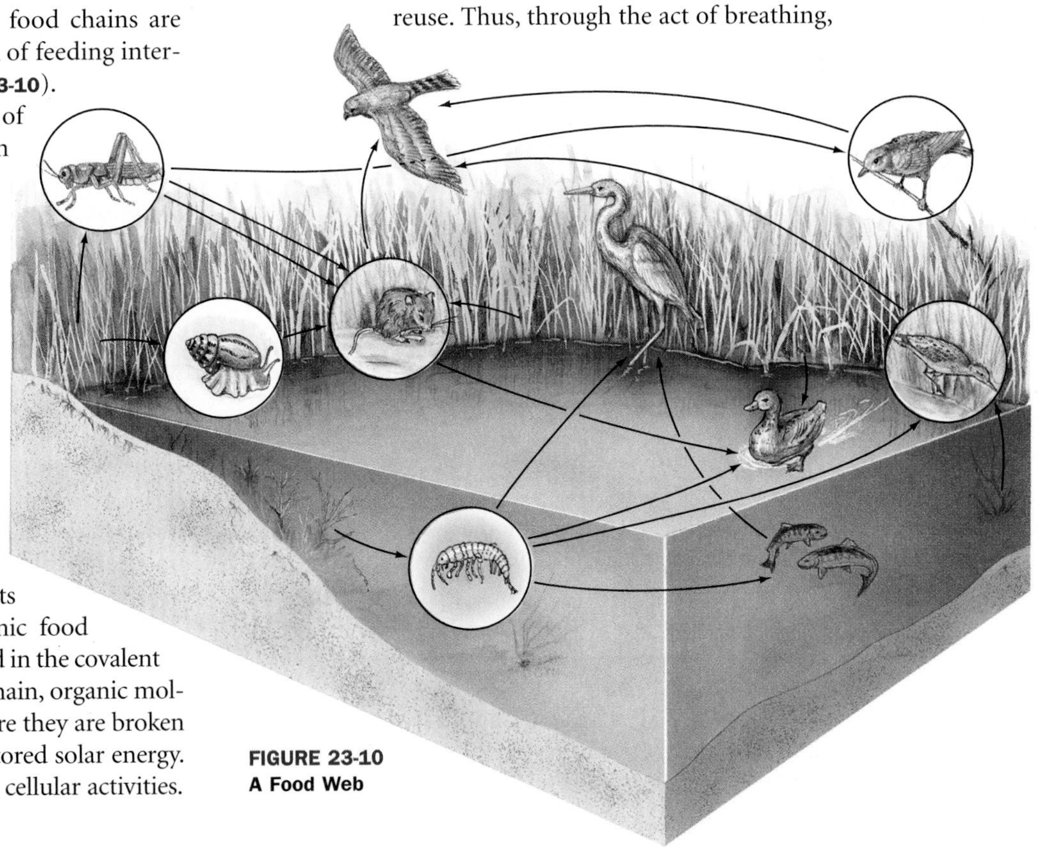

dead material—either animal wastes (feces) or the remains of plants and animals (**FIGURE 23-9**).

As in most things in nature, decomposer and grazer food chains are tightly linked and highly interdependent (**FIGURE 23-9**). For example, waste products from the grazer food chain enter the decomposer food chain. Nutrients liberated by the decomposer food chain enter the soil and water and are reincorporated into plants at the base of the grazer food chain. The survival of all living things depends on this linkage.

Food chains exist only on the pages of textbooks; in a community of living organisms, all food chains are part of a much more complex network of feeding interactions known as **food webs** (**FIGURE 23-10**). Food webs present a complete picture of the feeding relationships in any given ecosystem.

Energy and Nutrients Both Flow Through Food Webs, But in Very Different Ways

Although we tend to think of food chains simply as a display of feeding relationships, they are much more. From an ecological perspective, food chains are conduits for the flow of energy and the cycling of nutrients through the environment. Let's begin with energy.

Solar energy is captured by plants and algae and used to produce organic food molecules. Energy from the Sun is stored in the covalent bonds of these molecules. In the food chain, organic molecules pass from plants to animals, where they are broken down in mitochondria, which release stored solar energy. This energy is used to power numerous cellular activities.

FIGURE 23-10
A Food Web

you play an important role in the global recycling system that makes life possible.

Nutrients also reenter the environment through the decomposition of dead organisms. When a plant or animal dies, it decomposes. Decomposition is caused by bacteria and fungi. These organisms release enzymes that break down organic matter. Although these microorganisms absorb many nutrients released during this process, some nutrients escape, entering the soil and water for reuse. And, of course, when a bacterium dies, it too breaks down, releasing nutrients into its environment.

Finally, many nutrients reenter the environment through the decomposition of animal waste. The feces of a rhinoceros, for example, are broken down by bacteria, which liberate carbon dioxide, nitrogen, and numerous minerals.

One way or another, all nutrients eventually make their way back to the environment for recycling. Each new generation of organisms therefore relies on the recycling of material in the biosphere. The atoms in your body, for instance, have been recycled many times since the beginning of life on Earth. Who knows, some of those atoms may have been in the very first cell that lived in the shallow seas.

The Organisms of a Food Chain Exist on Different Trophic Levels

Ecologists classify the organisms in a food chain according to their position, or **trophic level** (TROE-fic; literally, "feeding" level). The producers comprise the base of the grazer food chain and are therefore members of the first trophic level. The grazers are members of the second trophic level. Carnivores that feed on grazers are members of the third trophic level, and so on.

Most terrestrial food chains are limited to three or four trophic levels. In fact, longer terrestrial food chains are quite rare. The reason for this is that food chains generally do not have a large enough producer base to support many levels of consumers. Put another way, in the longer food chains, less food is available for the top-level consumers. Why?

Plants absorb only a small portion of the sunlight that strikes the Earth (only 1%–2%), which they use to produce organic matter, or biomass. Technically, **biomass** is the dry weight of living material in an ecosystem. The biomass at the first trophic level is the raw material for the second trophic level. The biomass at the second trophic level is the raw material for the third trophic level, and so on.

Not all of the biomass produced by plants is converted to grazer biomass. At least three reasons account for the incomplete transfer of biomass from one trophic level to the next. First, some of the plant material such as the roots is not eaten. Second, not all of the material that the grazers eat is digested. Third, some of the digested material is broken down to produce energy and heat and therefore cannot be used to build biomass in the grazers. As a rule, only 5%–20% of the biomass at any one trophic level can be passed to the next. (The amount varies depending on the organisms involved in the food chain.)

When plotted, the biomass at the various trophic levels forms a pyramid, the **biomass pyramid** (**FIGURE 23-11**). Because biomass contains energy (stored in the covalent bonds), the biomass pyramid can be converted into a graph of the chemical energy in the various trophic levels. This graph is called an **energy pyramid**.

In most food chains, the number of organisms also decreases with each trophic level, forming a **pyramid of numbers**. Knowledge of ecological pyramids will help you understand why people in many less developed countries generally subsist on a diet of grains (corn, rice, or wheat) rather than meat. **FIGURE 23-12** shows energy pyramids for two food chains. In the grain → human food chain on the right, 20,000 kilocalories of grain can feed 10 people for a day. In the grain → steer → human food chain, the 20,000 kilocalories fed to the cow produces only 2000 kilocalories of meat, barely enough to feed one person for a day (assuming a 10% transfer of biomass). Thus, the shorter the food chain, the more food is available to top-level consumers.

This simple rule has profound implications for the human race. The human population increases by about 80–90 million people a year, and feeding these people poses an enormous challenge. How can new residents be fed most efficiently?

The most efficient food source will be crops such as corn, rice, and wheat that are fed directly to people. It is obviously far less efficient to feed corn and other grains to cattle and other livestock, which are then slaughtered for human consumption. Vegetarianism, say supporters, is not only good for your health but better for the environment, because it requires less grain production than a nonvegetarian diet. People, however, require an adequate intake of protein. Protein can be supplied by fish, meat, and a mixture of legumes. (Chapter 24 has more on feeding the world's hungry.)

Nutrient Cycles Consist of Two Phases, the Environmental and the Organismic

As noted earlier, nutrients flow from the environment through food webs, then are released back into the environment. This circular flow constitutes a **nutrient cycle**, also known as a **biogeochemical cycle**.

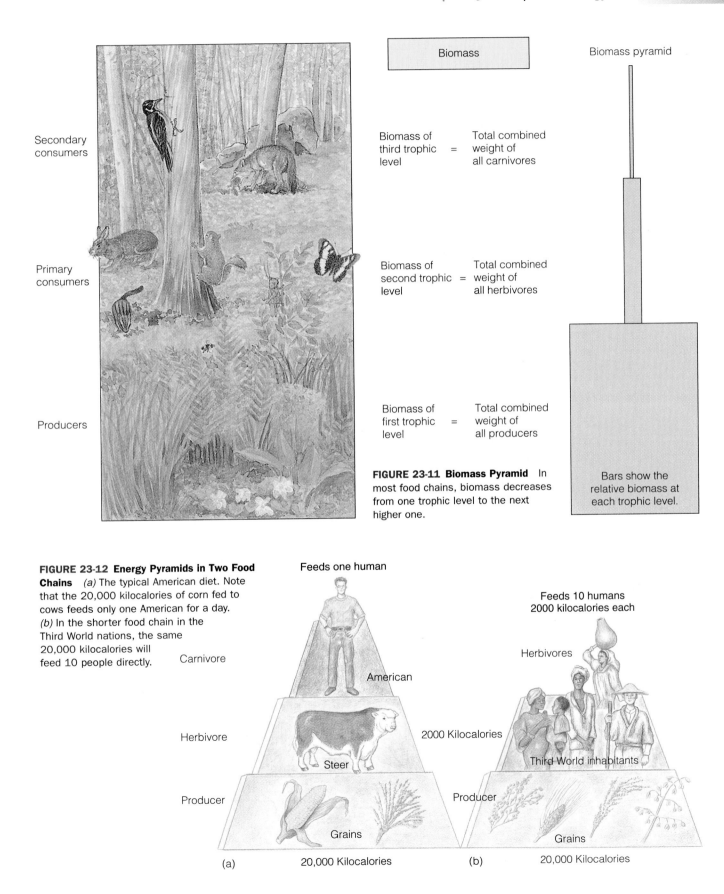

Secondary consumers

Primary consumers

Producers

Biomass

| Biomass of third trophic level | = | Total combined weight of all carnivores |

| Biomass of second trophic level | = | Total combined weight of all herbivores |

| Biomass of first trophic level | = | Total combined weight of all producers |

FIGURE 23-11 Biomass Pyramid In most food chains, biomass decreases from one trophic level to the next higher one.

Biomass pyramid

Bars show the relative biomass at each trophic level.

FIGURE 23-12 Energy Pyramids in Two Food Chains *(a)* The typical American diet. Note that the 20,000 kilocalories of corn fed to cows feeds only one American for a day. *(b)* In the shorter food chain in the Third World nations, the same 20,000 kilocalories will feed 10 people directly.

Feeds one human

Feeds 10 humans 2000 kilocalories each

Carnivore

Herbivore

Producer

American

Steer

Grains

Herbivores

2000 Kilocalories

Third World inhabitants

Producer

Grains

(a)   20,000 Kilocalories

(b)   20,000 Kilocalories

Nutrient cycles can be divided broadly into two phases: the environmental and organismic. In the **environmental phase**, a nutrient exists in the air, water, or soil or sometimes in two or more of them simultaneously. In the **organismic phase**, nutrients are found in the biota—the plants, animals, and microorganisms.

Dozens of global nutrient cycles operate continuously to ensure the availability of chemicals vital to all living things, present and future. Unfortunately, however, a great many human activities disrupt nutrient cycles. These activities can profoundly influence the survival of a species. This section looks at two of the most important nutrient cycles, the carbon and nitrogen cycles, and the ways they are being altered. (Solutions to the problems are described in Chapter 24.) Remember that the cycles mentioned here are only two of several dozen nutrient cycles essential to life.

The Carbon Cycle.

The carbon cycle is illustrated in **FIGURE 23-13** in a simplified form. We begin with free carbon dioxide. In the environmental phase of the cycle, carbon dioxide resides in two reservoirs, which scientists often call *sinks*: the atmosphere and the surface waters (oceans, lakes, and rivers). **FIGURE 23-13** shows that atmospheric carbon dioxide is absorbed by plants and photosynthetic organisms and thus enters the organismic phase of the cycle. These organisms (autotrophs) convert carbon dioxide into organic food materials, which travel along the food chain from one trophic level to the next. Carbon dioxide reenters the environmental phase via cellular energy production (cellular respiration) of the organisms in the grazer and decomposer food chains.

For tens of thousands of years, most of our ancestors lived in harmony with nature. With the advent of the Industrial Revolution, however, human beings began to interfere with natural processes on a large scale. One of the victims of our technological development has been the global carbon cycle. The widespread combustion of fossil fuels (which releases carbon dioxide) and rampant deforestation (which reduces carbon dioxide uptake) have seriously overwhelmed the cycle.

For many years before the Industrial Revolution, global carbon dioxide production equaled carbon dioxide absorption by plants and algae. Today, approximately 6 billion tons of carbon—in the form of carbon dioxide—is added to the atmosphere each year. Three quarters of the carbon dioxide comes from the combustion of fossil fuels like the gasoline in our cars; the remaining quarter stems from deforestation. How does deforestation add carbon dioxide to the atmosphere? Trees absorb enormous amounts of carbon dioxide for photosynthesis. As forests are cleared, the amount of atmospheric carbon dioxide they absorb declines. Making matters worse, many forests are burned after cutting, further adding to the carbon dioxide levels in the atmosphere.

In the past 100 years, global atmospheric carbon dioxide levels have increased 25%. In the atmosphere, carbon dioxide traps heat escaping from Earth and reradiates it to the Earth's surface. As carbon dioxide levels increase, global temperature may rise dramatically. Such a rise could have devastating effects on global climate. It could shift rainfall patterns, destroy agricultural production in many regions, and wipe out thousands of species. A rising global temperature might cause glaciers and the polar ice caps to melt, raising the sea level and flooding many

FIGURE 23-13 The Carbon Cycle

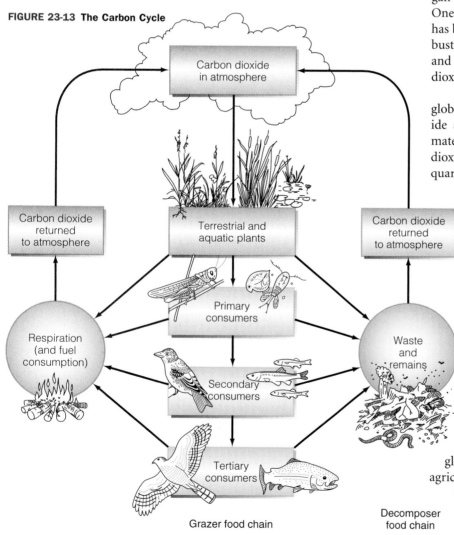

Carbon dioxide in atmosphere

Carbon dioxide returned to atmosphere

Carbon dioxide returned to atmosphere

Terrestrial and aquatic plants

Respiration (and fuel consumption)

Primary consumers

Waste and remains

Secondary consumers

Tertiary consumers

Grazer food chain

Decomposer food chain

low-lying coastal regions. Fortunately, there are many cost-effective strategies for greatly reducing our dependency on fossil fuel. The most notable are energy efficiency and renewable fuels—solar energy and wind, for example. (Global warming is detailed in Chapter 24.)

The Nitrogen Cycle.

Nitrogen is an element essential to many important biological molecules, including amino acids, DNA, and RNA. The Earth's atmosphere contains enormous amounts of it, but atmospheric nitrogen exists as nitrogen gas (N_2), which is unusable to all but a few organisms. To be incorporated into living organisms, atmospheric nitrogen must first be converted to a usable form—either nitrate or ammonia.

The conversion of nitrogen to ammonia is known as **nitrogen fixation**. As **FIGURE 23-14** shows, nitrogen fixation partly occurs in the roots of leguminous plants (peas, beans, clover, alfalfa, vetch, and others) in small swellings called **root nodules**. Inside the nodules live symbiotic bacteria that convert atmospheric nitrogen to ammonia. Ammonia is also produced by cyanobacteria that live in the soil. Once ammonia is produced, other soil bacteria convert it to nitrite and then to nitrate. Nitrates are incorporated by plants and used to make amino acids and nucleic acids. All consumers therefore ultimately receive the nitrogen they require from plants.

Nitrate in soil is also produced indirectly from the decay of animal waste and the remains of plants and animals. As shown on the right side of **FIGURE 23-14**, this process returns ammonia to the soil for reuse. Ammonia is converted to nitrite, then to nitrate and reused. Some nitrate, however, may be converted to nitrite and then to nitrous oxide (N_2O) by denitrifying bacteria (**FIGURE 23-14**). Nitrous oxide is converted to nitrogen and released into the atmosphere.

Humans alter the nitrogen cycle in at least four ways: (1) by applying excess nitrogen-containing fertilizer on farmland, much of which ends up in waterways; (2) by disposing of nitrogen-rich municipal sewage in waterways; (3) by raising cattle in feedlots adjacent to waterways; and (4) by burning fossil fuels, which release a class of chemicals known as nitrogen oxides into the atmosphere. The first three activities increase the concentration of nitrogen in the soil or water, upsetting the ecological balance. Nitrogen oxides released into the atmosphere by power plants, automobiles, and other sources are converted to nitric acid, which falls with rain or snow. Be-

sides changing the pH of soil and aquatic ecosystems, nitric acid also adds nitrogen to surface waters and may be responsible for 25% of the nitrogen pollution in some coastal waters in the United States.

Nitrogen, like phosphate, is a plant nutrient. It stimulates the growth of aquatic plants and causes rivers and lakes to become congested with dense mats of vegetation, making them unnavigable. Sunlight penetration to deeper levels is also impaired by the growth of plants, causing oxygen levels in deeper waters to decline. In the autumn, when aquatic plants die and decay, oxygen levels can fall further, killing aquatic life.

Section 23-3

Ecosystem Homeostasis

In this book, I have repeatedly pointed out that human health requires internal balance. A breakdown of internal homeostasis can lead to

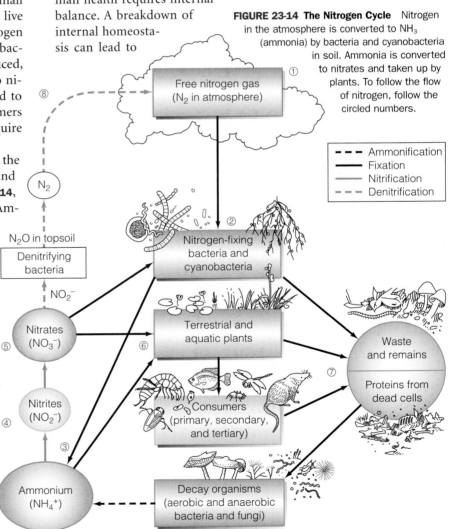

FIGURE 23-14 The Nitrogen Cycle Nitrogen in the atmosphere is converted to NH_3 (ammonia) by bacteria and cyanobacteria in soil. Ammonia is converted to nitrates and taken up by plants. To follow the flow of nitrogen, follow the circled numbers.

- - - - Ammonification
———— Fixation
———— Nitrification
- - - - Denitrification

disease and even death. Human health is also dependent on environmental homeostasis to maintain conditions conducive to life. I have used the word *environment* to include the physical, chemical, social, and psychological environments we are part of. As this chapter shows, the physical and chemical environments depend on the operation of homeostatic mechanisms like the carbon and nitrogen cycles operating in the biosphere to maintain proper conditions.

Homeostatic mechanisms operate to maintain ecosystem balance. **Ecosystem balance** is not a steady state, but a kind of dynamic equilibrium. In balanced ecosystems, for example, populations grow and decline in natural cycles (they are dynamic). But from year to year, they remain more or less the same (they are in a state of equilibrium).

Ecosystem Homeostasis or Balance Is the Result of Many Interacting Factors

Ecosystem homeostasis or balance results in large part from forces that act on individual populations within an ecological system (**FIGURE 23-15**). The first set of forces consists of those that cause populations to grow. I refer to them as **growth factors**. Favorable weather, ample food supplies, and a high reproductive rate are three of many factors that cause populations to increase. The second set of factors includes those that cause populations to decline, the **reduction factors**. Adverse weather, lack of food, disease, and predation, for instance, can cause populations to decline. Reduction factors collectively constitute **environmental resistance**—so named because they resist population growth.

Numerous growth and reduction factors operate simultaneously in ecosystems. As **FIGURE 23-15** shows, each set of factors has two components: abiotic and biotic. To understand how they work, consider a simplified example.

In the midwestern United States, grasses and other plants grow well during wet years. Because of the abundance of food, mice and other rodents that live there thrive and their populations increase. This increase, in turn, results in larger populations of hawks and owls, which feed on rodents. The more food that is available, the more owl and hawk young that will survive.

In this simplified chain of events, favorable weather conditions shift ecosystem balance. Conditions are partly restored, however, by the increase in the hawk and owl populations. Increased predatory pressure from these birds tends to decrease the mouse population. As the mouse population falls, the number of predators decreases.

Balance can also be restored by abiotic conditions. A harsh winter, for example, might cause the mouse population to plummet. Because the number of mice decreases, the owl and hawk populations fall.

This example illustrates some mechanisms by which slight ecosystem imbalances are corrected. Owls and hawks are biotic reduction factors that help maintain ecosystem stability. In contrast, adverse winter weather is an abiotic reduction factor that can offset the

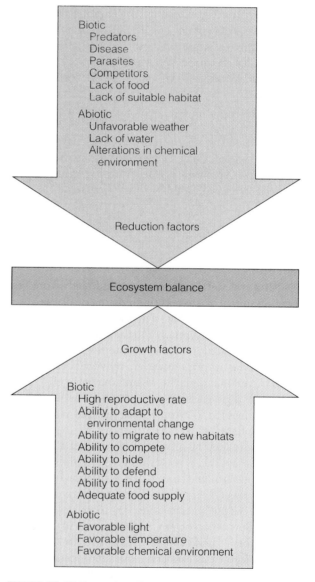

FIGURE 23-15 Ecosystem Balance Ecosystem balance is a complex equilibrium brought about by biotic and abiotic growth and reduction factors. From *Environmental Science*, 4th edition, by Daniel D. Chiras. Copyright © 1994 by The Benjamin/Cummings Publishing Company. Reprinted by permission.

(a)

FIGURE 23-16 Aftermath of an Eruption *(a)* This lifeless landscape was created by an enormous volcanic eruption at Washington's Mount Saint Helens. Thousands of acres were destroyed. *(b)* Life soon began to take root in the volcanic ash, but complete recovery may take decades.

(b)

rise in the rodent population and even reduce the size of the hawk and owl populations.

Changes in abiotic and biotic conditions occur with great regularity in ecosystems. Minor fluctuations are of little consequence, however, as the simplified example above shows. As a rule, when minor changes occur, ecosystems appear to remain fairly resilient—capable of bouncing back. Unfortunately, in most instances, human intrusions are so severe that natural systems cannot offset them.

Succession Is the Progressive Development of Biological Communities, Either on Barren Ground or in Damaged Ecosystems

The ability of ecosystems to recover from minor changes gives them a measure of stability. The same is true in the human body. But not all changes in ecosystems (or our bodies) can be quickly reversed. Thus, ecosystem balance can be drastically upset by natural forces such as volcanoes, or by human forces (**FIGURE 23-16**). In such instances, ecosystems may recover, but the recovery is quite slow. A deciduous forest that is cleared, then planted with crops, for example, will return to forest on its own if it is abandoned. However, a full recovery may require 70 years or more. This process of restoration, described more fully shortly, is an example of secondary succession and is a means by which natural systems are restored following severe disruption. In a sense, secondary succession is nature's healing process.

Succession, in general, is a process of sequential change in which one community is replaced by another until a mature, or climax, ecosystem is formed. A mature ecosystem is one that has reached a state of long-term dynamic balance. It is characterized by a stable level of species diversity (explained below). Two types of succession exist: primary and secondary.

Primary succession occurs where no biotic community previously existed—for example, when deep-sea volcanoes form islands. On volcanic islands in the tropics, a rich paradise can form on the barren rock, but it will take tens of thousands of years. Seeds for plants may be carried to the islands by waves or may be dropped by birds. Over time, the plants take root, then spread to cover the entire island. Birds may settle on the island as well. In the ensuing years, new species may evolve from the old. The process of primary succession on rock exposed by the retreat of glaciers is shown in **FIGURE 23-17**. Take a few minutes to study this figure.

Secondary succession occurs when a biotic community that had previously existed is destroyed by natural forces or human actions. **FIGURE 23-18** (page 533) shows secondary succession in an abandoned eastern farm field. As illustrated, crabgrass first invades the abandoned field. Crabgrass is well suited to life in open, sunny fields. Unlike many other plants, it can tolerate hot, sunny locations because it is capable of fixing atmospheric carbon dioxide when other plants have closed down their pores to conserve water.

In the next 2 years, grasses and other herbaceous plants take root from seeds in the soil. Seeds may also be blown in from neighboring fields or carried in by animals. Over the next three decades, pine seedlings take root and begin to grow. As they get larger, sun-loving pine trees produce shade—so much, in fact, that shade-tolerant hardwood trees begin to take root. Hardwoods eventually grow so large that they "shade out" many of the pine trees. In ecological terms, they outcompete the pines. Seventy years after this process begins, a mature, or climax, forest is formed, consisting mostly of hardwoods and an occasional pine. This is a climax forest, which exists in a state of dynamic equilibrium. It will remain more or less the same if it is not disturbed by fire, timber cutting, or some catastrophic event.

Secondary succession in the abandoned farm field illustrates some key principles of succession. One of those is that colonizers such as crabgrass and transitional species such as pines thrive in their new habitat because environmental resistance is low and the conditions necessary for growth (growth factors) are favorable. These same species, however, gradually alter the environment, creating conditions suitable for the growth of other species. The new species, in turn, alter environmental conditions so much that they eliminate the species that made their growth possible.

Secondary succession occurs much more rapidly than primary succession because soil is already present. In primary succession, soil must be formed from rock, gravel, or sand, and this can take 100–1000 years.

Just as damage to the human body cannot always be repaired, damage to ecosystems, if severe enough, also may not be remedied by secondary succession. In Vietnam, for example, thousands of acres of tropical forests were destroyed by allied forces who sprayed Agent Orange, a herbicide, on them from planes and helicopters in an effort to reduce the likelihood of ambush. Broadleaf trees are quite susceptible to Agent Orange and died quickly. Today the forests, however, have been invaded by a hardy brush that some scientists think may prevent mature forests from reestablishing.

▮ Ecosystem Stability May Be Related to Species Diversity

Ecosystem stability results from the interplay of growth and reduction factors, which are part of nature's homeostatic mechanisms. Some ecologists believe that ecosystem stability may also result from species diversity. Generally speaking, **species diversity** is a measure of the number of different species in a given community or ecosystem. The more species there are, the more diverse an ecosystem is said to be.

Ecologists who believe that species diversity leads to ecosystem stability support this contention with the following facts. First, extremely stable ecosystems such as tropical rain forests are characterized by a remarkably high species diversity. By various estimates, tropical rain forests cover about 10% of the Earth's land surface, but are home to about 60% of the world's species. In contrast, some ecosystems such as the tundra contain populations that tend to oscillate widely. Ecologists sometimes attribute this instability to their low species diversity. A second fact that supports the notion that species diversity is related to ecosystem stability is that intentionally simplifying an ecosystem (removing species) tends to make it unstable. Farm fields are a good example. Large fields containing one crop are known as **monocultures.** Monocultures are much more vulnerable to pests and disease than **heterocultures**—that is, fields containing several different crops. And both monoculture and heteroculture crops are more vulnerable than nature's heteroculture

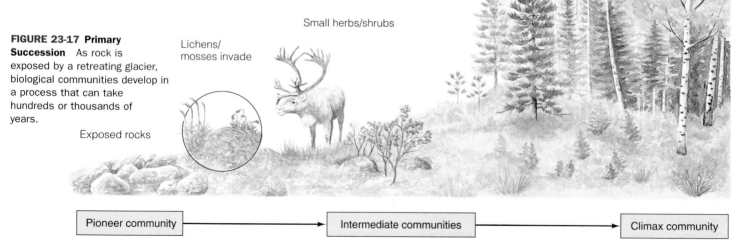

FIGURE 23-17 Primary Succession As rock is exposed by a retreating glacier, biological communities develop in a process that can take hundreds or thousands of years.

Exposed rocks

Lichens/ mosses invade

Small herbs/shrubs

Heath mat

Jack pine black spruce and aspen

Balsam fir, paper birch and white spruce

Pioneer community → Intermediate communities → Climax community

FIGURE 23-18 Secondary Succession   On an abandoned eastern farm (top left), nature begins the process of restoring a biotic community.

grasslands, which contain many more species of plants and animals.

To understand why scientists think diversity leads to stability, consider the simple models shown in **FIGURE 23-19**. As illustrated, in the simplified ecosystem, there are few species and few connections. Thus, the loss of one species—for example, a producer—in a simplified ecosystem would be far more noticeable than the loss of a producer in the much more diverse one where several producers are present.

Not all ecologists agree with the preceding ideas. Some say that diverse ecosystems such as tropical rain forests are stable because of their uniform climate, not because of their species diversity. Instability in the tundra could be a result of the volatile climate, not of the low level of diversity.

Despite differences of opinion, some generalizations can be drawn from the evidence. The most important is that intentionally reducing species diversity makes ecosystems unstable. That is important advice for farmers and foresters, who tend to convert diverse ecosystems into much simpler ones.

Scientists debate the concept of ecosystem stability with vigor these days because some recent research has cast doubt on the concept of a naturally maintained equilibrium. Some researchers believe that populations and ecosystems rarely, if ever, return to equilibrium

FIGURE 23-19 Diversity and Stability   *(a)* In a simplified ecosystem, the loss of one species generally has a more profound impact on the food web than the loss of one species in *(b)* a more complex ecosystem.

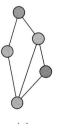

(a)

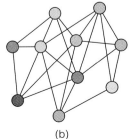

(b)

once disturbed. The debaters agree on one thing though: A return to normal equilibrium depends in large part on the nature and severity of the disturbance. Evidence suggests that small perturbations like changes in rainfall or short-term drought are resisted or adequately compensated for. More severe alterations such as deforestation of the tropics may render a system unable to recover.

Unfortunately, the topic is so complex and so easily misconstrued that the debate is more confusing than edifying. What can be said at this time is that our understanding of ecosystem homeostasis is a long way from complete. As the critical thinking rules outlined in Chapter 1 suggest, sometimes a little uncertainty is necessary.

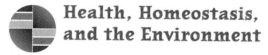

Health, Homeostasis, and the Environment

How We Alter the Environment

Over and over again in this section, I've shown that human health is dependent on a healthy environment. Environmental disturbances such as pollution affect homeostasis, which can then lead to disease. Creating an enduring human presence, therefore, depends on our achieving a better balance—a healthy balance. In order to understand how we can achieve such a balance, the topic of the next chapter, we must first look at the ways we impact our environment.

Humans impact the environment in two principal ways: (1) by altering the abiotic components and (2) by altering the biotic components.

Consider some examples of human tampering with abiotic factors—chemical and physical components. Chemical pollutants from factories, power plants, lawns, and even gardens enter streams and lakes. Here they alter the chemical environment, often killing fish and other aquatic organisms. Hot water from power plants and factories is yet another pollutant released in large quantities into lakes and streams where it alters the physical environment of these bodies of water, often killing aquatic life. Changing the physical and chemical environment has an impact on plants, nonhuman animals, and microorganisms, but it can also affect humans.

Human populations also directly impact the biotic components of the environment. Overfishing, for example, eliminates species of valuable aquatic food

FIGURE 23-20 Foreign Competitor Introduced to the United States Kudzu grows wild in the South, covering fields and homes.

chains. In the Pacific Northwest, for example, overfishing of salmon reduces food available for seals and other sea mammals. Another important impact occurs when foreign species are introduced into an ecosystem. In their new home, these species often outcompete native species of plants and animals. An example is shown in **FIGURE 23-20**. The plant is known as *kudzu*. Intentionally introduced in the South to control soil erosion, kudzu literally takes over fields, growing over trees and covering abandoned homes.

It is important to point out that all organisms have an impact on their environment. That is to say, all living things alter the abiotic and biotic components of the ecosystem in which they live. But the impact of human populations is different from that of other organisms, for several reasons. First, our advanced technological development has given us incredible power to alter the face of the Earth. Second, the human population has reached unprecedented numbers. Today, around 6 billion people live on the planet. The cumulative impact of all these individuals is enormous and may be endangering the biosphere. Chapter 24 examines the evidence behind this assertion and describes ways to reduce human impact and build a sustainable future. The key to success, many experts believe, is restoring global ecosystem balances and building a relationship with nature that does not endanger the homeostatic mechanisms that maintain the planet's health and the health and well-being of the millions of species that make the Earth their home.

PRINCIPLES OF ECOLOGY: ECOSYSTEM STRUCTURE

1. Most of us see our lives as apart from nature. But in truth we depend heavily on our environment for food, oxygen, clothing, materials, and much else.

2. *Ecology* is the study of ecosystems. It examines the relationship of organisms to their environment and the many interactions between the abiotic and biotic components of ecosystems.

3. The living "skin" of the planet is called the *biosphere.* It extends from the bottom of the oceans to the top of the highest mountains.

4. The biosphere is a *closed system* in which materials are recycled over and over. The only outside contribution is sunlight, which powers virtually all biological processes.

5. The Earth's surface is divided into large biological regions, or *biomes*, each with a characteristic climate and characteristic plant and animal life.

6. The oceans can also be divided into biological regions, known as *aquatic life zones*.

7. An *ecosystem* consists of a community of organisms, its environment, and all of the interactions between them.

8. Organisms are adapted to a range of conditions in the ecosystem in which they live. This is called their *range of tolerance.*

9. Human activities can alter the abiotic and biotic conditions in an ecosystem, causing considerable harm.

10. In any ecosystem, one factor tends to limit growth and is therefore called a *limiting factor.*

11. A group of organisms of the same species living in a specific region constitutes a *population.* Two or more populations occupying that region form a *community.*

12. The physical space a species occupies is called its *habitat.* A species' *niche* includes its habitat, its position in the food chain, its range of tolerance, and so on.

ECOSYSTEM FUNCTION

13. Virtually all life on Earth depends on the *producers*, organisms that synthesize organic materials from sunlight, carbon dioxide, and water. The major producers are the plants, photosynthetic protists (single-celled organisms like algae), and photosynthetic bacteria.

14. Organisms dependent on producers and other organisms for food are called *consumers.* Four types of consumers are present: *herbivores, carnivores, omnivores,* and *detritivores.*

15. All organisms are part of food chains. A *food chain* represents a feeding relationship in an ecosystem. Food chains that begin with plants consumed by grazers (herbivores) are known as *grazer food chains.* Those that begin with animal waste or the remains of plants, animals, and microorganisms are known as *decomposer food chains.*

16. Food chains are simplified components of larger networks, called *food webs*, that represent a truer picture of the feeding relationships in an ecosystem.

17. Food chains and food webs are a conduit for the one-way flow of energy through an ecosystem; they provide avenues for the cycling of minerals and other nutrients through the ecosystem.

18. The position of an organism in a food chain is called its *trophic level.* In a grazer food chain, plants (producers) are on the first trophic level; herbivores are on the second; carnivores are on the third.

19. Nutrients necessary for life are involved in *nutrient cycles.* Nutrients in the environment enter the organismic phase through producers. The nutrients are then shunted through the food chain (organismic phase of the cycle) and eventually reenter the environment (environmental phase), where they are generally available for reuse.

20. Humans can interrupt nutrient cycles, locally and globally. For example, the global *carbon cycle* is flooded with carbon dioxide from the worldwide consumption of fossil fuels. Deforestation also increases atmospheric carbon dioxide.

21. The *nitrogen cycle* is flooded with nitrogen from human sources (fertilizer, sewage, and animal waste) in towns, cities, and rural communities.

ECOSYSTEM HOMEOSTASIS

22. Human health is dependent on homeostasis at two levels: within the environment and within ourselves. Healthy physical, chemical, social, and psychological environments are crucial to a healthy body.

23. The environment contains numerous homeostatic mechanisms. *Ecosystem balance* is the net result of the interplay of biotic and abiotic *growth* and *reduction factors.*

24. *Environmental resistance* is the sum of the reduction factors.

25. Ecosystems can recover from minor disturbances rather easily, but more significant changes may require more time. In some cases, damage may be so severe that recovery is impossible.

26. The term *succession* refers to a series of changes in an ecosystem in which one community replaces another until a mature, or climax, ecosystem is produced. Primary succession occurs where no community previously existed. Secondary succession occurs where a community was destroyed by natural or human events.

27. Ecosystem stability may also result from high species diversity or from favorable abiotic factors such as climate.

HEALTH, HOMEOSTASIS, AND THE ENVIRONMENT: HOW WE ALTER THE ENVIRONMENT

28. Humans change the environment by altering the abiotic and the biotic components.

29. All organisms have an impact on their environment. The impact of human populations, however, is different from that of other organisms because of our technological prowess and because our population has reached unprecedented numbers. Even small acts are now beginning to add up and may be putting many other species and our own population at risk.

Critical Thinking

THINKING CRITICALLY— ANALYSIS

This Analysis corresponds to the Thinking Critically scenario that was presented at the beginning of this chapter.

At this point in your education, it may be hard to discern fact from fiction in the debate over the environment. But you do know from reading this book that the environment is extraordinarily important to our well-being and our long-term future. The environment is not just another issue. It is a make-it-or-break-it issue. What we do to the environment we do to ourselves.

But can environmental goals be achieved without harming the economy? If you listen to the news or read the newspaper on a regular basis, you've been exposed to a common misconception—notably, that environment and economics embody fundamentally conflicting goals. What you do to protect the environment will automatically harm the economy. This myth has been repeated so many times, in fact, that it may take some time to dislodge.

Fortunately, numerous examples show that we can meet, even in many cases exceed, both environmental and economic goals with alternative measures. These measures, in fact, often make extraordinary economic sense. For instance, by using energy more efficiently, we can greatly reduce air pollution and other environmental impacts. Energy efficiency measures generally cost a lot less than new sources of energy. Cheaper energy as a rule results in a stronger economy. Numerous other examples also exist, each showing that environmental protection, rather than being an impediment to business, can represent a net gain. The missing element is often creativity and a willingness to examine alternative strategies.

EXERCISING YOUR CRITICAL THINKING SKILLS

1. The following statements express the view of some citizens in this country: Technology can solve all problems. We don't need to worry about global warming, overpopulation, and other environmental problems because scientists will find technical answers to address them.

   This view is often labeled as pure technological optimism. Do you agree or disagree with the viewpoint expressed above? Why or why not? Give specific examples that support your case. Is bias entering into your opinion, and are your views based on solid facts?

2. You work for your state's Department of Natural Resources and are asked to comment on the proposed introduction of a deer from a remote part of the Soviet Union. The rationale behind the introduction is that it would enhance hunting in the state, and this would improve the economy of several rural areas. How would you evaluate the proposal? On what ecological criteria would you rely as you determine whether it is a good idea?

TEST OF CONCEPTS

1. Humans are a part of nature. Do you agree or disagree with this statement? Support your answer.
2. Define the term *ecology*, and give examples of its proper and improper use.
3. The Earth is a closed system. What are the implications of this statement?
4. Define the following terms: biosphere, biome, aquatic life zone, and ecosystem.
5. Define the term *range of tolerance*. Using your knowledge of ecology, give some examples of ways in which humans alter the abiotic and biotic conditions of certain organ-

isms, and describe the potential consequences of such actions.
6. Describe ways in which humans alter their own range of tolerance for various environmental factors.
7. What is a limiting factor? Give some examples.
8. Define the following terms: habitat, niche, producer, consumer, trophic level, food chain, and food web.
9. Explain why the biomass at one trophic level is lower than the biomass at the next lower trophic level.
10. Outline the flow of carbon dioxide through the carbon cycle, and de-

scribe ways in which humans adversely influence the carbon cycle.
11. Using what you have learned about ecology, describe why it is important to protect natural ecosystems and other species.
12. With the knowledge you have gained, explain why it is beneficial to set aside habitat to protect an endangered species.
13. Looking back over the principles you have learned in this chapter, write a set of guidelines for human society that would enable us to live sustainably on the Earth.

TOOLS FOR LEARNING

Tools for Learning is an on-line student review area located at this book's web site HumanBiology (www.jbpub.com/humanbiology). The review area provides a variety of activities designed to help you study for your class:

www.jbpub.com/humanbiology

Chapter Outlines. We've pulled out the section titles and full sentence sub-headings from each chapter to form natural descriptive outlines you can use to study the chapters' material point by point.

Review Questions. The review questions test your knowledge of the important concepts and applications in each chapter. Written by the author of the text, the review provides feedback for each correct or incorrect answer. This is an excellent test preparation tool.

Flash Cards. Studying human biology requires learning new terms. Virtual flash cards help you master the new vocabulary for each chapter.

Figure Labeling. You can practice identifying and labeling anatomical features on the same art content that appears in the text.

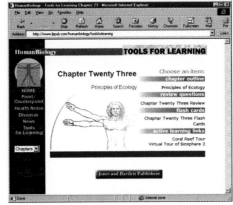

Active Learning Links. Active Learning Links connect to external web sites that provide an opportunity to learn basic concepts through demonstrations, animations, and hands-on activities.

CHAPTER 24

ENVIRONMENTAL ISSUES
POPULATION, POLLUTION, AND RESOURCES

Thinking Critically

A business magazine article notes that "on the issue of global warming the scientific community is divided." In support of this assertion, it quotes two prominent scientists. One says that he's "convinced the world is in a human-induced warming phase" and another argues that "there's simply not enough evidence to support such a conclusion." The article goes on to say that, because of the uncertainty among the scientific community, it makes no sense to launch a global effort to reduce carbon dioxide emissions, a position supported by the U.S. government under former presidents Ronald Reagan and George Bush and some prominent business interests, especially the oil and coal industry. Can you detect any problem in this reportage? What critical thinking rules were helpful to you in this examination? ▪

Floating barriers help restrain the monstrous oil spill of the Exxon *Valdez* in Alaska's Prince William Sound in 1989.

THE THREAT TO THE ENVIRONMENT IS COMmonly referred to as the "environmental crisis." To many people, the phrase "environmental crisis" seems like an exaggeration. But a close examination of trends in population growth, pollution, and resource use and depletion yields a disturbingly different view, and many experts who have studied the trends believe that the problems are severe enough to warrant the term "crisis."

This chapter looks at several major environmental issues. It focuses on three principal areas—population, pollution, and resource use—and discusses pressing environmental issues in each category. This chapter also discusses a variety of solutions for individuals, corporations, and governments and the ways in which human society can steer onto a sustainable course—that is, a path that ensures long-term ecological stability on the planet, a condition essential for a healthy human existence.

Overshooting the Earth's Carrying Capacity

Environmental problems of significant import occur in both the rich, industrialized nations and the poor, less-developed nations. Although the problems vary from one nation to another, they all are signs of a common root cause: human society is exceeding the Earth's carrying capacity. **Carrying capacity** is the number of organisms an ecosystem can support indefinitely—that is, the number it can sustain. Carrying capacity for all organisms, including humans, is determined by at least three factors: (1) food production, (2) resource supply, and (3) the environment's ability to assimilate pollution. Let's look at each one.

In Many Places, the Human Population Is Exceeding Food Production

As the global human population grows, many nations are finding it more and more difficult to meet rising demands for food. Starvation abounds in the developing nations, and millions of people perish each year from malnutrition and starvation or from diseases worsened by hunger, a sure sign of populations living beyond the local carrying capacity (**FIGURE 24-1**).

Many Resources Are in Short Supply and Will Be Depleted in the Near Future

Human populations require many resources such as fuel, fiber, and building materials. Those resources that are finite such as oil, natural gas, and minerals are known as **nonrenewable resources**. Resources that replenish themselves via natural biological and geological processes are called **renewable resources**. Wind, hydropower, trees, and fishes are examples.

Today, some important nonrenewable resources such as oil are on the decline, and some nonrenewable resources are just about used up. Renewable resources such as tropical forests are being depleted faster than they are being replanted. Such rapid depletion provides partial proof for the assertion that human populations are living beyond the Earth's carrying capacity. The rapid destruction of renewables also indicates that we're destroying the carrying capacity itself—that is, systematically eroding the life-support system of the planet.

Pollution from Human Activities Exceeds the Environment's Assimilative Capacity

The final determinant of carrying capacity is the capability of the environment to assimilate and degrade pollutants. In natural ecosystems, wastes are usually diluted to harmless levels and are recycled in nutrient cycles, so they can be reused. Animal feces, for instance, are broken down by soil bacteria. These bacteria release carbon dioxide, nitrogen, and other nutrients into the environment for reuse. In nature, nutrient cycles ensure a steady supply of resources for all generations. Unfortunately, many human activities interrupt these vital cycles. Too much animal waste in an aquatic environment, for example, may shift the concentration of nitrogen and phosphorus, starting a series of changes that results in rapid growth of algae and aquatic plants; algae and plants decay in the fall, consuming oxygen in the process and killing fish.

Human populations produce enormous amounts of wastes that

FIGURE 24-1 Food Shortages Overpopulation, drought, political turmoil, and mismanagement of farmland are the chief causes of hunger and starvation. Children die by the thousands every day.

FIGURE 24-2 Air Pollution Studies show that air pollutants in some locales can increase respiratory disease such as emphysema and lung cancer. Air pollution is aesthetically unappealing, and levels that do not directly affect human health can still cause considerable environmental damage.

overwhelm natural systems. The U.S. population represents only 6% of the world population but produces about 25% of the world's pollution. The environment can dilute some of this waste to harmless levels. It can break down other wastes, rendering them harmless, but in many locations, the environment's assimilative capacity is being severely overtaxed. In lakes, rivers, and oceans, pollution is taking a toll on fish and wildlife. Even though efforts have been made to reduce the amount of pollution our society produces, continued economic expansion and growing numbers of people often offset these gains.

The stress on the assimilative capacity of the atmosphere is evident around most cities with populations over 50,000. These pockets of air and water pollution pose a threat to human health and to a great many species that make this planet their home (**FIGURE 24-2**). Today, over 130 million Americans (nearly half of our population) live in cities whose air is deemed harmful to their health.

The environmental crisis is clearly multifaceted. In all instances, the problem is basically the same: We humans are exceeding the Earth's carrying capacity. One of the main reasons for this is overpopulation.

Section 24-2

Overpopulation: Problems and Solutions

Overpopulation can be explained in six words: too many people, reproducing too rapidly. The world population is nearly 6 billion and is increasing at a rate of 1.5% per year. Although the growth rate may seem small, it translates into 220,000 people being added to the world population *every day*. If the current rate continues, world population could reach 10 billion people

by the year 2030. To understand this problem requires a little background information.

Overpopulation Is a Problem in Virtually All Countries, Rich and Poor

The annual **growth rate** of the world population is calculated by the following formula:

$$\text{growth rate} = \text{birth rate} - \text{death rate}$$

In 1997, the world birth rate was 24 per 1000. In other words, 24 children were born each year for every 1000 people living on the Earth. The global death rate was 9 per 1000. To calculate the growth rate, you simply subtract the death rate from the birth rate. The global growth rate was therefore equal to 24/1000 − 9/1000, or 15 per 1000. This means that 15 new residents are added to the planet each year for every 1000 people alive. To convert the growth rate (15/1000) to a percentage, simply multiply by 100. Thus, the growth rate is equal to 15/1000 × 100, or 1.5%.

Growth rates are important measures of population dynamics, but they can be deceiving. For this reason, **demographers** (scientists who study populations) often convert growth rates into **doubling time**, the time it takes a population to double. Doubling time is calculated by the following equation:

$$\text{doubling time} = \frac{70}{\text{growth rate (\%)}} = \frac{70}{1.5} = 47 \text{ years}$$

The figure 70 in this equation is a demographic constant. The global growth rate is an average of all countries and thus masks regional growth differences. For example, the U.S. growth rate for the last 10 years has averaged about 0.9%. In other words, 9 new residents are added to the population each year for every 1000 people alive.[1] If this growth rate continues, our present population of 268 million people will double in 78 years. In contrast, in Africa the growth rate is 2.6%, yielding an alarming doubling time of 26 years!

Today, the most rapid growth is occurring on three continents: Africa, Asia, and Latin America (**TABLE 24-1**). In Europe, the population is actually shrinking.

Most people view overpopulation as a problem of the developing nations. Actually, overpopulation is a problem in all countries. It is as serious in the United

---

[1]Population growth in any country is the result not only of the difference between birth rate and death rate, but also of the net migration—that is, the number of people entering or leaving the country. In the United States, about 40% of our annual growth results from legal and illegal immigration.

TABLE 24-1

Growth Rate and Doubling Time

Region	Growth Rate (%)	Doubling Time (Years)
World	1.5	47
Developed countries	0.1	564
Less-developed countries	1.8	38
Africa	2.6	26
Asia	1.6	44
North America	0.6	117
Latin America	1.8	38
Europe	−0.1	—
Oceania	1.1	63

States as it is in Bangladesh. The reason for this apparent irony is the fairly high standard of living in the rich, industrialized countries. Citizens of these nations have an enormous impact on the environment. In fact, each baby born in the United States will use 20–40 times as many resources as a baby born to a family in India. This means the 268 million Americans alive in 1997 caused as much environmental damage as 5 to 10 billion people in the Third World!

The Human Population Is Growing Exponentially

The human population has not always been so large, nor has it always grown so rapidly. As **FIGURE 24-3A** shows, it was not until the last 200 years that global human population began to skyrocket, and then only because of better sanitation, improvements in medicine, and advances in technology. As you may have noticed, the graph of human population growth in **FIGURE 24-3A** is J-shaped. This is an **exponential curve**.

What is exponential growth? To begin, consider an example. Suppose your parents invested $1000 in a savings account at 10% interest the day you were born. Suppose also that the interest your money earned was applied to the balance, so that it too earned interest. If this were the case, your bank account would be said to be growing exponentially. **Exponential growth** occurs anytime a value, such as population size, grows by a fixed percentage if the interest (increase) is applied to the base amount.

Exponential growth is deceptive. In this example, the growth in your account seems small at first. At 10% interest, the $1000 will double in 7 years, yielding $2000 (**FIGURE 24-3B**). It will double again in 7 years, yielding $4000. The next doubling, occurring when you are 21 years old, will give you $8000. By age 42, the account will have grown to $64,000. By the time you were 49, your account would hold $128,000. At age 56, you'd have over $250,000. If you waited 7 more years, the account would grow to over $500,000. In 7 more years, at 70 years of age, you'd be a millionaire.

What is deceptive about exponential growth is that it begins slowly, then seems to go wild, even though the rate of growth is constant throughout the entire period. In this example, it took 49 years for your account to grow from $1000 to $128,000. In the next 21 years, however, your account grew by $900,000. Why?

Although your account doubled every 7 years, it was not until the base amount was very large that the doublings amounted to much. Thus, once the base amount

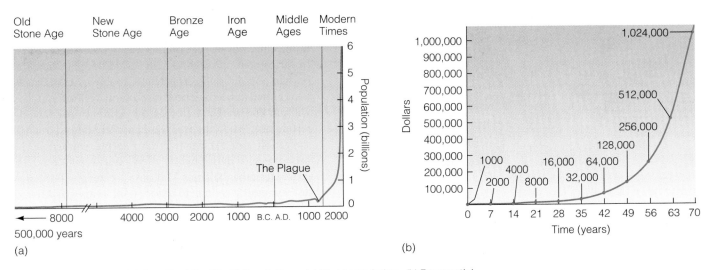

FIGURE 24-3 Exponential Growth of the World Population *(a)* World population. *(b)* Exponential growth of a bank account starting with $1000 at 10% interest.

TABLE 24-2

Global Population Growth

Population Size	Year	Time Required to Double
1 billion	1850	All of human history
2 billion	1930	80 years
4 billion	1975	45 years
8 billion (projected)	2017	42 years

reached a certain critical level, each doubling yielded what appeared to be incredible gains—even though the account grew at a steady 10% per year. The human population has taken over 3 million years to reach its current size of 5.6 billion, but because of exponential growth, it could increase by another 5.6 billion in the next 43 years, if current growth continues (**TABLE 24-2**).[2]

Pollution and resource demand will also grow exponentially as the human population skyrockets, and therein lies another concern. Even though supplies of many resources seem large, exponential growth could deplete them in a very short time. Just to show you how insidious exponential growth is, consider a resource with a billion-year life span at the current rate of consumption. If the rate of demand increases 5% per year, that billion-year resource will last only about 500 years.

---

[2]Note that the human population has not been growing at 1.6% per year for the 3 million years.

FIGURE 24-4 Poverty Abounds About three-fifths of the world's people live in poverty. Nearly one of every five people on this planet does not have enough to eat. Many live in makeshift shelters like these in Rio de Janeiro.

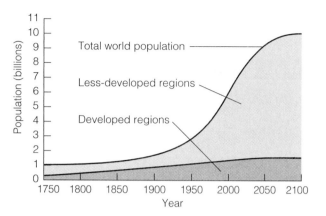

FIGURE 24-5 Human Population Growth The graph shows the relative size of the populations in the less-developed regions and the developed regions and projected growth. Note that 90% of the growth is occurring in the Third World. Today, about one of every five people in the world lives in an industrialized country. By 2100, if current growth rates continue, the ratio will be about 1 to 10.

Many People in Developing Nations Are Already Suffering Enormously as a Result of Overpopulation

In many countries, large segments of the population are living in extreme poverty. Many of these people do not have enough to eat. People live in makeshift homes or on the streets (**FIGURE 24-4**). An estimated 700–800 million people are malnourished or severely undernourished. About 2 billion people are on the edge of poverty, with barely enough to eat. Thus, nearly three of every five people on the planet are living in horrible conditions.

Because humans are living beyond the carrying capacity of the environment now, further growth will very likely worsen problems. Unfortunately, little is being done to stem the swelling tide of humanity. Each year, in fact, 80 million new residents are added to the world population. Approximately 9 of every 10 of these people are born in the poorer developing nations, where hunger and poverty are often a way of life (**FIGURE 24-5**). In African nations such as Kenya and Ethiopia where the human population is expected to double in 25 years, efforts aimed at keeping up the substandard food supplies will put enormous strains on the economy and the environment. Making improvements in diet to reduce starvation and persistent hunger seems impossible. In this struggle, many believe that nature will restore a more equitable balance. Millions of people will die unless something is done to reduce population growth and increase food supply.

Solving World Hunger and Living Sustainably on the Planet Require Many Actions

There is no easy answer to world hunger. Solving such a complex issue and learning to live sustain-

TABLE 24-3

Some Solutions to Alleviate World Hunger

Reduce population growth.

Reduce soil erosion.

Reduce desertification.

Reduce farmland conversion.

Improve yield through better crop strains.

Improve yield through better soil management.

Improve yield through fertilization.

Improve yield through better pest control.

Reduce spoilage and pest damage after harvest.

Use native animals for meat production.

Tap farmland reserves available in some countries.

ably on the Earth require a multifaceted response (**TABLE 24-3**).

Slowing the Rate of Growth.

One of the most important steps is to reduce population growth. Most world leaders agree that greatly reducing the rate of increase will help developing countries meet the demand for food and will improve the human condition. Reducing the rate of growth will also help industrial countries reduce their demand for resources and reduce pollution, habitat destruction, and a host of other environmental problems. Population control measures will also reduce expenditures for social services in the developing nations. Every birth averted through family planning, studies suggest, saves a developing nation between $15 and $200 per year in social services.

Reducing Numbers Through Attrition.

Judging from the state of the environment, today's population of nearly 6 billion people is well beyond the carrying capacity of our planet. Slowing the growth of human population, therefore, will not be sufficient to build a sustainable human presence. Over the next 50 years, it will very likely be necessary to reduce the size of the human population. This will permit human society to live within the Earth's carrying capacity. The principal means of reducing population size is through attrition—reducing the birth rate so that it falls below the death rate.

Reducing the Total Fertility Rate.

Reducing the population size of a country, indeed the world, is not as difficult as many might believe, nor as

undesirable. Already, dozens of countries in Europe have reached a stage of extremely slow growth, no growth, and even "negative" growth (shrinkage), among them Hungary, Denmark, Italy, and Austria. The decline in population growth is achieved, in large part, because couples are having smaller families. Globally, birth rates also appear to be falling dramatically, a very good sign.

The number of children a woman has in her lifetime is called the **total fertility rate**. For a population to remain stable in countries such as ours, the total fertility rate must be maintained at 2.1.[3] This means that each woman of reproductive age has, on average, 2.1 children. This level of fertility is called the **replacement-level fertility**. It is the number of children that will replace a couple when they die. A replacement-level fertility rate of 2.1 means that every 10 couples in a population must have 21 children to maintain a steady population size. The additional child accounts for typical mortality.

When the total fertility rate is below the replacement-level fertility, a population will decline, but only if there is no net immigration—that is, there are no newcomers from other countries. In the United States, the total fertility rate has been below replacement-level fertility since 1972 (**FIGURE 24-6**). Despite this important development, the U.S. population continues to grow. Why?

One of the reasons for growth is the steady inflow of illegal and legal immigrants—over a million people each year. Growth also occurs because of the age structure of

---

[3]Assuming there is no immigration.

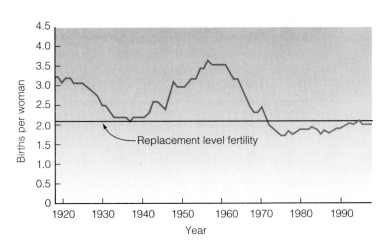

FIGURE 24-6 Total Fertility Rate in the United States   Since 1972, the total fertility rate in the United States has been below replacement-level fertility. Yet overall population growth continues because of immigration and because of the large number of women having children.

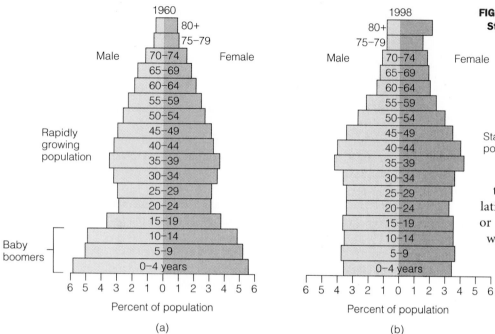

(a)

(b)

FIGURE 24-7 Population Profiles for the United States *(a)* In 1960, the U.S. population was growing; the broad base of the population profile reflects this fact. *(b)* In 1998, growth was slowing, and the population was beginning to stabilize; the profile is becoming more boxlike.

our population. **FIGURE 24-7** is a **population histogram**, or **population profile**, of the United States in 1960 and 1998. It shows the number or percentage of males and females in each age group. As you can see, the profile for 1960 is bottom-heavy. There are more people in the lower age groups than in the upper ones. Many of the people in the lower age groups are products of the baby boom era; that is, they were born soon after World War II.

After the war, America's prospects looked bright. Judging from the rise in total fertility to well over 3, people must have been extremely optimistic and happy to be done with the brutal war. The large postwar families created a demographic phenomenon known as the baby boom (**FIGURE 24-8**). Now the baby boomers are having children of their own, and even though they are having far fewer children than their parents, the reproductive-age group is larger than it was after the war. The current growth in the U.S. population therefore occurs not because women are having more children,

but because more women are having children.

If immigration quotas do not increase and if the total fertility rate remains the same, the U.S. population profile should become more boxlike, or stationary. At this point, the population would cease growing, reaching a steady state called **zero population growth**. However, Congress passed legislation in 1991 that greatly increases the number of legal immigrants and could again accelerate U.S. population growth. Current projections show that the U.S. population will increase from 268 to 390 million between the present and 2050.

Worldwide, the population profile today looks a lot like the U.S. population profile did shortly after the war—that is, the histogram is triangular, or expansive. Today, 33% of the world's people are under the age of 15. Soon, they will be entering the reproductive-age group and will start having families. Unless these children restrict their own reproduction, they will cause massive population growth in the next four decades.

Section 24-3

Resource Depletion: Eroding the Prospects of All Organisms

The human population depends on a variety of resources for its survival and well-being—among them, forests, soil, water, minerals, and oil. Today, however,

FIGURE 24-8 The Baby Boom Effect This figure follows the babies born from 1955 to 1959 through the year 2010. Population profiles provide a means of projecting trends in populations.

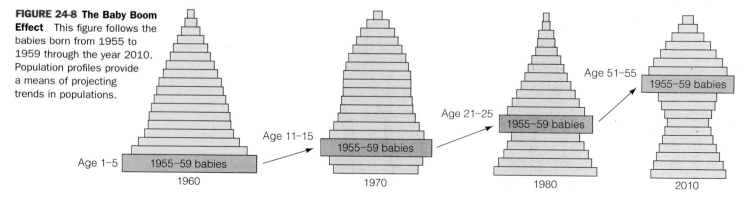

many of these resources are in danger. This section describes the problems and some of the solutions.

■ Humanity Is Destroying the World's Forests, but Massive Replanting Efforts Could Reduce or Even Stop Deforestation

At one time, tropical rain forests covered a region about the size of the United States (**FIGURE 24-9**). Today, about half of those forests are gone. Because of an exponential increase in population size and the resulting timber harvests "needed" to provide a host of products, most of the remaining tropical rain forests could be destroyed within your lifetime. Along with them, thousands of species, perhaps as many as a million, would vanish. Making matters worse, in the tropics, only 1 tree is replanted for every 10 trees that are cut down. In tropical Africa, this ratio is 1 to 29.

The decline of the world's forests is not limited to the tropics. In the United States, for example, 45% of the forested land present when the eastern seaboard was first colonized has been converted to other uses. In the Pacific Northwest, 90%–95% of the old-growth forests, consisting of 250-to-1000 year-old trees, have been cut.

India's forests decline by 1.5 million hectares (3.7 million acres) per year. On every continent, forests are in danger.

Because global deforestation exceeds reforestation and because population and demand for resources continue to climb, many scientists have little hope for the world's forests. Consequently, many environmentalists are urging sharp cutbacks in demand for timber and wood products and other strategies that will ease the pressure on forests. Smaller homes, paper recycling, widespread tree planting, and better forest management could all help avert a shortage of timber in the coming years and thereby protect remaining forests. Increasing the rate of paper recycling by 30% in the United States alone would save an estimated 350 million trees per year.

Saving forests is not just a matter of ensuring a steady supply of wood and wood products. Forest conservation also protects the habitat of wild species, helps purify air, and reduces carbon dioxide buildup in the atmosphere. It protects watersheds, reduces soil erosion, and protects

FIGURE 24-9 Tropical Rain Forests *(a)* This map shows the steady decline in tropical forests and projected remaining trees in the year 2000. *(b)* Deforestation: a desperate effort to stave off poverty results in an environmental disaster. These farmers live near the Andasibe reserve in Madagascar, an island where the per capita income is less than $250 a year. Clearing the rain forest allows these impoverished farmers to plant rice and graze cattle. But the environmental price may be too steep. In Madagascar alone, 80% of the rain forest has been destroyed.

(b)

(a)

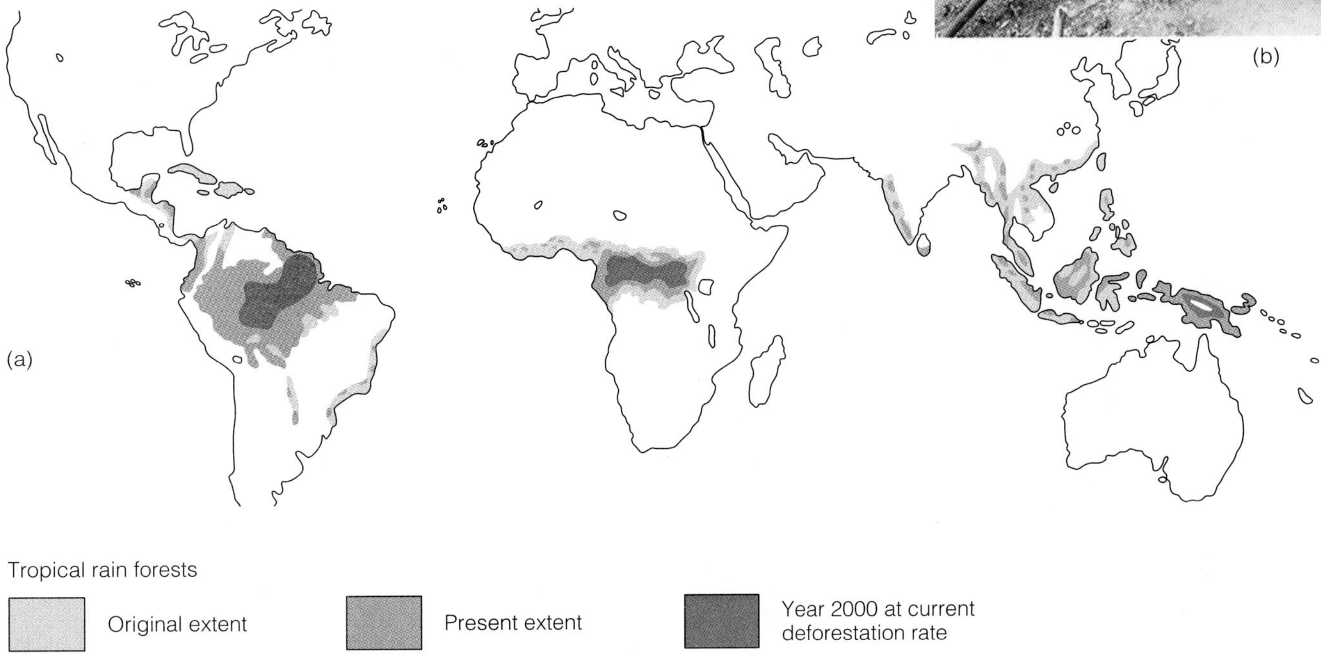

Tropical rain forests

 Original extent

 Present extent

Year 2000 at current deforestation rate

recreational opportunities as well. Tropical forests are a potential source of new medicines and food plants that could help feed the world's people.

Destroying tropical forests also disrupts native cultures. Protecting forests is a way of protecting these people and their cultures. New efforts are now under way to save the forests by developing markets for sustainable products such as nuts and fruits. Several recent studies suggest that sustainable harvesting of the forest is far more profitable than timber harvesting and cattle ranching—and much better for the forest and its inhabitants.

Soil Erosion, Like Deforestation, Is Also a Worldwide Phenomenon

Although soil erosion occurs naturally, it is often greatly accelerated by human activities such as farming, construction, and mining (**FIGURE 24-10**). In the past 100 years, about one-third of the fertile topsoil on American farmland has been eroded away by water and wind, largely because of poor land management. Each year, an estimated 1 billion tons of topsoil are eroded by wind and water from U.S. farmland.

Worldwide, about 25 billion tons of soil is eroded from farmland each year. Just to give you a perspective, in a decade a 25 billion ton per year erosion rate would equal 250 billion tons. That is equivalent to all of the topsoil on half the U.S. farmland. Globally, annual erosion rates are 18–100 times greater than rates at which soil is reformed. (The average renewal time is 500 years, but it ranges from 200 to 1000 years.) The annual loss of topsoil causes farmers to retire millions of acres of once-productive farmland and, in many cases, to cut down forests or plow up grassland to replace what has been lost. Approximately 25% of the tropical rain

FIGURE 24-10 Soil Erosion Millions of acres of farmland are destroyed each year because of poor land management practices that lead to severe soil erosion.

FIGURE 24-11 Farmland Conversion Cities are often surrounded by excellent farmland soils that drain well and are flat and highly productive. Many of the features that make soils suitable for farmland also make them suitable for building. This scene, unfortunately, is all too common in expanding urban areas.

forests cut down each year are leveled to replace farmland destroyed by human activities.

Making matters worse, millions of acres of farmland are lost to urban and suburban sprawl, highway construction, and other human activities (**FIGURE 24-11**). In the United States, at least 3500 acres of rural land—actual or potential farmland, pasture, and rangeland—are lost every day. That's equivalent to 1.25 million acres a year, or a strip 0.3 mile wide extending from New York City to San Francisco. Overgrazing and poor land management are destroying millions of acres of farmland as well. Worldwide, an area the size of Ohio becomes desert each year, principally because of overgrazing by livestock and poor land management practices.

These figures paint a rather grim picture for the long-term future of food production here and abroad, especially when viewed in light of increasing populations. But this trend need not be our destiny. The loss of topsoil can be stopped, and soils can be replenished. However, worldwide conservation efforts are needed as are population growth measures, which help to reduce the conversion of farmland to other uses such as housing. Additional solutions shown in **TABLE 24-3** could alleviate world hunger and could also reduce the loss of productive farmland and rangeland. For these actions to be effective, society must begin work soon.

Many Areas of the World Are Facing Water Shortages or Will Soon Face Them as Population and Demand Increase

FIGURE 24-12 is a map of areas in the United States that will face a water shortage in the near future. Regional water shortages result because too many people are drawing on limited water supplies. Long-term prospects here and abroad appear bleak. Between 1975 and 2000, irrigated agriculture worldwide is expected to double to

FIGURE 24-12 Water-Short Regions in the United States

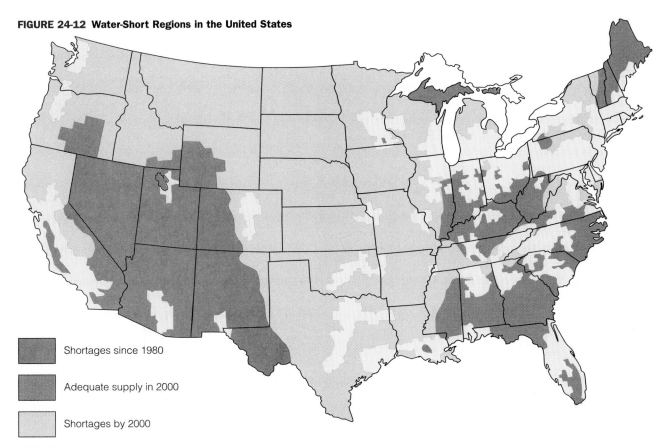

Shortages since 1980

Adequate supply in 2000

Shortages by 2000

Metropolitan areas with
population greater than 1 million

FIGURE 24-13 The Ogallala Aquifer   The water of the Ogallala is replenished very slowly and is currently being withdrawn much faster than it can be replaced. Farmers' wells are running dry, and eventually many irrigated farms will go out of business.

meet the rising demand for food. Industry's demand for water is expected to increase twentyfold. By the end of the century, water demand is expected to exceed supply in at least 30 countries.

FIGURE 24-13 shows the location of an enormous **aquifer**, a porous underground zone saturated with water. Known as the **Ogallala aquifer**, this zone supplies irrigation and drinking water for farms in Nebraska, Kansas, Colorado, Oklahoma, and Texas. Unfortunately, severe **groundwater overdraft**, withdrawals that exceed natural replenishment, may put an end to irrigated farming in these states.

Reducing the mining of groundwater is essential to meeting future needs. That will require strict conservation efforts, such as lining irrigation ditches with concrete or using pipes rather than open ditches to transport water to fields, to reduce evaporation. More efficient sprinklers, computerized systems that monitor soil moisture so that farmers know exactly how much irrigation water is required, and other measures can also help.

Many Essential Minerals Will Be Depleted in the Next Four Decades

Metals such as steel and aluminum are produced from the Earth's mineral deposits. More than 100 minerals, worth billions of dollars to the global economy, are traded on the world market each year. Several dozen of these minerals are so important to modern society that if they were suddenly no longer available or were unavailable at a reasonable price, the U.S. economy, like those of other industrialized countries, would come to a standstill.

At least 18 economically important minerals will fall into short supply in the next 40 years, even if countries expand their recycling programs. Silver, mercury, lead, sulfur, tin, tungsten, and zinc are among this group. Even if new discoveries and new technologies make it possible to extract five times the currently known reserves, this group will be 80% depleted by or before 2040.

Oil Supplies Are Limited, and Most Students Alive Today Will See the End of Oil in Their Lifetimes

Oil is the lifeblood of modern society. In the United States, for example, oil supplies 43% of our annual energy demand. But the supply of oil is finite. No one knows how much oil is left in Earth's crust, but this we do know: Oil reserves could be depleted within the next 40 to 100 years, depending on demand. If demand continues to increase as it has, oil supplies could be gone by 2038. Clearly, time is running out for oil. You will probably see the end of oil within your lifetime. So what do we do?

The first step in meeting future demand is to make current energy use much more efficient. Efficiency will no doubt occur as fuel supplies rise and citizens shift to more efficient modes of transportation. Waiting until that time, say many experts, wastes precious fuel and unnecessarily pollutes the planet.

Enormous supplies of energy exist today—in our waste. The efficient use of oil and oil products, such as gasoline, diesel fuel, and home heating oil, can help us stretch current oil supplies considerably while actually improving the economy. In fact, many of the leading economic powers, such as Japan and West Germany, are thriving in large part because they are so energy-efficient.

Energy-efficient investments such as storm doors, storm windows, and added insulation coupled with measures to cut air inflow (infiltration) are often inexpensive and can reduce home energy consumption by 30%–50%. More efficient cars could cut demand enormously. Even higher mileage is possible. The Japanese, for example, have a car that gets 98 miles per gallon on the highway and seats four people! Even newer models are being developed. The hybrid car, for example, uses a small gasoline engine to make electricity which powers four electric motors in the car's wheels. This car could get over 150 miles per gallon. Toyota released a similar model in 1998 in Japan that got 65 mpg.

Another important step in increasing energy efficiency is mass transit—buses and trains to transport commuters to and from work. Mass transit is four times more efficient than the automobile.

Energy efficiency is the cheapest and most cost-effective means of meeting future demand. Amory Lovins, a world-renowned energy expert, calculates that Americans could reduce oil demand by 80% without changing life-styles by using energy-efficient technologies currently on the market. Lovins's calculations also show that U.S. electrical demand could be cut by 75% by using energy-efficient light bulbs, motors, and other technologies that are currently available (**FIGURE 24-14**). These measures could save us hundreds of billions of dollars a year.

Alternative fuels can also help meet future demand. Bear in mind that the immediate need is for an alternative source of energy to replace oil, which is used chiefly in transportation and home heating. One example is ethanol. Produced from corn, wheat, and other crops, ethanol can power automobiles and trucks. Grown on special fuel farms, ethanol could virtually replace gasoline, but a shift to ethanol in the United States would not be without its costs. One of those might be a reduction in food output. Brazil is a leader in ethanol use: 98% of the new cars sold in Brazil are equipped to use it.

Renewable energy could provide enormous amounts of fuel in the years to come. Unknown to many, the renewable energy supply is enormous and, in some cases, can be tapped quickly and inexpensively. According to one estimate, nonrenewable energy reserves (coal, oil, natural gas, and so on) would provide the equivalent of 8.8 trillion barrels of oil. Renewable energy could provide 10 times that amount of energy every year!

FIGURE 24-14
Energy-Efficient Many new light bulbs use only 25% of the energy a standard light bulb requires. Although they cost more, they last as long as 10 standard bulbs and save $20–$40 in electricity over their lifetime.

Pollution: Fouling Our Nest

Like death and taxes, waste is an inescapable fact of life. All organisms produce it. Humans, however, are by far the most prolific generators of waste on the planet. Today, waste from human society is overwhelming many nutrient cycles, poisoning other species (and ourselves), and altering planetary homeostasis. This section recaps four of the most serious waste problems.

Global Warming Results from the Release of Carbon Dioxide and Other Greenhouse Gases

Carbon dioxide is produced during cellular respiration and the combustion of all organic materials—most importantly, fossil fuels. In normal concentrations in the atmosphere, carbon dioxide has a warming effect on the planet. Acting much like the glass in a greenhouse, it traps heat escaping from the Earth and radiates it back to the surface. Carbon dioxide is, therefore, also known as a **greenhouse gas.** A little bit of carbon dioxide is essential to life on Earth. In fact, without carbon dioxide in the atmosphere, the planet would be about 55°F (30°C) cooler than it is. Too much, however, may lead to overheating.

Over the past 100 years, global carbon dioxide levels have increased about 25%, principally as a result of industrialization (powered by the combustion of fossil fuels) and deforestation. Today, atmospheric carbon dioxide levels continue to rise at a rate of about 0.5% per year (**FIGURE 24-15A**).

Several other pollutants also contribute to global warming. Methane, chlorofluorocarbons, nitrous oxide, and even water vapor all radiate heat back to the Earth, causing the atmosphere to heat up. Methane is released from livestock and the manure they produce; humankind's nearly 1 billion cattle annually release about 73 million metric tons of this gas (providing another good reason to reduce or eliminate one's consumption of beef). Methane production has increased 435% in the last century. Incidentally, termites, which thrive in the deforested tropics, also excrete methane. Methane output from all sources is increasing at a rate of about 1% per year and is 20 times more effective at trapping heat than carbon dioxide.

Chlorofluorocarbons, or CFCs, are best known for their effect on the ozone layer and are released from refrigerators, air conditioners, and freezers.[4] One class of

---

[4]CFCs used in spray cans were banned in the United States and several other countries in the late 1970s.

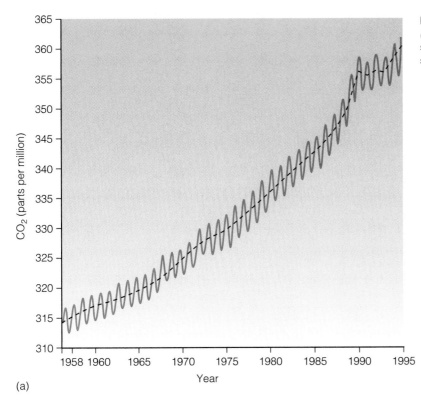

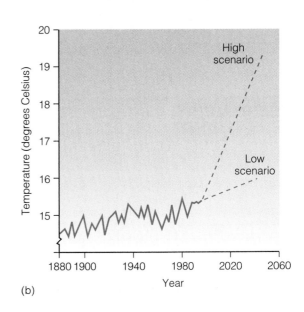

FIGURE 24-15 Global Carbon Dioxide and Temperature Trends (a) Carbon dioxide levels from before 1958 to 1995 showing steady rise. (b) Global temperature since 1880 with two possible scenarios.

CFCs was used to clean circuit boards used in computers and other electronic equipment. In the lower atmosphere, CFCs also trap heat, contributing to the greenhouse effect.

Because of the dramatic increase in the release of greenhouse gases, many atmospheric scientists believe that global temperature is on the rise and may rise dramatically in the coming decades, causing an extraordinary shift in climate. **FIGURE 24-15B** shows average global temperature over the past century and projected temperature.

The Impacts of Global Warming Could Be Ecologically Devastating.

According to latest projections based on the present rate of increase in greenhouse gases, global temperatures could be 2°C (4°F) hotter by 2100. The models suggest that global rainfall patterns will shift dramatically as a result of warming (**FIGURE 24-16**). Computer simulations of U.S. climate suggest that the Midwest and much of the western United States will be drier and hotter than they are today. If that happens, many midwestern farmers will be driven out of business. As rainfall declines in this agriculturally productive region, farming may intensify in the northern states. Overall agricultural productivity in the United States may fall, because northern soils are not as rich as those in the Midwest. Food shortages caused by the decline in U.S. agriculture in the Midwest and rising food prices could affect the economy in profound ways. Computer climate models predict that the southern United States and Pacific Coast may be wetter but hotter.

A rise in global temperature is expected to melt glaciers and the polar ice caps (**FIGURE 24-17**). Warmer temperatures would also expand the volume of the seas.

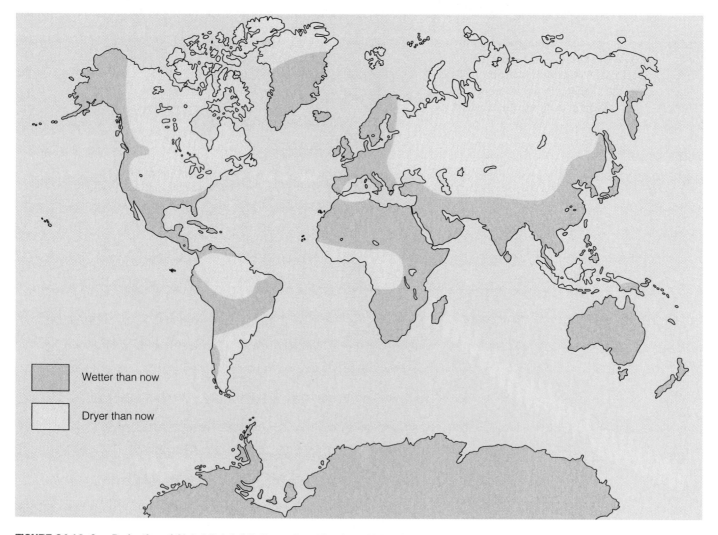

Wetter than now

Dryer than now

FIGURE 24-16 One Projection of Global Rainfall Patterns Resulting from Global Warming

FIGURE 24-17 Mount Rainier Glaciers This enormous glacier and others like it could melt as the Earth gets warmer, raising the sea level.

FIGURE 24-18 Miami Underwater? A rising sea level due to global warming could flood many coastal cities the world over.

Together, melting ice and expanding oceans would result in an increase in sea level. In the past 50 years, sea level has risen 10–12 centimeters (4–6 inches). By 2100, the computer models suggest that the sea level will rise 50 centimeters (2 feet) because of global warming. What impact would this have?

In the United States, approximately half of the population lives within 50 miles of the ocean, and many large cities such as Miami are located only a few feet above sea level (**FIGURE 24-18**). A rising sea level would flood many low-lying regions. Storms could cause more damage than they do now because high water would be able to move farther inland. Expensive dikes and levees would be needed to protect cities such as Miami and New Orleans. Other coastal cities would have to be rebuilt on higher ground, at a staggering cost.

A rising sea level would be particularly hard on Bangladesh and other Asian countries with extensive lowland rice paddies. Approximately 17% of Bangladesh would be reclaimed by the sea if sea level rose 2–3 feet.

Rising temperatures would also have an impact on many species. Some might adapt or find new habitat; many plants, however, would be wiped out by a relatively rapid increase in temperature, for temperature would probably increase far faster than many species could adapt. As a result, forests would dry out and die off. The incidence of forest fires may increase, further adding to global warming.

These predictions sound dire—and indeed they are. But a small number of scientists believe that we don't know enough about the atmosphere and the impact of greenhouse gases to know for certain whether the predictions will come true. Global climate is extremely complex and several factors could dampen the warming trend. Other factors could worsen it.

Solving Global Warming Requires Massive Action, and Soon.

While scientists and politicians debate global warming, the Earth appears to be getting hotter. In fact, 12 of the hottest years in the past 100 years have occurred from 1980 to 1996. Although warm years are not unusual, having so many record-breaking years in such a short period is highly unlikely. In 1997, these and other statistics spurred the world's nations to agree to measures that will decrease CO_2 emissions.

Sharp reductions in fossil-fuel consumption through energy efficiency, conservation, and the use of alternative sources of fuel will be required. Global reforestation is essential. For each family of four in the United States, 6 acres of fast-growing trees would have to be planted to offset the carbon dioxide the family will produce during its life. Additional strategies are shown in **TABLE 24-4**. We as individuals can also help. On average, each gallon of gasoline you consume in your automobile produces 5 pounds of carbon dioxide. Every hour you watch television results in the production of

TABLE 24-4

Measures to Reduce Global Climate Change

Reduce the rate of population growth.

Switch from coal- and oil-fired power plants to natural gas, which produces much less carbon dioxide per unit of electricity.

Implement the technologies that burn coal more efficiently.

Expand cogeneration—processes that trap waste heat and put it to good use.

Boost automobile efficiency.

Expand mass transit.

Develop alternative liquid fuels for the transportation sector.

Improve the efficiency of industry.

Make new and existing homes more energy-efficient through insulation, weather stripping, storm doors, and storm windows.

Build many new homes that use solar energy for space heating.

Reduce global deforestation.

Begin a massive global reforestation effort.

Phase out all CFCs and other CFC-damaging chemicals soon.

Reduce consumption of unnecessary items.

Expand recycling efforts.

TABLE 24-5

Carbon Dioxide Production from Common Activities

Electrical Appliances	Pounds of Carbon Dioxide Added to Atmosphere•
Color television	0.64 per hour
Steam iron	0.85 per hour
Vacuum cleaner	1.70 per hour
Air conditioner, room	4.00 per hour
Toaster oven	12.80 per hour
Ceiling fan	4.00 per day
Refrigerator, frost-free	12.80 per day
Waterbed heater	24.00 per day
with thermostat	12.80 per day
Clothes dryer	10.00 per load
Dishwasher	2.60 per load
Toaster	0.12 per use
Microwave oven	0.25 per 5-minute use
Coffeemaker	0.50 per brew

*At room temperature and sea level, every pound of carbon dioxide occupies 8.75 cubic feet, about half the size of a refrigerator.

SOURCE: Reprinted from the February/March 1990 issue of *National Wildlife.* Copyright © 1990 by the National Wildlife Federation.

TABLE 24-6

Individual Actions That Can Reduce Global Climate Change

Automobile energy savings
Buy energy-efficient vehicles.
Reduce unnecessary driving.
Car-pool, take mass transit, walk, or bike to work.
Combine trips.
Keep your car tuned and your tires inflated to the proper level.
Drive at or below the speed limit.

Home energy savings
Increase your attic insulation to R30 or R38.
Caulk and weather-strip your house.
Add storm windows and insulated curtains.
Install an automatic thermostat.
Turn the thermostat down a few degrees in winter, and wear warmer clothing.
Replace furnace filters when needed.
Lower water heater setting to 120–130°F.
Insulate water heater and pipes, install a water heater insulation blanket, and repair or replace all leaky faucets.
Take shorter showers.
Use cold water as much as possible.
Avoid unnecessary appliances.
Buy energy-efficient appliances.
Use low-energy light bulbs.

Reducing waste and resource consumption
Recycle at home and at work.
Avoid products with excessive packaging.
Reuse shopping bags.
Refuse bags for single items.
Use a diaper service instead of disposable diapers.
Reduce consumption of throwaways.
Donate used items to Goodwill, Disabled American Veterans, the Salvation Army, or other charities.
Buy durable items.
Give environmentally sensitive gifts.

0.64 pound of carbon dioxide while your frost-free refrigerator accounts for nearly 13 pounds daily (TABLE 24-5). We can play a role in reducing the problem by using energy much more efficiently: using mass transit, recycling, insulating our homes, and a great many other strategies (TABLE 24-6).

Large Portions of the World Are Threatened by Acid Deposition

In the Adirondack Mountains of New York, hundreds of lakes are dying. Fishes have vanished as the lakes turn acidic (FIGURE 24-19). A similar phenomenon is occurring in southeastern Canada and in Sweden and Norway, where dying lakes number in the thousands.

The acids falling from the skies are produced from two atmospheric pollutants, sulfur dioxide and nitrogen

FIGURE 24-19 Victims of Modern Society These fish were killed by acids from acid deposition.

which are responsible for **acid deposition.** Rain and snow wash acids from the sky. Fog and clouds also carry heavy loads of acids that are deposited on trees.

In the United States and Europe, acid deposition has been worsening for four decades, largely as a result of increased fossil-fuel combustion. The map in **FIGURE 24-20** shows that the region of the United States affected by acid deposition is growing larger and that the level of acidity, measured by pH, is increasing. Today, acid deposition is commonly encountered downwind from virtually all major population centers. It comes from power plants, motorized vehicles, factories, and even our own homes.

Acid deposition changes the pH of lakes and streams, killing fishes and other aquatic organisms. Acids on land dissolve toxic minerals such as aluminum from the soil and wash them into surface waters. Aluminum causes the gills of fishes to clog with mucus, suffocating them. Extensive acidification of lakes and rivers is putting resort owners out of business in the Northeast, upper Midwest, and southern Canada.

FIGURE 24-21 shows areas most susceptible to acid deposition. These regions are generally mountainous

dioxide. These pollutants arise chiefly from the combustion of fossil fuels: coal, oil, and natural gas. In the atmosphere, these gases, or **acid precursors**, combine with water and oxygen to form sulfuric and nitric acids,

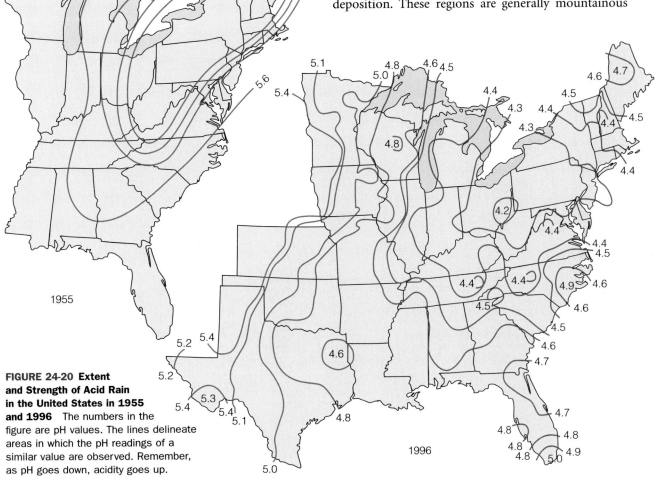

FIGURE 24-20 Extent and Strength of Acid Rain in the United States in 1955 and 1996 The numbers in the figure are pH values. The lines delineate areas in which the pH readings of a similar value are observed. Remember, as pH goes down, acidity goes up.

and contain soils with little capacity to neutralize acids. Consequently, acids that fall on the land quickly wash to nearby lakes and streams.

Acids also damage crops and trees, directly and indirectly. Direct damage results when acids fall on growing buds or leaves. Indirect damage occurs when acids alter soil chemistry, inhibiting the soil bacteria necessary for nutrient recycling. Acids also leach important minerals from the soil, resulting in slower plant growth.

Some scientists believe that massive forest diebacks occurring throughout the world may be the result of acidic rainfall and fog that often blankets forests (**FIGURE 24-22**). Finally, acid deposition also damages buildings, statues, and other structures. The estimated cost in the United States is about $5 billion per year.

Reducing the deposition of acids will require a dramatic reduction in the release of sulfur dioxide and nitrogen dioxide from power plants, factories, and automobiles. One way of reducing sulfur dioxide is the smokestack scrubber, a device that traps sulfur dioxide gas escaping from power plants, removing up to 95% of this pollutant. Installing and operating a scrubber is rather expensive, and some utilities have objected to this strategy. Using low-sulfur coal can also help, but eastern coal producers find this strategy objectionable because their coal is especially high in sulfur.

A far better strategy is energy efficiency. By using less fossil-fuel energy, we cut the combustion of polluting fuels. This action reduces both sulfur dioxide and nitrogen dioxide release and also reduces carbon dioxide, a major contributor to global warming. Energy efficiency therefore has multiple environmental benefits. Add the economic benefits, and it becomes an even more attractive strategy for environmental protection.

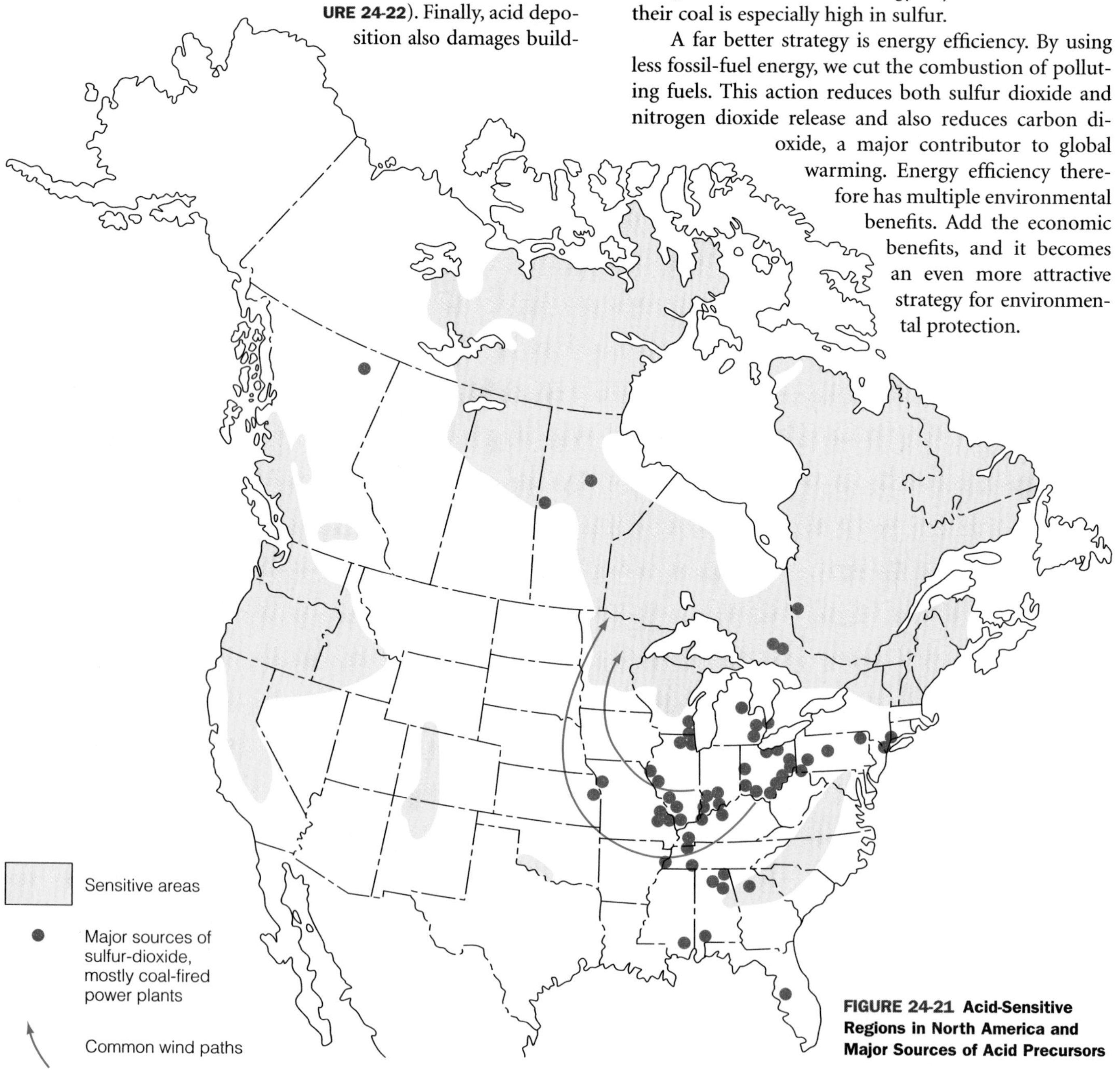

Sensitive areas

Major sources of sulfur-dioxide, mostly coal-fired power plants

Common wind paths

FIGURE 24-21 Acid-Sensitive Regions in North America and Major Sources of Acid Precursors

FIGURE 24-22 Forest Die-off Ghostly remains of trees killed by acid deposition in the Pisgah National Forest, North Carolina.

The Ozone Layer is Endangered by Human Pollutants

Encircling the Earth, 20–30 miles above its surface, is a region of the atmosphere, the **ozone layer**, containing a slightly elevated level of ozone gas. The ozone layer shields the Earth from harmful ultraviolet radiation from the Sun. Early in the evolution of life, in fact, the formation of the ozone layer probably "allowed" the colonization of land by plants and animals.

Today, however, the ozone layer is being destroyed by chemicals released into the atmosphere by modern society. The most potent destroyer of ozone is a class of chemicals known as the chlorofluorocarbons (CFCs).[5] Used as spray can propellants (not in the United States), refrigerants, blowing agents for plastic foam, and cleansing agents, CFCs are highly stable molecules. Scientists, in fact, selected them for use in spray cans in large part because of their lack of chemical reactivity. In the early 1970s, however, a U.S. scientist discovered that CFCs were broken down by sunlight. Another scientist showed that CFCs released from human activities gradually drift into the upper atmosphere. Here, the scientists hypothesized, the breakdown products react with and destroy ozone molecules.

Estimates of ozone depletion came and went over the years, but scientists were hard pressed to detect any measurable decline, in large part because ozone levels

fluctuate naturally from year to year. In 1988, however, a panel of atmospheric scientists reviewing 20 years of satellite data on ozone levels concluded that the ozone layer is indeed on the decline. What they found was that satellite measurements of ozone automatically logged into their computer had shown a consistent decline. It hadn't shown up because the computer was programmed to delete low levels on the belief that such readings represented instrument errors. When scientists reexamined the data, they found that over North America ozone levels had fallen 1.7%–3% (**FIGURE 24-23**). Much larger declines were evident at the poles. In fact, scientists found that each year a giant hole in the ozone layer about the size of the United States formed over Antarctica. Ozone levels in the hole declined by as much as 50%. In 1992, the ozone hole reached a record high—almost three times the size of the United States. Studies strongly suggested that the main reason for the decline in the ozone layer was the accumulation of chlorofluorocarbons.

In a show of solidarity on September 1988, 23 nations met in Montreal to sign an agreement to cut back

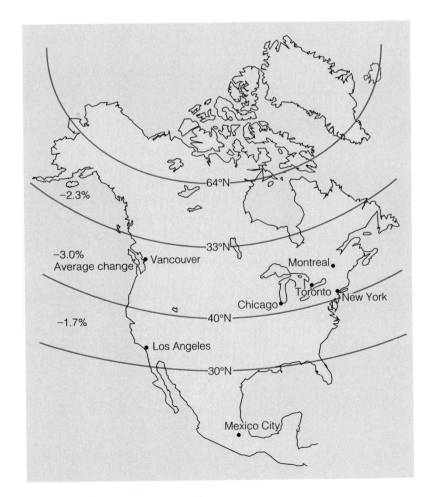

FIGURE 24-23 Decline in the Ozone Layer

---

[5] Nitric oxide from high-flying jet airplanes such as the supersonic transport can also destroy the ozone layer.

on the production of many ozone-destroying chemicals. By 1999, the treaty was to achieve a 50% reduction in CFC manufacture and release. Some critics objected, saying that a 50% cut, while important, would still result in a 10% decline in the ozone layer. Clearly, greater reductions were needed. Because of these concerns and the cooperation of the CFC industry, the Montreal treaty was revised. The new plan signed in London in 1990 called for a complete phase-out of many CFCs by 1999. After recent findings suggesting an even greater loss than previously suspected, a new treaty was signed in 1992. It calls for a faster phase-out of ozone-depleting chemicals and includes more chemicals than previous agreements. In 1997, CFC production had been nearly eliminated.

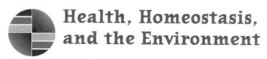

Health, Homeostasis, and the Environment

Building a Sustainable Future

The trends and problems discussed in the previous sections could spawn a crisis of epic proportions. What many people forget when they deal with such issues is that these problems could combine to produce impacts far greater than anticipated. For example, soil erosion, global warming, farmland conversion, acid deposition, ozone depletion, and population growth together could produce incredible food shortages.

Solving these problems will require tougher laws and regulations, reductions in population growth, more efficient technologies, and changes in human behavior. Many people, according to opinion polls, believe that broader societal changes are required. One suggestion is that we build a **sustainable society**, one that lives within the carrying capacity of the environment. Living within the limits of nature means achieving a population size and a way of life that do not exceed the planet's ability to supply food and other resources and to handle wastes. It means promoting human societies that manage their affairs in ways that protect the Earth's homeostatic mechanisms.

Building a sustainable society will require a shift to the patterns seen in nature. It will require that we become more efficient in our use of resources; recycle to the maximum extent possible; shift to renewable resources, especially energy; restore ecosystem damage; and stabilize, if not reduce, world population.

Building a sustainable society will also depend on a profound change in attitudes. The prevalent frontier notions of unlimited resources and human dominance over nature, say proponents, must be replaced with a new ethic, a **sustainable-Earth ethic**. This new ethic is based on three tenets: (1) the world has a limited supply of resources, which must be shared with all living things; (2) humans are a part of nature and subject to its rules; and (3) nature is not something to conquer but, rather, a force with which we must learn to cooperate. New attitudes will not come easily, but they are essential as we attempt to strike a balance with nature.

Building a sustainable society will require a lifetime of commitment on the part of businesses, governments, and individuals. Given the course of modern society, it means taking a 180-degree turn. By comparison, the moon landing will seem like a weekend fix-it project.

Some changes required to create an enduring human presence may come as a result of legislative actions—new laws and regulations—that force us to rely more heavily on conservation, recycling, and renewable resources and to restore damaged ecosystems and reach a stable, sustainable population size. Technological innovations will also help us become a sustainable society. For example, a new line of compact fluorescent light bulbs is now available. These bulbs fit into a regular light socket and use one-fourth as much energy as a regular incandescent bulb (see **FIGURE 24-14**). Photovoltaics, thin silicon wafers that produce electricity from sunlight, could also help us make the transition to solar energy. Water-sensing computers can help farmers prevent overwatering. Improvements in automobile efficiency can help us cut back on fossil-fuel consumption and clean up the air. But new technologies are not the only, or even the most important, answer. Many existing technologies simply need to be installed. Insulation, weather stripping, flow restrictors for shower heads and faucets, and efficient appliances already on the market can make tremendous inroads into waste.

Many proponents also believe that we will eventually need to reduce consumption and alter our lifestyles. Taking shorter showers, shaving without the water running, turning off lights, and hundreds of other small actions on the part of individuals, when combined with similar actions by millions of other people, can result in significant cuts in resource demand.

The days of profligate resource use are quickly coming to an end. Ending the waste and pollution—and soon—is crucial if we are to protect the environment and ourselves. The balance of nature is, after all, the balance that sustains us. We risk upsetting it at our own peril. Protecting the planet is the ultimate form of health-care.

SUMMARY

1. The economy of nature and the human economy are at odds. Today, that "conflict" manifests itself in a global environmental crisis characterized by overpopulation, resource depletion, and pollution.

OVERSHOOTING THE EARTH'S CARRYING CAPACITY

2. Although environmental problems vary from one nation to another, they are all the result of human populations exceeding the carrying capacity of the environment.

3. *Carrying capacity* is the number of organisms an ecosystem can support indefinitely. It is determined by food and resource supplies and by the capacity of the environment to assimilate or destroy waste products of organisms.

OVERPOPULATION: PROBLEMS AND SOLUTIONS

4. *Overpopulation* occurs anytime a population overshoots its carrying capacity. It is manifest in shortages of food, lack of other resources, or excessive pollution—sometimes all three.

5. The world population is nearly 6 billion and is growing at a rate of 1.5% per year, a doubling time of 47 years.

6. The most rapid growth in human population is occurring in the developing world in parts of Africa, Asia, and Latin America. Resource depletion and food shortages are the most common problems in these regions.

7. Many experts believe that the industrialized countries are also overpopulated. Judging from the quality of our air, water, and soils, we are clearly exceeding the Earth's carrying capacity.

8. Because of our excessive resource demand, in fact, an average American has 20–40 times as much impact on the environment as a resident of a developing country.

9. The human population is growing exponentially. *Exponential growth* occurs anytime a value, such as population size, grows by a fixed percentage with the "interest" (increase) being applied to the base amount.

10. Exponential growth of population, resource demand, and pollution are at the heart of all environmental problems.

11. The human population problem can be summed up in six words: too many people, reproducing too rapidly. The size of our population and its rapid rate of growth result in resource shortages, excessive pollution, and poverty.

12. A decline in *total fertility rate*, the number of children a woman will have over her lifetime, to *replacement-level fertility*, the rate at which parents replace themselves, is required to reduce the rate of population growth.

RESOURCE DEPLETION: ERODING THE PROSPECTS OF ALL ORGANISMS

13. The human population requires a variety of renewable and nonrenewable resources for survival, and many of these resources are being depleted.

14. Worldwide, forests are being cut faster than they can regenerate. Forests are important sources of wood and wood products, but they also provide wildlife habitat, protect watersheds, regenerate oxygen, and remove pollutants such as carbon dioxide.

15. To prevent the destruction of the world's forests, tree planting, paper recycling, and other strategies are needed.

16. Worldwide, soils are being eroded from rangeland and farmland at an unsustainable rate, and millions of acres of farmland are being destroyed by human encroachment. The destruction of productive soils threatens the long-term prospects for food production. Soil conservation and population control measures can help ensure an adequate supply of soil, but serious efforts must begin soon.

17. Because of regional overpopulation, many areas suffer water shortages. The rise in population and the rise in demand are likely to cause further shortages over the coming decades. Reducing the growth of the human population and strict water conservation are needed.

18. At least 18 economically important metals could fall into short supply in the next 20 years. The more efficient use of metals, recycling, and a reduction in demand are needed to help offset inevitable shortages.

19. Oil supplies, like mineral supplies, are also finite. Given the most optimistic projects of oil supplies, the world will run out of oil by 2038.

20. By reducing population growth, using oil much more efficiently, and seeking clean, renewable alternative fuels, modern society can make a smooth transition to a sustainable energy supply.

POLLUTION: FOULING OUR NEST

21. All organisms produce waste, but humans are by far the most prolific generators of waste on the planet. Today, our waste is overwhelming nutrient cycles, poisoning other species (and ourselves), and destroying the planet's homeostatic mechanisms.

22. One of the most serious threats from pollution comes from carbon dioxide. Global atmospheric concentrations of carbon dioxide have increased 25% in the past century, in large part due to the combustion of fossil fuels and deforestation.

23. Carbon dioxide is a *greenhouse gas*, trapping heat in the Earth's atmosphere. A little carbon dioxide is beneficial, but too much will increase the planet's surface temperature, altering its climate, shifting rainfall patterns and agricultural zones, flooding low-lying regions, and destroying many species that cannot adapt to the sudden change in temperature.

24. Chlorofluorocarbons and methane are also greenhouse gases that are on the increase.

25. Many scientists believe that global warming has begun. To slow it down or stop it, they recommend massive reforestation projects and dramatic improvements in the efficiency of fossil-fuel combustion. Alternative fuels and reductions in population growth can also help.

26. Sulfur dioxide and nitrogen dioxide are two gaseous pollutants released from

power plants, factories, automobiles, and other sources. In the atmosphere, they are converted to sulfuric and nitric acid, respectively.

27. Acids fall from the sky in wet and dry deposition. Acids alter the pH of lakes and streams, killing fish and other organisms. They also leach toxic metals from soils that kill fishes. They destroy trees and crops and deface buildings, costing society billions of dollars a year.

28. Pollution-control devices called *scrubbers*, which are now in use on some power plants, can remove sulfur dioxide from smokestack gases. Stopping acid rain requires a multifaceted approach, but conservation, use of cleaner coal, and pollution control devices are the three prominent strategies.

29. The ozone layer encircles the Earth, trapping ultraviolet light. It is being destroyed by chlorofluorocarbons, or CFCs, and other pollutants.

30. Fears over the decline in ozone led to several international agreements to reduce CFC production by 100% within this decade.

HEALTH, HOMEOSTASIS, AND THE ENVIRONMENT: BUILDING A SUSTAINABLE FUTURE

31. Solving our problems will not only require tougher laws and tighter regulations, but also a profound change in the way we live and conduct business.

32. A sustainable society is built on five operating principles: conservation, recycling, renewable resources, restoration, and population control.

33. A sustainable society is based on respect for nature and a willingness to cooperate with natural processes.

Critical Thinking

THINKING CRITICALLY—ANALYSIS

This Analysis corresponds to the Thinking Critically scenario that was presented at the beginning of this chapter.

Critical thinking rules encourage us to question sources of information and their conclusions. In this example, we find that the author of this article is grossly in error when he claims that the scientific community is divided on the issue. About 99% of the nation's 700 atmospheric scientists believe that global warming is a reality; only a handful embrace the opposite view. The evidence the author introduces to support his assertion—a quote from each side of the issue—is terribly misleading.

This type of reporting is quite common in newspapers, television, and magazines. Although it is intended to give a balanced view of issues, in reality it provides an extremely unbalanced view.

What would have been a more accurate conclusion? Perhaps that not all scientists agree that global warming is occurring, but that the vast majority do. It's important to note that even though the majority of the world's atmospheric scientists agree that global warming is happening, this doesn't mean they're right.

Digging a little deeper, finding out more, often throws conclusions into question.

EXERCISING YOUR CRITICAL THINKING SKILLS

One of the most dramatic changes on the planet in the last two decades has been the steady march of the Sahara, the largest desert on Earth. Studies published in the 1970s and 1980s documented an impressive southward expansion of the desert at a rate of 5 kilometers (3 miles) per year and attributed the problem to drought, overgrazing, and agricultural land abuse in semiarid lands bordering the vast Sahara. This projection, however, was based on measurements of the southward spread in a few isolated locations, assuming that they represented the entire desert.

More recent satellite observations of vegetation over the entire continent of Africa, however, show that the Sahara advances and retreats like a tide, largely in response to rainfall. In the period from 1980 to 1984, for example, the desert's southern boundary moved 240 kilometers south. Between 1984 and 1985, the southward migration reversed itself by 110 kilometers. In 1987, the boundary between desert and semiarid lands shifted northward again by 55 kilometers, and in 1988 it shifted northward by 100 kilometers. In 1989 and 1990, however, the desert boundary shifted 77 kilometers southward.

Although the southern border of the desert in 1990 was 130 kilometers farther south than in 1980, some researchers believe that the shift does not reflect a long-term trend but, rather, differences in year-to-year rainfall. Clearly, continued tracking is necessary.

Many critics of the theory of global warming will find this report encouraging and use it to argue that desertification caused by climatic shift is not occurring. Critically analyze the conclusion of the study. Can you arrive at an alternative explanation?

Note: One alternative interpretation is that although the Sahara may ebb and flow, it may still be marching southward as drought, overgrazing, and other poor land management practices take their toll. Although heavier rainfalls may result in vegetative recovery, it is possible that the recovery is temporary. Furthermore, if stress continues, the recuperative ability of the ecosystem may be overwhelmed.

TEST OF CONCEPTS

1. In what ways are the economy of nature and the human economy working in opposite directions? In your estimation, is this a serious problem, and, if so, how can it be reduced or eliminated?

2. Human populations in both the industrialized and nonindustrialized countries are overshooting the Earth's carrying capacity. Do you agree or disagree with this statement? Be sure to describe what is meant by overshooting the carrying capacity.

3. The global human population is growing at a rate of 1.8% per year, so there's nothing to worry about. Do you agree or disagree with this statement? Support your position.

4. Why is overpopulation as much a problem in the United States as it is in Bangladesh?

5. The U.S. population fell below replacement-level fertility in 1972, yet it continues to increase. Why?

6. Describe key trends in the use of forests, soils, water, minerals, and oil suggesting that humanity is on an unsustainable course. List and discuss solutions to each of the problems.

7. Given trends in natural resource use, many experts believe that global population growth must stop. Do you agree or disagree? Why? Is your position supported by scientific fact or based more on a general feeling?

8. Describe the cause and impacts of each of the following: global warm-

ing, acid deposition, stratospheric ozone depletion, and hazardous wastes.

9. Make a list of solutions for each of the problems you identified in Question 8. How could conservation (efficiency), recycling, renewable resources, and population control factor into the solutions?

10. Explain what is meant by the following statement: A sustainable society is based on a design from nature.

11. Outline the operational and ethical principles of a sustainable society. Using your critical thinking skills, determine how they differ from the principles of modern society.

TOOLS FOR LEARNING

Tools for Learning is an on-line student review area located at this book's web site HumanBiology (www.jbpub.com/humanbiology). The review area provides a variety of activities designed to help you study for your class:

Chapter Outlines. We've pulled out the section titles and full sentence sub-headings from each chapter to form natural descriptive outlines you can use to study the chapters' material point by point.

Review Questions. The review questions test your knowledge of the important concepts and applications in each chapter. Written by the author of the text, the review provides feedback for each correct or incorrect answer. This is an excellent test preparation tool.

Flash Cards. Studying human biology requires learning new terms. Virtual flash cards help you master the new vocabulary for each chapter.

Figure Labeling. You can practice identifying and labeling anatomical features on the same art content that appears in the text.

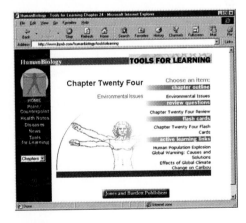

Active Learning Links. Active Learning Links connect to external web sites that provide an opportunity to learn basic concepts through demonstrations, animations, and hands-on activities.

APPENDIX A

PERIODIC TABLE OF ELEMENTS

Period	Group IA	Group IIA	Group IIIA	Group IVA	Group VA	Group VIA	Group VIIA	Group VIII			Group IB	Group IIB	Group IIIB	Group IVB	Group VB	Group VIB	Group VIIB	Group 0
Period 1	1 H 1.008																	2 He 4.003
Period 2	3 Li 6.941	4 Be 9.012											5 B 10.81	6 C 12.01	7 N 14.01	8 O 16.00	9 F 19.00	10 Ne 20.18
Period 3	11 Na 22.99	12 Mg 24.31											13 Al 26.98	14 Si 28.09	15 P 30.97	16 S 32.06	17 Cl 35.45	18 Ar 39.95
Period 4	19 K 39.10	20 Ca 40.08	21 Sc 44.96	22 Ti 47.90	23 V 50.94	24 Cr 52.00	25 Mn 54.94	26 Fe 55.85	27 Co 58.93	28 Ni 58.70	29 Cu 63.55	30 Zn 65.38	31 Ga 69.72	32 Ge 72.59	33 As 74.92	34 Se 78.96	35 Br 79.90	36 Kr 83.80
Period 5	37 Rb 85.47	38 Sr 87.62	39 Y 88.91	40 Zr 91.22	41 Nb 92.91	42 Mo 95.94	43 Tc (98)	44 Ru 101.1	45 Rh 102.9	46 Pd 106.4	47 Ag 107.9	48 Cd 112.4	49 In 114.8	50 Sn 118.7	51 Sb 121.8	52 Te 127.6	53 I 126.9	54 Xe 131.3
Period 6	55 Cs 132.9	56 Ba 137.3	57 La 138.9	72 Hf 178.5	73 Ta 180.9	74 W 183.9	75 Re 186.2	76 Os 190.2	77 Ir 192.2	78 Pt 195.1	79 Au 197.0	80 Hg 200.6	81 Tl 204.4	82 Pb 207.2	83 Bi 209.0	84 Po (209)	85 At (210)	86 Rn (222)
Period 7	87 Fr (223)	88 Ra (226.0)	89 Ac (227)	104 Unq	105 Unp	106 Unh	107 Uns		109 Une									

Lanthanides (rare earth metals)

58 Ce 140.1	59 Pr 140.9	60 Nd 144.2	61 Pm (145)	62 Sm 150.4	63 Eu 152.0	64 Gd 157.3	65 Tb 158.9	66 Dy 162.5	67 Ho 164.9	68 Er 167.3	69 Tm 168.9	70 Yb 173.0	71 Lu 175.0

Actinides

90 Th 232.0	91 Pa (231)	92 U 238.0	93 Np (244)	94 Pu (242)	95 Am (243)	96 Cm (247)	97 Bk (247)	98 Cf (251)	99 Es (252)	100 Fm (257)	101 Md (258)	102 No (259)	103 Lr (260)

KEY

16	— Atomic number
S	— Symbol of element
32.06	— Atomic mass

 Metals

 Nonmetals

Metalloids

Noble Gases

APPENDIX B

THE METRIC SYSTEM

Most Common English Units and the Corresponding Metric Units

Measure	English Unit	Metric Unit
Weight	tons	metric tons
	pounds	kilograms
	ounces	grams
Length	miles	kilometers
	yards	meters
	inches	centimeters
Square Measure	acres	hectares
	square miles	square kilometers
Volume	quarts and gallons	liters
	fluid ounces	milliliters

Converting English Units to Metric Units

Measure	English Unit	Metric Unit
Weight	1 ton (2000 pounds)	0.9 metric tons
	1 pound	0.454 kilograms
	1 ounce	28.35 grams
Length	*1 mile	1.6 kilometers
	1 yard	0.9 meters
	*1 inch	2.54 centimeters
Square Measure	1 acre	0.4 hectares
	1 square mile	2.59 square kilometers
Volume	1 quart	0.95 liters
	*1 gallon	3.78 liters
	1 fluid ounce	29.58 milliliters

*Most useful conversions to know.

Converting Metric Units to English Units

Measure	Metric Unit	English Unit
Weight	*1 metric ton	2204 pounds
	1 metric ton	1.1 tons
	*1 kilogram	2.2 pounds
	1 gram	28.35 ounces
Length	*1 kilometer	0.6 miles
	1 meter	1.1 yards
	1 centimeter	0.39 inches
Square Measure	*1 hectare	2.47 acres
	1 square kilometer	0.386 square miles
Volume	1 liter	1.057 quarts
	1 liter	0.26 gallons
	1 milliliter	0.0338 fluid ounces

*Most useful conversions to know.

GLOSSARY

Abiotic factors Physical and chemical components of an organism's environment.

Accommodation Change in the shape of the lens caused by contraction or relaxation of the smooth muscle of the ciliary body. Through accommodation, the lens adjusts the degree to which incoming light rays are bent, permitting objects to be focused on the retina.

Acetylcholine Neurotransmitter substance in the central and peripheral nervous systems of humans.

Acetylcholinesterase Enzyme that destroys the neurotransmitter acetylcholine in the synaptic cleft.

Achondroplasia Genetic disease that results from an autosomal dominant gene. Individuals with the disease have short legs and arms, but a relatively normal body size.

Acid Any chemical substance that adds hydrogen ions to a solution, such as sulfuric acid.

Acid deposition Deposition of sulfuric and nitric acids in the atmosphere onto the Earth's surface. Damages buildings, lakes, streams, crops, and forests. Acids come from sulfur dioxide and nitrogen dioxide produced during the combustion of fossil fuels.

Acid precursors Sulfur dioxide and nitrogen dioxide gases that combine with water and oxygen to form sulfuric and nitric acids in the atmosphere.

Acrosome Enzyme-filled cap over the head of a sperm. Helps the sperm dissolve its way through the corona radiata and zona pellucida of the ovum.

Actin microfilaments Contractile filaments made of protein and found in cells as part of the cytoskeleton. Especially abundant in the microfilamentous network beneath the plasma membrane and in muscle cells.

Action potential Recording of electrical change in membrane potential when a neuron is stimulated.

Activation energy The energy needed to force the electron clouds of reactants together before the formation of products; also, the energy needed to break internal chemical bonds.

Active immunity Immune resistance gained when an antigen is introduced into the body either naturally or through vaccination.

Active site The region of an enzyme molecule that binds substrates and performs the catalytic function of the enzyme.

Active transport Movement of molecules across membranes using protein molecules and energy supplied by ATP. Moves molecules and ions from regions of low to high concentration.

Adaptation Genetically based characteristic that increases an organism's chances of passing on its genes.

Adaptive radiation Process in which one species gives rise to many others that occupy different environments. Also known as divergent evolution.

Adenine One of the purine bases found in nucleotides such as ATP.

Adenosine diphosphate (ADP) Precursor to ATP; consists of adenine, ribose, and two phosphates; *see also* Adenosine triphosphate.

Adenosine triphosphate (ATP) A molecule composed of ribose sugar, adenine, and three phosphate groups. The last two phosphate groups are attached by "high-energy bonds" that require considerable energy to form but release that energy again when broken. ATP serves as the major energy carrier in cells.

Adenylate cyclase Enzyme bound to the inner surface of the plasma membrane. Linked to plasma membrane hormone receptors. Responsible for the conversion of ATP to cyclic AMP (a second messenger).

Adipose tissue Type of loose connective tissue containing numerous fat cells. Important storage area for lipids.

Adolescence Period of human life from puberty until adulthood. Characterized by sexual maturity.

Adrenal cortex Outer portion of the adrenal gland. Produces a variety of steroid hormones, including cortisol and aldosterone.

Adrenal gland Endocrine gland located on top of the kidney. Consists of two parts: adrenal cortex and medulla, each with separate functions.

Adrenal medulla Inner portion of the adrenal glands. Produces epinephrine (adrenalin) and norepinephrine (noradrenalin).

Adrenalin (epinephrine) Hormone secreted under stress. Contributes to the fight-or-flight response by increasing heart rate, shunting blood to muscles, increasing blood glucose levels, and other functions.

Adrenocorticotropic hormone (ACTH) Polypeptide hormone produced by the anterior pituitary. Stimulates the cells of the adrenal cortex, causing them to synthesize and release their hormones, especially glucocorticoids.

Aerobic exercise Exercise, such as swimming, that does not deplete muscle oxygen. Excellent for strengthening the heart and for losing weight.

Age-structure diagram (population histogram) Graphical representation of the number or percentage of males and females in various age groups in a population.

Agglutination Clumping of antigens that occurs when antibodies bind to several antigens.

Aging Inevitable and progressive deterioration of the body's function, especially its homeostatic mechanisms.

AIDS Acquired immune deficiency syndrome. Fatal disease caused by the HIV virus, which attacks T helper cells, greatly reducing the body's ability to fight infection.

Albinism Genetic disease resulting in a lack of pigment in the eyes or the eyes, skin, and hair. An autosomal recessive trait.

Aldosterone Steroid released by the adrenal cortex in response to a decrease in blood pressure, blood volume, and osmotic concentration. Acts principally on the kidney.

Alga (algae, plural) Heterogeneous group of aquatic plants consisting of three major groups: green, red, and brown. Important producer essential to aquatic food chains.

Alleles Alternative form of a gene.

Allergen Antigen that stimulates an allergic response.

Allergy Extreme overreaction to some antigens, such as pollen or foods. Characterized by sneezing, mucous production, and itchy eyes.

Allosteric inhibition Enzyme regulation in which an inhibitor molecule binds to an enzyme at a site away from the active site, changing the shape or charge of the active site so that it can no longer bind substrate molecules.

Allosteric site Region of an enzyme where products of metabolic pathways bind, changing the shape of the active site. In some enzymes this prevents substrates from binding to the active site; in others, it allows them to bind. Thus, allosteric sites can either turn on or turn off enzymes.

Alpine tundra A relatively cold, treeless region similar to the Arctic tundra but found on the tops of mountains.

Altitudinal biomes Regions of a mountain, each characterized by a distinct group of plants and animals. Results from the varying climatic conditions at different altitudes.

Alveoli Tiny, thin-walled sacs in the lung where oxygen and carbon dioxide are exchanged between the blood and the air.

Ameboid motion Cellular locomotion common in single-celled organisms and some cells in the human body. The cells send out slender cytoplasmic projections that attach to the substrate "ahead" of the cell. The cytoplasm flows into the projections, or pseudopodia, advancing the organism.

Amniocentesis Procedure whereby physicians extract cells and fluid from the amnion surrounding the fetus. The cells are examined for genetic defects, and the fluid is studied biochemically.

Amnion Layer of cells that separates from the inner cell mass of the embryo and eventually forms a complete sac around the fetus.

Amniotic fluid Liquid in the amniotic cavity surrounding the embryo and fetus during development. Helps protect the fetus.

Ampulla Enlarged area of each semicircular canal that houses receptor cells for movement.

Amylase Enzyme in saliva that helps break down starch molecules.

Anabolic reaction A reaction in which substances are formed; for example, the synthesis of glucose during photosynthesis is an anabolic reaction.

Anabolic steroids Synthetic androgen hormones that promote muscle development.

Analogous structures Anatomical structures that function similarly but differ in structure; for example, the wing of a bird and the wing of an insect.

Anaphase Phase of mitosis during which the chromatids of each chromosome begin to uncouple and are pulled in opposite directions with the aid of the mitotic apparatus.

Androgen-binding protein Cytoplasmic receptor protein that binds to and concentrates testosterone within the Sertoli cell of the seminiferous tubule. Production of ABP is stimulated by FSH.

Androgens Male sex steroids such as testosterone produced principally by the testes.

Anemia Condition characterized by an insufficient number of red blood cells in the blood or insufficient hemoglobin. Often caused by insufficient iron intake.

Aneuploidy Describes a genetic condition in which there is an abnormal number of chromosomes.

Aneurysm Ballooning of the arterial wall caused by a degeneration of the tunica media.

Anoxia Lack of oxygen.

Antagonistic Refers to hormones or muscles that exert opposite effects.

Anterior pituitary Major portion of the pituitary gland, which is controlled by hypothalamic hormones. Produces seven protein and polypeptide hormones.

Anthropoids Monkeys, the great apes, and humans.

Antibodies Proteins produced by immune system cells that destroy or inactivate antigens, including pollen, bacteria, yeast, and viruses.

Anticodon Sequence of three bases found on the transfer RNA molecule. Aligns with the codon on messenger RNA and helps control the sequence of amino acids inserted into the growing protein.

Anticodon loop Part of the transfer RNA molecule that bears three bases that bind to the three bases of the codon on messenger RNA.

Antidiuretic hormone (ADH) Hormone released by the posterior pituitary. Increases the permeability of the distal convoluted tubule and collecting tubules, increasing water reabsorption.

Antigens Any substance that is detected as foreign by an organism and elicits an immune response. Most antigens are proteins and large molecular weight carbohydrates.

Anvil (incus) One of three bones of the middle ear that helps transmit sound waves to the receptor for sound in the inner ear.

Aorta Largest artery in the body; carries the oxygenated blood away from the heart and delivers it to the rest of the body through many branches.

Appendicular skeleton Bones of the arms, legs, shoulders, and pelvis. Contrast with axial skeleton.

Aquatic life zones Ecologically distinct regions in fresh water and salt water.

Aqueous humor Liquid in the anterior and posterior chambers of the eye.

Aquifer Porous underground zone containing water.

Arctic tundra Massive biome north of taiga characterized by low precipitation and cold temperatures.

Arteries Vessels that transport blood away from the heart.

Arteriole Smallest of all arteries; usually drains into capillaries.

Arthropods Phylum of invertebrate animals with jointed appendages and chitinous exoskeletons such as insects, arachnids, and crustaceans. Many arthropods inhabit freshwater and marine ecosystems, but the vast majority live on land.

Arthroscope Device used to examine internal injuries, such as in joints.

Asexual reproduction Reproductive strategy common in single-celled organisms, such as the amoeba. Reproduction occurs by cell division.

Association cortex Area of the brain where integration occurs.

Association neurons Nerve cells that receive input from many sensory neurons and help process them, ultimately carrying impulses to nearby multipolar neurons.

Aster Array of microtubules found in the cell in association with the spindle fibers during cell division.

Asthma Respiratory disease resulting from an allergic response. Allergens cause histamine to be released in the lungs. Histamine causes the air-carrying ducts (bronchioles) to constrict, cutting down airflow and making breathing difficult.

Astigmatism Unequal curvature of the cornea (sometimes the lens) that distorts vision.

Atom Smallest particles of matter that can be achieved by ordinary chemical means, consisting of protons, neutrons, and electrons.

Atomic mass unit Unit used to measure atomic weight. One atomic mass unit is 1/12 the weight of a carbon atom.

Atomic weight Average mass of the atoms of a given element, measured in atomic mass units.

Atrioventricular bundle Tract of modified cardiac muscle fibers that conduct the pacemaker's impulse into the ventricular muscle tissue.

Atrioventricular node (AV node) Knot of tissue located in the right ventricle. Picks up the electrical signal arriving from the atria and transmits it down the atrioventricular bundle.

Atrioventricular valves Valves between the atria and ventricles.

Auditory (eustachian) tube Collapsible tube that joins the nasopharynx and middle ear cavities and helps equalize pressure in the middle ear.

Auricle (or pinna) Skin-covered cartilage portion of the outer ear.

Autocrines Chemical substances produced by cells, which affect the function of the cells producing them.

Autoimmune reaction Immune response directed at one's own cells.

Autonomic nervous system That part of the nervous system not under voluntary control.

Autosomal dominant trait Trait that is carried on the autosomes and is expressed in heterozygotes and homozygote dominants.

Autosomal recessive trait Trait that is carried on the autosomes and is expressed only when both recessive genes are present.

Autosomes All human chromosomes except the sex chromosomes.

Autotrophs Organisms such as plants that, unlike animals, are able to synthesize their own food.

Axial skeleton The skull, vertebral column, and rib cage. Contrast with appendicular skeleton.

Axon Long, unbranched process attached to the nerve cell body of a neuron. Transports bioelectric impulses away from the cell body.

Basal body Organelle located at the base of the cilium and flagellum. Consists of nine sets of microtubules arranged in a circle. Each "set" contains three microtubules.

Basilar membrane Membrane that supports the organ of Corti in the cochlea.

Batesian mimicry Strategy in which one species evolves a warning coloration similar to a species with active defense mechanisms such that the imitating species benefits from the outward similarity.

Behavior An individual's response to environmental stimuli such as other individuals of the same species, members of other species, or some aspect of the physical environment.

Benign tumor Abnormal cellular proliferation. Unlike a malignant tumor, the cells in a benign tumor stop growing after a while, and the tumor remains localized.

Beta cells Insulin-producing cells of the islets of Langerhans in the pancreas.

Bile Fluid produced by the liver and stored and concentrated in the gallbladder.

Bile salts Steroids produced by the liver, stored in the gallbladder, and released into the small intestine where they emulsify fats, a step necessary for enzyme digestion.

Binary fission Bacterial cellular division.

Bioelectric impulse Nerve impulse resulting from the influx of sodium ions along the plasma membrane of a neuron.

Biomass The dry weight of living material in an ecosystem.

Biomass pyramid Diagram of the amount of biomass at each trophic level in an ecosystem or, more commonly, a food chain.

Biome Terrestrial region characterized by a distinct climate and a characteristic plant and animal community.

Biorhythms (biological cycles) Naturally fluctuating physiological process.

Biosphere Region on Earth that supports life. Exists at the junction of the atmosphere, lithosphere, and hydrosphere.

Biotic factor Biological components of ecosystems.

Bipedal Refers to the ability to walk on two legs.

Birth control Any method or device that prevents conception and birth.

Birth control pill Pill generally taken to inhibit ovulation. The most commonly used pills contain both synthetic estrogen and progesterone.

Birth defects Physical or physiological defects in newborns. Caused by a variety of biological, chemical, and physical factors.

Blastocyst Hollow sphere of cells formed from the morula. Consists of the inner cell mass and the trophoblast.

Blood Specialized form of connective tissue. Consists of white blood cells, platelets, red blood cells, and plasma.

Blood clot Mass of fibrin containing platelets, red blood cells, and other cells. Forms in walls of damaged blood vessels, halting the efflux of blood.

B-lymphocytes Type of lymphocyte that transforms into a plasma cell when exposed to antigens.

Bone (organ) Structure comprised of bone tissue. Provides internal support, protects organs, and helps maintain blood calcium levels.

Bone (tissue) Tissue consisting of a calcified extracellular material with numerous cells (osteocytes) embedded in it.

Bowman's capsule Cup-shaped end of the nephron that participates in glomerular filtration.

Bowman's space Cavity between the inner and outer layer of Bowman's capsule.

Brain stem Part of the brain that consists of the medulla and pons. Houses structures, such as the breathing control center and reticular activating systems, that control many basic body functions.

Braxton-Hicks contractions Contractions that begin a month or two before childbirth. Also known as false labor.

Breathing center Aggregation of nerve cells in the brain stem that controls breathing.

Breech birth Delivery of a baby feet first.

Bronchi Ducts that convey air from the trachea to the bronchioles and alveoli.

Bronchioles Smallest ducts in the lungs. Their walls are largely made of smooth muscle that contracts and relaxes, regulating the flow of air into the lungs.

Bronchitis Infection or persistent irritation of the bronchi characterized by cough.

Buffer A chemical substance that helps resist changes in acidity.

Bulimia Eating disorder characterized by recurrent binge eating followed by vomiting.

Calcitonin (thyrocalcitonin) Polypeptide hormone produced by the thyroid gland that inhibits osteoclasts and stimulates osteoblasts to produce bone, thus lowering blood calcium levels.

Canaliculi Tiny canals in compact bone that provide a route for nutrients and wastes to flow to and from the osteocytes.

Cancer Any of dozens of diseases, all characterized by the uncontrollable replication various body of cells.

Capillaries Tiny vessels in body tissues whose walls are composed of a flattened layer of cells that allow water and other molecules to flow freely into and out of the tissue fluid.

Capillary bed Branching network of capillaries supplied by arterioles and drained by venules.

Capsid Protein coat of a virus.

Capsomere Globular proteins that make up the capsid of viruses.

Carbohydrate Organic compound consisting of carbon, hydrogen, and oxygen. A structural component of plant cells, it is used principally as a source of energy in animal cells.

Carbonic anhydrase Enzyme found in red blood cells that catalyzes the conversion of carbon dioxide to carbonic acid.

Carcinogens Cancer-causing agents.

Cardiac muscle Type of muscle found in the walls of the heart; it is striated and involuntary.

Carnivores Meat eaters; organisms that feed on other animals.

Carrier proteins Class of proteins that transport smaller molecules and ions across the plasma membrane of the cell. Are involved in facilitated diffusion.

Carriers Individuals who carry a gene for a particular trait that can be passed on to their children, but who do not express the trait.

Carrying capacity The number of organisms an ecosystem can support on a sustainable basis.

Cartilage Type of specialized connective tissue. Found in joints on the articular surfaces of bones and other locations such as the ears and nose.

Catabolic reaction A reaction in which molecules are broken down; for example, the breakdown of glucose is a catabolic reaction.

Catalyst Class of compounds that speed up chemical reactions. Although they take an active role in the process, they are left unchanged by the reaction. Thus, they can be used over and over again. Catalysts lower the activation energy of a reaction. *See also* Enzyme.

Cataracts Disease of the eye resulting in cloudy spots on the lens (and sometimes the cornea) that cause cloudy vision.

Cell body Part of the nerve cell that contains the nucleus and other cellular organelles; the center of chemical synthesis.

Cell culture Glass bottle or shallow dish containing nutrient medium and designed to permit cells to grow in the laboratory under controlled conditions.

Cell cycle Repeating series of events in the lives of many cells. Consists of two principal parts: interphase and cellular division.

Cell division Process by which the nucleus and the cytoplasm of a cell are split, creating two daughter cells. Consists of mitosis and cytokinesis.

Cellular respiration The complete breakdown of glucose in the cell, producing carbon dioxide and water. Composed of four separate but interconnected parts: glycolysis, the intermediate reaction, the citric acid cycle, and the electron transport system.

Central nervous system The brain and spinal cord.

Centriole Organelle consisting of a ring of microtubules, arranged in nine sets of three. Structurally identical to basal bodies, but associated with the spindle apparatus. Gives rise to the basal body in ciliated cells.

Centromere Region on each chromatid that joins with the centromere of its sister chromatid.

Cerebellum Structure of the brain that lies above the pons and medulla. It has many important functions including synergy.

Cerebral cortex Outer layer of each cerebral hemisphere, consisting of many multipolar neurons and nerve cell fibers.

Cerebral hemisphere Convoluted mass of nervous tissue located above the deeper structures, such as the hypothalamus and limbic system. Home of consciousness, memory, and sensory perception; originates much conscious motor activity.

Cervical cap Birth control device consisting of a small cup that fits over the tip of the cervix.

Cervix Lowermost portion of the uterus; it protrudes into the vagina.

Cesarean section Delivery of a baby via an incision through the abdominal and uterine walls.

Chemical equilibrium The point at which the "forward" reaction from reactants to products proceeds at the same rate as the "backward" reaction from products to reactants, so that there is no net change in chemical composition.

Chemical evolution Formation of organic molecules from inorganic molecules early in the history of the Earth.

Chemiosmosis Production of ATP in the chloroplast's and the mitochondrion's electron transport systems. H^+ flowing through pores in ATPase molecules drives the endergonic synthesis of ATP.

Chemosynthetic organisms Autotrophic cells that lack chlorophyll and acquire electrons from inorganic molecules.

Chlamydia Bacterium that causes nonspecific urethritis, a type of sexually transmitted disease.

Chlorofluorocarbons (CFCs) Chemical substances once used as spray can propellants and refrigerants. These substances drift to the upper stratosphere and dissociate. Chlorine released by CFCs reacts with ozone, thus eroding the ozone layer.

Chlorophyll A green pigment that acts as the primary light-trapping molecule for photosynthesis.

Cholecystokinin (CCK) Hormone produced by cells of the duodenum when chyme is present. Causes the gallbladder to contract, releasing bile.

Chordae tendineae Tendinous cords that anchor the atrioventricular valves to the inner walls of the ventricles.

Chorionic villus biopsy Medical procedure to detect genetic defects; involves removing a small portion of the villi and then examining it for chromosomal abnormalities.

Choroid Middle layer of the eye that absorbs stray light and supplies nutrients to the eye.

Chromatid Strand of the chromosome consisting of DNA and protein.

Chromatin Long, threadlike fibers containing DNA and protein in the nucleus.

Chromosome Structure found in the nucleus of cells that houses the genetic information. Consists of DNA and protein.

Chromosome-to-pole fibers Microtubules of the spindle that extend from the centriole to the chromosome where they attach. They play a crucial role in separating the double-stranded chromosomes during mitosis.

Chronic bronchitis Persistent irritation of the bronchi, which causes mucous buildup, coughing, and difficulty breathing.

Chyme Liquified food in the stomach.

Ciliary body Portion of the middle layer of the eye that is located near the lens. Contains smooth muscle that constricts, thus helping control the shape of the lens and permitting the eye to focus.

Circadian rhythm Biorhythm that occurs on a daily cycle.

Circulatory system Organ system consisting of the heart, blood vessels, and blood.

Circumcision Operation to remove the foreskin of the penis. Generally performed on newborns.

Cisterna Channels of the endoplasmic reticulum and Golgi.

Citric acid Six-carbon compound produced in the first reaction of the citric acid or Krebs cycle. Formed when oxaloacetate reacts with acetyl Coenzyme A.

Citric acid cycle A cyclic series of reactions in which the pyruvates produced by glycolysis are broken down to CO_2, accompanied by the formation of ATP and electron carriers. Occurs in the matrix of mitochondria.

Class Subgroup of a phylum.

Clitoris Small knot of tissue located where the labia minora meet. Consists of erectile tissue.

Cloning Technique of genetic engineering whereby many copies of a gene are produced.

Closed system System that receives no materials from the outside.

Coacervates Microscopic globules that selectively incorporate molecules from their environment. Coacervates existing on Earth over 3 billion years ago may have given rise to the earliest cells.

Cochlea Sensory organ of the inner ear that houses the receptor for hearing.

Codominant Refers to two equally expressed alleles.

Codon Three adjacent bases in the messenger RNA that code for a single amino acid.

Coelom Body cavity (a space between the body wall and internal organs) lined by mesoderm.

Coevolution Process in which two or more species act as selective forces on each other, resulting in anatomical, behavioral, and functional changes in each other.

Collecting tubules Tubules in the kidney into which nephrons drain. They converge and drain into the renal pelvis.

Color blindness Condition that occurs in individuals who have a deficiency of certain cones. The most common form involves difficulty in distinguishing between red and green.

Colostrum Protein-rich product of the breast, produced for two to three days immediately after delivery and prior to milk production.

Community All of the plants, animals, and microorganisms in an ecosystem.

Compact bone Dense bony tissue in the outer portion of all bones.

Competition Struggle by two or more individuals for the same limited resource.

Complement Group of blood proteins that circulate in the blood in an inactive state until the body is invaded by bacteria; then they help destroy the bacteria.

Complementary base pairing Unalterable coupling of the purine adenine to the pyrimidine thymine, and the purine guanine to the pyrimidine cytosine. Responsible for the accurate transmission of genetic information from parent to offspring.

Condom Birth control device, consisting of a thin latex rubber sheath or other material that is rolled onto the erect penis. Prevents sperm from entering the vagina and helps prevent the spread of sexually transmitted diseases.

Conduction deafness Loss of hearing that occurs when the conduction of sound waves to the inner ear is impaired. May be caused by ruptured eardrum or damage to ossicles.

Cone Type of photoreceptor that operates in bright light; is responsible for color vision.

Connective tissue One of the primary tissues. It contains cells and varying amounts of extracellular material and holds cells together, forming tissues and organs.

Connective tissue proper Name referring to loose and dense connective tissue; supports and joins various body structures.

Consumers Organisms that eat plants and algae (producers) and other consumers.

Contact inhibition Cessation of growth that results when two or more cells contact each other. A feature of normal cells but absent in cancer cells.

Continental shelf Gradually sloping ocean bottom next to continents.

Contraceptive Any measure that helps prevent fertilization and pregnancy.

Contractile vacuole Vacuole in amoeboid protists that collects excess water from the cytoplasm and, when full, contracts, voiding the water through a temporary opening in the plasma membrane.

Control group A group of subjects (plants, animals, etc.) that is used in an experiment; this group is identical to the experimental group in all regards except that it does not receive the experimental treatment.

Convergence Inward turning of the eyes to focus on a nearby object.

Coral reef Biologically rich life zones, found in relatively warm and shallow waters in the tropics or nearby regions consisting of calcium carbonate or limestone produced by calcarous red and green algae and by colonies of organisms called stony corals.

Corepressor Molecule that binds to a repressor protein, allowing it to bind to the operator site, which, in turn, shuts down the structural genes by blocking RNA polymerase.

Cornea Clear part of the wall of the eye continuous with the sclera; allows light into the interior of the eye.

Coronary bypass surgery Surgical technique used to reestablish blood flow to the heart muscle by grafting a vein to shunt blood around a clogged coronary artery.

Corpus luteum (CL) Structure formed from the ovulated follicle in the ovary; produces estrogen and progesterone.

Cortical granules Secretory vesicles lying beneath the plasma membrane of the oocyte that are released when a sperm cell contacts the oocyte membrane. They block additional sperm from fertilizing the ovum.

Cortisol Glucocorticoid hormone that increases blood glucose by stimulating gluconeogenesis. Also stimulates protein breakdown in muscle and bone.

Coupled reactions A pair of reactions, one exergonic and one endergonic, that are linked so that the energy produced by the exergonic reaction provides the energy to drive the endergonic reaction.

Cowper's gland Smallest of the sex accessory glands; empties into the urethra.

Cranial nerves Nerves arising from the brain and brain stem.

Creatine phosphate High-energy molecule in skeletal muscle.

Cristae Folds formed by the inner membrane of the mitochondrion.

Critical thinking Process in which one seeks to determine the validity of conclusions based on sound reasoning. This often requires one to examine hidden bias, the appropriateness of experimental design, and other factors.

Cro-Magnons Earliest known members of *Homo sapiens sapiens.*

Crossing over Exchange of chromatin by homologous chromosomes during prophase I of meiosis. Results in considerably more genetic variation in gametes and offspring.

Crystallin Protein inside the lens that may denature, causing cataracts.

Culture The ideas, customs, skills, and arts of a given people in a given time that can change or evolve over time.

Cushing's disease Disease that results from pharmacologic doses of cortisone usually administered for rheumatoid arthritis or allergies.

Cyanobacteria The most abundant of the photosynthetic bacteria; once called blue-green algae.

Cyclic AMP Nucleotide derived from ATP. Its synthesis is stimulated when protein and polypeptide hormones bind to the plasma membrane of cells. In the cytoplasm, it activates protein kinase, which, in turn, activates other enzymes.

Cyclosporine One of numerous drugs used to suppress graft rejection.

Cystic fibrosis Autosomal recessive disease that leads to problems in sweat glands, mucous glands, and the pancreas. Pancreas may become blocked, thus reducing the flow of digestive enzymes to the small intestine. Mucus buildup in the lungs makes breathing difficult.

Cytokinesis Cytoplasmic division brought about by the contraction of a microfilamentous network lying beneath the plasma membrane at the midline. Usually begins when the cell is in late anaphase or early telophase.

Cytoplasm Material occupying the cytoplasmic compartment of a cell. Consists of a semifluid substance, the cytosol, containing many dissolved substances, and formed elements, the organelles.

Cytoskeleton A network of protein tubules in the cytoplasmic compartment of a cell. Attaches to many organelles and enzyme molecules and thus helps organize cellular activities, increasing efficiency.

Cytotoxic cells Type of T cell (T-lymphocyte) that attacks and kills virus-infected cells, parasites, fungi, and tumor cells.

Daughter cells Cells produced during cell division.

Decibel Unit used to measure the intensity of sound.

Deciduous Term referring to trees that shed their leaves during the fall.

Decomposer food chain Series of organisms that feed on organic wastes and the dead remains of other organisms.

Defibrillation Procedure to stop fibrillation (erratic electrical activity) of the heart.

Deletion Loss of a piece of a chromosome.

Demographer Scientist who studies populations.

Dendrite Short, highly branched fiber that carries impulses to the nerve cell body.

Dense connective tissue Type of connective tissue that consists primarily of densely packed fibers, such as those found in ligaments and tendons.

Deoxyribonuclease Pancreatic enzyme that breaks RNA and DNA into shorter chains.

Depth perception Ability to judge the relative position of objects in our visual field.

Dermis Layer of dense irregular connective tissue that binds the epidermis to underlying structures.

Desert Biome characterized by low rainfall and a hot climate. Contains organisms well adapted to these conditions.

Detrivores Organisms that feed on animal waste or the remains of plants and animals.

Diabetes insipidus Condition caused by lack of the pituitary hormone ADH. Main symptoms are polydipsia (excessive drinking) and polyuria (excessive urination).

Diabetes mellitus Insulin disorder either resulting from insufficient insulin production or decreased sensitivity of target cells to insulin. Results in elevated blood glucose levels unless treated.

Dialysis Procedure used to treat patients whose kidneys have failed. Blood is removed from the body and pumped through an artificial filter that removes impurities.

Diaphragm (birth control) Birth control device consisting of a rubber cup that fits over the end of the cervix. Used in conjunction with spermicidal jelly or cream.

Diaphragm (muscle) Dome-shaped muscle that separates the abdominal and thoracic cavities.

Diaphysis Shaft of the long bones. Consists of an outer layer of compact bone and an inner marrow cavity.

Diastolic pressure The pressure at the moment the heart relaxes. The lower of the two blood pressure readings.

Differentiation Structural and functional divergence from the common cell line. Occurs during embryonic development.

Dihybrid cross Procedure where one plant is bred with another to study two traits.

Diplopia Double vision. May occur when the eyes fail to move synchronously.

Distal convoluted tubule Section of the nephron that connects the loop of Henle to the collecting tubule. Site of tubular reabsorption.

Divergent evolution Process in which organisms evolve in different directions due to exposure to different environmental influences.

Diverticulitis Expansion of the large intestine due to obstruction.

DNA polymerase Enzyme that helps align the nucleotides and join the phosphates and sugar molecules in a newly forming DNA strand.

Dominant Adjective used in genetics to refer to an allele that is always expressed in heterozygotes. Designated by a capital letter.

Dorsal root (of a spinal nerve) Inlet for sensory nerve fibers to the spinal cord.

Double helix Describes the helical structure formed by two polynucleotide chains making up the DNA molecule.

Doubling time Time it takes a population to double.

Down syndrome Genetic disorder caused by an additional chromosome 21 that results in distinctive facial characteristics and mental retardation. Also known as Trisomy 21.

Dryopithecus Genus of apelike creatures that is thought to have given rise to the gibbons, gorillas, orangutans, and chimpanzees.

Duodenum First portion of the small intestine; site where most food digestion and absorption takes place.

Dust cell Cell found in and around the alveoli; phagocytizes particulate matter that has entered the lung.

E. coli Common bacterium that lives in the large intestine of humans and other mammals. Digests leftover glucose and other materials from food. Used in much genetic research.

Ecological niche An organism's habitat and all of the relationships that exist between that organism and its environment.

Ecological system (ecosystem) System consisting of organisms and their environment and all of the interactions that exist between these components.

Ecology Study of living organisms and the web of relationships that binds them together in the economy of nature. The study of ecosystems.

Ecosystem balance Dynamic equilibrium in ecosystems. Maintained by the interplay of growth and reduction factors.

Ectoderm One of the three types of cells that emerges in human embryonic development. Gives rise to the skin and associated structures, including the eyes.

Edema Swelling resulting from the buildup of fluid in the tissues.

Effector General term for any organ or gland that is controlled by the nervous system.

Ejaculation Ejection of semen from the male reproductive tract.

Elastic arteries Arteries that contain numerous elastic fibers interspersed among the smooth muscle cells of the tunica media.

Elastic cartilage Type of cartilage containing many elastic fibers found in regions where support and flexibility are required.

Electron Highly energetic particle carrying a negative charge that orbits the nucleus of an atom.

Electron carrier A molecule that can reversibly gain and lose electrons. Electron carriers generally accept high-energy electrons produced during an exergonic reaction and donate the electrons to acceptor molecules that use the energy to drive endergonic reactions.

Electron cloud A region surrounding the nucleus of an atom where electrons orbit.

Electron transport system Series of protein molecules in the inner membrane of the mitochondrion that pass electrons from the citric acid cycle from one to another, eventually donating them to oxygen. The electrons come from the citric acid cycle. During their journey along this chain of proteins, the electrons lose energy, which is used to make ATP. *See also* Chemiosmosis.

Elements Purest form of matter; substances that cannot be separated into different substances by chemical means.

Emphysema Progressive, debilitating disease that destroys the tiny air sacs in the lung (alveoli), caused by smoking and air pollution.

Endergonic reaction A chemical reaction that requires an input of energy to proceed.

Endocrine glands Glands of internal secretion that produce hormones secreted into the bloodstream.

Endocrine system Numerous, small, hormone-producing glands scattered throughout the body.

Endocytosis Process by which cells engulf solid particles, bacteria, viruses, and even other cells.

Endoderm One of the three types of cells that emerges during embryonic development. Gives rise to the intestinal tract and associated glands.

Endolymph Fluid inside the semicircular canals that deflects the cupula, signaling rotational movement of the head and body.

Endometrium Uterine endothelium or lining.

Endoplasmic reticulum Branched network of channels found throughout the cytoplasm of many cells. Formed from flattened sheets of membrane derived from the nuclear membrane. See rough and smooth endoplasmic reticulum for functions.

Endosymbiotic evolution Theory that accounts for the development of the first eukaryotes. Says that free-living bacteria-like organisms were engulfed by other cells and became internal symbionts. Internal symbionts later became the organelles of eukaryotes.

Endothelium Single-celled lining of blood vessels.

End-product inhibition The inhibition of an enzyme by a product of the chemical reaction it catalyzes or by the product of a series of chemical reactions of which the enzyme is a part; may result from binding of the end product to the allosteric site or active site of the enzyme.

Energy The capacity to do work.

Energy carrier A molecule that stores energy in "high-energy" chemical bonds and releases the energy again to drive coupled endergonic reactions. ATP is the most common energy carrier in cells; *NAD* and *FAD* are others.

Energy pyramid Diagram of the amount of energy at various trophic levels in a food chain or ecosystem.

Enhancer Segment of DNA that increases the activity of nearby genes several hundred times.

Envelope Protective membrane of some viruses; lies outside the capsid.

Enzyme A protein catalyst that speeds up the rate of specific biochemical reactions.

Epidermis Outermost layer of the skin that protects underlying tissues from drying out and from bacteria and viruses.

Epididymal duct Duct within the epididymis; site where sperm are stored until ejaculation.

Epididymis Storage site of sperm. Located on the testis, it consists of a long, tortuous duct, the epididymal duct.

Epiglottis Flap of tissue that closes off the trachea during swallowing.

Epilimnion The warm-water layer of a lake located along the surface.

Epiphyseal plate Band of cartilage cells between the shaft of the bone and the epiphysis. Allows for bone growth.

Epiphysis Expanded end of the long bones.

Episiotomy Surgical incision that runs from the vaginal opening toward the rectum. Enlarges the vaginal opening, easing childbirth.

Epithelium One of the primary tissues. Forms linings and external coatings of organs.

Erectile tissue Spongy tissue of the penis that fills with blood during sexual excitement, making the penis turgid.

Erythropoietin Hormone produced by the kidney when oxygen levels decline. Stimulates red blood cell production in the bone marrow.

Esophagus Muscular tube that transports food to the stomach.

Essential amino acid One of nine amino acids that must be provided in the human diet.

Estuarine zone Estuary and coastal wetland. One of the richest coastal life zones.

Euchromatin Metabolically active chromatin.

Euglenoids Protists with characteristics of both plants and animals. They get their name from the best-known of their kind, *Euglena*.

Eukaryote Any cell containing a distinct nucleus and organelles. They are found in single-celled organisms of the kingdom Protista and all multicellular organisms of the kingdoms Plantae, Animalia, and Fungi.

Evolution Process that leads to structural, functional, and behavioral changes in species, making them better able to survive in their environment; also leads to the formation of new species. Results from natural genetic variation and environmental conditions that select for organisms best suited to their environment.

Exergonic reaction A chemical reaction that releases energy.

Exhalation Expulsion of air from the lungs.

Exocrine gland Gland of external secretion; empties its contents into ducts.

Exocytosis Process by which cells release materials stored in secretory vesicles. The reverse of endocytosis.

Exon Expressed segment of DNA.

Experiment Test performed to support or refute a hypothesis.

Exponential growth Type of growth that occurs when a value grows by a fixed percentage and the increase is applied to the base amount.

Extension Movement of a body part (limbs, fingers, and toes) that opens a joint.

External auditory canal Channel that directs sound waves to the eardrum.

External genitalia External portion of the female reproductive system consisting of the clitoris, labia minora, and labia majora.

External sphincter of the bladder Voluntary muscular valve that controls urine release under conscious control. Formed by a flat band of muscle that forms the floor of the pelvic cavity.

Extrinsic eye muscles Six muscles located outside the eye that are responsible for eye movement.

Facilitated diffusion Process in which carrier proteins shuttle molecules across plasma membranes. The molecules move in response to concentration gradients.

Feces Semisolid material containing undigested food, bacteria, ions, and water; produced in the large intestine.

Feedback mechanism A mechanism in which the product of one process regulates the rate of another process, either turning it on or shutting it off.

Fermentation Process occurring in eukaryotic cells in the absence of oxygen, during which pyruvic acid is converted to lactic acid. Also occurs in those prokaryotes that live in oxygen-free environments.

Fertilization Union of sperm and ovum.

Fiber Any of the indigestible polysaccharides in fruits, vegetables, and grains.

Fibrillation Cardiac muscle spasms occurring during heart attacks due to a loss of synchronized electrical signals.

Fibrin Fibrous protein produced from fibrinogen, a soluble plasma protein. Helps form blood clots.

Fibrinogen Protein in plasma that forms fibrin.

Fibroblast Connective tissue cell, found in loose and dense connective tissues that produces collagen, elastic fibers, and a gelatinous extracellular material; responsible for repairing damage created by cuts or tears to connective tissue.

Fibrocartilage Type of cartilage whose extracellular matrix consists of numerous bundles of collagen fibers. Principally found in the intervertebral disks.

Fight-or-flight response An automatic response of an organism to danger that enables it to flee or stand and fight; it is activated by the autonomic nervous system and results in an increase in heart rate and breathing and an increase in blood flow to the muscles.

First law of thermodynamics Law stating that energy can be converted from one form to another but is never created or destroyed.

First polar body Cast-off nuclear material produced during the first meiotic division during oogenesis.

Fitness Measure of reproductive success of an organism and, therefore, the genetic influence an individual has on future generations.

Flagellum Long, whiplike extension of the plasma membrane of certain protozoans and sperm cells in humans. Used for motility.

Flavine adenine dinucleotide (FAD) A molecule that carries high-energy electrons from one chemical reaction to another; found in the citric acid cycle.

Flexion Movement of a limb, finger, or toe that involves closing a joint.

Follicle (ovary) Structure found in the ovary. Each follicle contains an oocyte and one or more layers of follicle cells that are derived from the loose connective tissue of the ovary surrounding the follicle.

Follicle (thyroid) Structure found in the thyroid gland. Consists of an outer layer of cuboidal cells surrounding thyroglobulin, a proteinaceous material from which thyroxine is formed.

Follicle-stimulating hormone (FSH) Gonadotropic hormone from the anterior pituitary that promotes gamete formation in both men and women.

Food chain Series of organisms in an ecosystem in which each organism feeds on the organism preceding it.

Food vacuole Membrane-bound vacuole in a cell containing material engulfed by the cell.

Food web All of the connected food chains in an ecosystem.

Foreskin Sheath of skin that covers the glans penis.

Fossil Remains or imprints of organisms that lived on Earth many years ago, usually embedded in rocks or sediment.

Fovea centralis Tiny spot in the center of the macula of the eye that contains only cones. Objects are focused onto the fovea for sharp vision.

Fusion Joining of two atoms, which releases large amounts of energy.

Gallbladder Sac on the underside of the liver that stores and concentrates bile.

Gastrin Stomach hormone that stimulates HCl production and release by the gastric glands.

Gastroesophageal sphincter Ring of muscle located in the lower esophagus that opens when food arrives, allowing food to pass into the stomach, and then closes to keep food and stomach acid from percolating upward.

Gene Segment of the DNA that controls cell structure and function.

Gene flow Introduction of new genes into a population when new individuals join the population.

Gene pool All the genes of all of the members of a population or species.

Gene therapy The use of artificially produced genes to treat, even cure, diseases.

Genera Plural of genus.

Genetic engineering The artificial manipulation of genes in which certain genes from one organism are removed and transferred to another organism of the same or a different species. This permits scientists to transfer important genes to improve species, for example, to increase resistance to disease.

Genome All of the genes of an organism.

Genotype Genetic makeup of an organism.

Genus Subgroup of a family of organisms.

Geographic isolation Physical separation of a population by some barrier. Sometimes results in reproductive isolation and the formation of new species.

Germ cell Refers to the sperm or ovum (egg) and the cells from which they are derived; germ cells contain half the chromosomes of somatic cells.

Germinal epithelium Germ cells in the wall of the seminiferous tubule that give rise to sperm.

Gestation The period of pregnancy.

Glans penis Slightly enlarged tip of the penis.

Glaucoma Disease of the eye caused by pressure resulting from a buildup of aqueous humor in the anterior chamber.

Glomerular filtration Movement of materials out of the glomeruli into Bowman's capsule in the kidney.

Glomerulus Tuft of capillaries that make up part of the nephron; site of glomerular filtration.

Glucagon Hormone released by the pancreas that stimulates the breakdown of glycogen in the liver and the release of glucose molecules, thus increasing blood levels of glucose.

Glucocorticoids Group of steroid hormones produced by the adrenal cortex that stimulate gluconeogenesis.

Gluconeogenesis Synthesis of glucose from fatty acids and amino acids. Takes place in the liver where amino acids and fatty acids are stored.

Glycogenolysis Breakdown of glycogen, releasing glucose.

Glycolysis Metabolic pathway in the cytoplasm of the cell, during which glucose is split in half, forming two molecules of pyruvic acid. The energy released during the reaction is used to generate two molecules of ATP.

Glycoproteins Proteins that have carbohydrate attached to them.

Goiter Condition in which the thyroid gland enlarges due to lack of dietary iodide.

Golgi complex Organelle consisting of a series of flattened membranes that form channels. It sorts and chemically modifies molecules and repackages its proteins into secretory vesicles.

Golgi tendon organs Special receptors found in tendons that respond to stretch. Also known as neurotendinous organs.

Gonadotropin General term for FSH and LH, which are produced by the anterior pituitary and target male and female gonads.

Gonadotropin-releasing hormone or GnRH Hormone produced by the hypothalamus that controls the release of FSH (ICSH in males) and LH.

Gonorrhea Sexually transmitted disease caused by a bacterium.

Gray matter Gray, outermost region of the cerebral cortex.

Grazer Herbivorous organism.

Grazer food chain Food chain beginning with plants and grazers (herbivores).

Greenhouse gas Gas, such as carbon dioxide and chlorofluorocarbons, that traps heat escaping from the Earth and radiates it back to the surface.

Growth factor Any biotic or abiotic factor that causes a population to grow.

Growth hormone A protein hormone produced by the anterior pituitary that stimulates cellular growth in the body, causing cellular hypertrophy and hyperplasia. Its major targets are bone and muscle.

Growth rate (of a population) Determined by subtracting the death rate from the birth rate.

Habitat Place in which an organism lives.

Habituation Condition where sensory receptors stop generating impulses, even though a stimulus is still present.

Hammer (malleus) One of three bones of the middle ear. Abuts the tympanic membrane and helps transmit sound from the eardrum to the inner ear.

Helper cell Type of T-lymphocyte that stimulates the proliferation of T and B cells when antigen is present.

Heme group Subunit of the hemoglobin molecule. Consists of a porphyrin ring and a central iron ion to which oxygen binds.

Hemoglobin Protein molecules inside red blood cells; binds to oxygen.

Hemophilia Disease caused by a gene defect occurring on the Y chromosome. Results in absence of certain blood-clotting factors.

Herbicide Any chemical applied to crops to control weeds.

Herbivore Any organism that feeds directly on plants. Also known as a grazer.

Herpes One of the most common sexually transmitted diseases; caused by a virus.

Heterochromatin Inactive chromatin that is slightly coiled or compacted in the interphase nucleus.

Heterotrophic fermenter Evolutionarily probably one of the first cells. Absorbed glucose from the environment and broke it down by anaerobic glycolysis.

Heterotrophs Organisms such as animals that, unlike plants, are unable to synthesize their own food. Consume plants and other organisms.

Heterozygous Adjective describing a genetic condition in which an individual contains one dominant and one recessive gene in a gene pair.

High-density lipoproteins (HDLs) Complexes of lipid and protein that transport cholesterol to the liver for destruction.

Histamine Potent vasodilator released by certain cells in the body during allergic reactions.

Histone Globular protein thought to play a role in regulating the genes.

Homeostasis A condition of dynamic equilibrium within any biological or social system. Achieved through a variety of automatic mechanisms that compensate for internal and external changes.

Hominid First humanlike creatures.

Hominoids Subgroup of anthropoids.

Homo sapiens neanderthalensis The Neanderthals. Subspecies of *Homo sapiens.*

Homo sapiens sapiens Species of modern humans that emerged about 400,000 years ago.

Homologous structures Structures thought to have arisen from a common origin.

Homozygous Adjective describing a genetic condition marked by the presence of two identical alleles for a given gene.

Hormone Chemical substance produced in one part of the body that travels to another, typically through the bloodstream, where it elicits a response.

Human chorionic gonadotropin (HCG) Hormone produced by the embryo that stimulates the corpus luteum in the mother's body to produce estrogen.

Humoral immunity Immune reaction that protects the body primarily against viruses and bacteria in the body fluids via antibodies produced by plasma cells.

Hyperglycemia High blood glucose levels.

Hyperopia (farsightedness) Condition that occurs when the eyeball is too short or the lens is too weak, resulting in poor focus on nearby objects.

Hypertension High blood pressure.

Hypertonic Adjective describing a solution with a higher solute concentration than the cell's cytoplasm, causing the cell to shrivel.

Hypolimnion The coldest water of a lake, which lies below the thermocline.

Hypothalamus Structure in the brain located beneath the thalamus. It consists of many aggregations of nerve cells and controls a variety of autonomic functions aimed at maintaining homeostasis.

Hypothesis Tentative and testable explanation for a phenomenon or observation.

Hypotonic Adjective describing a solution with a solute concentration lower than the cell's cytoplasm, resulting in a swelling of the cell.

I gene Gene that controls blood type through the synthesis of glycoproteins that end up on the plasma membrane of the red blood cell.

Immune system Diffuse system consisting of trillions of cells that circulate in the blood and lymph and take up residence in the lymphoid organs, such as the spleen, thymus, lymph nodes, and tonsils, as well as other body tissues. Helps protect the body against foreign cells, such as bacteria and viruses, and protects against cancer cells.

Immunity Term referring to the resistance of the body to infectious disease.

Immunocompetence Process in which lymphocytes mature and become capable of responding to specific antigens.

Immunoglobulins Antibodies.

Implantation Process in which the blastocyst embeds in the uterine lining.

Impotence Inability of a male to achieve an erection.

Imprinting Attachment of young animals such as birds to their mothers or artificial substitutes during a critical period.

In vitro Term referring to any procedure carried out in a test tube or petri dish, such as *in vitro* fertilization.

Incomplete dominance Partial dominance. Occurs when an allele exerts only partial dominance over another allele, resulting in an intermediate trait.

Incontinence Inability to control urination.

Induced abortion Deliberate expulsion of a fetus or embryo.

Inducer Chemical substance that activates inducible genes.

Inducible operon Set of genes that remains inactive until needed. Activated by inducers.

Infectious mononucleosis White blood cell disorder caused by a virus. Characterized by a rapid increase in the number of monocytes and lymphocytes in the blood.

Inferior vena cava Large vein that empties deoxygenated blood from the body below the heart into the right atrium of the heart.

Infertility Inability to conceive; can be due to problems in either the male or the female or both partners.

Inflammatory response Response to tissue damage including an increase in blood flow, the release of chemical attractants, which draw monocytes to the scene, and an increase in the flow of plasma into a wound.

Inhalation Process of air being drawn into the lungs.

Inhibin Substance produced by the seminiferous tubules that inhibits the production of FSH by the anterior pituitary.

Inhibiting hormone Hormone from the hypothalamus that inhibits the release of hormones from the anterior pituitary.

Initiator codon Codon found on a messenger RNA strand that marks where protein synthesis begins.

Inner cell mass Cells of the blastocyst that become the embryo and amnion.

Insulin Hormone that stimulates the uptake of glucose by body cells, especially muscle and liver cells. Stimulates the synthesis of glycogen in liver and muscle cells.

Insulin-dependent diabetes Type of diabetes that can only be treated with injections of insulin. May be caused by an autoimmune reaction. Also known as early-onset diabetes.

Insulin-independent diabetes Type of diabetes that often occurs in obese people. In most patients, it can be controlled by diet. Also known as late-onset diabetes.

Integral protein Large protein molecules in the lipid bilayer of the plasma membrane.

Integration Process of making sense of various nervous inputs so that a meaningful response can be achieved.

Intercostal muscles Short, powerful muscles that lie between the ribs. Involved in inspiration and active exhalation.

Interferon Protein released from cells infected by viruses that stops the replication of viruses in other cells.

Interleukin 2 Chemical released by helper cells that activates T and B cells, stimulating cell division.

Internal sphincter (of the bladder) Involuntary muscular valve that relaxes reflexively, releasing urine. Formed by a smooth muscle in the neck of the bladder at the junction of the bladder and the urethra.

Interphase Period of cellular activity occurring between cell divisions. Synthesis and growth occur in preparation for cell division.

Interstitial cells Testosterone-producing cells located in the loose connective tissue between the seminiferous tubules of the testes.

Interstitial cell stimulating hormone (ICSH) Luteinizing hormone in males. Stimulates testosterone secretion.

Interstitial fluid Fluid surrounding cells in body tissues. Provides a path through which nutrients, gases, and wastes can travel between the capillary and the cells.

Intervertebral disks Shock-absorbing material between the bones of the spine.

Intrauterine device (IUD) Birth control device that consists of a small plastic or metal object with a string attached that is inserted into the uterus through the cervix to prevent implantation.

Intron Segment of DNA that is not expressed. Lies between exons (expressed segments).

Invertebrates Animals without a spinal column such as insects.

Ion Atom that has gained or lost one or more electrons. May be either positively or negatively charged.

Ionic bond Weak bond that forms between oppositely charged ions.

Iris Colored segment of the middle layer of the eye visible through the cornea.

Irritability Ability to perceive and respond to stimuli.

Islets of Langerhans Group of endocrine cells found in the pancreas that produce insulin and glucagon.

Isotonic Having the same solute concentration as a cell or body fluid.

Isotope Alternative form of an atom; differs from other atoms in the number of neutrons found in the nucleus.

Joint capsule Connective tissue that connects to the opposing bones of a joint and forms the synovial cavity. The inner layer of the joint capsule produces synovial fluid.

Kidney Organ that rids the body of wastes and plays a key role in regulating the chemical constancy of blood.

Kingdom A large grouping of organisms; scientists typically recognize five major kingdoms: Monera, Protista, Animalia, Plantae, and Fungi.

Klinefelter syndrome Genetic disorder that results from an XXY genotype.

Krebs cycle *See citric acid cycle.*

Labia majora Outer folds of skin of the external genitalia in women.

Labia minora Inner folds of the external genitalia in women.

Labor The process or period of childbirth.

Lactation Milk production in the breasts.

Laparoscope Instrument used to examine internal organs through small openings made in the skin and underlying muscle.

Laryngitis Inflammation of the lining of the larynx, resulting in hoarseness. Caused by bacterial and viral infection and also excessive use of the voice.

Larynx Rigid but hollow cartilaginous structure that houses the vocal cords and participates in swallowing.

Learning Process in which stored information can lead to changes in innate behavior.

Lens Transparent structure that lies behind the iris and in front of the vitreous humor. Focuses light on the retina.

Leukemia Cancer of white blood cells.

Leukocytosis An increase in the concentration of white blood cells, which often occurs during a bacterial or viral infection.

Life expectancy Average length of time a person will live.

Ligament Connective tissue structure that runs from bone to bone, located alongside and sometimes inside the joint. Offers support for joints.

Limbic system Array of structures in the brain that work in concert with centers of the hypothalamus. Site of instincts and emotions.

Limiting factor One factor that is most important in regulating growth in an ecosystem.

Lipase Enzyme that removes some of the fatty acids from the glycerol molecule, forming a monoglyceride. Produced by the salivary glands and the pancreas.

Lipid Commonly known as fats. Water-insoluble organic molecules that provide energy to body cells, help insulate the body from heat loss, and serve as precursors in the synthesis of certain hormones. A principal component of the plasma membrane.

Liposuction Surgical technique used to remove subcutaneous fat.

Liver Organ located in the abdominal cavity that performs many functions essential to homeostasis. It stores glucose and fats, synthesizes some key blood proteins, stores iron and certain vitamins, detoxifies certain chemicals, and plays an important role in digestion by producing bile.

Long bones Bones of the skeleton that form parts of the extremities.

Loose connective tissue Type of connective tissue that serves primarily as a packing material. Contains many cells among a loose network of collagen and elastic fibers. Often contains cells that help protect the body from foreign organisms.

Low-density lipoproteins (LDLs) Complexes of protein and lipid that transport cholesterol, depositing it in body tissues.

Lungs Two large saclike organs in the thoracic cavity where the blood and air exchange carbon dioxide and oxygen.

Luteinizing hormone (LH) Hormone produced by the anterior pituitary that stimulates gonadal hormone production. In men, LH stimulates the production of testosterone, the male sex steroid. In women, LH stimulates estrogen secretion.

Lymph Fluid contained in the lymphatic vessels. Similar to tissue fluid, but also contains white blood cells and may contain large amounts of fat.

Lymph node Small nodular organ interspersed along the course of the lymphatic vessels. Serves as a filter for lymph.

Lymphatic system Network of vessels that drains extracellular fluid from body tissues and returns it to the circulatory system.

Lymphocyte Type of white blood cell. *See also* B-lymphocyte and T-lymphocyte.

Lymphoid organs Organs, such as the spleen and thymus, that belong to the lymphatic system.

Lymphokine Chemical released by suppressor T cells that inhibits the division of B and T cells.

Lysosome Membrane-bound organelle that contains enzymes. Responsible for the breakdown of material that enters the cell by endocytosis. Also destroys aged or malfunctioning cellular organelles.

Lysozyme Enzyme produced in saliva that dissolves the cell wall of bacteria, killing them.

Macronutrients Nutrients required in relatively large amounts by organisms. Includes water, proteins, carbohydrates, and lipids.

Macrophage Phagocytic cell derived from monocytes that resides in loose connective tissues and helps guard tissues against bacterial and viral invasion.

Macula lutea Region of the retina located lateral to the optic disc where cones are most abundant.

Maculae Receptor organs in the saccule and utricle that play a role in position sense.

Malignant tumor Structure resulting from uncontrollable cellular growth. Cells often spread to other parts of the body.

Mammals Class of vertebrates with a single jaw bone, specialized teeth, hair, mammary glands, a four-chambered heart, and an advanced brain. Reproduction is the most complex of animals.

Marfan's syndrome Autosomal dominant genetic disorder that affects the skeletal system, the eye, and the cardiovascular system.

Marrow cavity Cavity inside a bone containing either red or yellow marrow.

Mast cell Cell found in many tissues, especially in the connective tissue surrounding blood vessels. Contains large granules containing histamine.

Matrix Extracellular material found in cartilage. Also the material in the inner compartment of the mitochondrion.

Matter Anything that has mass and occupies space.

Medulla Term referring to the central portion of some organs; for example, the adrenal medulla.

Megakaryocyte Large cell found in bone marrow that produces platelets.

Meiosis Type of cell division that occurs in the gonads during the formation of gametes. Requires two cellular divisions (meiosis I and meiosis II). In humans, it reduces the chromosome number from 46 to 23.

Meiosis I First meiotic division.

Meiosis II Second meiotic division.

Meissner's corpuscle Encapsulated sensory receptor thought to respond to light touch.

Membranous epithelium Refers to any sheet of epithelium that forms a continuous lining on organs.

Memory cells T or B cells produced after antigen exposure. They form a reserve force that responds rapidly to antigen during subsequent exposure.

Menopause End of the reproductive function (ovulation) in women. Usually occurs between the ages of 45 and 55.

Menstrual cycle Recurring series of events in the reproductive functions of women. Characterized by dramatic changes in ovarian and pituitary hormone levels and changes in the uterine lining that prepare the uterus for implantation.

Menstruation Process in which the endometrium is sloughed off, resulting in bleeding. Occurs approximately once every month.

Merkel disk Light touch receptor. Consists of dendrites that end on cells in the epidermis.

Mesoderm One of the three types of cells that emerge in human embryonic development. Lies in the middle of the forming embryo. Gives rise to muscle, bone, and cartilage.

Messenger RNA (mRNA) Type of RNA that carries genetic information needed to synthesize proteins to the cytoplasm of a cell.

Metabolic pathway Series of linked chemical reactions in which the product of one reaction becomes the reactant in another.

Metabolic water Water produced during cellular respiration by the addition of protons (hydrogen ions) and electrons to oxygen. Occurs in the electron transport system.

Metabolism The chemical reactions of the body, including all catabolic and anabolic reactions.

Metaphase Stage of cellular division in which chromosomes line up in the center of the cell.

Metarterioles Arterioles that serve as circulatory short cuts, connecting arterioles with venules in a capillary bed. Also known as thoroughfare channels.

Metastasis Spread of cancerous cells throughout the body, through the lymph vessels and circulatory system or directly through tissue fluid.

Microfilament Solid fiber consisting of contractile proteins that is found in cells in a dense network under the plasma membrane. Forms part of the cytoskeleton.

Micronutrients Nutrients required in small quantities. They include two broad groups, vitamins and minerals.

Microspheres Small globules consisting of protein that may have been precursors of the first cells. Also known as proteinoids.

Microtubules Hollow protein tubules in the cytoplasm of cells that form part of the cytoskeleton. Also form spindles.

Microvilli Tiny projections of the plasma membranes of certain epithelial cells that increase the surface area for absorption.

Middle ear Portion of the ear located within a bony cavity in the temporal bone of the skull. Houses the ossicles.

Mineralocorticoids Group of steroid hormones produced by the adrenal cortex. Involved in electrolyte or mineral salt balance.

Mitochondrion Membrane-bound organelle where the bulk of cellular energy production occurs in eukaryotic cells. Houses the citric acid cycle and electron transport system.

Mitosis Term referring specifically to the division of a cell's nucleus. Consists of four stages: prophase, metaphase, anaphase, and telophase.

Mitotic spindle Array of microtubules constructed in the cytoplasm during prophase. Microtubules of the mitotic spindle connect to the chromosomes and help draw them apart during mitosis.

Molecule A stucture formed by two or more atoms.

Monerans Kingdom containing prokaryotic organisms, bacteria.

Monocyte White blood cell that phagocytizes bacteria and viruses in body tissues.

Monohybrid cross Procedure in which one plant is bred with another to study the inheritance of a single trait.

Monosomy Genetic condition caused by a missing chromosome.

Morning sickness Nausea that often occurs in the first two to three months of pregnancy.

Morula Solid ball of cells produced from the zygote by numerous cellular divisions.

Motor unit Muscle fibers supplied by a single axon and its branches.

Mucus Thick, slimy material produced by the lining of the respiratory tract and parts of the digestive tract. Moistens and protects them.

Mullerian mimicry Phenomenon of several species, each with toxic characteristics, evolving similar color patterns.

Multipolar neuron Motor neuron found in the central nervous system. Contains a prominent, multiangular cell body and several dendrites.

Muscle fiber Long, unbranched, multinucleated cell found in skeletal muscle.

Muscle spindles Stretch receptors found in skeletal muscle. Also known as neuromuscular spindles.

Muscle tone Inherent firmness of muscle, resulting from contraction of muscle fibers during periods of inactivity.

Muscular artery Any one of the main branches of the aorta. Tunica media consists primarily of smooth muscle cells.

Mutation Technically, a change in the DNA caused by chemical and physical agents. Also refers to a wide range of chromosomal defects.

Myelin sheath Layer of fatty material coating the axons of many neurons in the central and peripheral nervous systems.

Myofibril Bundle of contractile myofilaments in skeletal muscle cells.

Myoglobin Cytoplasmic protein in muscle cells that binds to oxygen.

Myometrium Uterine smooth muscle.

Myopia (nearsightedness) Visual condition that results when the eyeball is slightly elongated or the lens is too strong. In the uncorrected eye, light rays from distant images come into focus in front of the retina.

Myosin Protein filament found in many cells in the microfilamentous network. Also found in muscle cells.

Myosin ATPase enzyme found in the myosin cross bridges that splits ATP during muscle contraction.

Naked nerve ending Unmodified dendritic ending of the sensory neurons. Responsible for at least three sensations: pain, temperature, and light touch.

Natural childbirth Childbirth without the use of drugs.

Natural selection Evolutionary process in which environmental abiotic and biotic factors "weed" out the less fit—those organisms not as well adapted to the environment as their counterparts.

Nephron Filtering unit in the kidney. Consists of a glomerulus and renal tubule.

Nerve Bundle of nerve fibers. May consist of axons, dendrites, or both. Carries information to and from the central nervous system.

Nerve deafness Loss of hearing resulting from nerve or brain damage.

Nervous tissue One of the primary tissues. Found in the nervous system and consists of two types of cells: conducting cells (neurons) and supportive cells.

Neural groove Ectodermal groove that forms early in embryonic development and runs the length of the embryo, later forming the neural tube.

Neural tube Tube of ectoderm that arises from the neural groove and will become the spinal cord.

Neuroendocrine reflex A reflex involving the endocrine and nervous systems.

Neuron Highly specialized cell that generates and transmits bioelectric impulses from one part of the body to another.

Neurosecretory neurons Specialized nerve cells of the hypothalamus and posterior pituitary that produce and secrete hormones.

Neurotransmitter Chemical substance released from the terminal ends (terminal boutons) of axons when a bioelectric impulse arrives. May stimulate or inhibit the next neuron.

Neutron Uncharged particle in the nucleus of the atom.

Neutrophil Type of white blood cell that phagocytizes bacteria and cellular debris.

Nicotinamide adenine dinucleotide (NAD) Electron acceptor molecule that shuttles energetic electrons from glycolysis, the transition reaction, and the citric acid cycle to the electron transport system.

Nitrogen fixation Process in which bacteria and a few other organisms convert atmospheric nitrogen to nitrate or ammonia, forms usable by plants.

Node of Ranvier Small gap in the myelin sheath of an axon; located between segments formed by Schwann cells. Responsible for saltatory conduction.

Nondisjunction Failure of a chromosome pair or chromatids of a double-stranded chromosome to separate during mitosis or meiosis.

Nonpoint source (of pollution) A source that does not release pollutants via an easily identifiable route. *See also* Point source.

Nonspecific urethritis (NSU) One of the most common sexually transmitted diseases. Caused by several different bacteria.

Noradrenaline (norepinephrine) Hormone produced by adrenal medulla and secreted under stress. Contributes to the fight-or-flight response.

Nuclear envelope Double membrane delimiting the nucleus.

Nuclear pores Minute openings in the nuclear envelope that allow materials to pass to and from the nucleus.

Nucleoli Temporary structures in the nuclei of cells during interphase. Regions of the DNA that are active in the production of RNA.

Nucleus (atom) Dense, center region of the atom that contains neutrons and protons.

Nucleus (cell) Cellular organelle that contains the genetic information that controls the structure and function of the cell.

Nutrient cycle Circular flow of nutrients from the environment through the various food chains back into the environment.

Olfactory membrane Receptor for smell; found in the roof of the nasal cavity.

Olfactory nerve Nerve that transmits impulses from the olfactory membrane to the brain.

Omnivores Organisms that feed on both plants and animals.

Oogenesis Production of ova.

Oogonium Germ cell in ovary that contains 46 double-stranded chromosomes. Forms primary oocytes.

Operator site Region of the DNA molecule adjacent to the structural genes that acts as a switch to turn the operon on or off.

Operon Functional unit of the DNA of bacteria. Consists of structural and regulatory genes.

Optic disk Site in the retina where the optic nerve exits. Also known as the blind spot.

Optic nerve Nerve that carries impulses from the retina to the brain.

Order Taxonomic term that refers to a subgroup of a class.

Organ Discrete structure that carries out specialized functions.

Organ of Corti Receptor for sound; located in the inner ear within the cochlea.

Organ system Group of organs that participate in a common function.

Organogenesis Organ formation during embryonic development.

Osmosis Diffusion of water across a selectively permeable membrane.

Osmotic pressure Force that drives water across a selectively permeable membrane. Created by differences in solute concentrations.

Ossicles Three small bones inside the middle ear that transmit vibrations created by sound waves to the organ of Corti.

Osteoarthritis Degenerative joint disease caused by wear and tear that impairs movement of joints.

Osteoblast Bone-forming cell; secretes collagen.

Osteoclast Cell that digests the extracellular material of bone. Stimulated by the parathyroid hormone.

Osteocyte Bone cell derived from osteoblasts that has been surrounded by calcified extracellular material.

Osteoporosis Degenerative disease resulting in the deterioration of bone. Due to inactivity in men and women and loss of the ovarian hormone estrogen in postmenopausal women.

Outer ear External portion of the ear.

Oval window Membrane-covered opening in the cochlea where vibrations are transmitted from the stirrup to the fluid within the cochlea.

Ovary Female gonad. Produces ova (eggs) and steroid hormones, estrogen and progesterone.

Overpopulation Condition in which a species has exceeded the carrying capacity of the environment.

Ovulation Release of the oocyte from ovary. Stimulated by hormones from the anterior pituitary.

Ovum (ova, plural) Germ cell containing 23 single-stranded chromosomes. Produced during the second meiotic division.

Oxaloacetate Four-carbon compound of the citric acid cycle. It is involved in the very first reaction of the cycle and is regenerated during the cycle.

Oxidation The loss of hydrogens or electrons from a substance.

Oxytocin Hormone from the posterior pituitary hormone. Stimulates contraction of the smooth muscle of the uterus and smooth-muscle-like cells surrounding the glandular units of the breast.

Ozone (O_3) Molecule produced in the stratosphere (upper layer of the atmosphere) from molecular oxygen. Helps screen out incoming ultraviolet light. *See also* Ozone layer.

Ozone layer Region of the atmosphere located approximately 12 to 16 miles above the Earth's surface where ozone molecules are produced. Helps protect the Earth from ultraviolet light.

Pacinian corpuscle Large encapsulated nerve ending that is located in the deeper layers of the skin and near body organs. Responds to pressure.

Pancreas Organ found in the abdominal cavity under the stomach, nestled in a loop formed by the first portion of the small intestine. Produces enzymes needed to digest foodstuffs in the small intestine and hormones that regulate blood glucose levels.

Pap smear Procedure in which cells are retrieved from the cervical canal to be examined for the presence of cancer.

Papillae Small protrusions on the upper surface of the tongue. Some papillae contain taste buds.

Paracrines Chemicals released by cells that elicit a response in nearby regions.

Parasitism Process in which an individual feeds on (usually without killing) another larger individual—the host.

Parasympathetic division (of the autonomic nervous system) Portion of the autonomic nervous system responsible for a variety of involuntary functions.

Parathyroid glands Endocrine glands located on the posterior surface of the thyroid gland in the neck. Produce parathyroid hormone.

Parathyroid hormone (PTH) Hormone that helps regulate blood calcium levels. Stimulates osteoclasts to digest bone, thus raising blood calcium levels. Also known as parathormone.

Passive immunity Temporary protection from antigen (bacteria and others) produced by the injection of immunoglobulins.

Penis Male organ of copulation.

Pepsin Enzyme released by the gastric glands of the stomach. Breaks down proteins into large peptide fragments.

Pepsinogen Inactive form of pepsin.

Perforin Chemical released by cytotoxic cells that destroys bacteria. Binds to plasma membrane of target cells, forming pores that make the target cells leak and die.

Perichondrium Connective tissue layer surrounding most types of cartilage. Contains blood vessels that supply nutrients to cartilage cells.

Periodic table of elements Table that lists elements by ascending atomic number. Also lists other vital statistics of each element.

Peripheral nervous system Portion of the nervous system consisting of the cranial and spinal nerves and receptors.

Peristalsis Involuntary contractions of the smooth muscles in the wall of the esophagus, stomach, and intestines, which propel food along the digestive tract.

Peritubular capillaries Capillaries that surround nephrons. They pick up water, nutrients, and ions from the renal tubule, thus helping maintain the osmotic concentration of the blood.

Permafrost Permanently frozen subsoil in the Arctic tundra.

Permanent threshold shift Permanent hearing loss caused by repeated exposure to noise. Results from damage to hair cells of the organ of Corti.

Pharynx Chamber that connects the oral cavity with the esophagus.

Phenotype Outward appearance of an organism.

Phonation Production of sound.

Phosphofructokinase Enzyme in the glycolytic pathway that catalyzes the conversion of glucose-6-phosphate to fructose-6-phosphate; regulated by ATP and citric acid.

Phosphoglyceraldehyde A three-carbon monosaccharide produced during glycolysis by the splitting of the glucose molecules.

Photoreceptors Modified nerve cells that respond to light. Located in the retina of humans and other animals.

Photosynthesis The series of chemical reactions in which the energy of light is used to synthesize high-energy organic molecules, usually carbohydrates, from low-energy inorganic molecules, usually carbon dioxide and water.

Phylum (phyla, plural) Major grouping of organisms. Animal kingdom contains about two dozen primary phyla, nearly all of which are represented today.

Phytoplankton Microscopic, free-floating organisms, principally algae, which capture sunlight energy, using it to produce carbohydrates from carbon dioxide dissolved in the water.

Pineal gland Small gland located in the brain that secretes a hormone thought to help control the biological clock.

Pituitary gland Small pea-sized gland located beneath the brain in the sella turcica. It produces numerous hormones and consists of two main subdivisions: anterior and posterior pituitary.

Placenta Organ produced from maternal and embryonic tissue. Supplies nutrients to the growing embryo and fetus and removes fetal wastes. Also produces hormones that help maintain pregnancy.

Plasma Extracellular fluid of blood. Comprises about 55% of the blood.

Plasma cell Cell produced from B-lymphocytes (B cells); synthesizes and releases antibodies.

Plasma membrane Outer layer of the cell. Consists of lipid and protein and controls the movement of materials into and out of the cell.

Plasmids Small circular strands of DNA found in bacterial cytoplasm separate from the main DNA.

Plasmin Enzyme in the blood that helps dissolve blood clots.

Plasminogen Inactive form of plasmin.

Platelet Cell fragment produced from megakaryocytes in the red bone marrow. Plays a key role in blood clotting.

Point source (of pollution) A discrete, easily identifiable source, usually releasing pollutants directly into waterways via pipes.

Polar body Discarded nuclear material produced during meiosis I and meiosis II of oogenesis.

Polar fibers Type of microtubule found in the spindle. Extend from one centriole to the center of the cell.

Polygenic inheritance Transmission of traits that are controlled by more than one gene.

Polyploidy Term referring to a genetic disorder caused by an abnormal number of chromosomes. Includes tetraploidy and triploidy.

Polyribosome Also known as polysome. Organelle formed by several ribosomes attached to a single messenger RNA. Synthesizes proteins used inside the cell.

Polysaccharide A carbohydrate molecule such as glycogen and starch which is made of many smaller molecules (monosaccharides).

Population Group of like organisms occupying a specific region.

Porphyrin ring Part of the chlorophyll molecule that absorbs sunlight.

Portal system Arrangement of blood vessels in which a capillary bed drains to a vein, which drains to another capillary bed.

Posterior chamber Posterior portion of the anterior cavity of the eye.

Posterior pituitary Neuroendocrine gland that consists of neural tissue and releases two hormones, oxytocin and antidiuretic hormone.

Precapillary sphincters Tiny rings of smooth muscle that surround the capillaries arising from the metarterioles.

Predation Process in which an individual kills and feeds on another smaller individual.

Premature birth Birth of a baby before 37 weeks of gestation.

Premenstrual syndrome (PMS) Condition that occurs in some women in the days before menstruation normally begins. Characterized by a variety of symptoms such as irritability, depression, fatigue, headaches, bloating, swelling and tenderness of breasts, joint pain, and tension.

Premotor area Region of the brain in front of the primary motor area. Controls muscle contraction and other less voluntary actions (playing a musical instrument).

Presbyopia Visual impairment caused by aging. Lens becomes stiffer, making it more difficult to focus on nearby objects.

Primary follicle Structure in the ovary consisting of a primary oocyte and a complete single layer of cuboidal follicle cells.

Primary motor area Ridge of tissue in front of a central groove (the central sulcus). Controls voluntary motor activity.

Primary oocyte Germ cell produced from oogonium in the ovary. Undergoes the first meiotic division.

Primary response Immune response elicited when an antigen first enters the body.

Primary sensory area Region of the brain located just behind the central sulcus. The point of destination for many sensory impulses traveling from the body into the spinal cord and up to the brain.

Primary spermatocyte Cell produced from spermatogonium in the seminiferous tubule. Will undergo first meiotic division.

Primary succession Process of sequential change in which one community is replaced by another. Occurs where no biotic community has existed before.

Primary tissue One of major tissue types, including epithelial, connective, muscle, and nervous tissue.

Primary tumor Cancerous growth that gives rise to cells that spread to other regions of the body.

Primates An order of the kingdom Animalia. Includes prosimians (premonkeys), monkeys, apes, and humans.

Primordial follicle Structure in the ovary that consists of a primary oocyte surrounded by a layer of flattened follicle cells. Gives rise to the primary follicle.

Primordial germ cells Cells that originate in the wall of the yolk sac and eventually become either spermatogonia or oogonia.

Principle of independent assortment Mendel's second law. Hereditary factors are segregated independently during gamete formation. Occurs only when genes are on different chromosomes.

Principle of segregation Mendel's first law, which states that hereditary factors separate during gamete formation.

Producers Generally refers to organisms that can synthesize their own foodstuffs. Major producers are the algae and plants that absorb sunlight and use its energy to synthesize organic foodstuffs from water and carbon dioxide.

Prokaryote A cell that has no nucleus and no organelles such as a bacterium. All prokaryotes are members of the Kingdom Monera.

Prolactin Protein hormone secreted by the posterior pituitary. In humans, it is responsible for milk production by the glandular units of the breast.

Promoter Region of the operon between the regulator gene and operator site. Binds to RNA polymerase.

Pronuclei Name of the ovum and sperm cell nuclei shortly after fertilization occurs. Each contains 23 chromosomes.

Prophase First phase of mitosis during which chromosomes condense, the nuclear membrane disappears, and the spindle forms.

Proprioception Sense of body and limb position.

Prosimians Premonkeys; tarsiers and lemurs.

Prostaglandins Group of chemical substances that have a variety of functions. Act on nearby cells.

Prostate gland Sex accessory gland that is located near the neck of the bladder and empties into the urethra. Produces fluid that is added to the sperm during ejaculation.

Protein A polymer consisting of many amino acids.

Proteinoids Spherical structures composed of small amino acid chains formed when amino acids are heated in air. May have been an early precursor of the first cells.

Protists Unicellular eukaryotic organisms such as protozoans and amoebae. Some are autotrophic and some are heterotrophic.

Proton Subatomic particle found in the nucleus of the atom. Each proton carries a positive charge.

Proto-oncogenes Genes in cells that, when mutated, lead to cancerous growth.

Puberty Period of sexual maturation in humans.

Pulmonary circuit (or circulation) Short circulatory loop that supplies blood to the lungs and transports it back to the heart.

Pulmonary veins Veins that carry oxygenated blood from the lungs to the left atrium.

Punctuated equilibrium Hypothesis explaining how evolutionary change occurs. States that long periods of relatively little change are broken up by briefer periods of relatively rapid evolution.

Pupil Central opening in the iris that allows light to penetrate deeper into the eye.

Purine Type of nitrogenous base found in DNA nucleotides. Consists of two fused rings.

Purkinje fiber Modified cardiac muscle fiber that conducts bioelectric impulses to individual heart muscle cells.

Pus Liquid emanating from a wound. Contains plasma, many dead neutrophils, dead cells, and bacteria.

Pyloric sphincter Ring of smooth muscle cells in the lower portion of the stomach where it joins the duodenum. Serves as a gate valve. Opens periodically after a meal, releasing spurts of chyme (liquified, partially digested food) into the small intestine.

Pyramid of numbers Diagram of the number of organisms at various trophic levels in a food chain or ecosystem.

Pyrimidine One of two types of nitrogen base found in DNA nucleotides. Consists of one ring.

Pyrogen Chemical released primarily from macrophages that have been exposed to bacteria and other foreign substances. Responsible for fever.

Radial keratotomy Procedure to correct nearsightedness. Numerous, small superficial incisions are made in the cornea, flattening it and reducing its refractive power.

Radioactivity Tiny bursts of energy or particles emitted from the nucleus of some unstable atoms. Results from excess neutrons in the nuclei of some atoms.

Radionuclide Radioactive isotope of an atom.

Range of tolerance Range of conditions in which an organism is adapted.

Receptor Any structure that responds to internal or external changes. Three types of receptors are found in the body: encapsulated, nonencapsulated (naked nerve endings), and specialized (e.g., the retina and semicircular canals).

Recessive Term describing an allele of a gene that is expressed when the dominant factor is missing.

Recombinant DNA technology Procedure in which scientists take segments of DNA from an organism and combine them with DNA from other organisms.

Recombination Process of crossing over during meiosis, resulting in new genetic combinations.

Red blood cells (RBCs) Enucleated cells in blood that transport oxygen in the bloodstream.

Red bone marrow Tissue found in the marrow cavity of bones. Site of blood cell and platelet production.

Reduction The addition of hydrogens or electrons to a molecule.

Reduction factor Any of the factors that cause populations to decline.

Reflex Automatic response to a stimulus. Mediated by the nervous system.

Refraction Bending of light.

Regulator gene Gene that codes for the synthesis of repressor protein in an operon.

Relaxin Hormone produced by the corpus luteum and the placenta. It is released near the end of pregnancy and softens the cervix and the fibrocartilage uniting the pubic bones, thus facilitating birth.

Releasing hormone Any of a group of hypothalamic hormones that stimulates the release of other hormones by the anterior pituitary.

Renal pelvis Hollow chamber inside the kidney. Receives urine from the collecting tubules and empties into the ureter.

Renal tubule That portion of the nephron where urine is produced.

Renewable resources Resources that replenish themselves via natural biological and geological processes, such as wind, hydropower, trees, fish, and wildlife.

Replacement-level fertility Number of children that will replace a couple when they die.

Repressible operon Operon whose genes remain active unless turned off. Found in bacteria and may be present in eukaryotes as well.

Repressor protein Protein produced by a regulator gene. Binds to a region of the DNA molecule (the operator site) adjacent to the structural genes. Blocks RNA polymerase from transcribing structural genes.

Reproductive isolation Condition in which two groups of similar organisms derived from the same parent stock lose the ability to interbreed. Often due to geographic isolation.

Reptiles Class of vertebrates in which fertilization occurs internally. Includes snakes, lizards, and turtles.

Resting potential Minute voltage differential across the membrane of neurons. Also known as the membrane potential.

Restriction endonuclease Enzyme used in recombinant DNA technology. Cuts off segments of the DNA molecule for cloning and splicing.

Reticular activating system (RAS) Region of the medulla that receives nerve impulses from neurons transmitting information to and from the brain. Impulses are transmitted to the cortex, alerting it.

Retina Innermost, light-sensitive layer of the eye. Consists of an outer pigmented layer and an inner layer of nerve cells and photoreceptors (rods and cones).

Retrovirus Special type of RNA virus that carries an enzyme enabling it to produce complementary strands of DNA on the RNA template.

Reverse transcriptase Enzyme that allows the production of DNA from strands of viral RNA.

Rheumatoid arthritis Type of arthritis in which the synovial membrane of the joint becomes inflamed and thickens. Results in pain and stiffness in joints. Thought to be an autoimmune disease.

Rhythm method Birth control method in which a couple abstains from sexual intercourse around the time of ovulation. Also known as the natural method.

Ribosomal RNA (rRNA) RNA produced at the nucleolus. Combines with protein to form the ribosome.

Ribosome Cellular organelle consisting of two subunits, each made of protein and ribosomal RNA. Plays an important part in protein synthesis.

RNA polymerase Enzyme that helps align and join the nucleotides in a replicating RNA molecule.

Rod Type of photoreceptor in the eye. Provides for vision in dim light.

Root nodule Swelling in the roots of certain plants (legumes) containing nitrogen-fixing bacteria.

Rough endoplasmic reticulum (RER) Ribosome-coated endoplasmic reticulum. Produces lysosomal enzymes and proteins for use outside the cell.

Saccule Membranous sac located inside the vestibule. Contains a receptor for movement and body position.

Salivary gland Any of several exocrine glands situated around the oral cavity. Produces saliva.

Saltatory conduction Conduction of a bioelectric impulse down a myelinated neuron from node to node.

Saprophyte Organism that releases enzymes that digest food materials externally. Smaller food molecules generated in the process are absorbed by the organism. Includes most fungi and nonphotosynthetic bacteria.

Sarcomere Functional unit of the muscle cell. Consists of the myofilaments, actin, and myosin.

Sarcoplasmic reticulum Term given to the smooth endoplasmic reticulum of a skeletal muscle fiber. Stores and releases calcium ions essential for muscle contractions.

Schwann cell Type of neuroglial cell or supportive cell in the nervous system. Responsible for the formation of the myelin sheath.

Science Body of knowledge on the workings of the world and a method of accumulating knowledge. *See also* Scientific method.

Scientific method Deliberate, systematic process of discovery. Begins with observation and measurement. From observations, hypotheses are generated and tested. This leads to more observation and measurement that supports or refutes the original hypothesis.

Sclera Outermost layer of the eye.

Scrotum Skin-covered sac containing the testes.

Sebum Oil excreted by sebaceous glands onto the surface of the skin.

Second law of thermodynamics Law stating that no energy conversion is ever 100% efficient.

Secondary response Generally, a powerful, swift immune system response occurring the second time an antigen enters the body. Much faster than the primary response.

Secondary sex characteristics Distinguishing features of men and women resulting from the sex steroids. In men, includes facial hair growth and deeper voices. In women, includes breast development and fatty deposits in the hips and other regions.

Secondary succession Process of sequential change in which one community is replaced by another. It occurs where a biotic community previously existed, but was destroyed by natural forces or human actions.

Secondary tumor Cancerous growth formed by cells arising from a primary tumor.

Secretin Hormone produced by the cells of the duodenum. Stimulates the pancreas to release sodium bicarbonate.

Secretory vesicles Membrane-bound vesicles containing protein (hormones or enzymes) produced by the endoplasmic reticulum and packaged by the Golgi complex of some cells. They fuse with the membrane, releasing their contents by exocytosis.

Selective permeability Control of what moves across the plasma membrane of a cell.

Semen Fluid containing sperm and secretions of the secondary sex glands.

Semicircular canal Sensory organ of the inner ear. Houses the receptors that detect body position and movement.

Semilunar valve Type of valve lying between the ventricles and the arteries that conduct blood away from the heart.

Seminal vesicles Sex accessory glands that empty into the vas deferens. Produce the largest portion of ejaculate.

Seminiferous tubule Sperm-producing tubule in the testis.

Sensitization Type of associative learning in which an individual learns to pay heightened attention to stimuli.

Sertoli cell Cell in the germinal epithelium of the seminiferous tubule. Houses spermatogenic cells as they develop.

Sex accessory gland One of several glands that produce secretions that are added to sperm during ejaculation.

Sex chromosomes X and Y chromosomes that help determine the sex of an individual. They also carry a few other traits.

Sex-linked trait Trait produced by a gene carried on a sex chromosome.

Sex steroid Steroid hormones produced principally by the ovaries (in women) and testes (in men). Help regulate secretion of gonadotropins and determine secondary sex characteristics.

Sexually transmitted diseases (venereal diseases) Infections that are transmitted by sexual contact.

Sickle-cell anemia Genetic disease common in African Americans that results in abnormal hemoglobin in red blood cells, causing cells to become sickle shaped when exposed to low oxygen levels. Sickling causes cells to block capillaries, restricting blood flow to tissues.

Sinoatrial node The heart's pacemaker. Located in the wall of the right atrium, it sends timed impulses to the heart muscle, thus synchronizing muscle contractions.

Skeletal muscle Muscle that is generally attached to the skeleton and causes body parts to move.

Skeleton Internal support of humans and other animals. Consists of bones joined together at joints.

Sliding filament mechanism Sliding of actin filaments toward the center of a sarcomere, causing muscle contraction.

Smooth endoplasmic reticulum (SER) Endoplasmic reticulum without ribosomes. Produces phosphoglycerides used to make the plasma membrane. Performs a variety of different functions in different cells.

Smooth muscle Involuntary muscle that lacks striations. Found around circulatory system vessels and in the walls of such organs as the stomach, uterus, and intestines.

Somatic cell A body cell such as those of bone, muscle, liver, brain, and blood.

Somite Block of mesoderm that gives rise to the vertebrae, muscles of the neck, and trunk.

Special sense Vision, hearing, taste, smell, and balance.

Speciation Formation of new species often as a result of geographic isolation.

Species Group of organisms that is structurally and functionally similar. When members of the group breed, they produce viable, reproductively competent offspring. Also a subgroup of a genus.

Specificity The property of an enzyme allowing it to catalyze only one or a few chemical reactions.

Spermatogenesis Formation of sperm in the seminiferous tubules.

Spermatogonia Sperm-producing cells in the periphery of the germinal epithelium of the seminiferous tubules.

Spermatozoan Sperm. Male germ cell.

Spinal nerve Nerve that arises from the spinal cord.

Sponges Phylum of simple, immobile animals that live in colonies, mostly in saltwater. They demonstrate somewhat more complexity than one-celled organisms but less than organisms with distinct tissues.

Spongy bone Type of bony tissue inside most bones. Consists of an irregular network of bone spicules.

Spores (bacterial) Resistant structures produced when environmental conditions become unfavorable. They house the circular chromosome and a tiny amount of cytoplasm and are encased in a thick cell wall.

Starch A polysaccharide found in plants and made of many glucose units.

Sterilization Procedure to render a man or woman sterile or infertile. In men, the method is generally a vasectomy; in women, it is usually tubal ligation.

Stirrup (stapes) One of three bones of the middle ear that conducts vibrations from the eardrum to the inner ear.

Stoma (plural, *stomata*) Adjustable openings in plant leaves. Most gas exchange between leaves and the air occurs through the stoma.

Stroma The semifluid medium of chloroplasts, in which the membranous grana are embedded; the site of the light-independent reactions.

Structural gene Any gene of an operon that codes for the production of enzymes and other proteins.

Subatomic particles Electrons, protons, and neutrons. Particles that can be separated from an atom by physical means.

Substrate Molecule that fits into the active site of an enzyme.

Succession Process of sequential change in which one community is replaced by another until a mature or climax ecosystem is formed. *See also* Primary succession and Secondary succession.

Sulcus Indented region or groove in the cerebral cortex between ridges.

Suppressor cell Cell of the immune system that shuts down the immune reaction as the antigen begins to disappear.

Surfactant Detergent-like substance produced by the lungs. Dissolves in the thin watery lining of the alveoli; helps reduce surface tension, keeping the alveoli from collapsing.

Suspensory ligament Zonular fibers that connect the lens to the ciliary body.

Sustainable society Society that lives within the carrying capacity of the environment.

Symbiosis Refers to a number of different types of biotic interaction within a community, including mutualism and commensalism.

Sympathetic division Division of the autonomic nervous system that is responsible for many functions, especially those involved in the fight-or-flight response.

Synapse Juncture of two neurons that allows an impulse to travel from one neuron to the next.

Synaptic cleft Gap between an axon and the dendrite or effector (e.g., gland or muscle) it supplies.

Synergy Coordination of the workings of antagonistic muscle groups.

Synovial fluid Lubricating liquid inside joint cavities. Produced by the synovial membrane.

Synovial membrane Inner layer of the joint capsule.

Syphilis Potentially serious, sexually transmitted disease caused by a bacterium.

Systemic circulation System of blood vessels that transports blood to and from the body and heart, excluding the lungs.

Systolic pressure Peak pressure at the moment the ventricles contract. The higher of the two numbers in a blood pressure reading.

Taiga The northern coniferous forests biome.

Taste bud Receptor for taste principally found in the surface epithelium and certain papillae of the tongue.

T cell *See* T-lymphocyte.

Telophase Final stage of mitosis in which the nuclear envelope reforms from vesicles and the chromosomes uncoil.

Temperate deciduous forest Biome that in the United States lies east of the Mississippi River and is characterized by broadleafed trees.

Temporary threshold shift Temporary loss of hearing after being exposed to a noisy environment.

Tendons Connective tissue structures that generally attach muscles to bones.

Teratogen Chemical, biological, or physical agent that causes birth defects.

Teratology Study of birth defects.

Terminal boutons Small swellings on the terminal fibers of axons. They lie close to the membranes of the dendrites of other axons or the membranes of the effectors, and transfer bioelectric impulses from one cell to another.

Terminator codon Codon found on each strand of messenger RNA that marks where protein synthesis should end.

Testes Male gonads. They produce sex steroids and sperm.

Testosterone Male sex hormone that stimulates sperm formation and is responsible for secondary sex characteristics, such as facial hair growth and muscle growth.

Theories Principles of science—the broader generalizations about the world and its components. Theories are supported by considerable scientific research.

Theory of Natural Selection A theory proposed by Charles Darwin to explain how evolution works; it states that the fittest organisms of a population survive and reproduce, thus passing their genes on to future generations. Over time, this results in a shift in the genetic makeup of the population.

Thermocline A layer of water between the epilimnion and hypolimnion; characterized by rapid temperature change.

Thoracic duct Duct carrying lymph to the circulatory system. Empties into the large veins at the base of the neck.

Thyroid gland U- or H-shaped gland located in the neck on either side of the trachea just below the larynx. Produces three hormones: thyroxin, triiodothyronine, and calcitonin.

Thyroid-stimulating hormone (TSH) Hormone produced by the pituitary gland. Stimulates production and release of thyroxine and triiodothyronine by the thyroid gland.

Thyroxin Hormone produced by the thyroid gland that accelerates the rate of mitochondrial glucose catabolism in most body cells and also stimulates cellular growth and development.

Tissue Component of the body from which organs are made. Consists of cells and extracellular material (fluid, fibers, and so on). See primary tissue.

T-lymphocyte Type of lymphocyte responsible for cell-mediated immunity. Attacks foreign cells, virus-infected cells, and cancer cells directly. Also known as T cell.

Total fertility rate Number of children a woman is expected to have during her lifetime.

Trachea Duct that leads from the pharynx to the lungs.

Transcription RNA production on a DNA template.

Transfer RNA (tRNA) Small RNA molecules that bind to amino acids in the cytoplasm and deliver them to specific sites on the messenger RNA.

Transformation Conversion of a normal cell to a cancerous one.

Transition reaction Part of cellular respiration in which one carbon is cleaved from pyruvic acid, forming a two-carbon compound, which reacts with Coenzyme A. The resulting chemical enters the citric acid cycle.

Translation Synthesis of protein on a messenger RNA template.

Translocation Process in which a segment of a chromosome breaks off but reattaches to another site on the same chromosome or another one.

Transpiration Process by which water evaporates from leaf surfaces. Creates a tension (reduced pressure) in the xylem that pulls columns of water from the roots to the stems.

Triiodothyronine Hormone produced by the thyroid gland. Nearly identical in function to thyroxin.

Triploidy Genetic disorder in which cells have 69 chromosomes instead of 46.

Trisomy Genetic condition characterized by the presence of one extra chromosome.

Trophic hormones Hormones that stimulate the production and secretion of other hormones. Also known as tropic hormones.

Trophic level Feeding level in a food chain.

Trophoblast Outer ring of cells of the blastocyst that form the embryonic portion of the placenta.

TSH-releasing hormone (TSH-RH) Hormone secreted by the posterior lobe of the pituitary gland. Stimulates thyroxin secretion by the thyroid gland.

T tubules (transverse tubules) Invaginations of the plasma membrane of skeletal muscle fibers that conduct an impulse to the interior of the cell.

Tubal ligation Sterilization procedure in women. Uterine tubes are cut, preventing sperm and ova from uniting.

Tubular reabsorption Process in which nutrients are transported out of the nephron into the peritubular capillaries.

Tubular secretion Process in which wastes are transported from the peritubular capillaries into the nephron.

Tumor Mass of cells derived from a single cell that has begun to divide. In malignant tumors, the cells divide uncontrollably and often release clusters of cells or single cells that spread in the blood and lymphatic systems to other parts of the body. Benign tumors grow to a certain size, then stop.

Tundra Northernmost biome with long, cold winters and a short growing season.

Turner syndrome Genetic disorder in which an offspring contains 22 pairs of autosomes and a single, unmatched X chromosome. Phenotypically female.

Tympanic membrane (eardrum) Membrane between the external auditory canal and middle ear that oscillates when struck by sound waves.

Type I diabetes Form of diabetes that occurs mainly in young people and results from an insufficient amount of insulin production and release. Brought on by damage to insulin-producing cells of the pancreas. Also known as early-onset diabetes.

Type II alveolar cell Cell found in the lining of the alveoli. Produces surfactant.

Type II diabetes Form of diabetes that occurs chiefly in older individuals (around age of 40) and results from a loss of tissue responsiveness to insulin. Also known as late-onset diabetes.

Umbilical artery One of two arteries in the umbilical cord that carries blood from the embryo to the placenta.

Umbilical vein Vein in the umbilical cord that carries blood from the placenta to the fetus.

Ureter Hollow, muscular tube that transports urine by peristaltic contractions from the kidney to the urinary bladder.

Urethra Narrow tube that transports urine from the urinary bladder to the outside of the body. In males, it also conducts sperm and semen to the outside.

Urinary bladder Hollow, distensible organ with muscular walls that stores urine. Drained by the urethra.

Urine Fluid containing various wastes that is produced in the kidney and excreted out of the urinary bladder.

Uterus Organ that houses and nourishes the developing embryo and fetus.

Utricle Membranous sac containing a receptor for body position and movement. Located inside the vestibule of the inner ear.

Vaccine Preparation containing dead or weakened bacteria and viruses that, when injected in the body, elicits an immune response. *See also* Active immunity.

Vagina Tubular organ that serves as a receptacle for sperm and provides a route for delivery of the baby at birth.

Vagus nerve Nerve that terminates in the stomach wall and stimulates HCl production by cells in the gastric glands.

Variation Genetically based differences in physical or functional characteristics within a population.

Varicose vein Vein whose wall balloons out because the flow of blood downstream is obstructed.

Vas deferens Duct that carries sperm from the testis to the urethra. Contracts during ejaculation.

Vasectomy Contraceptive procedure in men in which the vas deferens is cut and the free ends sealed to prevent sperm from entering the urethra during ejaculation.

Vasopressin Also known as anti-diuretic hormone, which in high concentrations increases blood pressure.

Vein Type of blood vessel that carries blood to the heart.

Vena cava One of two large veins that empty into the right atrium of the heart.

Venule Smallest of all veins. Empties into capillary networks.

Vertebrates Animals with a spinal column (backbone).

Villi Fingerlike projections of the lining of the small intestine that increase the surface area for absorption.

Virus Nonliving entity consisting of a nucleic acid—either DNA or RNA—core surrounded by a protein coat, the capsid. Viruses are cellular parasites, invading cells and taking over their metabolic machinery to reproduce.

Visible light Electromagnetic radiation visible to humans and other animals.

Vitamin Any of a diverse group of organic compounds. Essential to many metabolic reactions.

Vitreous humor Gelatinous material found in the posterior cavity of the eye.

Vocal cords Elastic ligaments inside the larynx that vibrate as air is expelled from the lungs, generating sound.

Watershed A region drained by a river.

White blood cells (WBCs) Cells of the blood formed in the bone marrow. Principally involved in fighting infection.

White matter The portion of the brain and spinal cord that appears white to the naked eye. Consists primarily of white, myelinated nerve fibers.

Yellow marrow Inactive marrow of bones in adults containing fat. Formed from red marrow.

Yolk sac Embryonic pouch formed from endoderm. Site of early formation of red blood cells and germ cells.

Zero population growth Condition in which a population stops growing.

Zona pellucida Band of material surrounding the oocyte.

Zonular fibers Thin fibers that attach the lens to the ciliary body.

Zygote Cell produced by a sperm and ovum during fertilization. Contains 46 chromosomes.

INDEX

Note: Page numbers followed by f indicate figures; those followed by t indicate tables.

Photo Credits

Custom Medical Stock Photo; **FIGURE 6-17**, Fred Marsik/ Visuals Unlimited; Scientific Discoveries 6-1, **FIGURE 1**, Bettmann Archive

■ Chapter 7

Chapter opener CNRI/Photo Researchers; **FIGURE 7-2A**, David M. Phillips/Visuals Unlimited; **FIGURE 7-2B**, Stanley Flegler/Visuals Unlimited; **FIGURE 7-3**, Stanley Flegler/ Visuals Unlimited; **FIGURE 7-6**, R. Calentine/Visuals Unlimited; **FIGURE 7-7**, G. Prance/Visuals Unlimited; **FIGURE 7-8A**, Science VU/Visuals Unlimited; **FIGURE 7-8B**, Peter K. Ziminski/Visuals Unlimited; **FIGURE 7-9**, John D. Cunningham/Visuals Unlimited; **FIGURE 7-10**, David M. Phillips/Visuals Unlimited; **TABLE 7-2**, John D. Cunningham/ Visuals Unlimited

■ Chapter 8

Chapter opener W. Johnson/Visuals Unlimited; **FIGURE 8-1**, Visuals Unlimited/G. Musil; **FIGURE 8-2**, David M. Phillips/ Visuals Unlimited; **FIGURE 8-10**, W. Johnson/Visuals Unlimited; **FIGURE 8-12**, Leonard L Rue III/Visuals Unlimited; **FIGURE 8-16A**, Science VU/Visuals Unlimited; **FIGURE 8-16B**, Science VU/Visuals Unlimited; Health Note 8-1, **FIGURE 1**, John D. Cunningham/Visuals Unlimited

■ Chapter 9

Chapter opener CNRI/Photo Researchers; **FIGURE 9-5A**, David M. Phillips/Visuals Unlimited; **FIGURE 9-5B**, David M. Phillips/Visuals Unlimited; **FIGURE 9-5C**, Don Fawcett/ Visuals Unlimited; **FIGURE 9-7**, R. Calentine/Visuals Unlimited; **FIGURE 9-8**, Science VU/Visuals Unlimited; **FIGURE 9-9**, Chet Childs/Custom Medical Stock; **FIGURE 9-14A**, SIU/ Visuals Unlimited; **FIGURE 9-14B**, SIU/Visuals Unlimited; **FIGURE 9-15**, SIU/Visuals Unlimited; Health Note 9-2, **FIGURE 1A**, American Cancer Society; Health Note 9-2, **FIGURE 1B**, James Stevenson/SPL/Photo Researchers

■ Chapter 10

Chapter opener CNRI/Photo Researchers; **FIGURE 10-1**, SIU/Visuals Unlimited; **FIGURE 10-10A**, NMSB/Custom Medical Stock; **FIGURE 10-10B**, Runk/Schoenberger/Medical Images, Inc.

■ Chapter 11

Chapter opener Cabisco/Visuals Unlimited; **FIGURE 11-4**, David M. Phillips/Visuals Unlimited; **FIGURE 11-6**, C. Raines/Visuals Unlimited; **FIGURE 11-7**, Howard Sochureck/ Visuals Unlimited; **FIGURE 11-11A**, Science VU/E.R. Lewis/ Visuals Unlimited; **FIGURE 11-11B**, T. Reese, D.W. Fawcett/ Visuals Unlimited; **FIGURE 11-24**, SIU/Visuals Unlimited; Health Note 11-1, **FIGURE 1**, Allan Clear/Impact Visuals/PNI

■ Chapter 12

Chapter opener Cabsico/Visuals Unlimited; **FIGURE 12-3A**, Cabisco/Visuals Unlimited; **FIGURE 12-3B**, Biophoto Associates/Photo Researchers; **FIGURE 12-8**, A.L. Blum/Visuals Unlimited; **FIGURE 12-10**, Bill Beatty/Visuals Unlimited; **FIGURE 12-20**, Beltone Electronics; **FIGURE 12-21**, Science VU/Cochlear Corp./Visuals Unlimited; Health Note 12-1, **FIGURE 1**, © Photodisc

■ Chapter 13

Chapter opener Science Photo Library/Photo Researchers; **FIGURE 13-1**, Jeffrey Howe/Visuals Unlimited; **FIGURE 13-7A**, SIU/Visuals Unlimited; **FIGURE 13-7B**, SIU/Visuals Unlimited; **FIGURE 13-8**, SIU/Visuals Unlimited; **FIGURE 13-9A**, SIU/ Visuals Unlimited; **FIGURE 13-9B**, SIU/Visuals Unlimited; **FIGURE 13-10**, Science VU/Visuals Unlimited; **FIGURE 13-13A**, Reprinted with permission from Calcified Tissue Research, 1967; **FIGURE 13-13B**, Reprinted with permission from Calcified Tissue Research, 1967; **FIGURE 13-13C**, Reprinted with permission from Calcified Tissue Research, 1967; **FIGURE 13-13D**, Reprinted with permission from Calcified Tissue Research, 1967; **FIGURE 13-15**, R. Calentine/Visuals Unlimited; **FIGURE 13-18**, John D. Cunningham/Visuals Unlimited; **FIGURE 13-20**, Bruce Bergs/Visuals Unlimited; Health Note 13-1, **FIGURE 1**, Yoav Levy/Phototake

■ Chapter 14

Chapter opener SIU/Visuals Unlimited; **FIGURE 14-7A**, AP/Wide World Photos; **FIGURE 14-7B**, AP/Wide World Photos; **FIGURE 14-8**, Reprinted with permission from American Journal of Medicine, 20 (1956); **FIGURE 14-14A**, R. Calentine/ Visuals Unlimited; **FIGURE 14-14B**, David M. Phillips/Visuals Unlimited; **FIGURE 14-15**, Ken Greer/Visuals Unlimited; **FIGURE 14-16**, John D. Cunningham/Visuals Unlimited; **FIGURE 14-20**, Science VU/Visuals Unlimited

■ Chapter 15

Chapter opener M. Schliwa/Visuals Unlimited; **FIGURE 15-1**, Stan W. Elems/Visuals Unlimited; **FIGURE 15-4**, Cabisco/ Visuals Unlimited; **FIGURE 15-5**, Custom Medical Stock Photography; **FIGURE 15-6**, Science VU/Visuals Unlimited; **FIGURE 15-7**, SIU/Visuals Unlimited; **FIGURE 15-8A**, Michael Abbey/Visuals Unlimited; **FIGURE 15-8B**, Michael Abbey/Visuals Unlimited; **FIGURE 15-8C**, Phototake/PNI; **FIGURE 15-8D**, John D. Cunningham/Visuals Unlimited; **FIGURE 15-8E**, John D. Cunningham/Visuals Unlimited; **FIGURE 15-8F**, John D. Cunningham/Visuals Unlimited; **FIGURE 15-11**, David M. Phillips/Visuals Unlimited; **FIGURE 15-12A**, Jack Bostrack/Visuals Unlimited; **FIGURE 15-12B**, John D. Cunningham/Visuals Unlimited; Scientific Discovery 15-1, **FIGURE 1**, Cold Spring Harbour Laboratory; Scientific

Discovery 15-1, **FIGURE 2**, Science VU/NIHLBL/Visuals Unlimited; Health Note 15-1, **FIGURE 1**, Cabisco/Visuals Unlimited

■ Chapter 16

Chapter opener Biophoto Associates/Photo Researchers; **FIGURE 16-4A**, The Bettmann Archive; **FIGURE 16-4B**, Malcolm Gutter/Visuals Unlimited; **FIGURE 16-8**, Joe McDonald/Visuals Unlimited; **FIGURE 16-9A**, Science VU/Daniel V. Schidlow/Visuals Unlimited; **FIGURE 16-9B**, Science VU/Daniel V. Schidlow/Visuals Unlimited; **FIGURE 16-10**, Jeffrey Reed/Medichrome; **FIGURE 16-11A**, Bruce Berg/Visuals Unlimited; **FIGURE 16-11B**, Bruce Berg/Visuals Unlimited; **FIGURE 16-12**, Dr. Ira Rosenthal, Department of Pediatrics, University of Illinois at Chicago; **FIGURE 16-15A**, Mark E. Gibson/Visuals Unlimited; **FIGURE 16-15B**, Christopher Arnesen/Allstock/PNI; **FIGURE 16-15C**, M. Long/Visuals Unlimited; **FIGURE 16-15D**, Marek W. Litwin/Visuals Unlimited; **FIGURE 16-15E**, John D. Cunningham/Visuals Unlimited; **FIGURE 16-17B**, Cabisco/Visuals Unlimited; **FIGURE 16-19**, Biophoto Associates/Photo Researchers; **FIGURE 16-21**, Ron Spomer/Visuals Unlimited; **FIGURE 16-23A**, Science VU/Valerie Lindgren/Visuals Unlimited; **FIGURE 16-23B**, Bernd Wittich/Visuals Unlimited; **FIGURE 16-25A**, Science VU/Valerie Lindgren/Visuals Unlimited; **FIGURE 16-25B**, Martin Rotker; **FIGURE 16-26A**, Science VU/Valerie Lindgren/Visuals Unlimited; **FIGURE 16-26B**, Dr. Ira Rosenthal/Department of Pediatrics, University of Illinois at Chicago; **FIGURE 16-27**, S.C.R. Reuman/Visuals Unlimited; Health Note 16-1, **FIGURE 1**, Bill Bachmann/Photo Network/PNI

■ Chapter 17

Chapter opener Science VU/Lawrence Livermore Laboratory/Visuals Unlimited; **FIGURE 17-12**, Ralph A. Slepecky/Visuals Unlimited; **FIGURE 17-16**, Dana Richter/Visuals Unlimited; **FIGURE 17-17**, Martha Cooper/Peter Arnold, Inc.; Scientific Discoveries 17-1, **FIGURE 1**, AP/Wide World Photos

■ Chapter 18

Chapter opener SIU/Visuals Unlimited; **FIGURE 18-1**, Tom McHugh/Photo Researchers; **FIGURE 18-3**, Huntington Potter and David Pressler/Department of Neurobiology, Harvard Medical School; **FIGURE 18-6**, Science VU/©Jackson Lab/Visuals Unlimited; **FIGURE 18-7**, Keith V. Wood/Science VU/Visuals Unlimited

■ Chapter 19

Chapter opener David M. Phillips/Visuals Unlimited; **FIGURE 19-3A**, Biophoto Associates/Photo Researchers; **FIGURE 19-3B**, Lester Bergman and Associates; **FIGURE 19-4**, John D. Cunningham/Visuals Unlimited; **FIGURE 19-5**, Fred Hossler/Visuals Unlimited; **FIGURE 19-6**, David M. Phillips/Visuals Unlimited; **FIGURE 19-17**, Carter-Wallace Company; **FIGURE 19-21**, Science-VU-ortho/Visuals Unlimited; **FIGURE 19-22A**, Cabisco/Visuals Unlimited; **FIGURE 19-22B**, Cabisco/Visuals Unlimited; **FIGURE 19-23**, SIU/Visuals Unlimited; **FIGURE 19-24**, SIU/Visuals Unlimited; **FIGURE 19-25**, M. Long/Visuals Unlimited; **FIGURE 19-26**, SIU/Visuals Unlimited; **FIGURE 19-29**, Phototake/PNI; **FIGURE 19-30**, Christopher Brown/Stock Boston

■ Chapter 20

Chapter opener Cabisco/Visuals Unlimited; **FIGURE 20-2**, David M. Phillips/Visuals Unlimited; **FIGURE 20-11A**, SIU/Visuals Unlimited; **FIGURE 20-11B**, SIU/Visuals Unlimited; **FIGURE 20-14**, Ka Botzis; **FIGURE 20-16**, Michael DeMocker/Visuals Unlimited

■ Chapter 21

Chapter opener Kjell Sandved/Visuals Unlimited; **FIGURE 21-3**, Sidney Fox/Science VU/Visuals Unlimited; **FIGURE 21-4**, Science VU-USM/Visuals Unlimited; **FIGURE 21-7**, Rick Wallace/Visuals Unlimited; **FIGURE 21-8**, A. Kerstitch/Visuals Unlimited; **FIGURE 21-9**, John D. Cunningham/Visuals Unlimited; **FIGURE 21-10A**, Dale Jackson/Visuals Unlimited; **FIGURE 21-10B**, Daniel D. Chiras; **FIGURE 21-11**, Science VU/Visuals Unlimited; **FIGURE 21-14A**, Nada Pecnick/Visuals Unlimited; **FIGURE 21-14B**, Alex Kerstitch/Visuals Unlimited; **FIGURE 21-14C**, John D. Cunningham/Visuals Unlimited; **FIGURE 21-16**, John Cancalos/Visuals Unlimited; **FIGURE 21-18**, R. Calentine/Visuals Unlimited; Health Note 21-1, **FIGURE 1**, Link/Visuals Unlimited

■ Chapter 22

Chapter opener Fred Espenak/NASA/Science Photo Library/Photo Researchers; **FIGURE 22-1**, Milton H. Tierney, Jr./Visuals Unlimited; **FIGURE 22-3A**, Michael Dick/Animals Animals; **FIGURE 22-3B**, Walt Anderson/Visuals Unlimited; **FIGURE 22-4A**, Walt Anderson/Visuals Unlimited; **FIGURE 22-4B**, John D. Cunningham/Visuals Unlimited; **FIGURE 22-5A**, Thomas C. Boyden/Visuals Unlimited; **FIGURE 22-5B**, Charles Rushing/Visuals Unlimited; **FIGURE 22-7**, Charles Rushing/Visuals Unlimited; **FIGURE 22-9**, National Museum of Kenya/VU; **FIGURE 22-10**, Cabisco/Visuals Unlimited; **FIGURE 22-12**, Science VU/Visuals Unlimited; **FIGURE 22-13A**, Henri Sommer/Visuals Unlimited; **FIGURE 22-13B**, Emily Strong/Visuals Unlimited; **FIGURE 22-13C**, Daniel D. Chiras; **FIGURE 22-13D**, Charles Sykes/Visuals Unlimited; **FIGURE 22-13E**, John Cancalosi/Visuals Unlimited; **FIGURE 22-14**, G. Prance/Visuals Unlimited; **FIGURE 22-15**, David Cavagnaro/Visuals Unlimited; **FIGURE 22-16**, Clyde H. Smith/Peter Arnold, Inc.

■ Chapter 23

Chapter opener Daniel D. Chiras; **FIGURE 23-3A**, Steve McCutcheon/Visuals Unlimited; **FIGURE 23-3B**, William J. Weber/Visuals Unlimited; **FIGURE 23-3C**, Albert J. Copely/Visuals Unlimited; **FIGURE 23-3D**, Ron Spomer/Visuals Unlimited; **FIGURE 23-3E**, John D. Cunningham/Visuals Unlimited; **FIGURE 23-5**, John N. Rinne/Visuals Unlimited; **FIGURE 23-6**, Richard Thom/Visuals Unlimited; **FIGURE 23-16A**, Larry Nielsen/Peter Arnold, Inc.; **FIGURE 23-16B**, Pete K. Ziminski/Visuals Unlimited; **FIGURE 23-20**, R. F. Ashley/Visuals Unlimited

■ Chapter 24

Chapter opener B. Nation/Sygma; **FIGURE 24-1**, Reuters/Bettmann; **FIGURE 24-2**, Jerome Wyckoff/Visuals Unlimited; **FIGURE 24-4**, Sylvan Wittwer/Visuals Unlimited; **FIGURE 24-9**, Frans Lanting/Minten Pictures; **FIGURE 24-10**, William Banaszewski/Visuals Unlimited; **FIGURE 24-11**, Science VU/Visuals Unlimited; **FIGURE 24-14**, Courtesy of General Electric; **FIGURE 24-17**, Martin G. Miller/Visuals Unlimited; **FIGURE 24-18**, Max Hunn/Visuals Unlimited; **FIGURE 24-19**, David S. Addison/Visuals Unlimited; **FIGURE 24-21**, Andre Jenny/Stock South/PNI